D1629025

Encyclopedia of Electrochemistry

Edited by A.J. Bard and M. Stratmann

Volume 6
Semiconductor Electrodes
and Photoelectrochemistry

Volume edited by S. Licht

Encyclopedia of Electrochemistry

Editors-in-Chief: Allen J. Bard, Martin Stratmann

Volume 1: Thermodynamics and Electrified Interfaces (Editors: Eliezer Gileadi, Michael Urbakh)
Volume 2: Interfacial Kinetics and Mass Transport (Editor: Ernesto Julio Calvo)
Volume 3: Instrumentation and Electroanalytical Chemistry (Editor: Pat Unwin)
Volume 4: Corrosion and Oxide Films (Editors: Martin Stratmann, Gerald S. Frankel)
Volume 5: Electrochemical Engineering (Editor: Digby D. Macdonald)
Volume 6: Semiconductor Electrodes and Photoelectrochemistry (Editor: Stuart Licht)
Volume 7: Inorganic Electrochemistry (Editors: William E. Geiger, Chris Pickett)
Volume 8: Organic Electrochemistry (Editor: Hans-J. Schäfer)
Volume 9: Bioelectrochemistry (Editor: George S. Wilson)
Volume 10: Modified Electrodes (Editors: Israel Rubinstein, Masamichi Fujihira)
Volume 11: Index

Encyclopedia of Electrochemistry

Edited by A.J. Bard and M. Stratmann

Volume 6
Semiconductor Electrodes and Photoelectrochemistry

Volume edited by S. Licht

Editors:

Prof. Dr. Allen J. Bard
Department of Chemistry
University of Texas
Austin, TX 78712
USA

Prof. Dr. Martin Stratmann
Max-Planck-Institut
für Eisenforschung
Max-Planck-Str. 1
40237 Düsseldorf
Germany

Prof. Stuart Licht
Department of Chemistry
Technion Israel Institute of Technology
Haifa, 32000
Israel

COVER: Courtesy of Prof. Dr. D. M. Kolb, University of Ulm, Germany

Library of Congress Card No: applied for
British Library Cataloguing-in-Publication Data. A catalogue record for this book is available from the British Library.

Die Deutsche Bibliothek-CIP-Cataloguing-in-Publication Data A catalogue record for this publication is available from Die Deutsche Bibliothek

Printed in the Federal Republic of Germany
Printed on acid-free paper.

Composition: Laser Words Private Limited, Chennai, India
Printing: betz-druck GmbH, Darmstadt
Bookbinding: J. Schäffer GmbH + Co.KG, Grünstadt
ISBN 3-527-30398-7

Preface

Electrochemistry plays a dominant role in a vast number of research and applied areas. Electrochemical reactions can overcome kinetic limitations, even at very low temperatures, by application of a potential and they are chemically and stereochemically specific, leading to chemical synthesis applications.

Electrochemical analyses are based on the high sensitivity of the reactions. The excellent spatial and temporal control is of importance in the emerging field of nanotechnology. Finally electrochemical reactions are known for a wide range of materials like metals, semiconductors, polymers, and biological systems leading to a large role in a number of rather diverse areas, such as preparative chemistry, analytical chemistry, energy storage, energy conversion, biochemistry, solid state chemistry, materials science, and microelectronics.

While studies of the transport of charged species and thermodynamic considerations dominated early studies, kinetic aspects of electrochemistry have become more important in electrochemical research with an increased understanding of the chemical and electronic structure of the solid/solution interface (accelerated by the application of numerous in situ and ex situ spectroscopic techniques, combined with electrochemical experiments). More recently the introduction of in situ scanning probe techniques has allowed one to follow electrochemical reactions on an atomic or molecular scale.

However, the most important development over the last few decades has been the spread of electrochemical concepts into very different areas of research and development. Based on research by electrochemists, electrochemistry is now used in many fundamental fields, such as the study of new organic and inorganic compounds and biological systems. In more applied areas, it is used to shape materials from the macroscopic to the microscopic scale, to analyze for chemical impurities, to understand and prevent the corrosion of materials, to probe the functioning of living cells, and to convert chemical energy into electricity. Scientists and engineers working in diverse areas need to locate and use electrochemical information.

It is the aim of this encyclopedia to fill this need by providing an up-to-date electrochemical source for engineers and scientists, as well as for students needing a starting point in their search for reliable information. The *Encyclopedia of Electrochemistry* is organized into 10 volumes plus 1 index volume, which concentrate on the major areas of electrochemistry, with each volume containing some 20 articles by experts. These are intended not only for electrochemists, but also for those in other fields who use electrochemistry. The *Encyclopedia of Electrochemistry* also includes the latest reviews of recent achievements in fundamental and applied electrochemistry. The extensive cross-referencing between volumes allows users to deepen their knowledge to the extent defined by their needs.

We hope that the *Encyclopedia of Electrochemistry* facilitates the broad use of electrochemistry in academia and industry but also provides a basis for further progress in this highly interesting and fast moving area of science.

December 2001

Allen J. Bard
Martin Stratmann

About the Editors

Allen J. Bard

Born December 18, 1933, Professor Bard received his early education in the public schools of New York City and attended The City College of New York (B.Sc., summa cum laude, 1955). He did his graduate work at Harvard University with J. J. Lingane (MA, 1956; PhD, 1958) in electroanalytical chemistry. In 1958 he joined the faculty of The University of Texas at Austin where he currently holds the Norman Hackerman/Welch Regents' Chair in Chemistry. His research interests have been in the application of electrochemical methods to the study of chemical problems and include investigations in electro-organic chemistry, photoelectrochemistry, electrogenerated chemiluminescence, and electroanalytical chemistry. He has published three books (*Electrochemical Methods*, with Larry Faulkner, *Integrated Chemical Systems*, and *Chemical Equilibrium*) and over 600 papers and chapters while editing the series *Electroanalytical Chemistry* (21 volumes) and the *Encyclopedia of the Electrochemistry of the Elements* (16 volumes) plus co-editing the monograph, *Standard Potentials in Aqueous Solution*. He is currently editor-in-chief of the *Journal of the American Chemical Society*. The ISI listing of the "50 most cited chemists from 1981–1997" ranks Professor Bard at number 13 (taken from a total of 627,871 chemists surveyed).

Martin Stratmann

Born 20 April 1954, Professor Stratmann studied chemistry at the Ruhr-Universität Bochum and received his diploma in 1980. He finished his PhD in 1982 at the Max Planck Institut für Eisenforschung in Düsseldorf with H.J. Engell on electrochemical studies of phase transformations in rust layers and spent his postdoctoral education with Ernest Yeager at the Case Western Reserve University. His professorship in physical chemistry followed in 1992 at the University of Düsseldorf with electrochemical studies on metal surfaces covered with ultrathin electrolyte layers. In 1994 he took over the Chair in Corrosion Science and Surface Engineering at the University of Erlangen and since 2000 has been a scientific member of the Max Planck Gesellschaft and director at the Max Planck Institut für Eisenforschung in Düsseldorf, heading a department of interface chemistry and surface engineering. Further he is a faculty member of the departments of Materials Science and Chemistry at the Ruhr-Universität Bochum. His research interests concentrate on corrosion related electrochemistry, in particular with emphasis on microscopic aspects and in situ spectroscopy, electrochemistry at buried metal/polymer interfaces – an area where he pioneered novel electrochemical techniques – atmospheric corrosion, adhesion and surface chemistry of reactive metal substrates. He has published more than 150 papers and is co-editor of *Steel Research and Materials and Corrosion*.

Stuart Licht

Professor Stuart Licht is an electrochemist with a principal focus on clean, renewable energy solutions, and with diverse research interests, which include solar energy, energy storage, unusual analytical methodologies, and fundamental physical chemistry. He received his BA and MSc from Wesleyan University, and his doctorate in 1986 from the Weizmann Institute of Science, followed by appointments as a Postdoctoral Fellow and Visiting Scientist at MIT. In 1988 he was appointed the first Carlson Professor of Chemistry at Clark University, and in 1995 was awarded a Gustella new Immigrant Professor at the Technion Israel Institute of Science, where he is currently Professor of Chemistry. He has contributed over 200 papers and patents ranging from novel efficient solar semiconductor/electrochemical processes, to unusual batteries, to elucidation of complex equilibria and quantum electron correlation theory. He has established the field of Fe(VI) charge storage, as well furthering the understanding of the battery, microelectrodes, photoelectrochemicals, and energy conversion processes. Stuart Licht has held National American Chemical Society offices, and chaired local sections of the ACS and the New England Section of the Electrochemical Society, and is Chairman of the Israel Section of the Electrochemical Society. Awards received for his research include the Herschel Rich Innovation Award, Discover Magazine Energy Award, Goldberg Research Prize, Industry Week's Technology of the Year Award, Alcoa Aluminum Foundation Prize, the Dov Elad Award for Excellence in Chemical Research, the Bradley Award for Excellence in Chemical Research and the Max Tishler Synthesis Prize. Born 24 July 1954, Professor Licht is a third generation chemist: the grandson of the industrial chemist Joseph Licht, and son of analytical chemist Truman S. Light.

Contents

1 Fundamentals of Semiconductor Electrochemistry and Photoelectrochemistry

Krishnan Rajeshwar
The University of Texas at Arlington, Arlington, Texas

1.1 Introduction and Scope

The study of semiconductor–electrolyte interfaces has both fundamental and practical incentives. These interfaces have interesting similarities and differences with their semiconductor–metal (or metal oxide) and metal–electrolyte counterparts. Thus, approaches to garnering a fundamental understanding of these interfaces have stemmed from both electrochemistry and solid-state physics perspectives and have proven to be equally fruitful. On the other hand, this knowledge base in turn impacts many technologies, including microelectronics, environmental remediation, sensors, solar cells, and energy storage. Some of these are discussed elsewhere in this volume.

It is instructive to first examine the historical evolution of this field. Early work in the fifties and sixties undoubtedly was motivated by application possibilities in electronics and came on the heels of discovery of the first transistor. Electron transfer theories were also rapidly evolving during this period, starting from homogeneous systems to heterogeneous metal-electrolyte interfaces leading, in turn, to semiconductor-electrolyte junctions. The 1973 oil embargo and the ensuing energy crisis caused a dramatic spurt in studies on semiconductor–electrolyte interfaces once the energy conversion possibilities of the latter were realized. Subsequent progress at both fundamental and applied levels in the late eighties and nineties has been more gradual and sustained. Much of this later research has been spurred by technological applicability in environmental remediation scenarios. Very recently, however, renewed interest in clean energy sources that are nonfossil in origin, has provided new impetus to the study of semiconductor–electrolyte interfaces. As we also learn to understand and manipulate these interfaces at an increasingly finer (atomic) level, new microelectronics application possibilities may emerge, thus completing the cycle that first began in the 1950s.

The ensuing discussion of the progress that has been made in this field mainly hinges on studies that have appeared since about 1990. Several review articles and chapters have appeared since then that deal with semiconductor–electrolyte interfaces [1–10]; aspects related to electron transfer are featured in several of these. This author is also aware of at least three books/monographs/proceedings volumes that have appeared since 1990 [11–13]. The reader is referred to the many authoritative accounts that exist before this time frame for a thorough coverage of details on semiconductor–electrolyte interfaces in general. Entry to this early literature may be found in the references cited earlier. In some instances, however, the discussion that follows necessarily delves into research dating back to the 1970s and 1980s.

To facilitate a self-contained description, we will start with well-established aspects related to the semiconductor energy band model and the electrostatics at semiconductor–electrolyte interfaces in the "dark". We shall then examine the processes of light absorption, electron-hole generation, and charge separation at these interfaces. The steady state and dynamic aspects of charge transfer are then briefly considered. Nanocrystalline semiconductor films and size quantization are then discussed as are issues related to electron transfer across chemically modified semiconductor–electrolyte interfaces. Finally, we shall introduce the various types of photoelectrochemical devices ranging from regenerative and photoelectrolysis cells to dye-sensitized solar cells.

1.2 Electron Energy Levels in Semiconductors and Energy Band Model

Unlike in molecular systems, semiconductor energy levels are so dense that they form, instead of discrete molecular orbital energy levels, broad energy bands. Consider a solid composed of N atoms. Its frontier band will have $2N$ energy eigenstates, each with an occupancy of two electrons of paired spin. Thus, a solid having atoms with odd number of valence electrons (e.g. Al with $[Ne]3s^2 3p^1$) will have a partially occupied frontier band in which the electrons are delocalized. On the other hand, a solid with an even number of valence electrons (e.g. Si having an electron configuration of $[Ne]3s^2 2p^2$) will have a fully occupied frontier band (termed a valence band, (VB)). The situation for Si is schematized in Fig. 1.

As Fig. 2 illustrates, the distinction between semiconductors and insulators is rather arbitrary and resides with the magnitude of the energy band gap (E_g) between the filled and vacant bands. Semiconductors typically have E_g in the 1 eV–4 eV range (Table 1). The vacant frontier band is termed a conduction band, (CB) (Fig. 2). We shall see later that E_g has an important bearing on the optical response of a semiconductor.

For high density electron ensembles such as valence electrons in metals, Fermi statistics is applicable. In a thermodynamic sense, the Fermi level, E_F (defined at 0 K

Tab. 1 Some elemental and compound semiconductors for photoelectrochemical applications

Semiconductor	***Conductivity type(s)***	***Optical band gap energy [eV]***[a]
Si	*n*, *p*	1.11
GaAs	*n*, *p*	1.42
GaP	*n*, *p*	2.26
InP	*n*, *p*	1.35
CdS	*n*	2.42
CdSe	*n*	1.70
CdTe	*n*, *p*	1.50
TiO_2	*n*	3.00(rutile) 3.2(anatase)
ZnO	*n*	3.35

[a]The values quoted are for the bulk semiconductor. The gap energies increase in the size quantization regime (see Sect. 7).

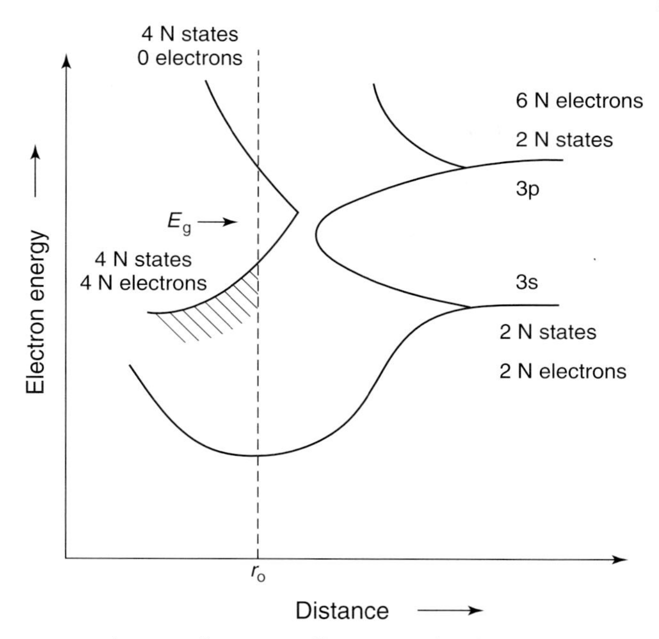

Fig. 1 Schematic illustration of how energy bands in semiconductors evolve from discrete atomic states for the specific example of silicon.

Fig. 2 Relative disposition of the CB and VB for a semiconductor (a) and an insulator (b). E_g is the optical band gap energy.

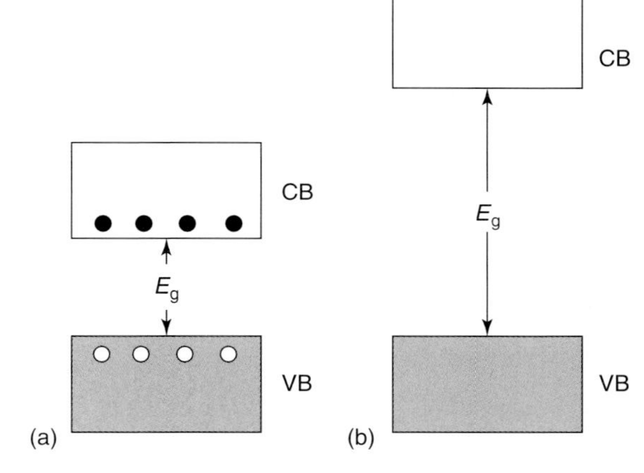

as the energy at which the probability of finding an electron is 1/2) can be regarded as the electrochemical potential of the electron in a particular phase (in this case, a solid). Thus, all electronic energy levels below E_F are occupied and those above E_F are likely to be empty.

Electrons in semiconductors may be regarded as low-density particle ensembles such that their occupancy in the valence and CBs may be *approximated* by the

Boltzmann function [14, 15]:

$$n_e \approx N_o \exp\left[-\frac{E_o - E_F}{kT}\right] \quad (1)$$

Now we come to another important distinction between metals and semiconductors in that two types of electronic carriers are possible in the latter. Consider the thermal excitation of an electron from VB to CB. This gives rise to a free electron in the CB and a vacancy or hole in the VB. A localized chemical picture for the case of Si shows that the hole may be regarded as a missing electron in a chemical bond (Fig. 3). There is a crude chemical analogy here with the dissociation of a solvent such as water into H_3O^+ and OH^-. In either case, equal numbers of oppositely charged species are produced. Thus, Eq. (1) becomes:

$$n_i \approx N_c \exp\left[-\frac{E_F - E_c}{kT}\right] \quad (2)$$

$$p_i \approx N_v \exp\left[-\frac{E_v - E_F}{kT}\right] \quad (3)$$

In Eqs. (2–3), N_c and N_v are the effective density of states (in cm^{-3}) at the lower edge and top edge of CB and VB, respectively. These expressions can be combined with the recognition that $n_i = p_i$ to yield

$$n_i^2 \approx N_o \exp\left[-\frac{E_c - E_v}{kT}\right] \approx N_o \exp\left[-\frac{E_g}{kT}\right] \quad (4)$$

To provide a numerical sense of the situation, N_c and N_v are typically both approximately 10^{19} cm^{-3} so that the constant N_o ($N_c N_v$) in Eq. (1) is about 10^{38} cm^{-3}. For a semiconductor such as Si (with $E_g = 1.11$ eV, Table 1), n_i will be about 10^{10} cm^{-3} at 300 K according to Eq. (4). This rough calculation lends credence to the original rationale for the use of Boltzmann statistics for the electron energy distribution in semiconductors (see preceding section).

The preceding case refers to the semiconductor in its intrinsic state with very low carrier concentrations under ambient conditions. The Fermi level, E_F, in this case lies approximately in the middle of the energy band gap (Fig. 4a). This simply reflects the fact that the probability of electron occupancy is very high in VB and very low in CB and does not imply an occupiable energy level at E_F itself.

In extrinsic semiconductors the carrier concentrations are perturbed such that $n = p$. Again the analogy with the addition of an acid or base to water is quite instructive here. Consider the case when donor impurities are added to a neutral semiconductor. Since the intrinsic carrier concentrations are so low (sub-parts per trillion), even additions in parts per billion levels can have a profound electrical effect. This process is known as doping of the semiconductor. In this particular case, the Fermi level shifts toward the CB edge (Fig. 4b). When the donor level is

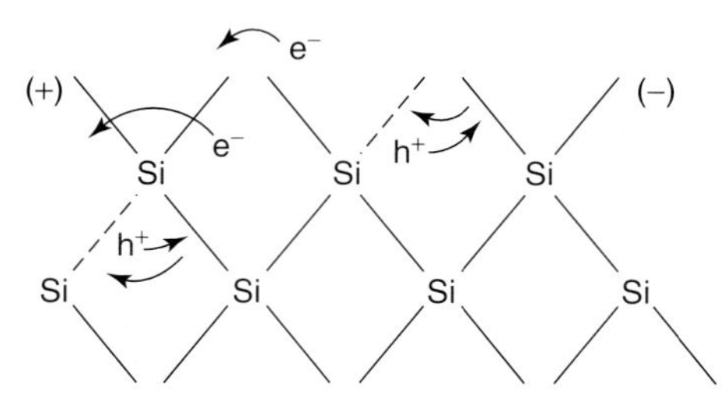

Fig. 3 A localized picture of electron-hole pair generation (see also Fig. 2a) in silicon.

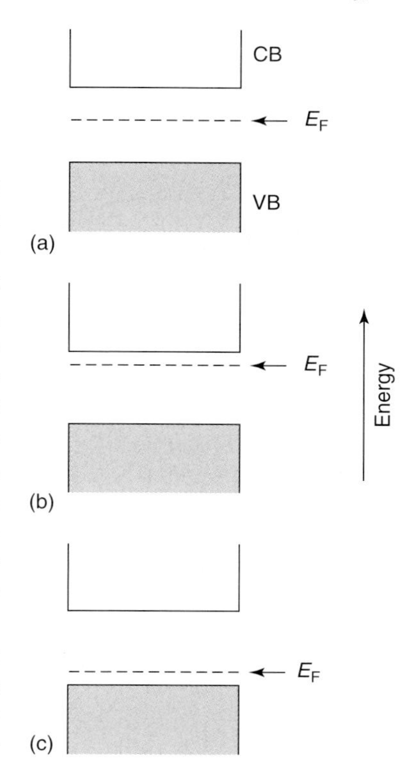

Fig. 4 Relative disposition of the Fermi level (E_F) for an intrinsic semiconductor (a), for an *n*-type semiconductor (b), and a *p*-type semiconductor (c).

within a few kT in energy from the CB edge, appreciable electron concentrations are generated by the donor ionization process (at ambient temperatures) such that now $n \gg p$. This is termed n-type doping, and the resultant (extrinsic) semiconductor is termed n-type. By analogy, p-type semiconductors have $p \gg n$. The terms minority and majority carriers now become appropriate in these cases. For a p-doped semiconductor case, the Fermi level now lies close to the VB edge (Fig. 4c). The movement of E_F with dopant concentration can also be rationalized via the Nernst formalism [6].

Doping can be accomplished by adding altervalent impurities to the intrinsic semiconductor. For example, P (a Group 15 or VA element) will act as a donor in Si (a Group 14 or IVA element). This can be rationalized on chemical terms by noting that P needs only four valence electrons for tetrahedral bonding (as in the Si lattice) – the fifth electron is available for donation by each P atom. The donor density, N_D nominally is approximately 10^{17} cm^{-3}. Thus, assuming that $n \simeq N_D$ (complete ionization at 300 K), p will be only approximately 10^3 cm^{-3} [recall that the product $n_i p_i$ is $\sim 10^{20}$ cm^{-6} (see preceding section)], bearing out the earlier qualitative assertion that $n \gg p$.

Impurity addition, however, is not the only doping mechanism. Nonstoichiometry in compound semiconductors such as CdTe (Table 1) also gives rise to n- or p-type behavior, depending on whether Cd or Te is in slight excess, respectively. The defect chemistry in these solid chalcogenides controls their conductivity and doping in a complex manner, a discussion of which is beyond the scope of this section. Excellent treatises are available on this topic and on the solid-state chemistry of semiconductors in general [16–22].

The foregoing discussion strictly refers to semiconductors in single-crystal form. Amorphous and polycrystalline counterparts present other complications caused by the presence of defects, trap states, grain boundaries, and the like. For this reason we orient the subsequent discussion mainly toward single crystals, although comparisons with less ideal cases are made where appropriate. The distinction between metal and semiconductor electrodes is important when we consider the electrostatics across the corresponding solid–liquid interfaces; this distinction is made in the following section.

1.3 The Semiconductor–Electrolyte Interface at Equilibrium

1.3.1 The Equilibration Process

The electrochemical potential of electrons in a redox electrolyte is given by the Nernst expression

$$E_{redox} = E^{o}{}_{redox} + \frac{RT}{nF} \ln \left[\frac{c_{ox}}{c_{red}} \right] \quad (5)$$

In Eq. (5), c_{ox} and c_{red} are the concentrations (roughly activities) of the oxidized and reduced species, respectively, in the redox couple. The parameter ($E_{redox} = \mu_{e,redox}$) as defined by Eq. (5) can be identified with the Fermi level ($E_{F,redox}$) in the electrolyte. This was the topic of debate some years back [23], although this premise now appears to be well founded. The task now is to relate the electron energy levels in the solid and liquid phases on a common basis.

The semiconductor solid-state physics community has adopted the electron energy in vacuum as reference, whereas electrochemists have traditionally used the standard hydrogen electrode (SHE) scale. While estimates vary [23–25], SHE appears to lie at −4.5 eV with respect to the vacuum level. We are now in a position to relate the redox potential E_{redox} (as defined with reference to SHE) with the Fermi level $E_{F,redox}$ expressed versus the vacuum reference (Fig. 5a)

$$E_{F,redox} = -4.5 \text{ eV} - e_o E_{redox} \quad (6)$$

When a semiconductor is immersed in this redox electrolyte, the electrochemical potential (Fermi level) is disparate across the interface. Equilibration of this interface thus necessitates the flow of charge from one phase to the other and a "band bending" ensues within the semiconductor phase. The situation before and after contact of the two phases is illustrated in Fig. 5(b) and (c) for an n-type and p-type semiconductor, respectively. After contact, the net result of equilibration is that $E_F = E_{F,redox}$ and a "built-in" voltage, V_{SC} develops within the semiconductor phase, as illustrated in the right hand frames of Fig. 5(b) and (c).

It is instructive to further examine this equilibration process. Consider again an n-type semiconductor for illustrative purposes (Fig. 5b). The electronic charge needed for Fermi level equilibration in the semiconductor phase originates from the donor impurities (rather than from bonding electrons in the semiconductor lattice). Thus, the depletion layer that arises as a consequence within the semiconductor contains positive charges from these ionized donors. The Fermi level in the semiconductor ($E_{F,n}$) moves "down" and the process stops when the Fermi level is the same on either side of the interface. The rather substantial difference in the density of states on either side dictates that $E_{F,n}$ moves farther than the corresponding level, $E_{F,redox}$ in the electrolyte. A particularly lucid account of this initial charge transfer is contained in Ref. 6.

The band-bending phenomenon, shown in Fig. 5(b) and (c), is by no means unique to the semiconductor–electrolyte interface. Analogous electrostatic adjustments occur whenever two dissimilar phases are in contact (e.g. semiconductor-gas, semiconductor–metal). An important point of distinction from the corresponding metal case now becomes apparent. For a metal, the charge, and thus the associated potential drop, is concentrated at the surface penetrating at most a few Å into the interior. Stated differently, the high electronic conductivity of a metal cannot support an electric field

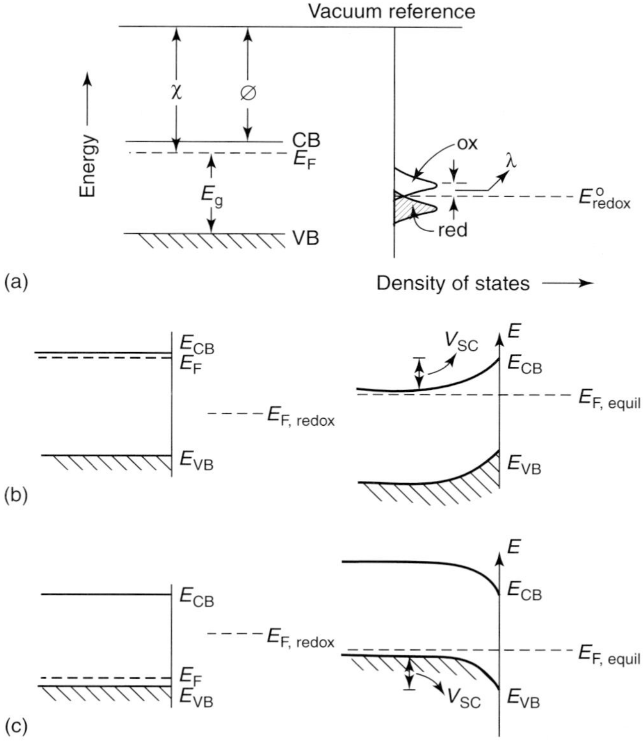

Fig. 5 (a) Energy levels in a semiconductor (left-hand side) and a redox electrolyte (right-hand side) shown on a common vacuum reference scale. χ and ϕ are the semiconductor electron affinity and work function, respectively. (b) The semiconductor-electrolyte interface before (LHS) and after (RHS) equilibration (i.e. contact of the two phases) shown for a *n*-type semiconductor. (c) As in (b) but for a *p*-type semiconductor.

within it. Thus, when a metal electrode comes into contact with an electrolyte, almost all the potential drop at the interface occurs within the Helmholtz region in the electrolyte phase. On the other hand, the interfacial potential drop across a semiconductor-electrolyte junction (see following) is partitioned both as V_{SC} and as V_H leading to a simple equivalent circuit model comprising two capacitors (C_{SC} and C_H) in series (Fig. 6). Further refinements of the equivalent circuit description are given later but the point to note is the rather variant behavior of a metal and a semiconductor at equilibrium with a redox electrolyte.

1.3.2
The Depletion Layer

There is a characteristic region within the semiconductor within which the charge would have been removed by the equilibration process. Beyond this boundary, the ionized donors (for a *n*-type semiconductor), have their compensating

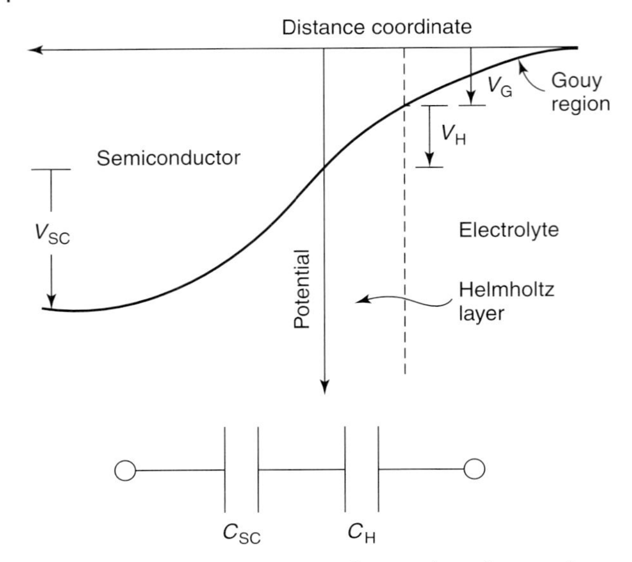

Fig. 6 Electrostatics at a semiconductor–electrolyte interface. A highly simplified equivalent circuit for the interface at equilibrium is shown at the bottom. The Gouy layer is neglected in the latter case (see text).

charge (electrons), and the semiconductor as a whole is electrically neutral. This layer is the space charge region or the depletion layer, so termed because the layer is depleted of the majority carriers. The potential distribution in this interfacial region can be quantified by relating the charge density and the electric field strength as embodied by the Poisson equation [14, 26]. Under restrictive conditions, more fully discussed in Refs. 6 and 14, we obtain a particularly simple expression

$$V(x) = -\left(\frac{e_o N_D}{2\varepsilon_s}\right) x^2 \quad \times (o \leq x \leq W) \tag{7}$$

In Eq. (7), e_o is the electronic charge and ε_s is the static dielectric constant of the semiconductor. The potential distribution is mapped in Fig. 6. We are now in a position to quantify the parameter V_{SC}:

$$V_{SC} = -\left(\frac{e_o N_D}{2\varepsilon_s}\right) W^2 \tag{8}$$

where W is the depletion layer width.

Further reflection shows how the magnitude of W should depend on the semiconductor parameter N_D. Consider two cases of a semiconductor, one that is lightly doped (say $N_D \sim 10^{16}$ cm^{-3}) and another that is heavily doped ($N_D \sim 10^{18}$ cm^{-3}). Obviously in the former case, the charge needed for Fermi level equilibration has to come from deeper into the solid and so the magnitude of W will be larger. This suggests a strategy for chemical control of the electrostatics at the semiconductor–electrolyte interface [6]. Nominal dimensions of W are in the 10–1000 nm range. This may be compared with the corresponding Helmholtz layer width, typically 0.4–0.6 nm. With the capacitor-in-series model (see earlier section), we can see that the semiconductor space charge layer is usually the determinant factor in the total capacity of the interface. Once again, the contrast with the corresponding metal–electrolyte interface is striking. Only when the semiconductor is degenerately doped (leading to rather large space charge layer charge, Q_{SC} and "thin" depletion layer widths) or when its surface is in accumulation does the situation become akin to the metal–electrolyte interface (see following).

1.3.3 Mapping of the Semiconductor Band-edge Positions Relative to Solution Redox Levels

Considerations of interfacial electron transfer require knowledge of the relative positions of the participating energy levels in the two (semiconductor and solution) phases. Models for redox energy levels in solution have been exhaustively treated elsewhere [27, 28]. Besides the Fermi level of the redox system (Eq. 6), the thermal fluctuation model [27, 28] leads to a Gaussian distribution of the energy levels for the occupied (reduced species) and the empty (oxidized species) states, respectively, as illustrated in Fig. 5(a). The distribution functions for the states are given by

$$D_{ox} = \exp\left[-\frac{E - E_{F,redox} - \lambda^2}{4\,kT\,\lambda}\right] \quad (9a)$$

$$D_{red} = \exp\left[-\frac{E - E_{F,redox} + \lambda^2}{4\,kT\,\lambda}\right] \quad (9b)$$

In Eqs. (9a) and (9b), λ is the solvent reorganization energy.

Now consider the relative disposition of these solution energy levels with respect to the semiconductor band edge positions at the interface. The total potential difference across this interface (Fig. 6) is given by

$$V_t = V_{SC} + V_H + V_G \quad (10)$$

In Eq. (10), V_t is the potential as measured between an ohmic contact on the rear surface of the semiconductor electrode and the reference electrode (Fig. 6). The problematic factors in placing the semiconductor and solution energy levels on a common basis involve V_H and V_G. In other words, theoretical predictions of the magnitude of V_{SC} (and how it changes as the redox couple is varied) are hampered by the lack of knowledge on the magnitude of V_H and V_G. A degree of simplification is afforded by employing relatively concentrated electrolytes such that V_G can be ignored.

As with metals, the Helmholtz layer is developed by adsorption of ions or molecules on the semiconductor surface, by oriented dipoles, or especially in the case of oxides, by the formation of surface bonds between the solid surface and species in solution. Recourse to band edge placement can be sought through differential capacitance measurements on the semiconductor–redox electrolyte interface [29].

In the simplest case as more fully discussed elsewhere [14, 15, 29], one obtains the Mott-Schottky relation (for the specific instance of a n-type semiconductor) of the semiconductor depletion layer capacitance (C_{SC}), again by invoking the Poisson equation

$$1/C_{SC}^2 = \frac{2}{N_D e_o \varepsilon_s}\left[(V - V_{fb}) - \frac{kT}{e_o}\right] \quad (11)$$

In Eq. (11), V_{fb} is the so-called flat band potential, that is the applied potential (V) at which the semiconductor energy bands are "flat", leading up to the solution junction. Several points with respect to the applicability of Eq. (11) must be noted.

The Mott-Schottky regime spans about 1 V in applied bias potential for most semiconductor–electrolyte interfaces (i.e.

in the region of depletion layer formation of the semiconductor space charge layer, see preceding section) [15]. The simple case considered here involves no mediator trap states or surface states at the interface such that the equivalent circuit of the interface essentially collapses to its most rudimentary form of C_{SC} in series with the bulk resistance of the semiconductor. Further, in all the earlier discussions, it is reiterated that the redox electrolyte is sufficiently concentrated that the potential drop across the Gouy (diffuse) layer in the solution can be neglected. Specific adsorption and other processes at the semiconductor–electrolyte interface will influence V_{fb}; these are discussed elsewhere [29, 30] as are anomalies related to the measurement process itself [31]. Figure 7 contains representative Mott-Schottky plots for both n- and p-type GaAs electrodes in an ambient temperature molten salt electrolyte [32].

Once V_{fb} is known (from measurements), the Fermi level of the semiconductor at the surface is defined. It is then a simple matter to place the energies corresponding to the conduction and VBs at the surface (E_{CB} and E_{VB}, respectively) if the relevant doping levels are known. The difference between E_{CB} and E_{VB} should approximately correspond to the semiconductor band gap energy, E_g (see Figs. 4 and 7). Alternatively, if V_{fb} is measured for one given state of doping of the semiconductor (n- or p-doped), the other band edge position can be fixed from

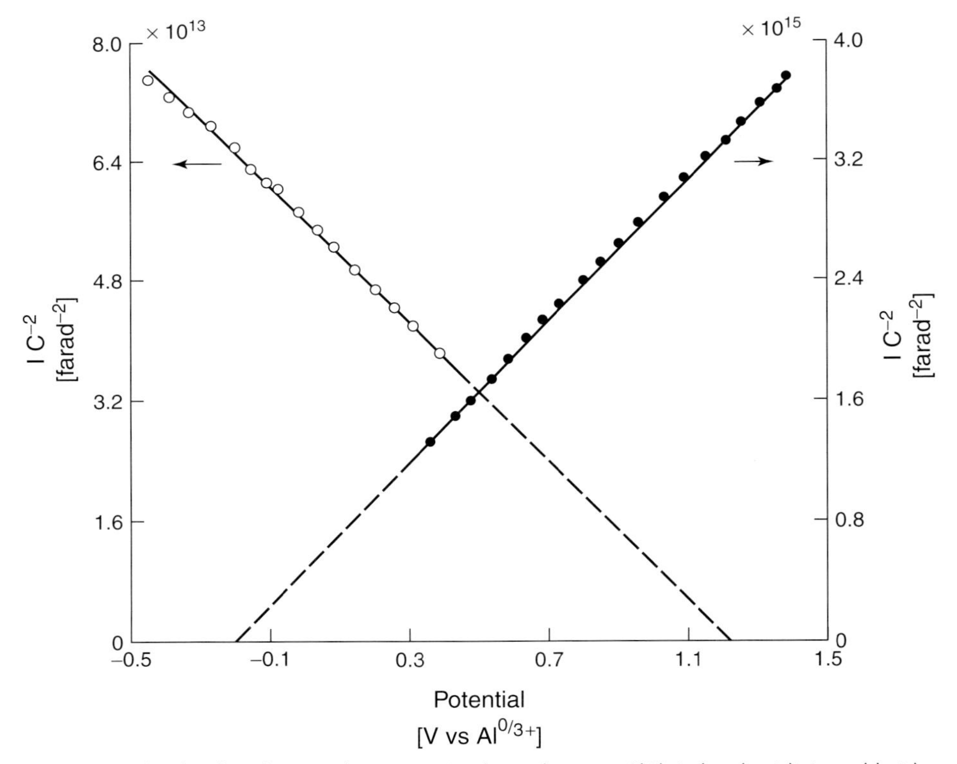

Fig. 7 Mott-Schottky plots for n- and p-type GaAs electrodes in an $AlCl_3$/n-butylpyridinium chloride molten-salt electrolyte. (Reproduced with permission from Ref. 32.)

knowledge of E_g. It is important to stress that the semiconductor surface band edge positions (as estimated from V_{fb} measurements) comprises all the terms in Eq. (10) and reflects the situation in situ for a given set of conditions (solution pH, redox concentration, etc.) of the semiconductor-redox electrolyte. The situation obviously becomes complex when the charge distribution and mediation at the interface changes either via surface states and illumination or both. These complications are considered later. Figure 8 contains the relative dispositions of the surface band edges mapped for a number of semiconductors in aqueous media.

Having located the semiconductor band edge positions (relative either to the vacuum reference or a standard reference electrode), we can also place the Fermi level of the redox system, $E_{F,redox}$, on the same diagram. Energy diagrams such as those in Fig. 8 are important in considerations of charge transfer as we shall see later. In anticipation of this discussion, it is apparent that the three situations illustrated in Fig. 9 for an n-type semiconductor–electrolyte interface entail the participation of the semiconductor CB, VB, and even states in its band gap in charge exchange with the solution species. Here again is a point of departure from the metal case; viz., for a semiconductor, hole, electron, and surface state pathways must all be considered.

Let us return to the band bending process at the interface. For a given semiconductor, the expectation is that as the redox Fermi level is moved more positive ("down" on the energy diagram),

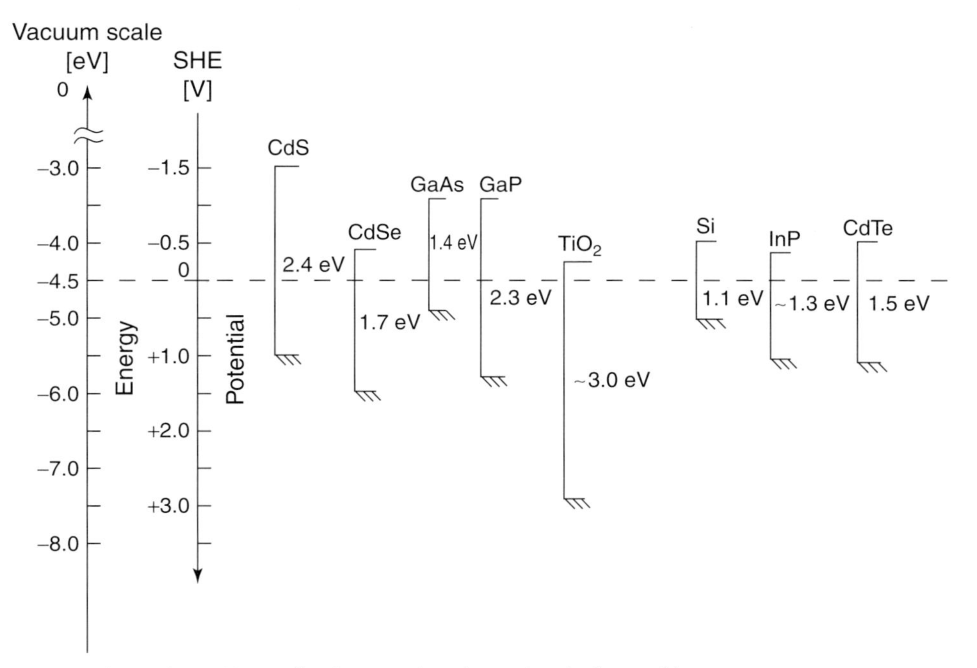

Fig. 8 Relative dispositions of various semiconductor band edge positions shown both on the vacuum scale and with respect to the SHE reference. These band edge positions are for an aqueous medium of pH ~1.

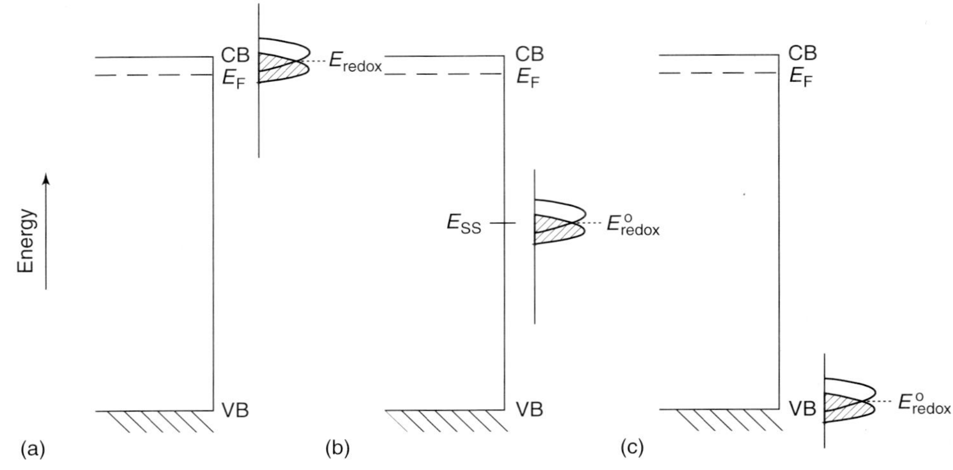

Fig. 9 Three situations for a *n*-type semiconductor–electrolyte interface at equilibrium showing overlap of the redox energy levels with the semiconductor CB (a), with surface states (b), and with the semiconductor VB (c). A discrete energy level is assumed for the surface states as a first approximation.

V_{SC} should increase concomitantly. This is the ideal (band edge "pinned") situation. In other words [23]

$$\frac{d\Delta V_{SC}}{dE_{redox}} = 1 \quad (12)$$

Equation (12) reflects the fact that the change in band bending faithfully tracks the redox potential change. A measure of the former is the open-circuit photopotential (see following). Figure 10 shows that this ideal situation indeed is realized for selected semiconductor–electrolyte interfaces [33]. As further discussed later, the analogy with the corresponding metal-semiconductor junctions (Schottky barriers) is direct [5, 34–36].

Complications arise when there are surface states that mediate charge exchange at the interface. When their density is

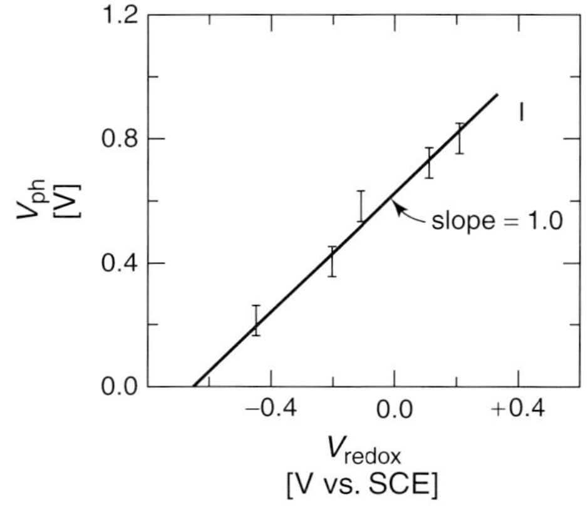

Fig. 10 Plot of the open-circuit photovoltage for amorphous Si-methanol interfaces containing a series of one-electron redox couples. (Reproduced with permission from Ref. 33.)

high [37], they act as a "buffer", in that in the extreme case, carriers in the semiconductor energy bands are completely excluded from the equilibration process.

1.3.4 Surface States and Other Complications

Surface states arise because of the abrupt termination of the crystal lattice at the surface; obviously the bonding arrangement is different from that in the bulk. Consider our prototype semiconductor, Si. The tetrahedral bonding characteristic of the bulk gives way to coordinative unsaturation of the bonds for the Si surface atoms. This unsaturation is relieved either by surface reconstruction or bonds with extraneous (e.g. solvent) species [29, 38–40]. The surface bonding results in a localized electronic structure for the surface that is different from the bulk. The energies of these localized surface orbitals nominally lie in the forbidden band gap region. The corresponding states are thus able (depending on their energy location) to exchange charge with the conduction or VBs of the semiconductor and/or the redox electrolyte [29].

Unlike the case illustrated in Fig. 10, changes in the solution redox potential have been observed to cause no change in the magnitude of V_{SC}. This situation is termed Fermi level pinning; in other words, the band edge positions are unpinned in these cases so that the movement of E_{redox} is accommodated by V_H rather than by V_{SC}. As mentioned earlier, it appears [37] that surface state densities as low as 10^{13} cm^{-2} ($\sim$1% of a monolayer) suffice to induce complete Fermi level pinning in certain cases. Of course, intermediate situations are possible. Thus, the ideal case manifests a slope of 1 in a plot of V_{SC} (or an equivalent parameter) versus E_{redox} (see Fig. 10). On the other hand, complete pinning results in a slope of zero. Intermediate cases of Fermi level exhibit slopes between 0 and 1 [41]. As stated earlier, there is a direct analogy here with the metal/semiconductor junction counterparts [42, 43]:

$$\phi_B = S\phi_m + \text{const} \qquad (13)$$

In Eq. (13), ϕ_B is the so-called Schottky barrier height, ϕ_m is the metal work function, and S is a dimensionless parameter. Attempts have been made to relate S to semiconductor properties [44–48].

To further complicate matters, the nonideal behavior of semiconductor–electrolyte interfaces as noted earlier is exacerbated when the latter are irradiated. Changes in the occupancy of these states cause further changes in V_H, so that the semiconductor surface band edge positions are different in the dark and under illumination. These complications are considered later. The surface states as considered earlier are shallow (with respect to the band edge positions) and can essentially be considered as completely ionized at room temperature. However, for many oxide semiconductors, the trap states may be deep and thus are only partially ionized. Specifically, they may be disposed with respect to the semiconductor Fermi level such that they are ionized only to a depth that is small relative to W [49]. The physical manifestation of such deep traps as observed in the AC impedance behavior of semiconductor–electrolyte interfaces has been discussed [14, 49].

Finally, within the Mott-Schottky approximation (Eq. 11), large values of ε_s or N_D can lead to the ratio V_H/V_{SC} becoming significant. Figure 11 contains estimates of this ratio for several values of N_D for a semiconductor with a large ε_s value (TiO_2, $\varepsilon_s = 173$) mapped as a function of the total

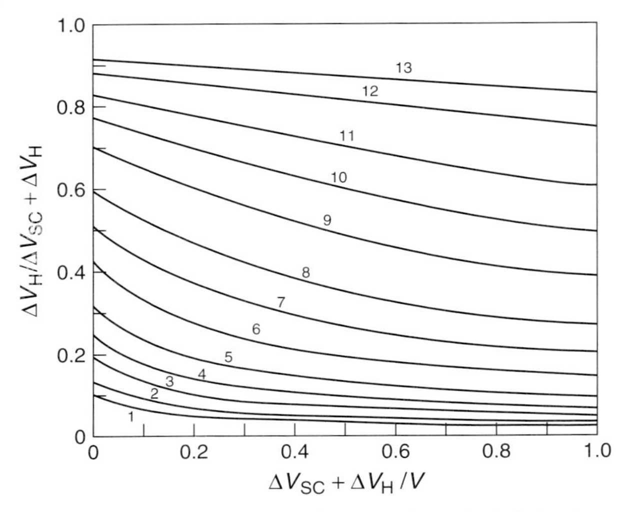

Fig. 11 The ratio of the potential drop in the Helmholtz to the total potential change computed as a function of the total potential change. A static dielectric constant of 173 (typical of TiO_2) and a Helmholtz capacitance of 10 μF/cm^2 were assumed and the doping density was allowed to vary from 10^{16} cm^{-3} (curve 1) to 10^{20} cm^{-3} (curve 13). (Reproduced with permission from Ref. 50.)

potential drop across the interface [50]. Clearly V_H can become a sizable fraction of the total potential drop (approaching the situation for metals) under certain conditions. It has been shown [51] that the Mott-Schottky plots will still be linear but the intercept on the potential axis is shifted from the V_{fb} value.

1.4
Charge Transfer Processes in the Dark

1.4.1
Current-potential Behavior

Let us return to the equilibrium situation of an n-type semiconductor in contact with a redox electrolyte and reconsider the situation in Fig. 9(a). This is shown again in Fig. 12(a) to underscore the fact that the interface is in a state of *dynamic* equilibrium. That is, the forward and reverse (partial) currents exactly balance each other and there is no *net* current flow across the interface. In fact, the situation here is similar to that occurring for a metal–redox electrolyte interface at the rest potential. We can write down expressions for the net current using a kinetics methodology as in Ref. 6 with some minor changes in notation:

$$i_c = -e_o A k_{et} c_{ox}(n_s - n_{so}) \qquad (14)$$

In Eq. (14), k_{et} is the rate constant for electron transfer, c_{ox} is the concentration of empty (acceptor) state in the redox electrolyte, n_s and n_{so} are the surface concentrations of electrons, the subscript "o" in the latter case denoting the equilibrium situation. Thus, as long as the semiconductor–electrolyte interface is not perturbed by an external (bias) potential, $n_s \equiv n_{so}$ and the net current is zero. The voltage

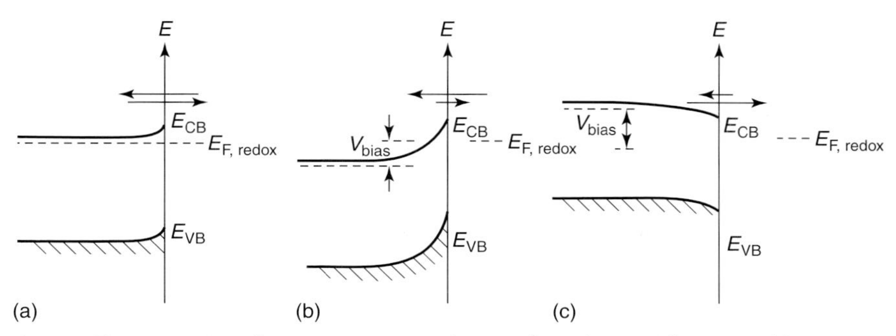

Fig. 12 Three situations for a *n*-type semiconductor–electrolyte interface at equilibrium (a), under reverse bias (b), and under forward bias (c). The size of the arrows denotes the magnitudes of the current in the two (i.e. anodic and cathodic) directions.

dependence of the current is embodied in the ratio, n_S/n_{SO}, which can be regarded as a measure of the extent to which the interface is driven away from equilibrium. It must be noted that n_{SO} is *not* the bulk concentration of majority carriers (n) in the semiconductor because of the potential drop across the space charge layer [6, 35].

$$n_{SO} = n \exp\left(-\frac{e_o V_{SC}}{kT}\right) \tag{15}$$

A few words about the units of the terms in Eq. (14) are in order at this juncture. The term i/e_oA may be regarded as a flux (J) in units of number of carriers crossing per unit area per second [1, 3, 8]. The concentration terms are in cm^{-3}; thus k_{et} has the dimensions of cm^4s^{-1} because of the second-order kinetics nature stemming from the two multiplied concentration terms in Eq. (14) [1, 3, 8].

Consider now the application of a bias potential to the interface. Intuitively when it is such that $n_S > n_{SO}$, a reduction current (cathodic current) should flow across the interface such that the oxidized redox species are converted to reduced species (Ox → Red). On the other hand, when $n_{SO} > n_S$, the current flow direction is reversed and an anodic current should flow. Once again the situation here is somewhat similar to the metal case. The major difference resides in the vastly different state densities in the solid and the existence of an energy gap region. The two nonequilibrium situations are shown in Figs. 12(b) and 12(c), respectively. Away from equilibrium, we have the analogous Boltzmann expression counterpart to Eq. (15)

$$n_S = n \exp\left[-\frac{e_o(V_{SC} + V)}{kT}\right] \tag{16}$$

leading, in turn, to

$$i_c = -e_o A k_{et} c_{ox} n_{SO} \times \left[\exp\left(-\frac{e_o V}{kT}\right) - 1\right] \tag{17}$$

The assumption is inherent in the preceding discussion that *all* of the applied bias (V) drops across the space charge layer such that we are modulating only the majority carrier population at the surface (and not the potential drop across the Helmholtz layer). In other words, the band edge positions are pinned or there is no Fermi level pinning (see Sect. 1.3.4).

Analogous expressions may be developed for majority carrier flow for a *p*-type semiconductor in contact with a redox electrolyte, with the important caveat that the VB is involved in this process instead.

Equation (17) suggests that the cathodic current is exponentially dependent on potential for $V < 0$. This is the so-called forward-bias regime. On the other hand, when $V > 0$ (reverse-bias regime) the current is essentially independent of potential and, importantly, is of opposite sign. Simply put, in this case, the electron flow direction (i.e. anodic) is from the occupied redox states into the semiconductor CB (Fig. 12c). It should not, thus, be surprising that this process is independent of potential. Both bias regimes are contained in curve 1 in Fig. 13.

Of particular interest to this discussion is the "preexponential" term in Eq. (17):

$$i_o = e_o A k_{et} c_{ox} n_{so} \qquad (18)$$

Analogous to the metal case, we can call this term the exchange current; it is the current that flows at equilibrium when the partial cathodic and anodic components exactly balance one another. Of particular interest is the dependence of i_o on n_{so}. Also, variations in c_{ox} will affect the magnitude of i_o. Both these trends can be readily rationalized. Finally, i_o will increase with doping because of the "thinness" of the resultant barrier at the surface.

When $E_{F,redox}$ is moved "down," that is more positive, the band bending increases, V_{SC} increases and thus n_{so} decreases. A similar alteration in c_{ox} affects $E_{F,redox}$ through the Nernst expression. In both instances, we are influencing the Fermi level at the interface at equilibrium. Thus, in a sense, i_o is a quantitative measure of the extent of rectification of a given interface; that is, a smaller i_o value translates to better rectification.

The reverse bias current remains at a very low value because of the lack of minority carriers (i.e. holes for n-type semiconductor) in the dark. Alternatively, injection of electrons from occupied redox levels (also an anodic current) has to thermally surmount the surface barrier [5, 34, 35]. Under extreme reverse bias, however, this barrier becomes "thin" and electrons can tunnel through it, leading to an abrupt increase in the anodic current. This process was studied even in the early days of

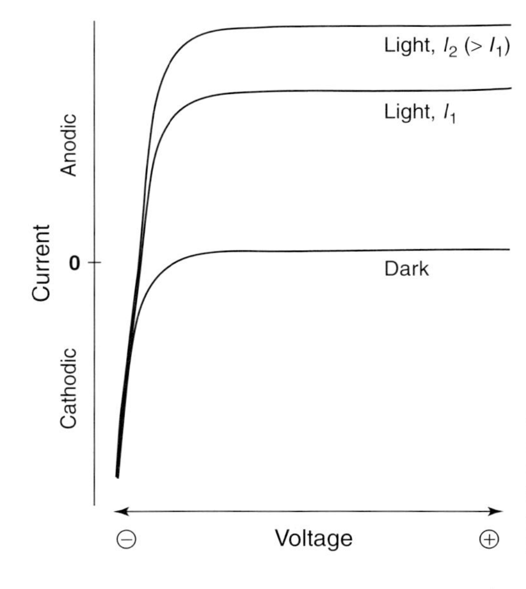

Fig. 13 Current-potential curves for a n-type semiconductor in the dark (curve 1) and under band gap illumination (curves 2 and 3). Two levels of photon fluxes are shown in the latter case.

semiconductor electrochemistry [52] and a detailed discussion is found in a book chapter [14]. Ultimately the junction "breaks down" (at the so-called Zener limit). This dark current flow is not shown in Fig. 13 (curve 1).

Returning to the forward-bias (cathodic) current flow, Eq. (17) bears some analogy to the famous Tafel expressions in electrochemical kinetics. Thus, ignoring the unity term within the square brackets, Eq. (17) predicts a Tafel slope of 60 mV per decade at 298 K. In many instances [53, 54], such a slope indeed is observed. In many cases, however, the slopes are higher than the "ideal" value [14, 55–59].

The causes for this anomalous behavior are still not fully understood. It appears likely that many factors are involved: surface film formation, varying potential drop across the Helmholtz region caused, for example, by surface state charging, and so on. Even crystallographic orientations appear to be important [59]. These aspects have been discussed by other authors [14, 55, 60].

We have so far considered only (majority carrier) processes involving the CB (again assuming for illustrative purposes a n-type semiconductor). Consider the interfacial situation depicted in Fig. 9(c). The energy states of the redox system now overlap with the VB of the semiconductor such that *hole injection* in the "dark" is possible. When the band bending is large, the injected holes remain at the surface and attack the semiconductor itself, causing the latter to undergo corrosion. If the bias potential is such that the band bending is modest and the holes recombine with electrons (either via the surface states or in the space charge region itself), a cathodic current flows that is carried by the majority carriers in the bulk. This recombination current pathway is schematized in Fig. 14 and is further discussed in the next section. Hole injection has been extensively studied especially for III–V (Group 13–15) semiconductors such as GaAs and GaP because of the relevance of this process to electroless etching and device fabrication technology. This topic has been reviewed [61–64].

The invokement of either the CB or the VB of the semiconductor in charge exchange in the dark with solution redox species is not always straightforward. This is particularly true for multielectron redox processes to be discussed later. Movement of the semiconductor band edge positions (i.e. band edge unpinning) relative to the redox energy levels also presents a complicating situation (see following). Some cases (e.g. $Eu^{2+/3+}$ in contact with GaAs electrodes) are interesting in that the same

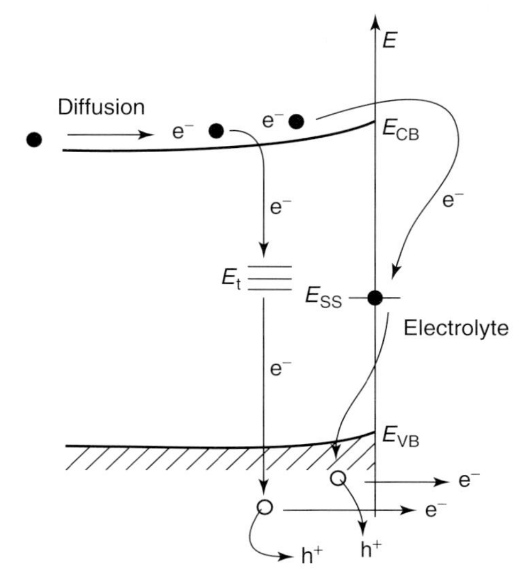

Fig. 14 Hole injection into the VB of a n-type semiconductor from an oxidant (e.g. Fe^{3+}) and the injection or recombination pathway. Both surface state–mediated and depletion layer trap–mediated routes are shown for the recombination.

couple can interact with both bands [55, 65]. Thus, the oxidation of Eu^{2+} is a VB process (occurring at *p*-GaAs but not at *n*-GaAs in the dark), whereas the reduction of Eu^{3+} (a facile process that reportedly occurs at rates close to the thermionic emission limit, Ref. 55) is mediated by CB electrons [65]. The $[Cr(CN)_6]^{3-/4-}$ redox system behaves in a similar manner with respect to GaAs [66].

Electroluminescence, (EL), is a versatile probe for studying such carrier injection processes. Thus, hole injection into the VB of a *n*-type semiconductor leads to cathodic EL, whereas electron injection into the CB of a *p*-type semiconductor leads to anodic EL [67]. Examples of studies of cathodic EL are commonplace [68–70]; however, anodic EL is not very common because the energy requirement for the redox couple has a very negative redox potential. Nonetheless, anodic EL has been reported for the *p*-InP-$[Cr(CN)_6]^{4-}$ interface [71]. Radical intermediates can also cause EL as discussed later for multielectron redox processes. EL is treated in more depth in another chapter.

This finally brings us to the comparability of the current-potential behavior of *n*-type and *p*-type samples of a given semiconductor. It may be noted that for a redox process occurring via one of the bands (e.g. VB), the cathodic currents (electron transfer from VB to Ox) are expected to be equal for *n*- and *p*-type materials. This idea has been pursued using the so-called quasi-Fermi level concept [55, 72, 73]. This model has been demonstrated quantitatively by studying the anodic decomposition of GaAs and the oxidation of redox species such as Cu^{+} and Fe^{2+} at *n*- and *p*-type GaAs electrodes [72, 73].

1.4.2
Dark Processes Mediated by Surface States or by Space Charge Layer Recombination

Surface states were considered earlier (Sect. 1.3.4) from an electrostatic perspective. Here we examine their dynamic consequences. There are two principal charge transfer routes involving surface states. Consider again an *n*-type semiconductor; the forward-bias current can either involve direct exchange of electrons between the semiconductor CB and Ox states in solution (Fig. 12b) or can be mediated by surface states (Fig. 15). The second

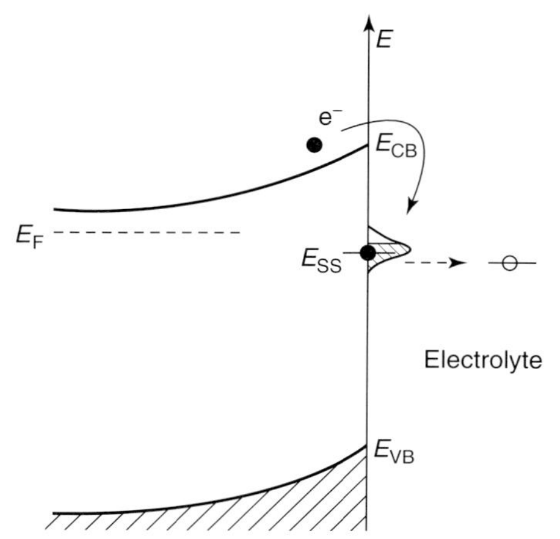

Fig. 15 Surface state–mediated electron injection from the CB of a *n*-type semiconductor into the electrolyte.

route involves hole injection into the semiconductor VB again from Ox states in solution (Figs. 9c and 14). The recombination current is mediated either by surface states or via space charge layer recombination. We will consider first the CB process.

Initial evidence for the intermediacy of surface states came from dark current measurements on n-TiO_2 and n-$SrTiO_3$ in the presence of oxidizing agents such as $[Fe(CN]_6{}^{3-}$, Fe^{3+}, and $[IrCl_6]^{2-}$ [74, 75]. Similar early evidence that the charge transfer process was more complex than direct transfer of electrons from the semiconductor CB also came from AC impedance spectroscopy measurements on n-ZnO, n-CdS, and n-CdSe in contact with $[Fe(CN)_6]^{3-}$ species [76, 77].

The electrochemical impedance for surface state–mediated charge transfer has been computed recently [78]. The key results are summarized in Fig. 16. Figure 16(a) contains the proposed equivalent circuit for the process and features a parallel connection of the impedance for the Faradaic process $[Z_F(\omega)]$ (ω = angular frequency, $2\pi f$) and the capacitance of the semiconductor depletion layer, C_{SC}. The

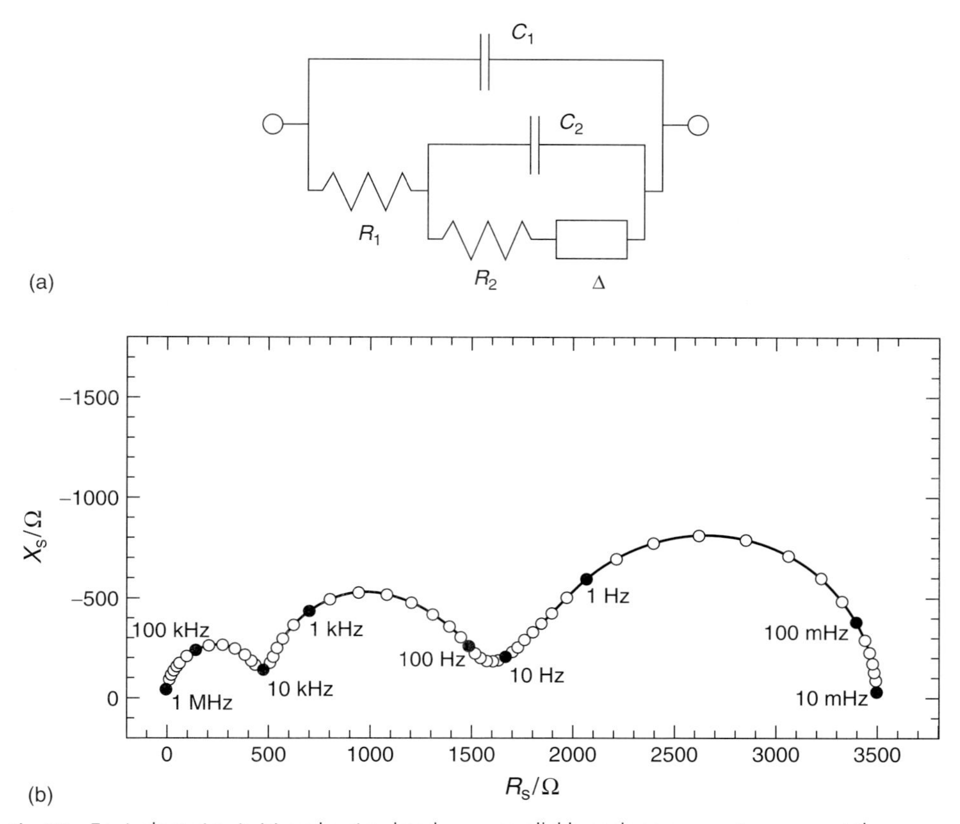

Fig. 16 Equivalent circuit (a) and a simulated Nyquist plot (b) for the charge-transfer pathway illustrated in Fig. 15. The capacitance C_1 represents that of the space charge layer and the parallel branch components represent the Faradaic charge-transfer process. Refer to the original work for further details. (Reproduced with permission from Ref. 78.)

former also involves a diffusion impedance (Δ) of the Warburg type (see following). The complex plane (Nyquist) plot predicted for the circuit is illustrated in Fig. 16(b). The theoretically predicted AC impedance response was compared with experiments on *n*-GaAs rotating disk electrodes in sulfuric acid media [79]. The equivalent circuit in Fig. 16(a) was also compared with previous versions proposed by other authors [80–83]. These alternate versions differ in their assumption of no variations in potential drop across the Helmholtz region (i.e. infinite C_H) and no concentration polarization in the electrolyte phase (infinite diffusion coefficient for the redox species). Also discussed is the application of AC impedance spectroscopy for studying the kinetic reversibility of majority carrier charge transfer via the CB of a *n*-type semiconductor [82].

AC impedance spectroscopy also has seen extensive utility in the study of the hole injection or recombination process depicted in Fig. 14. An equivalent circuit for this process is illustrated in Fig. 17; it does resemble the circuit in Fig. 16(a), except for the Warburg component [84]. Early studies [85–88] utilized the recombination resistance parameter, R_r, that was extracted from model fits of the measured AC impedance data. This parameter was seen to be inversely related to the hole injection current, thus signifying that it is indeed related to the recombination process. However, the challenge is to differentiate whether recombination is mediated via surface states or whether it occurs in the depletion layer. Thus, the parameter R_r alone cannot afford this information and both the real and the imaginary parts of this additional impedance must be considered. Subsequent studies have addressed this aspect [85, 89–93]. The admittance corresponding to recombination at the surface [92] and in the space charge layer [93] was calculated from first principles. These computations show that the recombination *capacitance* increases monotonically with decreasing band bending in the latter case, whereas it shows a peak in the former case as a function of potential.

Experimental studies [91] show that in the case of *n*-GaAs electrodes in contact with Ce^{4+} as the hole injection agent, surface recombination prevails. On the other hand, with *n*-GaP electrodes, recombination in the depletion layer must also be taken into account. Other discussions of the use of AC impedance spectroscopy for the study of hole injection or recombination are contained in Refs. 78 and 84.

The consequences of potential drop variations across the Helmholtz layer in the hole injection process have been examined by a variety of techniques [94, 95]. For example, chemical reaction of the GaAs surface with iodine results in a downward shift of the semiconductor band edge positions such that the reduction of iodine is mediated by CB electrons [95]. When sufficient negative charge accumulates at

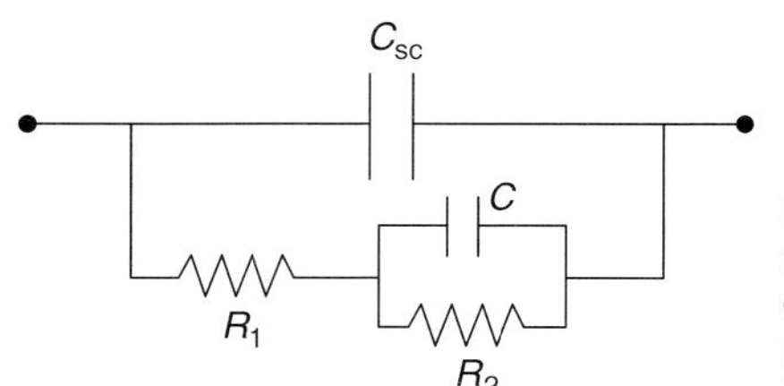

Fig. 17 Equivalent circuit representation of the injection or recombination process. (Reproduced with permission from Ref. 84.)

the surface, the potential is redistributed between the semiconductor spacecharge layer and the Helmholtz region. Now iodine is reduced by hole injection as gauged by EL and AC impedance measurements [95].

1.4.3 Rate-limiting Steps in Charge Transfer Processes in the Dark

The assumption is implicit in the discussion in Sect. 1.4.1 (leading to Eq. 18) that charge transfer kinetics at the semiconductor–electrolyte interface is the rate-limiting step. Fundamentally, we have to differentiate majority carrier *capture* and minority carrier *injection* processes in the dark. In the former case, transit through the semiconductor itself or charge exchange with the surface states can be potentially rate-limiting. In the latter case, there are three steps involved: hole injection into the semiconductor VB, charge exchange between the recombination center and the semiconductor CB, and diffusion of majority carriers (electrons) from the neutral region. Finally, mass transport processes in the electrolyte phase itself can be a limiting factor in the overall current flow. We shall examine carrier capture and injection processes in turn.

The vast majority of outer-sphere, nonadsorbing redox systems to date have yielded values for k_{et} in the 10^{-17} – 10^{-16} $cm^4\ s^{-1}$ range [3, 8]. These include n-Si-CH_3OH [96, 97], n-InP-CH_3OH [98], $GaInP_2$-coated n-GaAs-acetonitrile [99], and p-GaAs-HCl [100] interfaces. The redox couples in these studies have mostly involved metallocenes that show low proclivity to adsorb on the semiconductor surface. In these cases, the rate-determining step in the overall current flow undoubtedly lies in the electron transfer event at the interface itself. However, values for k_{et} approximately three orders of magnitude higher have also been reported for similar interfaces, namely, n-GaAs-acetonitrile-containing cobaltocenium $[Co(Cp)_2{}^+]$ acceptors [99, 101]. Similarly, high values were reported for p-GaAs-acetonitrile interfaces with ferricenium and cobaltocenium redox species [102]. In these latter cases, alternative mechanisms (e.g. thermionic emission, see following) must be invoked in a rate-limiting role. Quartz crystal microbalance measurements have yielded evidence for adsorption of redox species (and consequently high *local* substrate concentration) in some of these "anomalous" instances [101].

In the majority carrier capture process, if the interfacial charge transfer kinetics are facile, the transport of majority carriers through the space charge region can play a rate-limiting role. The thermionic emission theory [34] assumes that every electron that reaches the semiconductor surface, and has the appropriate energy to overcome the potential barrier there, will cross the interface with a tunnel probability of unity. However, if the interfacial kinetics are sluggish, some of the electrons will be reflected at the interface. In this case, the exchange current i_o is no longer described by Eq. (18) but by Eq. (19) [34]

$$i_o = AA^* \left(\frac{m^*}{m_e}\right) T^2 \left(\frac{n_s}{n}\right) \tag{19}$$

In Eq. (19), A is the electrode area, A^* is the Richardson constant (120 A K^{-2} cm^{-2}), m^*/m_e is the relative effective electron mass in the CB, and T is the absolute temperature.

In many of the reported instances [53, 55, 103], the current calculated from Eq. (19) is much higher than that measured experimentally, signaling that interfacial charge-transfer kinetics are limiting the overall rate. On the other hand, in the n-GaAs-acetonitrile-$Co(Cp)_2{}^+$ case [101], AC impedance spectroscopy data appear to support the assumption that thermionic emission is the current-limiting transport mechanism.

Another factor that enters into this discussion is the mobility of the majority carriers. It has been argued [14] that in the case of low mobility materials (e.g. $\mu_n \sim 1\ cm^2\ V^{-1} s^{-1}$), carrier transport from the semiconductor bulk to the interface itself can become limiting. Clearly a multitude of factors are important in majority carrier capture: k_{et}, acceptor concentration in the electrolyte and carrier mobility.

What about the minority carrier injection process depicted in Fig. 14? Here, contrasting with the process considered earlier, the hole injection step is usually very fast (see following). Then the current is limited by diffusion or recombination described by the Shockley equation [104]

$$i_o = \frac{e_o A D_p n_i^2}{n L_p} \tag{20a}$$

for bulk recombination, and

$$i_o = 0.5\, e_o A W \sigma \nu_{th} N_t n_i \tag{20b}$$

for recombination within the semiconductor space charge region. In Eqs. (20a) and (20b), D_p is the diffusion coefficient for holes, L_p is the hole diffusion length, n_i is the intrinsic carrier concentration, σ is an average capture cross-section for electrons and holes, ν_{th} is the thermal velocity of charge carriers, and N_t is the areal density of recombination (trap) centers in the middle of the energy gap (Fig. 14).

The diffusion or recombination mechanism results in considerable overpotential for (cathodic) current flow in the dark (again assuming a n-type semiconductor for illustration). Such a rate-limiting process was found to describe the charge transfer at n-GaAs in 6 M HCl containing Cu^+ as the hole injecting species [55, 73].

Whatever the limiting mechanism, ultimately the current becomes limited by concentration polarization, that is, by the transport of redox species from the bulk electrolyte to the semiconductor surface. The situation in this regard is no different from that at metal electrode–electrolyte interfaces. As in the latter case, hydrodynamic voltammetry is best suited to study mass transport. AC impedance spectroscopy can be another useful tool in this regard [105].

In the former case, the data can be processed via Levich plots of current vs. $\omega^{1/2}$ (ω = angular frequency). If processes other than solution mass transport become rate-limiting, then the plot will show a curvature and the current will even become independent of the electrode rotation rate. In this case, inverse Levich (or Koutecky-Levich) plots of $1/i$ vs. $\omega^{-1/2}$ can be used for further analyses. Such analyses have been done, for example, for n-GaAs-acetonitrile-$Co(Cp)_2{}^+$ interfaces [101] and n- and p-GaAs electrodes in contact with the $Cu^{+/2+}$ redox couple in HCl electrolyte [55, 73].

The diffusion impedance at semiconductor electrodes has been considered recently [105]. This author described the applicability of AC impedance spectroscopy for the study of electron capture and hole injection processes at n-GaAs-H_2O/C_2H_5OH-methyl viologen, p-InP-aq. KOH-$Fe(CN)_6{}^3$, n-GaAs-H_2SO_4-Ce^{4+}, and n-InP-aq. KOH-$Fe(CN)_6{}^{3-}$ interfaces. In the case of electron capture processes, a Randles-like equivalent circuit was found to be applicable [105]. On the other hand, no Warburg component was present in the hole injection case when the reverse reaction was negligible (Fig. 17). For a nonideal semiconductor–electrolyte contact (see Sect. 1.3.4), a Warburg impedance appeared in the electrochemical impedance of an injection reaction as well, as exemplified by the n-InP-$Fe(CN)_6^{3-}$ case [105].

1.5 Light Absorption by the Semiconductor Electrode and Carrier Collection

1.5.1 Light Absorption and Carrier Generation

The optical band gap of the semiconductor (Sect. 2) is an important parameter in defining its light absorption behavior. In this quantized process, an electron-hole pair is generated in the semiconductor when a photon of energy $h\nu$ (ν = frequency and $h\nu > E_g$) is absorbed. Optical excitation thus results in a delocalized electron in the CB, leaving behind a delocalized hole in the VB; this is the band-to-band transition. Such transitions are of two types: direct and indirect. In the former, momentum is conserved and the top of VB and the bottom of CB are both located at $k = 0$ (k is the electron wave vector). The absorption coefficient (α) for such transitions is given by [106]

$$\alpha = A'(h\nu - E_g)^{1/2} \qquad (21)$$

In Eq. (21), A' is a proportionality constant. Indirect transitions involve phonon modes; in this case Eq. (21) takes the form

$$\alpha = A'(h\nu - E_g)^2 \qquad (22)$$

A given material can exhibit a direct or indirect band-to-band transition depending on its crystal structure. For example, Si single crystals have an indirect transition located at 1.1 eV (Table 1). On the other hand, amorphous Si is characterized by a direct optical transition with a larger E_g value (shorter wavelengths). Both types of transitions can also be seen in the same material, for example GaP [107].

Within the present context, the important point to note is that the absorption depths (given by $1/\alpha$) are vastly different for direct and indirect transitions. While in the former case absorption depths span the 100–1000 nm range, they can be as large as 10^4 nm for an indirect transition [9].

Optical transitions in semiconductors can also involve localized states in the band gap. These become particularly important for semiconductors in nanocrystalline form (see following). Sub–band gap transitions can be probed with photons of energy below the threshold defined by E_g.

1.5.2 Carrier Collection

The number of carriers collected (in an external circuit, for example) versus those optically generated defines the quantum yield (Φ) – a parameter of considerable interest to photochemists. The difficulty here is to quantify the amount of light actually absorbed by the semiconductor as the cell walls, the electrolyte, and other components of the assembly are all capable of either absorbing or scattering some of the incident light. Unfortunately, this problem has not been comprehensively tackled, unlike in the situation with photocatalytic reactors involving semiconductor particulate suspensions, where such analyses are available [108–111]. Pending these, an *effective* quantum yield can still be defined.

Returning to the carrier collection problem, consider Fig. 18 for an n-semiconductor–electrolyte interface. As can be seen, the electron-hole pairs are optically generated, both in the field-free and in the space charge regions within the semiconductor. Recombination of these

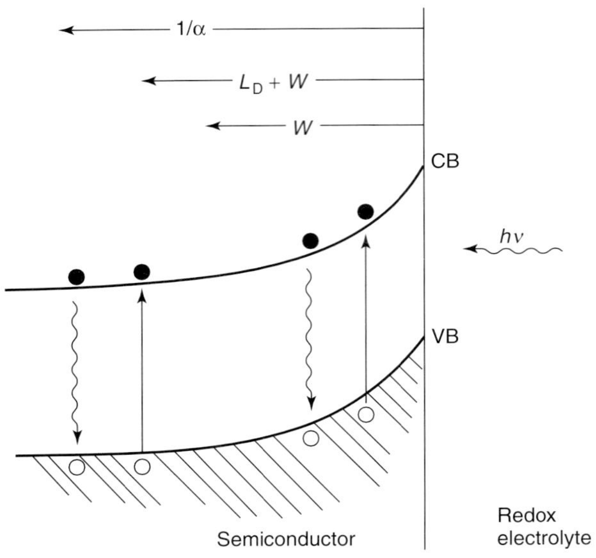

Fig. 18 Photogeneration of electron-hole pairs in the field-free region and depletion layer for a *n*-type semiconductor–electrolyte interface. The characteristic regions defined by the depletion layer (W), Debye length (L_D), and the light penetration depth ($1/\alpha$) are also compared.

carriers must be considered in the bulk, in the space charge layer, and on the semiconductor surface (the latter in contact with the redox electrolyte). We are assuming here that light is incident from the electrolyte side. Rear illumination geometry can be profitably employed and is considered later for the nanocrystalline film case.

The direction of the electric field at the interface (Fig. 6, Sect. 1.3.2) is such that the minority carriers (holes in this case) are swept to the surface and the electrons are driven to the rear ohmic contact. How fast the holes are drained away (by Faradaic reactions involving the redox electrolyte) will dictate how the Fermi levels compare with the equilibrium situation discussed earlier. The departure from equilibrium has been quantified in terms of the quasi-Fermi level concept discussed later.

The extent of collection of minority carriers from the region beyond the depletion layer is dictated by the diffusion process. A diffusion length, L, can be defined

$$L_p = \sqrt{D_p \tau_p} = \sqrt{kT \mu_p \tau_p} \quad \text{(for } n\text{-type semiconductor)} \qquad (23)$$

The subscripts in Eq. (23) remind us that we are dealing with minority carrier collection; μ_p is the hole mobility and τ_p is the hole life-time. The characteristic length L_p defines the region within which electron-hole pair generation is fully effective. Pairs generated at depths longer than the Debye length, L_D ($L_D = W + L$) will simply recombine. Thus, the effective quantum yield for a given interface will depend on the relative magnitudes of L_D and the light penetration depth, $1/\alpha$ (Fig. 18) [112, 113].

An expression for the flux of photogenerated minority carriers arriving at the surface was originally given for a solid-state junction [114] and subsequently adapted to semiconductor-liquid junctions [115]. The major weakness of these early models hinges on their underlying assumption for the boundary condition that the surface concentration of minority carriers is zero. As pointed out elsewhere [14], this is a demanding condition necessitating very high magnitudes for the interfacial charge transfer rate constant, k_{et} (see previous section). A modicum of improvement to the basic Gärtner model was found [116] by defining a flux rather than a concentration expression for the holes and a characteristic length where the bulk diffusion current transitions into a drift current. However, this treatment still assumes that every hole entering the depletion layer edge exits this region and out into the electrolyte phase.

The Gärtner equation [114] can be written in normalized form [113]

$$\Phi \equiv \frac{j_{ph}}{I_o} = 1 - \frac{\exp(-\alpha W)}{1 + \alpha L_p} \qquad (24)$$

In Eq. (24), Φ is the effective quantum yield (see previous section), given by the ratio of the photocurrent density (i_{ph}/A), j_{ph}, to the incident light flux, I_o. Recasting Eq. (24) in the form

$$-\ln(1 - \Phi) = \ln(1 + \alpha L_p) + \alpha W \qquad (25)$$

and recalling that W is proportional to ${V_{SC}}^{1/2}$ (Eq. 8), and $V_{SC} = V - V_{fb}$, a test of the rudimentary model would lie in a plot of the LHS of Eq. (25) against $(V_{fb} - V)^{1/2}$. Such plots are shown in Fig. 19 at four selected wavelengths for the p-GaP-H_2SO_4 electrolyte interface [117].

While adherence to the Gärtner model is satisfactory for large values of V_{SC} (i.e. large band bending, see Fig. 6), the model fails close to the flat band situation. Interestingly, this problem is exacerbated as the semiconductor excitation wavelength becomes shorter (Fig. 19). Thus, another weakness of the basic Gärtner model [114, 115] is the neglect of surface recombination. At the flat band situation, this model still predicts finite current flow arising from the diffusive flow of minority carriers toward and out of the interface (Fig. 20).

A variety of refinements have been made to take into account the surface recombination effect [117–132]. The earliest of these [119, 120, 123] involve some simplifying assumptions:

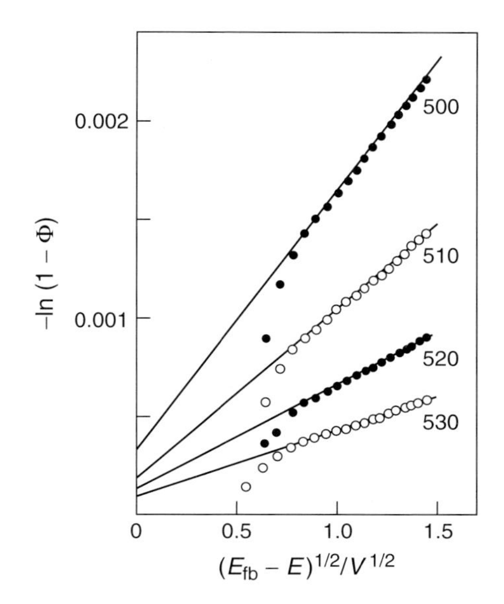

Fig. 19 Gärtner plots (see Eq. 25) for the p-GaP −0.5 M H_2SO_4 interface. The numbers on the plots refer to the excitation wavelength; E_{fb} is the flat band potential and E is the bias potential. (Note that this notation is different from that employed in the text.) (Reproduced with permission from Ref. 117.)

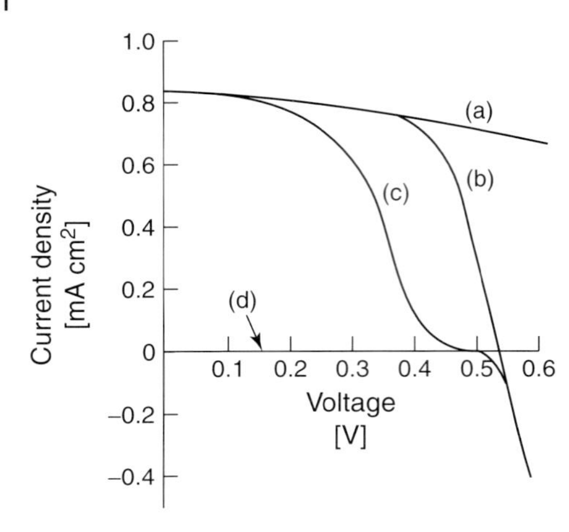

Fig. 20 Comparison of calculated current voltage profiles in the dark (curve d) and under illumination (curves a–c). Curve a is obtained from the basic Gärtner model. Curve (b) considers surface recombination and curve (c) considers both surface recombination and recombination in the space charge layer. These simulations are for a *n*-type semiconductor–electrolyte interface. (Reproduced with permission from Ref. 138.)

1. There is no recombination in the depletion layer. That is, all the holes optically generated in the bulk and within the depletion layer (Fig. 18) are swept to the surface without loss;
2. The steady state concentration of the optically generated minority carriers does not perturb the potential distribution in the dark (Fig. 6); and
3. There is a quasi-thermodynamic distribution of minority carriers within the depletion layer. This translates to a constant product term np across this region.

Surface recombination in the vast majority of these treatments invoke the Hall-Shockley-Read model [133, 134]. Defining the Gärtner limiting expression (Eq. 24) as Φ_G, we obtain [14]

$$\Phi_{ss} = \Phi_G \Bigg/ \left(1 + \frac{D_p \exp V_{SC}}{L_D(k_t + S_t)}\right) \quad (26)$$

In Eq. (26), we have two new parameters, k_t and S_t. These are the first-order rate constant for hole transfer (units of cm s^{-1}, see following) and the surface recombination velocity, S_t. In the combined situation of high L_D, high k_t, and very low (or zero) S_t, Eq. (26) collapses to the Gärtner limit.

At this juncture, it is worth noting that only one trap state at the surface has been assumed till now; further it is assumed that this surface state functions both as a carrier recombination site and as a charge-transfer pathway (Fig. 21). Both these assumptions are open to criticism. An alternative model invoking two distinct types of surface states – one active in recombination and the other capable of mediating charge transfer – has been considered [135]. Nonetheless, the most serious flaw in the above treatments lie in the neglect of carrier recombination in the depletion layer itself (as distinct from recombination at the surface). Reexamination of Eqs. (24) and (26) shows that the larger the Debye length, L_D, and the depletion layer width, W, the higher the quantum yield. However, recombination in the space charge layer must become significant at some value of W, thus providing a further limit to carrier collection.

Recombination within the space charge region is a nontrivial problem to treat from a computational perspective [14]. The methodology of Sah, Noyce, and Shockley [136] has been used by several authors [126, 127, 128, 131, 138–140]. Figure 20 illustrates the sensitivity of the

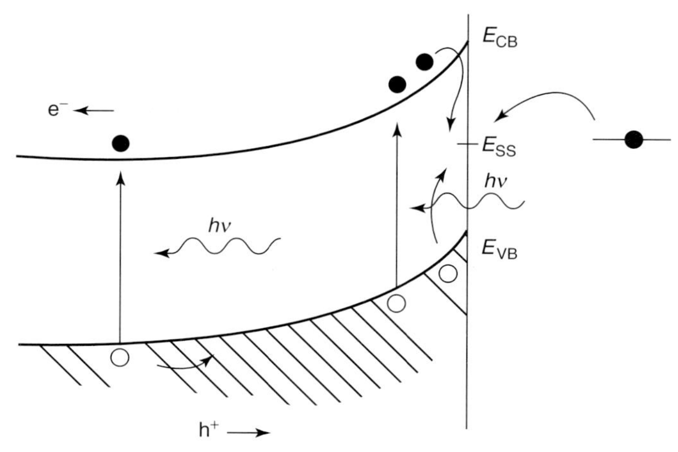

Fig. 21 Surface state mediation of both minority carrier (i.e. hole) transfer and recombination for a *n*-type semiconductor electrolyte interface.

current-potential profiles at the semiconductor–electrolyte interface to this recombination mode [137].

Other models taking the above nonidealities to varying extent have been proposed; a detailed discussion of these lies beyond the scope of this section [141–147]. However, it is worth noting here that in some instances involving high-quality semiconductor–electrolyte interfaces the rate-determining recombination step does indeed appear to lie in the bulk semiconductor [1, 148]. Silicon photoelectrodes in methanolic media containing fast, one-electron, outer-sphere redox couples were studied in these cases.

1.5.3 Photocurrent-potential Behavior

The current-voltage characteristics of an illuminated semiconductor electrode in contact with a redox electrolyte can be obtained by simply adding together the majority and minority current components. The majority carrier component is given by the diode equation (Eq. 17) while the minority carrier current (i_{ph}) is directly proportional to the photon flux (Eq. 24). Thus, the net current is given by:

$$i = i_{ph} - i_o \left[\exp\left(-\frac{e_o V}{kT}\right) - 1\right] \quad (27)$$

The minus sign in Eq. (27) underscores the fact that the majority carrier component flows opposite (or "bucks") the minority carrier current flow. This photocurrent component is shown as curves 2 and 3 in Fig. 13.

Equation (27) shows that the diode equation is offset by the i_{ph} term; this is exactly what is seen in Fig. 13. The plateau photocurrent is proportional to the photon flux, I_o, as illustrated for two different values of the incident light intensity in Fig. 13. At the open-circuit condition of the interface, $i = 0$ (and neglecting the unity term within the square brackets relative to the exponential quantity), Eq. (27) leads to

$$V_{oc} \simeq \frac{kT}{e_o} \ln \frac{i_{ph}}{i_o} \quad (28)$$

Equation (28) underlines two important trends: First, V_{oc} increases logarithmically with the photon flux (with a slope of ~60 mV at 298 K). Second, V_{oc} decreases with an increase of i_o (again logarithmically). This underlines the importance of ensuring that the majority carriers do not "leak" through the interface. Because of the diode nature of the interface, from a device perspective, the semiconductor surface must be designed to have fast minority carrier transfer kinetics (and thus high i_{ph}), but must be blocking to the flow of majority carriers (from the CB for a n-type semiconductor) into the redox electrolyte. This challenge is similar to what the solid-state device physicists face, but relative to metals (with a high density of acceptor states), *chemical* control of redox electrolytes offers a powerful route to performance optimization of liquid-based interfaces as also pointed out by previous authors [1, 6, 8, 149–154].

Referring back to Fig. 13, the current-potential curves under illumination of the semiconductor simply appear shifted "up" relative to the dark $i - V$ counterpart. This, however, is the ideal scenario. Anomalous photoeffects (APEs) are often observed that manifest as a cross-over of light and dark current-voltage curves, as illustrated in Fig. 22. Thus, the superposition principle [149–154] is not obeyed in this instance. The dashed line in Fig. 22 is produced by translating the photocurrent-voltage data by j_{SC}, the short-circuit current density. If the superposition principle is held, this dashed curve would have overlaid the dark current-voltage curve. Thus, this "excess" (forward bias) current embodies the APE, and the failure of superposition is quantified as the voltage difference (ΔV) between the dark $j - V$ data and the dashed line.

What are the ramifications of the cross-over or the APE? First, mathematical modeling of carrier transport in a junction-based solar photovoltaic system, according to

$$j = j_{SC} - j_{bk}(V) \tag{27a}$$

is not valid in the presence of this effect. (In Eq. (27a), j_{bk} is the "bucking" current density in the forward-bias regime, see previous section.) That is, a fully linearized system of differential equations and boundary conditions cannot be used to model the interface carrier transport. Second, computation and modeling of the open-circuit voltage for such devices by

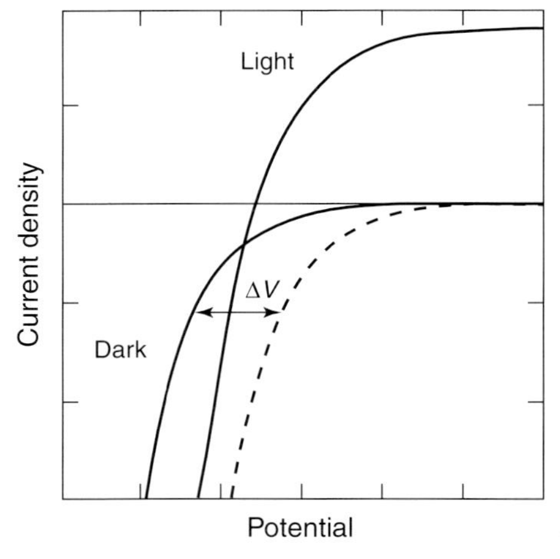

Fig. 22 Anomalous photoeffect (APE) showing cross-over of the dark and light current-voltage curves again for a n-type semiconductor-based interface. The dashed line is obtained as described in the text.

simply equating a constant photocurrent flux, j_{ph} against the dark (recombination) current, j_o is no longer possible (see Eq. (28) and the accompanying earlier discussion).

Third, and perhaps practically of most significance, the ΔV component represents a loss pathway in the photovoltage deliverable by the given device. Thus, the (open-circuit) photovoltage is V_{oc} instead of $V_{oc} + \Delta V$ in the ideal case in the absence of the APE. Therefore, it is important to quantify and understand the molecular and chemical origins of this effect. This has not been done so far, at least to this writer's knowledge, for semiconductor–electrolyte interfaces.

Of course, the cross-over effect is not confined to such interfaces. It is interesting that a recent textbook [155], dealing primarily with solid-state solar cell devices, makes only a fleeting reference to the underlying origin of the APE. Reference was made in this book to the cross-over of experimental dark and light $j - V$ characteristics for a Cu_2S-CdS solid-state heterojunction solar cell but its origin was not explored. A light-induced junction modification has also been reported for the (Cd, Zn)S-$CuInSe_2$ solid-state system [156, 157]. The cross-over effect appears to have been treated in even lesser depth in some classical textbooks on semiconductor devices [104, 155].

Probably the first reported instance of observation of an APE was in 1977 for a n-TiO_2-NaOH electrolyte interface [158]. The APE was observed in the saturation region of cathodic current flow and was induced by sub–band gap irradiation of the photoanode. A peak in the spectrum of the photoresponse at 800 nm (the corresponding photon energy being lower than the 3.0 eV band gap of TiO_2) was used by the authors to invoke a surface state–mediated electron transfer to O_2 (in the electrolyte) as the origin of the photoeffect. Surface states were again invoked to explain a cathodic photoeffect at n-CdS-aqueous polysulfide interfaces [159]. This photoeffect was only observed for the (0001) single-crystal face of n-CdS but not for the (0001) orientation. A subsequent study of photoelectrochemical effects at selenium films reported an anomalous *anodic* photocurrent at potentials positive of the flat band location for the p-type film [160]. This effect was assigned to a hole injection process via a tunneling mechanism. An increase of the tunneling probability under illumination was accommodated by a shrinking of the space charge layer at the interface. Photoenhancement of the forward current flow was observed again for n-CdS, in this instance in contact with a $[Fe(CN)_6]^{3-/4-}$ redox electrolyte [161]. This effect was observed only with a mechanically damaged surface, and disappeared after it had been etched with concentrated HCl.

Subsequent work [162] describes suppression of the cathodic photocurrent for n-CdS-$[Fe(CN)_6]^{3-/4-}$ interfaces mechanical polishing of the electrode. As with an earlier study [163], the spectral dependence of the photocathodic effect implicated sub–band gap states. The suppression was explained by two alternative models involving a compensated insulating layer or by Fermi-level pinning. Illumination was claimed to result in a dramatic increase of the (suppressed) cathodic current, which interestingly enough was observed only for n-CdS films but not for crystals including n-CdTe, n-CdSe, n-GaAs, n-ZnO, n-TiO_2 and n-ZnSe. On the other hand, a subsequent paper describes a photocathodic effect for n-CdSe-sulfide interfaces in which an interfacial layer of selenium was implicated [163].

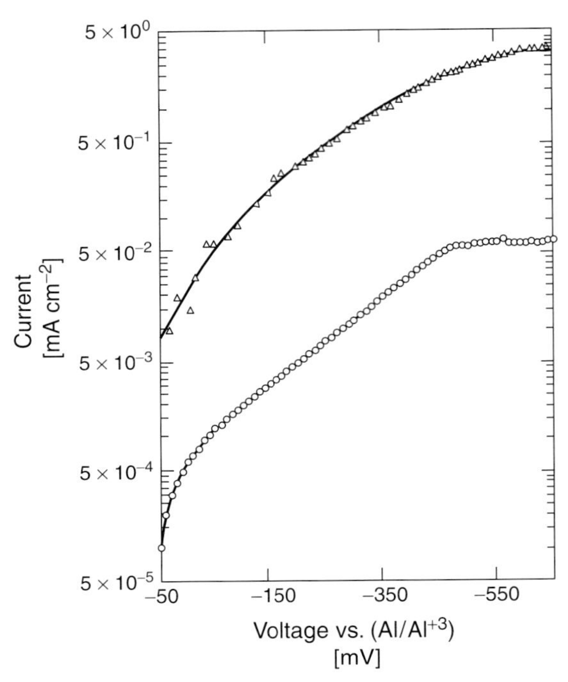

Fig. 23 Experimental data embodying APE for the n-GaAs-$AlCl_3$/n-butylpyridinium chloride molten-salt electrolyte interface. Refer to the text for details. (Reproduced with permission from Ref. 167.)

More recent studies on n-CdS [164, 165] and n-TiO_2 photoanodes [166] implicate the formation of photoconductive layers in the APEs. Thus, the foregoing review suggests the following

1. APE is a very general phenomenon that has been observed for solid-state junctions for n- and p-type semiconductors alike, and for a wide variety of semiconductor materials.
2. The reported results and trends are often contradictory. It is quite possible that the experimental conditions in these studies were quite variant, thus precluding direct comparison of the results.
3. The mechanistic reasons given for the APE are possibly many, and generalizations may not be warranted. Clearly, more research is needed on this topic.

Figure 23 contains an example of the APE for the n-GaAs-$AlCl_3$/n-butylpyridinium chloride molten salt electrolyte interface [167]. The bottom curve in Fig. 23 is the measured dark current-voltage profile. The top curve was obtained from the photocurrent-voltage data (under irradiation of the semiconductor). Clearly, if the superposition principle held, the two curves would have coincided with one another.

APEs have also been observed for nanocrystalline and chemically modified films, as discussed in a subsequent section.

Light-induced changes in the electrostatics at the semiconductor–electrolyte interface are conveniently probed by capacitance-voltage measurements in the dark and under illumination of the semiconductor electrode. If charge trapping at the interface plays a decisive role (whatever be the mechanism), the voltage drop across the illuminated interface is altered, and consequently the semiconductor band edge positions are shifted. This, of course, is the Fermi-level pinning situation that was encountered earlier (Sect. 1.3.4). Examples of studies addressing this aspect may be found, for example on $p-$GaAs [124] and CdTe [168, 169]–based aqueous electrolyte interfaces.

1.5.4
Dynamics of Photoinduced Charge Transfer

So far the discussion has centered on the steady state aspects of carrier generation and collection at semiconductor–electrolyte interfaces. As with their metal electrode counterparts, a wealth of information can be gleaned from perturbation-response type of measurements. An important difference, however, lies in the vastly different timescale windows that are accessible in the two cases. The critical RC time-constant of the cell in a transient experiment is given by

$$\tau_{\mathrm{cell}} = C(R_{\mathrm{m}} + R_{\mathrm{el}}) \qquad (29)$$

In Eq. (29), R_{m} is the measurement resistor (across which the current or photocurrent is measured) and R_{el} is the electrolyte resistance. The term C is the capacitance, which, in the metal case, is the Helmholtz layer capacitance, C_{H}. (Once again, the Gouy region is ignored here.) For semiconductor–electrolyte interfaces, we have seen that two layers are involved in a series circuit configuration with corresponding capacitances of C_{SC} and C_{H} (Fig. 6). Because $C_{\mathrm{H}} \gg C_{\mathrm{SC}}$, $C \simeq C_{\mathrm{SC}}$. This assumption is usually justified because $C_{\mathrm{H}} \simeq 10^{-5}$ F cm^{-2} and $C_{\mathrm{SC}} = 10^{-8} - 10^{-0}$ F cm^{-2}. If the composite resistance $(R_{\mathrm{m}} + R_{\mathrm{el}})$ is 100 ohm, then τ_{cell} for metal electrodes is $\sim 10^{-3}$ s and that for the semiconductor case is $10^{-6} - 10^{-7}$ s.

What are the processes important in a dynamic interrogation of the semiconductor–electrolyte interface?

1. Carrier generation within the semiconductor,
2. diffusion of minority carriers from the field-free region to the space charge layer edge,
3. transit through this layer,
4. charge transfer across the interface, and
5. carrier recombination via surface states or via traps in the space charge layer.

Other phenomena such as thermalization also are important as discussed later in the context of hot carrier effects. The time constant (τ_{cell}) of the cell and the measurement circuitry has complicated matters further and have caused some confusion in the interpretation of transient data. If a potentiostat is not used, then this time constant is given by Eq. (29).

One can envision three types of perturbation: an infinitesimally narrow light pulse (a Dirac or δ-functional), a rectangular pulse (characteristic of chopped or interrupted irradiation), or periodic (usually sinusoidal) excitation. All three types of excitation and the corresponding responses have been treated on a common platform using the Laplace transform approach and transfer functions [170]. These perturbations refer to the temporal behavior adopted for the excitation light. However, classical AC impedance spectroscopy methods employing periodic potential excitation can be combined with steady state irradiation (the so-called PEIS experiment). In the extreme case, both the light intensity and potential can be modulated (at different frequencies) and the (nonlinear) response can be measured at sum and difference frequencies. The response parameters measured in all these cases are many but include the photocurrent, voltage, luminescence, or microwave conductivity. Clearly, semiconductor–electrolyte interfaces present a rich, albeit demanding landscape for probing non–steady state phenomena.

The dynamics of charge transfer across semiconductor–electrolyte interfaces are considered in more detail elsewhere in this volume.

1.5.5 Hot Carrier Transfer

Short wavelength photons (of energy much greater than E_g) create "hot" carriers. If, somehow, thermalization of these carriers can be avoided, photoelectrochemical reactions that would otherwise be impossible with the "cooled" counterparts, that is, at very negative potentials for n-type semiconductors, would be an intriguing possibility. The key issue here is whether the rate of electron transfer across the interface can exceed the rate of hot electron cooling. The observation of hot carrier effects at semiconductor–electrolyte interfaces is a controversial matter [3, 7, 11, 171] and practical difficulties include problems with band edge movement at the interface and the like [4]. Under certain circumstances (e.g. quantum-well electrodes, oxide film-covered metallic electrodes), it has been claimed that hot carrier transfer can indeed be sustained across the semiconductor–electrolyte interface [7, 172, 173].

1.6 Multielectron Photoprocesses

This section has thus far considered redox electrolytes comprising one electron oxidizing or reducing agents. Multielectron redox processes, however, are important in a variety of scenarios. Consider the reduction of protons to H_2 (HER) – a technologically important electrochemical process that has also been extensively studied from a mechanistic perspective on metallic electrodes.

Photoelectrolytic processes such as HER can be carried out on semiconductor electrodes. One can envision a HER mechanism on a p-type semiconductor of the sort:

$$p-\mathrm{SC} \xrightarrow{h\nu} \mathrm{e}^-_{\mathrm{CB}} + h^+_{\mathrm{VB}} \quad (30)$$

$$\mathrm{e}^-_{\mathrm{CB}} + \mathrm{S} \longrightarrow \mathrm{S}^- \quad (30a)$$

$$\mathrm{S}^- + \mathrm{H}^+ \longrightarrow \mathrm{S} + \mathrm{H}^\bullet \quad (30b)$$

$$\mathrm{e}^-_{\mathrm{CB}} + \mathrm{H}^+ \longrightarrow \mathrm{H}^\bullet \quad (30c)$$

$$\mathrm{H}^+ + \mathrm{H}^\bullet + \mathrm{e}^-_{\mathrm{CB}} \longrightarrow \mathrm{H}_2 \quad (30d)$$

$$\mathrm{H}^+ + \mathrm{H}^\bullet + \mathrm{S}^- \longrightarrow \mathrm{H}_2 + \mathrm{S} \quad (30e)$$

$$\mathrm{h}^+_{\mathrm{VB}} + \mathrm{H}^\bullet \longrightarrow \mathrm{H}^+ \quad (30f)$$

$$\mathrm{h}^+_{\mathrm{VB}} + \mathrm{S}^- \longrightarrow \mathrm{S} \quad (30g)$$

In this above scheme, S denotes a surface state and both direct (Reactions 30c and 30d) and indirect (i.e. surface state–mediated) (Reactions 30b and 30e) radical and H_2-generating pathways are shown. Reactions 30f and 30g represent recombination routes involving the reaction intermediates.

Admittedly, this scheme is daunting in its complexity, and the kinetics implications are as yet unclear. Early studies on p-GaP, p-GaAs, and other Group III–V (13–15) semiconductors reported onset of cathodic photocurrents (attributable to HER) only at potentials far removed (ca. 0.6 V) from V_{fb} [174]. This was attributed to Steps 30b and 30g in the preceding scheme. More recent work [175] has shown that the HER at illuminated p-InP-electrolyte contacts is accompanied by a photocorrosion reaction, leading to indium formation on the semiconductor surface.

Interestingly, surface states themselves were chemically identified with $H^{\bullet}_{ads}$ (adsorbed hydrogen atom intermediates) in the aforementioned study [166]. These species have also been implicated in accumulation layer formation and anodic EL at *n*- and *p*-GaAs-electrolyte interfaces [176–178].

Another interesting characteristic of many multiequivalent redox systems is the phenomenon of photocurrent multiplication. This phenomenon may be illustrated for two systems utilizing illuminated *n*-type and *p*-type semiconductors respectively

n-type

$$n\text{-SC} \xrightarrow{h\nu} e^- + h^+ \quad (31)$$

$$HCOOH + h^+ \longrightarrow COOH^{\bullet} + H^+ \quad (31a)$$

$$COOH^{\bullet} \longrightarrow CO_2 + H^+ + e^- \quad (31b)$$

p-type

$$p\text{-SC} \xrightarrow{h\nu} h^+ + e^- \quad (31)$$

$$O_2 + H^+ + e^- \longrightarrow HO_2^{\bullet} \quad (31c)$$

$$HO_2^{\bullet} + H^+ \longrightarrow H_2O_2 + h^+ \quad (31d)$$

Thus, the key feature of photocurrent multiplication is a majority carrier injection step (Reactions 31b or 31d) from a reaction intermediate (usually a free radical) into the semiconductor CB or VB, respectively. In the preceding examples, each photon generates two carriers in the external circuit, affording a quantum yield (in the ideal case) of 2. This is the "current-doubling" process.

Practically, however, quantum yields somewhat lower than 2 are usually measured because Steps 31b or 31d compete with the further photooxidation or photoreduction of these intermediates, respectively. This is true especially at high photon flux values. Even multiplication factors as high as 4 are possible as in the photodissolution of *n*-Si in NH_4F media [179–182].

Photocurrent multiplication has been observed for a variety of semiconductors including Ge [180], Si [179–182], ZnO [183–189], TiO_2 [190–193], CdS [194, 195], GaP [196], InP [197], and GaAs [198–200]. These studies have included both *n*- and *p*-type semiconductors, and have spanned a range of substrates, both organic and inorganic. As in the Si case, this phenomenon can also be caused by photodissolution reactions involving the semiconductor itself. The earlier studies have mainly employed voltammetry, particularly hydrodynamic voltammetry [193]. As more recent examples [2, 9, 10] reveal, intensity-modulated photocurrent spectroscopy, (IMPS), is also a powerful technique for the study of photocurrent multiplication.

This leads us to another important category of multielectron photoprocesses involving the semiconductor itself. While photocorrosion is a nuisance from a device operation perspective, it is an important component of a device fabrication sequence in the microelectronics industry. Two types of wet etching of semiconductors can be envisioned [201]. Both occur at open-circuit but one involves the action of chemical agents that cause the simultaneous rupture and formation of bonds. Several aspects of photoetching have been reviewed [62–64, 202, 203] including reaction mechanisms, morphology of the etched surfaces, and etching kinetics in the dark and under illumination. General rules for the design of anisotropic photoetching systems have also been formulated [204]. Photoelectrochemical etching is considered in more detail elsewhere in this volume.

1.7 Nanocrystalline Semiconductor Films and Size Quantization

1.7.1 Introductory Remarks

From a materials perspective, the field of semiconductor electrochemistry and photoelectrochemistry has evolved from the use of semiconductor single crystals to polycrystalline thin films and, more recently, to nanocrystalline films. The latter have been variously termed membranes, nanoporous or nanophase films, mesoporous films, nanostructured films, and so on; they are distinguished from their polycrystalline electrode predecessors by the crystallite size (nm versus μm in the former) and by their permeability to the electrolyte phase. These films are referred to as "nanocrystalline in the following sections. These features render three-dimensional geometry to nanocrystalline films as opposed to the "flat" or two-dimensional (planar) nature of single crystal or polycrystalline counterparts.

What are the virtues of these emerging photoelectrode materials? The first is related to their enormous surface area. Consider that the 3D structure is built up of close-packed spheres of radius, r. Then ignoring the void space, the specific area, A_S (area/volume) is given by $3/r$ [205]. For $r = 10$ nm, A_S is on the order of 10^6 cm^{-1}, and for a 1 cm^2 film of 1 μm thickness, this value corresponds to an *internal* surface area of ~100 cm^2 (i.e. a "surface roughness factor" of 100). Clearly, this becomes important if we want the electrolyte redox species to be adsorbed on the electrode surface (see following). Alternatively, a large amount of sensitization dye can be adsorbed onto the support semiconductor although this dye sensitization approach is not considered in this chapter. By way of contrast, the amount of species that can be confined in a monolayer on a corresponding flat surface would be negligibly small.

As we shall see later, electron transport in nanocrystalline films is necessarily accompanied by charge-compensating cations because the holes are rapidly injected into the flooded electrolyte phase. This provides opportunities for studying ion transport processes in mesoporous media that are coupled to electron motion. Ion insertion also has practical consequences as in energy storage device applications [206].

Surface state densities on the order of ~10^{12} cm^2 are commonplace for semiconductor electrodes of the sort considered in previous sections of this chapter. These translate to equivalent volume densities of ~10^{18} cm^{-3} for nanocrystalline films. Such high densities enhance light absorption by trapped electrons in surface states, giving rise to photochromic and electrochromic effects [198–200] (see following). Unusually high photocurrent quantum yields are also observed with sub–band gap light with these photoelectrode materials. Corresponding sub–band gap phenomena are rather weak and difficult to detect with single-crystal counterparts.

1.7.2 The Nanocrystalline Film–Electrolyte Interface and Charge Storage Behavior in the Dark

Understanding of the electrostatics across nanocrystalline semiconductor film-electrolyte junctions presents interesting challenges, particularly from a theoretical perspective. Concepts related to space charge layers, band bending, flat band potential, and the like (Sect. 3) are not applicable here because the crystallite dimensions comprising these layers are comparable to (or even smaller than) nominal depletion layer widths.

The rather complete interpenetration of the electrode and electrolyte phases must mean that the Helmholtz double layer extends throughout the interior surface of the nanoporous network, much like a supercapacitor [9, 210] situation. Finally, unlike in the cases treated earlier, the semiconductor (especially if it is a metal oxide) is not heavily doped such that free majority carriers are not present in appreciable amounts. This is indicated, for example, by the sensitivity of the conductivity of nanocrystalline TiO_2 layers to UV light – the conductivity is strongly enhanced on UV exposure, similar to a photoconductive effect. This effect has been interpreted, in terms of trap filling with recombination times considerably slower than the trapping processes under reverse bias [211, 212]. The light sensitivity also is diagnostic of the fact that the low electronic conductivity in the dark is not due to high interparticle resistances (i.e. in the "neck" regions), but rather is indicative of the low electron concentrations.

The electron concentration can be increased by forward-biasing the nanocrystalline electrode–electrolyte interface potentiostatically. The interface is driven thus into the accumulation regime for the majority carriers, and if a transparent rear contact (e.g. F-doped, SnO_2 or Sn-doped indium oxide) is used, the resultant blue (or bluish-black) coloration of the film can be spectroscopically monitored [208, 209, 213]. Whether the optical response arises from CB electrons or from electrons trapped in surface states is not entirely clear. It has been claimed [214] that the absorption spectrum of the latter differs significantly from CB electrons. Electrons in surface states can be chemically identified with Ti^{3+} defect sites that can be detected, for example, by electron paramagnetic resonance spectroscopy [215, 216].

In either case, the resultant negative charge generated by electron accumulation at the internal surfaces has to be balanced by cations (from the electrolyte phase) for charge compensation. Such ion insertion reactions have been studied using techniques such as voltammetry, reflectance or absorption spectroscopy, chronoamperometry, and electrochemical quartz-crystal microgravimetry [213, 217–22]. Both aqueous and aprotic electrolytes have been deployed for these studies.

Unlike in the single-crystal cases treated earlier, placement of the semiconductor energy band positions at the interface via Mott-Schottky analyses is not straightforward for nanocrystalline films. Abrupt changes in slope and other nonidealities [215, 227, 223] have been observed, for example, in the Mott-Schottky plots for TiO_2 films and attributed to the influence of the conductive glass that is normally employed to support these films. This behavior is especially prevalent under reverse bias. The onset of majority carrier optical absorption (in the visible and near IR range) under forward-bias instead has been profitably employed to place the CB positions of TiO_2 in aqueous media [208].

Impedance spectroscopy and electrochemical dye desorption experiments have been employed [224] to study the electrical characteristics of TiO_2 nanocrystalline films in the dark. This study as well as the others cited earlier demonstrate how the conductivity changes (as a result of electron injection from the support electrode) can cause the porous or nanocrystalline layer to manifest itself electrically, such that the active region moves away (i.e. outward) from the support as the forward-bias voltage is increased. The potential distribution has also been analyzed depending on whether the depletion layer width exceeds or is smaller than the typical dimension of the structural units in the nanocrystalline network [223].

1.7.3
Photoexcitation and Carrier Collection: Steady State Behavior

Figure 24 contains a schematic representation of the nanocrystalline semiconductor film–electrolyte interface at equilibrium (Fig. 24a) and the corresponding situation under band gap irradiation of the semiconductor (Fig. 24b) [9]. Because the diffusion length of the photogenerated carriers is usually larger than the physical dimensions of the structural units, holes and electrons can reach the impregnated electrolyte phase before they are lost via bulk recombination. This contrasts the situation with the single-crystal cases discussed earlier.

If, as is the case with TiO_2 nanocrystalline films, the holes are rapidly scavenged by the electrolyte redox (specifically Red) species, collection of the photogenerated electrons at the rear contact becomes the determinant factor in the quantum yield. Thus, the quasi-Fermi level for holes remains close to $E_{F,redox}$ and that for electrons, $E_{F,e}$ moves "up", as depicted in Fig. 24b. Illumination thus induces an electron flux, $J_n(x)$ through the nanocrystalline phase. Under steady state conditions, the photocurrent density (j_{ph}) is equal to $e_o J_n(x = d)$. The driving force for electron diffusion through the network of nanocrystallites has been calculated from first principles [225]. It has been found that the driving force is approximately kT/e_o divided by the thickness of the network. Importantly, this free-energy gradient is found to be independent of the incident photon flux.

It is important to reiterate that the charge separation in a nanocrystalline semiconductor–electrolyte interface does not depend on a built-in electric field at the junction as in the single-crystal

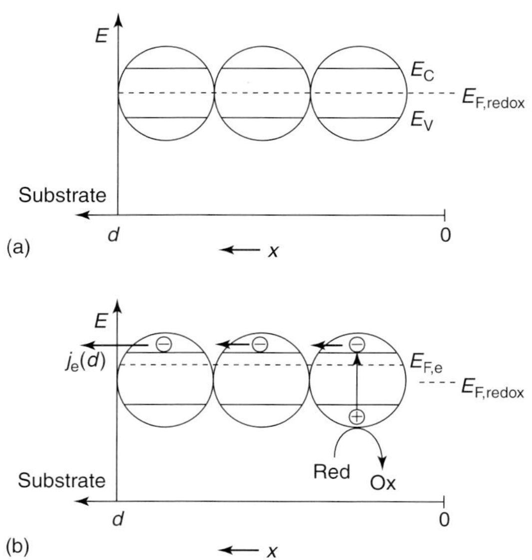

Fig. 24 Schematic representation of a nanocrystalline semiconductor–electrolyte interface in the dark (a) and under illumination from the electrolyte side (b). E_c and E_v correspond to E_{CB} and E_{VB} in our notation. (Reproduced with permission from the authors of Ref. 9.)

case. Instead, the differential kinetics for the reactions of photogenerated electrons and holes with electrolyte redox species account for the charge separation (and the generated photovoltage). The molecular factors underlying the sluggish scavenging of electrons at the nanocrystalline film-electrolyte boundary (by the redox species) are as yet unclear. Clearly, the competition between *surface* recombination of these electrons (with the photogenerated holes) and collection at the rear contact dictates the magnitude of the quantum yield that is experimentally measured for a particular junction.

Photocurrent losses have been recorded for electrolytes dosed with electron acceptors such as O_2 and iodine [226]. Nanocrystalline TiO_2 electrodes with thicknesses ranging from 2 μm to 38 μm were included in this study. In the presence of these electron-capture agents, electron collection (i.e. photocurrent) at the rear contact was seriously compromised. On the other hand, as high as 10% of the photons were converted to current for a 38 μm thick film in a N_2-purged solution [226].

The result was obtained with front-side illumination geometry. As one would intuitively expect, carrier collection is most efficient close to the rear contact. Indeed, marked differences have been observed for photoaction spectra with the two irradiation (i.e. through the electrolyte side vs. through the transparent rear contact) geometries for TiO_2, CdS, and CdSe nanocrystalline films [227, 228]. Obviously, the relative magnitudes of the excitation wavelength and the film thickness critically enter into this variant behavior.

In the vast majority of cases, the iodide/triiodide redox couple has been employed (presumably because of its success in shuttling the photooxidized dye in the sensitization experiments) although other redox electrolytes [e.g. $SCN^-/(SCN)_2^-$; 228] have been employed as well. For the chalcogenide films, sodium selenosulfite was employed [227]. It must be noted that, aside from losses caused by the surface recombination and back-reactions, an additional loss component from the increase in film resistance must also be recognized, especially as the film thickness is increased. The resistance loss manifests as a deterioration in the photovoltage and fill factor.

In the discussion to this point, we have not considered trapping or release of the photogenerated electrons as they undergo transit to the rear contact. However, electrons trapped in localized interfacial states induce a countercharge in the Helmholtz double-layer, as discussed in the preceding discussion. The resultant voltage drop can introduce a nonnegligible field component into the diffusional process. The time-dependence of the electron density, $n(x,t)$ is given by [9]

$$\frac{\partial n(x,t)}{\partial t} = \eta \alpha I_o e^{-\alpha x} + D_n \frac{\partial^2 n(x,t)}{\partial x^2} - \frac{n(x,t) - n_o}{\tau} \qquad (32)$$

In Eq. (32), η is the electron injection efficiency, D_n is the diffusion coefficient of electrons, and τ is the pseudo first-order lifetime of electrons determined by back-reaction with Ox.

Even if the migration component is negligible (but see following), solution of Eq. (32) presents difficulties because of the possible dependence of D_n on n and x. Similarly, τ may depend on these two variables also. Nonetheless, the steady state solution of Eq. (32) has been obtained [229] by assuming that D and τ are constant and that $\eta = 1$. Under these conditions, the photocurrent is predicted to be independent of voltage – a rather physically implausible situation. In the forward-bias regime, η is expected to decrease and the back-reaction of the photogenerated electrons (with Ox) can no longer be neglected.

This brings us to the rear support-film interface. What sort of barrier exists at this junction? Are the electron exchange kinetics voltage-dependent at this interface? The effect of changing the work function of the substrate on the current-voltage curves (in the dark and under illumination) has been investigated for TiO_2 nanocrystalline films [230]. The onset potential for the photocurrent is found to be the same regardless of whether SnO_2, Au, or Pt is used to support the film. A Fermi level pinned *rear* interface was used to explain the results.

In general, the voltammograms for nanocrystalline electrodes are similar to what is observed for their single-crystal counterparts. An example of a photovoltammogram for CdS is contained in Fig. 25. The fact that a S-type profile is observed culminating in a photon flux-limited plateau regime is rather surprising given that (1) the film is rather insulating and (2) the electrolyte permeates into the network and possibly contacts the rear support electrode. The transition in the profiles from spiked at potentials near V_{on} (photocurrent onset) to more rectangular at positive bias potentials (not shown in Fig. 25) must mean that the voltage does exert an effect on carrier transit through the network. No satisfactory explanation appears to exist at present to resolve this apparent anomaly.

1.7.4
Photoexcitation and Carrier Collection: Dynamic Behavior

In this section, we briefly consider the response of nanocrystalline semiconductor–electrolyte interfaces to either pulsed or periodic photoexcitation. Several points are noteworthy in this respect: (1) The photocurrent rise-time in response to an illumination step is nonlinear. Further, the response is *faster* when the light intensity is higher. (2) The decay profiles exhibit features on rather slow timescales extending up to several seconds. (3) The photocurrent decay transients exhibit a peaking behavior. The time at which this peak

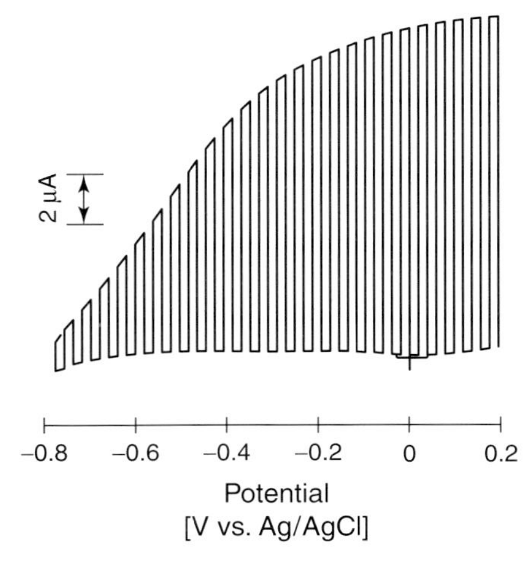

Fig. 25 Photovoltammogram under interrupted illumination of a nanocrystalline CdS-sodium sulfite electrolyte interface in the reverse-bias regime.

occurs varies with the square of the film thickness, d. (4) A similar pattern is also seen in IMPS data where the transit time, τ is seen to be proportional to d^2.

These observations have been interpreted within the framework of two distinct models, one involving trapping or detrapping of the photogenerated electrons [231, 232] and the other based on electron diffusion (or field-assisted diffusion) not attenuated by electron localization [233, 234]. The millisecond transit times also mean that the transit times are very long compared to equilibration of majority carriers in a bulk semiconductor or electron-hole pair separation within the depletion layer of a flat electrode. The slow transport is rationalized by a weak driving force and by invoking percolation effects [223].

It is interesting that the response patterns differ for different nanocrystalline electrodes [223]. For example, while trapping or detrapping effects appear to be relatively unimportant for GaP, the response for TiO_2, especially at low photon fluxes, is governed by electron trapping or detrapping kinetics. This accounts for the faster response at higher photon fluxes (see preceding section).

1.7.5 Size Quantization

When electronic particles such as electrons and holes are constrained by potential barriers to regions of space that are comparable to or smaller than their de Broglie wavelength, the corresponding allowed energy states become discrete (i.e. quantized) rather than continuous. This manifests in the absorption (or emission) spectra as discrete lines that are reminiscent of atomic (line) transitions; these sharper features often appear superimposed on a broader envelope. Another manifestation for semiconductors is that the energy band gap (E_g) increases, or equivalently, the absorption threshold exhibits a blue shift. The critical dimension for size quantization effects to appear in semiconductors depends on the effective mass (m^*) of the electronic charge carriers. For $m^* \sim 0.05$, the critical dimension is about 300 Å; it decreases approximately linearly with increasing m^* [7].

Size quantization effects and quantum dot photoelectrochemistry are discussed in more detail elsewhere in this volume.

1.8 Chemically Modified Semiconductor–Electrolyte Interfaces

1.8.1 Single Crystals

Much of the research in the early 1980s on chemically modified semiconductor–electrolyte interfaces was directed toward protecting them from photocorrosion; this body of work has been reviewed [226]. Parallel efforts also went into improving minority carrier transfer at the interface by chemisorbing metal ions such as Ru^{3+} on the semiconductor surface. Chemical agents such as sulfide ions are known to passivate the semiconductor against surface recombination [6]. A study [22a] on electron transfer dynamics at p-GaAs-acetonitrile interfaces where the semiconductor surface was sulfide-passivated exemplifies this fact. In general, the mechanistic issue of whether these chemical agents improve minority carrier charge transfer by minimizing surface recombination or by a true catalytic action has not been completely resolved [1].

Yet another tactic involves perturbing the electrostatics at the semiconductor–electrolyte interface by adsorbing

Tab. 2 A summary of approaches[a] to chemical modification of semiconductor–electrolyte interfaces

Modification agent(s)	*Semiconductor(s)*	*Modification objective*[b]	*Sample Reference(s)*
Ru^{3+}	*n*-GaAs	A	239, 240
Co^{3+}	*n*-GaAs	A	241
Os^{3+}	*n*-GaAs	A	241
Ag^{+}	*p*-InP	A	242
S^{2-}	*p*-GaAs	A	236
HS^{-}	CdS, CdSe	B	243
Thiolates	CdS, CdSe	B	244
Dithiocarbamate	CdS, CdSe	B	245, 246
Lewis acids	CdS, CdSe	B	247, 248
Lewis bases	CdS, CdSe	B	249
Cl^{-}	*n*-GaAs	B	238
Benzoic acid derivatives	n^{+}-GaAs	B	250
Noble metals	*n*-TiO_2	C	251
Noble metals	*p*-InP	C	239, 252–254
RuO_2[c]	*n*-CdS	C	255, 256
RuO_2[c]	*n*-Si	C	257
Pt[c]	*p*-Si	C	257, 258
Pt[c]	*n*-CdS	C	259
Noble metals	*n*-CdS	C	254
Noble metals[d]	*p*-InP	C	260

[a] Approaches to photoanode stabilization based on polymer films containing redox functionalities have been reviewed elsewhere, e.g. Refs. 6 and 226.
[b] A: minority carrier transfer catalysis and or surface state passivation; B: electrostatic modification; C: catalysis of multielectron photoprocesses (refer to text).
[c] In these cases, the semiconductor electrode also contained a coating, either polymeric or indium tin oxide.
[d] The chemically modified photocathode was used in conjunction with *n*-$MoSe_2$ (or *n*-WSe_2) in a two-photoelectrode cell configuration.

(or even chemically attaching) electron donors or acceptors on the semiconductor surface [237]. In favorable cases, this increases the band bending at the interface by thus introducing a fixed countercharge of opposite polarity (negative for a *n*-type semiconductor) at the junction. Chloride ion adsorption on the *n*-GaAs surface from ambient temperature $AlCl_3$/*n*-butylpyridinium chloride melts [30, 238] is a case in point; this process serves to improve the junction and the photovoltage that it delivers. Of course, such "fixed-charge" effects have long been known to the solid-state device physics and gas phase catalysis communities. Other agents that have been deployed for chemical tuning of the interfacial energetics at the semiconductor–electrolyte interface are listed in Table 2.

Native semiconductor surfaces are fairly inactive from a catalysis perspective. Thus, noble metal or metal oxide islands have been implanted on photoelectrode surfaces as electron storage centers to drive multielectron redox processes such as HER, photooxidation of H_2O, and photooxidation of HCl, HBr, or HI. Examples of this sort of chemical modification strategy are also contained in Table 2.

The advent of self-assembled monolayer (SAM) films on electrode surfaces has rendered a high degree of molecular order and predictability to the chemical modification approach. In particular, the use of these insulating, molecular spacers enables interrogation of critical issues in electron transfer such as the influence of chemical bonding and distance between the support electrode and the redox moieties on the rate constant for electron transfer. Many such studies on gold-confined SAMs have appeared recently [261–263]. Corresponding studies on *semiconductor* surfaces (particularly Group II–V compounds such as GaAs and InP [264–266] and elemental semiconductors such as Si [267]) have also begun to appear.

Alkanethiol-based or alkylsiloxane-based SAMs have been profitably employed in all these instances to probe the distance effect in electron transfer dynamics. The thiol-based SAMs have the virtue that the spacer length can be predictably altered simply by varying the number of methylene units in the chain. The distance dependence of k_{et} is embodied in the parameter β, the decay coefficient. For a critical discussion of the subtleties involved in the extraction and interpretation of this parameter, we refer to Ref. 262. A value of 0.49 ± 0.07 has been reported for this parameter for *n*-InP-alkanethiol-ferrocyanide interfaces [266]. This value is smaller than its counterpart for corresponding films on gold surfaces, which range from $\sim$0.6 to 1.1 per methylene unit. The reason for this difference is not entirely clear, although several hypotheses were advanced by the authors [266].

1.8.2 Nanocrystalline Semiconductor Films and Composites

Dye sensitization of nanocrystalline semiconductor films certainly represents one popular approach to chemical modification of the interface. However, this topic is covered in detail elsewhere in this volume. Other examples, from a non–dye sensitization perspective, are less common but two recent studies are noted [268, 269]. One utilizes the surface affinity of TiO_2 toward suitably derivatized viologens to construct chemically modified nanocrystalline films suitable for displays, electrochromic (smart) windows, sensors, and the like [268]. In the other study [269], the TiO_2 film surface was modified with phosphotungstic acid (PWA). This compound belongs to a family of polyoxometallates that exhibit interesting electron- and proton-transfer or storage properties and also high thermal stability [269]. Thus, these modified films would be applicable in areas such as catalysis, sensors, electronics, and even medicine.

These TiO_2-PWA films represent a logical bridge connecting single-phase semiconductor films and multicomponent composite systems. Of course, highly evolved multicomponent assemblies occur in nature and there is no better example than the plant photosynthetic system. The plant photosynthetic architecture contains synergistic components (e.g. light-harvesting antennae, membranes) each with a well-defined and complementary function, to convert the incident photon energy, to move electrons vectorially, and to store the reaction products. The design and implementation of artificial analogs have proved to be a daunting task, both from a synthetic and characterization perspective. While this topic is covered elsewhere in this series of volumes, we briefly discuss in what follows, some simple multicomponent assemblies based on semiconductors.

Early examples in the 1980s were aimed at the design of composite systems for photoelectrolytic generation of H_2. Thus, Nafion and SiO_2 were used as supports for coprecipitated ZnS and CdS for photoassisted HER from aqueous sulfide media [270]. Subsequent work has addressed the mechanistic role of the support in the photoassisted HER [271]. Vectorial electron transfer was demonstrated in bipolar TiO_2/Pt or CdSe/CoS photoelectrode panels arranged in series arrays for the photodecomposition of water to H_2 and O_2 [272, 273].

More recently, matrix-semiconductor composites, that is, films comprising of semiconductor particles that are dispersed in a nonphotoactive continuous matrix have been developed. Examples of matrix candidates are metals and polymers [274–279]. Occlusion electrosynthesis is a versatile method for preparing such composite films as exemplified by the Ni/TiO_2 and Ni/CdS family [280–282].

Matrix-semiconductor composite films have two virtues from a photoelectrochemical perspective. First, their components can be separately chosen and optimized for a specific function. Thus, the matrix component can be chosen to have good adsorption tendency toward a targeted substrate. The semiconductor component then functions in the role of photogenerating charge carriers either for reducing or oxidizing this sequestered substrate. This photocatalytic strategy has been recently demonstrated both for organic substrates (methanol and formate ion) [283, 284] and an inorganic substrate (sulfite) [285]. The net result in either case is an enhanced photocatalytic performance of the composite because of the high *local* concentration of the substrate resulting from the matrix adsorption process. In principle, high surface area supports of the sort that are normally used in the gas-phase catalysis community can also be used in conjunction with TiO_2 [286, 287]. These would include materials such as Al_2O_3, SiO_2, or diatomaceous earth. The resultant composite films, however, cannot be used as electrodes because of their poor electronic conductivity.

The second important feature of a metal-semiconductor composite approach is that the metal can function as a template for chemical or electrochemical derivatization to afford a film comprising molecular redox-semiconductor (or even semiconductor-semiconductor) contacts. Figure 26 generically illustrates the occlusion electrosynthesis approach for preparing M/TiO_2 composite films and a subsequent derivatization with ferri/ferrocyanide to afford the corresponding metal hexacyanoferrate (MHCF)/TiO_2 counterparts [288]. These chemically modified films exhibit interesting "bipolar" photoelectrochemical behavior [289] and photoelectrochromic properties [290].

1.9 Types of Photoelectrochemical Devices

As Fig. 27 illustrates, there are basically three types of photoelectrochemical devices for solar energy conversion. The first type is regenerative in nature and the species that are photooxidized at the n-type semiconductor electrode are simply re-reduced at the counterelectrode (Fig. 27a). Instead of an electrocatalytic electrode [291, 292] where the counterelectrode reaction occurs in the dark (this is the situation schematized

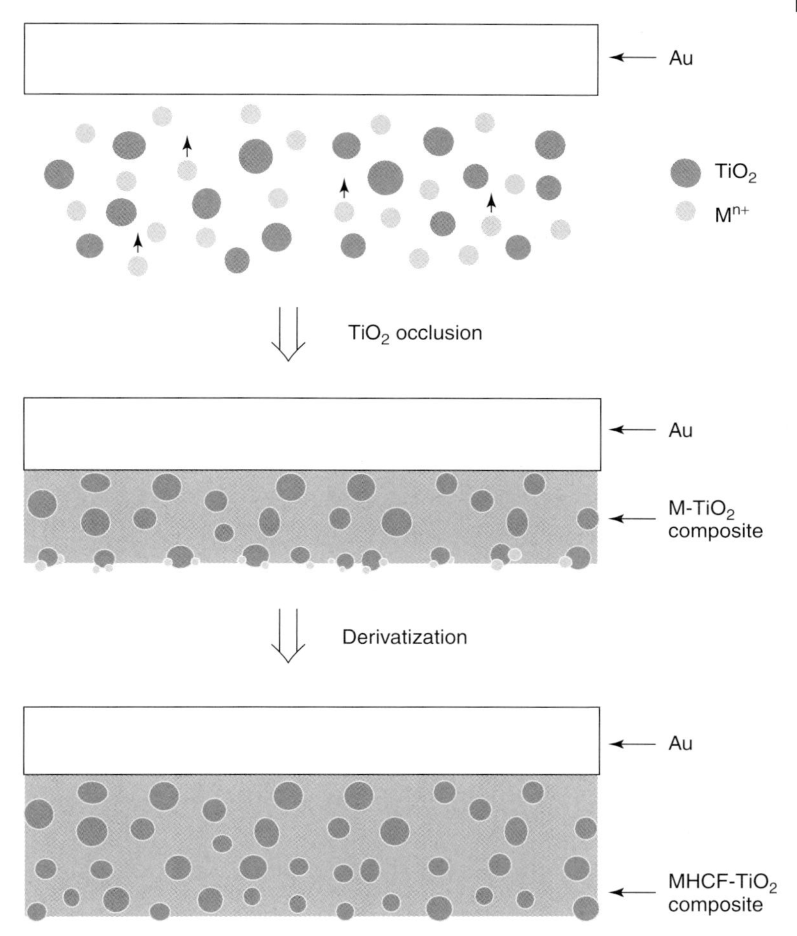

Fig. 26 Schematic illustration of the occlusion electrosynthesis approach for the preparation of M/TiO_2 (M = metal) composite films and subsequent chemical derivatization to yield the MHCF/TiO_2 counterparts. Refer to the text for further details.

in Fig. 27a), a *p*-type semiconductor photo-cathode may also be deployed in a tandem regenerative cell. In all these cases, the cells operate in the photovoltaic mode where the input photon energy is converted into electricity.

Interesting enough, it is the second type of device, namely a photoelectrolytic cell (Fig. 27b), that first caught the attention of a scientific and technological community in the 1970s that was searching for alternative energy sources to fossil-derived fuels. Thus in a landmark paper, Fujishima and Honda [293] demonstrated that sunlight could be used to drive the photoelectrolysis of water using an *n*-TiO_2 photoanode and a Pt counterelectrode. Unfortunately, the requirements for efficiently splitting water are rather stringent, as discussed elsewhere in this volume.

In the third type of energy conversion device, the initial photoexcitation does not

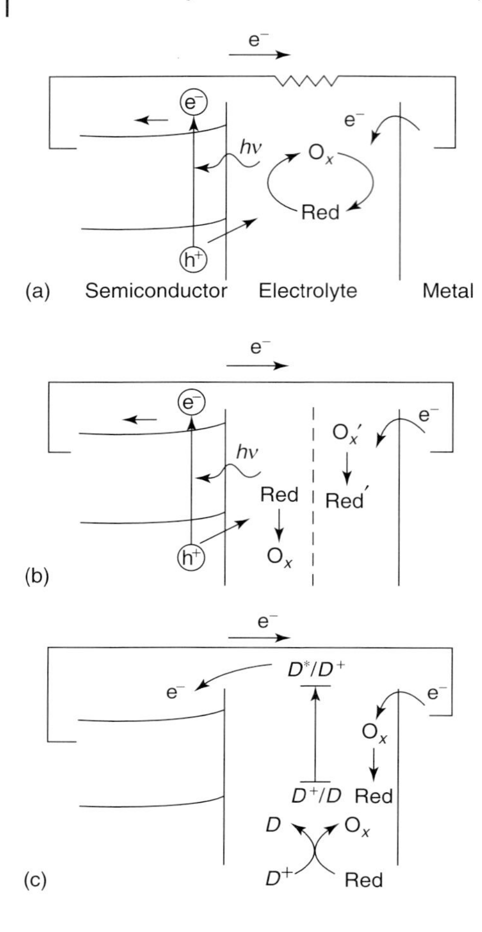

Fig. 27 Types of photoelectrochemical devices for solar energy conversion. (a), (b), and (c) depict regenerative, photoelectrolytic, and dye-sensitized configurations, respectively. As in the remainder of this chapter, an *n*-type semiconductor is assumed in these cases for specificity.

occur in the semiconductor (unlike in the device counterparts in Figs. 27a and b) but occurs instead in a visible light-absorbing dye (Fig. 27c). Subsequent injection of an electron from the photoexcited dye into the semiconductor CB results in the flow of a current in the external circuit. Sustained conversion of light energy is facilitated by regeneration of the reduced form of the dye via a reversible redox couple (e.g. iodide/triiodide) [294]. Therefore, this device, as its counterpart in Fig. 27(a), also operates in a photovoltaic mode, or perhaps more appropriately, in a *photogalvanic* mode.

Other variants of the three types of device operation may be envisioned for semiconductor-liquid junctions. Thus, in the photoelectrolytic mode, the cell reaction clearly is driven (by light) in the contra-thermodynamic direction, that is, $\Delta G > 0$. However, there are many instances, involving, for example, the photooxidation of organic compounds in which light merely serves to accelerate the reaction rate. Thus these cells operate in the photocatalytic mode. In fact, aqueous suspensions comprising irradiated semiconductor particles may be considered to be an assemblage of short-circuited microelectrochemical cells operating in the photocatalytic mode.

Finally, a storage electrode may be incorporated even in a regenerative photoelectrochemical cell of the sort schematized in Fig. 27(a). Thus, when the sun is shining, this storage electrode is "charged"; in the dark, energy may be tapped (as from a battery) from this storage electrode [295–298].

Further details of these device types as well as *nonenergy*-related applications of photoelectrochemical cells (such as in environmental remediation) may be found in the chapters that follow in this volume.

1.10 Conclusion

In this introductory chapter, we have discussed the electrostatics of the semiconductor–liquid interface considering both single crystals as well as their nanocrystalline counterparts. The charge transfer dynamics across both these types of interfaces have been described in the dark and under photoexcitation of the semiconductor. Finally, the various types of photoelectrochemical devices for solar energy conversion are introduced. Subsequent chapters in this volume provide further elaboration of some of these topics considered herein.

Acknowledgments

Research in the author's laboratory on semiconductor electrochemistry and photoelectrochemistry since 1995 is funded, in part, by the Office of Basic Energy Sciences, the US Department of Energy. A number of talented and dedicated coworkers and colleagues have been involved in collaborative research with the author over the past twenty years; their names appear in the publications cited from this laboratory. I also thank the University of Texas at Arlington for providing the facilities and infrastructure. I am grateful to Prof. S. Licht for comments on an earlier version of the manuscript. Last but not least, I thank Ms. Gloria Madden and Ms. Rita Anderson for assistance in the preparation of this chapter.

References

1. N. S. Lewis, *Acc. Chem. Res.* **1990**, *23*, 176.
2. L. M. Peter, *Chem. Rev.* **1990**, *90*, 753.
3. N. S. Lewis, *Annu. Rev. Phys. Chem.* **1991**, *42*, 543.
4. C. A. Koval, J. N. Howard, *Chem. Rev.* **1992**, *92*, 411.
5. A. Kumar, W. C. Á. Wilisch, N. S. Lewis, *Crit. Rev. Solid State Mater. Sci.* **1993**, *18*, 327.
6. M. X. Tan, P. E. Laibinis, S. T. Nguyen et al., *Prog. Inorg. Chem.* **1994**, *41*, 21.
7. A. J. Nozik, R. Memming, *J. Phys. Chem.* **1996**, *100*, 13061.
8. N. S. Lewis, *J. Phys. Chem.* **1998**, *102*, 4843.
9. L. M. Peter, D. Vanmaekelbergh, *Adv. Electrochem. Sci. Eng.* **1999**, *6*, 77.
10. L. M. Peter in *Comprehensive Chemical Kinetics*, (Eds.: R. G. Compton, G. Hancock), Elsevier, Amsterdam, 1999, pp. 223–280.
11. R. J. D. Miller, G. McLendon, A. J. Nozik et al., in *Surface Electron-transfer Processes*, VCH Publishers, New York, 1995.
12. K. Rajeshwar, L. M. Peter, A. Fujishima et al., Eds., in *Photoelectrochemistry Proc.*, The Electrochemical Society, Pennington, NJ, 1997, Vol. 20.
13. N. Sato in *Electrochemistry at Metal and Semiconductor Electrodes*, Elsevier, Amsterdam, 1998.
14. A. Hamnett in *Comprehensive Chemical Kinetics* (Eds.: R. G. Compton), Elsevier, Amsterdam, 1987, p. 61, Vol. 27.
15. J. S. Newman in *Electrochemical Systems*, 2nd edition, Prentice-Hall, Englewood Cliffs, NJ, 1991, p. 496.
16. R. A. Smith in *Semiconductors*, Cambridge University Press, Cambridge, 1964.
17. F. A. Kröger, H. J. Vink in *Solid State Physics* (Eds.: F. Seitz, D. Turnbull), Academic Press, New York, 1956, Vol. 3.
18. F. A. Kröger in *The Chemistry of Imperfect Crystals*, North Holland, Amsterdam, 1964.
19. N. B. Hannay, Ed., in *Semiconductors*, Reinhold, New York, 1959.
20. A. R. West in *Solid State Chemistry and Its Applications*, John Wiley, New York, 1984.
21. P. A. Cox in *The Electronic Structure and Chemistry of Solids*, Oxford University Press, Oxford, England, 1987.
22. R. Hoffmann in *Solids and Surfaces*, VCH, New York and Weinheim, 1988.
23. K. Uosaki, H. Kita in *Modern Aspects of Electrochemistry* (Eds.: R. E. White, J. O'M. Bockris, B. E. Conway), Plenum, New York, 1986, pp. 1–60, Vol. 18.
24. S. Trasatti in *The Absolute Electrode Potential: An Explanatory Note*, IUPAC Commission I.3 (Electrochemistry), 1984.
25. H. Reiss, *J. Electrochem. Soc.* **1978**, *125*, 937.
26. Yu. V. Pleskov, Yu. Ya. Gurevich in *Semiconductor Photoelectrochemistry*, Consultants Bureau, New York, 1986.
27. H. Gerischer, *Adv. Electrochem. Eng.* **1961**, *1*, 139.
28. H. Gerischer in *Physical Chemistry: An Advanced Treatise* (Eds.: H. Eyring, D. Henderson, W. Jost), Academic Press, New York, 1970, Vol. 9A.

29. S. R. Morrison in *The Chemical Physics of Surfaces*, Plenum, New York and London, 1977.
30. P. Singh, R. Singh, R. Gale et al., *J. Appl. Phys.* **1980**, *51*, 6286.
31. K. Rajeshwar in *Molten Salt Techniques* (Eds.: R. J. Gale, D. G. Lovering), Plenum, New York and London, 1984, pp. 221–252, Vol. 2.
32. R. Thapar, K. Rajeshwar, *Electrochim. Acta* **1983**, *28*, 198.
33. C. M. Gronet, N. S. Lewis, G. W. Cogan et al., *J. Electrochem. Soc.* **1984**, *131*, 2873.
34. E. H. Rhoderick in *Metal-Semiconductor Contacts*, Clarendon Press, Oxford, 1980.
35. S. Kar, K. Rajeshwar, P. Singh et al., *Sol. Energy* **1979**, *23*, 129.
36. G. Horowitz, P. Allongue, H. Cachet, *J. Electrochem. Soc.* **1984**, *131*, 2563.
37. A. J. Bard, A. B. Bocarsly, F. R. F. Fan et al., *J. Am. Chem. Soc.* **1980**, *102*, 3671.
38. A. Many, Y. Goldstein, N. B. Grover in *Semiconductor Surfaces*, North Holland, Amsterdam, 1965.
39. G. A. Somorjai in *Introduction to Surface Chemistry and Catalysis*, John Wiley, New York, 1994.
40. A. W. Adamson, A. P. Gast in *Physical Chemistry of Surfaces*, John Wiley, New York, 1997.
41. R. L. Van Meirhaeghe, F. Cardon, W. P. Gomes, *J. Electroanal. Chem.* **1985**, *188*, 287.
42. J. Bardeen, *Phys. Rev.* **1947**, *71*, 717.
43. A. M. Cowley, S. M. Sze, *J. Appl. Phys.* **1965**, *36*, 3212.
44. S. Kurtin, T. C. McGill, C. A. Mead, *Phys. Rev. Lett.* **1969**, *22*, 1433.
45. T. C. McGill, *J. Vac. Sci. Technol.* **1974**, *11*, 935.
46. L. J. Brillson, *Phys. Rev. Lett.* **1978**, *40*, 260.
47. L. J. Brillson, *J. Vac. Sci. Technol.* **1979**, *16*, 1137.
48. W. E. Spicer, P. W. Chye, P. R. Skeath et al., *J. Vac. Sci. Technol.* **1979**, *16*, 1422.
49. G. Nogami, *J. Electrochem. Soc.* **1982**, *129*, 2219.
50. K. Uosaki, H. Kita, *J. Electrochem. Soc.* **1983**, *130*, 895.
51. R. de Gruyse, W. P. Gomes, F. Cardon et al., *J. Electrochem. Soc.* **1975**, *125*, 711.
52. J. C. Tranchart, L. Hollan, R. Memming, *J. Electrochem. Soc.* **1978**, *125*, 1185.
53. S. R. Morrison, *Surf. Sci.* **1969**, *15*, 363.
54. R. A. L. Vanden Berghe, F. Cardon, W. P. Gomes, *Surf. Sci.* **1973**, *39*, 368.
55. D. Meissner, R. Memming, *Electrochim. Acta* **1992**, *37*, 799.
56. H. Gerischer, *Ber. Bunsen-Ges. Phys. Chem.* **1965**, *69*, 578.
57. R. Memming, G. Schwandt, *Electrochim. Acta* **1968**, *13*, 1299.
58. M. J. Madou, F. Cardon, W. P. Gomes, *Ber. Bunsen-Ges. Phys. Chem.* **1977**, *81*, 1186.
59. H. H. Goossens, W. P. Gomes, *J. Electrochem. Soc.* **1991**, *138*, 1696.
60. D. Vanmaekelbergh, L. S. Yun, W. P. Gomes et al., *J. Electroanal. Chem.* **1987**, *221*, 187.
61. J. J. Kelly, J. E. A. M. van den Meerakker, P. H. L. Notten et al., *Philips Tech. Rev.* **1988**, *44*, 61.
62. P. H. L. Notten, J. E. A. M. van den Meerakker, J. J. Kelly in *Etching of III-V Semiconductors: An Electrochemical Approach*, Elsevier, Oxford, 1991.
63. H. H. Goossens, W. P. Gomes, *Electrochim. Acta* **1992**, *37*, 811.
64. W. P. Gomes, H. H. Goossens in *Adv. Electrochem. Sci. Eng.* (Eds.: H. Gerischer, C. W. Tobias), VCH, Weinheim, 1994, Vol. 3.
65. D. Meissner, C. Sinn, R. Memming et al., in *Homogeneous Photocatalysis* (Eds.: E. Pelizetti, N. Serpone, D. Reidel), Dordrecht, 1986.
66. J. E. A. M. van den Meerakker, *J. Electroanal. Chem.* **1988**, *243*, 161.
67. B. Pettinger, H-R. Schöppel, H. Gerischer, *Ber. Bunsen-Ges. Phys. Chem.* **1976**, *80*, 849.
68. K. Uosaki, *Trends Anal. Chem.* **1990**, *9*, 98.
69. K. Uosaki, H. Kita, *Ber. Bunsen-Ges. Phys. Chem.* **1984**, *91*, 447.
70. A. Manivannan, A. Fujishima, *J. Lumin.* **1988**, *42*, 43.
71. G. Oskam, E. A. Meulenkamp, *J. Electroanal. Chem.* **1992**, *326*, 213.
72. R. Reincke, R. Memming, *J. Phys. Chem.* **1992**, *96*, 1310.
73. R. Reineke, R. Memming, *J. Phys. Chem.* **1992**, *96*, 1317.
74. J. Vandermolen, W. P. Gomes, F. Cardon, *J. Electrochem. Soc.* **1980**, *127*, 324.
75. P. Salvador, C. Gutierrez, *J. Electrochem. Soc.* **1984**, *131*, 326.
76. V. A. Tyagai, G. Ya. Kolbasov, *Surf. Sci.* **1971**, *28*, 423.

77. W. P. Gomes, F. Cardon, *Ber. Bunsen-Ges. Phys. Chem.* **1970**,*74*, 431.
78. Z. Hens, *J. Phys. Chem. B* **1999**, *103*, 122.
79. Z. Hens, W. P. Gomes, *J. Phys. Chem. B* **1999**, *103*, 130.
80. F. Cardon, *Physica* **1972**, *57*, 390.
81. G. Horowitz, *J. Electroanal. Chem.* **1983**, *159*, 421.
82. D. Vanmaekelbergh, *Electrochim. Acta* **1997**, *42*, 1135.
83. Z. Hens, W. P. Gomes, *Electrochim. Acta* **1998**, *43*, 2577.
84. W. P. Gomes, D. Vanmaekelbergh, *Electrochim. Acta* **1996**, *41*, 967.
85. P. A. Allongue, H. Cachet, *J. Electrochem. Soc.* **1985**, *132*, 45.
86. J. E. A. M. Van der Meerakker, J. J. Kelly, P. H. L. Notten, *J. Electrochem. Soc.* **1985**, *132*, 638.
87. K. Schröder, R. Memming, *Ber. Bunsen-Ges. Phys. Chem.* **1985**, *89*, 385.
88. D. Vanmaekelbergh, W. P. Gomes, F. Cardon, *Ber. Bunsen-Ges. Phys. Chem.* **1986**, *90*, 431.
89. D. Vanmaekelbergh, W. P. Gomes, F. Cardon, *Ber. Bunsen-Ges. Phys. Chem.* **1985**, *89*, 994.
90. D. Vanmaekelbergh, W. P. Gomes, F. Cardon, *J. Electrochem. Soc.* **1987**, *134*, 891.
91. D. Vanmaekelbergh, R. P. ter Heide, W. Kruijt, *Ber. Bunsen-Ges. Phys. Chem.* **1989**, *93*, 1103.
92. D. Vanmaekelbergh, F. Cardon, *J. Phys. D: Appl. Phys.* **1986**, *19*, 643.
93. D. Vanmaekelbergh, F. Cardon, *Semicond. Sci. Technol.* **1988**, *3*, 124.
94. J. J. Kelly, P. H. L. Notten, *Electrochim. Acta* **1984**, *29*, 589.
95. J. E. A. M. van den Meerakker, *Electrochim. Acta* **1985**, *30*, 435.
96. A. M. Fajardo, N. S. Lewis, *Science* **1996**, *274*, 969.
97. A. M. Fajardo, N. S. Lewis, *J. Phys. Chem. B* **1997**, *101*, 11136.
98. K. E. Pomykal, N. S. Lewis, *J. Phys. Chem. B* **1997**, *101*, 2476.
99. A. Meier, S. S. Kocha, M. C. Hanna et al., *J. Phys. Chem. B* **1997**, *101*, 7038.
100. I. Uhlendorf, R. Reincke-Koch, R. Memming, *J. Phys. Chem.* **1996**, *100*, 4930.
101. A. Meier, D. C. Selmarten, K. Siemoneit et al., *J. Phys. Chem. B* **1999**, *103*, 2122.
102. Y. Rosenwaks, B. R. Thacker, R. K. Ahrenkiel et al., *J. Phys. Chem.* **1992**, *96*, 10096.
103. H. Gerischer, *J. Phys. Chem.* **1991**, *95*, 1356.
104. S. M. Sze in *Physics of Semiconductor Devices*, John Wiley, New York, 1981.
105. Z. Hens, W. P. Gomes, *J. Phys. Chem. B* **1997**, *101*, 5814.
106. J. L. Pankove in *Optical Processes in Semiconductors*, Prentice Hall, Englewood Cliffs, NJ, 1971.
107. D. Vanmaekelbergh, B. H. Erne, C. W. Cheung et al., *Electrochim. Acta* **1995**, *40*, 686.
108. I. Rosenberg, J. R. Brock, A. Heller, *J. Mol. Catal.* **1992**, *96*, 3423.
109. J. Valladares, J. R. Bolton in *Photocatalytic Purification and Treatment of Water and Air* (Eds.: D. F Ollis, H. Al-Ekabi), Elsevier, Amsterdam, 1993, p. 111.
110. M. I. Cabrera, O. M. Alfano, A. E. Cassano, *J. Phys. Chem.* **1996**, *100*, 20043.
111. C. A. Martin, M. A. Baltanas, A. E. Casano, *Environ. Sci. Technol.* **1996**, *30*, 2355.
112. K. Rajeshwar, *Spectroscopy* **1993**, *8*, 16.
113. L. M. Peter in *Comprehensive Chemical Kinetics* (Eds.: R. G. Compton), Elsevier, Amsterdam, 1989, pp. 353–383, Vol. 29.
114. W. W. Gärtner, *Phys. Rev.* **1959**, *116*, 84.
115. M. A. Butler, *J. Appl. Phys.* **1977**, *48*, 1914.
116. V. A. Tyagai, *Russ. J. Phys. Chem.* **1965**, *38*, 1335.
117. J. Li, L. M. Peter, *J. Electroanal. Chem.* **1984**, *165*, 41.
118. L. M. Peter, J. Li, R. Peat, *J. Electroanal. Chem.* **1984**, *165*, 29–41.
119. R. H. Wilson, *CRC Crit. Rev. Solid State Mater. Sci.* **1980**, *10*, 1.
120. R. H. Wilson, *J. Appl. Phys.* **1977**, *48*, 4297.
121. K. Rajeshwar, *J. Electrochem. Soc.* **1980**, *129*, 1003.
122. J. F. McCann, D. Haneman, *J. Electrochem. Soc.* **1982**, *129*, 1134.
123. R. Memming, *Surf. Sci.* **1964**, *1*, 88.
124. J. J. Kelly, R. Memming, *J. Electrochem. Soc.* **1982**, *129*, 730.
125. S. U. M. Khan, J. O'M. Bockris, *J. Phys. Chem.* **1984**, *88*, 2504.
126. F. E. Guibaly, K. Colbow, B. L. Funt, *J. Appl. Phys.* **1981**, *52*, 3480.
127. F. E. Guibaly, K. Colbow, *J. Appl. Phys.* **1982**, *53*, 1737.
128. F. E. Guibaly, K. Colbow, *J. Appl. Phys.* **1983**, *54*, 6488.
129. J-N. Chazalviel, *J. Electrochem. Soc.* **1982**, *129*, 963.

130. M. Nishida, *J. Appl. Phys.* **1980**, *51*, 1669.
131. F. E. Guibaly, K. Colbow, *Can. J. Phys.* **1981**, *59*, 1682.
132. D. L. Ullman, *J. Electrochem. Soc.* **1981**, *128*, 1269.
133. S. Ramakrishna, S. K. Rangarajan, *J. Electroanal. Chem.* **1991**, *308*, 49.
134. W. Shockley, W. T. Read, *Phys. Rev.* **1952**, *87*, 835.
135. R. N. Hall, *Phys. Rev.* **1982**, *87*, 387.
136. S. J. Fonash in *Solar Cell Device Physics*, Academic Press, San Diego, 1981, Chapter 6, pp. 262–326.
137. C. T. Sah, R. N. Noyce, W. Shockley, *Proc. IRE* **1957**, *45*, 1228.
138. J. Reichman, *Appl. Phys. Lett.* **1980**, *36*, 574.
139. W. I. Albery, P. N. Bartlett, A. Hamnett et al., *J. Electrochem. Soc.* **1981**, *128*, 1492.
140. W. J. Albery, P. N. Bartlett, *J. Electrochem. Soc.* **1983**, *130*, 1699.
141. H. Reiss, *J. Electrochem. Soc.* **1978**, *125*, 937.
142. J. Thomchick, A. M. Buoncristiani, *J. Appl. Phys.* **1980**, *51*, 6265.
143. W. Lorenz, M. Handschuh, *J. Electroanal. Chem.* **1980**, *111*, 181.
144. W. Lorenz, C. Aegerter, M. Handschuh, *J. Electroanal. Chem.* **1987**, *221*, 33.
145. M. Handschuh, W. Lorenz, *J. Electroanal. Chem.* **1988**, *251*, 1.
146. P. Lemasson, A. Etcheberry, J. Gautron, *Electrochim. Acta* **1982**, *27*, 607.
147. S. Ramakrishna, S. K. Rangarajan, *J. Electroanal. Chem.* **1994**, *369*, 289.
148. M. L. Rosenbluth, C. M. Lieber, N. S. Lewis, *Appl. Phys. Lett.* **1984**, *45*, 423.
149. S. Licht, *Nature* **1987**, *330*, 148.
150. S. Licht, D. Peramunage, *Nature* **1990**, *345*, 330.
151. S. Licht, N. Myung, *J. Electrochem. Soc.* **1995**, *142*, 840.
152. S. Licht, N. Myung, *J. Electrochem. Soc.* **1995**, *142*, 845.
153. S. Licht, F. Forouzan, *J. Electrochem. Soc.* **1995**, *192*, 1539.
154. S. Licht, D. Peramunage, *J. Phys. Chem. B* **1996**, *100*, 9082.
155. H. J. Hovel in *Semiconductors and Semimetals, Solar Cells*, Academic Press, New York, 1975, Vol. 11.
156. R. R. Potter, J. R. Sites, *IEEE Trans. Electron Devices* **1984**, *ED-31*, 571.
157. R. R. Potter, J. R. Sites, *Appl. Phys. Lett.* **1983**, *43*, 843.
158. H. Morisaki, M. Hariya, K. Yazawa, *Appl. Phys. Lett.* **1977**, *30*, 7.
159. H. Minoura, M. Tsuiki, *Chem. Lett.* **1978**, 205.
160. W. Gissler, *J. Electrochem. Soc.* **1980**, *127*, 1713.
161. B. Vainas, G. Hodes, J. DuBow, *J. Electroanal. Chem.* **1981**, *130*, 391.
162. N. Müller, G. Hodes, B. Vainas, *J. Electroanal. Chem.* **1984**, *172*, 155.
163. R. Tenne, W. Giriat, *J. Electroanal. Chem.* **1985**, *186*, 127.
164. D. P. Amalnerkar, S. Radhakrishnan, H. Minoura et al., *Sol. Energy Mater.* **1988**, *18*, 37.
165. D. P. Amalnerkar, S. Radhakrishnan, H. Minoura et al., *J. Electroanal. Chem.* **1989**, *260*, 433.
166. S.-E. Lindquist, H. Vidarsson, *J. Mol. Catal.* **1986**, *38*, 131.
167. P. Singh, R. Singh, K. Rajeshwar et al., *J. Electrochem. Soc.* **1981**, *128*, 1145.
168. D. Lincot, J. Vedel, *J. Electroanal. Chem.* **1987**, *220*, 179.
169. D. Lincot, J. Vedel, *J. Phys. Chem.* **1988**, *92*, 4103.
170. L. M. Peter in *Photocatalysis and the Environment* (Ed.: M. Schiavello), Kluwer Academic Publishers, Dordrecht, 1988, p. 243, Vol. 237.
171. B. Ba, H. Cachet, B. Fotouhi et al., *Electrochim. Acta* **1992**, *37*, 309.
172. Y-E. Sung, F. Galliard, A. J. Bard, *J. Phys. Chem. B* **1998**, *102*, 9797.
173. Y-E. Sung, A. J. Bard, *J. Phys. Chem. B* **1998**, *102*, 9806.
174. M. P. Dare-Edwards, A. Hamnett, J. B. Goodenough, *J. Electroanal. Chem.* **1981**, *119*, 109.
175. J. Schefold, *J. Phys. Chem.* **1992**, *96*, 8692.
176. K. Uosaki, H. Kita, *J. Am. Chem. Soc.* **1986**, *108*, 4294.
177. S. Kaneko, K. Uosaki, H. Kita, *J. Phys. Chem.* **1986**, *90*, 6654.
178. K. Uosaki, Y. Shigematsu, S. Kaneko et al., *J. Phys. Chem.* **1989**, *93*, 6521.
179. M. Matsumura, S. R. Morrison, *J. Electroanal. Chem.* **1983**, *144*, 113.
180. M. Matsumura, S. R. Morrison, *J. Electroanal. Chem.* **1983**, *147*, 157.
181. H. J. Lewerenz, J. Stumper, L. M. Peter, *Phys. Rev. Lett.* **1988**, *61*, 1989.
182. L. M. Peter, A. N. Borazio, H. J. Lewerenz et al., *J. Electroanal. Chem.* **1990**, *290*, 229.

183. S. R. Morrison, T. Freund, *J. Chem. Phys.* **1967**, *47*, 1543.
184. W. P. Gomes, T. Freund, S. R. Morrison, *Surf. Sci.* **1968**, *13*, 201.
185. W. P. Gomes, T. Freund, S. R. Morrison, *J. Electrochem. Soc.* **1968**, *115*, 818.
186. K. Micka, H. Gerischer, *J. Electroanal. Chem.* **1972**, *38*, 397.
187. J.-S. Lee, T. Kato, A. Fujishima et al., *Bull. Chem. Soc. Jpn.* **1984**, *57*, 1179.
188. G. H. Schoenmakers, D. Vanmaekelbergh, J. J. Kelly, *J. Phys. Chem.* **1996**, *100*, 3215.
189. G. H. Schoenmakers, D. Vanmaekelbergh, J. J. Kelly, *J. Chem. Soc., Faraday Trans.* **1997**, *93*, 1127.
190. M. Miyake, H. Yoneyama, H. Tamura, *Chem. Lett.* **1976**, 635.
191. N. Hykaway, W. M. Sears, H. Morisaki et al., *J. Phys. Chem.* **1986**, *90*, 6663.
192. D. J. Fermin, E. A. Ponomarev, L. M. Peter, in *Photoelectrochemistry, Proc.*, (Eds.: K. Rajeshwar, L. M. Peter, A. Fujishima et al.), The Electrochemical Society, Pennington, NJ, 1997, Vol. 97–20.
193. G. Nogami, J. H. Kennedy, *J. Electrochem. Soc.* **1989**, *136*, 2583.
194. W. J. Albery, N. L. Dias, C. P. Wilde, *J. Electrochem. Soc.* **1987**, *134*, 601.
195. P. Herrasti, L. M. Peter, *J. Electroanal. Chem.* **1991**, *305*, 241.
196. J. Li, L. M. Peter, *J. Electroanal. Chem.* **1985**, *182*, 399.
197. B. H. Erné, D. Vanmaekelbergh, I. E. Vermeir, *Electrochim. Acta* **1993**, *38*, 2559.
198. J. Li, R. Peat, L. M. Peter, *J. Electroanal. Chem.* **1986**, *200*, 333.
199. R. Peat, L. M. Peter, *Electrochim. Acta* **1986**, *31*, 731.
200. R. Peat, L. M. Peter, *J. Electroanal. Chem.* **1986**, *209*, 307.
201. D. Vanmaekelbergh, J. J. Kelly, *J. Phys. Chem.* **1990**, *94*, 5406.
202. W. P. Gomes, S. Lingier, D. Vanmaekelbergh, *J. Electroanal. Chem.* **1989**, *269*, 237.
203. R. Tenne in *Semiconductor Micromachining: Fundamentals Electrochemistry and Physics* (Eds.: S. A. Campbell, H.-J. Lewerenz), John Wiley, New York, 1994, pp. 139–175, Vol. 1.
204. J. van de Ven, H. J. P. Nabben, *J. Electrochem. Soc.* **1990**, *137*, 1603.
205. G. A. Somorjai in *Chemistry in Two Dimensions: Surfaces*, Cornell University Press, Ithaca, NY, 1981.
206. S. Y. Huang, L. Kavan, I. Exnar et al., *J. Electrochem. Soc.* **1995**, *142*, L142.
207. J. G. Highfield, M. Grätzel, *J. Phys. Chem.* **1988**, *92*, 464.
208. B. O'Regan, M. Grätzel, D. Fitzmaurice, *Chem. Phys. Lett.* **1991**, *183*, 89.
209. B. O'Regan, M. Grätzel, D. Fitzmaurice, *J. Phys. Chem.* **1991**, *95*, 10525.
210. B. E. Conway, *J. Electrochem. Soc.* **1991**, *138*, 1539.
211. R. Könenkamp, R. Henninger, P. Hoyer, *J. Phys. Chem.* **1993**, *97*, 7328.
212. R. Könenkamp, R. Henninger, *Appl. Phys. A* **1994**, *58*, 87.
213. I. Bedja, S. Hotchandani, P. V. Kamat, *J. Phys. Chem.* **1996**, *100*, 19489.
214. G. Boschloo, D. Fitzmaurice, *J. Phys. Chem. B* **1999**, *103*, 2228.
215. F. Cao, G. Oskam, P. C. Searson et al., *J. Phys. Chem.* **1995**, *99*, 11974.
216. R. F. Howe, M. Grätzel, *J. Phys. Chem.* **1985**, *89*, 4495.
217. L. A. Lyon, J. T. Hupp, *J. Phys. Chem.* **1995**, *99*, 15718.
218. H. Lindström, S. Södergren, A. Solbrand et al., *J. Phys. Chem. B* **1997**, *101*, 7710.
219. H. Lindström, S. Södergren, A. Solbrand et al., *J. Phys. Chem. B* **1997**, *101*, 7717.
220. S. Lunell, A. Stashans, L. Ojamäe et al., *J. Am. Chem. Soc.* **1997**, *119*, 7374.
221. L. A. Lyon, J. T. Hupp, *J. Phys. Chem. B* **1999**, *103*, 4623.
222. R. van de Krol, A. Goossens, J. Schoonman, *J. Electrochem. Soc.* **1997**, *144*, 1723.
223. J. J. Kelly, D. Vanmaekelbergh, *Electrochim. Acta* **1998**, *43*, 2773.
224. A. Zaban, A. Meier, B. A. Gregg, *J. Phys. Chem. B* **1997**, *101*, 7985.
225. D. Vanmaekelbergh, P. E. de Jongh, *J. Phys. Chem. B* **1999**, *103*, 747.
226. H. Rensmo, H. Lindström, S. Södergren et al., *J. Electrochem. Soc.* **1996**, *143*, 3173.
227. G. Hodes, I. D. J. Howell, L. M. Peter, *J. Electrochem. Soc.* **1992**, *139*, 3136.
228. A. Hagfeldt, U. Björkstén, S.-E. Lindquist, *Sol. Energy Mater. Sol. Cells* **1992**, *27*, 293.
229. S. Södergren, A. Hagfeldt, J. Olsson et al., *J. Phys. Chem.* **1994**, *98*, 5552.
230. A. Shiga, A. Tsujiko, T. Ide et al., *J. Phys. Chem. B* **1998**, *102*, 6049.
231. K. Schwarzburg, F. Willig, *Appl. Phys. Lett.* **1991**, *58*, 2520.
232. P. E. de Jongh, D. Vanmaekelbergh, *Phys. Rev. Lett.* **1996**, *77*, 3427.

233. A. Solbrand, A. Henningson, S. Södergren et al., *J. Phys. Chem. B* **1999**, *103*, 1078.
234. A. Solbrand, H. Lindström, H. Rensmo et al., *J. Phys. Chem. B* **1997**, *101*, 2514.
235. K. Rajeshwar, *J. Appl. Electrochem.* **1985**, *15*, 1.
236. Y. Rosenwaks, B. R. Thacker, R. K. Ahrenkiel et al., *J. Phys. Chem.* **1992**, *96*, 10096.
237. S. Licht, V. Marcu, *J. Electroanal Chem.* **1986**, *210*, 197.
238. P. Singh, K. Rajeshwar, *J. Electrochem. Soc.* **1981**, *128*, 1724.
239. A. Heller, *Acc. Chem. Res.* **1981**, *14*, 154.
240. B. A. Parkinson, A. Heller, B. Miller, *Appl. Phys. Lett.* **1978**, *33*, 521.
241. B. J. Tufts, I. L. Abrahams, L. G. Casagrande et al., *J. Phys. Chem.* **1989**, *93*, 3260.
242. A. Heller, H. J. Leamy, B. Miller et al., *J. Phys. Chem.* **1983**, *87*, 3239.
243. K. Uosaki, Y. Shigematsu, H. Kita et al., *Anal. Chem.* **1989**, *61*, 1980.
244. M. J. Natan, J. W. Thackeray, M. S. Wrighton, *J. Phys. Chem.* **1986**, *90*, 4089.
245. J. W. Thackeray, M. J. Natan, P. Ng et al., *J. Am. Chem. Soc.* **1986**, *108*, 3570.
246. J. J. Hickman, M. S. Wrighton, *J. Am. Chem. Soc.* **1991**, *113*, 4440.
247. C. J. Murphy, A. B. Ellis, *J. Phys. Chem.* **1990**, *94*, 3082.
248. J. Z. Zhang, A. B. Ellis, *J. Phys. Chem.* **1992**, *96*, 2700.
249. C. J. Murphy, G. C. Lisensky, L. K. Leung et al., *J. Am. Chem. Soc.* **1990**, *112*, 8344.
250. S. Bastide, R. Butruille, D. Cahen et al., *J. Phys. Chem. B* **1997**, *101*, 2678.
251. C-M. Wang, A. Heller, H. Gerischer, *J. Am. Chem. Soc.* **1992**, *114*, 5230.
252. E. Aharon-Shalom, A. Heller, *J. Electrochem. Soc.* **1982**, *129*, 2865.
253. A. Heller, E. Aharon-Shalom, W. A. Bonner et al., *J. Am. Chem. Soc.* **1982**, *104*, 6942.
254. D. E. Aspnes, A. Heller, *J. Phys. Chem.* **1983**, *87*, 4919.
255. K. Rajeshwar, M. Kaneko, A. Yamada et al., *J. Phys. Chem.* **1985**, *89*, 806.
256. K. Rajeshwar, M. Kaneko, *J. Phys. Chem.* **1985**, *89*, 3587.
257. L. Thompson, J. DuBow, K. Rajeshwar, *J. Electrochem. Soc.* **1982**, *129*, 1934.
258. G. Hodes, L. Thompson, J. DuBow et al., *J. Am. Chem. Soc.* **1983**, *105*, 324.
259. A. W.-H. Mau, C-B. Huang, N. Kakuta et al., *J. Am. Chem. Soc.* **1984**, *106*, 6537.
260. C. Levy-Clement, A. Heller, W. A. Bonner et al., *J. Electrochem. Soc.* **1982**, *129*, 1701.
261. C. E. D. Chidsey, *Science* **1991**, *251*, 919.
262. J. F. Smalley, S. W. Feldberg, C. E. D. Chidsey et al., *J. Phys. Chem.* **1995**, *99*, 13141.
263. S. B. Sachs, S. P. Dudek, R. P. Hsung et al., *J. Am. Chem. Soc.* **1997**, *119*, 10563.
264. C. W. Sheen, J. X. Shi, J. Martensson et al., *J. Am. Chem. Soc.* **1992**, *114*, 1514.
265. Y. Gu, D. H. Waldeck, *J. Phys. Chem.* **1996**, *100*, 9573.
266. Y. Gu, D. H. Waldeck, *J. Phys. Chem. B* **1998**, *102*, 9015.
267. A. Haran, D. H. Waldeck, R. Naaman et al., *Science* **1994**, *263*, 948.
268. A. Hagfeldt, L. Walder, M. Grätzel, *Proc. Soc. Photo-Opt. Instrum. Eng.* **1995**, *2531*, 60.
269. H. Lindstrom, H. Rensmo, S.-E. Lindquist et al., *Thin Solid Films* **1998**, *323*, 141.
270. N. Kakuta, K. H. Park, M. F. Finlayson et al., *J. Phys. Chem.* **1985**, *89*, 732.
271. A. Ueno, N. Kakuta, K. H. Park et al., *J. Phys. Chem.* **1985**, *89*, 3828.
272. E. S. Smotkin, A. J. Bard, A. Campion et al., *J. Phys. Chem.* **1986**, *90*, 4604.
273. E. S. Smotkin, S. Cervera-March, A. J. Bard et al., *J. Phys. Chem.* **1987**, *91*, 6.
274. S. Kuwabata, N. Takahashi, S. Hirao et al., *Chem. Mater.* **1993**, *5*, 437.
275. C. S. C. Bose, K. Rajeshwar, *J. Electroanal. Chem.* **1992**, *333*, 235.
276. Y. Son, N. R. de Tacconi, K. Rajeshwar, *J. Electroanal. Chem.* **1993**, *345*, 135.
277. N. R. de Tacconi, Y. Son, K. Rajeshwar, *J. Phys. Chem.* **1993**, *97*, 1042.
278. N. R. Avvaru, N. R. de Tacconi, K. Rajeshwar, *Analyst* **1998**, *123*, 113.
279. S. Ito, T. Deguchi, K. Imai et al., *Electrochem. Solid State Lett.* **1999**, *2*, 440.
280. M. Zhou, W-Y. Lin, N. R. de Tacconi et al., *J. Electroanal. Chem.* **1996**, *402*, 221.
281. M. Zhou, N. R. de Tacconi, K. Rajeshwar, *J. Electroanal. Chem.* **1997**, *421*, 111.
282. N. R. de Tacconi, H. Wenren, K. Rajeshwar, *J. Electrochem. Soc.* **1997**, *144*, 3159.
283. N. R. de Tacconi, H. Wenren, D. McChesney et al., *Langmuir* **1998**, *14*, 2933.
284. N. R. de Tacconi, J. Carmona, K. Rajeshwar, to be published.
285. N. R. de Tacconi, M. Mrkic, K. Rajeshwar, *Langmuir* **2000**, *16*, 8426.
286. C. Anderson, A. J. Bard, *J. Phys. Chem. B* **1997**, *101*, 2611.

287. N. Takeda, M. Ohtani, T. Torimoto et al., *J. Phys. Chem. B* **1997**, *101*, 2644.
288. N. R. de Tacconi, K. Rajeshwar, R. O. Lezna, *Electrochim. Acta* **2000**, *45*, 3403.
289. N. R. de Tacconi, J. Carmona, K. Rajeshwar, *J. Phys. Chem. B* **1998**, *101*, 10151.
290. N. R. de Tacconi, J. Carmona, W. L. Balsam et al., *Chem. Mater.* **1998**, *10*, 25.
291. G. Hodes, J. Manassen, D. Cahen, *J. Appl. Electrochem.* **1977**, *7*, 182.
292. G. Hodes, J. Manassen, D. Cahen, *J. Electrochem. Soc.* **1977**, *127*, 544.
293. A. Fujishima, K. Honda, *Nature* **1972**, *238*, 37.
294. B. O'Regan, M. Grätzel, *Nature* **1991**, *353*, 737.
295. J. Manassen, G. Hodes, D. Cahen, *J. Electrochem. Soc.* **1977**, *124*, 532.
296. S. Licht, G. Hodes, R. Tenne et al., *Nature* **1987**, *326*, 863.
297. S. Licht, B. Wang, T. Soga et al., *Appl. Phys. Lett.* **1999**, *74*, 4055.
298. S. Licht, *J. Phys. Chem. B* **2001**, *105*, 6281.

2
Experimental Techniques

2.1
Photoelectrochemical Systems Characterization

John J. Kelly, Zeger Hens and Daniel Vanmaekelbergh
Utrecht University, Utrecht, The Netherlands

Zeger Hensalso
Laboratorium voor Fysische Chemie, Gent, Belgium

2.1.1
Introduction

This chapter focuses on the characterization of photoelectrochemical systems. Before the essential features of such systems are described in Sect. 2.1.2, the scope of the chapter is first defined on the basis of a brief history of the field.

2.1.1.1 Scope of the Chapter

Semiconductor electrochemistry developed as a discipline with the development of semiconductor device technology [1–3]. Wet chemical processing including etching was important for the fabrication of early silicon and germanium devices. With the increasing sophistication and miniaturization in silicon technology, wet processes were replaced by dry, mainly physical, methods. As a result, interest in semiconductor electrochemistry declined.

However, two developments in the 1970s led to a huge revival of interest and to the "birth" of photoelectrochemistry. In 1971, Fujishima and Honda showed that it was possible to photolyze water in an electrochemical cell with a TiO_2 photoanode and a Pt cathode without an external source [4]. Subsequently, a number of groups [5–8] described regenerative photoelectrochemical solar cells based on narrow band gap semiconductors and a redox couple. They showed that with the proper choice of reducing species it was possible to achieve stability of the photoanode during operation of the cell over longer periods. In further studies, Licht and coworkers developed efficient systems that couple solar energy conversion with energy storage and show improved energy conversion and stability [9–14]. At the same time, the growth of the optoelectronics industry, based largely on III–V materials, revealed a need for new processing methods [15], including etching and metallization. Wet chemical and electrochemical methods proved very successful in this field. In addition, electrochemical methods could be used in a simple way to make and characterize materials for a whole range of applications. The work described so far relates to single crystalline and, to a lesser extent, polycrystalline materials.

Interest in porous photoelectrochemical systems was stimulated by the report in 1991 of O'Regan and Grätzel [16] of a novel solar cell based on a dye-sensitized nanoparticulate TiO_2 photoanode. This work raised a whole range of interesting fundamental issues with regard to charge carrier dynamics, transport and interfacial transfer in porous semiconductor matrices permeated by an electrolyte solution. The discovery in 1992 [17] of the striking optical properties of nanoporous silicon, obtained by (photo)anodic etching, led to a reappraisal of the photoelectrochemistry of this material and to studies of other porous crystalline semiconductors [18].

Research on size quantization in colloidal systems was responsible for the development by Bard and coworkers of a new field in photoelectrochemistry, that of nanodot electrodes [19, 20]. An ordered or disordered monolayer or sub-monolayer of nanometer-sized semiconductor particles is attached to a conducting substrate either directly or via a self-assembled organic monolayer. The monolayer acts as a spacer, allowing the distance between the dot and the substrate to be varied. Absorption of light by the semiconductor dots gives rise to processes similar in many ways to those observed in bulk electrodes. However, because of the size quantization and the distinctive electrode geometry, striking new effects are found.

In this chapter, the characterization of the three types of systems described earlier have been considered: single crystal and polycrystalline bulk electrodes (Sect. 2.1.3), dye-sensitized and quantum dot electrodes (Sect. 2.1.4), and macroporous and nanoporous semiconductors (Sect. 2.1.5). Some essential features, common to all these systems, are introduced in the following subsection. The obvious way to characterize a photoelectrochemical system is by electrochemical methods, and this approach will constitute the focus of the chapter. A brief and very general introduction to (photo)electrochemical characterization is given in Sect. 2.1.2. This is not intended to be a conclusive review; for this a whole volume would be required. The application of these methods to the three systems is described in the subsequent sections. Where relevant, other nonelectrochemical methods of characterization are mentioned.

2.1.1.2 Special Features of Photoelectrochemical Systems

Photoelectrochemical systems rely on the properties of a semiconductor electrode or of an electrode provided with a "sensitizer" layer consisting of, for example, dye molecules or quantum dots. This working electrode (WE) forms part of an electrochemical cell that also contains a counter electrode (CE) and, in many cases, a reference electrode (RE). Absorption of light by the semiconductor or the sensitizer layer gives rise to a photocurrent (PC) and/or a photovoltage, which can be measured in the external circuit. Conversely, the passage of current through the interface of the working electrode with the electrolyte solution can lead to light emission. In this section, some important aspects of such photoelectrochemical systems are reviewed. These topics are dealt with later in the chapter.

When a semiconductor is brought into electrical contact with an electrolyte solution containing a redox couple (Red/Ox^+), equilibrium is established by exchange of charge between the two phases [21–23]. Figure 1(a) shows an example of an n-type semiconductor with a redox couple whose Fermi energy (which is related to the redox potential $V_{Red/Ox}$) is located in the

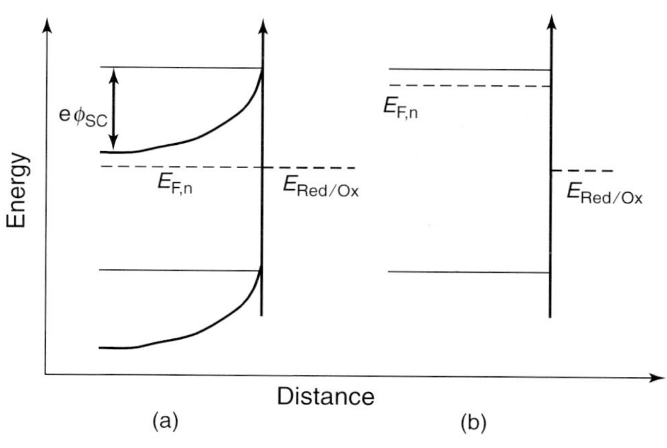

Fig. 1 A scheme of the energetics at an *n*-type semiconductor electrode in contact with a redox system in an electrolyte solution. (a) The situation under conditions of electronic equilibrium. The electrochemical potential of the electrons is the same in both phases, i.e. the electron Fermi-level in the semiconductor $E_{F,n}$ has the same value as the Fermi-level of the electrons in the redox system $E_{Red/Ox}$. (b) Case in which the energy bands in the semiconductor are flat; this situation, corresponding to maximum photovoltage, is reached under strong illumination at open circuit. W_{SC} is the width of the depletion layer and $e\phi_{SC}$ is the band-bending.

band gap of the solid. Electrons have been transferred from the conduction band (CB) of the semiconductor to solution creating a space charge layer (a depletion layer of width W_{SC}) within the semiconductor with a band bending $e\phi_{SC}$. In equilibrium, the Fermi level is constant throughout the system. There are various ways in which equilibrium can be disturbed, for example, by illumination, by an externally applied potential, or by charge carrier injection from solution.

A photovoltaic (two-electrode) cell operates without an external voltage source [22, 23]. The working electrode is illuminated. Two limiting cases can be distinguished. If the electrodes are short-circuited, then the electrons and holes, generated by supra–band gap light, are separated by migration within the depletion layer and by diffusion. The holes react at the interface, oxidizing the reduced species

$$\mathrm{Red} + \mathrm{h_{VB}}^{+} \longrightarrow \mathrm{Ox}^{+} \tag{1}$$

The electrons are detected as photocurrent in the external circuit. This "short-circuit photocurrent" depends on the competition between hole transfer across the interface (the Faraday reaction) and electron-hole recombination in the bulk semiconductor and at the surface of the electrode. Under open-circuit conditions no photocurrent can flow in the external circuit. The Fermi level in the semiconductor is raised with respect to that in solution and a photovoltage is established. In the limiting case at high light intensity, the energy bands become flat (Fig. 1b). The maximum photovoltage is determined by the difference in Fermi level under flat band and equilibrium conditions. Clearly, such a system can supply electrical power (Sect. 2.1.2.2).

Unlike photovoltaic cells, most photoelectrochemical systems do not operate without an external source. In a more general approach, the potential of the working electrode can be varied with respect to the equilibrium potential V_{eq} (or to the potential of a RE) by means of an external voltage source (e.g. a potentiostat) connected between WE and CE [21, 24]. The current density (indicated here by j) is measured both in the dark and under illumination as a function of the applied potential V. This is described in more detail in Sect. 2.1.2.2.

A strong oxidizing agent ($Ox_S{}^+$), that is one with electron acceptor levels corresponding to the valence band of the semiconductor, can extract electrons from the band thus creating holes

$$Ox_S{}^+ \longrightarrow Red_S + h_{VB}{}^+ \qquad (2)$$

In an n-type semiconductor, these holes can recombine with majority carriers (electrons) from the CB [25]. Electroluminescence (EL) is expected from a semiconductor with a direct band gap for conditions in which the surface electron concentration is significant. Similarly, electron injection into a p-type semiconductor can give rise to light emission [25].

Macroporous semiconductor electrodes resemble in many respects the bulk electrodes described earlier. However, there are clear differences between porous and nonporous systems that become even more pronounced in nanoporous electrodes. Porous systems are considered in Sect. 2.1.5.

Recently, insulating nanocrystals have been attached to conducting substrates (metal or indium tin oxide) by van der Waals interactions or covalent bonding. When such a system is used as a working electrode in a photoelectrochemical cell, a photocurrent (in the μA range) can be measured [19, 20, 26]. Clearly, a monolayer of nanocrystals can absorb a sufficient fraction of the incident light to give to a measurable photocurrent. The absorption of a photon in a nanocrystal leads to a photoexcited state; an electron is promoted from a lower energy level (HOMO or valence band) to a higher energy level (LUMO or CB) (Fig. 2). A photocurrent will be observed if the electron and hole are separated effectively before the system relaxes to the ground state. In Fig. 2, the electron is transferred to the

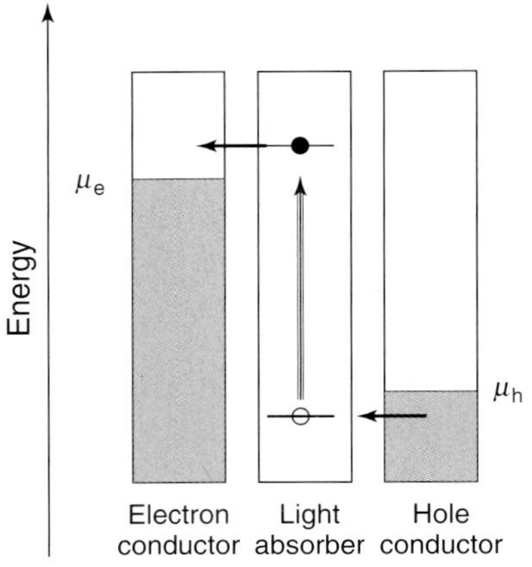

Fig. 2 Scheme representing the general principle of a photoelectrochemical system. Electrons are photoexcited in the absorber. The electron and hole are selectively transferred to an electron conductor (usually a metal or a semiconductor) and to a hole conductor (a redox system in a liquid electrolyte). The photovoltage is the difference between the electrochemical potentials in the electron and hole conducting phases; thus $\mu_e - \mu_h$.

metal, the hole to an oxidizable species in solution. This leads to an upward shift of the electrochemical potential of the electrons μ_e and a downward shift of the electrochemical potential of the holes μ_h (and thus to a photovoltage $V_{photo} = (\mu_e - \mu_h/e)$. Such a photoelectrochemical system shows a strong analogy with molecular (e.g. dye) systems adsorbed on metal or semiconductor electrodes.

The photoelectrochemical solar cell of O'Regan and Grätzel [16] combines the special features of semiconductor porosity and sensitization described earlier. This system consists of nanometer-sized TiO_2 crystallites interconnected to form a three-dimensional porous assembly. Dye molecules acting as the light absorber are chemisorbed on the internal surface. The efficiency of light absorbance is close to unity as a result of the fact that photons encounter an adsorbed dye molecule many times. The electrolyte solution, containing an I^-/I_3^- redox system, permeates the pores of the system. The TiO_2 assembly acts as an electron conductor and the electrolyte as the "hole conductor". Because of the interpenetration of the electron and hole conducting phases on a nanometer scale, back transfer of the electron to I_3^- is the main source of recombination.

The three systems described earlier have a number of features in common. To observe photocurrent, light obviously has to be absorbed by the system. The spatial separation of photogenerated electrons and holes must be more effective than their recombination. This requires efficient kinetics for electron transfer across the solid/solution interface. These are topics that are addressed in the remainder of the chapter. Electrode reactions in photoelectrochemical systems may be quite complicated. To illustrate the general approach to the elucidation of reaction mechanisms two complex systems are considered. Finally, charge carrier transport in porous semiconductor systems is a topic deserving special attention.

2.1.2 Photoelectrochemical Characterization Methods

For scientific research on photoelectrochemical systems, the photoelectrode is generally the working electrode in a three-electrode electrochemical cell. Using this set up, the system can be investigated by perturbing it and recording the system's response. In general, such perturbation-response methods may be classified according to the time dependence of the perturbation (steady state versus time-resolved) or to the physical nature of the perturbation and the response. In this section, the principles of steady state and time-resolved methods in general are first discussed. Later on, two different perturbation-response techniques used to study semiconductor-based photoelectrochemical systems are discussed. Finally, methods in which illumination or minority carrier injection gives rise to light emission from the electrode are considered.

2.1.2.1 Steady State and Time-resolved Methods

Steady state methods are basically simple: one applies a time-independent perturbation x to a system and records – if it exists – the time-independent response $y(x)$ of the system. This may be repeated for different levels of the perturbation. In this way, a functional relation between perturbation and response is determined, which should provide information on the system studied. Consider, for example, an electrical resistor through which a current is passed (perturbation). Instantaneously, the system responds by maintaining a potential difference across the resistor. The linear relation between both quantities allows the characterization of the system.

A major drawback of steady state techniques is that no information is obtained on the dynamic properties of a system. Take for instance an electronic circuit, which simulates a simple resistor except that the potential difference is established between its poles after some time delay. Clearly, this circuit would emerge from a steady state analysis as a simple resistor: the system dynamics, which is a basic characteristic of any system, can be revealed only by recording the response from the moment the perturbation is applied (time-resolved measurements).

Many time-resolved methods do not record the transient response as outlined in the earlier example. In the case of linear systems, all information on the dynamics may be obtained by using sinusoidally varying perturbations $x(t)$ (harmonic modulation techniques) [27], a method far less sensitive to noise. In this section, the complex representation of sinusoidally varying signals is used, that is, $x(t) = \mathrm{Re}[\widetilde{X}(\omega)\exp(i\omega t)]$, where $i = \sqrt{-1}$. The quantity $\widetilde{X}(\omega)$ contains the amplitude and the phase information of the sinusoidal signal, whereas the complex exponential $\exp(i\omega t)$ expresses the time dependence. A harmonically perturbed linear system has a response that is – after a certain transition time – also harmonic, differing from the perturbation only by its amplitude and phase (i.e. $y(t) = \mathrm{Re}[\widetilde{Y}(\omega)\exp(i\omega t)]$). In this case, all the information on the dynamics of the system is contained in its transfer function $H(\omega)$, which is a complex function of the angular frequency, defined as [27, 28]

$$H(\omega) = \frac{\widetilde{Y}(\omega)}{\widetilde{X}(\omega)} \tag{3a}$$

Time-resolved measurements on linear systems may be represented in many ways. For example, in the Nyquist representation, the transfer function $H(\omega)$ is plotted as a point in a two-dimensional plane having coordinates [$\mathrm{Re}(H)$ and $\mathrm{Im}(H)$], for each frequency measured. In specific cases, the transfer function may be represented also by an equivalent electrical circuit. This is a combination of lumped circuit elements (resistor, capacitor, etc.) having the same perturbation-response behavior as the system studied.

In the case of a nonlinear system, a similar approach using harmonic perturbations is possible if a "small-signal" perturbation $x(t) = \mathrm{Re}[\Delta\widetilde{X}(\omega)\exp(i\omega t)]$, superimposed on a time-independent "bias" perturbation, is applied to the system. If the signal level of the perturbation is sufficiently small, a linear dependence of the response on the perturbation can be achieved (i.e. $y(t) = \mathrm{Re}[\Delta\widetilde{Y}(\omega)\exp(i\omega t)]$). Clearly, the transfer function defined in Eq. (3a) becomes a differential quantity:

$$H(\omega) = \frac{\Delta\widetilde{Y}(\omega)}{\Delta\widetilde{X}(\omega)} \tag{3b}$$

At low frequencies, $H(\omega)$ is equal to the slope $\partial y/\partial x$ of the steady state response $y(x)$. The time-resolved electrochemical techniques discussed in Sects. 2.1.2.2 and 2.1.2.3 pertain to this class of small-signal modulation techniques.

In general, measuring the transfer function of a system under study using harmonic modulation techniques is straightforward. Interpreting the experimental data, however, is not. As will be demonstrated by experimental examples in Sects. 2.1.3 and 2.1.5, time-resolved methods become most powerful if the experimental impedance can be analyzed using a dynamic model for the system studied.

2.1.2.2 Current Density Versus Potential Techniques

Because most applications of (photo)electrochemical systems involve the transfer of electrons across an interface (Sect. 2.1.1), current density-potential techniques are commonly used in (photo)electrochemistry. In this case, the difference in electrochemical potential of electrons across the interface of interest (accessible via the working electrode – reference electrode potential difference) and the current density through this interface are used as the perturbation and the response (or vice versa). Two approaches can be distinguished. When (quasi) steady state signals are used, one speaks of current density versus potential measurements whereas harmonically modulated signals, superimposed on a bias, are involved in electrochemical impedance spectroscopy (EIS). We introduce these two approaches on the basis of the kinetics of the simple system shown in Fig. 1.

If the potential of the n-type electrode, whose energy band diagram is shown in Fig. 1(a), is made negative with respect to the equilibrium potential V_{eq}, then, the band-bending $e\phi_{SC}$ decreases until finally flat band condition ($V = V_{fb}$) is reached (Fig. 1b). If the potential is made more negative than the flat band value ($V < V_{fb}$), then majority carrier "accumulation" occurs at the surface. The decrease in band-bending on going from Figs. 1(a) to 1(b) is accompanied by an increase in the electron concentration at the surface. As a result, a net cathodic current flows across the interface due to the reduction of the oxidized species

$$\mathrm{Ox}^{+} + \mathrm{e_{CB}}^{-} \longrightarrow \mathrm{Red} \qquad (4)$$

Here, we assume that electron transfer only occurs via the CB and not via surface states. As in a Schottky diode, j generally increases exponentially with (decreasing) potential (Fig. 3a). The form of the dark current-potential curve, however, depends on the mechanism and kinetics of the charge-transfer reaction. At high overpotential, corresponding to a large deviation from equilibrium, the reaction expressed by Eq. (4) may become limited by mass transport in solution, that is, the cathodic current becomes potential-independent (this is not shown in Fig. 3).

If the potential of the electrode of Fig. 1 is made more positive than V_{eq}, the band-bending increases with respect to that at

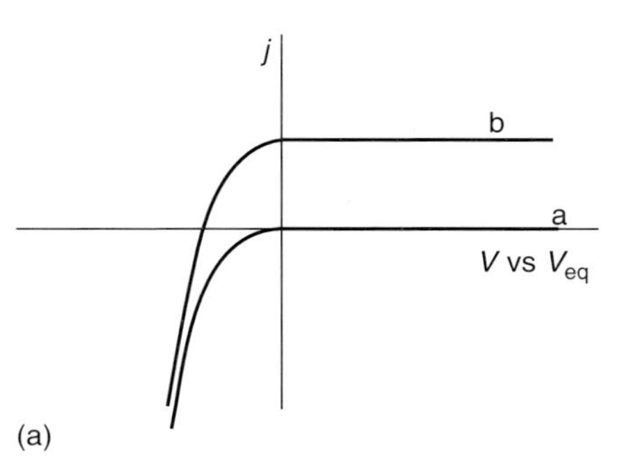

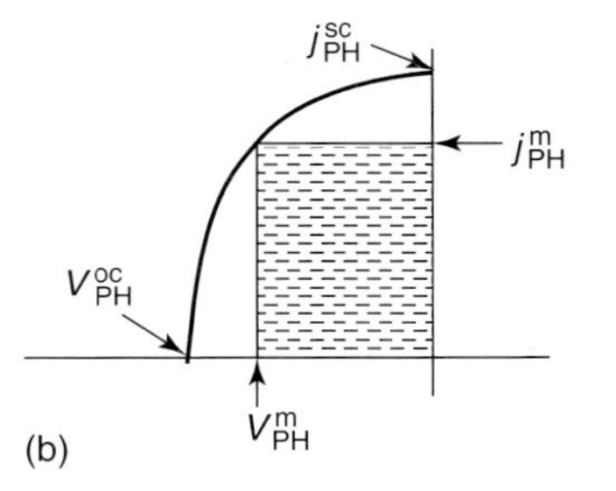

Fig. 3 (a) Schematic representation of the current-potential curves for an n-type semiconductor (see Fig. 1) in the dark (a) and under illumination with supra-band gap light (b). (b) The part of curve (b) relevant to photovoltaic applications. The open-circuit photovoltage V_{PH}^{oc} and the short-circuit photocurrent j_{PH}^{sc} are indicated, as is the rectangle defining j_{PH}^{m} and V_{PH}^{m}.

equilibrium and strong depletion, or even inversion, may result. This corresponds to the blocking current range of the diode ($V > V_{eq}$). A small potential-independent anodic current results from electron injection from the reduced species across the barrier into the conduction band.

Under illumination, a photocurrent is observed under depletion conditions if the photogenerated electron and hole are spatially separated before recombination can occur. The hole reacts at the surface (Eq. 1) and the electron is collected at the counter electrode. For the simplest case in which no recombination occurs at the surface, Gärtner [29] derived an expression for the photocurrent density j_{PH} taking into account the absorbed photon flux Φ, the absorption coefficient α, and the minority carrier diffusion length L_{min}

$$j_{PH} = e\Phi\left[1 - \frac{\exp(-\alpha W_{SC})}{1 + \alpha L_{min}}\right] \quad (5a)$$

The potential dependence of the photocurrent is determined in the model by the dependence of the depletion layer thickness on the band-bending

$$W_{SC} = \left(\frac{2\varepsilon\varepsilon_0\phi_{SC}}{eN_D}\right)^{1/2} \quad (5b)$$

where N_D is the donor density, ε is the dielectric constant of the semiconductor, and ϕ_{SC} equals $V - V_{fb}$ (see Sect. 2.1.3.1). In essence, this simple model states that an electron/hole pair will contribute to the photocurrent if it is generated within a distance from the electrolyte interface within which the electron and hole can be separated by migration and diffusion, that is, the penetration depth of the light $1/\alpha$ is less than $L_{min} + W_{SC}$. Apart from the applied potential (which determines W_{SC}), the efficiency of charge separation depends on the quality of the semiconductor (L_{min}) and on α (which is a function of the wavelength). Light absorption just above the band edge depends strongly on whether the semiconductor has a direct band gap (for which α is large and the penetration depth is small) or an indirect band gap (for which α is small and $1/\alpha$ is large). The importance of such factors is illustrated in Sect. 2.1.5.2.

Recombination of electrons and holes via surface states competes with hole transfer to solution and thus reduces, and may even suppress, the photocurrent. To obtain a limiting photocurrent, a stronger band-bending is required. Obviously, more extended models are needed to describe the photocurrent-potential characteristics in this "less favorable" case [24].

A schematic curve for the dependence of the total current density (dark current + PC) on the potential is shown in Fig. 3(a). It is clear that the form of this curve on the anodic side will depend on the parameters as discussed earlier. The part of the curve relevant to photovoltaic applications is shown in Fig. 3(b). The short-circuit photocurrent density j_{PH}^{sc} and the open-circuit photovoltage V_{PH}^{oc} are indicated. It should be noted that this part of the curve can be mapped without an external voltage source by measuring the photocurrent through and the voltage across a load resistor (between WE and CE) whose resistance is varied from zero to a very large value. The conversion efficiency η of a photovoltaic cell is defined as $\eta = P_m/P_o$ where P_o is the power of the incident radiation and P_m, the maximum power output of the cell, is given by

$$P_m = j_{PH}^m \times V_{PH}^m \quad (6a)$$

The values of j_{PH}^m and V_{PH}^m depend on the shape of the current-potential curve under illumination (i.e. on both the dark current and photocurrent-potential characteristics). For a given temperature and light intensity $j_{\mathrm{PH}}^{\mathrm{m}}$ and $V_{\mathrm{PH}}^{\mathrm{m}}$ are chosen to give a rectangle of maximum area in the current-potential plot (see shaded area in Fig. 3b). An important parameter for the characterization of a solar cell is the fill factor (FF) defined as

$$\mathrm{FF} = \frac{j_{\mathrm{PH}}^{\mathrm{sc}} V_{\mathrm{PH}}^{\mathrm{oc}}}{j_{\mathrm{PH}}^{\mathrm{m}} V_{\mathrm{PH}}^{\mathrm{m}}} \tag{6b}$$

In steady state measurements, one generally applies a fixed potential to the working electrode and measures the steady state current through the cell. Alternatively, the potential is scanned at a fixed rate and the current is measured continuously. As in metal electrochemistry, information about the kinetics of surface reactions and the hydrodynamics of the system can be obtained by varying the potential scan rate. The rotating disk electrode is a useful tool for studying the role of mass transport. The rotating ring-disk electrode (RRDE) has two important applications in semiconductor electrochemistry. In studies of competitive reactions at the semiconductor disk, the reaction products can be analyzed at the ring of the RRDE (see Sect. 2.1.3.3). Alternatively, reactive species can be generated at the metal disk of a RRDE and their electrochemistry studied at the semiconducting ring [25]. Important parameters in photoelectrochemistry are the intensity and the spectral distribution of the light used for photocurrent and photovoltage studies. In photovoltaics, steady state measurements can be used to obtain FF and solar energy conversion efficiencies [24]. Such results may give insight into reaction mechanisms.

From the earlier discussion, it is clear that the j versus the V relation of an electrochemical system is nonlinear. Therefore, the electrochemical admittance (or its inverse, the electrochemical impedance) is defined as the transfer function relating a small signal variation of the working electrode potential and of the current density through the working electrode/electrolyte interface:

$$Y_{\mathrm{EC}}(\omega) = \frac{\Delta \tilde{j}(\omega)}{\Delta \tilde{V}(\omega)} \tag{7}$$

In the case of a semiconductor-based photoelectrochemical system, the measurement of the electrochemical admittance serves two purposes. As is explained in Sect. 2.1.3.1, it allows on the one hand the in situ determination of the energetics of the (bulk) semiconductor surface. On the other hand, it makes the dynamics of various (photo)electrochemical processes experimentally accessible. Clearly, EIS is also possible using an illuminated semiconductor, an experimental method sometimes referred to as PEIS. Finally, it should be noted that although the electrochemical admittance is determined experimentally (the applied electrode potential is used as the perturbation), the electrochemical impedance is generally plotted as the result of an EIS measurement.

2.1.2.3 **PC versus Light Intensity Techniques**

Typical to the study of photoelectrochemical systems are measurement techniques that use the light flux Φ incident on the working electrode as a perturbation and the resulting PC j_{PH} as the system's response [30]. In this case, the electrode potential can be used as an additional experimental variable. Clearly, the incident light flux cannot be quantified directly. Therefore, a reference signal proportional to this light flux is generally used, for example, the voltage generated by the light flux on a reference photodiode. This technique, which for steady state conditions has no particular name, has been applied several times with success, for example, to identify (light-intensity dependent) photocurrent multiplication processes (see Sect. 2.1.3.3). The optoelectrical transfer function relates an incident sinusoidally modulated light flux and the resulting modulated photocurrent density. This quantity is defined as [30]

$$Y_{OE}(\omega) = \frac{\Delta \tilde{j}_{PH}(\omega)}{e\Delta\tilde{\Phi}(\omega)} \tag{8}$$

Again, it should be stressed that the optoelectrical transfer function is defined using small-signal perturbations, superimposed on time-independent bias signals. Measurement of the optoelectrical impedance is often referred to as intensity modulated photocurrent spectroscopy (IMPS).

2.1.2.4 Luminescence-based Techniques

Photoluminescence (PL), like photocurrent, is a technique in which a light flux incident on the working electrode acts as a perturbation. In this case, the response of the system is followed by measuring the intensity of the emitted light. As in photocurrent measurements, the electrode potential can be used as an additional variable. As explained in Sect. 2.1.2.2, photocurrent is obtained when, under depletion conditions, electrons and holes created by illumination are effectively separated by migration and diffusion (Fig. 4). PL, on the other hand, is expected when the photogenerated carriers recombine radiatively, that is, at potentials approaching the flat band potential V_{fb} (see Sect. 2.1.3.1). Emission may result either from direct band-band recombination or from indirect recombination via a band gap state [25]. In the simplest case in which surface recombination can be disregarded, the potential dependence of the emission intensity I_{PL} is described by the Gärtner equation:

$$I_{PL} = \kappa\Phi\left[\frac{\exp(-\alpha W_{SC})}{1+\alpha L_{min}}\right] \tag{9}$$

where Φ is the incident photon flux and κ is the ratio of the rate of radiative recombination to the total recombination rate. In principle, it is possible to obtain values for L_{min}, W_{SC}, and α from the potential dependence of I_{PL}. The Gärtner model has been extended by Gerischer and coworkers [31, 32] to account for surface recombination. In this case, surface recombination rates can be obtained from photoluminescence measurements.

While photoluminescence has been mainly studied under steady state conditions or with transient techniques, harmonic-modulation measurements are possible. Beckmann and Memming [33]

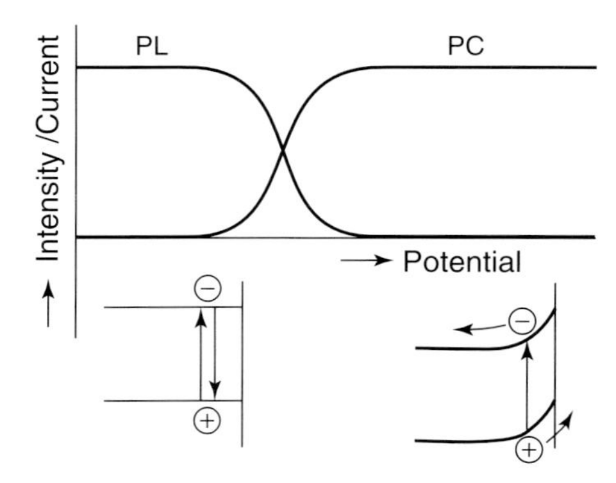

Fig. 4 Schematic representation of the potential dependence of the PC and the PL intensity of an *n*-type semiconductor in an indifferent electrolyte solution. Energy band diagrams are shown for the illuminated semiconductor under flat band and depletion conditions.

studied the photoluminescence of n-type GaP by perturbing the system with a small sinusoidal modulation of the potential during a potential scan and measuring the response with lock-in techniques.

EL is observed when minority carriers, injected into a semiconductor from solution, recombine radiatively with majority carriers [25]. This is illustrated schematically in Fig. 5 for an n-type semiconductor. At positive potentials, the surface electron concentration is very low. Holes injected by the oxidizing agent in solution are held at the electrode surface by the electric field of the depletion layer. The surface holes generally cause oxidation and dissolution of the semiconductor. Clearly, in this potential range neither current nor light emission is observed. As the potential is made negative and the band-bending decreases, the injected holes recombine with electrons supplied via the external circuit. This results in a cathodic current and, if recombination is radiative, in light emission. In general, the cathodic current becomes potential-independent at negative potentials as a result of mass-transport limitations in solution. The emission intensity should, therefore, become constant. However, this is often not the case. A more complicated potential-dependence points to changes in surface chemistry that influence nonradiative surface recombination. The observation of electroluminescence in an electrochemical system clearly shows the involvement of minority carriers in the charge-transfer processes.

As far as we are aware, experiments on electroluminescence involving a sinusoidal perturbation of the applied potential have not been performed but large-signal potential step and pulse measurements have been reported. We shall return briefly to these aspects in Sect. 2.1.3.2.5.

2.1.3
Bulk Systems

In this section, we consider some important aspects of photoelectrochemical systems based on single-crystal and polycrystalline electrodes. In Sect. 2.1.3.1, the use of EIS is described for the determination of the energetics of the semiconductor/electrolyte interface. Section 2.1.3.2 deals with the dynamics of electron/hole recombination at the semiconductor surface and the importance of electrochemical and optoelectrical impedance techniques for such studies. Finally, a somewhat

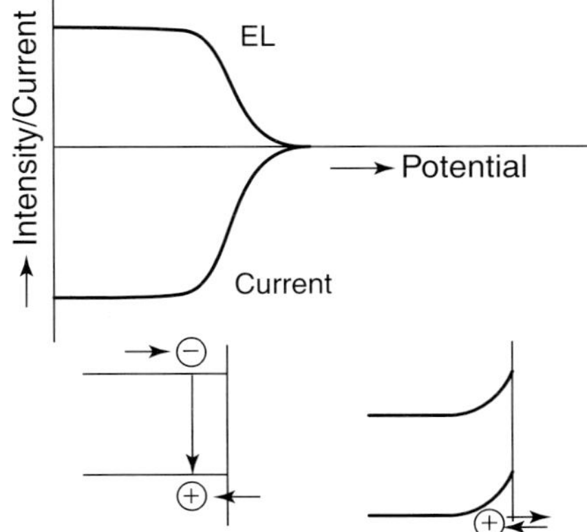

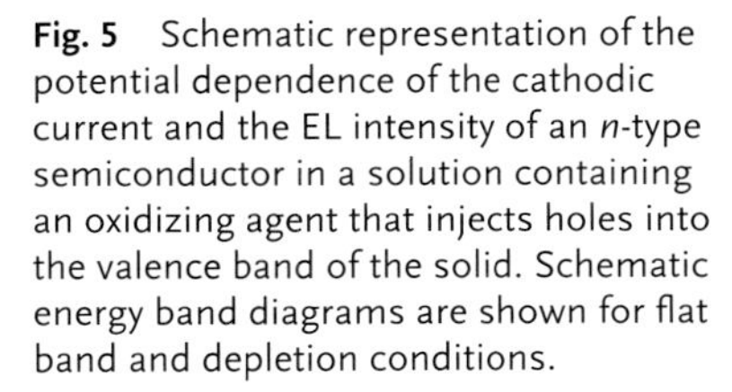
Fig. 5 Schematic representation of the potential dependence of the cathodic current and the EL intensity of an n-type semiconductor in a solution containing an oxidizing agent that injects holes into the valence band of the solid. Schematic energy band diagrams are shown for flat band and depletion conditions.

arbitrary choice of complex reactions is used to illustrate the general approach to the study of reaction mechanisms and reaction kinetics (see Sect. 2.1.3.3).

2.1.3.1 In Situ Energetics of Semiconductor/Electrolyte Interfaces

An aspect of great importance to the electrochemical properties of a (bulk) semiconductor/electrolyte (s/e) interface is the energetic position of the upper edge of the valence band and the lower edge of the CB of the semiconductor at the interface. According to the Gerischer-model for charge transfer at bulk semiconductor electrodes, the energetics of the s/e interface plays an essential role in determining the rate and the mechanism of an electrode process [34–36]. Capacitance measurements – hence, electrochemical impedance measurements – constitute the most widely used in situ method to determine the energetics of s/e interfaces [37].

As described earlier, the s/e interface is electrified, that is, charge is separated at the interface. According to Gauss' law, the charge separation is accompanied by an inner-potential difference between the semiconductor and the electrolyte. Consequently, in the absence of any electrochemical process, the s/e interface acts as a capacitor, the capacitance of which may be calculated from the electrochemical admittance ($Y_{EC} = i\omega C$).

Most simply, the system is modeled by assuming depletion or accumulation of free charge carriers at the semiconductor side of the interface. This charge is neutralized by an ionic counter-charge at the electrolyte side of the interface (Helmholtz layer) [36, 38]. From this picture, two (differential) capacitances may be defined. The first, C_{SC}, relates a change of the charge Q_{SC} accumulated in the semiconductor to a change of the potential drop ϕ_{SC} across the semiconductor ($C_{SC} = \partial Q_{SC}/\partial \phi_{SC}$). The second, C_{EL}, relates analogously the charge Q_{EL} in and the potential difference ϕ_{EL} across the electrolyte side of the interface ($C_{EL} = -\partial Q_{EL}/\partial \phi_{EL}$). Assuming that the total potential drop ϕ across the interface equals the sum $\phi_{SC} + \phi_{EL}$, it may be easily shown that the interfacial capacitance C corresponds to the capacitance of the series connection of the capacitors C_{SC} and C_{EL} [36, 37, 39].

In Fig. 6, the inverse square of the capacitance of an n-InP|1.2 M HCl solution interface is plotted as a function of the potential applied to the n-InP electrode. The capacitance has been calculated by modeling the interface as a parallel connection of a capacitor and a resistor in series with the cell resistor. Clearly, both quantities are linearly related. This result can be

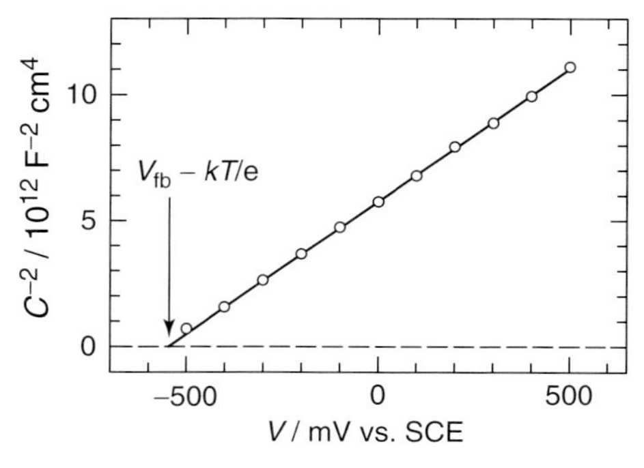

Fig. 6 Mott-Schottky curve determined at an n-InP electrode in a 1.2 M HCl aqueous solution. The interfacial capacitance is determined from the electrochemical impedance measured at 8.2 kHz using a parallel connection of a resistor and a capacitor in series with the cell resistor.

understood if majority charge carriers are depleted from the semiconductor surface. In that case, the capacitance C_{SC} is given by the Mott-Schottky equation, which for an n-type semiconductor reads [36, 37, 39]:

$$C_{SC} = \left(\frac{eN_D\varepsilon\varepsilon_0}{2}\right)^{1/2}\left(\phi_{SC} - \frac{k_BT}{e}\right)^{-1/2} \tag{10}$$

where T is the absolute temperature and the constants e and k_B indicate the elementary electrical charge and the Boltzmann constant, respectively. The capacitance C_{SC} can be identified directly with the interfacial capacitance if $C_{SC} \ll C_{EL}$. Although exact values of C_{EL} are scarce in the literature, this inequality is generally assumed to be fulfilled for moderately doped semiconductors under depletion conditions [36, 37, 39, 40]. As a consequence, a change of the potential drop ϕ across the interface results mainly in a change of ϕ_{SC} (i.e. ϕ_{EL} is approximately constant). Hence, ϕ_{SC} may be written as the difference between the applied electrode potential V and the so-called flat band potential V_{fb} [37]:

$$\phi_{SC} = V - V_{fb} \tag{11}$$

At the flat band potential, there is no potential drop across the semiconductor and, hence, the semiconductor energy bands are flat from the bulk up to the semiconductor surface (see Fig. 1b). Moreover, because ϕ_{EL} is approximately constant, the energy bands are fixed at the semiconductor surface. Hence, the position of the band edges at the surface may be calculated once the flat band potential is known.

A problem often encountered when calculating the flat band potential from Mott-Schottky data is the frequency dependence of both the slope and the extrapolation point of the C^{-2} versus V curve. Obviously, such so-called frequency dispersion hampers a proper determination of V_{fb} although reliable values are generally obtained at high measuring frequencies (>10 kHz). The origin of this nonideal behavior is not well understood [37, 41]. However, to check the reliability of Mott-Schottky measurements, the capacitance should be measured in a broad frequency range [42–44].

2.1.3.2 Electron-hole Recombination Dynamics at the s/e Interface

2.1.3.2.1 **Introduction** The dynamics of electron-hole recombination at the semiconductor surface has been extensively studied both at illuminated and at dark s/e interfaces [45–53]. For recombination, minority charge carriers should be present at the interface. For n-type semiconductors, holes may be supplied to the surface by illumination under depletion conditions using supra-band gap light (see Sect. 2.1.2.2). Alternatively, holes may be injected into the valence band by a strong oxidizing agent in the electrolyte solution. If a depletion layer exists, the injected holes accumulate at the semiconductor surface.

As it gives a nice and relatively simple illustration of the use of various characterization methods, we will discuss the subject of electron-hole recombination dynamics in the next sections, taking the n-GaAs photoanode as the main example. More complicated topics – related to interfacial transfer of photogenerated charge carriers – are discussed in Section 2.1.3.3.

2.1.3.2.2 **Current Density Versus Potential Measurements** Figure 7 shows a current density versus potential curve for an illuminated n-GaAs electrode ($\lambda = 480$ nm) in a 0.1 M H_2SO_4 solution. As reported for various photoanodes in the literature,

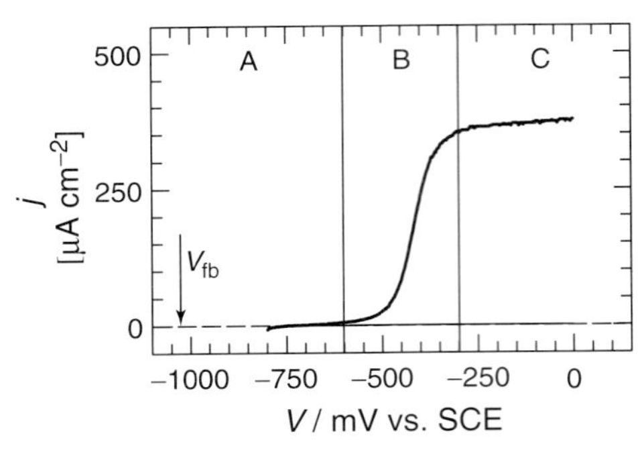

Fig. 7 Current density versus potential curve, recorded at an illuminated n-GaAs electrode in a 0.1 M H_2SO_4 aqueous solution. Indicated are the flat band potential (as determined in the dark) and the different potential ranges (see text).

the j versus V plot shows – at potentials more positive than the flat band potential – three different regions, indicated as A, B, and C in Fig. 7 [45, 47, 49, 50, 54]. In region A, that is closest to V_{fb}, no current passes through the s/e interface, despite the minority charge carrier flux towards the surface. On the other hand, a photocurrent plateau appears in region C – at the most positive potentials. The intermediate region B shows a transition between these two extreme situations. Clearly, the occurrence of a photocurrent plateau is in accordance with the Gärtner-equation: in this potential region, all photogenerated holes reach the semiconductor surface and participate in a charge-transfer reaction (the oxidation of the semiconductor). On the other hand, since $\alpha^{-1} \ll W_{SC}$ in the case of n-GaAs, the absence of photocurrent in region A is not accounted for by the Gärtner-equation. This indicates a loss of photogenerated holes, which can be attributed to electron-hole recombination at the semiconductor surface.

2.1.3.2.3 The Electrochemical Impedance of Surface Recombination

Figure 8 shows the impedance spectrum of the illuminated n-GaAs|0.1 M H_2SO_4 interface, as recorded in the potential region A (no steady state photocurrent). One can see a small semicircle at high frequencies. For various illumination intensities, the diameter of the semicircle fitting the data at high frequencies equals approximately $kT/e|j_{PH}|$ [45–47, 49]. In addition, it was shown that upon illumination, a capacitive peak appears in the C^{-2} versus V plot of the n-GaAs|0.1 M H_2SO_4 interface [45, 46, 51]. The peak value proved to be a function of the frequency and the photocurrent density as measured in region C [51]. This behavior is markedly different from the purely capacitive impedance (vertical line in the Nyquist plane and straight Mott-Schottky plot) expected for a blocking s/e interface (see Sect. 2.1.3.1).

The appearance of both the semicircle and the capacitive peak were accounted for by Vanmaekelbergh and coworkers, by considering recombination of photogenerated holes with CB electrons at the semiconductor surface [51, 55–57]. The recombination mechanism assumed by these authors consists of the successive capture of an electron in an empty surface state and of a hole in an occupied surface state. Taking the rates of the electron (hole) capture steps to be first order in the CB electron density n_S (valence band hole density p_S) and the density of empty (filled) surface states, an electrochemical impedance corresponding to the equivalent circuit shown in Fig. 9 was calculated for this recombination mechanism. The

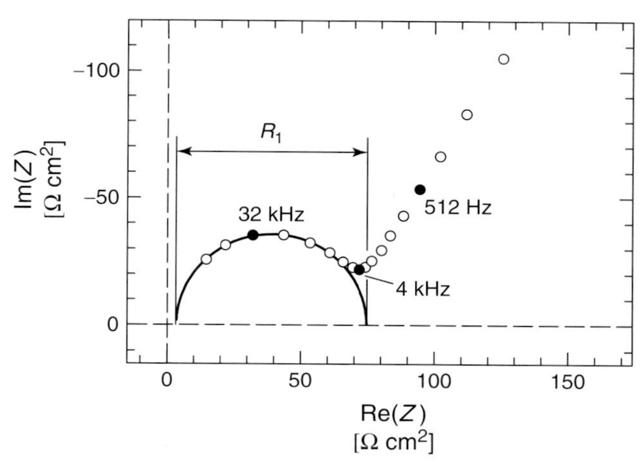

Fig. 8 High-frequency electrochemical impedance spectrum, obtained at an illuminated *n*-GaAs electrode in a 0.1 M H_2SO_4 aqueous solution. Bias potential: – 600 mV versus SCE (i.e. potential region A); limiting photocurrent density: 380 μA cm^{-2}. The resistance R_1 equals 71 Ω cm^2, that is, $1.06 \times kT/ej_{PH}$.

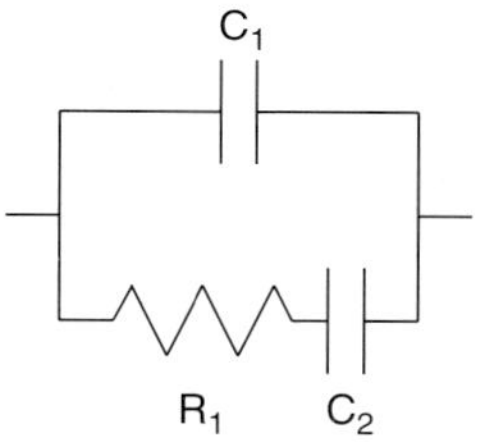

Fig. 9 Equivalent circuit obtained for surface recombination at a semiconductor/electrolyte interface. Explicit expressions of the different circuit elements are given in the text.

impedance of the three different circuit elements read [51]:

$$R_1 = \frac{kT}{e|j_{REC}|} \tag{12}$$

$$R_2 = \frac{kT}{e|j_{REC}|}\frac{\beta_n n_S}{\beta_p p_S}\frac{1}{1-|j_{REC}/j_{PH}|} \tag{13}$$

$$C_1 = \frac{e|j_{REC}|}{kT}\frac{1}{\beta_n n_S} \tag{14}$$

where j_{REC} equals the current density associated with recombination and β_n (β_p) denote the rate constant for electron (hole) capture.

Because the photocurrent density adds to the current density that (dis)charges the space charge layer of the semiconductor, the overall impedance of the s/e interface consists of the parallel connection of the surface-recombination impedance and the space charge layer capacitor. At high frequencies, the recombination impedance reduces to the resistor R_1. Hence, the high-frequency impedance of the interface corresponds to the parallel connection of R_1 and C_{SC}. This parallel combination accounts for the experimentally observed high-frequency semicircle (note the correspondence between the experimental diameter and R_1). In addition, the features of the capacitive peak in the Mott-Schottky plot could be qualitatively explained by considering the parallel connection of the recombination impedance and the space charge layer capacitor [51].

Clearly, electron-hole recombination is not limited to illuminated semiconductors. Minority charge carriers also may be injected by an oxidizing or reducing agent in the electrolyte solution (see Sects. 2.1.1.2 and 2.1.2.4). Also in this case, the parallel connection of the surface recombination impedance and the space charge capacitor provides an accurate description of the experimental impedance [45]. For instance, the impedance spectrum of the n-GaAs|Ce^{4+} system – Ce^{4+} is a well-known hole-injecting agent for n-GaAs – shows a capacitive semicircle with a diameter equal to $kT/e|j|$ at high frequencies [45, 58]. It was also demonstrated that a capacitive peak, exhibiting the same functional dependence on frequency and current density as obtained with the n-GaAs photoanode, is present in a Mott-Schottky plot measured with the n-GaAs|Ce^{4+} system [45].

Both examples attribute convincingly the loss of photogenerated holes at the n-GaAs photoanode polarized in potential region A to electron-hole recombination at the electrode surface. Because similar results have been obtained at the n-CdTe photoanode [50], one could think of the surface-recombination impedance as a general fingerprint of surface-recombination steps in an overall reaction mechanism. This is, however, not the case. For the n-InP photoanode, the features of the surface recombination are absent although recombination was shown to occur at the semiconductor surface [54]. In this particular case, this discrepancy was resolved by assuming that recombination does not occur at fixed recombination centers but rather at intermediates of the anodic decomposition of the semiconductor [53]. In addition, it was shown that the typical semicircle resulting from the parallel connection of the space charge layer capacitor and a resistor with resistance $kT/e|j|$ is not uniquely related to surface-recombination processes. Any reaction step, the rate of which is proportional to the density of majority charge carriers at the interface, may contribute this feature to the overall impedance [59–61]. Hence, for direct transfer of majority charge carriers or for surface-state mediated transfer, this semicircle also may appear in the overall impedance spectrum [59]. These latter examples demonstrate the need for a reliable dynamic model of the charge-transfer reaction for the interpretation of impedance data.

2.1.3.2.4 **The Optoelectrical Transfer Function of Surface Recombination** If the n-GaAs|0.1 M H_2SO_4 system, polarized in potential region A, is suddenly exposed to a constant illumination, a photocurrent transient decaying from the value as measured in potential region C to zero is recorded [30]. This photocurrent decay is related to the increase from zero recombination at the moment of the exposure (maximum photocurrent) to complete recombination (no photocurrent). The same picture arises from the optoelectrical transfer function (Fig. 10), which corresponds to a semicircle ranging from $Y_{OE} = 1$ at high frequencies to $Y_{OE} = 0$ at low frequencies [62]. As for the transient at short times, recombination is ruled out at high modulation frequencies leading to a maximum value of the modulated photocurrent, whereas at low frequencies recombination is fully operative, causing the disappearance of the photocurrent.

This impedance could be analyzed using the same dynamic model that describes the surface-recombination impedance. If the

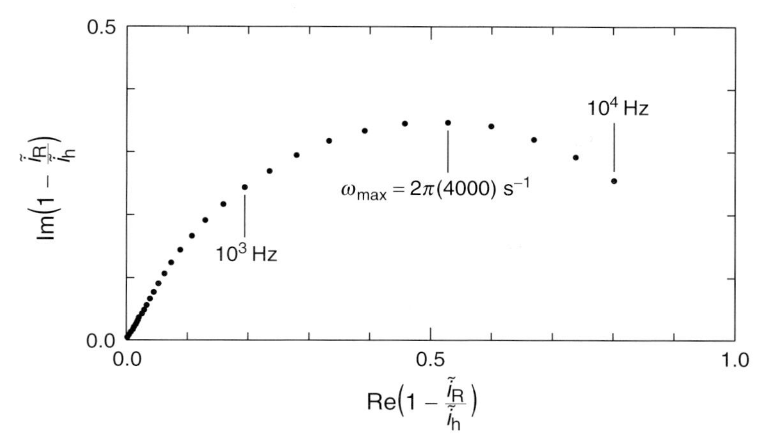

Fig. 10 Plot of the optoelectrical transfer function measured at an illuminated *n*-GaAs electrode in a 1 M H_2SO_4 aqueous solution. Bias potential: – 600 mV versus SCE (i.e. potential region A); limiting photocurrent density: 800 μA cm^{-2} (from Ref. 62).

response time of the electrochemical cell is sufficiently fast, the resulting optoelectrical impedance, valid for potential region A, reads according to Vanmaekelbergh and coworkers [62]:

$$Y_{OE} = 1 - \frac{\beta_n n_S}{i\omega + \beta_n n_S} \quad (15)$$

Apart from the flattening of the semicircle, this transfer function accounts well for the experimental impedance. Moreover, from the characteristic frequency of the semicircle, the rate constant of electron capture may be calculated if the electron density at the semiconductor surface is known. From the example given in Fig. 10, a value of $\beta_n = 10^{-6}$ $cm^3\ s^{-1}$ is obtained [62]. Analogous to the case of the *n*-GaAs photoanode, loss of photogenerated holes could be attributed to surface recombination by IMPS in the case of the illuminated *n*-CdS and *n*-InP electrodes [62, 54].

Summarizing, we can conclude that both examples demonstrate that impedance techniques are powerful tools for the determination of reaction mechanisms, especially if they are combined with a mathematical model of the system studied. Clearly, EIS and IMPS yield partly the same information. However, they are also complementary: IMPS enables the determination of the rate constant of majority charge carrier capture whereas the density of majority charge carriers at the interface is accessible only via EIS. Moreover, EIS allows one to investigate analogous processes at dark electrodes.

2.1.3.2.5 **Recombination Studied by Luminescence** On normalizing the photoluminescence intensity I_{PL} in Eq. (9) with respect to the maximum intensity I_{max} (for $W_{SC} = 0$) one obtains

$$\frac{I_{PL}}{I_{max}} = \exp(-\alpha W_{SC}) \quad (16)$$

Because the thickness of the depletion layer can be obtained as a function of potential from impedance measurements (see Sect. 2.1.3.1) Eq. (16) can be used to check the validity of the Gärtner

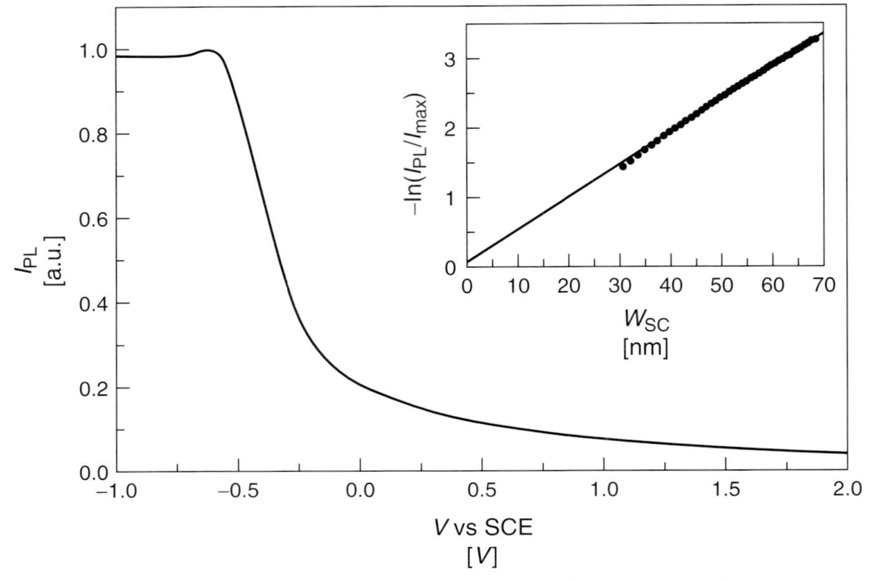

Fig. 11 The intensity of the PL measured at 2.2 eV (the "yellow" emission) from an n-type GaN electrode as a function of potential. The excitation energy was 4 eV. In the inset the results are plotted according to Eq. (16) (from Ref. 25).

approach and to determine the absorption coefficient [63] (see Fig. 11). Alternatively, if α is known accurately then Eq. (16) gives information about the potential distribution within the semiconductor, as shown by Ellis and coworkers for GaAs [63, 64]. If α values are either available or determined for different wavelengths, then the minority carrier diffusion length [$L_{\min}$ in Eq. 9] can be obtained.

In many cases, the photoluminescence intensity of n-type electrodes at negative potential does not attain a constant value as would be expected from Eq. (16) [25]. Hysteresis is also frequently encountered in cyclic scanning experiments [25]. Such effects are due to potential-dependent changes in the surface chemistry. Under accumulation conditions in aqueous solution, hydrogen is evolved and may be incorporated into the electrode. In addition, the semiconductor may undergo cathodic reduction. These processes introduce surface or near-surface states that provide pathways for nonradiative recombination. Gerischer and coworkers [31, 32] have studied such systems using an extended Gärtner model.

There are numerous reports on the potential dependence of the electroluminescence for n-type electrodes under (quasi) steady state conditions and a limited number for p-type electrodes [25]. Such studies were mainly used for diagnostic purposes, that is, to investigate the role of minority carriers in electrode processes.

While there have been reports of time-resolved photoluminescence [65–68] and electroluminescence experiments [69–71] for the study of recombination, trapping and detrapping of photogenerated charge carriers, potential modulation has scarcely been used. An illustration of the possibilities of this approach is provided by work of Decker and coworkers [72] who used a frequency-dependent train of potential pulses to study the effect of surface chemistry on the radiative recombination of holes injected into *n*-type GaAs from $Fe(CN)_6^{3-}$ in solution. An unexpected potential-dependence of the emission was attributed to a change in the surface termination from hydroxide to hydride coverage on going from positive to negative potentials. A time constant of 0.1 ms was estimated for this transition. This transformation was subsequently observed by Erné and coworkers using in situ infrared spectroscopy [73]. It is clear that potential-modulated luminescence techniques deserve wider attention.

2.1.3.3 **Complex Charge-transfer Processes**

In this section we use two types of complex charge-transfer reaction to illustrate the general approach to the elucidation of reaction mechanisms at single crystal electrodes. These reactions are photocurrent doubling at *n*-type and *p*-type semiconductors and the (photo)anodic oxidation of the semiconductor itself.

2.1.3.3.1 **PC-doubling Reactions** PC doubling refers to a type of charge-transfer reaction in which both bands of the semiconductor are involved, thus emphasizing the distinctive features of semiconductor electrochemistry. The first examples of such reactions relate to the photoanodic oxidation of species such as formate and tartrate at wide band gap *n*-type electrodes [74, 75]. A photon generates an electron-hole pair in the semiconductor. The electron and hole are separated by the electric field of the depletion layer. The electron is detected as photocurrent in the external circuit. The hole oxidizes a species from solution producing an intermediate that is capable of injecting an electron into the CB. This electron also contributes to the photocurrent. As a result, for each photon absorbed in the system two charge carriers are registered as current. This corresponds to a "quantum efficiency" of two. For formate oxidation at *n*-type ZnO, Morrison and coworkers [74, 75] suggested a mechanism of the type

$$HCOO^- + h_{VB}{}^+ \longrightarrow H^+ + COO^{\bullet-} \quad (17)$$

$$COO^{\bullet-} \longrightarrow CO_2 + e_{CB}{}^- \quad (18)$$

Results obtained by Honda and coworkers [76] showed that the reaction is more complicated. The ZnO electrode is, in fact, dissolved during the current-doubling reaction. The stoichiometry found in their work is given by

$$ZnO + 2HCOO^- + 2h_{VB}{}^+ \longrightarrow Zn^{2+} + H_2O + CO_2 + 2e_{CB}{}^- \quad (19)$$

This result calls into question mechanisms in which the reducing agent reacts directly with valence-band holes. Honda and coworkers proposed that formate is oxidized by atomic oxygen formed in the photoanodic oxidation of ZnO.

$$ZnO + 2h_{VB}{}^+ \longrightarrow Zn^{2+} + O_{ads} \quad (20)$$

A subsequent study of the dependence of the kinetics of the photocurrent-doubling reaction on the light intensity supports this idea [77]. The following mechanism was suggested

$$O_{ads} + HCOO^- \longrightarrow OH^\bullet + COO^{\bullet-} \quad (21)$$

$$OH^\bullet + HCOO^- \longrightarrow H_2O + COO^{\bullet-} \quad (22)$$

The stoichiometry of this scheme agrees with that reported by Honda and coworkers. This result could be checked by studying the kinetics of two types of competing process. Iodide ions present in solution compete for reactive intermediates and thus influence the quantum efficiency. In the potential range between the onset and saturation of photocurrent, electron-hole recombination competes with the charge-transfer reactions. The study of these competing processes by EIS and IMPS supports the mechanism indicated by Eqs. (21) and (22) [77]. One of the problems encountered in the investigation of such complex mechanisms is the identification of (reactive) intermediates. For the ZnO/formate system Harbor and Hair [78] used ESR spin-trapping experiments to show that the $COO^{-\bullet}$ radical anion is indeed formed.

Quantum efficiencies higher than one have been observed for a range of reducing agents at ZnO [74, 75]. The mechanisms have been studied less thoroughly. In the case of the oxidation of methanol, it is clear that the mechanism differs from that of formate (and tartrate) [77]. Photoanodic current doubling also has been observed for other n-type semiconductors such as TiO_2 [79] and CdS [80]. Bogdanoff and Alonso-Vante [81] have described an interesting study of the competitive photoanodic oxidation of formic acid and water at TiO_2 using differential electrochemical mass spectroscopy. In contrast to ZnO, TiO_2 is a stable photoanode. On the basis of on-line mass detection, the authors conclude that formic acid is oxidized by hydroxyl radicals produced by the photoanodic oxidation of water, a reaction somewhat similar to that of oxygen radicals at ZnO (see Eq. 21). In the case of formate oxidation on CdS, there is no evidence for corrosion of the semiconductor. A direct reaction of the current-doubling agent with a valence-band hole was suggested for the first reaction step [80].

In 1969, Memming reported photocurrent doubling for the reduction of H_2O_2 and $S_2O_8^{2-}$ at p-type GaP. In this case, two holes are detected as photocurrent for each photon absorbed. This result was explained by a two-step mechanism [82]. The first step involves reduction of the oxidizing agent (e.g. H_2O_2) by a photogenerated electron

$$H_2O_2 + e_{CB}^- \longrightarrow OH^- + OH^\bullet \quad (23)$$

and the hydroxyl radical intermediate injects a hole into the valence band

$$OH^\bullet \longrightarrow OH^- + h_{VB}^+ \quad (24)$$

Since this first report, photocurrent doubling has been found for a whole range of two-electron oxidizing agents at various semiconductors including Si, SiC, CdTe, and III–V materials [83]. A striking example of photocurrent "multiplication" is the reduction of iodate at p-GaAs [83]. In a wide range of light intensity, a quantum efficiency of three is observed, whereas at low light intensity there is evidence for an efficiency of six. This would mean that five of the six intermediates formed on reducing IO_3^- to I^- could inject a hole into the valence band of GaAs. In a number of these systems, the oxidizing agent causes chemical dissolution of the solid. In the case of GaAs, studies of etching and current doubling have led to the conclusion that the various processes involved (electron capture, hole injection, and chemical etching) are linked via a common intermediate formed by the chemisorption of the oxidizing agent on the surface [15, 84, 85].

Obviously, information about the mechanisms of such reactions can be obtained by studying the photocurrent as a function of system parameters (potential, light intensity, concentration, hydrodynamics, etc.). However, such measurements do not yield information about the reaction kinetics. Peter and coworkers [86, 87] were the first to show that rate constants for majority carrier injection could be determined for such systems by IMPS. They studied current doubling for oxygen reduction at p-type GaAs and GaP. From a complex-plane representation of the optoelectrical impedance, rate constants in the range $10^4-10^5\ s^{-1}$ were calculated. This result suggests that hole injection is a thermally activated process. $HO_2^{\bullet}$, postulated as the injecting species, gives rise to a surface energy level; in the case of GaAs, this is located about 0.4 eV above the valence-band edge.

2.1.3.3.2 **Dissolution of Semiconductors**
Another class of complex reactions that has been widely studied is the oxidative dissolution of elemental and compound semiconductors. There are a number of reasons for the interest in these systems. The possible application of semiconductors in regenerative photoelectrochemical solar cells required the complete suppression of corrosion of the photoelectrode [5, 6, 22, 23]. On the other hand, with the development of the optoelectronics industry based on III–V materials there was a need for a more fundamental understanding of etching processes [15]. The revival of interest in porous semiconductors, triggered by the discovery of the unusual optical properties of porous Si, led to a general revival of interest in the mechanisms of porous etching [18].

Because the bonding states correspond to the valence band of a semiconductor, one expects holes to be important for the oxidation reaction [15, 24]. This is the case with most etching systems. The importance of holes is immediately clear from simple current-potential or cyclic voltammetric measurements. Generally, the p-type semiconductor dissolves anodically in the dark showing an exponential increase of current with increasing potential, whereas the n-type semiconductor can only be oxidized if minority carriers are generated by light under depletions conditions (see Fig. 7). The anodic oxidation of most semiconductors is a complicated process. The reaction of GaAs, for example, which can be represented schematically by

$$GaAs + 6h_{VB}{}^{+} \longrightarrow Ga(III) + As(III) \tag{25}$$

requires six valence-band holes to form trivalent gallium and arsenic species. Chemical reactions are obviously also involved. The final products depend on the nature of the electrolyte solution. The information that can be obtained from cyclic voltammetry is clearly rather limited.

Kinetic studies in which oxidation of a reducing agent present in solution competes with oxidation of the semiconductor have yielded a wealth of information, allowing quite detailed dissolution mechanisms to be proposed [88]. Generally, in these studies an RRDE is used; the products formed at the semiconductor disk are detected electrochemically at a noble metal ring.

Semiconductors can be etched under open-circuit conditions with an oxidizing agent that is capable of being reduced by extracting electrons from the valence band, that is, creating holes in the band (see Sect. 2.1.3.2.3) [15]. Information on hole injection can be obtained from electroluminescence studies on n-type electrodes. Electrochemical studies of such "electroless" etching systems have shown that, during dissolution, intermediates are formed with energy levels in the band gap [89]. From such states, electrons can be thermally excited into the CB. If this occurs in an n-type semiconductor under depletion conditions then the injected electron can be detected as current in the external circuit. By measuring this anodic current as a function of injection rate and temperature, one gets extensive information about both chemical and electrochemical processes [88, 89]. Gomes and coworkers have shown that electron injection and RRDE competition studies are complementary [90]. The contribution of electron injection to the total anodic oxidation rate is, for most etching systems, small compared to that of holes.

Electron injection from reaction intermediates of the oxidation of n-type semiconductors can be observed as quantum efficiency larger than unity in photocurrent-potential measurements. There are two striking examples in the literature: the photoanodic dissolution of n-type silicon in HF solution [91, 92] and of n-type InP in HCl solution [54]. In these cases the quantum efficiency at low light intensity is exceptionally high, four for silicon and two for InP. In the case of Si, this means that only one photon (and thus one hole) is required to dissolve each silicon atom; three electrons are injected into the conduction band

$$\mathrm{Si} + \mathrm{h_{VB}}^{+} \longrightarrow \mathrm{Si(IV)} + 3\mathrm{e_{CB}}^{-} \quad (26)$$

At higher light intensity, the quantum efficiency drops indicating that reaction steps involving valence-band holes take over from electron injection steps. A quantum efficiency of two for InP means that three of the six oxidation steps require minority carriers and thus photons. As for silicon, the quantum efficiency decreases (from 2 to 1) as the light intensity is increased. As described in the previous section, IMPS can be used very effectively to study the mechanisms of "photocurrent multiplication" reactions. The method has proved particularly successful for the silicon and InP systems, providing information about the sequence of the reaction steps and the magnitude of the rate constants for majority carrier injection [54, 92].

Because electron injection is detected during dissolution of n-type semiconductors under illumination, it seems likely that it should also occur during dissolution of p-type semiconductors in the dark. There are two ways in which this can be checked. If the injected electrons recombine radiatively with majority carriers (the holes), light is emitted. EL corresponding to band-band recombination has been observed during anodic dissolution of p-type InP in HCl solution (Fig. 12) [93]. The quantum efficiency for light emission, that is, the number of photons emitted per electron passed through the external circuit, was very low (approximately 10^{-6}). This probably is due to a low rate of electron injection and a high rate of nonradiative recombination at the etching surface. Another approach that allows one to distinguish quantitatively between electron and hole contributions to the current is the p-n junction configuration [94, 95]. This technique has been used for the study of the anodic dissolution of silicon in alkaline solution. This is an unusual system showing a novel coupling of chemical and electrochemical steps [95]. Both p-type and n-type silicon can be oxidized electrochemically in the dark. In both cases, passivation occurs if a limiting current is exceeded. The anodic peak current observed as n-type silicon passivates is comparable to that of p-type silicon. Anodic current from an n-type electrode in the dark can, under normal circumstances, only result from electron injection into the CB. Because there is no electron donor present in solution, electron injection must occur from a surface species. This species very likely arises as a result of the chemical-etching reaction

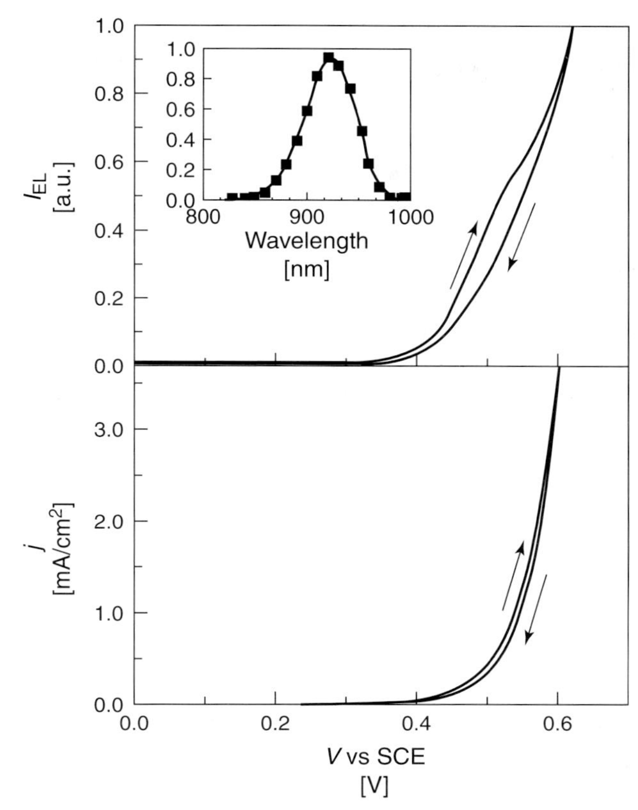

Fig. 12 The potential dependence of the EL intensity I_{EL} and of the current density j for a p-type InP electrode, dissolving anodically in a 1 M HCl solution. The inset shows the EL spectrum (from Ref. 93).

of silicon by water. Anodic current with p-type silicon can be due to a hole reaction or, as in the case of n-type silicon, to electron injection. Measurements with a p-n junction electrode showed the latter to be the case [95]; most of the anodic current can be attributed to minority carrier injection.

Electrical impedance measurements are essential in semiconductor dissolution studies. Under depletion conditions, a change in applied potential will usually give rise to a change in the potential drop within the semiconductor. Impedance measurements combined with Mott-Schottky theory can be used to determine the flat band potential (see Sect. 2.1.3.1). This then allows one to calculate the band-bending for any value of the applied potential, if the Helmholtz potential remains constant (i.e. the band edges are pinned). The band-bending determines the concentration of majority carriers at the surface (see Eqs. 25 and 26). In a p-type electrode these are holes, which are essential for the dissolution reaction. In an n-type electrode, the band-bending determines the surface electron concentration and thus, the rate of recombination with photogenerated or injected holes; this process competes with the oxidation of the semiconductor by holes.

The assumption of pinned band edges is very often not valid. Lincot and Vedel [50] in an early study of the photoanodic dissolution of *n*-type CdTe used EIS to show that the Fermi level becomes pinned in a wide potential range positive with respect to V_{fb}. This means that in this range the band-bending within the semiconductor remains constant while the potential changes across the Helmholtz layer. In their analysis, Lincot and Vedel consider the rate constants for change transfer to be exponentially dependent on the Helmholtz potential.

In many early studies of anodic oxidation of *p*-type semiconductors it was tacitly assumed that the reaction was controlled solely by the surface hole concentration, that is, by the potential drop across the space charge layer of the solid. The role of the Helmholtz potential was neglected. For *p*-type electrodes, anodic oxidation occurs at potentials close to the flat band value and under accumulation conditions. In addition, there will be a high density of surface states as a result of the breaking of surface bonds. In this case, a change in the applied potential is likely to be distributed partly or even completely over the Helmholtz layer. The potential dependence of the anodic current then is due to (in part) the changes in the rate constants for hole capture resulting from changes in the Helmholtz potential. Electrical impedance spectroscopy is necessary to decide whether etching is under space charge layer or Helmholtz-layer control. The latter was shown to be the case for the dissolution of *p*-GaAs in acidic electrolytes [96]. Vanmaekelbergh and Searson in a study of the dissolution of *p*-type silicon in HF solution showed that EIS also can be used to get information about electron injection processes [97].

Electrochemical measurements of the type described earlier give indirect evidence about dissolution processes. More direct chemical information can be obtained from in-situ spectroscopies, in particular from IR and Raman methods. Chazalviel and coworkers have showed the power of this approach in studies on silicon and GaAs [73, 98, 99]. Electrochemical and spectroscopic techniques are macroscopic methods giving a view of the whole electrode surface. To study semiconductor dissolution at the microscopic (atomic) level, one needs techniques such as scanning tunneling microscopy (STM) and atomic force microscopy (AFM). The anodic and chemical dissolution of silicon has been studied in very elegant work by Allongue and coworkers [100–102].

2.1.4
Sensitizer-based Photoelectrochemical Systems

Dye molecules adsorbed on an electrode form, in principle, the simplest class of photoelectrochemical system. The basic scheme presented at the end of Sect. 2.1.1.2 is directly applicable for understanding the mechanism of photocurrent generation. The dye molecules act as the light-absorbing species. If injection of a photogenerated charge carrier into the electrolyte or electrode competes effectively with relaxation of the excited state, a photocurrent might be observed in the external circuit. Singlet-to-triplet crossing often leads to a relatively long-lived excited state, which allows electron-hole separation to occur by an electron transfer process. The photochemistry and photoelectrochemistry of dye-sensitized electrodes has been studied extensively, and the research is still continuing. Dye-sensitization of photochemical reactions is a key topic in photographic research [103, 104]. At present, there is a considerable effort being devoted to dye-sensitized porous photoelectrochemical solar cells [16, 105]. The fundamental research is focused on the chemical bonding between the light-absorbing molecules and the electrode surface, the mechanism of light absorption, and the dynamics of charge injection and recombination in the dyes. In this research, scanning probe methods [106, 107] and time-resolved optical spectroscopy [108, 109] are used together with photoelectrochemical characterization [24]. The reader is referred to a more specialized review for details [24].

Electrodes to which insulating light-absorbing nanocrystals, instead of dyes, are attached form a relatively new class of system. With colloidal solution chemistry, a large variety of insulating nanocrystals can be prepared [110–113]. Well-known examples are II–VI compounds (CdS, CdSe, and ZnO), III–V compounds (InP, InAs, and GaN), and transition-metal oxides. In addition, there are various methods in solid-state chemistry and electrochemistry for preparing nanocrystals directly on surfaces. For instance, CdS, CdSe, and PbS nanocrystals can be electrochemically deposited on gold electrodes [114]. Electrochemical oxidation of Si leads to porous silicon that may contain a large number of Si nanocrystals [115]. Nanocrystalline colloidal systems have been extensively characterized with optical spectroscopy [110–113]. There are two effects that are essential for understanding the electronic and optical properties of insulating nanocrystals. First, quantum confinement of the electron waves in the nanocrystal leads to discrete electron states at the top of the valence band and bottom of the CB and to an increase of the band gap energy with respect to that of a macroscopic crystal. The energy-level spectrum shifts from that of a classical insulator to that of a molecule with a reduction of the dimensions in the 20–1-nm range. Second, a considerable fraction of the atoms of a nanocrystal lies at the surface. This leads to surface-electron states. States of energy in the band gap can have a strong influence on the optical properties of nanocrystals. It is clear that the surface chemistry is extremely important. Organic and inorganic molecules can passivate surface states, thus removing them from the optical gap. Functionalized capping molecules play an important role in providing stability against coagulation and in allowing the attachment of nanocrystals to solid (electrode) surfaces (see the following text).

Early reports have shown photoelectrochemical activity when, for example, CdS and PbS nanocrystals are attached to a metal electrode in a sub-monolayer array [19, 20, 116–120]. Clearly, photoexcitation of the nanocrystals can lead to a long-lived state, which allows one of the charge carriers to be transferred from the nanocrystal before recombination occurs (see Sect. 2.1.1.2).

A macroscopic PbS crystal has a band gap of 0.41 eV. Because of the small effective mass of the electrons and holes ($m_{e,eff} = m_{h,eff} = 0.09 \times m_e$), strong size-quantization occurs. The absorption spectrum of an aqueous suspension (of polyvinyl alcohol-capped) PbS nanocrystals, 6.5 nm in diameter (see TEM picture) is shown in Fig. 13. The HOMO-LUMO optical transition occurs at 2.1 eV, and two other absorption peaks are seen at 3.2 and 4.3 eV. When a gold electrode is immersed in this colloidal solution, PbS nanocrystals

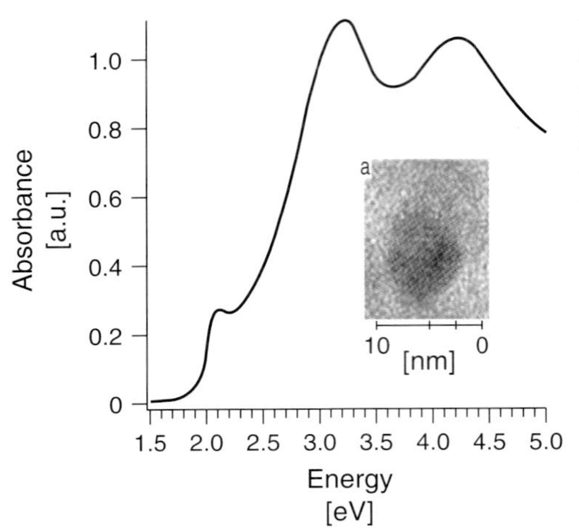

Fig. 13 Absorption spectrum of an aqueous suspension of nanocrystalline, size-quantized PbS particles (capped with polyvinylalcohol). A HR-TEM image of a typical PbS nanocrystal is shown in the insert; the diameter of the nanocrystals is about 6.5 nm (from Ref. 122).

are attached and form a monolayer array, see Fig. 14 [121, 122].

The photoelectrochemical activity of a Au/Q-PbS electrode, illuminated with chopped green light, is shown in Fig. 14 for three different aqueous electrolyte solutions. With a 1 M KCl solution (case a), photocurrent transients are observed when the electrode potential is more positive then −0.4 V (versus SCE). Illumination gives a cathodic current that decreases to zero with increasing time; turning off the light induces an anodic current. Clearly, illumination leads first to the transfer of an electron from gold to the photoexcited hole in the PbS nanocrystals (a "cathodic" transfer), followed by transfer of the photoexcited electron to empty states in the gold (an "anodic" transfer). Energy relaxation in the gold due to electron transfer competes with electron-hole recombination in Q-PbS. For this to occur, the electrochemical potential in the gold electrode must be located between the energy levels corresponding to the photoexcited electron and hole. Recombination transients are suppressed when the Fermi level is above 0.4 eV (versus SCE), which indicates that the long-lived excited state contains an electron located at 0.4 eV versus SCE, thus considerably below the LUMO of Q-PbS. The much faster kinetics of the photoexcited hole transfer indicates that the hole in the long-lived excited state is still delocalized, that is, it occupies the HOMO. It was thus concluded that the excited state in Q-PbS leading to photoinduced electron transfer consists of a trapped electron and a delocalized hole in the HOMO (denoted as $Q(\mathrm{e_{trap}}, h_{\mathrm{HOMO}})$). The following scheme describes the decay dynamics of the long-lived state in Q-PbS:

$$Q \xrightarrow{h\nu} Q^* \xrightarrow{\text{relaxation}}$$

$$Q(\mathrm{e_{trap}}, h_{\mathrm{HOMO}}) \xrightarrow{\text{relaxation}} Q + \text{heat} \tag{27}$$

$$Q(\mathrm{e_{trap}}, h_{\mathrm{HOMO}}) \xrightarrow{\text{transfer from Au}} Q^-(\mathrm{e_{trap}}) \tag{28}$$

$$Q^-(\mathrm{e_{trap}}) \xrightarrow{\text{transfer to Au}} Q \tag{29}$$

Interestingly, when tartrate is added as an electron donor to the solution, an anodic photocurrent is observed in the potential range positive with respect to −0.4 V

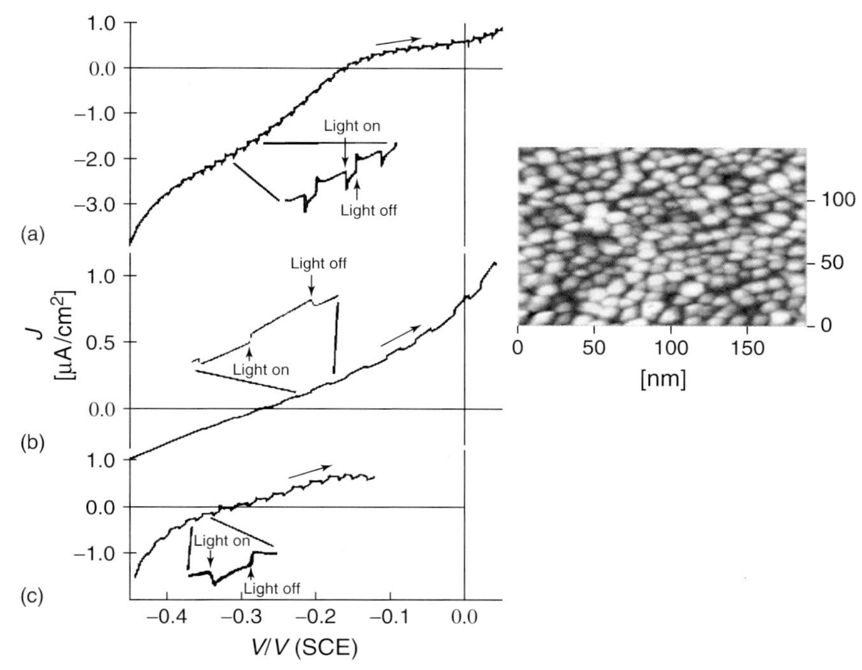

Fig. 14 Current-potential curves, under chopped light (514 nm), measured with a Q-PbS covered gold electrode in a photoelectrochemical cell with three aqueous electrolytes. The insert gives an STM picture of a part of the Q-PbS covered electrode showing a disordered monolayer coverage. (a) 1 M KCl; (b) 1 M KCl + 0.2 M tartrate (electron donor); (c) 1 M KCl + 0.01 M $K_3Fe(CN)_6$ (electron acceptor) (Ref. 122).

(Fig. 14, case b). This means that electron donation from tartrate can compete effectively with electron donation from the gold electrode [i.e. Eq. (28)]:

$$Q(e_{trap}, h_{HOMO}) + \text{Red} \xrightarrow{\text{transfer from tartrate}} Q^-(e_{trap}) + \text{Ox}^+ \quad (30)$$

In contrast, a cathodic photocurrent is observed when $Fe(CN)_6{}^{3-}$ is added to the solution (Fig. 14, case c). Thus, electron donation to the oxidizing species in solution competes effectively with electron donation to gold; it can be clearly seen in Fig. 14, case c that the electrochemical activity in the latter solution does not stop when the electrode Fermi level is increased above 0.4 eV. This is logical, because the cathodic photocurrent corresponds to electron transfer to the oxidized species, not to the electrode.

Photoelectrodes, consisting of CdS nanocrystals that are directly attached to an electrode (gold or conducting oxide) or indirectly via a dithiol molecular linker have been studied extensively [19, 20, 116–120]. Here, we will focus on work that demonstrates the power of small-amplitude methods in photoelectrochemical characterization [26, 123, 124]. By measuring the photoelectrochemical activity of gold/Q-CdS electrodes, as the potential of the gold electrode was varied, it could be concluded that photoinduced electron transfer between the gold surface and the quantum dots is due to a long-lived state in Q-CdS. This state, denoted as $[Q(\mathrm{e_{LUMO}}, h_{\mathrm{trap}})]$, consists of an electron in the LUMO and a hole trapped in a level about 0.6 eV below the LUMO:

$$Q \xrightarrow{h\nu} Q^* \xrightarrow{\text{relaxation}}$$

$$Q(\mathrm{e_{LUMO}}, h_{\mathrm{trap}}) \xrightarrow{\text{relaxation}} Q + \text{heat} \quad (31)$$

Time-resolved photobleaching experiments on Q-CdS colloidal solutions showed the existence of a long-lived state (lifetime 50 ms) that, very probably, corresponds to $Q(\mathrm{e_{LUMO}}, h_{\mathrm{trap}})$ [125].

Figure 15 shows the optoelectrical transfer function (see Sect. 2.1.2.4), measured with near UV-light from an argon laser for a 1 M KCl solution (a, b) and a KOH-tartrate solution (c, d). In a and c, the transfer function is plotted in the complex plane with the frequency as a parameter, in b and d, the modulus of the transfer function is plotted versus the modulation frequency ω. In the KOH-tartrate solution (c, d), an anodic vector is observed at sufficiently low frequency. This means that $\partial j_{\mathrm{PH}}/\mathrm{e}\partial\Phi > 0$ (see Sect. 2.1.2.2). This agrees with the observation of a steady state anodic photocurrent, which increases with increasing light intensity. The anodic vector shrinks with increasing modulation frequency ω and, eventually, becomes zero for $\omega > 5 \times 10^4\ \mathrm{s^{-1}}$. Two semicircles with characteristic frequencies $\omega_{\mathrm{c}} = 6\ \mathrm{s^{-1}}$ and $\omega_{\mathrm{b}} = 700\ \mathrm{s^{-1}}$ are observed (measurements at 25 °C). Measurements in the temperature range 0–60 °C showed that ω_{b} is temperature-independent, while ω_{c} increases with increasing temperature. Hence, ω_{c} corresponds to the rate of electron donation from tartrate to the hole trapped in Q-CdS, while ω_{b} is the rate of photoinduced electron transfer from the LUMO in Q-CdS to empty states in the gold electrode:

$$Q(\mathrm{e_{LUMO}}, h_{\mathrm{trap}}) \xrightarrow{\omega'_{\mathrm{b}}} Q^+(h_{\mathrm{trap}}) \quad (32)$$

$$Q^+(h_{\mathrm{trap}}) + \mathrm{Red} \xrightarrow{\omega_{\mathrm{c}}} Q + \mathrm{Ox}^+ \quad (33)$$

With a 1 M KCl solution, the low-frequency limit $\partial j_{\mathrm{PH}}/\mathrm{e}\partial\Phi$ of the optoelectrical transfer function is zero; this agrees with the fact that there is no steady state photocurrent. The high-frequency limit of the optoelectrical transfer function is also zero. Starting from $10^5\ \mathrm{s^{-1}}$, an anodic vector develops with decreasing modulation frequency. This corresponds to electron transfer from the Q-CdS LUMO to gold. The rate ω_{b} of this process cannot be clearly distinguished from the discharge frequency of the photoelectrochemical cell. Electron transfer from the LUMO to the gold is probably faster than $3 \times 10^4\ \mathrm{s^{-1}}$. At lower frequencies, the modulus of the optoelectrical transfer function shrinks to zero. This is because of the electron back-transfer from gold to the empty level in Q-CdS following the faster LUMO-to-gold electron transfer:

$$Q^+(h_{\mathrm{trap}}) \xrightarrow{\omega_{\mathrm{a}}} Q \quad (34)$$

The rate of this transfer (ω_{a}) is $6 \times 10^3\ \mathrm{s^{-1}}$ and is independent of temperature. Thus, at sufficiently low modulation frequency, both the processes represented by Eqs. (32) and (34) are in phase with the modulated light intensity, and the resulting photocurrent is zero. When no redox system is present, relaxation of the long-lived excited state in Q-CdS, $Q(\mathrm{e_{LUMO}}, h_{\mathrm{trap}})$, occurs via consecutive steps involving electron transfer from the LUMO to the gold, and from the gold to the trapped hole. This relaxation

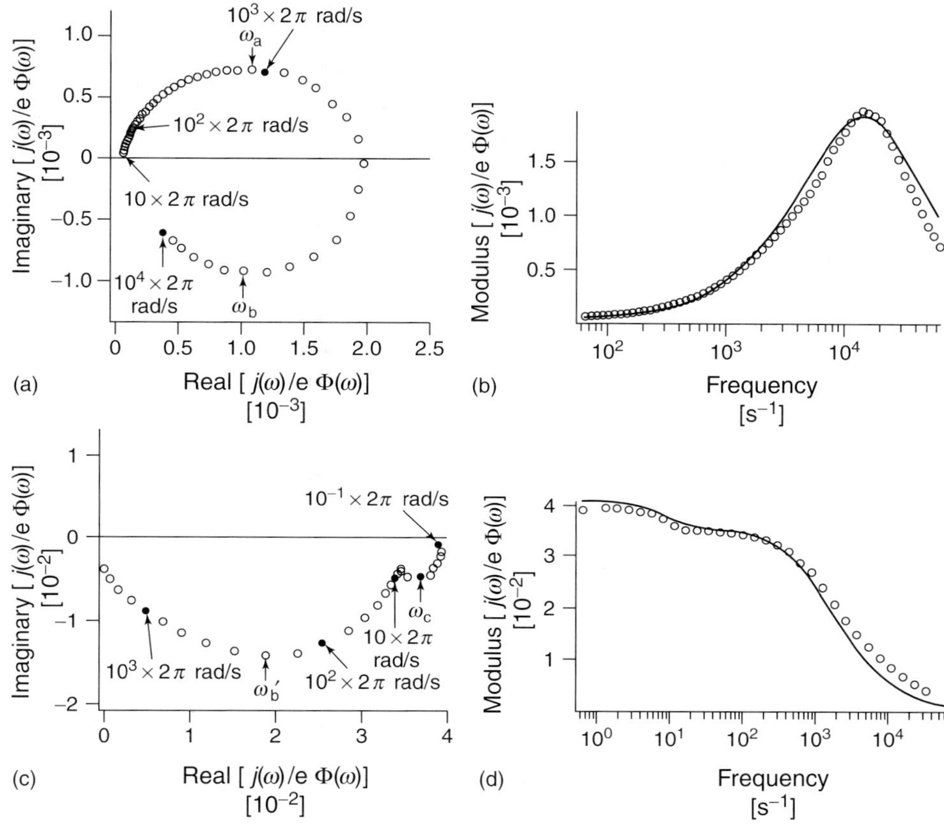

Fig. 15 The optoelectrical transfer function measured for a Q-CdS covered gold electrode in an electrochemical cell. The attached quantum dots have an average diameter of 5 nm. (a,b). Results obtained in an aqueous solution (pH = 7) with 1 M KCl (297 K). In (a), a plot of the transfer function is shown in the complex plane with the modulation frequency as a parameter. The meaning of the transfer function and of the characteristic frequencies (ω_a, ω_b) is discussed in the text. In (b), the modulus of the transfer function is plotted vs. the modulation frequency. (c,d) Results obtained in an aqueous solution (pH = 14) with 0.5 M tartrate as an electron donor. (c) Plot of the transfer function in the complex plane with the modulation frequency as a parameter. (d) Plot of the modulus of the transfer function versus the modulation frequency (from Ref. 123).

competes effectively with relaxation of the excited state in the particle.

It can be concluded that measurement of the optoelectrical transfer function in a photoelectrochemical cell is a powerful technique for studying the mechanism and kinetics of photoinduced electron transfer. The technique has been further exploited to determine the rates of photoinduced electron transfer in gold/spacer/Q-CdSe assemblies [121, 126, 127]. Cyclohexylidene disulfides form well-ordered and rigid molecular monolayers on gold (111) surfaces because of strong Au-S bonding and intermolecular van der Waals interactions [121]. Q-CdSe quantum dots were linked covalently at the other end of the molecule via the S-termination. The rates of photoinduced electron transfer from the LUMO of Q-CdSe to gold ($k_{L,Au}$), and from gold to the trapped hole in Q-CdSe ($k_{Au,T}$) were obtained from the optoelectrical transfer function. An exponential decrease of the electron transfer rate with increasing length of the spacer molecule is found, with a decay parameter $\beta = 0.5$ Å^{-1}. This low value (the value in vacuum is typically four times larger) indicates a strong through-bond electronic coupling in the cyclohexylidene spacer molecules. This agrees with the result of quantum chemical calculations and with the strong S–S coupling shown by photoelectron spectroscopy [126, 127].

The measurement of changes in the optical properties of nanocrystals attached to an electrode caused by variations in the electrode potential forms a challenging but interesting characterization method. Recently, a bleaching of the HOMO-LUMO transition of CdS nanocrystals has been observed when the electrode Fermi level was in resonance with the LUMO [128]. This led the authors to conclude that the photoexcited state in CdS contains an electron in the LUMO (and a trapped hole). This is in agreement with the results obtained by photoelectrochemical characterization of gold/Q-CdS electrodes (see earlier).

2.1.5 Porous Photoelectrochemical Systems

2.1.5.1 Introduction

Here, we define a porous solid as a phase that contains empty spaces that are interconnected. Thus, there is a single-solid phase that can be permeated with a second phase. Porous metals have been extensively studied, and are widely applied in heterogeneous catalysis, storage batteries, fuel cells, and super-capacitors [41]. Porous insulating or semiconducting phases have been considered as photochemical devices for the light-stimulated oxidation of organic waste components [heterogeneous photocatalysis] [110–113, 129]. Porous photoelectrochemical systems have been studied extensively only in the last ten years [16, 105, 130–137]. Their fabrication is more demanding than that of photochemical systems, because electrical work is delivered in an external circuit upon illumination (see Sect. 2.1.2.2). In a porous photoelectrochemical system, the solid phase is insulating or semiconducting, whereas the permeated phase is an electrolyte solution. The current in the solid phase is due to the motion of electrons or holes; the current in the solution is due to the motion of an oxidized or reduced species.

A porous photoelectrochemical system can be prepared by the deposition of colloidal particles on a conducting substrate. It is essential that the particles are electrically connected and that there is electrical contact between the particles and the conducting substrate. Furthermore, the pore system should form a single-permeated phase. A well-known example of such a system is the particulate TiO_2 photoelectrode, which forms the basis of the photoelectrochemical solar cells proposed by O'Regan and Grätzel and other groups [16, 130–137]. A SEM picture of a TiO_2 network, consisting of interconnected spheres with a diameter of 30 nm, is shown in Fig. 16(a); electrical contact between the particles and between the conducting substrate and the particles was achieved by slight sintering at 450 °C.

An alternative route exists for the preparation of porous semiconductors. Many n-type single crystals, such as GaP, GaAs,

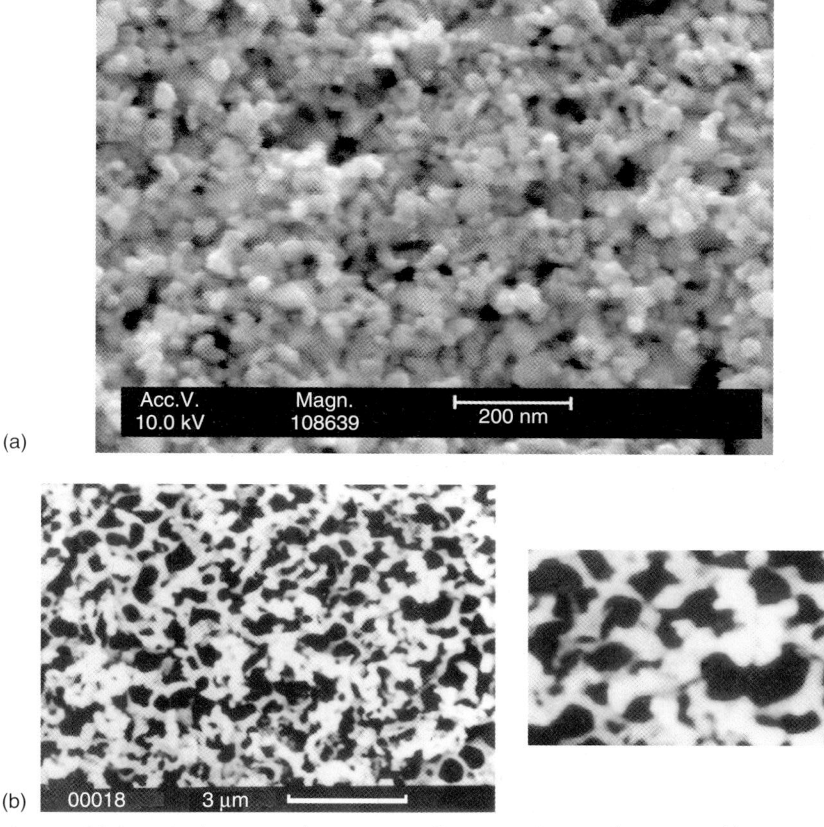

Fig. 16 (a) A particulate network consisting of 25 nm TiO_2 particles prepared by deposition from a colloidal solution followed by slight sintering; (b) A porous GaP crystalline network prepared by anodic etching of an *n*-type GaP crystal at positive potential. The structural units have dimensions in the 100–300 nm range.

Si, SiC, and TiO_2 can be transformed into a porous network by anodic etching under conditions of severe band-bending [18, 115, 138–144]. Under such conditions, surface electrons located at the top of the valence band or in band gap states can tunnel through the gap into the CB; the surface localized holes generated in such a way are consumed in anodic dissolution of the material. The rate of inter-band tunneling (Zener breakdown [145]) is strongly dependent on the electric field and the presence of surface defects. This would explain why anodic dissolution is so strongly nonuniform over the surface and eventually leads to the formation of a porous semiconductor. However, many questions remain regarding porous photoelectrochemical etching. A well-studied example is macroporous GaP [139, 143]. When *n*-type GaP is subjected to a potential of 5 V or more (versus SCE) a highly porous network is formed (Fig. 16b) with dimensions of the structural units and the pores in the 100-nm range. The GaP network retains its crystallinity [146, 147].

A "single-crystal GaP sponge" is different from a particulate network in that there are no grain boundaries and the connection between the network and the bulk GaP matrix is ideal.

2.1.5.2 **Special Properties of Porous Photoelectrochemical Systems**

Porous photoelectrochemical systems consist of an insulating or semiconducting solid network permeated with a conducting electrolyte solution; the dimensions of the solid structures and pores are in the 1–500-nm range. A typical semiconductor/electrolyte interface has a width of between 0.5 nm (the Helmholtz layer in a concentrated electrolyte solution) and 100 nm (typical depletion layer in a semiconductor). Thus, the width of the solid/electrolyte interfacial layer can be much smaller but also larger than the dimensions of the solid structures and the pores. Therefore, porous semiconductors can show remarkable charge-storage properties. Other relevant length scales are the wavelength of visible light (400–700 nm) and the diffusion length of charge carriers before recombination. These length scales also can be in the same range as the dimensions in the porous structure; this leads to striking optical and electrodynamic properties. Extensive research is being performed in these fields, and a comprehensive review is beyond the scope of this section. Instead, the electrostatic, optical, and electrodynamic properties of porous semiconductor (electrodes) are briefly discussed and the reader is referred to more detailed publications.

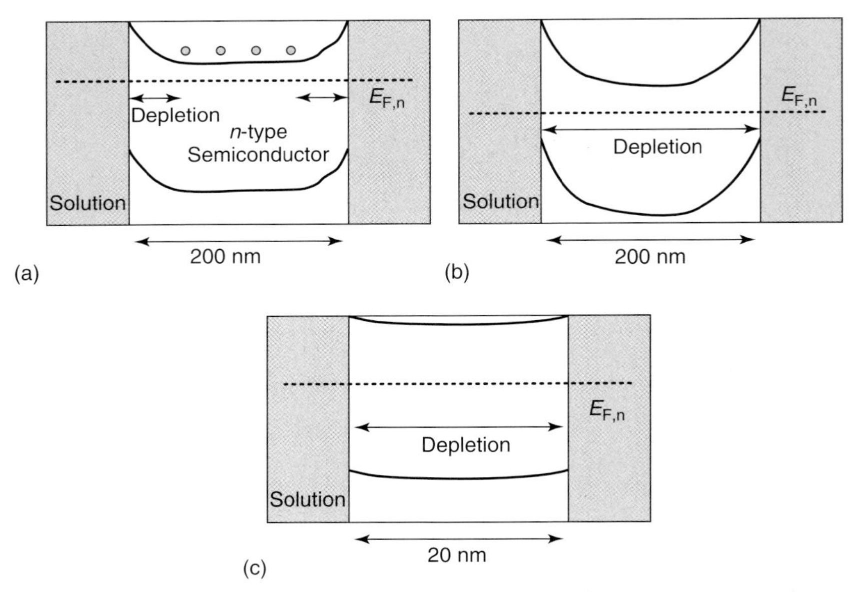

Fig. 17 Schemes of the semiconductor/electrolyte interface for a macroporous and a nanoporous electrode. (a) An *n*-type macroporous electrode under moderate depletion: structural units contain a depleted region and a bulk region (free electrons in the nondepleted region). (b) A macroporous *n*-type electrode at a strongly positive potential: the entire porous electrode is depleted of free electrons. (c) A nanoporous electrode in which depletion occurs without band bending.

2.1.5.2.1 Penetration of the Interfacial Layer in a Porous Semiconductor Electrode

In Fig. 17(a), the energetics of a typical interfacial region between an n-type semiconductor and an electrolyte solution is shown (see also Sect. 2.1.2.2). Electronic equilibrium exists between the semiconductor and a redox system present in the solution: the electrochemical potential of electrons μ_e in the solid is equal to that in the liquid phase, and does not change with the spatial coordinate x, perpendicular to the solid/liquid interface. The electrochemical potential of the electrons is also equal to the electron Fermi level, denoted as $E_{F,n}$ and can be written as

$$E_{F,n}(x) - E_F^{ref} = -e[\varphi(x) - \varphi^{ref}] + kT \ln[n(x)/n^{ref}] \qquad (35)$$

where $n(x)$ is the electron concentration in the CB of the semiconductor (i.e. the free-electron concentration), and $-e\varphi(x)$ the potential energy of an electron.

It is clear from Fig. 17(a) that a depletion layer for free electrons is present near the solid/solution interface. From Eq. (35) it follows that

$$n(x) = n_{bulk} e^{-[E_c(x) - E_{c,bulk}]/kT} \qquad (36)$$

Here, $E_C(x)$ and $E_{C,bulk}$ is the energy of the CB edge at position x in the depletion layer and in the bulk, respectively; $E_C(x = 0) - E_{C,bulk}$ is the band bending. The concentration of free electrons in the bulk, n_{bulk}, is determined by the density of dopant atoms. The depletion layer of an n-type electrode is positively charged, because the ionized dopant atoms are not fully compensated by free electrons. The counter charge is located on the electrolyte side, very close to the interface (Helmholtz layer, width 0.5 nm).

Consider now electronic equilibrium for a macroporous electrode. The structural units in macroporous GaP (doping density 10^{17} cm^{-3}) have typical dimensions of about 150 nm [138, 139, 146, 147]. This means that, if the band bending of the electrode is smaller than one 1 eV, W_{SC} (50 nm, see Sect. 2.1.2.2) is smaller than half the width of a structure: the inside edge of the depletion layer is in the porous network; there is still a semiconducting bulk region in the porous solid (see Fig. 17a). Therefore, the total surface area of the inside edge of the depletion layer is very large and so is the interfacial capacitance that is approximately equal to the internal surface area times C_{SC}. However, if the band-bending is more than 3 eV, W_{SC} is larger than half the width of typical structures in macroporous GaP; the entire porous GaP structure is depleted of electrons, the inside edge of the depletion layer is located in the bulk substrate, outside the porous film (Fig. 17b). The surface area of the inside edge of the depletion layer then corresponds to the macroscopic (geometrical) area of the electrode.

Erné and coworkers measured the interfacial capacitance of macroporous GaP electrodes as a function of the electrode potential [138, 139]. It was found that the capacitance is large for sufficiently small band-bending (interfacial layer in the porous solid) and decreases to the value of a nonporous interface at larger band-bending. Similar effects have been found with macroporous SiC and Si electrodes [18, 141]. In fact, the interfacial capacitance is a measure for the surface area of the macroporous network, with the width of the depletion layer, W_{SC}, as a measuring stick.

Finally, in nanoporous networks, the structural units have dimensions in the range 25–1 nm. For instance, the TiO_2 particulate electrode forming the basis of a dye-sensitized solar cell consists of nanocrystals of 10–25 nm (Fig. 16b). It is clear that, under conditions of free-electron depletion, band bending is nearly absent. (Fig. 17c). When electrons are supplied to the nanoporous system (by injection from a photoexcited dye or from a conducting substrate), the difference between E_F and E_C becomes smaller in the entire nanocrystal and the conductivity of the system increases.

Under conditions of electron accumulation, the interfacial capacitance C of a semiconductor/electrolyte contact tends to that of the Helmholtz layer (see Sect. 2.1.3.1 with $C_{SC} \geq C_H$). The width of the interfacial double layer is reduced to about 0.5 nm; hence, it follows the internal surface of a porous electrode. As a result, the overall interfacial capacitance of a nanoporous system can be huge, being determined by the product of the total internal surface area of the system and the Helmholtz-capacitance per unit geometric surface area [148, 149].

2.1.5.2.2 **Charge Storage in a Porous Semiconductor Electrode** In a bulk single-crystal electrode with a flat semiconductor/electrolyte interface, the electrochemical potential can be changed considerably by a relatively small change in the number of electrons present in the semiconductor. This is due to the relatively small interfacial capacitance per unit of geometric surface area. In contrast, the entire three-dimensional structure is interfacial in the case of macroporous and nanoporous systems interpenetrated with a conducting electrolyte. Consequently, the capacitance per geometric area can be very large, which means that a relatively large number of electrons are needed for a change in the electrochemical potential. This is exemplified by the considerable variation in microwave absorption induced by a given change in the Fermi level observed at macroporous GaP [150]. The large dynamic range in the total number of free electrons in a semiconducting network can be used as a tool for photoelectrochemical characterization. For instance, the optical absorbance by free carriers in nanoporous TiO_2 electrodes has been used to detect changes in the free carrier concentration in this system due to voltage modulation and modulation of the photoexcitation rate [151].

Clearly, these considerations are of importance for systems in which porous semiconductors or insulators are used [super-capacitors, (chemical) sensors, and electrochromic devices] [152, 153].

2.1.5.2.3 **Charge Storage in a Quantum Dot System** Hoyer and coworkers [154, 155] reported that the electrochemical potential of a porous particulate ZnO electrode (with ZnO dots of 5-nm diameter) shifts to higher energy with increasing electron density n in a much more pronounced way than predicted from Eq. (35). This is caused by two physical phenomena that become important with very small particles (quantum dots). First, as a result of size-quantization, the energy levels of the CB (and valence band) become discrete and separated by considerable energy gaps (typically in the range of 0.1 eV). Even in assemblies in which the nanocrystals are covalently linked, size-quantization may persist. Second, the charging energy per particle, $e^2/C_{particle}$, (typically 0.01–0.1 eV) can form an important contribution to the electrochemical potential of nanometer-sized particles. Investigating nanoparticulate ZnO electrodes similar to those used by Hoyer, Meulenkamp [156] reported that the relationship between the electrochemical potential and the electron density depends on the nature of the electrolyte solution; this clearly shows the importance of the charging energy. Size-quantization and single-dot charging energy play an important role in electron transport in metal–nanodot–metal double barrier tunnel junctions [157–161]. Study of electron transport in two-dimensional or three-dimensional assemblies consisting of nanometer-sized particles is still in its infancy; one may expect that single-dot charging (leading to Coulomb-blockade) and size-quantization will result in interesting and novel transport phenomena.

2.1.5.2.4 **Light Scattering in Macroporous Semiconductors** The dimensions of the structural units and pores in macroporous semiconducting and insulating networks are often in the 100-nm range. This is the same range as the wavelength of visible and UV light. Because of the structural variation of the refractive index on the wavelength scale, visible light can be strongly scattered in macroporous networks [162–165]. In macroporous GaP, for instance, the propagation of red (sub–band gap) light is strongly attenuated [146, 147]. The importance for photoelectrochemical systems lies in the fact that the effective absorption length of supra–band gap light in macroporous systems is reduced considerably with respect to that in single crystals. For example, the penetration depth $1/\alpha$ of green light in a bulk GaP single crystal is about 10 μm, whereas macroporous GaP networks with a thickness of only 2 μm completely absorb green light [138, 139]. Visible and near-UV light is not scattered in nanoporous systems because the structural variation in the refractive index occurs on a scale much smaller than the wavelength of light. In such a case, the effective absorption coefficient of the light can be estimated from effective medium theory.

2.1.5.2.5 **Electron-hole Photoexcitation by Sub-band Gap Light** In a dye-sensitized porous photoelectrode, an electron from the dye is photoexcited into the CB by a photon of energy considerably below the band gap of the semiconductor. The dye molecules are anchored on the internal surface of the porous semiconductor. Light absorption is very effective because of multiple interactions of a single photon with the dye molecules. Similarly, a porous semiconductor without dye molecules may absorb sub-band gap light, and this may lead to photogeneration of free electrons and holes. The mechanisms of free carrier generation with sub-band gap light in macroporous GaP photoelectrodes have been investigated in detail [166, 167]. Surface-localized electrons involved in two-photon transitions and in a coupled optical-thermal transition were found to give rise to significant sub-band gap photocurrent in this system.

2.1.5.2.6 **Effective Electron-hole Separation** The ability of porous photoelectrochemical systems to separate effectively electrons and holes is widely known since the presentation of the dye-sensitized particulate TiO_2 solar cell [16, 105, 130–137]. In this system, the photocurrent quantum yield (the number of electrons counted in the external circuit as photocurrent divided by the number of absorbed photons) is close to unity. This means that electron-hole pair recombination is essentially absent. Efficient separation of photogenerated electrons and holes was demonstrated with several other photoelectrochemical systems [105, 130–137]. Photovoltaic devices based on permeated hole-conducting and electron-conducting polymer phases also show an enhanced photocurrent quantum yield [168–170]. The origin of this desirable feature can be demonstrated by comparing the photocurrent quantum yield of a GaP bulk single crystal with a macroporous GaP photoelectrode (see Fig. 18). GaP absorbs light of energy between 2.2 and 2.7 eV by an indirect transition (hence, weakly, the absorption depth of green light is 10 μm). Because of the diffusion length L_{min} of minority carriers in n-GaP is relatively small (about 50 nm), the penetration depth $1/\alpha$ is much larger than the width of the retrieval region ($L_{\mathrm{min}} + W_{\mathrm{SC}}$). This results in a photocurrent quantum yield of about 0.01 [$\cong \alpha(L_{\mathrm{min}} + W_{\mathrm{SC}})$]; that is, 99% of the absorbed photons are converted into heat by recombination in the bulk. A macroporous GaP electrode, on the other hand, shows a photocurrent quantum yield of unity in a large potential range. The reason for this spectacular enhancement is illustrated in Fig. 17(b): in a macroporous GaP electrode, all minority carriers (holes in the valence band) are photogenerated within a distance from the interface that is smaller than the diffusion length L_{min}. Minority carriers can thus reach the surface without recombining. If surface recombination is slower than transfer of the hole to the liquid electrolyte phase, the quantum yield will be close to one, in agreement with the experimental result. A considerable enhancement of the photocurrent quantum yield has been observed in several porous photoelectrochemical [16, 18, 105, 130–139] and photovoltaic systems [168–170].

2.1.5.2.7 **Luminescence from Porous Electrodes**

As in bulk systems, photogeneration of charge carriers in a porous electrode or injection of minority carriers from solution can lead to light emission [25]. In a macroporous system in which the depletion layer can follow the contours of the porous matrix, one does not expect significant differences between bulk and porous electrodes with regard to the potential dependence of the emission. If, however, the porosity is high and the dimensions of the structures become very small (e.g. <5 nm) then special effects may be expected. These are indeed found as, for example, with nanoporous silicon.

One of the most striking properties of nanoporous silicon is its strong

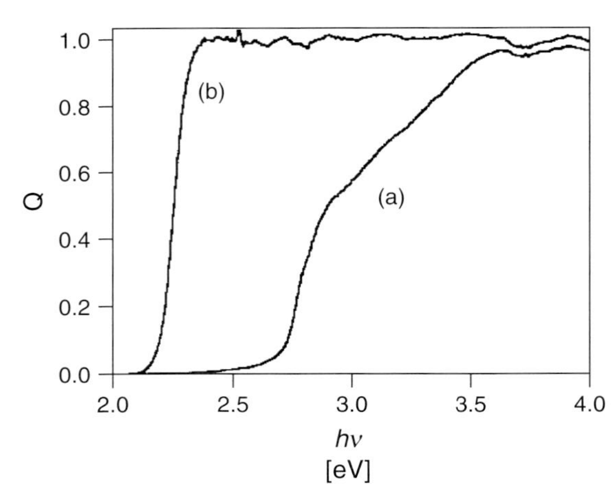

Fig. 18 Comparison of the photocurrent quantum yield [$Q = j_{\mathrm{PH}}/e(1 - R)\Phi$] versus the wavelength of the incident light, measured with a nonporous n-type single-crystal electrode of GaP (a), and with a macroporous GaP electrode (b). With a nonporous electrode the quantum yield is very low for light absorbed in the indirect transition ($h\nu < 2.7$ eV). In contrast, for a porous electrode the quantum yield is unity for light of energy above the gap ($h\nu > 2.2$ eV) (from Ref. 139).

photoluminescence in the visible spectral range [171]. Of the various explanations given for this phenomenon, the most widely accepted is that of size quantization. As a result of the confinement of charge carriers within nanometer-sized structures in the porous matrix, the effective band gap is widened while the ratio of radiative to nonradiative recombination is considerably enhanced [171]. In situ luminescence measurements provide information about the physical and chemical properties of porous silicon and about charge-transfer reactions at the silicon/solution interface [25].

In contrast to the photoluminescence from a single-crystal electrode (see Fig. 11), the emission from a porous *n*-type silicon electrode is constant at positive potentials and decreases only in the range negative with respect to the flat band potential (Fig. 19) [172, 173]. Because of the absence of an electric field in the porous layer and the strong confinement of the carriers, the electron and hole are not separated in the potential range corresponding to depletion in a bulk electrode. As at a single-crystal electrode, hole injection from a strongly oxidizing species such as $SO_4^{-\bullet}$ (generated electrochemically by the reduction of $S_2O_8^{2-}$ at the electrode) gives rise to visible electroluminescence in porous *n*-type silicon [172, 173]. The emission increases in the range in which the photoluminescence decreases [172, 173] (see Fig. 19). An interesting aspect of the electroluminescence is the voltage tunability of the colour. The emission maximum shifts to shorter wavelength as the potential is scanned to negative values, until finally the emission

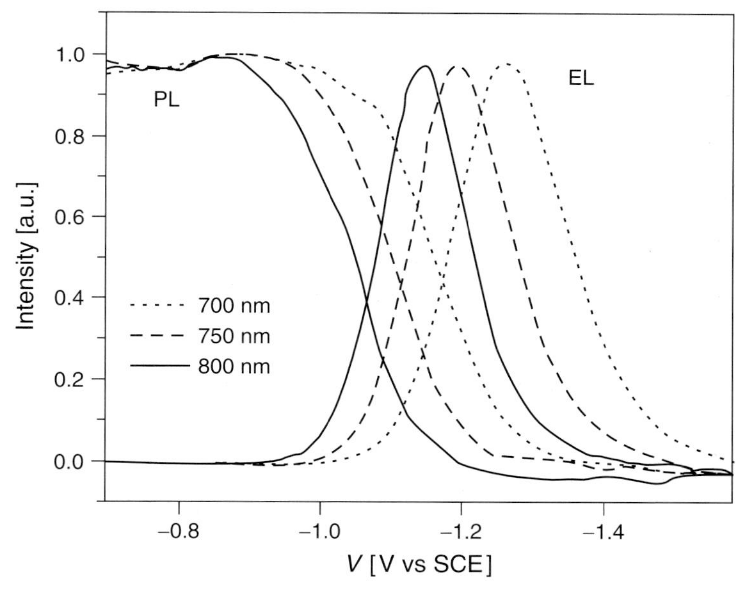

Fig. 19 The potential dependence of the emission intensity from a porous *n*-type silicon electrode in H_2SO_4 solution. Three emission wavelengths are shown. For the PL measurements an argon-ion laser was used as excitation source. EL was excited by reduction of peroxydisulfate, added to the H_2SO_4 solution. Note, the potential scale is reversed in this figure (from Ref. 173).

is quenched. EL can be expected from a quantized structure if it is populated by an electron supplied from the bulk silicon. Because of the larger band gap of porous silicon there is a mismatch of the CB edges. As a result the Fermi energy of the bulk silicon must be raised to a level close to the conduction band edge of the quantized structure. This will occur for larger particles at less negative potential because the mismatch of the CB edges is small in this case. As a result, long wavelength emission will be turned on first. Gradually as the Fermi level is raised further, that is, as the potential is made more negative, the smaller particles will participate and the emission maximum will shift to shorter wavelength. Quenching of EL and PL has been attributed to Auger recombination. This can explain why the rise in EL on going to negative potentials is coupled to the quenching of PL. To give EL, an electron is required in a particle to create a hole via $S_2O_8^{2-}$ reduction (compare with Eqs. 23 and 24). On the other hand, photoexcitation of a particle already occupied by an electron leads to Auger recombination; that is, the PL is quenched. At more negative potentials, the supply of an electron to a particle in which an electron-hole pair is present, leads to Auger quenching of the EL [172, 173].

A wide range of luminescence effects have been reported for porous silicon; these have been reviewed by Kelly and coworkers [25]. During anodic oxidation, porous p-type silicon shows electroluminescence [174, 175], similar to that described for p-type InP in Sect. 2.1.3.3.2. In the case of porous silicon, however, a very strong light emission is observed in the visible spectral range (because of size quantization). On excitation of the bulk substrate with near-infrared light, porous n-type silicon can be photoanodically oxidized; this process is also accompanied by strong emission of visible light [176]. These results provide insight into the mechanism of anodic oxidation of the porous semiconductor.

2.1.5.3 Electron Transport

2.1.5.3.1 Electron Diffusion, Collection, and Recombination A unique feature of a porous photoelectrochemical system is the permeation of the solid semiconducting network by an electrolyte solution on a scale smaller than L_{min} (Fig. 17b). As a result, one of the photogenerated charge carriers can be transferred to the solution. In the following, we will assume that the hole is transferred to the electrolyte solution and oxidizes a reduced species. Photogenerated electrons are left in the

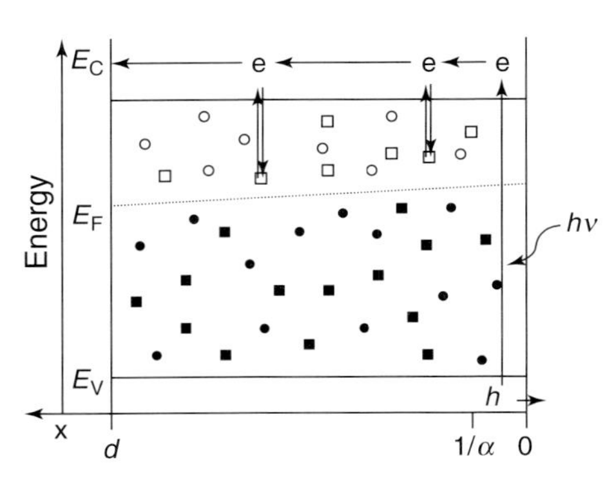

Fig. 20 Energy scheme for photogeneration and diffusion of electrons in a porous photoelectrode under steady state conditions. The light is incident from the electrolyte side ($x = 0$) and is absorbed in a region of width $1/\alpha$. Photogenerated electrons diffuse toward the collecting back contact ($x = d$) caused by a gradient in the electrochemical potential E_F (dashed line). Transport is attenuated by multiple trapping/detrapping.

solid network and diffuse, because of a gradient of the electrochemical potential, through the network over a considerable distance toward the metal contact, where they are collected (Fig. 20). The thickness, d, of a porous electrode and, hence, the length of the electron pathway is between 1 and 100 μm.

The transport of electrons in porous semiconductors is of interest to a wide audience, not only to electrochemists. On one hand, porous semiconducting and insulating networks can be considered as disordered systems, which show a strong resemblance to amorphous semiconductors [177, 178]. In porous semiconductors, diffusing electrons can be scattered not only by the lattice but also by the surface of the matrix and by grain boundaries. Scattering is a friction phenomenon reducing the free-electron mobility μ. In addition, porous networks have a large interfacial area. Therefore, a huge volume density of interfacial electronic states, distributed in the band gap, can be expected. The volume density of band gap states can be much larger than that of a macroscopic crystal and comparable to that of an amorphous semiconductor [177, 178]. A diffusing electron can be trapped in a state in the gap and hence, become temporarily localized. The electron is promoted back into the CB by thermal excitation (trapping/detrapping). It is clear that electron scattering and multiple trapping/detrapping are different physical processes; scattering reduces the electron flux in the system by reducing the mobility of the diffusing free electrons, whereas trapping decreases the electron flux by reducing the density of free electrons.

Many recent experimental results show that electron transport through a porous semiconducting network is a slow process [179–183]. For instance, the average time that electrons need to travel through the system before collection, that is, the transit time $\tau_{tran}(d)$ is in the millisecond to second range [179–183, 187, 188]. As a result, photogenerated electrons can be lost before collection, by transfer to the oxidized species in the solution, a process characterized by a time constant τ_{rec}. Electron back-transfer forms an important recombination process in dye-sensitized photoelectrochemical systems [16, 179–183].

2.1.5.3.2 **Characterization of Electron Diffusion and Back-transfer by Light Intensity Modulated Techniques** In Sect. 2.1.2, it was shown that time-resolved methods are required to obtain information on the kinetics and dynamics of a system. The measurement of the photocurrent response upon a small modulation of the light intensity is a very effective method to study electron dynamics in a porous system [184–186, 187, 188]. Neglecting trapping/detrapping of diffusing electrons and assuming the electrochemical potential gradient in the porous system to be independent of the spatial coordinate x (see Fig. 20), Vanmaekelbergh and coworkers demonstrated that the optoelectrical transfer function may be written as a function of τ_{tran} en τ_{rec} [189, 190]:

$$\frac{\Delta \tilde{j}_{PH}(\omega)}{e(1-R)\Delta\tilde{\Phi}(\omega)} = \frac{1-e^{-i\omega\tau_{tran}(d)}e^{-\tau_{tran}(d)/\tau_{rec}}}{i\omega\tau_{tran}(d)+[\tau_{tran}(d)/\tau_{rec}]} \quad (37)$$

Plots of the optoelectrical transfer function in the complex plane are presented in Fig. 21. Attention is drawn to two limiting cases. If the transit time of photogenerated electrons $\tau_{tran}(d)$ through the porous network is much smaller than the electron lifetime τ_{rec}, photogenerated electrons will

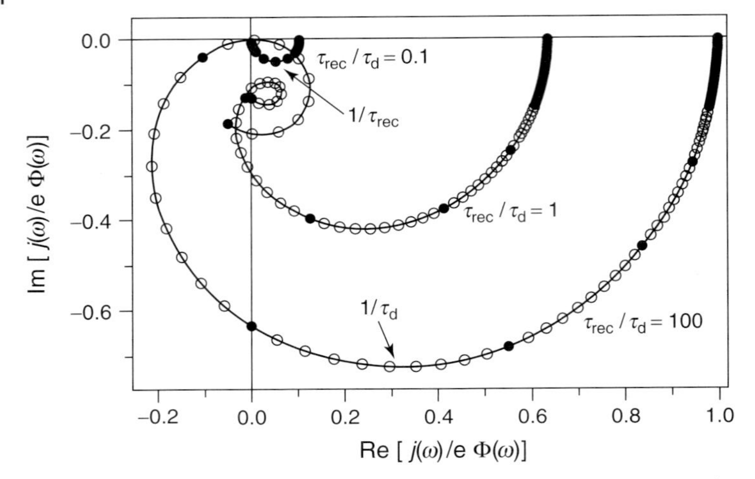

Fig. 21 Plots of the calculated optoelectrical transfer function (Eq. 37) in the complex plane for three ratios of the electron lifetime τ_{rec} to transit time τ_d. For $\tau_{rec}/\tau_d = 100$, all electrons are collected, the transfer function is typical for electron transport; the characteristic frequency gives the transit time. For $\tau_{rec}/\tau_d = 0.1$, only 10% of the photogenerated electrons are collected: the characteristic frequency gives the recombination lifetime.

reach the collecting contact without being lost; the collection efficiency is nearly one, and thus also the low-frequency limit of the transfer function. In this case (Fig. 21), the optoelectrical transfer function describes electron propagation. The shape of the function is determined by electron diffusion only and enters, surprisingly, the third quadrant of the complex plane at high frequencies. The average transit time through the porous network follows from the angular frequency ω_m at which the imaginary part of the optoelectrical transfer function shows its first minimum (characteristic frequency, starting from zero frequency):

$$\tau_{tran}(d) = \frac{3}{\omega_m} \tag{38}$$

It is easily inferred from Eq. (38) that the factor $e^{-\tau_{tran}(d)/\tau_{rec}}$ damps the oscillating function $e^{-i\omega\tau_{tran}(d)}$. As a result, the optoelectrical transfer function becomes increasingly semicircular if $\tau_{tran}(d)/\tau_{rec}$ increases toward and above unity. When recombination is dominant ($\tau_{tran}(d) \gg \tau_{rec}$), the optoelectrical transfer function reduces to

$$\frac{\Delta\tilde{j}(\omega)}{e(1-R)\Delta\tilde{\Phi}(\omega)} = \frac{\tau_{rec}/\tau_{tran}(d)}{i\omega\tau_{rec}+1} \tag{39}$$

This transfer function corresponds to a semicircle in the complex plane (Fig. 21). The recombination lifetime follows from the characteristic frequency of the transfer function: $1/\omega_m = \tau_{rec}$; the transit time can then be extracted from the low-frequency limit given by

$$\frac{\partial j_{photo}}{(1-R)e\,\partial\phi} = \frac{\tau_{rec}}{\tau_{tran}(d)} \tag{40}$$

Including trapping/detrapping in the model that leads to the transfer function expressed by Eq. (37) does not alter the optoelectrical transfer function itself. However, the transit time that is obtained from ω_m is increased by a factor $[1 + \Sigma(E_{F,n})]$. The value of the trapping parameter $\Sigma(E_{F,n})$ (with respect to 1) determines to what extent multiple trapping or detrapping attenuates electron diffusion: electron transport is trap-limited if $\Sigma(E_{F,n}) \gg 1$. It was found that the trapping parameter is given by the ratio of the overall volume density of states near the Fermi level and the average density of free electrons in the porous system for a given position of the electron Fermi level $[n(E_{F,n})]$:

$$\Sigma = \frac{kT \sum_{i} s_{v,i}(E_{F,n})}{n(E_{F,n})} \tag{41}$$

In Eq. (41), $\sum_i s_{v,i}(E)$ denotes the volume density of band gap states per unit energy, at a given energy E. In a porous system, this quantity can be easily between 10^{18} and 10^{20} $cm^{-3}eV^{-1}$; it is expected that Σ is considerably above unity. This means that transport will be trap-limited. In a porous photoelectrochemical system, $n(E_{F,n})$ increases linearly with increasing background light intensity Φ. Hence, we predict that the trapping parameter Σ and thus, the attenuation of electron transport caused by trapping will strongly decrease with increasing light intensity. This has been observed with several porous photoelectrochemical systems [184–186, 187, 188] (see following text). The fact that the transit time decreases with increasing background light intensity also shows that attenuation of electron transport is due to trapping, not to scattering. Furthermore, the experimental relationship between the transit time $\tau_{tran}(d)$ and the background light intensity Φ can be used to map the density of state function $kT \sum_i s_{v,i}(E)$ in a certain energy range of the band gap.

2.1.5.3.3 **Experimental Results on Electron Transport** Electron transport and recombination (corresponding to electron back-transfer to the oxidized species I_3^- in the electrolyte) in nanoporous TiO_2 photoelectrodes has been studied extensively [184–186, 187, 188, 191–194, 195]. It has been found that the transit time through such an electrode depends strongly on the background light intensity; several research groups found that $\tau_{tran}(d) \propto \Phi^{-0.7}$. In other words, the effective diffusion coefficient $D_n(E_{F,n})$ is proportional to $\Phi^{0.7}$. Values between 10^{-8} and 10^{-4} $cm^{-2}s^{-1}$ were found for $D_n(E_{F,n})$ for typical light intensities between 10^{10} and 10^{15} cm^{-2} s^{-1}. These very low values and their dependence on the light intensity (hence, on the position of the electron Fermi level in the band gap) show that electron diffusion is strongly attenuated by trapping. In the framework presented earlier (trap-limited electron transport), the effective diffusion coefficient reads

$$D_n(E_{F,n}) = \frac{\mu\, n(E_{F,n})}{e \sum_{i} s_{v,i}(E_{F,n})} \tag{42}$$

where μ is free-electron mobility. A density-of-state function, which increases exponentially with increasing energy in the band gap has been derived from the earlier relationship. The photovoltage response upon modulation of the light intensity (with a given background light intensity) has been used to study the recombination kinetics in dye-sensitized TiO_2 solar cells [191–195]. It was found that the back-transfer of the electrons from the TiO_2 network to I_3^- is extremely slow. This contrasts with the fast reduction of this species at Pt, where chemisorbed I_3^- plays an essential role. A detailed study of the modulated photovoltage as a function of the background light intensity revealed that $\tau_{rec} \propto 1/n(E_{F,n})$ (this means that the rate of back-transfer is second order in the concentration of electrons) [196]. Peter and coworkers proposed a mechanism for this multistep electron transfer process in which the radical anion $I_2^{-\bullet}$ plays an essential role.

Electron diffusion through macroporous GaP electrodes has also been studied. Macroporous GaP is an ideal model system for several reasons: (1) the porous network is prepared by anodic etching of a single crystal; this gives a very good reproducibility in the preparation of the samples, (2) the porous network is a

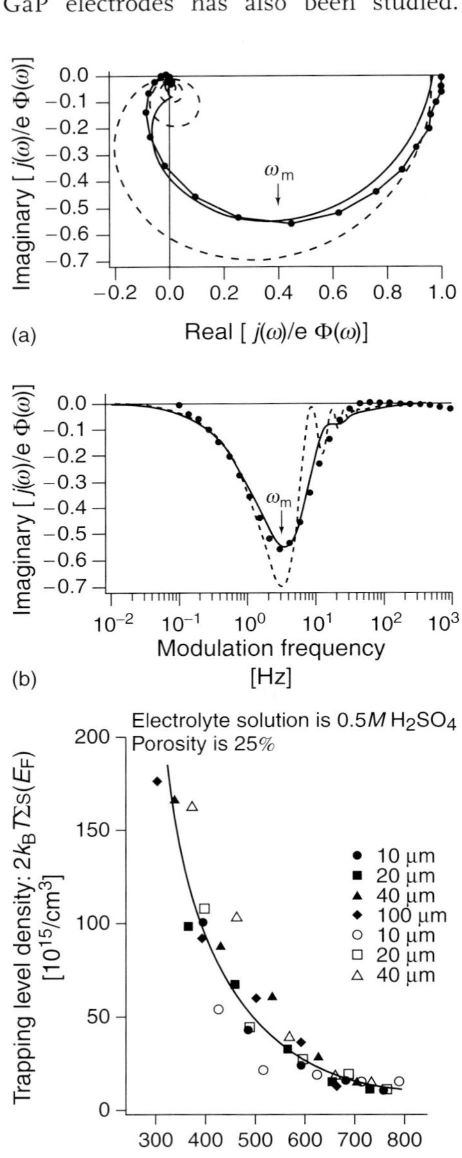

Fig. 22 (a) Complex-plane representation of the measured optoelectrical transfer function (points) for a porous GaP electrode (40-µm thick) permeated with an aqueous H_2SO_4 solution. The background light intensity is 2×10^{14} cm^{-2} s^{-1}. (b) Plot of the imaginary part (points) as a function of the modulation frequency. Full and dashed lines in (a) and (b) represent calculated plots: the dashed plot corresponds to Eq. (37); the full plot takes into account a Gaussian distribution around the average transit time. (c) Density of state function versus the energy in the band gap for macroporous GaP. The data are obtained from the trap-limited transit time ($3/\omega_m$) measured with different background light intensities, (see text).

single crystal without grain boundaries, and (3) the thickness of the porous film and hence, the length of the electron pathway can be varied at will between 20 and 200 μm. This enables an additional check of the theoretical understanding of electron transport. The optoelectrical transfer function measured with a 40-μm porous GaP photoelectrode is shown in Fig. 22 [188]. The transfer function is presented in the complex plane; the evolution of the imaginary component as a function of the modulation frequency is also shown. It is clear that the results are in agreement with the model presented earlier (Eq. 37) if a Gaussian distribution around the average transit time of the electrons is assumed. At a background light intensity Φ of 2×10^{14} cm^{-2}s^{-1}, the transit time $\tau_{tran}(d) = 3/\omega_m$ is 1 s, that is, a factor 10^5 larger than that expected if one assumes that the mobility of the diffusing electrons is the same as in bulk GaP. Thus, electron transport is strongly attenuated by the effect of multiple trapping. The fact that $\tau_{tran}(d)$ decreases strongly with increasing background light intensity shows that trapping/detrapping in band gap states close to the electron Fermi level attenuates diffusion. A detailed analysis of the results obtained with porous GaP electrodes of different thickness led to a mapping of the density-of-state function in the upper half of the band gap (Fig. 22) [188].

References

1. A. Uhlir, *Bell Systems Techn. J.* **1956**, *35*, 333.
2. H. Gerischer, F. Beck, *Z. Phys. Chem. N.F.* **1957**, *13*, 389.
3. D. R. Turner, *J. Electrochem. Soc.* **1958**, *105*, 402.
4. A. Fujishima, K. Honda, *Bull. Chem. Soc. Jpn.* **1971**, *44*, 1148.
5. A. B. Ellis, S. W. Kaiser, M. S. Wrighton, *J. Am. Chem. Soc.* **1976**, *98*, 1635.
6. A. B. Ellis, J. M. Bolts, M. S. Wrighton, *J. Electrochem. Soc.* **1977**, *124*, 1603.
7. G. Hodes, J. Manassen, D. Cahen, *Nature* **1976**, *261*, 403.
8. B. Miller, A. Heller, *Nature* **1976**, *262*, 680.
9. S. Licht, G. Hodes, R. Tenne et al., *Nature* **1987**, *330*, 148.
10. B. Wang, S. Licht, T. Soga et al., *Solar Energy Mater. Sol. Cells* **2000**, *64*, 311.
11. S. Licht, H. Tributsch, B. Wang et al., *J. Phys. Chem. B* **2000**, *104*, 8920.
12. S. Licht, *J. Phys. Chem.* **1986**, *90*, 1096.
13. S. Licht, D. Peramunage, *Nature* **1990**, *345*, 330.
14. S. Licht, O. Khaselev, T. Soga et al., *Electrochem. Sol. State Lett.* **1998**, *1*, 20.
15. P. H. L. Notten, J. E. A. M. van den Meerakker, J. J. Kelly, in *Etching of III–V Semiconductors, An Electrochemical Approach*, Elsevier Advanced Technology, Oxford, 1991.
16. B. O'Regan, M. Grätzel, *Nature* **1991**, *353*, 737.
17. L. T. Canham, *Appl. Phys. Lett.* **1990**, *57*, 1046.
18. J. J. Kelly, D. Vanmaekelbergh, Porous-etched Semiconductors: Formation and Characterization, in *The Electrochemistry of Nanostructures: Preparation and Properties*, Chap. 4 (Eds.: G. Hodes), Wiley-VCH, Weinheim, 2001.
19. S. Ogawa, F. F. Ran, A. J. Bard, *J. Phys. Chem. B* **1995**, *99*, 11182.
20. S. Ogawa, K. Hu, F. F. Ran et al., *J. Phys. Chem. B* **1997**, *101*, 5707.
21. S. R. Morrison, in *Electrochemistry at Semiconductor and Oxidized Metal Electrodes*, Plenum Press, New York, 1980.
22. J. Gobrecht, H. Gerischer, *Solar Energy Mater.* **1979**, *2*, 131.
23. R. Memming, *Electrochim. Acta* **1980**, *25*, 77.
24. R. Memming, in *Semiconductor Electrochemistry*, Wiley-VHC, Weinheim, 2000.
25. J. J. Kelly, E. S. Kooij, E. A. Meulenkamp, *Electrochim. Acta* **1999**, *45*, 561.
26. E. P. A. M. Bakkers, E. Reitsma, J. J. Kelly et al., *J.Phys. Chem. B* **1999**, *103*, 2781.
27. J. R. Macdonald in *Impedance Spectroscopy, Emphasizing Solid Materials and Systems*, John Wiley & Sons, New York, 1987.
28. R. Unbehauen in *Systemtheorie 1, Allgemeine Grundlagen, Signale und lineare Systeme im Zeit- und Frequenzbereich* R. Oldenbourg Verlag, München, 1997.

29. W. Gärtner, *Phys. Rev.* **1959**, *116*, 84.
30. L. M. Peter, *Chem. Rev.* **1990**, *90*, 753.
31. B. Smandek, G. Chmiel, H. Gerischer, *Ber. Bunsen-Ges. Phys. Chem.* **1989**, *93*, 1094.
32. G. Chmiel, H. Gerischer, *J. Phys. Chem.* **1990**, *94*, 1612.
33. K. H. Beckmann, R. Memming, *J. Electrochem. Soc.* **1969**, *116*, 368.
34. H. Gerischer, *Z. Phys. Chem.* **1960**, *26*, 325.
35. H. Gerischer, *Z. Phys. Chem.* **1961**, *27*, 48.
36. H. Gerischer, Semiconductor Electrochemistry, in *Physical Chemistry, an Advanced Treatise*, Vol. IXA/Electrochemistry, Academic Press, New York, 1970.
37. W. P. Gomes, F. Cardon, in *Progress in Surface Science* (Eds.: S. G. Davison), Pergamon Press, 1982, p. 155, Vol. 12.
38. J. O' M. Bockris, A. K. N. Reddy in *Modern Electrochemistry*, Plenum Press, New York, 1977, Vol. 2.
39. Y. V. Pleskov, Y. Y. Gurevich in *Semiconductor Photoelectrochemistry*, Consultants Bureau, New York, 1986.
40. R. Degryse, W. P. Gomes, F. Cardon et al., *J. Electrochem. Soc.* **1975**, *122*, 711.
41. T. Pajkossy, *J. Electroanal. Chem.* **1994**, *364*, 111.
42. J. F. McCann, S. P. S. Baldwall, *J. Electrochem. Soc.* **1982**, *129*, 551.
43. Z. Hens, W. P. Gomes, *Phys. Chem. Chem. Phys.* **1999**, *1*, 3607.
44. Z. Hens, W. P. Gomes, *Phys. Chem. Chem. Phys.* **1999**, *1*, 3617.
45. K. Schröder, R. Memming, *Ber. Bunsen-Ges. Phys. Chem.* **1985**, *89*, 385.
46. D. Vanmaekelbergh, W. P. Gomes, *Ber. Bunsen-Ges. Phys. Chem.* **1985**, *89*, 994.
47. J. E. A. M. van den Meerakker, J. J. Kelly, P. H. L. Notten, *J. Electrochem. Soc.* **1985**, *132*, 638.
48. J. E. A. M. van den Meerakker, *Electrochim. Acta* **1985**, *30*, 435.
49. D. Vanmaekelbergh, W. P. Gomes, F. Cardon, *Ber. Bunsen-Ges. Phys. Chem.* **1986**, *90*, 431.
50. D. Lincot, J. Vedel, *J. Electroanal. Chem.* **1987**, *220*, 179.
51. D. Vanmaekelbergh, W. P. Gomes, F. Cardon, *J. Electrochem. Soc.* **1987**, *134*, 891.
52. D. Vanmaekelbergh, R. P. ter Heide, W. Kruijt, *Ber. Bunsen-Ges. Phys. Chem.* **1989**, *93*, 1103.
53. Z. Hens, W. P. Gomes, *J. Phys. Chem. B* **2000**, *104*, 7725.
54. B. H. Erné, D. Vanmaekelbergh, I. E. Vermeir, *Electrochim. Acta* **1993**, *38*, 2559.
55. D. Vanmaekelbergh, F. Cardon, *J. Phys. D: Appl. Phys.* **1986**, *19*, 643.
56. D. Vanmaekelbergh, F. Cardon, *Semicond. Sci. Technol.* **1988**, *3*, 124.
57. D. Vanmaekelbergh, F. Cardon, *Electrochim. Acta* **1992**, *37*, 837.
58. Z. Hens, W. P. Gomes, *J. Phys. Chem. B* **1997**, *101*, 5814.
59. Z. Hens, W. P. Gomes, *J. Electroanal. Chem.* **1997**, *437*, 77.
60. Z. Hens, *J. Phys. Chem. B* **1999**, *103*, 122.
61. Z. Hens, W. P. Gomes, *J. Phys. Chem. B* **1999**, *103*, 130.
62. D. Vanmaekelbergh, A. R. de Wit, F. Cardon, *J. Appl. Phys.* **1993**, *73*, 5049.
63. A. A. Burk, P. B. Johnson, W. S. Hobson et al., *J. Appl. Phys.* **1986**, *59*, 1621.
64. A. B. Ellis, Luminescent Properties of Semiconductor Electrodes, in *Chemistry and Structure of Interfaces, New Laser and Optical Techniques*, Chapter 6 (Eds.: R. B. Hall, A. B. Ellis), VCH, Deerfield Beach (FL), 1986.
65. B. A. Balko, G. L. Richmond, *J. Phys. Chem.* **1993**, *97*, 9002.
66. O. Kruger, Ch. Jung, *Ber. Bunsen-Ges. Phys. Chem.* **1994**, *98*, 1022.
67. J. F. Kauffman, C. S. Liu, M. W. Karl, *J. Phys. Chem.* **1998**, *102*, 6766.
68. G. N. Ryba, C. N. Kenyon, N. S. Lewis, *J. Phys. Chem.* **1993**, *97*, 13814.
69. A. Manivannan, K. Hashimoto, A. Fujishima, *J. Phys. Chem.* **1992**, *96*, 3766.
70. A. Manivannan, K. Itoh, K. Hashimoto et al., *J. Electrochem. Soc.* **1990**, *137*, 3121.
71. J. Ouyang, F.-R. Fan, A. J. Bard, *J. Electrochem. Soc.* **1989**, *136*, 1033.
72. F. Decker, F. Prince, P. Motisuke, *J. Appl. Phys.* **1985**, *57*, 2900.
73. B. H. Erné, F. Ozanam, J.-N. Chazalviel, *Phys. Rev. Lett.* **1998**, *80*, 4337.
74. S. R. Morrison, T. Freund, *J. Chem. Phys.* **1967**, *47*, 1543.
75. W. P. Gomes, T. Freund, S. R. Morrison, *J. Electrochem. Soc.* **1968**, *115*, 818.
76. A. Fujishima, T. Kato, E. Maekawa et al., *Bull. Chem. Soc. Jpn.* **1981**, *54*, 1671.
77. G. H. Schoenmakers, D. Vanmaekelbergh, J. J. Kelly, *J. Chem. Soc., Faraday Trans.* **1997**, *93*, 1127.
78. J. R. Harbor, M. L. Hair, *J. Phys. Chem.* **1979**, *83*, 652.

79. G. Nogami, J. H. Kennedy, *J. Electrochem. Soc.* **1989**, *136*, 2583.
80. P. Herrasti, L. Peter, *J. Electroanal. Chem.* **1991**, *305*, 241.
81. P. Bogdanoff, N. Alonso-Vante, *J. Electroanal. Chem.* **1994**, *379*, 415.
82. R. Memming, *J. Electrochem. Soc.* **1969**, *116*, 785.
83. J. J. Kelly, B. P. Minks, N. A. M. Verhaegh et al., *Electrochim. Acta* **1992**, *37*, 909.
84. B. P. Minks, G. Oskam, D. Vanmaekelbergh et al., *J. Electroanal. Chem.* **1989**, *273*, 119.
85. B. P. Minks, D. Vanmaekelbergh, J. J. Kelly, *J. Electroanal. Chem.* **1989**, *273*, 133.
86. R. Peat, L. M. Peter, *Electrochim. Acta* **1986**, *31*, 731.
87. R. Peat, L. M. Peter, *J. Electroanal. Chem.* **1986**, *209*, 307.
88. W. P. Gomes, H. H. Goossens, Electrochemistry of III–V Compound Semiconductors: Dissolution Kinetics and Etching in *Advances in Electrochemical Science and Engineering*, (Eds.: H. Gerischer, C. W. Tobias), VCH, Weinheim, 1994, Vol. 3.
89. D. Vanmaekelbergh, J. J. Kelly, *J. Electrochem. Soc.* **1989**, *136*, 108.
90. W. P. Gomes, S. Lingier, D. Vanmaekelbergh, *J. Electroanal. Chem.* **1989**, *269*, 237.
91. M. Matsumura, S. R. Morrison, *J. Electroanal. Chem.* **1983**, *147*, 159.
92. L. M.Peter, A. M. Borazio, H. J. Lewerenz et al., *J. Electroanal. Chem.* **1990**, *290*, 229.
93. G. H. Schoenmakers, R. Waagenaar, J. J. Kelly, *J. Electrochem. Soc.* **1995**, *142*, L60.
94. S. Cattarin, L. M. Peter, D. J. Riley, *J. Phys. Chem. B* **1997**, *101*, 4071.
95. X. Xia, C. M. A. Ashruf, P. J. French et al., *J. Phys. Chem. B* **2001**, *105*, 5722.
96. B. H. Erné, D. Vanmaekelbergh, *J. Electrochem. Soc.* **1997**, *144*, 3385.
97. D. Vanmaekelbergh, P. C. Searson, *J. Electrochem. Soc.* **1994**, *141*, 697.
98. B. H. Erné, F. Ozanam, M. Stchakovsky et al., *J. Phys. Chem. B* **2000**, *104*, 5961.
99. B. H. Erné, F. Ozanam, M. Stchakovsky et al., *J. Phys. Chem. B* **2000**, *104*, 5974.
100. P. Allongue, H. Brune, H. Gerischer, *Surf. Sci.* **1992**, *275*, 414.
101. P. Allongue, V. Costa Gieling, H. Gerischer, *J. Electrochem. Soc.* **1993**, *140*, 1009.
102. P. Allongue, *Phys. Rev. Lett.* **1996**, *77*, 1986.
103. W. West, B. H. Carrol in *The Theory of Photographic Processes*, (Eds.: C. E. K. Mees, T. H. James), McMillan, New York, 1966.
104. R. W. Berriman, P. B. Gilman, *Photogr. Sci. Eng.* **1973**, *17*, 235.
105. J. J. Kelly, D. Vanmaekelbergh, *Electrochim. Acta* **1998**, *43*, 2773.
106. J. W. Gerritsen, E. J. G. Boon, G. Janssens et al., *Appl. Phys. A* **1998**, *66*, S79.
107 M. Kawasaki, H. Ishii, *J. Imaging Sci. Technol.* **1995**, *39*, 210.
108. T. Hannappel, B. Burfeindt, W. Storck et al., *J. Phys. Chem. B* **1997**, *101*, 6799.
109. Y. Tachibana, J. E. Moser, M. Graetzel et al., *J. Phys. Chem.* **1996**, *100*, 20056.
110. A. Henglein, *Top. Curr. Chem.* **1988**, *143*, 115.
111. H. Weller, *Adv. Mater.* **1993**, *5*, 88.
112. A. P. Alivisatos, *J. Phys. Chem.* **1996**, *100*, 13226.
113. A. N. Shipway, E. Katz, I. Willner, *Chem. Phys. Chem.* **2000**, *1*, 18.
114. B. Alperson, H. Demange, I. Rubinstein et al., *J. Phys. Chem. B* **1999**, *103*, 4943.
115. R. L. Smith, S. D. Collins, *J. Appl. Phys.* **1992**, *71*, R1.
116. V. L. Colvin, A. N. Goldstein, A. P. Alivisatos, *J. Am. Chem. Soc.* **1992**, *114*, 5221.
117. M. Miyake, H. Matsumoto, M. Nishizawa et al., *Langmuir* **1997**, *13*, 742.
118. T. Nakanishi, B. Ohtani, K. Uosaki, *J. Phys. Chem. B* **1998**, *102*, 1571.
119. K. Hu, M. Brust, A. J. Bard, *Chem. Mater.* **1998**, *10*, 1160.
120. S. Drouard, S. G. Hickey, J. D. Riley, *Chem. Commun.* **1999**, *1*, 67.
121. E. P. A. M. Bakkers in *Charge transfer between semiconductor nanocrystals and a metal*, Ph. D. Thesis, University of Utrecht, 2000.
122. E. P. A. M. Bakkers, J. J. Kelly, D. Vanmaekelbergh, *J. Electroanal. Chem.* **2000**, *482*, 48.
123. E. P. A. M. Bakkers, E. Reitsma, J. J. Kelly, D. Vanmaekelbergh, *J. Phys. Chem. B* **1999**, *103*, 2781.
124. S. G. Hickey, D. J. Riley, *Electrochim. Acta* **2000**, *45*, 3277.
125. W. J. Albery, G. T. Brown, J. R. Darwent et al., *J. Chem. Soc. Faraday Trans. 1* **1985**, *81*, 1999.
126. E. P. A. M. Bakkers, A. W. Marsman, L. W. Jenneskens et al., *Angew. Chem. Int. Ed.* **2000**, *39*, 2297.
127. E. P. A. M. Bakkers, A. L. Roest, A. W. Marsman et al., *J. Phys. Chem. B* **2000**, *104*, 7266.

128. S. G. Hickey, J. D. Riley, E. Tull, *J. Phys. Chem. B* **2000**, *104*, 7623.
129. M. R. Hoffmann, S. T. Martin, W. Choi et al., *Chem. Rev.* **1995**, *95*, 69.
130. G. Hodes, *J. Electrochem. Soc.* **1992**, *139*, 3136.
131. R. Grünwald, H. Tributsch, *J. Phys. Chem. B* **1997**, *101*, 2564.
132. S. Y. Huang, *J. Phys. Chem. B* **1997**, *101*, 2576.
133. D. Liu, R. W. Fessenden, G. L. Hug et al., *J. Phys. Chem. B* **1997**, *101*, 2583.
134. R. Argazzi, C. A. Bignozzi, T. A. Heimer et al., *J. Phys. Chem. B* **1997**, *101*, 2591.
135. H. Rensmo, K. Keis, H. Lindstroem et al., *J. Phys. Chem. B* **1997**, *101*, 2598.
136. A. Wahl, J. Augustynski, *J. Phys. Chem. B* **1998**, *102*, 7820.
137. W. Kubo, K. Murakoshi, T. Kitamura et al., *Chem. Lett.* **1998**, *12*, 1241.
138. B. H. Erné, D. Vanmaekelbergh, J. J. Kelly, *Adv. Mater.* **1995**, *7*, 739.
139. B. H. Erné, D. Vanmaekelbergh, J. J. Kelly, *J. Electrochem. Soc.* **1996**, *143*, 305.
140. G. Oskam, A. Natarajan, P. C. Searson, F. M. Ross, *Appl. Surf. Sci.* **1997**, *119*, 160.
141. J. van de Lagemaat, M. Plakman, D. Vanmaekelbergh et al., *Appl. Phys. Lett.* **1996**, *69*, 2801.
142. T. Sugira, T. Yoshida, H. Minoura, *Electrochem. Solid-State Lett.* **1998**, *1*, 175.
143. A. O. Konstantinov, C. I. Harris, E. Janzén, *Appl. Phys. Lett.* **1994**, *65*, 2699.
144. A. O. Konstantinov, A. Henry, C. I. Harris et al., *Appl. Phys. Lett.* **1995**, *66*, 2250.
145. C. Zener, *Proc. R. Soc. London* **1934**, *145*, 523.
146. F. J. P. Schuurmans, D. Vanmaekelbergh, J. Van de Lagemaat et al., *Science* **1999**, *284*, 141.
147. F. J. P. Schuurmans, M. Megens, D. Vanmaekelbergh et al., *Phys. Rev. Lett.* **1999**, *83*, 2183.
148. L. M. Peter, D. J. Riley, R. I. Wielgosz, *Appl. Phys. Lett.* **1995**, *66*, 2355.
149. L. M. Peter, R. I. Wielgosz, *Appl. Phys. Lett.* **1996**, *69*, 806.
150. W. H. Lubberhuizen, D. Vanmaekelbergh, E. Van Faassen, *Journal of Porous Materials* **2000**, *7*, 147.
151. B. Enright, B. G. Redmond, D. Fitzmaurice, *J. Phys. Chem.* **1994**, *98*, 6195.
152. M. Gratzel, *Nature* **2001**, *409*, 575.
153. D. Cummins, G. Boschloo, M. Ryan et al., *J. Phys. Chem. B* **2000**, *104*, 11449.
154. P. Hoyer, R. Eichberger, H. Weller, *Ber. Bunsen-Ges. Phys. Chem.* **1993**, *97*, 630.
155. P. Hoyer, H. Weller, *J. Phys. Chem.* **1995**, *99*, 14096.
156. E. A. Meulenkamp, *J. Phys. Chem. B* **1999**, *103*, 7831.
157. D. L. Klein, P. L. McEuen, J. E. B. Katari et al., *Appl. Phys. Lett.* **1996**, *68*, 2574.
158. D. L. Klein, R. Roth, A. K. L. Lim et al., *Nature* **1997**, *389*, 699.
159. U. Banin, Y. Cao, D. Katz et al., *Nature* **1999**, *400*, 542.
160. O. Millo, D. Katz, Y. Cao et al., *Phys. Rev. B* **2000**, *61*, 16773.
161. E. P. A. M. Bakkers, D. Vanmaekelbergh, *Phys. Rev. B* **2000**, *62*, R7743.
162. J. D. Joannopoulos, P. R. Villeneuve, S. Fan, *Nature* **1997**, *386*, 143.
163. U. Grüning, V. Lehmann, C. M. Engelhardt, *Appl. Phys. Lett.* **1995**, *66*, 3254.
164. J. E. G. J. Wijnhoven, W. L. Vos, *Science* **1998**, *281*, 802.
165. Y. A. Vlasov, N. Yao, D. J. Norris, *Adv. Mater.* **1999**, *11*, 165.
166. D. Vanmaekelbergh, L. van Pieterson, *Phys. Rev. Lett.* **1998**, *80*, 82.
167. D. Vanmaekelbergh, M. A. Hamstra, L. van Pieterson, *J. Phys. Chem. B* **1998**, *102*, 7997.
168. J. J. M. Halls, C. A. Walsh, N. C. Greenham et al., *Nature* **1995**, *376*, 498.
169. G. Yu, J. Gao, J. C. Hummelen et al., *Science* **1995**, *270*, 1789.
170. U. Bach, D. Lupo, P. Comte et al., *Nature* **1998**, *395*, 583.
171. A. G. Cullis, L. T. Canham, P. D. J. Calcott, *J. Appl. Phys.* **1997**, *82*, 909.
172. E. A. Meulenkamp, L. M. Peter, D. J. Riley et al., *J. Electroanal. Chem.* **1995**, *392*, 97.
173. L. M. Peter, D. J. Riley, R. I. Wielgosz et al., *Thin Solid Films* **1996**, *276*, 123.
174. A. Bsiesy, F. Gaspard, R. Herino et al., *J. Electrochem. Soc.* **1991**, *138*, 3450.
175. M.Ligeon, F. Muller, R. Herino et al., *J. Appl. Phys.* **1993**, *74*, 1265.
176. E. S. Kooij, A. R. Rama, J. J. Kelly, *Surf. Sci.* **1997**, *370*, 125.
177. R. A. Street, Cambridge Solid State Science Series, in *Hydrogenated Amorphous Silicon*, (Eds.: R. W. Cahn, E. A. Davis, I. M. Ward), 1991.
178. H. Scher, M. F. Shlesinger, J. T. Bendler, *Phys. Today* **1991**, *44*, 26.

179. L. M. Peter, D. Vanmaekelbergh in *Advances in Electrochemical Science and Engineering*, (Eds.: R. C. Alkire, D. M. Kolb), Wiley VCH, Weinheim, 1999, Vol. 6.
180. K. Schwarzburg, F. Willig, *Appl. Phys. Lett.* **1991**, *58*, 2520.
181. P. Hoyer, H. Weller, *J. Phys. Chem.* **1995**, *99*, 14096.
182. A. Solbrand, H. Lindstroem, H. Rensmo et al., *J. Phys. Chem. B* **1997**, *101*, 2514.
183. F. Cao, G. Oskam, G. J. Meyer et al., *J. Phys. Chem.* **1996**, *100*, 17021.
184. P. E. de Jongh, D. Vanmaekelbergh, *Phys. Rev. Lett.* **1996**, *77*, 3427.
185. L. Dloczik, O. Ileperuma, I. Lauermann et al., *J. Phys. Chem. B* **1997**, *101*, 10281.
186. P. E. de Jongh, D. Vanmaekelbergh, *J. Phys. Chem. B* **1997**, *101*, 2716.
187. D. Vanmaekelbergh, F. Iranzo Marín, J. van de Lagemaat, *Ber. Bunsen-Ges. Phys. Chem.* **1996**, *100*, 616.
188. A. L. Roest, D. Vanmaekelbergh, *Phys. Rev. B* **2000**, *62*, 16926.
189. D. Vanmaekelbergh, P. E. de Jongh, *Phys. Rev. B* **2000**, *61*, 4699.
190. P. E. de Jongh in *Photoelectrochemistry of nanoporous semiconductor electrodes*, Ph.D. Thesis,. University of Utrecht, 1999.
191. A. C. Fisher, L. M. Peter, E. A. Ponomarev et al., *J. Phys. Chem. B* **2000**, *104*, 949.
192. J. van de Lagemaat, N. G. Park, A. J. Frank, *J. Phys. Chem. B* **2000**, *104*, 2044.
193. N. Kopidakis, E. A. Schiff, N. G. Park et al., *J. Phys. Chem. B* **2000**, *104*, 3930.
194. J. van de Lagemaat, A. Frank., *J. Phys. Chem. B* **2000**, *104*, 4292.
195. N. W. Duffy, L. M. Peter, R. M. G. Rajapakse et al., *J. Phys. Chem. B* **2000**, *104*, 8916.
196. G. Schlichthörl, S. Y. Huang, J. Sprague et al., *J. Phys. Chem. B* **1997**, *101*, 8141.

2.2
Deposition of (Multiple Junction) Semiconductor Surfaces

Tetsuo Soga
Nagoya Institute of Technology, Nagoya, Japan

2.2.1
Introduction

A limited fraction of incident solar photons have sufficient (greater than band gap) energy to initiate charge excitation and separation within a semiconductor. Wide band gap semiconductor solar cells are capable of generating a high photovoltage but have a low limiting photocurrent caused by the low fraction of short wavelength light in the solar spectrum. Smaller band gap cells can utilize a larger fraction of the incident photons but generate lower photovoltage. Multijunction devices, also referred to as tandem, multiple window, split spectrum, and cascade solar cells, can overcome this limitation [1]. A variety of multiple semiconductor layers have been fabricated and explored for photovoltaic applications. These include GaInP on GaAs, GaAs on Si, GaAs on GaSb, InP on GaInAs, GaAs on GaInAsP, etc [2]. Through a detailed specific example of one of these important multiple semiconductor systems consisting of various GaAs layers on various Si layers, this paper describes preparation and optimization of multijunction systems that are capable of efficient photon absorption, charge separation, and charge transfer.

This section reviews the growth process of III–V compound semiconductor on Si substrate grown by metalorganic chemical vapor deposition (MOCVD). The nucleation of GaP and GaAs on Si substrate, the dislocation generation mechanism, antiphase domain (APD) structure of GaP on Si substrate, the method to reduce dislocation density, the method to passivate the defects in GaAs on Si, and the application to photovoltaic solar cell are presented. Although highly mismatched systems such as GaAs-on-Si, AlGaAs-on-Si, and GaN-on-Si are interesting for device applications, the study on the crystal growth of GaP on Si substrate is interesting for the fundamental understanding of the III–V compounds on Si substrate. It is expected that the understanding of the growth mechanism will lead us to achieve the growth of low-dislocation-density III–V compounds on Si substrate.

The deposition of III–V compounds has recently been actively performed. Especially, the crystal growth on Si substrate has attracted attention since high-quality GaAs layers were successfully grown on Si substrates in 1984 [3–7]. Although various devices such as lasers [8] and solar cells [9] have been fabricated on Si substrate, the device characteristics are not satisfactory because the existence of a high density of the threading dislocations in the epitaxial layer. The reduction of dislocation density is an important issue to obtain a high-performance compound semiconductor device on an Si substrate. The dislocations are generated due to lattice mismatch, thermal expansion mismatch, the crystal structure difference, the generation of APD, surface contamination, and so on.

The threading dislocations in the epitaxial layer on Si substrate are classified into (1) dislocations that originate from the dislocations in the Si substrate, (2) dislocations generated by the coalescence of the islands at the initial stage, (3) dislocations generated by the lattice mismatch, and (4) dislocations generated by the thermal stress during the cooling stage from the growth temperature.

Among these four types of dislocations, we do not need to take into account item (1) because the dislocation density of Si is generally very low. In order to reduce the dislocation density according to items (2), (3), and (4), many efforts have been made. The efforts to change the initial growth mode from three-dimensional (3D) to two-dimensional (2D) have been made to avoid the dislocations generated by the coalescence of the islands at the initial stage of the growth [10]. Although many methods using strained layer superlattice (SLS) buffer layer [3, 11, 12], rapid thermal annealing [13], thermal cycle annealing (TCA) [14, 15], and so on have been adopted to reduce the dislocation density of item (3), the dislocation density is still on the order of 10^6 cm^{-2}. The low temperature growth has been investigated to decrease the generation of the dislocation by the thermal stress [16, 17]. It is expected that the number of dislocations generated by the thermal stress is reduced when the growth temperature is low. Although the dislocation density of GaAs-on-Si on the order of 10^4 cm^{-2} has been obtained at the growth temperature of 350 °C, thermal stability is the problem.

2.2.2 MOCVD

The epitaxial growth was performed using low-pressure or atmospheric pressure MOCVD. The former consists of lamp-heated and the latter consists of rf-heated horizontal reactor with load lock chamber. The substrate was put on the SiC-coated carbon susceptor and the temperature was controlled by the thermocouple inserted into the susceptor. Source gases for Al, Ga, Zn, As and Se are trimethylalluminum (TMA), trimethylgallium (TMG), diethylzinc (DEZ), AsH_3 and H_2Se, respectively. The Si substrate orientation was (001) 2–4° tilted toward [110] direction. Si substrate was rinsed in organic solvents, followed by the repetition of the oxidation by $H_2SO_4 : H_2O = 4 : 1$ and the removal of the oxides by 25% HF solution. After loading the substrate into the reactor, the substrate was heated at 1000 °C for 10 min at hydrogen atmosphere. The V/III ratio was varied from 100 to 6400 with changing the PH_3 flow rate, keeping the TMG flow rate constant for the growth of GaP on Si substrate. The gas pressure was varied from 76 to 380 torr. The epitaxial layer thickness was varied from 20 nm to 3.7 μm. The growth temperature was kept constant at 900 °C. 3D growth was not observed under these growth conditions for the case of GaP growth on GaP substrate. The samples were examined using Nomarski optical microscopy, cross-sectional transmission electron microscopy (TEM), and so on. GaAs was grown on Si substrate by the two-step growth method with 10-nm-thick GaAs buffer layer grown at 400 °C.

2.2.3 Deposition of Gallium Phosphide on Silicon Substrate

2.2.3.1 Nucleation of GaP on Si Substrate

In general, the growth mode is divided into three categories, namely, 2D type, Volmer-Weber (3D) type, and Stranski-Krastanov (2D + 3D) type. In the case of the 2D mode, the dislocations are generated when the layer thickness exceeds the critical thickness [18]. On the other hand, the dislocations and stacking faults are generated at the coalescence of the islands formed at the initial stage of the growth in the case of the 3D mode [19]. Therefore, the defect density will be significantly

Tab. 1 Material parameters of III–V compound semiconductors and Si

	GaP	*GaAs*	*InP*	*Si*
Lattice constant (Å)	5.45	5.65	5.87	5.43
Linear thermal expansion coefficient ($\times 10^{-6}$K^{-1})	5.9	6.8	4.6	2.6
Crystal structure	ZB	ZB	ZB	Diamond

Note: ZB: Zinc Blend.

reduced if a 2D growth can be realized from the beginning of the growth.

The 3D growth mode has origins not only in the basic material property differences between the epitaxial layer and the substrate (e.g. lattice constant mismatch, polar or nonpolar effect, etc.) but also in the growth conditions.

The material properties for various III–V compounds and Si are shown in Table 1. As shown in this table, the lattice constant of GaP is closest to that of Si. Therefore, the effect of lattice mismatch on the growth mode is expected to be minimized and other effects such as the surface migration effect, surface contamination, polar or nonpolar structure are emphasized.

Figure 1 shows the growth mode of GaP-on-Si for various gas pressure and the V/III ratio [20]. All the surface morphologies of samples were classified into three-types, namely, island-type, mixture-type (mixture of island and layer-type), and layer-type. It is indicated that the growth mode of GaP changes from island-type to layer-type with increasing V/III ratio. The V/III ratio at which the growth mode changes from island-type to layer-type decreases with increasing gas pressure. A very high V/III ratio of 3200 is necessary to obtain a GaP layer without island-type crystal at the gas pressure of 76 torr [8]. Island-type growth is clearly demonstrated, and it is observed that the islands are not connected by a GaP layer between themselves at the edge of the island. This means that the growth mode of GaP on Si substrate under these growth conditions is not Stranski-Krastanov type but Volmer-Weber type. Faceting was observed at the boundary either on (111) or (211) type planes.

The island formation is interesting because these islands are not formed in

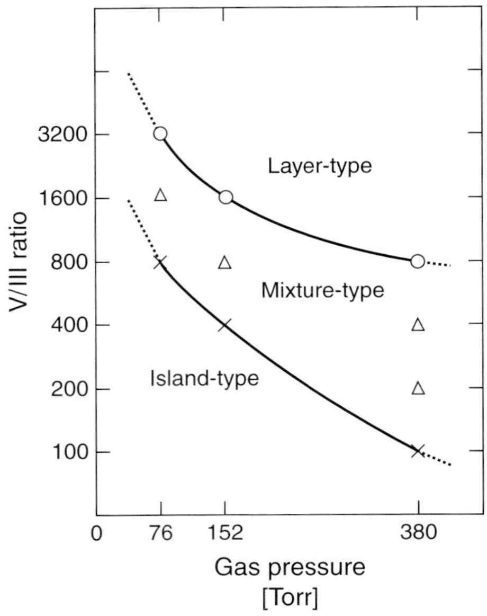

Fig. 1 Growth mode of GaP on Si substrate for various V/III ratio and gas pressure.

the case of homoepitaxy; it is particular for the heteroepitaxy. In the case of the homoepitaxial growth, the source gases are usually incorporated into the step edge or terrace of the misoriented substrate. However, the island spacing is several orders of magnitude larger than that of the average step distance of the misoriented substrate. Therefore, the nucleation site is not governed by the substrate steps. Furthermore, the residual oxide or impurity is not the nucleation site because the GaP island density changes drastically with the growth conditions [21].

In the nucleation process for GaP on Si substrate, the diffusion of the growth species through the boundary layer, the surface migration, and the nucleation at the nucleation sites should be taken into account. Comparing the heteroepitaxy of GaP-on-Si and the homoepitaxy, there should be no difference in the diffusion process if the growth conditions are the same. Therefore, it is deduced that the difference between the homoepitaxy and the heteroepitaxy on Si is caused by the difference in the migration on the Si substrate. The migration length of the migrating species on Si is considered to be longer than for homoepitaxy. This is due to the weak interaction between Si and Ga or P atoms. An example of strong atomic interaction is the growth of a III–V compound semiconductor containing Al on Si substrate. It has been reported that AlGaP [22], AlGaAs [23], and AlAs [24] layers grown on Si substrate are flat from the beginning of growth.

Before considering the migrating species for GaP on Si, the surface migration of Ga is discussed. Usually, the migration species for the deposition of Ga films on Si substrate is the cluster of Ga atoms. TMG is probably perfectly decomposed to Ga and a metal-radical at 900 °C [25]. Therefore, in the case of the deposition of Ga on Si, the migrating species are supposed to be Ga_x-type clusters. It is expected that the molecular mass of a cluster increases during the surface migration. Because the island density increases gradually with increasing V/III ratio, the migration species would be a Ga_xP_y-type cluster [21]. When the PH_3 flow rate is increased, the number of decomposed P atoms is increased, and a high density of Ga_xP_y-type cluster with large mass is easily formed. Clusters with large molecular mass are expected to migrate more slowly than those with smaller masses. When the cluster size exceeds the critical size, the clusters are deposited on the Si surface and the islands are formed.

Moreover, when the concentration of P atoms on the Si surface is high, P atoms absorbed on the Si surface are increased. It results in the formation of the flat layer because the P atoms absorbed on Si capture the migrating species.

To explain the gas pressure dependence of the growth mode, other factors in the reactor must be considered. A possible factor affected by the gas pressure is the flow velocity in the reactor. TMG is almost completely decomposed at the growth temperature [25]. In contrast, the decomposition of PH_3 varies with the flow velocity because the decomposition rate for PH_3 is not as fast as TMG [26]. These results are qualitatively explained as follows. It is evident that the flow velocity is increased when the pressure is decreased. The pyrolysis of PH_3 takes place in the heated region in the reactor. Therefore, the decrease of pressure makes the PH_3 decomposition difficult because the resident time for the source gases in the heated region becomes short, resulting in the atomic arrangement as shown in Fig. 2(a). So the higher V/III ratio is

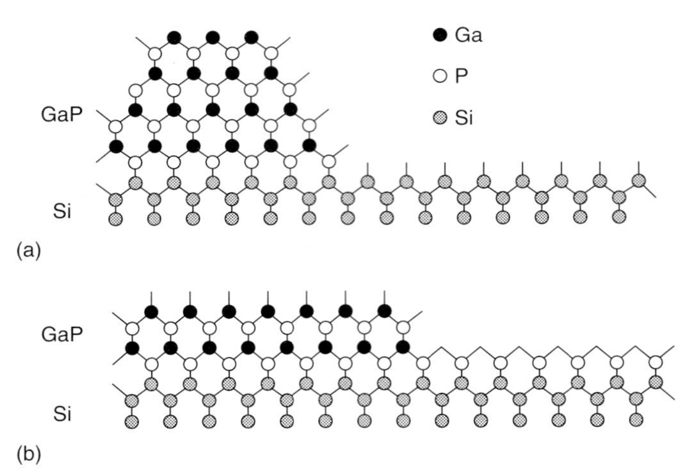

Fig. 2 Schematic atomic arrangement of 3D GaP on Si (a) and 2D GaP on Si (b).

necessary for lower growth pressure to produce a layer-type growth as shown in Fig. 2(b).

In the growth of GaP, a simple reaction involving PH_3 is considered to be the decomposition of PH_3 and the production of phosphorus molecules, that is,

$$PH_3 = P + 3H \tag{1}$$

Using simple reaction kinetics and solving the differential equations, that is,

$$\frac{d[PH_3]}{dt} = a[PH_3] \tag{2}$$

the partial pressures of PH_3 in these cases are given by

$$p_{PH_3} = p_{PH_{3_0}} \exp\left(-a\frac{x}{v}\right) \tag{3}$$

where p_{PH_3} is the partial pressure of PH_3 at the distance x from the entrance of the heated zone in the flow direction with the flow velocity v, $p_{PH_{3_0}}$ is the initial partial pressure of PH_3 or the supplied partial pressure into the reactor, and a is a constant including the reaction rate constant, the cross-sectional area of the reactor and other parameters. The flow velocity v is proportional to the total flow rate (F) and inverse of the growth pressure P_g, that is,

$$v = b\frac{F}{P_g} \tag{4}$$

and Eq. (3) is changed into

$$p_{PH_3} = p_{PH_{3_0}} \exp\left(-c\frac{P_g}{F}\right) \tag{5}$$

where c is a modified constant including a, b, and x. On the other hand, using these notations, V/III ratio is given by

$$[V/III] = \frac{p_{PH_{3_0}}}{p_{TMGa}} \tag{6}$$

The ratio of the phosphorus concentration([P]) to the gallium concentration ([Ga]), which is available to the growth on the substrate surface is expressed as,

$$[P/Ga] = [P]/[Ga] \tag{7}$$

where [P] and [Ga] are proportional to the concentration of the decomposed PH_3 and TMGa, respectively. [Ga] is supposed to be constant on the growing surface for various growth pressures because the

TMGa is almost decomposed at the growth temperature. Thus, substituting the partial pressures and the relation for the V/III ratio into Eq. (6), the above ratio is given by

$$[\mathrm{P/Ga}] = a\frac{(p_{\mathrm{PH}_{3_0}} - p_{\mathrm{PH}_3})}{p_{\mathrm{TMGa}}} = a[\mathrm{V/III}] \times \left\{1 - \exp\left(-c\frac{P_\mathrm{g}}{F}\right)\right\} \quad (8)$$

where a is the proportionality constant. Therefore, the initial growth mode of GaP, which is governed by [P]/[Ga] ratio is expressed using the V/III ratio, P_g and F. From these results, it can be said that at lower growth pressure, a higher V/III ratio is required for the layer growth.

2.2.3.2 Generation of Dislocation

In the heteroepitaxial growth, the generation of misfit dislocation and the stress relaxation are related to each other. The misfit dislocation generation and the stress relaxation of GaP layer on Si substrate grown under high V/III ratio (layer-type growth mode) are described. Because the defects associated with the coalescence of the islands are not generated, the observed dislocations are generated after the layer thickness exceeds the critical thickness.

From the cross-sectional TEM micrograph, it can be seen that the GaP surface is very flat, and defects such as dislocations or structural defects are not observed at all when the thickness is thinner than 90 nm [27]. This means that GaP grows on Si coherently with compressive stress at the initial stage. The TEM measurement for GaP on Si with various layer thickness shows that the dislocations at the interface are observed when the GaP layer thickness exceeds 90 nm.

In the TEM picture with the thickness of 3.7 μm, two kinds of dislocations with an extra-half plane in the Si substrate (A-type) and the GaP layer (B-type) are observed [28]. The schematic models of atomic arrangements for these structures are shown in Fig. 3. In general, the dislocation generation should take place at random in the isotropic crystal [29], that is, four kinds of A-type dislocations with Burgers' vector of 1/2[1 0 1], 1/2[0−11], 1/2[0 1 1] and 1/2[−101] can be generated. However, one salient feature came to be seen; instead of these four directional A-type dislocation, only two directional A-type dislocations with Burgers vector of 1/2[1 0 1] or 1/2[0−11] are observed. A possible interpretation for this generation of A-type dislocation with only two-kinds of Burgers' vectors instead of the four previously reported types of dislocations is the difference of the situation of the site, that is, dislocations are created at the step

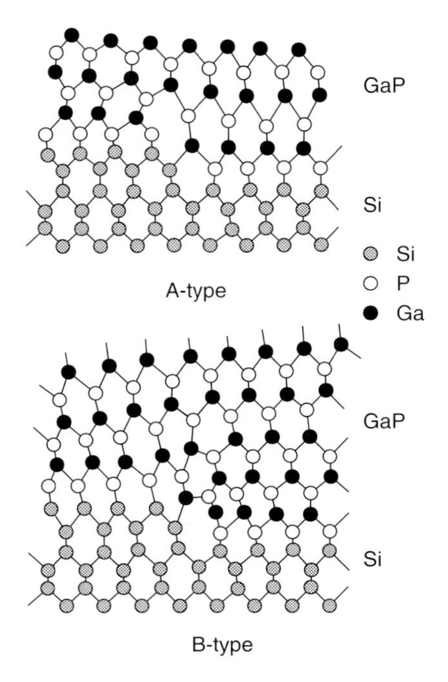

Fig. 3 Atomic arrangement for GaP on Si with A-type and B-type dislocation.

edge of the misoriented Si substrate rather than randomly isotropic generation. The direction of Burgers' vector for the B-type dislocations is the reverse of that of the A-type dislocation.

Assuming that the strain is completely accommodated at the growth temperature by 60° dislocations, the spacing between dislocations corresponds to 110 nm. However, the spacing between A-type dislocations is much smaller than the calculated value. This difference might be due to the uniformity of the Si substrate steps.

If the dislocations are generated by the lattice mismatch, an extra-half plane should occur in the Si substrate, which has a small lattice constant [29]. Because both the room temperature lattice constant and the thermal expansion coefficient of GaP are larger than those of Si, GaP should have the larger lattice constant than Si at the growth temperature. Hence, the lattice mismatch relaxation at the growth temperature is responsible for the generation of the A-type dislocation. The B-type dislocations cannot be explained merely on the grounds of lattice mismatch.

Because no dislocation is generated at the initial stage of the growth, the B-type dislocations should be generated during the growth or cooling down process. If the lattice strain of GaP is relaxed completely at the growth temperature by introducing A-type misfit dislocations, the tensile stress is produced in the GaP layer during the cooling process because of the difference of the thermal expansion coefficients of GaP and Si. The thermal expansion coefficient of GaP is about 2.5 times larger than that of Si. In order to relax the tensile stress in the GaP layer, the dislocations with the extra-half plane in the GaP layer should be introduced. Accordingly, it is proved that the dislocations with the extra-half plane in the Si substrate are formed by the lattice mismatch and that those in the GaP layer are formed during the cooling process to relax the thermal stress. This experiment supports the report that proved the generation of threading dislocations in GaP on Si during the cooling down process [30].

The stress applied to the GaP layer as a function of the thickness measured by X-ray diffraction is shown in Fig. 4 [20]. The dotted line shows the stress value for the thermal stress calculated by the bimetal model. A thin GaP layer has a compressive stress, and the stress changes to tensile with increasing thickness. Considering the thermal stress between the growth temperature and room temperature, the GaP layer

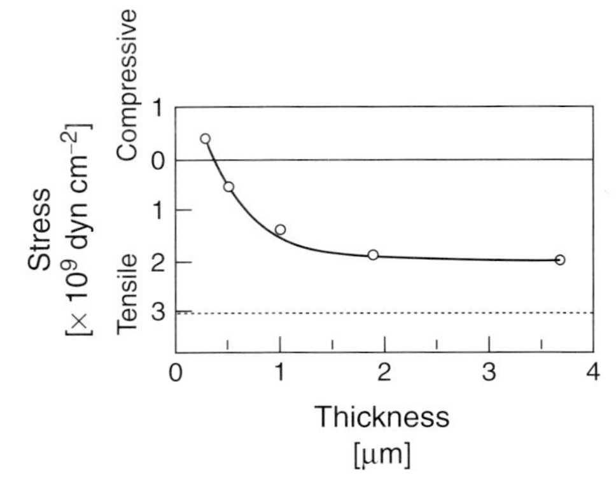

Fig. 4 Stress of GaP layer on Si substrate as a function of the thickness.

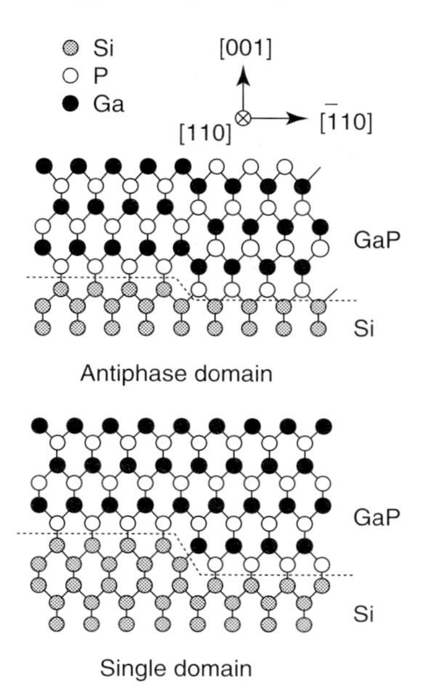

Fig. 5 APD structure and single domain structure at different step height.

must have compressive stress at growth temperature when the layer thickness is thinner than the critical thickness. Because the strain energy, which is caused by the misfit, increases with the thickness, the misfit dislocations are generated with increasing thickness. The change of stress in GaP on Si from compressive to tensile with the thickness is due to the increase of the misfit dislocation density at the growth temperature. The strain is smaller than the calculated thermal stress because the thermal stress is partly relaxed by the generation of misfit dislocations with the extra half plane in the GaP layer.

Although the dislocation is not generated at the initial stage of the growth, the dislocations with the extra-half plane in the Si substrate are generated at the growth temperature to relax the lattice mismatch. During the cooling down process, the thermal stress is relaxed by generating the misfit dislocation with the extra-half plane in the GaP layer. Below the dislocation frozen temperature, the thermal expansion mismatch produces the tensile stress in the epitaxial layer without generating new dislocations, resulting in the large tensile stress at room temperature.

2.2.3.3 Annihilation of APD Structure

The problem of the APD occurs at the Si surface step as a result of the polar or nonpolar structure. As shown in Fig. 5, the APDs are generated when the Si has a single (or odd) atomic step, whereas the single domains are formed when the Si surface is double (or even) atomic steps.

The typical APD is shown in Fig. 6. These were taken under dark field conditions using (002) (a) and (00$\bar{2}$) (b) reflections. In these figures, the contrasts of the domain region and matrix region are inverted by changing the reflection vector **g** from 002 to 00$\bar{2}$. The amplitudes of 002 and 00$\bar{2}$ reflections have been calculated to unequal for most thicknesses in the case of zincblende structure [31, 32]. Therefore, it is proved that the antiphase boundary is normal to the (001) plane near the Si substrate. In most cases, the APD is annihilated on changing the orientation of the boundary from the (001) normal to higher index planes so as to minimize the total energy.

The mechanisms for generation and annihilation of APDs are discussed. The Si surface steps are usually composed of single and double atomic steps. The initial growth mode of GaP on Si is 2D as it is grown under a high V/III ratio. In the initial stage of growth, the Si substrate is covered with P under a high PH_3 rate. Therefore, APD is introduced at the single atomic step position. In the case of

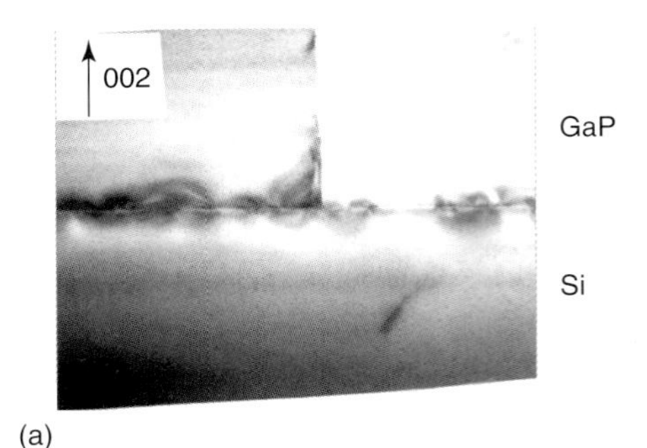

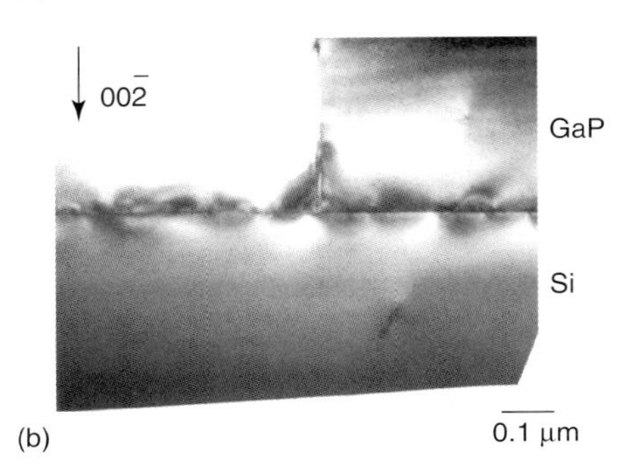

Fig. 6 Dark field TEM micrograph of APD annihilated during the growth. The reflection vector is (002) (a) and (00-2) (b).

If the size of APD introduced in the initial stage is small, the APD is annihilated in the early stage of growth. On the other hand, if the APD is large, a thick layer is necessary for all the APD to be annihilated. Therefore, APD remains at the surface. The observation that the APD is observed in GaP on 2° off (001) Si indicates the existence of single atomic steps. However, the density of single steps is lower than that of the exactly (001) Si substrate. On the other hand, all the steps change to double atomic steps in the cases of 4°-off and 6°-off substrates after the annealing process. This is inferred from the fact that the APD is not detected in these samples.

the growth on only (001) Si, the spacing between the steps is assumed to be large compared with the misoriented substrate. Therefore, the size of the APD is large. On the other hand, the size of the APD is small in the case of the misoriented substrate.

Calculation shows that the antiphase boundary for the (211) and (110) antiphase boundaries are energetically more likely to form than those for the (111) or (100) planes [33]. Therefore, the appearance of the (110) antiphase boundary is in good agreement with the calculations. The total energy is increased with increasing thickness. In order to reduce the total energy, the antiphase boundary changes its orientation to the low-energy index plane. The higher index plane of the APD is estimated to be the (211) plane from the angle [31]. This result also supports the calculation that the energy of the antiphase boundary for (211) is smaller than that of (110).

2.2.4 Deposition of Gallium Arsenide on Silicon Substrate

2.2.4.1 Nucleation

For lattice mismatched III–V semiconductors on Si, two kinds of misfit dislocations are observed; one is the pure-edge Lomer misfit dislocation, whose Burgers vector is parallel to the interface (type-I dislocation), and the other is the misfit dislocation, whose Burgers vector is 60° from the dislocation line (60° dislocation or type-II dislocation) [34]. Schematic illustrations of type-I and type-II dislocations are shown in Fig. 7(a) and 7(b), respectively.

A periodic array of misfit dislocations with an average spacing of about 8.1 nm is observed and the ratio of number of type-I to type-II dislocations is about 3 : 1 for GaAs on Si substrate. This means that the lattice mismatch is completely relaxed by the misfit dislocations. The majority of

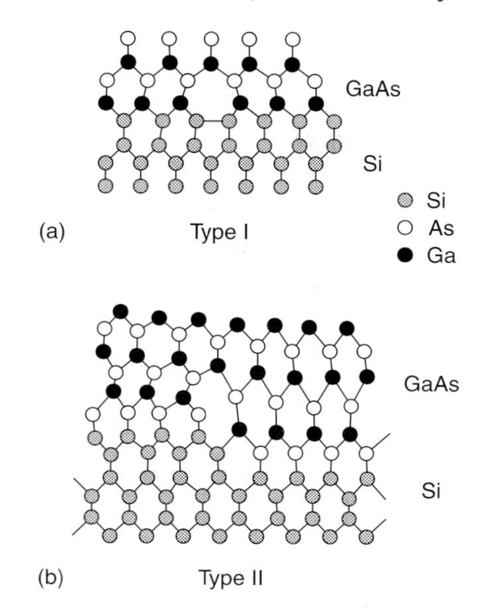

Fig. 7 Schematic illustration of type-I dislocation (a) and type-II dislocation (b).

the dislocations are type-I dislocation, and this is also the case for materials grown by molecular beam epitaxy [35].

On the other hand, in the case of GaP on Si, most dislocations are also type-II dislocations. The HRTEM micrographs show that most dislocations in GaAs on GaP substrates are type II.

This section discusses the generation mechanisms for type-I and type-II dislocations. Although the lattice mismatch for GaAs/Si and GaAs/GaP are almost the same, the types of dislocations are different. Therefore, the type of dislocation cannot be explained alone by lattice mismatch. Furthermore, the thermal stress in the epitaxial layer cannot explain the difference of the dislocation type because the thermal stress for GaAs/Si and GaP/Si is almost the same although the types of dislocations are different.

Some mechanisms have been proposed to explain the generation of type-I and type-II dislocations. The reported mechanisms for the generation of type-II dislocation are (1) the bending of the threading dislocation in the substrate parallel to the interface [36] and (2) the glide of dislocation from the surface, which forms a half-loop [29]. The reported mechanisms for the generation of type-I dislocation are (1) the reaction of two type-II dislocations [37] and (2) the dislocation climb of the pure edge dislocation from the surface [29].

Misfit dislocation generation can be explained by the earlier-mentioned mechanisms when the growth mode is 2D and the epitaxial layer is flat. On the other hand, no dislocation generation mechanism has been reported when 3D islands are formed at the initial stage of growth and the lattice mismatch is large. Therefore, another dislocation generation mechanism should be considered because the dislocation generation is modified by island formation. For example, misfit dislocations are generated in GaAs/Si heterostructures at the beginning of growth and before the coverage of Si with a GaAs layer has been completed [38, 39]. Therefore, it is deduced that the misfit dislocation generation is greatly affected by the initial growth mode.

The type of dominant dislocation, the lattice mismatch and the growth mode for GaAs on Si, GaP on Si, GaAs on GaP, and AlGaAs on Si are summarized in Table 2. The table shows that the type of dislocation is affected by the growth mode rather than the lattice mismatch; type-I dislocations are dominant for material systems with a 3D growth mode, and the type-II dislocations are dominant for the material systems with a 2D growth mode.

In the cases of GaP/Si and GaAs/GaP, the type-II dislocation generation probably

Tab. 2 Dominant type of dislocation, lattice mismatch, and growth modes

	Dislocation	*Lattice mismatch (%)*	*Growth mode*
GaAs on Si	Type-I	4.1	3D
AlGaAs on Si	Type-I	4.2	3D
GaP on Si	Type-II	0.37	2D
GaAs on GaP	Type-II	3.7	2D

is due to glide from the surface because 2D growth occurs at the beginning of the growth, and the dislocation density of the substrate is extremely low compared with that of the epitaxial layer. On the other hand, in the case of GaAs/Si and AlGaAs/Si, the dislocation generation cannot be explained solely by the mechanisms reported until now. If the type-I dislocations are generated after the reaction of two 60° dislocations, the Burgers vectors should satisfy the condition:

$$a/2[011] + a/2[10\bar{1}] \longrightarrow a/2[110]$$
$$(a : \text{lattice constant})$$
$$\text{(type II)} \qquad \text{(type II)} \qquad \text{(type I)}$$

However, the probability that all the type-II dislocations are changed to type-I dislocations is very low. Moreover, it is impossible to explain the generation of type-I dislocations by climb from the surface. The generation of type-I dislocations by the climb process is enhanced by increasing the point defect density. Therefore, the type-I dislocation should be observed in GaP on Si as the density of point defects is high in GaP grown on Si because of the high growth temperature. But, type-I dislocations are rarely observed in GaP on Si. Therefore, another mechanism should be considered to explain the generation of type-I dislocations.

If the 3D islands are formed at the initial stage of the growth, the size of the island increases with growth time, although the island density is constant. This means that the misfit dislocations are generated while the island size is increasing because the spacing between islands is much larger than that of the misfit dislocations. The type-I dislocation generation mechanism for 3D growth is shown in Fig. 8 in the case of GaAs on Si. When the island size is smaller than the critical size, the misfit strain is accommodated by elastic strain. The strain energy of the island is raised when the island size is increased. In order to relieve the lattice mismatch, misfit dislocations are generated at the edge of the island when the critical island size has been exceeded as shown in Fig. 8. Therefore, type-I dislocation generation is preferentially enhanced when the initial growth mode is

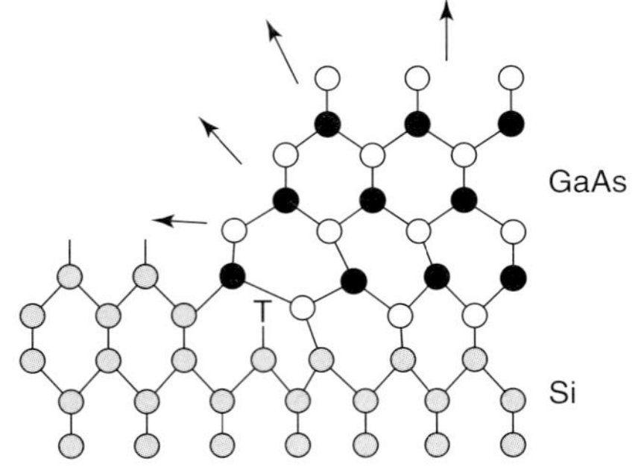

Fig. 8 Dislocation generation mechanism for GaAs on Si substrate.

3D, and type-II dislocations are dominant when the initial growth mode is 2D.

2.2.4.2 **Effect of TCA**

$In_{0.1}Ga_{0.9}As/GaAs$ SLS was inserted in the GaAs layer to reduce the dislocation density. The individual layer thickness and the total layer numbers of SLS are 20 nm and 10, respectively. The TCA was performed just before the SLS growth (TCA1) and after the SLS growth (TCA2). The upper and lower temperatures of TCA are 900 and 300 °C, respectively. Numbers of TCA were changed in this study. The intentional doping was not performed in GaAs on Si. The sample structure is shown in Fig. 9.

Figure 10 shows the dark spot defect (DSD) density as a function of the number of TCA1 for various number of TCA2. The DSD density of GaAs on Si without SLS is also shown. The DSD density decreases with increasing the number of TCA1 and TCA2. The lowest DSD density obtained in this study is 3.8×10^6 cm^{-2}. It is indicated that the TCA both before and after the SLS growth is more effective in reducing the DSD density. Even if the dislocation density is reduced by TCA1, dislocations are generated in the SLS by the lattice mismatch of SLS and GaAs. Although the individual layer thickness of SLS is thinner than the critical thickness, the total

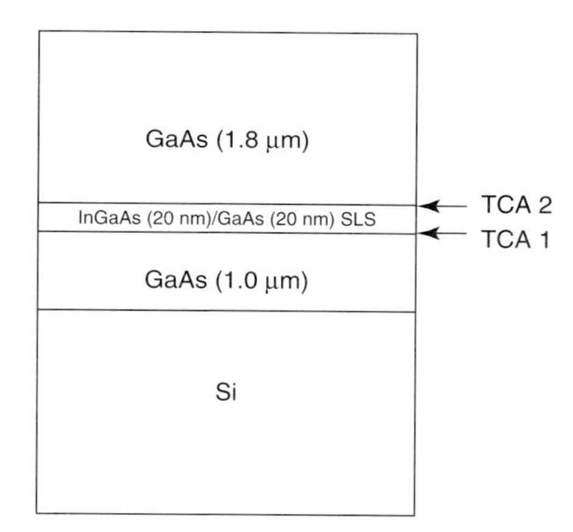

Fig. 9 GaAs on Si substrate using InGaAs/GaAs SLS buffer layer.

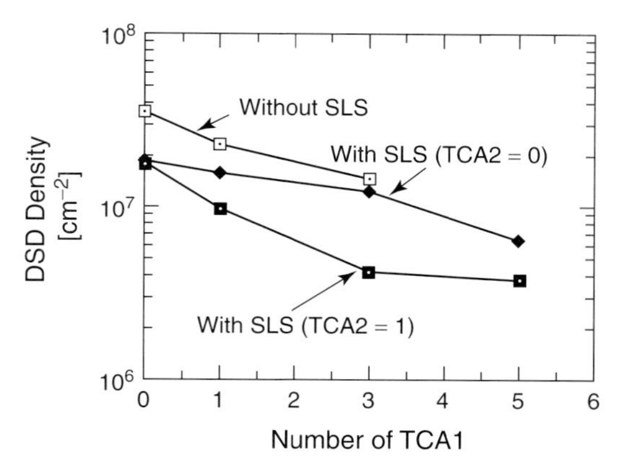

Fig. 10 DSD density of GaAs on Si as a function of TCA1 for various number of TCA2.

SLS thickness is thicker than the critical thickness [12]. It is suggested from the experimental results that the dislocations generated at SLS are bended by TCA2, resulting in the low dislocation density. Until now, the low etch pit density on the order of 10^6 cm^{-2} has been obtained using SLS and TCA for the total epitaxial layer thickness of more than 3.5 μm [40–43]. Few papers have been reported on the growth of GaAs on Si, with the dislocation density of 10^6 cm^{-2} at the epitaxial layer thickness of less than 3 μm.

The crystal quality of GaAs with 900 °C TCA and SLS is inferior to that on a GaAs substrate because a high density of dislocations is generated in the epitaxial layer, which degrades the minority carrier lifetime. The TCA temperature was optimized to improve the minority carrier lifetime. In this experiment, SLS buffer later was not used.

Figure 11 shows the DSD density of GaAs grown on Si substrate revealed by electron beam–induced current (EBIC) measurement for various TCA temperature. The DSD density decreases with increasing the TCA temperature gradually and is on the order of 10^6 cm^{-2} at 1000 °C. The crystal quality improvement by the relatively high TCA temperature is due to the enhancement of dislocation movement, large compressive stress, and the generation of point defects at higher temperature, which in turn reduce the dislocation density effectively.

Figure 12 shows the minority carrier lifetime of GaAs, $Al_{0.15}Ga_{0.85}As$ and $Al_{0.22}Ga_{0.78}As$ grown on Si for various TCA temperatures. The minority carrier lifetime is also improved with increasing the TCA temperature, which is supported by the decrease of DSD at high temperature. The minority carrier lifetime of GaAs grown on Si with 1000 °C TCA is 3.36 ns. Those for GaAs and AlGaAs for various Al compositions, are shown in Fig. 13. The lifetime of GaAs grown on GaAs substrate is also plotted for comparison [44–48]. Although it is impossible to compare the lifetime because the carrier concentration is not the same for all the samples, it is estimated that the lifetime of GaAs and AlGaAs on Si is approximately one order shorter than those grown on GaAs substrate.

2.2.4.3 Effect of Hydrogenation

Hydrogenation was carried out in a quartz tube, where a hydrogen plasma was excited by rf power via a copper coil encircling the quartz tube. The

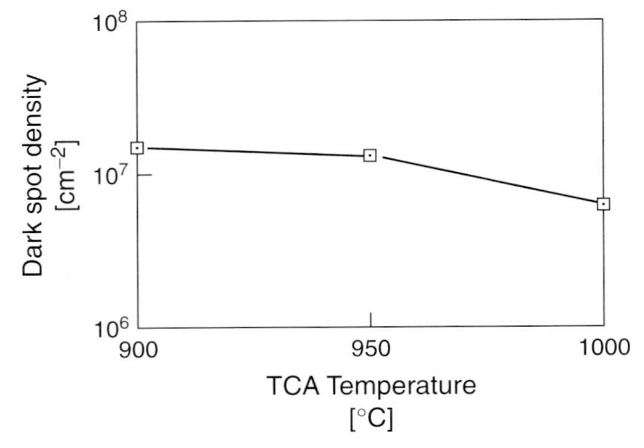

Fig. 11 DSD density of GaAs on Si as a function of TCA temperature.

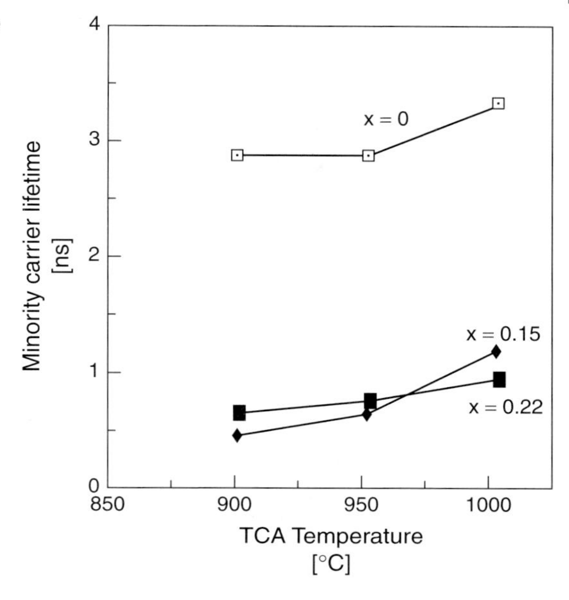

Fig. 12 Minority carrier lifetime of GaAs and AlGaAs for various TCA temperature.

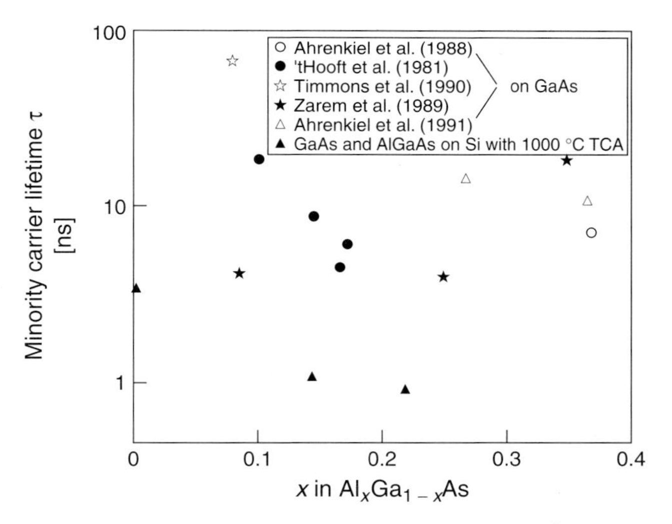

Fig. 13 Minority carrier lifetime of GaAs on Si with 1000 °C TCA.

plasma power, the treatment time, and the substrate temperature during the plasma treatment were 90 W, 2 hours, and 250 °C, respectively. In order to recover shallow level passivation and the damage induced by the plasma treatment, post annealing was performed in an $AsH_3 + H_2$ ambient at various temperatures ranging from 350 °C to 450 °C for 10 minutes. The TCA temperature was 900 °C and the SLS buffer layer was not used.

Carrier concentration profiles of unintentionally doped GaAs grown on Si substrates, before and after hydrogenation, are

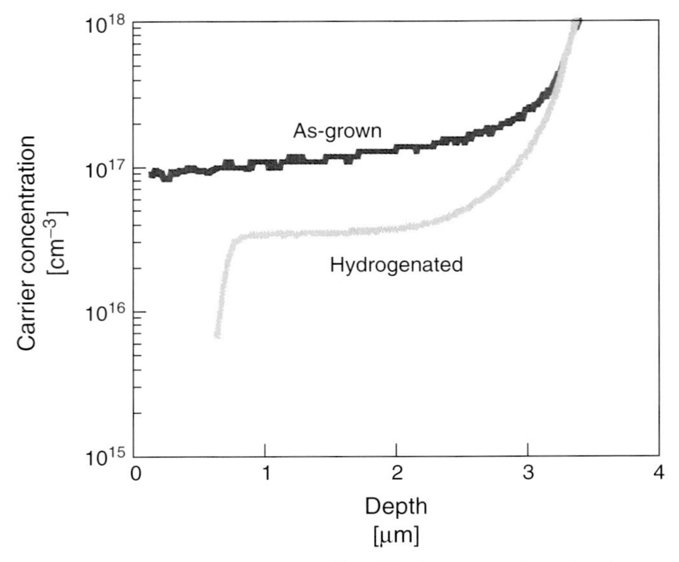

Fig. 14 Carrier concentration profile of GaAs on Si with and without hydrogenation.

shown in Fig. 14. Undoped GaAs-on-Si is n-type (1×10^{17} cm^{-3}) because of Si auto-doping during the growth [49]. For the hydrogenated sample, the carrier concentration is reduced to about 3×10^{16} cm^{-3} at depth exceeding 1 μm. This is due to the electrical passivation of the shallow levels. Because the major donor in GaAs grown on Si substrates is Si via auto-doping from the substrate, passivation will occur by the formation of SiH^0 complexes via the reaction

$$Si^+ + H^0 + e^- \longrightarrow SiH^0$$

The SiH^0 complex will then be dissociated by heat or applied electric fields. There is a kink in the concentration profile curve at a depth of nearly 0.8 μm, which corresponds to the plasma-induced damage [50], as the knee goes deeper with increased plasma treatment time. After a 10-minute annealing at 450 °C in $AsH_3 + H_2$ ambient, the donor electrical activities were completely restored to their initial levels.

Figure 15 shows the 4.2 K PL spectra of GaAs on Si for as-grown sample and hydrogenated sample, hydrogenated sample annealed at 450 °C. The major peaks are peak B corresponding to the heavy hole-associated free exciton and peak C corresponding to the carbon impurity-bound exciton. After the sample is treated by the hydrogen plasma, the full width at a half maximum (FWHM) of peak B narrows from 4.49 meV to 3.83 meV. This narrowing is due to the passivation of localized states. With 450 °C annealing where the shallow level is completely recovered, the FWHM is a little narrower than that of the as-grown sample.

Figure 16 shows the minority carrier lifetime derived from the time resolved photoluminescence decay curve. The minority carrier lifetime increases from 1.66 ns (as-grown) to 4.66 ns after the plasma treatment, and gradually decreases with increasing annealing temperature. It is difficult to judge the crystal quality only by the

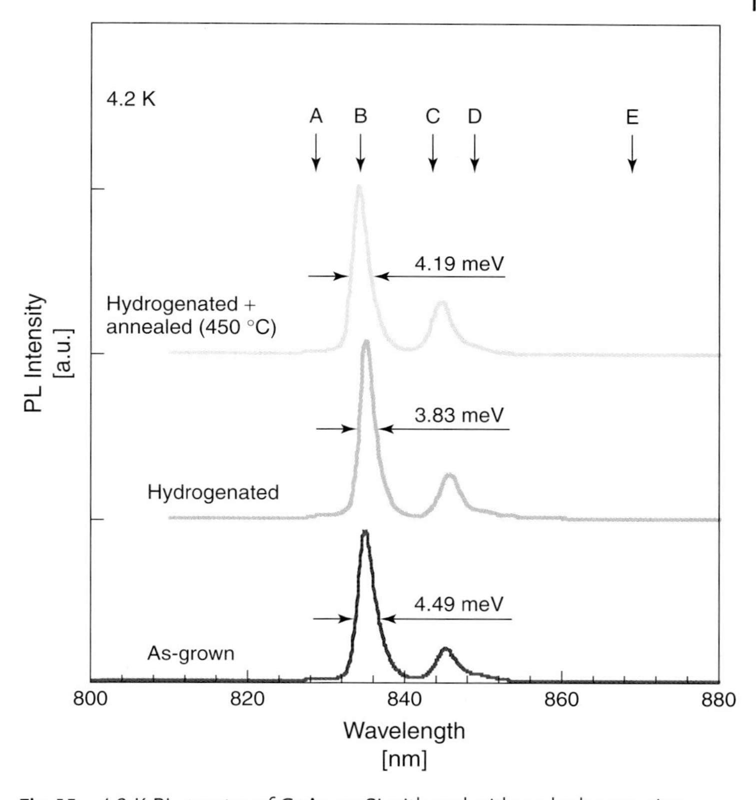

Fig. 15 4.2 K PL spectra of GaAs on Si with and without hydrogenation.

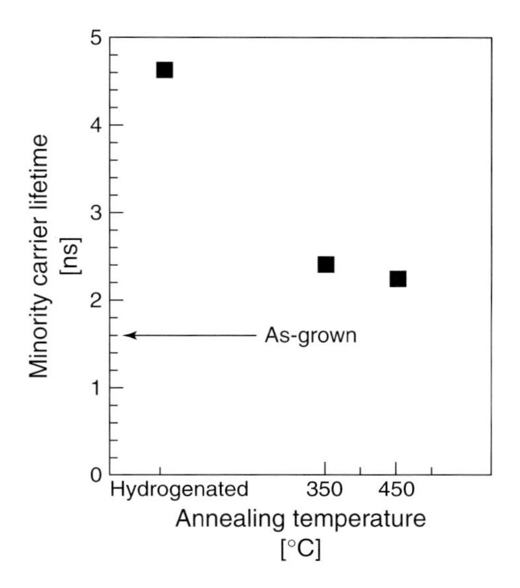

Fig. 16 Minority carrier lifetime of hydrogenated GaAs on Si with various postannealing temperature.

minority carrier lifetime because the lifetime is also affected by the shallow carrier concentration. In Fig. 14, the shallow carrier concentration of the 450 °C annealed sample is the same as that of the as-grown sample because with passivation the shallow level is completely restored. The minority carrier lifetime after hydrogen plasma treatment followed by annealing at 450 °C (2.27 ns) is longer than that of the as-grown sample (1.66 ns), suggesting that the defects generated by lattice mismatch and thermal expansion mismatch are electrically and optically passivated. The longer minority carrier lifetime of the hydrogenated sample (before annealing) is due to defect passivation and shallow level passivation. Species in the plasma include free radicals, ions, and electrons. Among these species, free radicals can effectively passivate the defects. Furthermore, it is well known that ions can damage the semiconductor surface. Therefore, during the hydrogenation process, defect passivation and damage formation take place at the same time. The minority carrier lifetime of the hydrogenated sample annealed at 450 °C is longer than that of the as-grown sample because the defects generated during the plasma treatment are passivated.

The DLTS spectra show that the peak of the Si-defect related level becomes smaller after hydrogen plasma treatment. It suggests that the Si-related defect level is passivated by hydrogenation. The passivation effect remains even after 450 °C annealing, where the shallow level is completely restored.

From these experiments it can be concluded that (1) the shallow level that has been passivated by hydrogenation is completely recovered by annealing at 450 °C in $AsH_3 + H_2$ ambient, (2) the deep levels are still passivated by hydrogen after annealing at 450 °C in $AsH_3 + H_2$ ambient, and (3) hydrogenation followed by 450 °C annealing produces a longer minority carrier lifetime at the same shallow carrier concentration.

2.2.4.4 Application to Photovoltaic Device

Figure 17 shows the schematic cross-sectional view of a GaAs/Si tandem solar cell. It consists of n^+-GaAs buffer layer,

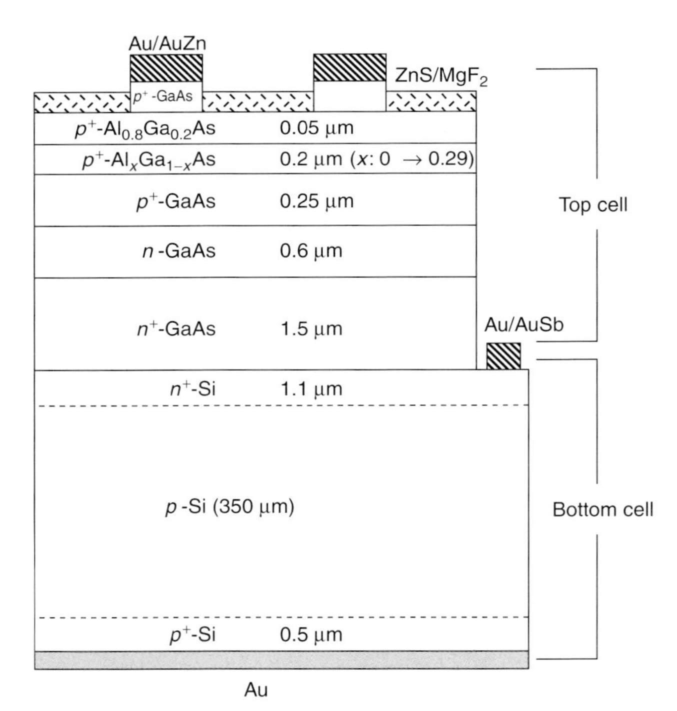

Fig. 17 Schematic cross-sectional view of three-terminal GaAs/Si tandem solar cell.

n-GaAs layer, p^+-GaAs, p^+-Al_xGa_{1-x}. As graded band emitter layer (x: 0 → 0.29), p^+-$Al_{0.8}Ga_{0.2}$. As window layer and p-GaAs cap layer. After the epitaxial growth, electrodes of Au-Zn/Au, Au-Sb/Au, and Au were formed on the p^+-GaAs layer, n^+-Si and p^+-Si, respectively. The surface of the cell was coated with a double-layer antireflection coating (ARC) using ZnS (49 nm) and MgF_2 (71 nm). The total area and active area of the solar cell are 25 mm^2 and 22.05 mm^2, respectively. The photovoltaic measurements were performed under AM0 and 1sun conditions at 27 °C.

The open-circuit voltage (V_{OC}) of GaAs solar cell fabricated on Si substrate as a function of TCA temperature is shown in Fig. 18 [51]. V_{OC} of GaAs solar cell fabricated on GaAs substrate is also indicated. It is known that V_{OC} is very sensitive to the minority carrier lifetime. V_{OC} of solar cell grown on Si is improved with increasing the TCA temperature, but 0.1–0.12 V smaller than that fabricated on GaAs substrate.

Table 3 shows short-circuit current (J_{SC}), V_{OC}, fill factor (FF) and conversion efficiency (η) of the solar cell with 1000 °C TCA and 900 °C TCA (shown in the parentheses). A total conversion efficiency of 22.1% has been achieved by combining the GaAs top cell (η = 17.7%) and Si bottom cell (η = 4.4%) in a three-terminal configuration. This is the highest efficiency for the GaAs/Si monolithic tandem solar cell ever reported.

Tab. 3 Photovoltaic properties of GaAs/Si tandem solar cell with 1000 °C TCA

	(mA/cm²) J_{sc}	*(V)* V_{oc}	*(%)* *FF*	*(%)* η
Top cell	34.8 (32.6)	0.90 (0.88)	76.8 (75.7)	17.7 (16.1)
Bottom cell	15.0 (14.8)	0.52 (0.52)	77.2 (78.3)	4.4 (4.5)
Total				22.1 (20.6)

Note: Values in ( ) show those with 900 °C TCA.

Although the conversion efficiency of the top cell is improved with increasing the TCA temperature from 900 °C to 1000 °C, the conversion efficiency of the Si bottom cell is reduced. This would be due to the degradation of minority carrier lifetime and the formation of the deep junction by the high-temperature heat treatment.

A two-terminal $Al_{0.15}Ga_{0.85}As$/Si tandem solar cell was fabricated in order to attain the photocurrent matching between the top cell and the bottom cell. The structure and the current-voltage

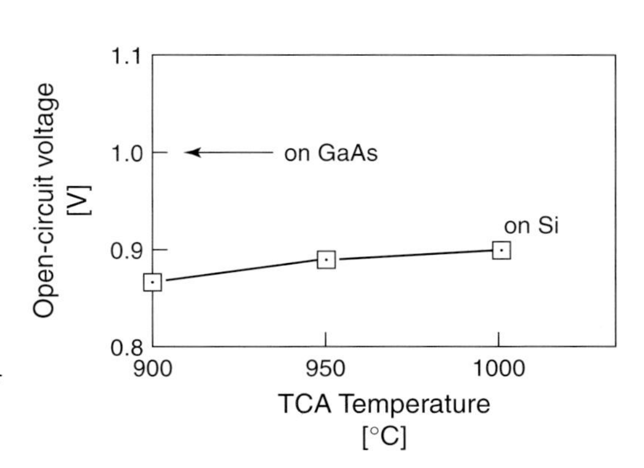

Fig. 18 Open-circuit voltage of GaAs solar cell on Si as a function of TCA temperature.

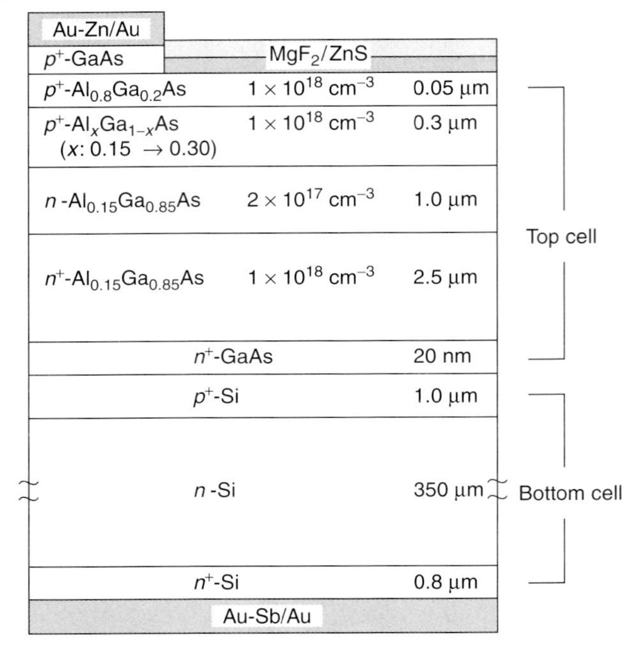

Fig. 19 Schematic cross-sectional view of two-terminal $Al_{0.15}Ga_{0.85}As/Si$ tandem solar cell.

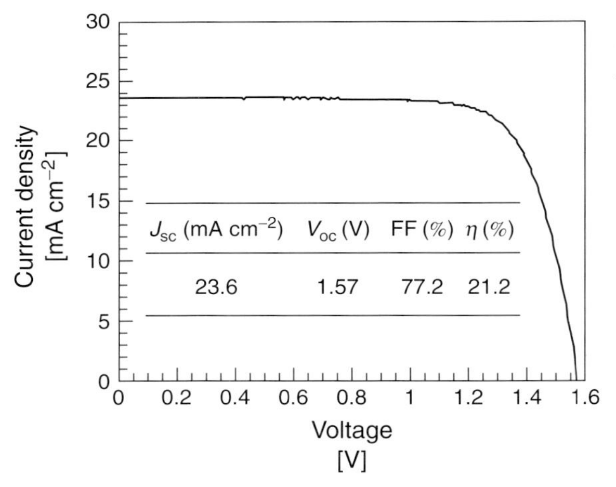

Fig. 20 Current-voltage characteristics of two-terminal $Al_{0.15}Ga_{0.85}As/Si$ tandem solar cell.

characteristics are shown in Figs. 19 and 20, respectively. It is observed that the short-circuit current of the top cell is perfectly matched to that of the bottom cell. A conversion efficiency of 21.2% utilizing the two-terminal configuration has been obtained [52].

It has been pointed out that a 2-terminal tandem solar cell with an efficiency higher than 30% can be obtained by using a top cell material with a band gap energy of 1.7–1.8 eV over an Si bottom cell. This structure resulted in photocurrent matching between the top cell and the bottom cell. However, in our study, the current matching was obtained by using the $Al_{0.15}Ga_{0.85}As$ top cell, of which the band gap energy is 1.61 eV. This is because the short-circuit current of the top cell is inferior to the ideal one. The main reason for the degradation is the short minority carrier lifetime caused by a high density of dislocation in the AlGaAs layer on Si. If it becomes possible to grow an AlGaAs layer on an Si substrate with a long minority carrier lifetime, comparable to that grown on GaAs substrate, a higher efficiency tandem solar cells can be obtained by increasing the Al composition so that the photocurrent matching between the top cell and the bottom cell is retained.

The improvement of the Si bottom cell is also important for the increase of the total conversion efficiency. The main problem is that the conversion efficiency is degraded after the crystal growth process [53]. The junction depth becomes deeper and the As atoms diffuse into the Si substrate during the growth. A low temperature growth process is necessary for the improvement of the bottom cell. Therefore, in the future, the technology to grow an AlGaAs layer with long minority carrier lifetime at low temperature should be investigated to increase the efficiency of the tandem cell.

2.2.5 Summary

The nucleation, dislocation generation, stress relaxation, and the annihilation of APD in the GaP/Si heteroepitaxial growth have been reviewed. The growth mode appears to be island-type for the low values of V/III ratio and the low gas pressure. When the V/III ratio or the gas pressure is increased, the growth mode changes from island-type to layer-type for thin GaP layer thickness. The two-type of misfit dislocations, which are generated by the lattice mismatch and the thermal expansion mismatch, are observed. A high density of APDs that propagate to the surface and are annihilated during growth has been observed. The nucleation of GaAs on Si and the effects of SLS buffer layer, TCA, and hydrogenation are described. The crystal quality has been improved by using SLS buffer layer, increasing the TCA temperature and using hydrogen plasma treatment. The GaAs layer grown on Si substrate has been applied to photovoltaic devices.

References

1. J. P. Benner, J. M. Olson, T. J. Coutts, in *Advances in Solar Energy*, (Ed.: K. W. Boer), American Solar Energy Society, Inc., 1992, p. 125, Vol. 7, Chapter 4.
2. M. A. Green, K. Emery, K. Bucher et al., *Progr. Photovolt.* **1999**, *11*, 31.
3. T. Soga, S. Hattori, S. Sakai et al., *Electron. Lett.* **1984**, *20*, 916–918.
4. M. Akiyama, Y. Kawarada, K. Kaminishi, *Jpn. J. Appl. Phys.* **1984**, *23*, L843–L845.
5. W. I. Wang, *Appl. Phys. Lett.* **1984**, *44*, 1149–1150.
6. B.-Y. Tsaur, G. M. Metze, *Appl. Phys. Lett.* **1984**, *45*, 535–537.
7. W. T. Masselink, T. Henderson, J. Klem et al., *Appl. Phys. Lett.* **1984**, *45*, 1309–1311.
8. T. Egawa, H. Tada, Y. Kobayashi et al., *Appl. Phys. Lett.* **1990**, *57*, 1179–1181.
9. T. Soga, T. Kato, M. Yang et al., *J. Appl. Phys.* **1995**, *78*, 4196–4199.
10. T. Soga, T. George, T. Suzuki et al., *Appl. Phys. Lett.* **1991**, *58*, 2108–2110.
11. R. Fischer, D. Neuman, H. Zabel et al., *Appl. Phys. Lett.* **1986**, *48*, 1223–1225.
12. Y. Watanabe, Y. Kadota, H. Okamoto et al., *J. Cryst. Growth* **1998**, *93*, 459–463.
13. N. Chand, R. People, F. A. Baicocchi et al., *Appl. Phys. Lett.* **1986**, *49*, 815–817.

14. M. Yamaguchi, A. Yamamoto, M. Tachikawa et al., *Appl. Phys. Lett.* **1988**, *53*, 2293–2295.
15. M. Yamaguchi, *J. Mater. Res.* **1991**, *6*, 376–384.
16. K. Nozawa, Y. Horikoshi, *Jpn. J. Appl. Phys.* **1991**, *30*, L668–L670.
17. H. Shimomura, Y. Okada, M. Kawabe, *Jpn. J. Appl. Phys.* **1992**, *31*, L628–L630.
18. F. R. N. Nabarro (Ed.), in *Dislocations in Solids*, North-Holland, Amsterdam, 1979, pp. 461–545.
19. K. Tamamura, K. Akimoto, Y. Mori, *J. Cryst. Growth* **1998**, *94*, 821–825.
20. T. Soga, T. Suzuki, M. Mori et al., *J. Cryst. Growth* **1993**, *132*, 134–140.
21. T. Soga, T. George, T. Jimbo et al., *Jpn. J. Appl. Phys.* **1991**, *30*, 3471–3474.
22. N. Noto, S. Nozaki, T. Egawa et al., *Mater. Res. Soc. Symp. Proc.* **1989**, *48*, 247–252.
23. T. Soga, T. George, T. Jimbo et al., *Appl. Phys. Lett.* **1991**, *58*, 1170–1172.
24. O. Ueda, K. Kitahara, N. Ohtsuka et al., *Mater. Res. Soc. Symp. Proc.* **1991**, *221*, 393–398.
25. J. Nishizawa, T. Kurabayashi, *J. Cryst. Growth* **1988**, *93*, 98–102.
26. G. B. Stringfellow, *J. Cryst. Growth* **1991**, *115*, 418–420.
27. T. Soga, T. Jimbo, M. Umeno, *Jpn. J. Appl. Phys.* **1993**, *32*, L767–L769.
28. T. Soga, T. Jimbo, M. Umeno, *Appl. Phys. Lett.* **1993**, *63*, 2543–2545.
29. J. W. Matthews, in *Epitaxial Growth Part B*, Academic, New York, 1975, Chapter 8.
30. M. Tachikawa, H. Mori, *Appl. Phys. Lett.* **1990**, *56*, 2225–2227.
31. T. Soga, H. Nishikawa, T. Jimbo et al., *Jpn. J. Appl. Phys.* **1993**, *32*, 4912–4915.
32. O. Ueda, T. Soga, T. Jimbo et al., *Defect Control in Semiconductor*, (Ed.: K. Sumino), Elsevier Science, Amsterdam, 1990, pp. 1141–1146.
33. P. M. Petroff, *J. Vac. Sci. Technol.* **1986**, *B4*, 874–876.
34. T. Soga, T. Jimbo, M. Umeno, *J. Cryst. Growth* **1994**, *145*, 358–362.
35. N. Otsuka, C. Choi, Y. Nakamura et al., *Mater. Res. Soc. Symp. Proc.* **1985**, *67*, 85–90.
36. J. W. Matthews, A. E. Blakeslee, *J. Cryst. Growth* **1974**, *27*, 118–123.
37. K. Rajan, M. Denhoff, *J. Appl. Phys.* **1987**, *62*, 1710–1715.
38. D. Gerthsen, D. K. Biegelsen, F. A. Ponce et al., *J. Cryst. Growth* **1990**, *106*, 157–163.
39. T. Yao, H. Kakao, H. Kawanami et al., *J. Cryst. Growth* **1989**, *95*, 107–111.
40. T. Soga, T. Jimbo, M. Umeno, *Appl. Phys. Lett.* **1990**, *56*, 1443–1445.
41. T. Soga, S. Nozaki, N. Noto et al., *Jpn. J. Appl. Phys.* **1989**, *28*, 2441–2445.
42. N. Hayafuji, S. Ochi, M. Miyashita et al., *J. Cryst. Growth* **1989**, *28*, 2441–2445.
43. T. Soga, H. Nishikawa, T. Jimbo et al., *J. Cryst. Growth* **1991**, *107*, 479–483.
44. R. K. Ahrenkiel, D. J. Dunlavy, T. Hanak, *Sol. Cells* **1998**, *24*, 339–347.
45. R. K. Ahrenkiel, B. M. Keyes, T. C. Shen et al., *J. Appl. Phys.* **1991**, *69*, 3094–3099.
46. H. A. Zarem, J. A. Lebens, K. B. Nordstrom et al., *Appl. Phys. Lett.* **1989**, *55*, 2622–2624.
47. M. L. Timmons, T. S. Colpitts, R. Venkaasubramanian et al., *Appl. Phys. Lett.* **1990**, *56*, 1850–1852.
48. G. W. 'tHooft, C. Van Opdorp, H. Veenvliet et al., *J. Cryst. Growth* **1981**, *55*, 173–179.
49. S. Nozaki, J. J. Murray, A. T. Wu et al., *Appl. Phys. Lett.* **1989**, *55*, 1674–1676.
50. N. Das Gupta, R. Riemenschneider, H. L. Hartnagel, *J. Electrochem. Soc.* **1993**, *140*, 2038–2043.
51. T. Soga, M. Kawai, K. Otsuka et al., *2nd World Conference and Exhibition on Photovoltaic Solar Energy Conversion*, European Commission, Joint Research Centre, Ispra, Italy, **1998**, pp. 3737–3740.
52. T. Soga, K. Baskar, T. Kato et al., *J. Cryst. Growth* **1997**, *174*, 579–584.
53. M. Yang, T. Soga, T. Egawa et al., *Sol. Energy Mater. Sol. Cells* **1994**, *35*, 45–51.

2.3 Grafting Molecular Properties onto Semiconductor Surfaces

Rami Cohen, Gonen Ashkenasy, Abraham Shanzer, David Cahen
Weizmann Institute of Science, Rehovot, Israel

2.3.1 Surface Electronic Properties

The chemistry of the crystal surface is very different from that of the bulk. Although the atoms are arranged in a well-ordered structure in the bulk, on the surface they may adopt a different position from that corresponding to the perfect crystal lattice because of the absence of part of the neighboring atoms. In that case, the surface is said to be relaxed or reconstructed [1]. A second difference between nearly every surface and the bulk is associated with the chemical environment of the bonding atoms. Under ambient conditions, the surface is covered by one or more atomic layers of adsorbates such as oxygen (mostly as oxide or hydroxide), carbon or hydrocarbons and is not atomically clean (i.e. the atomic composition is not the same as that in the bulk). Thus, it is not surprising that nearly always the "real" surfaces are disordered and lack the periodicity of the bulk. Correspondingly, the positions and distribution of energy levels and their occupation by electrons at the surface generally differ from those associated with the bulk.

The importance of this situation becomes obvious when we try to use these crystals, especially semiconductor crystals, in a device. In this case, surface defects and imperfections play a dominant role in the electron transport in and out of the device and in this way influence its performance. It is to be remembered that all contacts of semiconductors to the outside world and the junction between two different semiconductors involve their surfaces. Even one of the simplest of electronic devices, the homojunction diode, requires outside metal–semiconductor contacts. Therefore, understanding and control over the electronic properties of semiconductor surfaces are essential for constructing devices and for fine-tuning their performance.

Figure 1 shows the energy-band diagram of an n-type semiconductor and the electrical properties associated with the surface. This chapter mostly deals with n-type semiconductors and definitions are given only for this type. The difference between the energetics of the bulk and that of the surface is demonstrated by the relatively high density of energy levels inside the semiconductor band gap. These levels are referred to as surface states and their importance stems from the fact that they are involved in most electronic loss mechanisms such as charge-trapping and recombination.

It was shown [4], using quantum mechanical calculations, that surface states can localize electrical charges in contrast to bulk states where electrons are delocalized. Because of this localization the surface becomes charged with respect to the bulk and an electric potential difference is created between the surface and the bulk, the so-called built-in potential (V_S). This is shown in the band diagrams as bending of the valence and conduction bands, a feature called "surface band-bending". In relation to the bulk, the surface states of an n-type semiconductor localize negative charges and the opposite is true for p-type material. Because it is easier to measure the band-bending than the surface states, in many cases the surface band-bending is used as an indicator for changes in the density of surface states. Because the capture

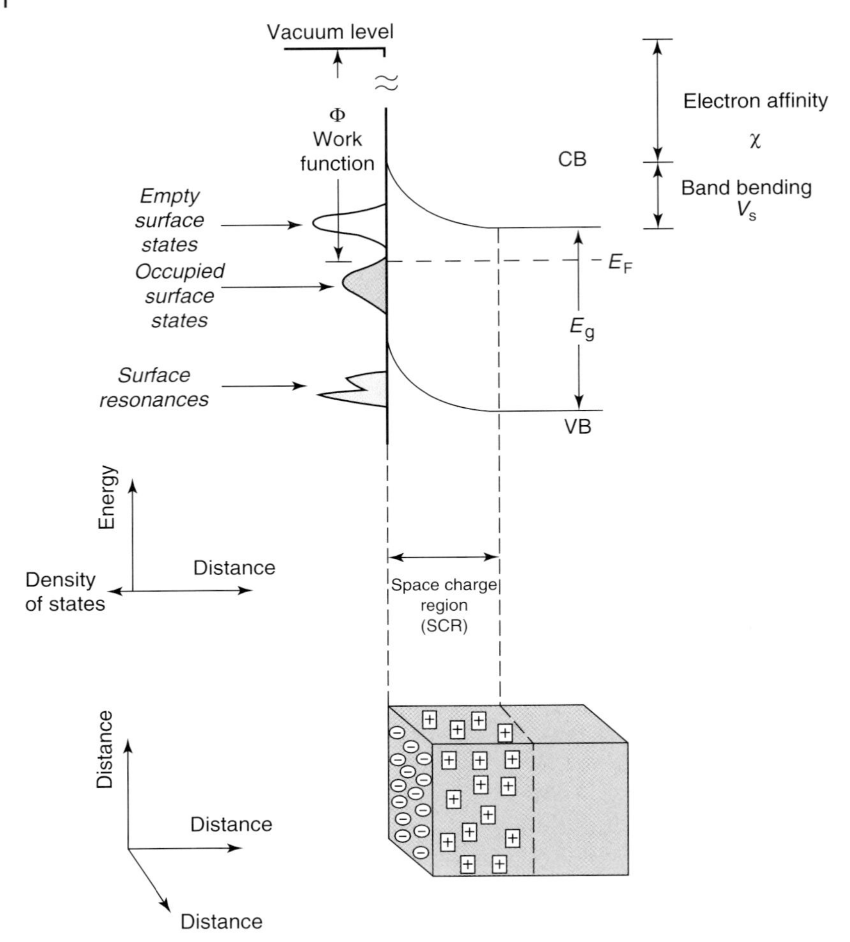

Fig. 1 (a) One-electron energy level diagram of an *n*-type doped semiconductor. VB represents the valence band, comprising largely filled, closely spaced energy levels. CB is the conduction band, which consists of largely empty, closely spaced energy levels. E_F is the semiconductor Fermi level and E_g is the band gap energy. The surface states are occupied until sufficient space charge is created corresponding to an electrostatic field, which prevents further electrons from going to the surface. It is this field that causes the shift in the band energies, i.e. V_S. (b) Side view of the crystal. The surface localizes electrons at the surface states, and thus is negatively charged. Figure 1 also shows the presence of surface energy levels that lie outside the band gap. These states are referred to as surface resonances and unlike surface states they are degenerate with the bulk states and can mix with them [2, 3]. A surface resonance has a varying degree of localization in the surface region, although it should be noted that there is no absolute definition for how strong a surface localization should be in order for it to be defined as a surface resonance.

cross-sections of electrons and holes are higher on the surface than in the bulk, surface states also serve as fast recombination centers of electrons and holes. This phenomenon can critically affect the electron-transfer efficiency in electronic devices and thus their performance. The recombination rate of electrons and holes at the surface is expressed in units of velocity (cm s^{-1}) and is called surface recombination velocity (SRV or s). It reflects the rate at which minority carriers (holes in n-type materials) are consumed at the surface. In many cases, SRV is determined by the surface states close to the middle of the band gap where the chances for electron-hole recombination are the highest [5].

Other surface electrical properties indicated in Fig. 1 are the work function, Φ, and the electron affinity, χ. In semiconductors, electron affinity is defined as the difference in energy between the local vacuum level [6] and the bottom of the conduction band (CB) at the surface. The (local) work function is the minimum energy required for an electron to escape into vacuum (just outside the range of the crystal potential) from the Fermi level. Although this definition is straightforward for metals, for semiconductors it should be borne in mind that there are mostly no real energy levels at the Fermi level. The work function is determined by the electron affinity, the band-bending, and the energy difference between the Fermi level and the conduction/valence band in the bulk.

2.3.2
Requirements of Molecular Surface Treatments

Because the surface charge and dipole are determined by the surface chemistry, it is the surface chemistry that to a large extent determines the surface electronic properties. Therefore, to control the surface electrical properties and to make the semiconductor useful for electronic devices, we look for treatments that will interact with the surface and modify it chemically. The requirements of molecular surface treatments, which are also named passivation treatments, are the following:

1. Remove surface and interface states or at least eliminate them in the energy interval of the band gap. Because the presence of surface states is conducive for charge-trapping and recombination effects, moving them out of the band gap is necessary for the construction of high-speed semiconductor devices like the ones that are based on GaAs and InP.
2. Tuning the surface electron affinity. Control over the electron affinity is essential for band edge engineering of interfaces such as those in photovoltaic solar cells and for controlling the barrier height for electron loss to the surroundings. The second effect is demonstrated for the GaAs/(Al,Ga)As system where a band offset of 0.4 eV [7] did not block electron loss from the GaAs into the passivating (Al,Ga)As layer.
3. Strong bonding that can withstand device-processing. Ideally, the molecules will chemisorb on the surface. It should be noted that the binding groups themselves, while interacting with the surface, could modify the surface chemistry and thus the surface energetics on binding.
4. Chemical protection from adsorbates, especially atmospheric adsorbates. In those cases where surface oxidation is detrimental, there is the daunting task of protection from O_2.

2.3.3
Strategies to Control Surface Electronic Properties

Strategies for tuning surface electronic properties, work function, electron affinity, band-bending, and SRV have to take into account the origin of chemical and physical properties. Use of the versatility of molecular chemistry can help control these properties in a predetermined fashion.

2.3.3.1 Controlling the Band-bending (V_S) and Surface Recombination Velocity (SRV)

As noted earlier, band-bending near the surface results from charge localization on the surface states. Thus, control over the band-bending requires a mechanism that will control the density and occupation of the surface states inside the band gap. Figure 2 indicates a method to modify the band-bending by shifting surface-state energies with respect to the band edges. If the surface states are shifted out of the band gap, above the CB minimum or below the valence band (VB) maximum (i.e. converted into surface resonances), they will couple to the band continuum and their charge will be delocalized in the band. As a result, V_S will be modified. Another way to change the V_S is by changing the surface-state energy inside the band gap. The states are occupied at energies below the Fermi level and are empty at those above it (this holds, strictly speaking, only at 0 °K). Therefore by modifying the surface states so that their energies will change from above (below) to below (above) the Fermi level, the surface states can be occupied (emptied) and the

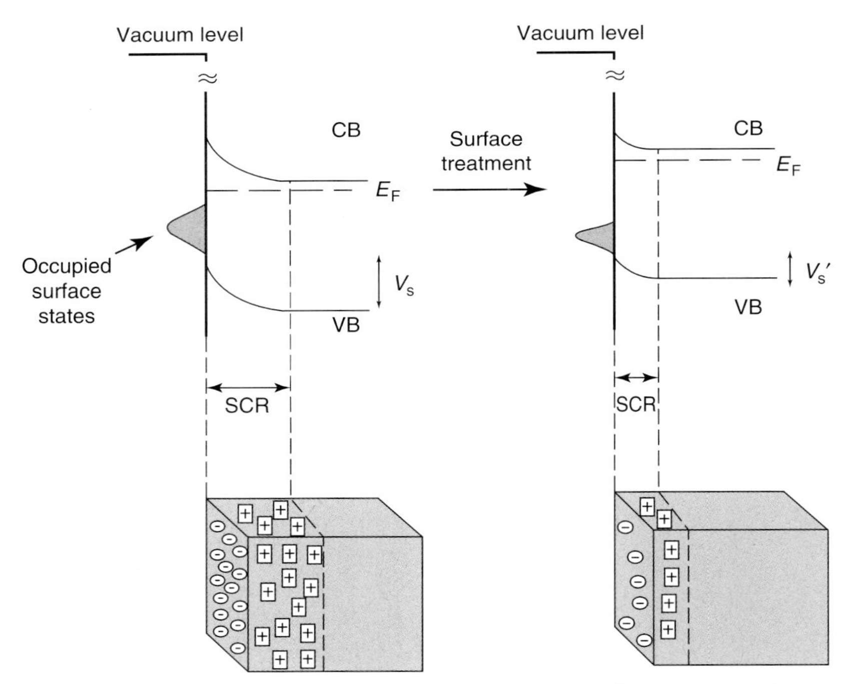

Fig. 2 Left to right: Reduction of the density of surface states of *n*-type semiconductor, as a result of surface treatment, which leads to decrease in width of the space charge region (SCR) and a decrease in the surface band-bending from V_S to V_S'.

surface charge changed. Change in the distribution of energies of the surface states can also affect the SRV. Because the SRV depends on the density of surface states with energies around the midgap, shifting of the surface-state energies away from the midgap can lead to reduction in the probability of charge recombination.

The strategy to modify surface states, and thus the V_S and SRV, is based on interaction of chemically grafted molecules with these states. The key is to find molecules that will modify the semiconductor surface chemistry in a way that involves the surface states. In this respect, the origin of the surface states should be considered. Intrinsic surface states originate from the termination of the crystal bulk and the breaking of chemical bonds at the surface, whereas extrinsic surface states originate from crystal imperfections, such as missing surface atoms, line defects, or from a change in the chemical environment caused by adsorbed extrinsic atoms.

2.3.3.2 Controlling the Electron Affinity, χ

By definition, χ depends on the energy difference between the vacuum level and the bottom of the CB at the surface. Therefore, any treatment that influences the surface potential will modify χ.

Modification of the surface potential can be achieved by utilizing polar molecules that will bind to the surface and change its potential. Figure 3 shows schematically the manner in which a polar molecule can modify the surface χ. The surface potential will be reduced if the molecular dipole is pointing toward the surface and will increase if the dipole is directed in the opposite direction. This approach of using dipolar molecules was applied to tune the χ of metals [8–11] and semiconductors [8, 12–14]. It should be mentioned

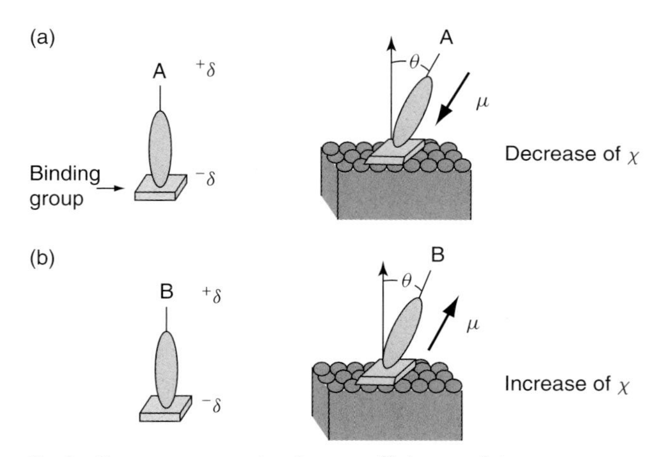

Fig. 3 Strategy to tune the electron affinity, χ, of the semiconductors by adsorption of polar molecules. All the molecules have a similar binding group that allows strong binding to the surface and a different end group (A, B) to tune the molecular polarity (i.e. dipole moment). A molecular dipole moment pointing toward the surface (a) leads to decrease in χ, whereas a dipole pointing away from the surface (b) increases the electron affinity. μ and θ are the molecular dipole moment and tilt angle, respectively.

that because short-range atomic forces determine the energy positions of the surface states inside the band gap, they need not be affected by the presence of the macropotential. Therefore, looking at Fig. 1, a polar molecule modifies the energy of the bands with respect to the vacuum level, but not necessarily those of the surface states with respect to the band edges or the surface-state (and band) energy level densities.

The change in the electron affinity caused by the molecule's dipole moment can be described in terms of a parallel plate capacitor, using the well-known Helmholtz equation

$$\Delta V = N \times \mu \times \frac{\cos\theta}{\varepsilon\varepsilon_o} \quad (1)$$

where ΔV is the potential drop caused by the dipole layer, μ is the dipole moment, N the dipole density per unit area, θ the angle between the dipole and the surface normal, ε is the relative dielectric permittivity of the film, and ε_o the permittivity of free space [15]. If values for N, $\cos\theta$, and ε are known, ΔV (and thus $\Delta\chi$) can be calculated. The equation also demonstrates that controlling χ can be achieved by tuning the dipole direction (toward/from the surface), its magnitude and its orientation with respect to the surface normal. It has to be noted that the concept of a dielectric constant for a monolayer is problematic. It can be related to the molecular polarizability using the Clausius-Mosotti relation (cf. discussion in [14]).

2.3.3.3 Controlling the Work Function, Φ

As shown in Fig. 1, the work function is the energy difference between the Fermi and vacuum levels and depends on the electron affinity and band-bending. Thus, by controlling either the electron affinity or band-bending, we can tune the work function

$$\Delta\Phi = \Delta\chi \pm \Delta V_s \quad (2)$$

with $+$ for n-type and $-$ for p-type.

2.3.4 Strategy for Molecule Selection

The chemical strategy to control the surface energetics using organic molecules is illustrated schematically in Fig. 4. The idea is to incorporate several molecular properties simultaneously in one molecule and to allow systematic modification of one specific property, independent of others. This approach provides

1. a simple tool to investigate the relation of the macroscopic properties of the semiconductors and the molecular properties of the adsorbed molecules;
2. a simple tool to enable the development of models for surface engineering.

A different approach, which is more suitable for nonmolecular, extended bonded, electrically conducting inorganic materials, is to use different molecular layers, each of which has a different function. For example, the first layer can electrically passivate the surface and a second layer can give long-term chemical protection [7].

2.3.5 Organic and Inorganic Molecular Surface Treatments

In general, we can classify the chemical surface treatments into two classes, organic and inorganic. Table 1 summarizes the advantages and disadvantages of each group. Although, as noted in the

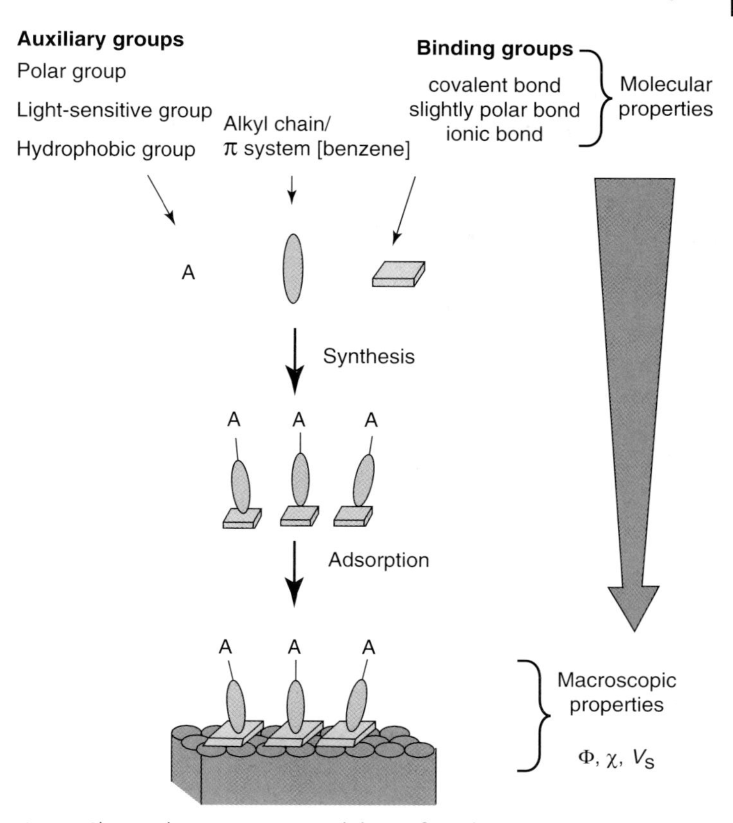

Fig. 4 Chemical strategy to control the surface electrical properties using multifunctional molecules.

table, the chemical stability of organic treatments is generally lower than that of inorganic ones, Dorsten and coworkers [16] demonstrated that sulfidization of GaAs using octadecylthiol gave rise to a sulfide layer that was as stable as that which could be obtained with corresponding inorganic treatments. Using the flexibility of organic synthesis, one can incorporate several functional groups in a single molecule, each performing a different role and ideally independent of each other. The versatility of the organic treatments can also be used to study the relationship between modification of the surface chemistry and surface electronic effects.

2.3.5.1 **Inorganic Surface Treatments**

Table 2 presents several examples of inorganic surface treatments used to modify the surface electronic properties. The first is oxidation of silicon (Si), a cornerstone of today's electronic device industry [17]. The treatment moves most of the surface states out of the Si band gap [2, 18] and enables the worldwide use of silicon for device manufacturing. On III–V and II–VI group semiconductors, the chemistry of the semiconductor's native oxides is more complicated than on silicon, and plain oxidation of the surface cannot lead to chemically stable, nearly defect-free surfaces [7]. For example, on

Tab. 1 Advantages and disadvantages of organic and inorganic treatments

	Organic treatments	*Inorganic treatments*
Advantages	* Structural versatility and flexibility	* Chemical stability
	* Can incorporate several properties simultaneously (polarity, hydrophobicity, light sensitivity)	* Thermal stability
	* Allow systematic modification of one specific property, often without effect on other properties	* Strong binding/interaction
Disadvantages	* Limited chemical and thermal stability	* Limited chemical flexibility

Tab. 2 Examples of substrates for which inorganic chemical surface modifications have been developed

Semiconductor	*Treatment*	*References*
Silicon	Oxygen	2, 18
	Hydrogen	18, 28
	Bromine	29
	As	30
GaAs	Sulfide and Selenide	20, 24, 27, 31, 32, 33–37
	Ruthenium	38
	Phosphine (PH_3)	39
	P_2O_5 /NH_4OH	40, 41
	Chlorine	42
	Cesium	43
	Sb	44, 24
	H_2S	26
	Iodine	45
InP	CdS	46
	Ruthenium	47, 48
CdTe (polycrystalline)	Hydrogen plasma	49
	Ruthenium	50, 51
	Reactive metal interlayer	52
	Oxygen anneal	53
(Hg, Cd)Te	Sulfide	54
	Hydrogen	55, 56
$CuInSe_2$ (polycrystalline)	Oxygen anneal	53, 57

GaAs(100) and (110) [2] surfaces in the ambient, high surface state densities pin the Fermi level and generate a high SRV. Covering GaAs(100) with (Al,Ga)As was found to be effective in terms of reducing surface electron-hole recombination, but could not block loss of electrons to the passivating (Al,Ga)As layer. Another approach to modify the surface chemistry of GaAs is the use of inorganic sulfides. These treatments were found to remove part of the surface oxide and form sulfides of Ga and As [19]. Studies of metal-insulator semiconductor (MIS) and Schottky diode structures of GaAs, treated with $(NH_4)_2S$ and Na_2S [20, 21], revealed that aqueous sulfide treatments induce only minor changes in the net surface state density. Their main effect is reduction of the density of the trap states, which are farther in energy from the band edges than the shallow, doping levels. This means that these treatments modify the positions of the surface-state energy levels [19, 22]. This observation agrees with results from other studies where sulfide treatment was claimed to repin the surface Fermi level at a different energy [22–25]. A different approach was introduced by Shen and coworkers [26] who used plasma H_2S treatment. In contrast to the aqueous sulfide treatments, this treatment leads to significant reduction of the surface-state density (up to three orders of magnitude) on sulfidization. The main drawback of the inorganic sulfide treatments is the instability of the sulfur-GaAs bond in the ambient (vs oxygen attack, in particular) [19, 27].

According to Lunt [58], the efficacy of a selenide treatment applied to GaAs is expected to be stronger than that of a sulfide one because of the electron-deficient nature of the GaAs binding site. Indeed, several studies [59, 60, 31] support this hypothesis and reveal a strong reduction in the band-bending of GaAs on treatment with Se-containing reagents. The stable phase was found to be Ga_2Se_3 [32], which has a close lattice match to GaAs, and therefore creates an almost strain-free layer.

Another class of surface treatments is based on halogens. Halogens are mostly used for surface etching and were found to dissociate upon adsorption [19]. The efficacy of halogen treatment for modifying the surface electronic properties was found to depend on the morphology and composition of the surface. In the case of GaAs, halogens showed high reactivity toward the Ga-rich surface [61, 62].

Parkinson and coworkers demonstrated improvement in the open-circuit voltage (V_{oc}) and fill factor of n-GaAs/K_2Se-K_2Se_2-KOH/C photoelectrochemical solar cells on treatment with Ru (III) [38]. The improvement in the cell performance was ascribed to the shift of the surface-state energies, which was thought to lead to a reduction in the main power-loss mechanism, electron-tunneling through the surface states. Treatment with Ru was also found to improve the performance, specially the stability, of InP-based photoelectrochemical solar cells [47]. This improvement was related to an increased barrier height [48].

2.3.5.2 **Organic Surface Treatments**

Table 3 summarizes examples of organic surface treatments used for changing the surface electronic properties. As a first example we consider self-assembly of organo-silanes on oxidized silicon. These self-assembled monolayers are known to be stable and well ordered [63, 64]. Modification of the monolayer properties can be achieved in a predetermined fashion by changing the chemical structure of the

Tab. 3 Several organic chemical treatments used for surface modifications

Semiconductor	*Treatments*	*References*
Silicon, single crystal and porous	Self assembly of organosilanes	65
	Acids and bases	69
	Amines	66, 70
	Common organic solvents	71
	Other organic adsorbates	67, 128
GaAs	Organic disulfides	72
	Porphyrins for NO Detection	73, 129
	Dicarboxylic acids	72, 74, 75–77
	Sulfides and thiols	58, 68, 78, 79, 80
	Dithiocarbamate	81
	CH_3CSNH_2	81, 82
	Benzoic acids	14, 83, 130
	Alcohols	84
	Ethyl iodide	85
	Dimethylcadmium	86
	Dimethylzinc	86
	Trimethylgallium	86
	Diethylzinc	87
	$HS(CH_2)Si(OCH_3)$	88
InP	Dicarboxylic acids	76
	Alkanethiol	89
	Various other molecules	90
CdTe	Benzoic acids	12
	Dicarboxylic acids	75
CdSe	Dicarboxylic acids	76
	Dithiocarbamate	91
	Thiols	92
	Anionic sulfur donor	93
	Ethylenediaminetetraacetic acid	94
	Dialkyl chalcogenides	95
	Olefins	96
	Aniline derivatives	97
	Amines	98, 99
	Boranes	100
	Fullerenes	101
	Carbonyl compounds	102
	TCNQ derivatives	103
	Benzoic acids	13
	Silapentanes and chlorinated Silanes	104
	Cyanide	105
	TOPO	106
	Porphyrins for O_2 Detection	131, 132
TiO_2	Benzoic acids	107
ITO	Benzoic and other carboxylic acids	108–110
$CuInSe_2$	Dicarboxylic acids	111
	Benzoic and hydroxamic acids	12
	Disulfides	8

self-assembling molecules. On the basis of that, a systematic modification of the surface electron affinity and band-bending of *n*-type silicon was demonstrated by using self-assembly of substituted quinoline chromophores [65]. The chemical scheme included two steps:

- self-assembly of organo-silanes that add organic functionality to the surface, and
- grafting of a series of substituted quinoline chromophores that add the variable (in this case polar) functionality.

On porous silicon, which has a high surface-to-volume ratio and thus a large number of accessible surface states, a substantial change in PL was recorded on exposure to amines and different solvents [66, 67]. On GaAs, organic sulfides and thiols were reported to induce PL changes that were attributed to changes in both the SRV and the V_S [58, 68].

Adsorbing a series of benzoic acids with systematically varying dipole moment led to a different response. Changes in contact potential differences (CPD) were correlated to changes in the electron affinity without observable effect on the surface band-bending [14]. Substituted dicarboxylic acids, which were found to bind by a two-site mechanism, rather than a one-site one, such as the benzoic acids [112], modified both the band-bending and the electron affinity of the GaAs [72, 74]. This difference in the electrical effect on treatment with carboxylic and dicarboxylic acids can be explained by the different binding group and the change in the molecular energy levels [75]. On CdTe, CPD measurements showed that organic benzoic and dicarboxylic acids could modify the electron affinity and/or the surface band-bending [12, 75]. On the wider band gap semiconductor, CdSe, PL data that were obtained after adsorption of several series of organic molecules (including aniline and carbonyl compounds) were interpreted in terms of changes in the band-bending [96–99, 102], whereas CPD changes induced by adsorption of benzoic acids and aniline derivatives were rationalized in terms of changes in electron affinity [13]. The different conclusions from luminescence intensity measurements and CPD measurements were ascribed to the fact that the PL measurements were done in solution, whereas the CPD measurements were performed in air. This difference also reflects the effects of interaction with the surrounding medium.

To provide information on molecule-surface interaction and on the feasibility of chemical treatments for use in electronic devices, like sensors, attempts have been made to establish a correlation between molecular properties and changes in surface properties. Table 4 summarizes works in which such a correlation was found.

2.3.5.3.1 **Correlation of Molecular Parameter with Changes in the Surface Electron Affinity** As mentioned earlier, control of the electron affinity or surface potential can be achieved by modifying the molecular dipole moments of the adsorbed molecules. Figure 5 shows the change in the electron affinity of CdTe, CdSe, and GaAs, which was deduced from CPD measured by a Kelvin probe, upon grafting a series of substituted benzoic acids onto the semiconductor surface. The linear relation between the change in the electron affinity and the dipole moment of the substituted benzoic acid is clearly observed. The dipole moments of the substituted phenyl groups reflect the electron withdrawing or donating power of the substituents. Bruening et al. used electrochemical and CPD measurements in

Tab. 4 Transfer of molecular properties to the macroscopic properties of semiconductors

Type of correlation	*Sample*	*References*
Change in electron affinity, χ, with dipole moment of the benzoyl substituent and cammett substituent constants	Silicon (single crystal)	65
	GaAs	14, 77, 130
	CdTe	12, 75
	CdSe	13
	$CuInSe_2$	12, 8
	TiO_2	107
	ITO	109
Change in built-in potential, V_s, with ionization potential, proton affinity, electron-accepting power; LUMO energy of the molecules	Silicon	65, 113
	CdTe	75
	CdSe	96, 97, 114, 99
Change in photoluminescence, PL, with electrochemical redox potential	CdSe	103
Change in surface recombination velocity (SRV), s, with ionization potential	CdSe	115
Change in PL with Hammett parameters	CdSe, CdS	102
Change in barrier height with number of carbon atoms in chain	GaAs (thiols)	80
Change in barrier height with molecular dipole moment	GaAs, Si	77, 128, 130

combination with ellipsometry and FTIR data to study the effects of binding a series of cyclic disulfides with systematically varying dipole moment and different degrees of hydrophobicity to Au and $CuInSe_2$ [8]. They found that the magnitude and direction of the change in electron affinity depend on the surface coverage, the orientation of the molecular dipole relative to the surface normal, and the mode of binding. Knowing the molecular surface coverage and tilt angle and thus the residual molecular dipole moment, as well as the atomic electronegativity difference at the molecule-surface interface allowed theoretical estimates of the experimentally observed change in the electron affinity, that is, it gave predictive power to Eq. (1).

In further work Wu et al. tested Eqn. (1) and used a series of molecules similar to that shown in Fig. 7, but completely conjugated. The main effect was found to be a smaller influence of the dipoles which can be understood from the increase in effective dielectric constant of the molecular layer [130].

2.3.5.3.2 **Correlation of Molecular Parameters with Changes in the Band-Bending** In general, the change in V_S can be viewed by two mechanisms,

- molecule-induced surface oxidation/reduction (generalized acid–base reaction) where the molecules either accept or donate electrons [117] according to the difference in the oxidation/reduction potentials, or alternatively by
- frontier orbital interaction mechanism, which involves the energy levels of the molecular frontier orbitals and of the semiconductor surface states [118, 119].

In the second mechanism, we consider the interaction between the highest occupied molecular orbital (HOMO)

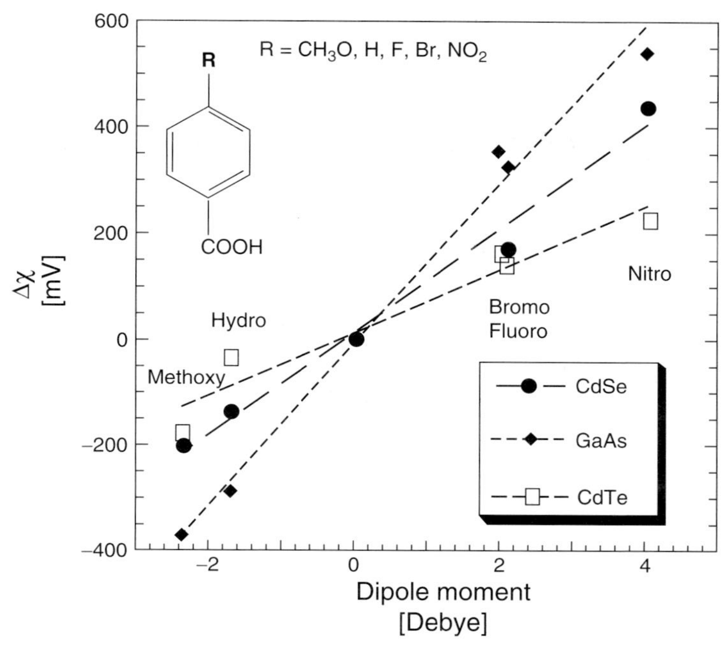

Fig. 5 Linear correlation of the change in the electron affinity of *n* CdTe, CdSe, and GaAs, as function of the dipole moment of benzoic acids [116], adsorbed on the semiconductors [12–14].

level of the surface or the molecule and the lowest unoccupied molecular orbital (LUMO) level of the other. It is an extension of the well-known frontier orbital interaction between the energy levels of two molecules forming a complex, as shown in Fig. 6(a). On interaction, the HOMO level is stabilized to lower energy, whereas the LUMO level is destabilized and pushed up in energy.

Table 4 summarizes studies in which the molecular ionization potential, which is related to the HOMO level, and the molecular electron affinity, related to the molecular LUMO level, are correlated to the changes in V_S. This correlation fits the orbital interaction mechanism.

Figure 6 shows several scenarios for the interaction of a given molecular LUMO level with surface states at different energies. In the first case, shown in Fig. 6(b), the molecule's LUMO level interacts with occupied surface states, which are below the Fermi level and therefore assume the role of the HOMO level. On interaction, the surface-state energies increase (with respect to vacuum), that is, they are "pushed" down in energy toward the VB, whereas the molecular LUMO level is "pushed" up in energy toward the CB. As a result, electrons that were formerly localized on the surface states may now occupy surface resonances with only partial localization at the surface. Therefore, they do not anymore (or little) contribute to the net surface charge. Reduction of V_S is thus expected. A different result is expected if the surface states are close to the semiconductor midgap, as shown in the second case (Fig. 6c). Because in this case the surface states are well removed from the band edge, only part of them will turn into

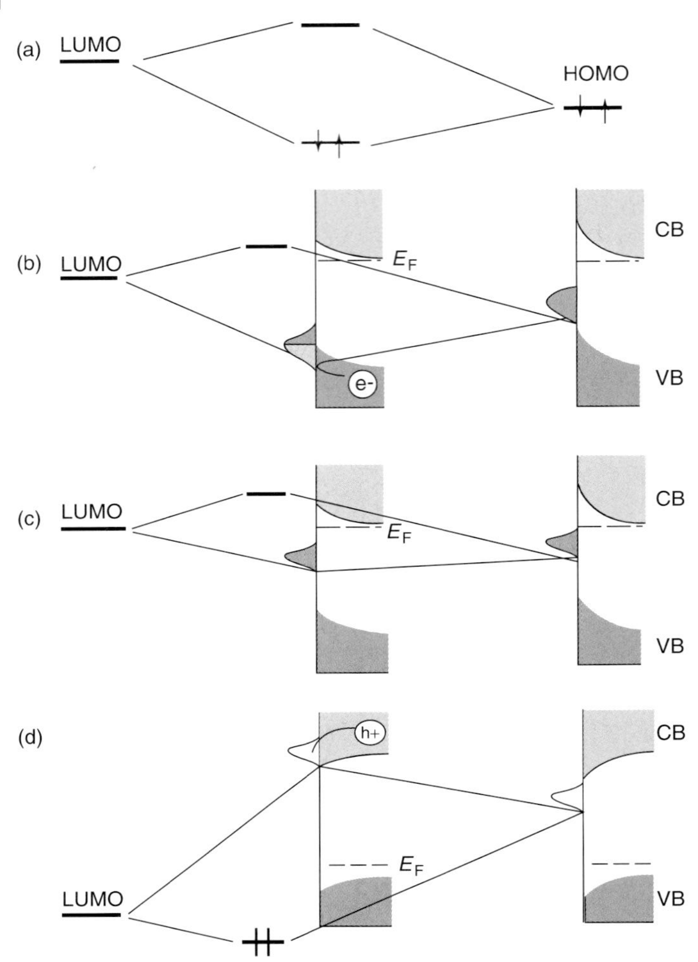

Fig. 6 Molecular orbital (HOMO–LUMO) interaction of two molecules (a) and of a molecule with semiconductor surface states (b–d). Different results are obtained after interaction with shallow acceptor states (occupied surface states close to the VB) (b), deep acceptor states (occupied states close to midgap) (c), and shallow donor states (close to the CB) (d). In general, the donor HOMO level is slightly stabilized by the interaction, whereas the acceptor LUMO level is slightly destabilized.

surface resonances and the net effect on the total surface state density, and thus on V_S, is expected to be moderate, compared to the first case (Fig. 6b). The SRV, on the other hand, is expected to change significantly because it depends critically on the density of the midgap (deep) surface states [5].

In the third scenario, shown in Fig. 6(d), we consider the interaction of a molecular LUMO level with empty surface states of a p-type semiconductor. Because the LUMO level of the molecule is well below the Fermi level, electron transfer from the surface to the molecule is expected and the molecule plays the role of the HOMO level in the interaction. As in the other cases, the surface states that were pushed in energies to more than the CB minimum will turn into surface resonances that will lead to a reduction in V_S.

The energy levels of the molecule after interaction should also be considered. These levels are modified (i.e. changed in energy, relative to the band edges) on interaction with the surface states, as was demonstrated [120, 133] for the interaction of thiophenol derivatives with Cu(111). If, on interaction, the molecular LUMO level is at an energy that corresponds to somewhere within the band gap and below E_F, the state will act as an occupied surface state and will increase (decrease) the net surface charge of n-(p-)type surfaces. On the other hand, if the molecular LUMO level is pushed below the VB maximum, then the state acts as a surface resonance and only a moderate effect on the band-bending is expected. It should be noted that although our discussion in this article is limited to the molecular LUMO level, analogous interactions of molecular HOMO levels with surface states can also be energetically favorable [119] and lead to modification of surface states and/or SRV.

Figure 7 shows an example of the change in V_S of an etched n-CdTe single crystal as a function of the LUMO energy level of a series of dicarboxylic acid derivatives [75]. The systematic modification of the molecular LUMO level was achieved by changing the molecule's substituents from electron-donating to electron-accepting groups. The change in V_S was suggested to stem from two contributions:

- a constant change of 170 mV because of the binding group and
- a second contribution that could be correlated with the molecules' LUMO energy level and that increased with decreasing energy separation between the molecule's LUMO state- and the surface state-energy level [75].

This second contribution is denoted schematically in Fig. 7 and can be attributed to extended coupling of the molecules' energy level and the semiconductor surface states, as the molecules' LUMO energies become closer in energy to those of the surface states. The predictive power of the frontier orbital interaction scheme, noted earlier, was demonstrated for n-CdSe, n-GaAs, n-InP, and p-GaAs crystals when changes in the V_S and in the SRV of the crystals on interaction with a given dicarboxylic acid molecule (DCDC and DHDC) could be explained [112]. On the basis of the studies mentioned earlier, it was found that the ability of the chemical treatment to modify the electronic properties of semiconductor surfaces depends on the following parameters:

1. The molecule's frontier orbital energy level and the difference in energy between that level and those of the interacting surface states. The smaller the energy distance, the stronger is the molecule-surface coupling and the larger can be the induced change in the surface electronic properties.
2. The surface-states' energy levels and densities. In surfaces where the dominant surface states are close to the band edges, the main effect on interaction is a change in band-bending. On surfaces with dominant states close to midgap, the main effect on interaction

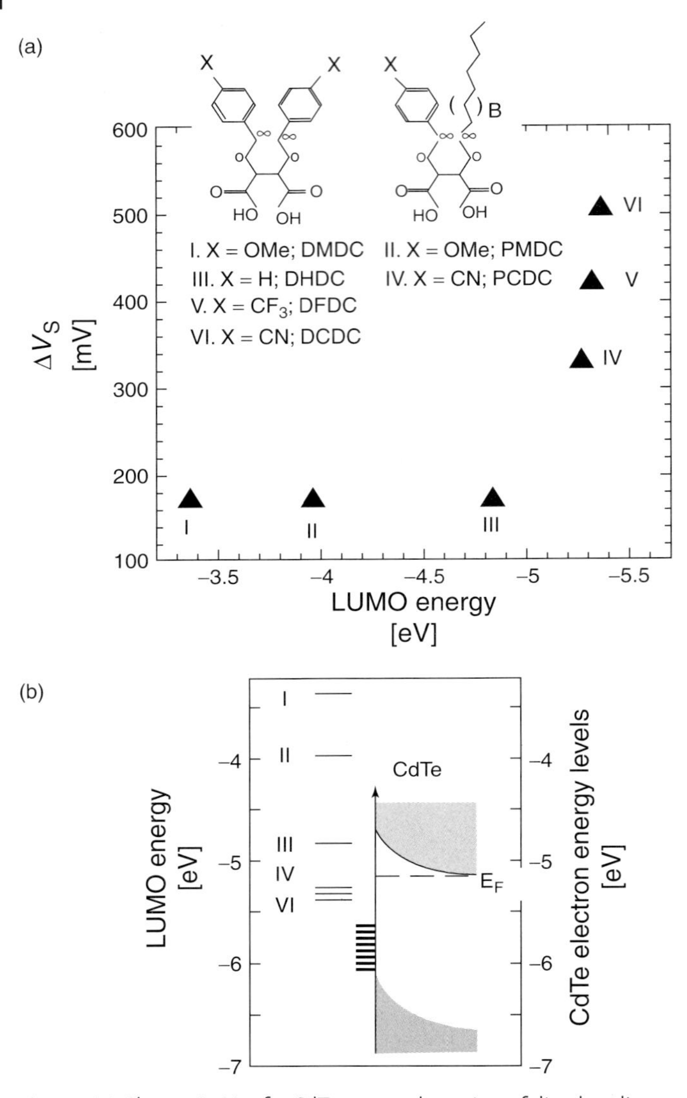

Fig. 7 (a) Change in V_s of n-CdTe upon adsorption of dicarboxylic acid derivatives as a function of the benzoyl substituents' LUMO energy [75]. ΔV_s is the change in V_s relative to the etched surface. (b) Energy diagram of bare CdTe and of LUMO of isolated molecule before adsorption. The CdTe energy levels were experimentally measured by CPD, and the Fermi level was calculated from the doping density.

is a change in surface recombination velocity.

According to this approach, designing molecular surface treatments should start by mapping the surface-state energies and then choosing a molecule with energy levels that are close to the target surface states and that can successfully interact with them. In practical cases, semiconductor surfaces may possess several populations of surface states and thus molecular selectivity will be required.

Further progress in this direction depends on

- better knowledge of molecular energy levels on an absolute scale (to allow comparison with semiconductor energy levels);
- consideration of other factors that might affect molecule-surface coupling, such as, for example, the shape of the orbitals, that is, symmetry considerations.

2.3.6 Examples of Molecular Control over Optoelectronic Devices

The progress in device-processing techniques and the ability to design new multifunctional molecules, whose properties can be modified systematically, opens new directions for construction of molecular devices. Given below are several examples of the application of organic molecules for controlling the performance of optoelectronic devices based on inorganic, nonmolecular materials.

The first system we consider is a Schottky diode and its derivatives. The electrical characteristics of this diode depend on the interface states and dipoles and can thus reflect changes induced by adsorbed molecules. Petty, Roberts, and coworkers [121, 122] used the Langmuir-Blodgett (LB) technique to prepare a Schottky diode of Au/Cd-stearate/CdTe on both *n*- and *p*-CdTe. Such a molecular film increased both the barrier height and the open-circuit voltage of this MIS (metal-insulator semiconductor) solar cell structure. The fill factor was found to depend on the thickness of the LB chain and was the highest for one monolayer (2.5 nm). These findings are in accordance with the optimum insulator thickness found for silicon MIS solar cells of 1 to 2 nm. Allara and coworkers [80] studied the effect of self-assembled monolayer (SAM) alkane thiols with different chain lengths on the Schottky barrier height of the Au/*n*-GaAs system. They found that while the barrier height increased only very slightly, the increase correlated linearly with the number of the carbon atoms in the alkane chain. They attributed this change to an increase of the negative charge at the binding group (i.e. sulfur atom) as the thickness of the SAM increased. However, an opposite effect on the barrier height was noted for the thiol-passivated Cu/*n*-GaAs system, which suggests that other mechanisms should also be considered.

Although tuning the Schottky barrier properties by way of changing the length of the molecules is essentially similar to the effect of changing the thickness of any dielectric film between the metal and semiconductor, recent examples [77, 107–109, 123, 128, 130] show that true molecular effects can be obtained.

Vilan et al. demonstrated that molecules adsorbed at the Au/*n*-GaAs interface can tune the energetics of this metal–semiconductor diode, because of their chemical character, namely, by systematic substitution on aromatic rings [77]. Using a series of dicarboxylic acid derivatives whose dipole was varied systematically, they produced diodes in which the effective barrier height is tuned by the molecular dipole moment. Qualitatively, compared to the unmodified junction, molecules with a dipole pointing toward the surface increased the forward current in the current–voltage ($I-V$) curve, whereas those with a dipole pointing away from the surface decreased it. The interesting thing here is that the molecules do not form a perfect monolayer and there is high probability of pinholes in the layer. Despite this, the molecules strongly affect the diode characteristics. These results reveal that it is not necessary to use molecules that conduct current through them to affect device electrical characteristics and that molecules that electrostatically interact with the surface can also be applied.

The results of Wu et al. [130], who prepared and measured Au/molecule/n-GaAs diodes with the earlier mentioned series of conjugated molecules agree with these conclusions.

Selzer and Cahen [128] showed that a complementary configuration can be used, as well. They adsorbed ligands on the metal rather than on the semiconductor side of the junction. In that case the metal work function, rather than the semiconductor's electron affinity is changed. They used the molecules of Ref. [8], which are similar to those shown in Fig. 7(a), but with cyclic disulfide, instead of dicarboxylic acid binding groups, so as to allow chemisorption onto Au. The molecularly modified metal was then used to prepare Au/molecule/SiO_x/Si diodes. Results are naturally opposite to those obtained with GaAS/Au diodes [77], because the substituted phenyl groups point in opposite directions in the two cases.

Molecular layers will probably also find their way in optoelectronic devices. Krueger et al. used a series of benzoic acids [12–14] to modify the current–voltage characteristics of the inorganic–organic TiO_2/spiro-MeOTAD (an amorphous organic hole conductor) heterojunction [107]. These changes were correlated with the changes in the work function of the modified TiO_2 and were attributed to variations in the built-in voltage at the TiO_2/spiro-MeOTAD interface. A rough correlation between the change of the work function and the current–voltage characteristics was demonstrated also by Cambell and colleagues [123], who used a self-assembled layer of conjugated thiol molecules to manipulate the Schottky barrier height between a Cu electrode and an organic electronic material used for light-emitting diode (LED) systems.

Zuppiroli, Nuesch, and coworkers demonstrated [108, 109, 134] that pretreatment of the transparent conductor indium tin oxide (ITO) with a few derivatized organic molecules can significantly improve (decrease) the turn-on characteristic (field) that denotes the field above which light emission is observed in the ITO/poly(paraphenylene)/Al LED system. The turn-on field was reduced from 200 MV m^{-1} to 100 MV m^{-1} after molecule adsorption. The decrease in the turn-on field was correlated with the dipole moment of the carboxylic group. The molecules modified (increased) the work function of the ITO. In this way they increased the hole injection efficiency in the device. Moreover, the device durability was increased. The stability under DC operation was increased from 15 minutes to 2.5 hours after the molecular treatment.

Friend and coworkers [124] showed that molecular surface treatments could also be applied in the new field of organic-based optoelectronic circuits. Improvement in charge mobility of a field effect transistor (FET)-LED device was demonstrated by using hexamethyldisilazane before

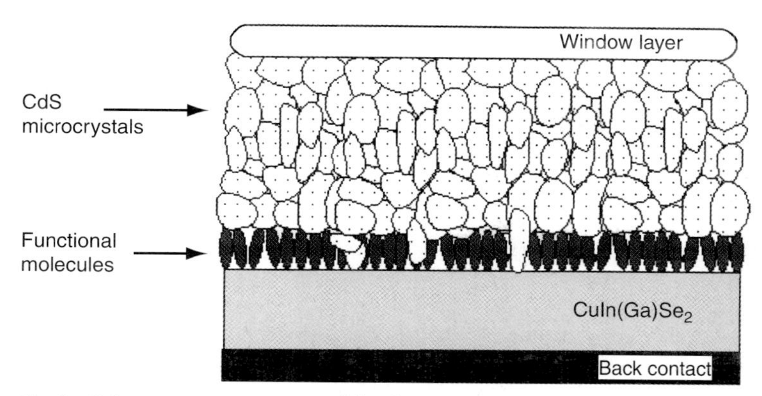

Fig. 8 Schematic view (not to scale!) of molecular control over $Cu(In,Ga)Se_2/CdS$ heterojunction thin film, polycrystalline solar cell. The molecules are at the interface between the $CuInSe_2$ film and the CdS microcrystals.

deposition of the conjugated polymer. The molecular treatment was suggested to promote phase segregation, that is, structural order at the interface. The earlier-mentioned studies reveal the potential of molecular treatments for tuning the charge transfer and charge transport across (semi)conductor/(semi)conductor interfaces.

A different potential application of a molecular treatment is their use as a blocking layer against charge leakage. For example, Langmuir-Blodgett (LB) films of 22-tricosanoic, $CH_2{=}CH(CH_2)_{20}{-}COOH$ [67] and hexadecanol [68] were shown to have charge-blocking properties on GaAs-based devices that are superior to those shown with them on (Al,Ga)As ones. Using capacitance measurements of the metal-(molecular insulator)-semiconductor (MIS) structure indicated reduction of the interface trap density [83] and of the barrier characteristics [84].

Solar cells are another field of applications for functional molecules. This idea is schematically presented in Fig. 8 for a heterojunction solar cell, where the molecules are located at the interface between the n- and p-type surface. Gal et al. [111] used an experimental scheme similar to Fig. 8 to show the effect of organic acids on the electrical characteristics of polycrystalline $CuIn(Ga)Se_2$(CIGSe)/CdS solar cells (see Fig. 9). The changes in the $I-V$ characteristics could be correlated with the molecules' dipole moments that modify the band line-up at the interface, rather than as a direct effect on the surface V_S. Specially in view of the later work [77, 107, 128, 130], these results indicate that it is not essential to form an ordered layer in order to modify device characteristics. However, what is required is a molecule that will modify the energetics at the interface so that any charge carrier passing from one side of the junction to the other will be influenced by it.

2.3.7 Summary

Fine-tuning of semiconductor surface electrical properties can be achieved by grafting multifunctional organic molecules onto the surface. In such molecules, one function takes care of the binding

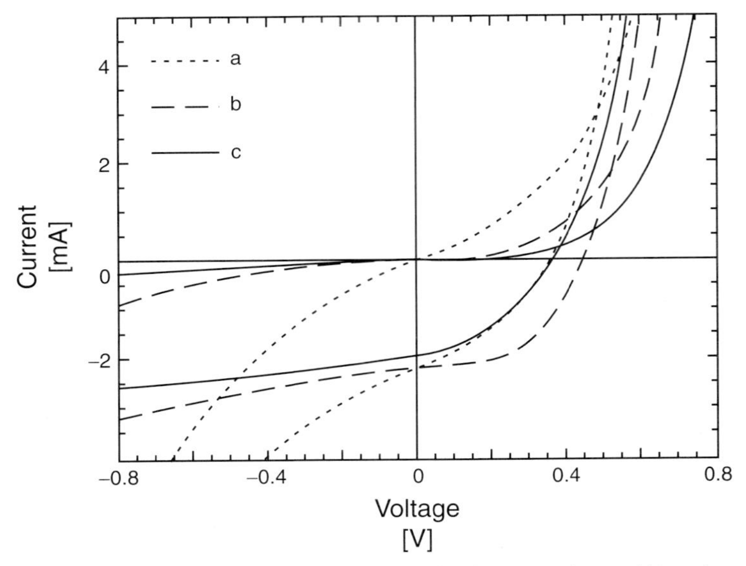

Fig. 9 Current–voltage characteristics of molecular, treated (a and b) and untreated (c) Au/CdS/Cu(In,Ga)Se_2/Mo solar cells in dark and under illumination (light intensity: ~1.5 AM). The molecules used, shown schematically in Fig. 7, are DMDC and DCDC for a and b, respectively. The molecules were deposited on the polycrystalline Cu(In,Ga)Se_2 surface before the wet chemical deposition of the CdS [111].

to the semiconductor surface, whereas another one conveys a desired property to that surface and thus to the semiconductor. The way in which this can happen is conveniently studied by using a series of molecules in which the desired property is varied in a systematic fashion.

Implementation of molecular tools provides new opportunities for hybrid organic–inorganic systems. The increasing interest in, and importance of, nanoscale technologies and the use of nanocrystals, where surface-related effects often dominate bulk properties [106, 125–127], adds to the importance of controlling and engineering surface properties. Correlating the molecular properties of the grafted molecules to the changes in the macroscopic semiconductor properties can extend our understanding of molecule-surface interaction mechanisms and help design molecular tools for controlling semiconductor surfaces. Recent work shows that this can be carried over to interfaces of semiconductors in devices where molecules will directly or indirectly influence active device characteristics. In this way, even small modifications in the molecular structure can induce significant and predictable changes in device behavior.

Acknowledgments

For our own contributions mentioned in this chapter we thank Leeor Kronik, Ellen Moons, Merlin Bruening, Stephane Bastide, Dori Gal, Ayelet Vilan, Tamar Moav and Rahel Lazar, as well Ron Naaman and his group, for fruitful collaborations. We acknowledge the pivotal contributions of the late Dr. Jacqueline Libman to the early parts of our work and Art Ellis (Univ. of Wisconsin, Madison) for collaboration and inspiration for the molecule-surface interaction mechanism. Partial support from the US-Israel Binational- and the Israel Science Foundations, the Minerva Foundation (Munich), the Fussfeld Fund and the German BMBF, through its bilateral Energy Research Program with the Israel Ministry of Science is gratefully acknowledged.

References

1. E. H. Rhoderick, R. H. Williams, *Metal-Semiconductor Contacts*, 2nd edition, Oxford, Clarendon, 1988.
2. H. Lüth, *Surfaces and Interfaces of Solid Materials*, 3rd edition, Springer, Germany, 1998.
3. G. V. Hansson, R. I. G. Uhrberg, Photoelectron Spectroscopy of Surface States on Semiconductor Surfaces, *Sur. Sci. Rep.* **1988**, *9*, 197–292.
4. A. Many, Y. Goldstein, N. B. Grover, *Semiconductor Surfaces*, Amsterdam, North-Holland, 1965.
5. D. V. Lang, C. H. Henry, Nonradiative Recombination at Deep Levels in GaAs and GaP by Lattice-Relaxation Multiphonon Emission, *Phys. Rev. Lett.* **1975**, *35*, 1525–1528.
6. H. Ishii et al., Energy Level Alignment and Interfacial Electronic Structures at Organic/Metal and Organic/Organic Interfaces, *Adv. Mater.* **1999**, *11*(8), 605–625.
7. A. M. Green, W. E. Spicer, Do We Need a New Methodology for GaAS Passivation? *J.Vac. Sci. Technol. A* **1993**, *11*, 1061–1069.
8. M. Bruening et al., Simultaneous Control of Surface Potential and Wetting of Solids with Chemisorbed Multifunctional Ligands, *J. Am. Chem. Soc.* **1997**, *119*, 5720–5728.
9. S. D. Evans et al., Self-Assembled Monolayers of Alkanethiols Containing a Polar Aromatic Group: Effect of the Dipole Position on Molecular Packing, Orientation, and Surface Wetting Properties, *J. Am. Chem. Soc.* **1991**, *113*, 4121–4131.
10. S. D. Evans, A. Ulman, Surface Potential Studies of Alkyl-Thiol Monolayers Adsorbed on Gold, *Chem. Phys. Lett.* **1990**, *170*(5, 6), 462–466.
11. I. H. Campbell et al., Controlling Schottky Energy Barriers in Organic Electronic Devices using Self-assembled Monolayers, *Phys. Rev. B* **1996**, *54*, R1432–R14 324.
12. M. Bruening et al., Polar Ligand Adsorption Controls Semiconductor Surface Potentials, *J. Am. Chem. Soc.* **1994**, *116*, 2972–2977.
13. M. Bruening et al., Controlling The Work Function of CdSe by Chemisorption of Benzoic Acid Derivatives and Chemical Etching, *J. Phys. Chem.* **1995**, *99*, 8368–8373.
14. S. Bastide et al., Controlling the Work Function of GaAs by Chemisorption of Benzoic Acid Derivatives, *J. Phys. Chem.*, **1997**, *101*, 2678–2684.
15. D. M. Taylor, O. N. Oliveira, H. Morgan, Models for Interpreting Surface Potential Measurements and their Application to Phospholipid Monolayers, *J. Coll. Int. Sci.* **1990**, *139*(2), 508–518.
16. J. F. Dorsten, J. E. Maslar, P. W. Bohn, Near-Surface Electronic Structure in GaAs (100) Modified with Self-assembled Monolayers of Octadecylthiol, *Appl. Phys. Lett.* **1995**, *66*, 1755–1757.
17. S. M. Sze, *Physics of Semiconductor Devices*, 7th edition, 2nd Chapter, John Wiley & Sons, USA, 1981.
18. W. Mönch, *Semiconductor Surfaces and Interfaces*, 2nd edition, in *Surface Science* (Ed.: E Gerhard), Springer, Berlin, 1995, pp. 274–284.
19. F. Seker et al., Surface Chemistry of Prototypical Bulk II–VI and III–V Semiconductors and Implications for Chemical Sensing, *Chem. Rev.* **2000**, *100*, 2505–2536.
20. J.-F. Fan, H. Oigawa, Y. Nannichi, The Effect of $(NH_4)_2S$ Treatment on the Interface Characteristics of GaAs MIS Structures, *Jap. J. Appl. Phys.* **1988**, *27*, L1331–L1333.
21. J.-F. Fan, H. Oigawa, Y. Nannichi, Metal-Dependent Schottky Height with the $(NH_4)_2S_x$ Treated GaAs. *Jap. J. Appl. Phys.* **1988**, *27*(11), L2125–L2127.
22. D. Liu et al., Deep level transient spectroscopy study of GaAs surface states treated with inorganic sulfides, *Appl. Phys. Lett.* **1988**, *53*, 1059–1061.

23. H. Hasegawa et al., Control of Fermi Level Pinning and Recombination Processes at GaAs Surfaces by Chemical and Photochemical Treatments, *J. Vac. Sci. Technol. B* **1988**, *6*, 1184–1192.
24. S. Hildebrandt et al., Optical Study of Band-Bending and Interface Recombination at Sb, S, and Se Covered Gallium Arsenide Surfaces, *Appl. Surf. Sci.* **1993**, *63*, 153–157.
25. R. S. Besser, C. R. Helms, Effect of Sodium Sulfide Treatment on Band Bending in GaAs, *Appl. Phys. Lett.* **1988**, *52*, 1707–1709.
26. H. Shen, W. Zhou, J. Pamulapati, Photoreflectance Study of H_2S Plasma-Passivated GaAs Surfaces, *Appl. Phys. Lett.* **1999**, *74*, 1430–1432.
27. V. L. Berkovits et al., Chemical and Photochemical Processes in Sulfide Passivation of GaAs(100), In Situ Optical Study and Photoemission Analysis, *J. Vac. Sci. Technol. A* **1998**, *16*, 2528–2538.
28. M. Stutzmann, C. P. Herrero, Hydrogen Passivation of Shallow Acceptors in Silicon, *Phys. Scripta.* **1989**, *T25*, 276–282.
29. M. J. Bedzyk, W. M. Gibson, J. A. Golovchenko, X-ray Standing Wave Analysis for Bromine Chemisorbed on Silicon, *J. Vac. Sci. Technol.* **1982**, *20*, 634–637.
30. R. I. G. Uhrberg et al., Electronic Structure, Atomic Structure, and the Passivated Nature of the Arsenic-Terminated Si (111) Surface, *Phys. Rev. B* **1987**, *35*, 3945–3951.
31. B. A. Kuruvilla et al., Passivation of GaAs (100) Using Selenium Sulfide, *J. Appl. Phys.* **1993**, *73*, 4384–4387.
32. J. Sun et al., Chemical Bonding and Electronic Properties of SeS_2-Treated GaAs(100), *J. Appl. Phys.* **1999**, *85*, 969–977.
33. E. Yablonovitch et al., Band-Bending, Fermi Level Pinning and Surface Fixed Charge on Chemically Prepared GaAs Surfaces, *Appl. Phys. Lett.* **1989**, *54*, 555–557.
34. E. Yablonovitch et al., Nearly Ideal Electronic Properties of Sulfide-Coated GaAs Surfaces, *Appl. Phys. Lett.* **1987**, *51*(6), 439–441.
35. C. J. Sandroff et al., Dramatic Enhancement in the Gain of a GaAs/AlGaAs Heterostructure Bipolar Transistor by Surface Chemical Passivation, *Appl. Phys. Lett.* **1987**, *51*(1), 33–35.
36. C. J. Sandroff et al., Electronic Passivation of GaAs Surfaces Through the Formation of Arsenic – Sulfur Bonds, *Appl. Phys. Lett.* **1989**, *54*(4), 362–364.
37. J. S. Herman, F. L. Terry, Hydrogen Sulfide Plasma Passivation of Gallium Arsenide, *Appl. Phys. Lett.* **1992**, *60*, 716–717.
38. B. A. Parkinson, A. Heller, B. Miller, Effects of Cations on the Performance of the Photoanode in the N-GaAs/K_2Se-K_2Se_2-KOH/C Semiconductor Liquid Junction Solar Cell, *J. Electrochem. Soc.* **1979**, *126*(6), 954–960.
39. P. Viktorovitch et al., Improved Electronic Properties of GaAs Surfaces Stabilized with Phosphorous, *Appl. Phys. Lett.* **1991**, *58*(21), 2387–2389.
40. H. H. Lee, R. J. Racicot, S. H. Lee, Surface Passivation of GaAs, *Appl. Phys. Lett.* **1989**, *54*(8), 724–726.
41. L. Ferrari et al., Optical and Spectroscopic Characterization of GaAs Passivated Surfaces, *Surf. Sci.* **1995**, *331–333*, 447–452.
42. D. Troost et al., Pinning of the Fermi level Close to the Valence-Band Top by Chlorine Adsorbed on Cleaved GaAs (110) Surfaces, *J. Vac. Sci. Technol.* **1987**, *B5*(4), 1119–1124.
43. H. J. Clemens, J. Von Wienskowski, W. Mönch, On the Interaction of Cesium with GaAs(110) and Ge(111) Surfaces: Work Function Measurements and Adsorption Site Model, *Surf. Sci.* **1978**, *78*, 648–666.
44. P. Skeath et al., Bonding of Antimony on GaAs (110): A Prototypical System for Adsorption of Column-V Elements on III–V Compounds, *Phys. Rev. B* **1983**, *27*, 6246.
45. K. Jacobi, G. Steinert, W. Ranke, Iodine Etching of the GaAs(-1-1-1) Surface Studied by LEED, AES, and Mass Spectroscopy, *Surf. Sci.* **1976**, *57*, 571–579.
46. K. Vaccaro et al., Indium Phosphide Passivation Using Thin Layers of Cadmium Sulfide, *Appl. Phys. Lett.* **1995**, *67*(4), 527–529.
47. D. N. Bose, Y. Ramprakash, S. Basu, Characterization of n-InP Surfaces Before and After Surface Modification, *Mater. Lett.* **1989**, *88*(9), 364–368.
48. D. N. Bose, J. N. Roy, S. Basu, Improved Schottky Barrier on n-InP by Surface Modification, *Mater. Lett.* **1984**, *2*(5B), 455–457.
49. A. J. Nelson, S. P. Frigo, R. A. Rosenberg, Chemistry and Electronic Structure of the

H_2 Plasma Passivated Surface of CdTe, *J. Appl. Phys.* **1994**, *73*(3), 1632–1637.
50. D. N. Bose, S. Basu, K. C. Mandal, Characterization of Chemically Modified CdTe Surfaces, *Thin Solid Films* **1988**, *164*, 13–19.
51. D. N. Bose et al., XPS Investigation of CdTe Surfaces: Effect of Ru Modification, *Semicon. Technol.* **1989**, *4*, 866–870.
52. J. L. Shaw et al., Chemically Controlled Deep Level Formation and Band-Bending at Metal-CdTe Interfaces, *Appl. Phys. Lett.* **1988**, *53*(18), 1723–1725.
53. D. Cahen, R. Noufi, Surface Passivation of Polycrystalline, Chalcogenide Based Photovoltaic Cells, *Solar Cells* **1991**, *30*(1–4), 53–59.
54. Y. Nemirovsky, D. Rosenfeld, Surface Passivation and 1/f Noise Phenomena in HgCdTe Photodiodes, *J. Vac. Sci. Technol.* **1990**, *A8*(2), 1159–1167.
55. Y. F. Chen et al., Hydrogen Passivation in $Cd_{1-x}Zn_xTe$ Studied by Photoluminescence, *Appl. Phys. Lett.* **1991**, *58*(5), 493–495.
56. Y. F. Chen, W. S. Chen, Influence of Hydrogen Passivation on the Infrared Spectra of $Hg_{0.8}Cd_{0.2}Te$, *Appl. Phys. Lett.* **1991**, *59*(6), 703–705.
57. D. Cahen, R. Noufi, Defect Chemical Explanation for the Effect of Air Anneal on $CdS/CuInSe_2$ Solar Cell Performance, *Appl. Phys. Lett.* **1989**, *54*(6), 558–560.
58. S. R. Lunt et al., Chemical Studies of the Passivation of GaAs Surface Recombination Using Sulfides and Thiols, *J. Appl. Phys.* **1991**, *70*(12), 7449–7467.
59. C. J. Sandorff et al., Enhanced Electronic Properties of GaAs Surfaces Chemically Passivated by Selenium Reactions, *J. Appl. Phys.* **1990**, *67*(1), 586–588.
60. S. A. Chambers, V. S. Sundaram, Structure, Chemistry, and Band-Bending at Se-Passivated GaAs(001) Surfaces, *Appl. Phys. Lett.* **1990**, *57*, 2342–2344.
61. W. C. Simpson, J. A. Yarmoff, Fundamental studies of halogen reactions with III–V semiconductor surfaces, *Ann. Rev. Phys. Chem.* **1996**, *47*, 527–554.
62. Y. Liu, A. J. Komrowski, A. C. Kummel, Side–selective reaction of Br_2 with second layer Ga atoms on the As-rich GaAs(001) 2×4 surface, *Phys. Rev. Lett.* **1998**, *81*, 413–416.
63. R. Maoz, J. Sagiv, On the Formation and Structure of Self-Assembling Monolayers, *J. Coll. Int. Sci.* **1984**, *100*(2), 465–496.
64. A. Ulman, *An Introduction to Ultrathin Organic Films*, Academic Press, New York, 1991.
65. R. Cohen et al., Molecular Electronic Tuning of Si Surfaces, *Chem. Phys. Lett.* **1997**, *279*, 270–274.
66. B. Sweryda et al., A Comparison of Porous Silicon and Silicon Nanocrystallite Photoluminescence Quenching with Amines, *J. Phys. Chem.* **1996**, *100*(32), 13 776–13 780, and reference therein.
67. J. M. Lauerhaas, M. J. Sailor, Chemical Modification of the Photoluminescence Quenching of Porous Silicon, *Science* **1993**, *261*, 1567–1568.
68. S. R. Lunt, P. G. Santangelo, N. S. Lewis, Passivation of GaAs Surface Recombination with Organic Thiols, *J. Vac. Sci. Technol. B* **1991**, *9*(4), 2333–2336.
69. J. K. M. Chun et al., Proton Gated Emission from Porous Silicon, *J. Am. Chem. Soc.* **1993**, *115*, 3024–3025.
70. R. R. Chandler et al., Steric Considerations in the Amine-Induced Quenching of Luminescent Porous Silicon, *J. Phys. Chem.* **1995**, *99*(21), 8851–8855.
71. J. M. Lauerhaas et al., Reversible Luminescence Quenching of Porous Si by Solvents, *J. Amer. Chem. Soc.* **1992**, *114*(5), 1911–1912.
72. A. Vilan, Chemical Modification of the electronic Properties of GaAs Surface, M.Sc thesis, Weizmann Institute of Science, Rehovot, 1996.
73. D. G. Wu et al., Novel NO biosensor based on surface derivatization of GaAs by "hinged" Fe porphyrin, *Angew. Chem. Int. Ed.* **2000**, *39*, 4496–4500.
74. R. Cohen et al., Frontier Orbital Model of Semiconductor Surface Passivation: Dicarboxylic Acids on *n*- and *p*-GaAs, *Adv. Mater.* **2000**, *12*, 33–37.
75. R. Cohen et al., Controlling Electronic Properties of CdTe by Adsorption of Dicarboxylic Acid Derivatives: Relating Molecular Parameters to Band-Bending and Electron Affinity Changes, *Adv. Mater.* **1997**, *9*, 746–749.
76. R. Cohen et al., Molecular Control over Semiconductor Surface Electronic Properties: Dicarboxylic Acids on CdSe, CdTe,

GaAs and InP, *J. Am. Chem. Soc.* **1999**, *121*, 10 545–10 553.
77. A. Vilan, A. Shanzer, D. Cahen, Molecular Control over Au/GaAs Diodes, *Nature* **2000**, *404*, 166–168.
78. C. D. Bain, A New Class of Self-Assembled Monolayers: Organic Thiols on Gallium Arsenide, *Adv. Mater.* **1992**, *4*(9), 591–594.
79. W. C. Sheen et al., A New Class of Organized Self-Assembled Monolayers: Alkane Thiols on GaAs (100), *J. Am. Chem. Soc.* **1992**, *114*, 1514–1515.
80. O. S. Nakagawa et al., GaAs Interfaces with Octadecyl Thiol Self-Assembled Monolayer: Structural and Electrical Properties, *Jap. J. Appl. Phys.* **1991**, *30*, 3759–376.
81. K. Asai et al., Electronic Passivation of GaAs Surfaces by Electrodeposition of Organic Molecules Containing Reactive Sulfur, *J. Appl. Phys.* **1995**, *77*, 1582–1586.
82. E. D. Lu et al., A Sulfur Passivation of GaAs Surface by an Organic Molecular, CH_3CSNH_2 Treatment, *Appl. Phys. Lett.* **1996**, *69*, 2282–2284.
83. M. Tabib-Azar, A. S. Dewa, W. H. Ko, Langmuir-Blodgett Film Passivation of Unpinned *n*-Type Gallium Arsenide Surfaces, *Appl. Phys. Lett.* **1988**, *52*, 206–208.
84. V. J. Rao, V. S. Kulkarni, Interface Barrier Height in *n*-GaAs(100)/Langmuir-Blodgett Film Structures, *Thin Solid Films* **1991**, *198*, 357–362.
85. N. K. Singh et al., Etching Reactions of C_2H_5I on GaAs(100), *Sur. Sci.* **1998**, *409*, 272–282.
86. P. J. Lasky et al., The Adsorption and Thermal Reaction of Dimethylcadmim, Dimethylzinc and Trimethylgallium on GaAs(110), *Surf. Sci.* **1996**, *364*, 312–324.
87. H. T. Lam, N. Venkateswaran, J. M. Vohs, The Reactions of Diethylzinc on Gallium-Rich and Arsenic-Rich Reconstructions of GaAs(100), *Surf. Sci.* **1998**, *401*, 34–46.
88. T. Hou et al., Passivation of GaAs (100) with an Adhesion Promoting Self-Assembled Monolayer, *Chem. Mater.* **1997**, *9*, 3181–3186.
89. Y. Gu, D. H. Waldeck, Electron Tunneling at the Semiconductor-Insulator-Electrolyte Interface. Photocurrent Studies of the *n*-InP-Alkanethiol-Ferrocyanide System, *J. Phys. Chem. B* **1998**, *102*, 9015–9028.
90. M. Sturzenegger, N. S. Lewis, An X-Ray Photoelectron Spectroscopic and Chemical Reactivity Study of Routes to Functionalization of Etched InP Surfaces, *J. Am. Chem. Soc.* **1996**, *118*, 3045–3046.
91. J. W. Thackeray et al., Interactions of Diethyldithiocarbamate with *n*-type Cadmium Sulfide and Cadmium Selenide: Efficient Photoelectrochemical Oxidation to the Disulfide and Flat-Band Potential of the Semiconductor as a Function of Adsorbate Concentration, *J. Am. Chem. Soc.* **1986**, *108*, 3570–3577.
92. M. J. Natan, J. W. Thackeray, M. S. Wrighton, Interaction of Thiols with *n*-Type Cadmium Sulfide and *n*-Type Cadmium Selenide in Aqueous Solutions: Adsorption of Thiolate Anion and Efficient Photoelectrochemical Oxidation to Disulfides, *J. Phys. Chem.* **1986**, *90*, 4089–4098.
93. J. J. Hickman, M. S. Wrighton, Face-Specific Interactions of Anionic Sulfur Donors with Oriented Crystals of (0001) CdX (X = Se, S) and Correlation with Electrochemical Properties, *J. Am. Chem. Soc.* **1991**, *113*, 4440–4448.
94. T. Uchihara et al., Effect of Ethylenediaminetetraacetic Acid on the Photocatalytic Activities and Flat-Band Potentials of Cadmium Sulfide and Cadmium Selenide, *J. Phys. Chem.* **1990**, *94*, 415–418.
95. J. K. Lorenz, T. F. Kuech, A. B. Ellis, Cadmium Selenide Photoluminescence as a Probe for the Surface Adsorption of Dialkyl Chalcogenides, *Langmuir* **1998**, *14*, 1680–1683.
96. G. J. Meyer et al., Semiconductor-Olefin Adducts. Photoluminescent Properties of Cadmium Sulfide and Cadmium Selenide in the Presence of Butenes, *J. Am. Chem. Soc.* **1989**, *111*, 5146–5148.
97. C. J. Murphy et al., Photoluminescence-Based Correlation of Semiconductor Electric Field Thickness with Adsorbate Hammett Substituent Constants. Adsorption of Aniline Derivatives onto Cadmium Selenide, *J. Am. Chem. Soc.* **1990**, *112*, 8344–8348.
98. G. C. Lisensky et al., Electro-Optical Evidence for Chelate Effect at Semiconductor Surfaces, *Science* **1990**, *248*, 840–843.
99. G. J. Meyer, G. C. Lisensky, A. B. Ellis, Evidence for Adduct Formation at the

Semiconductor-Gas Interface. Photoluminescent Properties of Cadmium Selenide in the Presence of Amines, *J. Am. Chem. Soc.* **1988**, *110*, 4914–4918.
100. D. R. Neu, J. A. Olson, A. B. Ellis, Photoluminescence as a probe of the Adsorption of Gaseous Boranes onto the Surface of Cadmium Selenide Crystals, *J. Phys. Chem.* **1993**, *97*, 5713–5716.
101. J. Z. Zhang, M. J. Geselbracht, A. B. Ellis, Binding of Fullerenes to Cadmium Sulfide and Cadmium Selenide Surfaces, Photoluminescence as a Probe of Strong, Lewis Acidity-Driven, Surface Adduct Formation, *J. Am. Chem. Soc.* **1993**, *115*, 7789–7793.
102. K. D. Kepler et al., Surface-Bound Carbonyl Compounds as Lewis Acids. Photoluminescence as a Probe for the Binding of Ketones and Aldehydes to Cadmium Sulfide and Cadmium Selenide Surfaces, *J. Phys. Chem.* **1995**, *99*(43), 16 011–16 017.
103. J. Z. Zhang, A. B. Ellis, Adsorption of TCNQ Derivatives onto the Surface of Cadmium Selenide Single Crystals. Quenching of Semiconductor Photoluminescence by a Family of Strong-Acids, *J. Phys. Chem.* **1992**, *96*, 2700–2704.
104. R. J. Brainard et al., Modulation of Cadmium Selenide Photoluminescence Intensity by Adsorption of Silapentanes and Chlorinated Silanes, *J. Phys. Chem. B* **1997**, *101*, 11 180–11 184.
105. S. Licht, D. Peramunage, Rational Electrolyte Modification of *n*-CdSe/([KFe(CN)$_6$]3-/2-) Photoelectrochemistry, *J. Electrochem. Soc.* **1992**, *139*, L23–L26.
106. J. K. Lorenz, A. B. Ellis, Surfactant-Semiconductor Interfaces: Perturbation of the Photoluminescence of Bulk Cadmium Selenide by Adsorption of Tri-*n*-octylphosphine Oxide as a Probe of Solution Aggregation with Relevance to Nanocrystal Stabilization, *J. Am. Chem. Soc.* **1998**, *120*, 10 970–10 975.
107. J. Krüger, U. Bach, M. Grätzel, Modification of TiO_2 Heterojunctions with Benzoic Acid Derivatives in Hybrid Molecular Solid-State Devices, *Adv. Mater.* **2000**, *12*, 447–451.
108. F. Nuesch et al., Derivatized Electrodes in the Construction of Organic Light Emitting Diodes, *Adv. Mater.* **1997**, *9*, 222–225.
109. L. Zuppiroli et al., Self-Assembled Monolayers as Interfaces for Organic Optoelectronic Devices, *Eur. Phys. J. B* **1999**, *11*, 505–512.
110. F. Nuesch et al., Chemical Potential Shifts at Organic Device Electrodes Induced by Grafted Monolayers, *Chem. Phys. Lett.* **1998**, *288*, 861–867.
111. D. Gal et al., Engineering the Interface Energetics of Solar Cells by Grafting Molecular Properties onto Semiconductors, *Proc. Indian Acad. Sci.* **1997**, *109*, 487–496.
112. A. Vilan et al., Real-Time Electronic Monitoring of Adsorption Kinetics: Evidence for Two-site Adsorption Mechanism of Dicarboxylic Acids on GaAs(100), *J. Phys. Chem. B* **1998**, *102*, 3307–3309.
113. Y. Kanemitsu, Slow Decay Dynamics of Visible Luminescence in Porous Silicon: Hopping of Carriers Confined on a Shell Region in Nanometer-size Si Crystallities, *Phys. Rev.* **1993**, *B48*, 12 357.
114. J. K. Lorenz, T. F. Kuech, A. B. Ellis, Cadmium Selenide Photoluminescence as a Probe for the Surface Adsorption of Dialkyl Chalcogenides, *Langmuir* **1998**, *14*, 1680–1683, and reference therein.
115. W. J. Dollard, M. L. Shumaker, D. H. Waldeck, Time-Resolved Studies of Charge Carrier Relaxation in Chemically Modified Semiconductor Electrodes: *n*-CdSe/Silane Interfaces, *J. Phys. Chem.* **1993**, *97*, 4141–4148.
116. A. L. McClellan, *Tables of Experimental Dipole Moments*, Freeman, San Francisco, 1963.
117. J. E. Huheey, *Inorganic Chemistry*, 3rd edition, Harper & Row, New-York, 1983, pp. 312–325.
118. G. L. Miessler, D. L. Tarr, *Inorganic Chemistry*, Prentice Hall, Englewood Cliffs, 1991.
119. R. Hoffman, *Solids and Surfaces: A Chemist's View of Bonding in Extended Structures*, VCH Publishers, New York, 1988, pp. 68–71.
120. T. Vondrak, C. J. Cramer, X.-Y. Zhu, The Nature of Electronic Contact in Self-Assembled Monolayers for Molecular Electronics: Evidence for Strong Coupling, *J. Phys. Chem. B* **1999**, *103*, 8915–8919.
121. M. C. Petty, G. G. Roberts, CdTe/Langmuir-Film MIS Structures, *Electron. Lett.* **1979**, *15*(12), 335–336.
122. I. M. Dharmadasa, G. G. Roberts, M. C. Petty, Cadmium Telluride/Langmuir Film

Photovoltaic Structures, *Electron. Lett.* **1980**, *16*(6), 201–202.
123. I. H. Campbell et al., Controlling Charge Injection in Organic Electronic Devices Using Self-Assembled Monolayers, *Appl. Phys. Lett.* **1997**, *71*, 3528–3530.
124. H. Sirringhaus, N. Tessler, R. H. Friend, Integrated Optoelectronic Devices Based on Conjugated Polymers, *Science* **1998**, *280*, 1741–1744.
125. L. Kronik et al., Surface States and Photovoltaic Effects in CdSe Quantum Dot Films, *J. Electrochem. Soc.* **1998**, *145*, 1748–1755.
126. A. Hagfeldt, M. Grätzel, Molecular Photovoltaics, *Acc. Chem. Res.* **2000**, *33*, 269–277.
127. D. Cahen et al., Nature of Photovoltaic Action in Dye-Sensitized Solar Cells, *J. Phys. Chem. B* **2000**, *104*, 2053–2059.
128. Y. Selzer, D. Cahen, Fine-tuning of Au/SiO_2/Si diodes by varying interfacial dipoles using molecular monolayers, *Adv. Mater.* **2001**, *13*, 508–511.
129. D. G. Wu et al., Direct Detection of Low Concentration NO in Physiological Solutions by a New GaAs-Based Sensor, *Chemistry Eur. J.* **2001**, *7*, 1743–1749.
130. D. G. Wu et al., Tuning of Au/*n*-GaAs Diodes with Highly Conjugated Molecules, *J. Phys. Chem. B* **2001**, in press.
131. G. Ashkenasy et al., Assemblies of "Hinged" Iron-Porphyrins as Potential Oxygen Sensors, *J. Amer. Chem. Soc.* **2000**, *122*, 1116–1122.
132. A. Ivanisevic et al., Linker-enhanced Binding of Metalloporphyrins to CdSe and Implications for O_2 detection, *Langmuir* **2000**, *16*, 7852–7858.
133. T. Vondrak et al., Interfacial Electronic Structure in Thiolate Self-assembled Monolayers; Implications for Molecular Electronics, *J. Amer. Chem. Soc.* **2000**, *122*, 4700–4707.
134. M. Carrara, F. Nüesch, L. Zuppiroli, Carboxylic acid anchoring groups for the construction of self-assembled monolayers on organic device electrodes, *Synth. Metals* **2001**, *121*, 1633–1634.

2.4 Capacitance, Luminescence, and Related Optical Techniques

Yoshihiro Nakato
Osaka University, Osaka, Japan

2.4.1 Capacitance

2.4.1.1 Space Charge Layer at a Semiconductor–Electrolyte Interface

An electrical double layer is formed at a semiconductor–electrolyte interface, similar to a metal–electrolyte interface [1–3]. The difference, however, is in the charge distribution between the two interfaces. At a metal–electrolyte interface, the charge on the metal side is localized just at the metal surface, whereas, at a semiconductor–electrolyte interface, the charge on the semiconductor side is distributed deep in the interior of the semiconductor, forming a wide space charge layer. In concentrated electrolyte solutions (~0.5 M and higher), the charge on the electrolyte side is localized at the (outer) Helmholtz layer for interfaces with either metals or semiconductors.

Figure 1 schematically illustrates (1) the charge distribution, (2) the charge-density distribution, (3) the electric field distribution, (4) the potential distribution, and (5) the band bending at the semiconductor–electrolyte interface, under an ideal condition that no surface charge nor surface dipole is present at the semiconductor. The semiconductor is assumed to be n-type, but a similar diagram can be drawn for a p-type semiconductor with the sign of electric charges inverted. The band bending in the semiconductor is simply caused by the potential distribution in it.

The n-type semiconductor has an electron donor in the crystal, doped as an impurity, such as phosphor atoms in n-Si, whereas the p-type semiconductor has an electron acceptor such as boron atoms in p-Si. The space charge in the semiconductor is composed of an ionized electron donor or acceptor. The space charge layer of a finite thickness is formed in the semiconductor when the density of the electron donor, N_D, or the density of the electron acceptor, N_A, is low, in a range from 10^{14} to 10^{18} cm^{-3}. The principle of electrical neutrality at the electrical double layer leads to an extended distribution of charges deep toward the interior of the semiconductor. The situation is rather similar to the formation of an extended charge distribution, called a Gouy layer, in the electrolyte in cases where the electrolyte concentration is low. The difference is that the space charges in the Gouy layer are composed of electrolyte ions and are mobile, whereas those in the semiconductor are composed of ionized donors or acceptors and therefore are spatially fixed and immobile.

2.4.1.2 Differential Capacitance of a Semiconductor–Electrolyte Interface

The potential distribution in the space charge layer of a semiconductor can be given by solving the Poisson equation for a given charge distribution [1–3]. For a semiconductor–electrolyte interface such as the one shown in Fig. 1, if the density of an electron donor, N_D, is constant throughout the semiconductor, the potential, $\phi(x)$, at a distance, x, from the surface is given as follows [1, 2]:

$$\phi(x) = \frac{qN_D}{\varepsilon_0\varepsilon_s}\left(Wx - \frac{1}{2}x^2\right) + \phi(0) \quad (0 \le x \le W) \qquad (1)$$

where q is the elementary charge, ε_0 the permittivity of vacuum, ε_s the dielectric constant of semiconductor, $\phi(0)$ the potential at the surface ($x = 0$), and W the

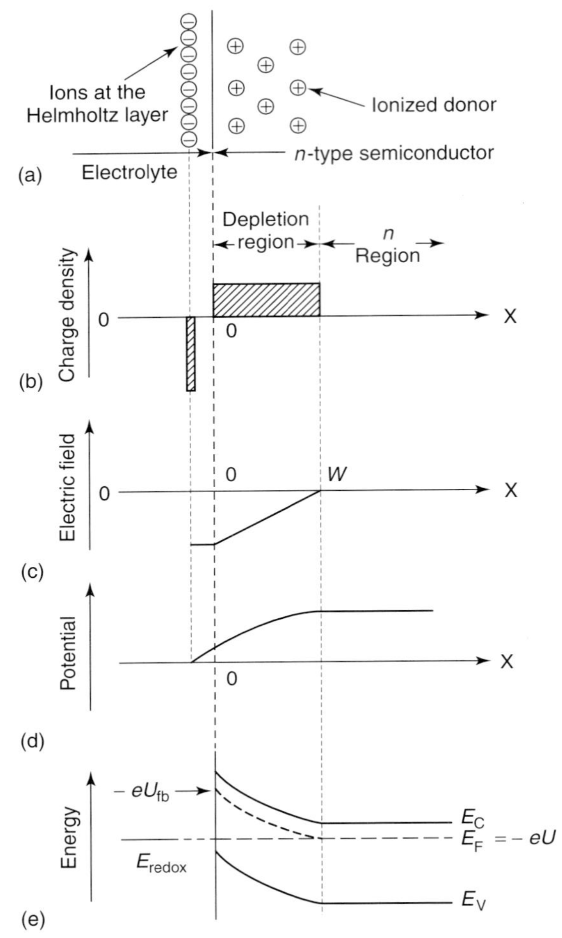

Fig. 1 Schematic illustration of (a) charge distribution, (b) charge-density distribution, (c) electric field distribution, (d) potential distribution, and (e) band bending at an *n*-type semiconductor–electrolyte interface under an ideal condition that neither surface charge nor surface dipole is present at the semiconductor.

width of the space charge layer. The $\phi(0)$ and W are given by

$$\phi(0) = \frac{qN_D W}{\varepsilon_0 \varepsilon_S}\delta \tag{2}$$

$$W = \left[\frac{2\varepsilon_0\varepsilon_S}{qN_D}\left(U - U_{fb} - \frac{kT}{q}\right)\right]^{1/2} \tag{3}$$

where δ is the thickness of the (outer) Helmholtz layer, U is the electrode potential, and U_{fb} the flat band potential. The space charge, Q_S, per unit area is given by

$$Q_S = qN_D W = [(2q\varepsilon_0\varepsilon_S N_D) \times (U - U_{fb} - kT/q)]^{1/2} \tag{4}$$

Thus, the differential capacitance of the space charge layer, C_d, per unit area is given as follows:

$$C_d = \frac{\partial Q_S}{\partial U} = \left[\frac{2}{q\varepsilon_0\varepsilon_S N_D} \times \left(U - U_{fb} - \frac{kT}{q}\right)\right]^{-1/2} \tag{5}$$

This equation can be rewritten as follows:

$$\frac{1}{C_d^2} = \frac{2}{q\varepsilon_0\varepsilon_S N_D}\left(U - U_{fb} - \frac{kT}{q}\right) \quad (6)$$

The plot of $1/C_d^2$ measured against U is called the *Mott–Schottky plot,* which can be used to determine U_{fb} and N_D (or N_A), as discussed later.

It has been assumed thus far that no surface charge nor surface dipole is present at the semiconductor. In general cases, both surface charges and surface dipoles are present in the semiconductor owing to adsorption equilibria of various ions between the electrolyte and the semiconductor surface as well as formation of polar bonds at the semiconductor surface. Such surface charges and surface dipoles cause a shift of the semiconductor band positions at the surface, as shown schematically in Fig. 2. The shift is expressed as a change in $\phi(0)$ or U_{fb} in the foregoing equations. However, the preceding equations themselves can be applied to such cases with the changed $\phi(0)$ or U_{fb}.

2.4.1.3 **Measurement of Differential Capacitance**

The capacitance of the semiconductor–electrolyte interface can be measured by use of a semiconductor electrode, in which the front side of the semiconductor is in contact with the electrolyte and the rear side is electrically connected with a metallic leading wire via an ohmic contact.

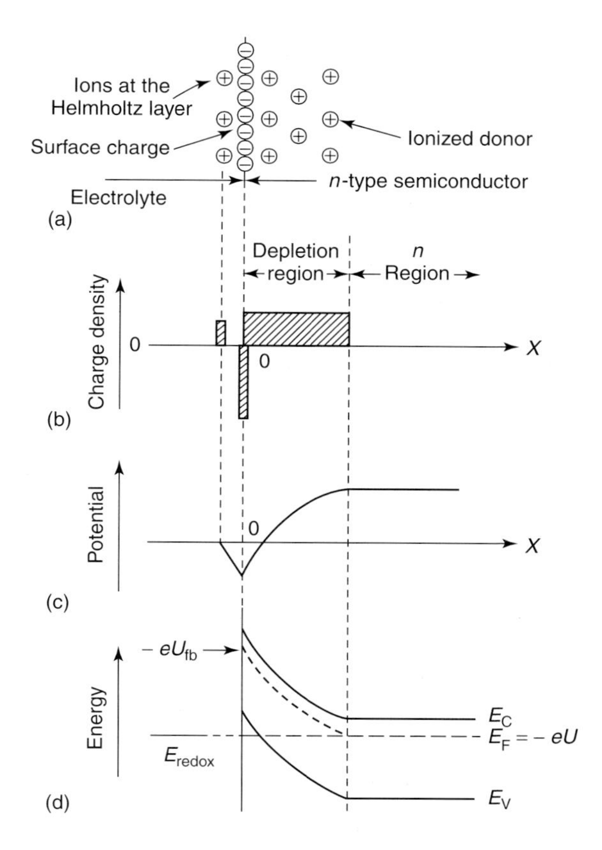

Fig. 2 Schematic illustration of (a) charge distribution, (b) charge-density distribution, (c) potential distribution, and (d) band bending at the semiconductor–electrolyte interface under a condition that negative surface charges are present.

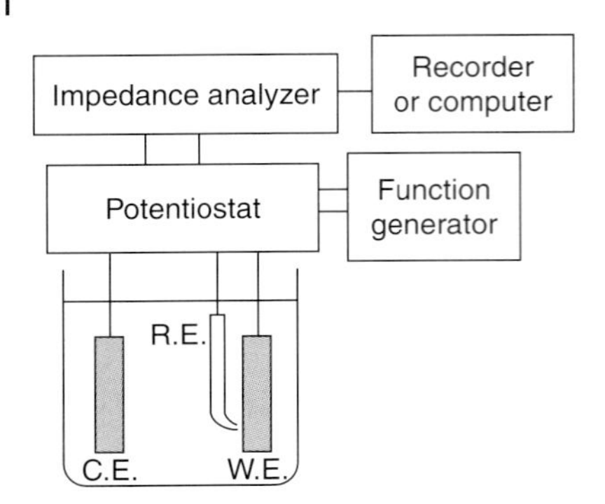

Fig. 3 Experimental setup for measurements of capacitance at the semiconductor–electrolyte interface. W.E.: working electrode (semiconductor electrode), C.E.: counter electrode, and R.E.: reference electrode.

The experimental setup for capacitance measurements is schematically shown in Fig. 3. The electrode potential U is regulated with a potentiostat. The differential capacitance is measured by superimposing an AC voltage with a small amplitude of about 10 mV and a frequency of a few Hz to 1 MHz on the electrode potential. One can use a commercial impedance analyzer to measure the differential capacitance, together with a personal computer to analyze obtained data automatically.

Recently, scanning capacitance microscopy (SCM) using a fine tip has been developed to investigate two-dimensional dopant profiling of semiconductor surfaces [4].

2.4.1.4 **Mott–Schottky Plots and Flat Band Potentials**

Equation 6 indicates that a plot of $1/C_{\mathrm{d}}^2$ against U gives a straight line with a slope of $(2/q\varepsilon_0\varepsilon_{\mathrm{s}}N_{\mathrm{D}})$, which is termed the Mott–Schottky plot, as mentioned earlier. The extrapolation of the straight line to $1/C_{\mathrm{d}}^2 = 0$ gives $(U_{\mathrm{fb}} + kT/q)$. Therefore, the plot can be used to determine the flat band potential U_{fb}. The donor density N_{D} (or the acceptor density N_{A}) can also be determined from the slopes of the plots. Figure 4 shows examples of Mott–Schottky plots, obtained for n-Si(111) and n-Si(100) electrodes in 7.1 M hydrogen iodide [5]. It should be noted that the slope of the straight line depends not only on N_{D} (or N_{A}) and ε_{s} but also on the true surface area (or surface roughness) of the semiconductor electrode.

The U_{fb} is one of the most important quantities for semiconductor electrodes because it determines the band edge positions at the semiconductor–electrolyte interface, which in turn, determine the energies of conduction-band electrons and valence-band holes reacting with the electrolyte solution. It is known that U_{fb} for most semiconductors, such as n- and p-GaAs, GaP, InP, n-ZnO, n-TiO_2, and n-SnO_2, in aqueous electrolytes is solely determined by the solution pH and shifts in proportion to pH with a ratio of -0.059 V/pH [1, 2]. This is explained by an adsorption equilibrium for H^+ or OH^- at the semiconductor–electrolyte interface, for example,

$$\mathrm{S_s\text{-}OH} + \mathrm{H_{aq}}^+ \rightleftharpoons \mathrm{S_s\text{-}OH_2}^+ \qquad (7)$$

where S_s-OH refers to surface OH group at the semiconductor.

The U_{fb} for n- and p-Si [6] and metal calcogenide semiconductors such as n-CdS, n-CdSe, and CdTe [2, 7] does not obey the foregoing rule, remaining nearly constant in a range of pH lower than about 6 for Si and about 10 for n-CdS. This is most probably because the semiconductor

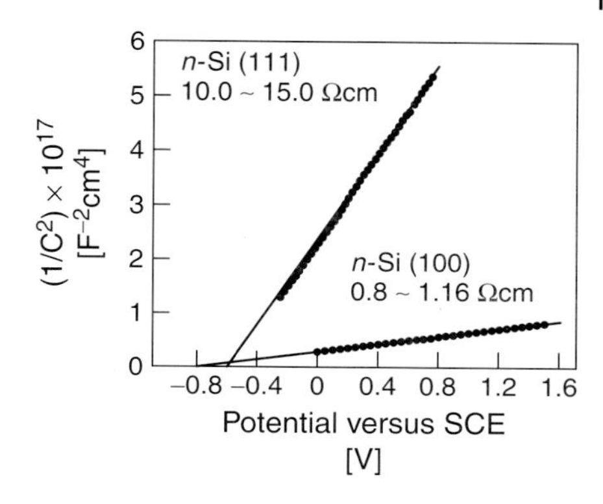

Fig. 4 Examples of Mott–Schottky plots obtained for n-Si(111) and n-Si(100) electrodes in aqueous 7.1 M HI.

surface has no OH group in this pH range and no adsorption equilibrium for H^+ or OH^- is attained. For metal calcogenide semiconductors such as n-CdS, the U_{fb} shifts by adsorption of HS^- and Cd^{2+} ions. It is also known that U_{fb} for some semiconductor electrodes shifts by a change in the surface termination bond [8, 9], as well as electrode illumination [10, 11] and the presence of a redox couple [10, 12]. The U_{fb} in nonaqueous electrolytes has been reported for some semiconductor electrodes [13, 14].

2.4.2 Luminescence from Semiconductor Electrodes

2.4.2.1 Photoluminescence and Electroluminescence

Illumination of a semiconductor electrode generates excited electrons in the conduction band and holes in the valence band. Some of them recombine with each other radiatively, resulting in emission of luminescence called photoluminescence (PL). The radiative recombination may occur directly between the conduction and valence bands (inter-band transition) or via certain impurity or defect levels within the band gap at which either electrons or holes, or both are trapped, as schematically illustrated in Fig. 5(a). Also, the radiative recombination may occur either in the semiconductor bulk or at the semiconductor surface (Fig. 5a).

On the other hand, EL is emitted when an n-type semiconductor electrode, for example, is negatively biased in an electrolyte solution containing a strong oxidant. Under this condition, holes are injected into the valence band by the oxidant, some of which recombine with electrons in the conduction band existing as the majority carrier in the n-type semiconductor, resulting in emission of luminescence called EL, as illustrated in Fig. 5(b). Similarly, EL is emitted when a p-type semiconductor electrode is positively biased in an electrolyte solution containing a strong reductant because some of the electrons, injected into the conduction band by the reductant, recombine with holes existing as the majority carrier in the p-type semiconductor. In both cases, EL may be emitted either via direct recombination between the conduction and valence bands or via an impurity level(s), and either via recombination in the semiconductor bulk or at the semiconductor surface. The situation is quite the same as that in PL emission (Fig. 5).

Figure 6 shows examples of PL and EL spectra from some semiconductor electrodes [15]. It is to be noted, however, that such spectra strongly depend on

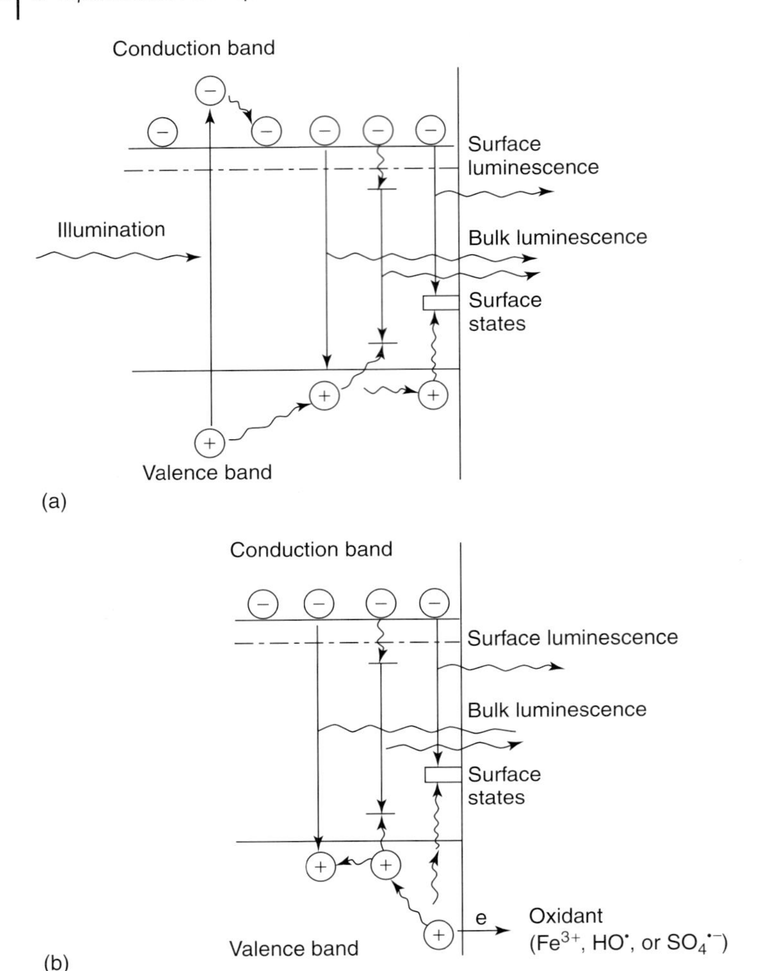

Fig. 5 Mechanisms of emission of (a) PL and (b) electroluminescence (EL) from a semiconductor electrode in an electrolyte solution.

semiconductor materials used (that is, the kind and the amount of impurities and/or defects) and the kind of surface treatments made.

2.4.2.2 Measurements of Luminescence from Semiconductor Electrodes

2.4.2.2.1 DC Methods The luminescence (PL or EL) from a semiconductor electrode can be measured by a simple conventional DC method as shown in Fig. 7, although no illumination equipment is necessary for EL measurements. The spectra in Fig. 6 are measured by this conventional DC method.

In the DC method, the luminescence from a semiconductor electrode is usually collected in a normal direction by use of a lens and led to a slit of a monochromator, followed by detection with an appropriate photomultiplier. The electrode potential of

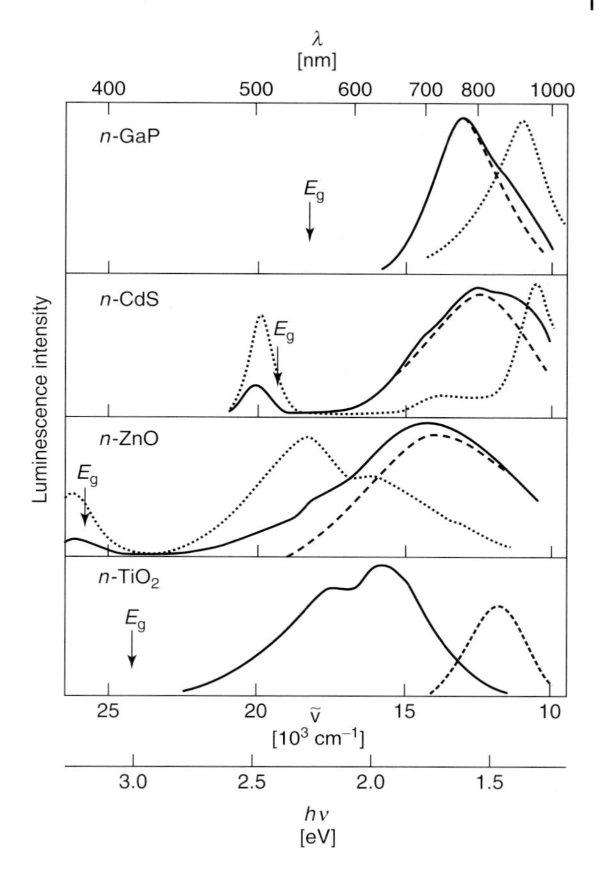

Fig. 6 Examples of PL and EL spectra of semiconductor electrodes. E_g: the band gap energy, $h\nu$: photon energy, λ: wavelength.

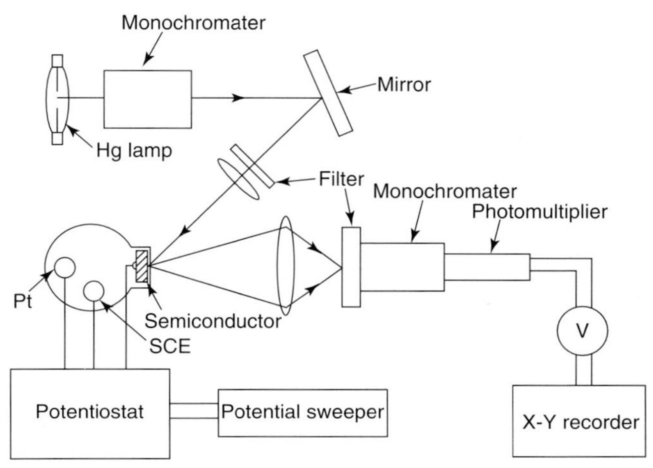

Fig. 7 Experimental setup for simple DC measurements of PL and EL from a semiconductor electrode.

the semiconductor electrode is regulated with a potentiostat. By this way, one can measure PL or EL spectra at constant electrode potentials, and the luminescence intensity versus potential curves as compared with simultaneously measured current versus potential curves. The use of a personal computer is convenient for data processing and storage.

For PL measurements, the semiconductor electrode is illuminated by monochromatic light, using a high-pressure mercury lamp, a nitrogen laser, an argon laser, and so forth, as the light source. Much care should be taken to minimize a contribution of scattered stray light from the illumination light to luminescence spectra. The detection limit of weak PL is in most cases determined by how much the intensity of the scattered stray light is reduced. High-intensity illumination with high-quality monochromatic light by use of a laser is effective to detect weak PL, but very high intensity illumination generates very high density electrons and holes at the semiconductor surface, which may alter the essential mechanism of carrier dynamics in the semiconductor. Very high intensity illumination may also generate much heat (temperature increase) at the semiconductor surface.

For EL measurements, the detection limit of weak EL is mainly determined by the detection sensitivity of an experimental apparatus used because no illumination is needed in this case. One can use various high-sensitivity detection methods such as a lock-in amplifier [16, 17], boxcar integration [17], and photon counting [17]. Care should, however, be taken when the time-averaging method is used because the luminescence intensity often changes with time even at a constant electrode potential, probably because the surface chemical structure of the semiconductor electrode may change. Recently, space-resolved two-dimensional detection of EL has been developed by use of a CCD camera [18].

2.4.2.2.2 **Pulsed Techniques** Time-resolved PL (and EL) measurements give much information on carrier dynamics in the semiconductor bulk and at the semiconductor surface. Figure 8 schematically illustrates a convenient method, called a time-correlated single photon counting method or a picosecond single photon timing method [19, 20, 21], by which weak PL can be measured at a high sensitivity with a time resolution on the order of picoseconds. A picosecond or femtosecond pulse laser with a pulse width of a few ps is used for excitation of the sample (semiconductor electrode). The luminescence from the sample is usually detected with a microchannel plate photomultiplier (MCP-PMT) having a small transient time spread. A part of the excitation laser pulse is split from the main beam and detected with a photodiode for use as the starting pulse. The luminescence signal, detected with the MCP-PMT, is used as the stopping pulse. This means that the sample was excited at the time of the starting pulse and a luminescent photon was emitted at the time of the stopping pulse. The starting and stopping pulses are led to a time-to-amplifier converter (TAC) to measure the time separation between them. Repeated measurements give a histogram of the number of luminescent photons versus time, which should agree with a luminescence decay curve.

PL measurements with a higher time resolution of several tens of femtoseconds can be achieved by means of femtosecond up-conversion spectroscopy [22]. The experimental setup is illustrated in Fig. 9. In this method, the luminescence intensity is measured by making use

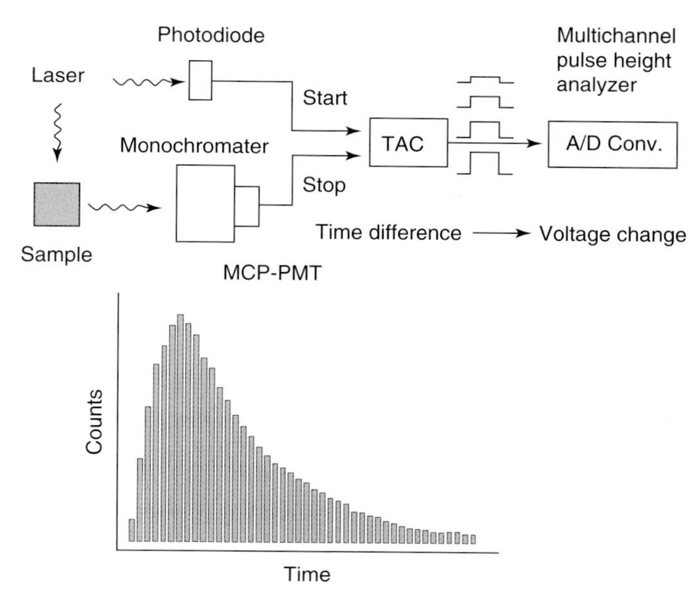

Fig. 8 Experimental setup for a time-correlated single-photon counting method or picosecond single-photon timing method, used for time-resolved luminescence measurements in a range of picoseconds.

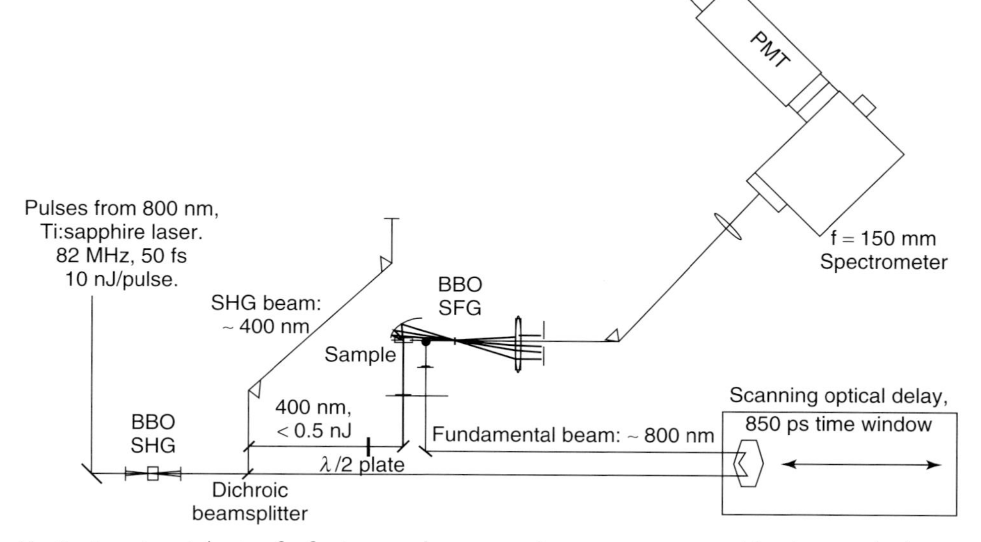

Fig. 9 Experimental setup for femtosecond up-conversion spectroscopy used for time-resolved luminescence measurements in a range of several tens femtoseconds.

of "gating" by another laser pulse. A femtosecond laser, such as a Ti-sapphire laser, is normally used as the light source for excitation. The emitted luminescence with a frequency ω_1 is mixed with a gating laser pulse having a frequency ω_2 in a nonlinear crystal such as BBO (β-barium borate) to give an up-conversion light pulse with a frequency $\omega_3 = \omega_1 + \omega_2$, which is measured with a photomultiplier. The up-conversion light pulse with ω_3 is generated only when the emitted luminescence and the gating laser pulse come to BBO simultaneously. Also, the intensity of the generated up-conversion light is in proportion to the product of the intensity of the emitted luminescence and that of the gating laser pulse. Thus, the luminescence decay curve in relative intensity is measured by giving various delays to the gating laser pulse.

Figure 10 shows examples of luminescence decays obtained by the femtosecond up-conversion spectroscopy [23]. An aqueous or methanol suspension of nanocrystalline TiO_2 particles, on which a cumarin dye is adsorbed, is used as the sample. The luminescence is emitted from the adsorbed dye, and its decay represents the rate of electron transfer from excited adsorbed dye to the conduction band of TiO_2.

2.4.2.3 **Bulk and Surface Luminescence**

The radiative recombination of electrons and holes can occur either in the semiconductor bulk (bulk PL or EL) or at the semiconductor surface (surface PL or EL), as mentioned earlier. In semiconductor electrochemistry, it is important to distinguish whether observed PL or EL is emitted from the semiconductor bulk or surface because surface PL and EL give more information on surface structures and processes at semiconductor electrodes, such as surface states, surface reaction intermediates, and so forth.

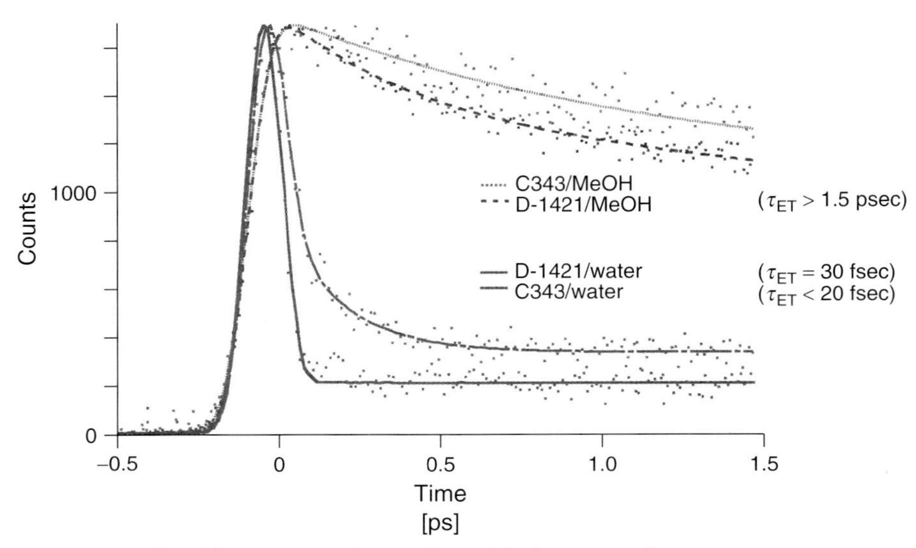

Fig. 10 Examples of luminescence decays obtained by femtosecond up-conversion spectroscopy. The luminescence is emitted from a cumarin dye adsorbed on nanocrystalline TiO_2 particles suspended in water or methanol (MeOH). C343: cumarin 343 and D-1421: 7-diethylaminocumarin-3-carboxylic acid.

Naturally, the intensity of surface PL is strongly influenced by the chemical composition of the electrolyte, in particular, by whether oxidants or reductants are included in it because electrons or holes trapped at surface states can easily react with oxidants or reductants in the electrolyte, resulting in luminescence quenching. The intensity of surface PL is also strongly dependent on surface pretreatments. However, it is to be noted that bulk PL is also influenced by these factors because photo-generated electrons and/or holes in the semiconductor bulk can diffuse to the surface and react with an oxidant or reductant in the electrolyte and also surface species (surface states) before they recombine. This leads to a decrease in the densities of electrons and holes in the semiconductor bulk and hence to a decrease in the intensity of bulk PL.

One effective way to distinguish experimentally the bulk and surface luminescence is to measure the electrode-potential dependence of the luminescence intensity. The surface luminescence is observed only under a forward bias with weak band bending or in the presence of an accumulation layer, namely, the surface luminescence is observed only when the surface densities of both electrons and holes are sufficiently high. On the other hand, the bulk luminescence is observed even in the presence of large band bending under which the surface density of the majority carrier is very low, if the penetration depth of incident illumination light is considerably larger than the width of the space charge layer.

The arguments made thus far show that both bulk and surface PL can be used to investigate surface processes at semiconductor electrodes. The measurements of bulk PL indicate that most of semiconductors emit bulk PL. They can be used to detect some surface processes and reactions [24]. The measurement of surface PL, if measured, is very powerful because its behavior gives direct information

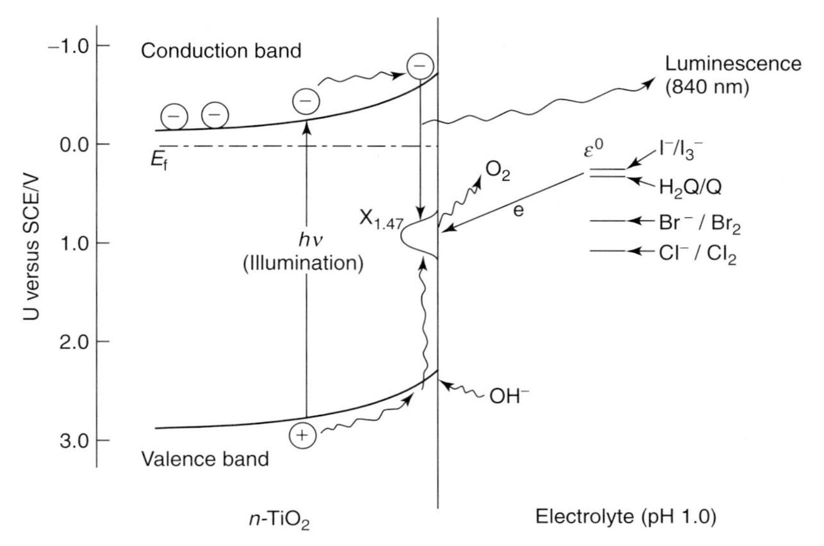

Fig. 11 Energy band diagram for explaining a PL band at 840 nm, emitted via a surface state connected with a surface reaction intermediate ($X_{1.47}$) of water photooxidation on n-TiO_2.

on surface structures and processes. The most important is to measure surface PL that is emitted via surface states connected with surface intermediates of electrode reactions. An example of such luminescence is a luminescence band peaked at 840 nm (Fig. 6), observed for a rutile-type n-TiO_2 electrode, which was activated beforehand by photoetching in aqueous H_2SO_4 under anodic bias. In situ PL measurements, combined with in situ photocurrent measurements, can lead to detailed studies on molecular mechanisms of electrode reactions. It is concluded [25] that the 840-nm band for rutile n-TiO_2 is emitted via a transition of an electron in the conduction band to a vacant level of a surface reaction intermediate ($X_{1.47}$, probably vacant 2p-level of surface Ti-O· radical) of water photooxidation, as shown in Fig. 11.

2.4.3
Other Optical Techniques

2.4.3.1 Time-resolved Laser Spectroscopy

2.4.3.1.1 Transient Absorption Spectroscopy Measurements of transient absorption spectra are much more difficult than the measurements of luminescence (PL) described thus far, especially for solid samples. However, time-resolved transient absorption spectroscopy has been used to study carrier dynamics in the

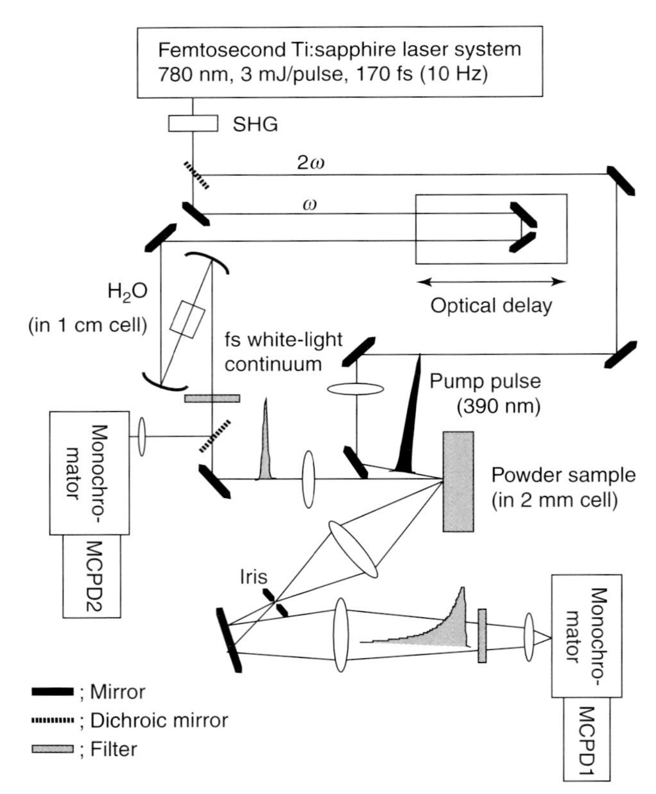

Fig. 12 Experimental setup for femtosecond (pump probe) transient absorption spectroscopy.

semiconductor bulk and at the semiconductor surface. Figure 12 shows an example of experimental setup for femtosecond (pump probe) transient absorption spectroscopy [26]. The principle is to generate an excited state of the sample (semiconductor electrode) by a laser pulse (pump pulse) and then measure an absorption spectrum of the excited state by another laser pulse (probe pulse), with varied delays in time from the pump pulse. A femtosecond laser, such as a Ti-sapphire laser, is used as the light source. In Fig. 12, a laser pulse with a frequency ω from the light source is converted with BBO to a light pulse with 2ω for use as the pump pulse. The original pulse with ω is separated with a mirror and used for the probe pulse. The latter pulse is led through an optical delay circuit to an H_2O cell to be converted into white light continuum for measurements of absorption spectra. The spectral intensity of the probe pulse is measured with a multichannel photodiode (MCPD).

Figure 13 shows an example of transient absorption spectra observed by direct excitation of nanocrystalline TiO_2 particles suspended in vacuum [27]. The broad absorption band in the red to near infrared region is assigned to light absorption of photo-generated electrons in the conduction band of TiO_2 particles. The rate of photoelectron transfer from adsorbed dyes to the conduction band of TiO_2 in vacuum and solution is also determined by use of measurements of transient absorption of injected electrons. It is reported that the photoelectron transfer occurs in a time range of 150–25 femtoseconds for chemically bound, strongly interacting dyes [28–31].

2.4.3.1.2 **Transient Grating Spectroscopy**
Transient grating spectroscopy is relatively easily handled compared with the transient absorption spectroscopy, and is often used to study carrier dynamics at semiconductor electrodes [32]. Figure 14 schematically shows the principle of transient grating spectroscopy. A femtosecond laser pulse for sample excitation is split into two beams, which are crossed again at the semiconductor surface to produce an optical striped interference pattern. The interference pattern produces a striped pattern of the densities of photo-generated electrons and holes near the semiconductor surface. The latter striped pattern gives rise to a striped pattern of optical refractive index near the semiconductor surface, which is monitored by measuring a diffraction pattern of a second probe laser

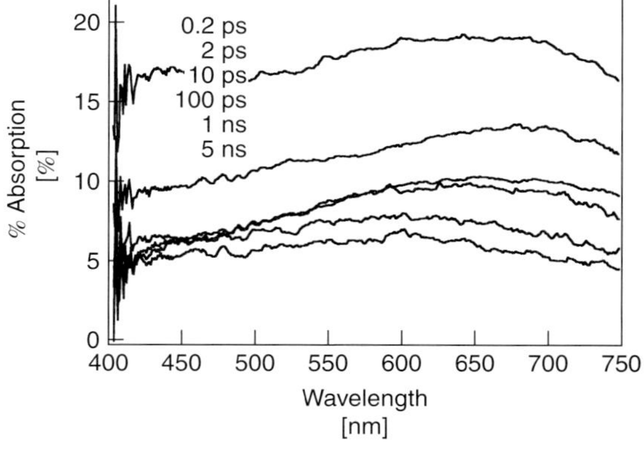

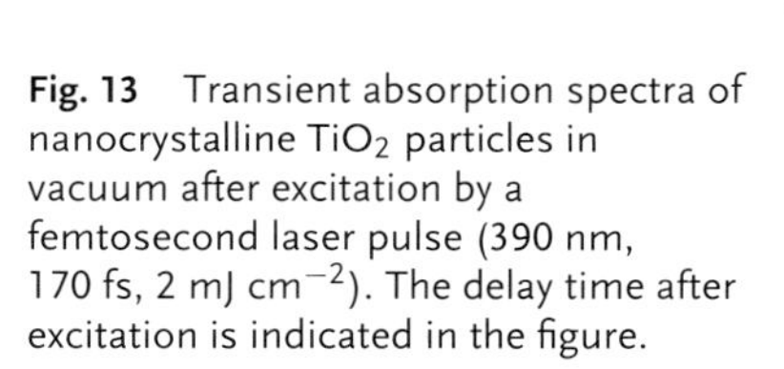
Fig. 13 Transient absorption spectra of nanocrystalline TiO_2 particles in vacuum after excitation by a femtosecond laser pulse (390 nm, 170 fs, 2 mJ cm^{-2}). The delay time after excitation is indicated in the figure.

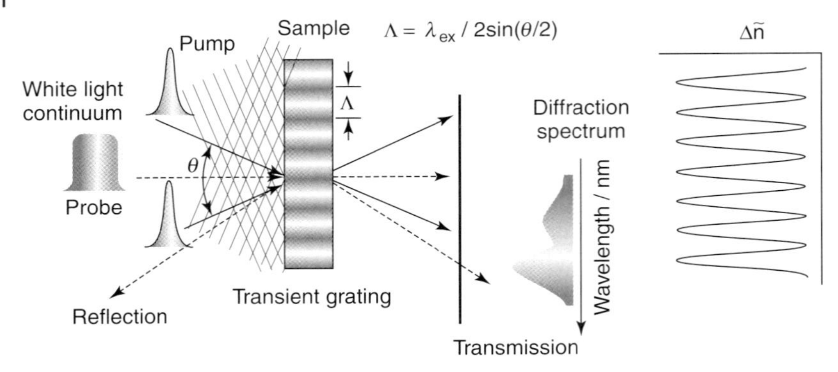

Fig. 14 A schematic view of a sample region in transient grating spectroscopy to explain its principle.

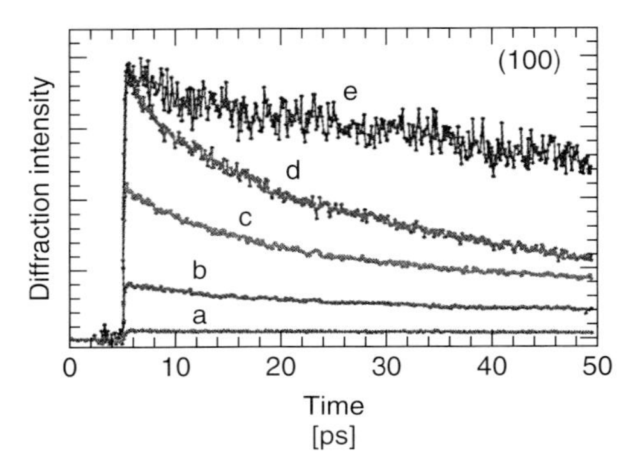

Fig. 15 Examples of transient grating signal decays, observed for an n-TiO_2 (100) electrode by excitation at 360 nm and probing at 670 nm. The excitation pulse intensity in a μW unit is (a) 1.4, (b) 4.1, (c) 6.1, (d) 9.7, and (e) 42.

pulse (white light continuum). The carrier dynamics near the semiconductor surface is observed as rise and decay curves for the diffraction pattern of the second laser probe pulse.

Figure 15 shows examples of decays of transient grating signal (the intensity of the diffraction pattern) observed for an n-TiO_2 (100) electrode by excitation at 360 nm and probing at 670 nm [33]. The decays are related with the rate of electron-hole recombination near the n-TiO_2 surface.

2.4.3.2 **In Situ Spectroscopic Investigation of Semiconductor Surfaces**

Various spectroscopic techniques have been used for in situ investigations of surface structures at semiconductor electrodes. In situ multiple internal reflectance Fourier Transform Infrared (FTIR) spectroscopy is widely used to investigate surface termination bonds and their reactions. Figure 16 shows a semiconductor (Si) wafer and an electrochemical cell for in situ multiple internal reflectance spectroscopy. By this method, various hydrogen termination bonds (Si−H, SiH_2, and SiH_3) at terraces and steps for hydrogen fluoride–etched or ammonium fluoride–etched Si (111) and (100) are clearly detected with good spectral resolution [34, 35].

In situ investigations of semiconductor surfaces are also done by means

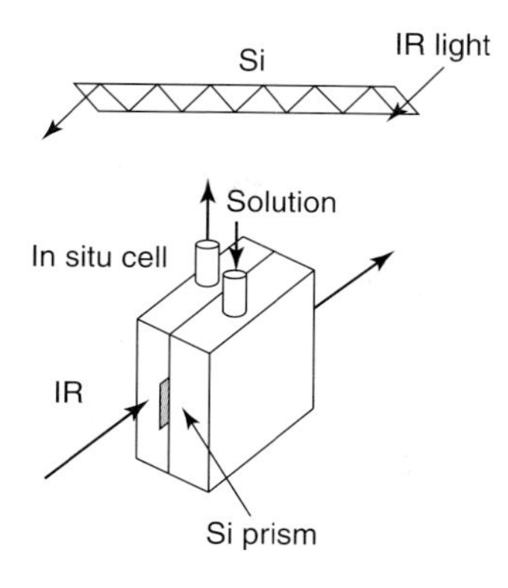

Fig. 16 A silicon wafer and an electrochemical cell for in situ multiple internal reflectance FTIR spectroscopy.

of second harmonic generation (SHG) spectroscopy, ultraviolet and infrared light mixing spectroscopy, (surface enhanced) laser Raman spectroscopy, X-ray diffraction (XRD) method, electron paramagnetic resonance (EPR) spectroscopy, and so forth. Space-resolved investigations are also done by means of surface near-field optical microscopy (SNOM), photoconductive atomic force microscopy, and so forth.

2.4.3.3 **Other Miscellaneous Techniques**

The electrolyte electroreflectance (EER) method is successfully used [36, 37] to determine the flat band potential (U_{fb}). In this method, the optical reflectance at the semiconductor electrode surface is measured under modulation of the electrode potential by superimposition of a small AC voltage. The modulation of the electrode potential causes modulation of the density of majority carriers near the semiconductor surface, which in turn, causes modulation in light reflectance.

In situ microwave photoconductivity measurements, combined with photocurrent measurements, are also successfully used [38] to investigate dynamics of photo-generated minority carriers and interfacial kinetics at semiconductor electrodes. The microwave signal is closely related with the density of free carriers, and modulated illumination enables us to investigate minority carrier dynamics.

Surface photovoltage (SPV) measurements have been used to investigate electronic structures at the semiconductor surfaces for semiconductor–vacuum and semiconductor–gas interfaces. This method is applied to semiconductor–liquid interfaces in case of insulating liquids [39].

2.4.4 **Summary**

Various methods to measure the capacitance and luminescence (PL and EL) at the semiconductor–electrolyte interface have been reviewed together with related optical techniques to investigate surface structures, optical properties, and carrier dynamics at the interfaces. Recent rapid progress in laser spectroscopy has enabled us to investigate very fast interfacial processes with a high time resolution of a few ten femtoseconds. Moreover, rapid progress in scanning probe microscopy, combined with optical techniques, has enabled us to investigate interfacial structures and processes with a high space resolution of an atomic (subnanometer) scale. Progress in these fields continues and it is thus highly probable that other new powerful methods will appear in the near future.

References

1. S. R. Morrison in *Electrochemistry at Semiconductor and Oxidized Metal Electrodes*, Plenum Press, New York, 1980.
2. Y. V. Pleskow, Y. Y. Gurevich, in *Semiconductor Photoelectrochemistry* (Translated by P. N. Bartlett), Consultants Bureau, New York, 1986.
3. S. M. Sze in *Physics of Semiconductor Devices*, 2nd edition, John Wiley & Sons, New York, 1981.
4. V. V. Zavyalov, J. S. McMurray, C. C. Williams, *Rev. Sci. Instrum.* **1999**, *70*, 158–164.
5. Unpublished result in a laboratory of Prof. Y. Nakato, Osaka University.
6. Y. Nakato, T. Ueda, Y. Egi et al., *J. Electrochem. Soc.* **1987**, *134*, 353–358.
7. H. Minoura, T. Watanabe, T. Oki et al., *Jpn. J. Appl. Phys.* **1977**, *16*, 865.
8. J. N. Chazalviel, *J. Electroanal. Chem.* **1987**, *233*, 37.
9. M. Fujitani, R. Hinogami, J. G. Jia et al., *Chem. Lett.* **1997**, 1041–1042.
10. A. J. Nozik, R. Memming, *J. Phys. Chem.* **1996**, *100*, 13061–13078.
11. A. J. McEvoy, M. Etman, R. Memming, *J. Electroanal. Chem.* **1985**, *190*, 225.
12. Y. Nakato, A. Tsumura, H. Tsubomura, *J. Electrochem. Soc.* **1981**, *128*, 1300–1304.
13. B. L. Wheeler, A. J. Bard, *J. Electrochem. Soc.* **1983**, *130*, 1680.
14. W. Jaegermann in *Modern Aspects of Electrochemistry* (Eds.: R. E. White, B. E. Conway, J. O'M. Bockris), Plenum, New York, 1996, pp. 136.
15. Y. Nakato, A. Tsumura, H. Tsubomura, *Chem. Phys. Lett.* **1982**, *85*, 387–390.
16. ftp://ftp.batnet.com/pub/wombats/srsys/ftp/lia.pdf
17. ftp://ftp.batnet.com/pub/wombats/srsys/ftp/sr.pdf
18. M. Oyama, M. Mitani, M. Washida et al., *J. Electroanal. Chem.* **1999**, *473*, 166–172.
19. D. V. O'Connor, D. Phillips in *Time-correlated Single Photon Counting*, Academic Press, New York, 1984.
20. I. Yamazaki, N. Tamai, H. Kume et al., *Rev. Sci. Instrum.* **1985**, *56*, 1187–1194.
21. N. Tamai, M. Ishikawa, N. Kitamura et al., *Chem. Phys. Lett.* **1991**, *184*, 398–403.
22. G. S. Beddard, T. Dout, G. Porter, *Chem. Phys.* **1981**, *61*, 17.
23. K. Murakoshi, S. Yanagida, M. Capel et al., *ACS Symp. Series 679 (Nanostructured Materials-Clusters, Thin Films, and Composites)*, Amer. Chem. Soc. 1997, Chap. 17, pp. 221–238.
24. K. Mecker, A. B. Ellis, *J. Phys. Chem. B* **1999**, *103*, 995–1001.
25. Y. Nakato, H. Akanuma, Y. Magari et al., *J. Phys. Chem. B* **1997**, *101*, 4934–4939.
26. T. Asahi, A. Furube, M. Ichikawa et al., *Rev. Sci. Instrum.* **1998**, *69*, 361.
27. A. Furube, T. Asahi, H. Masuhara, *J. Phys. Chem. B* **1999**, *103*, 3120.
28. Y. Tachibana, J. E. Moser, M. Graetzel et al., *J. Phys. Chem.* **1996**, *100*, 20056–20062.
29. T. Hannappel, B. Burfeindt, W. Storck et al., *J. Phys. Chem. B* **1997**, *101*, 6799–6802.
30. R. J. Ellingson, J. B. Asbury, S. Ferrere et al., *J. Phys. Chem. B* **1998**, *102*, 6455–6458.
31. J. B. Asbury, R. J. Ellingson, H. N. Ghosh et al., *J. Phys. Chem. B* **1999**, *103*, 3110–3119.
32. J. J. Kasinski, L. A. Comez-Jahn, K. J. Faran et al., *J. Chem. Phys.* **1989**, *90*, 1253–1269.
33. Unpublished result in a laboratory of Prof. N. Tamai, Kwansei-Gakuin University.
34. P. Jakob, Y. J. Chabal, *J. Chem. Phys.* **1991**, *95*, 2897–2909.
35. M. Niwano, T. Miura, N. Miyamoto, *J. Electrochem. Soc.* **1998**, *145*, 659–661.
36. M. Pujadas, P. Salvador, *J. Electrochem. Soc.* **1989**, *136*, 716–723.
37. L. J. Ferrer, H. Muraki, P. Salvador, *J. Phys. Chem.* **1986**, *90*, 2805–2807.
38. A. M. Chaparro, H. Tributsch, *J. Phys. Chem. B* **1997**, *101*, 7428–7434.
39. S. Bastide, D. Gal, D. Cahen, *Rev. Sci. Instrum.* **1999**, *70*, 4032–4036.

3
Semiconductor Nanostructure

3.1
Preparation of Nanocrystalline Semiconductor Materials

Gary Hodes, Yitzhak Mastai
Weizmann Institute of Science, Rehovot, Israel

3.1.1
Introduction

The field of nanophase materials is one that covers a wide and active area of research. The various properties of these materials, including mechanical, optical, electrical, and structural, are often very different from the same materials in the "bulk" phase. An example of this difference is the case of quantum dots (QDs) – nanoparticles that are sufficiently small that their electronic energy structure is changed from that of the bulk material. The size regime of semiconductor QDs varies from about 1 nm up to tens of nm in size, depending on the material properties.

Considerable effort has been made in recent years to synthesize nanocrystalline semiconductors with control over the size and size distribution and to characterize their properties. As a result of these studies, it is now possible to synthesize a variety of nanocrystalline semiconductors with varying degrees of control over the size, shape, and a narrow size distribution. Different strategies are used to prepare and keep the particles small and involve control of nucleation and growth of the crystallites. This can be achieved in a number of different ways and is described in the following section.

Electrodeposition normally leads to small particle size, largely because it is a low-temperature technique (molten salt electrolysis being an exception), thereby minimizing grain growth. Other factors also contribute to limitations of grain size and are discussed in this chapter. The low temperatures also minimize interdiffusion between electrodeposit and substrate or between different electrodeposited layers. In addition, electrodeposition allows a very high degree of control over the amount of material deposited through Faraday's law, which relates the amount of material deposited to the deposition charge. This feature is particularly desirable when isolated nanocrystals are to be deposited on a substrate.

Several different methods of electrodeposition, such as reduction of (chalcogen) oxide or zero valent (chalcogen) compounds in aqueous electrolytes, deposition from nonaqueous solvents, anodization, deposition from molten salts, and deposition by a multistage process, have been used. Electrodeposition offers a wide range of control of the nanocrystalline film owing to the many parameters involved (potential or current density, pulsed electrodeposition, solution composition, convection, and temperature).

Although this chapter is limited to electrodeposition of semiconductors, it is only fair to mention, even if briefly, some examples of electrodeposition of metal nanostructures. This is important because the principles and techniques used in electrodepositing metals are essentially the same as those used for depositing semiconductors – the main difference is that almost all studies on electrodeposition of nanocrystalline semiconductors involve compound semiconductors, with the added complications this entails. Examples include pulsed electrodeposition of metal multilayers [1, 2], porous membrane-templated electrodeposition of gold nanotubes [3], and Ni nanowires [4].

3.1.2 Electrochemical/Chemical Hybrid Deposition

A method for the synthesis of semiconductor QDs, called *the Electrochemical/Chemical method* (E/C), has been developed by Penner and his associates [5]. The E/C synthesis of semiconductors nanocrystals involves three steps:

1. In the first step, metal nanocrystals M^0 (e.g. Cd, Cu, Zn) are electrochemically deposited onto basal plane–oriented graphite electrode surfaces from aqueous solution of metal ions M^{n+}.
2. Next, the metal nanocrystals are oxidized, either chemically or electrochemically, to the metal oxide or hydroxide.
3. Finally, the oxide (hydroxide) is displaced with gaseous hydrogen chalcogenide or chalcogenide anion (e.g. S^{-2}) in solution to form nanocrystals of MX.

Step I – Electrodeposition of metal nanoparticles.

In the E/C synthesis, the first step in which metal nanocrystals are deposited on the substrate is critical. The semiconductor particles grow from the metal particles on a particle-by-particle basis (i.e. each metal particle is chemically transformed to the corresponding semiconductor). The size and size distribution, therefore, in the first step determine the final size of the semiconductor particles. For that reason, it is important to achieve an understanding of the growth mechanism of the metal nanocrystal deposition. Penner and associates studied the electrodeposition of various metal nanoparticles (Ag, Pt, Zn, Cu, Cd) mainly onto basal plane–oriented graphite and also onto Si electrodes [6–11]. The depositions were carried out from dilute aqueous solutions of metal ions using a potentiostatic pulse regime. A short (typically tens of ms) potential pulse was applied followed by open-circuit conditions.

It was demonstrated that the deposition of metal nanocrystals occurs via a Volmer-Weber (three-dimensional growth) mode. For example, in the deposition of platinum on graphite, platinum particles with a mean diameter of 5.2 nm were formed following the deposition of quantities equal to only 0.039 monolayers of platinum [10]. Volmer-Weber growth is favored at surfaces that are characterized by low interfacial free energy (like the graphite basal plane). Transmission electron microscopy (TEM) and selected area electron diffraction (ED) of the graphite surfaces revealed that the deposited Pt particles are not epitaxially oriented with the graphite surfaces. On the basis of electron microscopy and electrochemical studies, a two-step mechanism for the growth of Pt nanoparticles was proposed: "instantaneous" nucleation that takes part within a few ms, followed by diffusion-controlled growth from solution. A similar mechanism was observed in the deposition of other metal nanoparticles.

For short pulse durations (approximately 10 ms), particles of a few nm with a narrow size distribution were obtained. The particle size and size distribution both increased for increasingly greater pulse durations. In general for pulse plating, a high overpotential (or high current density) and short pulse duration favor reduced grain size caused by increased nucleation density and less time for growth.

This increasing size distribution with growth time contrasted with three-dimensional nucleation and growth of colloidal particles, where instantaneous nucleation followed by diffusion-limited growth leads to a narrow size distribution. This difference was explained by a nonhomogeneous distribution of initially nucleated particles on the substrate [12, 13]. Where many nuclei were formed close together, their diffusion layers interacted, resulting in a reduced rate of diffusion and slower growth. On the other hand, particles that nucleated relatively far from other nuclei retained independent and thinner diffusion layers and therefore grew faster. In other words, nanoparticles that nucleate close together end up being smaller than particles that nucleate far from neighbors.

The size distribution could be narrowed for greater amounts of deposit by employing a train of short deposition pulses ($\leq$10 ms) followed by a much longer off-time (at open circuit), rather than a single longer deposition pulse [14]. These shorter deposition pulses diffusionally decouple the growth of densely nucleated regions from sparsely nucleated regions, and by this means, they narrow the size distribution of the nanocrystallites.

Steps II and III – The conversion of metal nanoparticles to semiconductor nanoparticles.

The conversion procedure of metal nanoparticles to the corresponding semiconductor particles involves one or two steps. If the desired semiconductor is an oxide, such as ZnO [8], the conversion procedure involves one step, namely, the oxidation of the metal nanoparticles. Other anions require an additional step in which the oxide (or hydroxide) is displaced by other ions such as in the case of CdS [14–16] and CuI [6]. (Although the formation of an oxide is not essential to obtain the final semiconductor, oxide formation will normally occur unless steps are taken to prevent it, and it is more logical to deliberately oxidize the metal electrochemically.)

For CdS, the Cd nanoparticles are spontaneously oxidized in the plating solution (at pH = 6 and open circuit) to $Cd(OH)_2$. The displacement of OH by S to yield CdS can be carried out in two different ways: immersion of the $Cd(OH)_2$ in an aqueous sulfide solution (Na_2S at pH = 10) [15], or by exposure of the $Cd(OH)_2$ to gas phase H_2S at 300 °C [16]. In the synthesis from aqueous S^{2-}, CdS particles with a mean crystal diameter between 2 and 8 nm were prepared. The nanocrystals were oriented with the c-axis of the CdS-wurtzite unit perpendicular to the electrode surface (i.e. graphite basal plane). In the gas-phase conversion, CdS/S core-shell structures were obtained by decomposition of the H_2S to S on the CdS surface.

β-CuI was formed by first converting the Cu nanocrystals to Cu_2O followed by treatment in aqueous KI [6].

The various nanocrystalline semiconductors exhibited strong, size-dependent room temperature luminescence, although they were deposited directly on graphite [6, 8, 14]. The graphite, in contrast to metals, apparently does not efficiently quench the luminescence. Of particular importance is the observation that, under suitable preparation conditions, all the semiconductors emitted essentially band gap luminescence at room temperature with little or no deep sub–band gap band response that is normally characteristic of these materials. This implies that the nanocrystals were of very high quality.

3.1.3 Electrodeposition of Nanocrystals from Nonaqueous Solution

One of the simplest techniques used to electrodeposit semiconductors is cathodic deposition from nonaqueous solutions containing elemental chalcogen (S, Se) and a metal salt, first described by Baranski and Fawcett [17]. Two main mechanisms have been considered: deposition of metal (e.g. Cd) followed by chemical reaction with elemental chalcogen in solution and reduction of chalcogen to (poly)chalcogenide followed by ionic reaction between chalcogenide and metal cations. Which mechanism will dominate depends on the specific system (substrate, semiconductor, deposition conditions) and may change in the same deposition.

This technique was found to give nanocrystalline semiconductors which exhibited quantum size effects with typical lateral crystal size of 5 nm [18, 19]. The height of the crystals [measured by X-ray diffraction (XRD)] was often up to several times larger than the lateral dimensions (measured by TEM). In fact, the increased transparency to shorter wavelengths because of size quantization, together with good (photo)conductivity expected from "wires", was used to improve thin film $CdS/CuInS_2$ photovoltaic cells [20].

The nature of the anion of the metal salt is important in this technique. Salts of many anions are either insoluble or unstable in hot dimethyl sulfoxide (DMSO) containing dissolved chalcogen. Perchlorate and chloride are the most commonly used anions although other halides, methylsulfonate, and borofluoride have also been used. However, another reason that the anion is important, particularly relevant to the present discussion, is that it affects the crystal size and therefore the band gap. Films deposited from Cl^- solutions exhibit band gaps between 0.1 and 0.2 eV higher than those of the same compound deposited from ClO_4^- [21]. More recent and detailed studies indicate that that this difference is due to adsorption of the more strongly adsorbed Cl^- on the growing CdX surface, preventing further crystal growth [22].

Although both CdSe and CdS can be deposited in nanocrystalline form by this nonaqueous deposition, the essentially total insolubility of Te in DMSO prevents the use of the method for deposition of nanocrystalline tellurides (a small amount of Te can be codissolved with Se and mixed selenide-tellurides with small amounts of Te can be deposited; see following section). However, a related method to deposit CdTe has been described by Cocivera and associates [23, 24]. They reacted elemental Te with tri-*n*-butyl phosphine (TBP), which reacts with Te to form TBP telluride. This compound, together with a Cd salt dissolved in propylene carbonate, allowed cathodic electrodeposition of CdTe. The as-deposited films were reported to be X-ray amorphous, a fact that suggested that they might in fact be nanocrystalline [26]. (Cd,Hg)Te films grown by the same technique exhibited a crystal size of about 5 nm [25].

This technique was modified and simplified, in particular, by using a one-step technique to prepare the solution in DMSO [26]. The CdTe was indeed found to be nanocrystalline with a wide size distribution varying from several nm up to tens of nm. In addition, the films were generally nonstoichiometric with excess Te or Cd, depending on the deposition potential. It was difficult to deposit close to stoichiometric film. To improve the stoichiometry, reverse pulse deposition was used to strip excess Cd or Te during the anodic pulse. The pulse regime also decreased the crystal size to an extent depending on the pulse parameters (crystal size typically several nm) and improved the size distribution. The main factors limiting crystal growth were short pulse on-times and capping of the crystals with strongly adsorbing phosphine, mainly during the pulse off-time.

3.1.4 Size Control Using Semiconductor-substrate Lattice Mismatch

An important factor that can influence crystal size, particularly for the first layer of crystals, is the nature of the substrate. Deposition of CdSe from the above DMSO electrolyte using $Cd(ClO_4)_2$ onto films of evaporated Au on glass or mica resulted in crystals of about 4–5 nm in size. The distribution of the nanocrystals on the Au depended on current density and deposition temperature: high currents and low temperatures favored isolated crystals, whereas increasing aggregation occurred with higher temperature and decreased current [27].

ED [27] and HTEM [28] showed that the CdSe crystals were epitaxially deposited on the Au in a $\{111\}_{Au} \| \{00.2\}_{CdSe}$ and $\{110\}_{Au} \| \{11.0\}_{CdSe}$ orientation relationship. The epitaxy arises because of the good lattice match between the CdSe and Au in a 2 : 3 relationship (−0.6% mismatch). Beyond the first layer of crystals, the epitaxy is gradually lost and the crystal size grows for the perchlorate bath; for a $CdCl_2$ bath, the crystal size does not grow much because of capping, as discussed earlier. The mismatch strain, which gradually increases as the crystal grows, eventually leads to termination of growth. This can explain the relatively narrow size distribution obtained.

This hypothesis of strain-determined growth termination suggested that crystal size should be controllable by choice of the semiconductor and substrate lattice parameters: the larger the mismatch, the smaller the crystal size, and vice versa. A range of different electrode-posited semiconductor-substrate combinations has been investigated to test this.

The lattice parameter of the CdSe was varied with (assumed) little change in chemical interaction with the substrate by depositing an alloy of $CdSe_xTe_{1-x}$ [29]. From Vegard's law, a value of $x = 0.88$ (12% Te) should result in an increased lattice with a perfect match to Au. Although Te could not be dissolved in DMSO, small amounts of Te could be dissolved in the presence of dissolved Se. The concentration of Te in the electrolyte was not known (it was almost certainly <1 mM), but the amount of Te in the deposit was quantitatively varied by varying the deposition current density and temperature. Relatively high current densities resulted in severe depletion of the Te near the cathode, thereby favoring the more concentrated Se in the deposit. In the same way, higher deposition temperatures increased the concentration of Te at the cathode and therefore in the deposit.

The maximum Te concentration in the deposits was measured to be only about 5%. However, this was sufficient to see a major increase (by up to several times) in nanocrystal size with increase in Te concentration, whereas the epitaxy was, as expected, retained.

The lattice parameter was decreased (with expectation of reduced crystal size) by alloying the Cd with Zn [30]. $Cd_xZn_{1-x}Se$ was epitaxially deposited using very low concentrations of Cd in the deposition solution. The crystal size for $x = 0.78$ was between 2 and 3 nm – half the size of the corresponding CdSe deposits. The crystal height, measured by XRD, was 5 nm. The crystals are therefore short quantum columns rather than QDs.

CdS has a 4.5% mismatch with Au (2 : 3 ratio). If epitaxy was to occur for such a large mismatch, the crystal size should be very small. In this case, the properties of the deposit were dependent on the amount deposited [31]. For small amounts of deposit, the crystals are predominantly epitaxial (although not perfectly so as for CdSe on Au). The average crystal size grows, and the degree of orientation decreases as the amount of deposit increases. The crystal height for a nominally 5-nm thick (in reality, ≤ 3 nm) deposit was found to be 3 nm – similar to the lateral ones.

The mismatch may also be varied by changing the material of the substrate. Pd has the same fcc cubic structure as Au; the lattice spacing, however, is almost 5% smaller than that of Au, with the result that the mismatch between CdSe and Pd is +4.1%. This is close to the mismatch between CdS and Au but in the opposite direction, that is, the CdSe should be compressed instead of stretched. In contrast to CdSe on Au, which grows as isolated crystals on Pd, the CdSe covers the entire surface and, with further deposition, forms fairly large particles (10–20 nm), which become larger with increasing deposition time. Even more important, the CdSe deposit on Pd is both XRD and ED amorphous. However, careful examination of a HTEM image reveals the presence of many regions showing short-range order, often with hexagonal symmetry. An in-depth analysis of this short-range order showed the presence of irregular structures of ordered CdSe, on the order of 1 nm in size, surrounded by disordered CdSe. Furthermore, Fourier analysis of HTEM images showed some degree of preferential orientation of the ordered structures relative to the Pd lattice, but rotated 30° to the Pd [32]. Subsequent modeling of the superimposed CdSe-Pd lattices showed that this 30° rotation indeed resulted in an improved lattice match compared with the aligned lattices [33]. The very small ordered structures are in qualitative agreement with the mismatch strain argument that predicts very small crystals. However, it is likely that differences in chemical reactivity between CdSe and Au or Pd also play a role.

CdS has only a small mismatch with Pd – +0.17% (2 : 3 ratio). However, attempts to obtain thin electrodeposits of CdS on Pd were not successful. Small particles (approximately 2 nm) were obtained; however, the identity of these particles was not known – XPS analysis showed the almost total absence of Cd. In this case, chemical interaction between S and Pd probably dominates the surface.

In an attempt to separate chemical effects from lattice mismatch, we studied an alloy of 3% Cd with Au. Cd was chosen because of its strong interaction with Se. The change in lattice-spacing owing to the 3% of Cd is very small (+0.07%) and therefore any change in CdSe properties can be attributed to differences in chemical interaction. The lateral crystal size is approximately 5 nm – similar to that on pure Au. However, the crystal height is considerably smaller than this, as seen qualitatively by TEM contrast and surface coverage. Most surprising, however, is that the deposit is not only a Wurtzite structure but also a mixture of Wurtzite, Zincblende, and Rocksalt. Rocksalt CdSe is normally stable only at high pressures. It was suggested that surface tension forces, together with (or maybe because of) the strong interaction between the CdSe and the substrate, stabilized this high-pressure structure [34].

3.1.5 Electrodeposition of Layered Nanostructures

Many modern semiconductor devices comprise alternating layers of different materials forming superlattices and multiple quantum wells. One well-known example of such structures is the diode laser, a mass-produced device. This device depends on confinement of charges in the two-dimensional structures for enhanced laser output at lowered current thresholds. Such alternating semiconductor layers are usually manufactured either by chemical vapor deposition or by molecular beam epitaxy. The thickness of the layers can be closely controlled in both techniques. As mentioned earlier, electrodeposition also allows good control of thickness.

The group of Switzer has been active in electrodeposition of alternating layers of different metal oxides or metal/metal oxides. Ceramic superlattices of Pb_a-Tl_b-O_c/Pb_d-Tl_e-O_f with individual thickness of 5–10 nm were electrodeposited onto stainless steel from an aqueous solution of $TlNO_3$ and $Pb(NO_3)_2$ in 5 M (NaOH) at room temperature using either pulsed current or potential [35]. The deposition is based on oxidation of Tl(I) to Tl(III) and Pb(II) to Pb(IV); the higher oxidation states of these ions are readily hydrolyzed to the oxides. Because Tl_2O_3 is deposited preferentially (owing to a less anodic potential) compared to PbO_2, low current densities (or low anodic potentials) form essentially only Tl_2O_3 from mixed Tl/Pb solutions. If a solution is made with a higher Pb than Tl concentration, then as the current density (anodic potential) is increased the concentration of Pb in the resulting oxide increases as the low concentration Tl becomes diffusion-limited. The interplay of relative cation concentrations, mass transport conditions, and pulse characteristics (current/potential and time scales) imparts control of the stoichiometry, as well as thickness of the individual layers. The sharpness of the interface between layers can also be controlled: deposition where one of the constituent ions is in the diffusion-controlled regime results in a graded composition, whereas if deposition takes place in the kinetic control regime for both metal ions, a sharp compositional difference between layers is obtained [36]. STM was used to identify the individual layers [37, 38]. Both Tl_2O_3 and PbO_2 are degenerate semiconductors with band gaps of about 1.4 and 1.8 eV, respectively. Therefore, such layers could be used as quantum wells with electrons confined preferentially in the lower band gap Tl-rich layers.

Layers of Tl_2O_3 or Pb-Tl-O could be deposited on steel with strong (100) or (210) texturing, respectively, and superlattices of Pb-Tl-O could then be electrodeposited epitaxially on these prelayers with these same orientations [39].

Tl_2O_3 can be electrodeposited with a defect chemistry controllable through the deposition overpotential. Oxygen vacancies are produced at high values of overpotential, whereas Tl interstitials are favored at low overpotential. This was used to deposit alternating "defect chemistry superlattices" of several nm period by pulsing the potential during deposition [40]. Electrical and optical studies of Tl_2O_3 deposited at different overpotentials were carried out [41]. The resistivity of the degenerate oxide could be varied from several hundred μΩ-cm at low overpotential to 75 μΩ-cm over a range of <0.1 V. This change in resistivity was due to changes in the electron mobility (because the carrier density was constant at about 10^{21} cm^{-3}) caused by an increase in the concentration of charged interstitials. The optical band gaps of the high overpotential films were blue-shifted by as much as 1.1 eV from the normal value of 1.4 eV because of conduction band-filling (Burstein-Moss effect).

Another system studied by the same group comprises alternating Cu and Cu_2O layers formed by constant current electrolysis of aqueous Cu(II) + lactate solutions at a pH of 8–10 [42, 43]. Although a constant current is applied, spontaneous potential oscillations occur, leading to variations in the stoichiometry of the deposit from Cu to Cu_2O. The cause of these oscillations is not clear but appears to be related to formation and breakdown of surface compounds. The oscillation period (typically tens of seconds to minutes), and therefore the thickness of the different layers and their compositions, depends on the current density and, in particular, on the solution pH. Quartz crystal monitor studies suggested that Cu_2O is formed during positive potential spikes in the oscillations, whereas a composite of Cu and Cu_2O is deposited during the more negative plateau region of the oscillation [44].

At low current densities (≤ 1 mA cm^{-2}), no potential oscillations occur, and pure, strongly oriented (100) Cu_2O is formed with a crystal size > 1 μm. With increasing current density, the Cu content increases along with the appearance of the oscillations and the color changes from the red of the pure Cu_2O through green to gold. These layers are nanocomposites that comprise Cu_2O and metallic Cu nanocrystals. The nanocrystal size decreases as the current density is increased and varies from 5 to 10 nm for the Cu and from 4 to 20 nm for the Cu_2O in the composite. The potential oscillations suggest that the materials are modulated and this was observed in SEM micrographs [43].

Cu_2O is a p-type semiconductor with a band gap of about 2.1 eV, hence its red color. Absorption spectra of the composite films show a blueshift of the Cu_2O absorption edge by up to 0.4 eV (for 8 nm Cu_2O crystallites). On the basis of strong confinement, this shift is much larger than expected, and it may be because of the confinement in the nanocrystals beyond that due to simple physical size.

These structures are highly anisotropic with respect to their electrical conductivity. There is a difference of about 12 orders of magnitude in the conductivities of metallic Cu and Cu_2O. This difference is manifested in both the lateral (parallel to the layers) and perpendicular conductivities of the alternating layers. The lateral resistivity varied from 10^{-4} Ω-cm for very Cu rich samples to $>10^6$ Ω-cm for Cu-poor ones. In particular, at a Cu concentration >9.8 vol%, the resistivity decreased suddenly, typical of a percolation process. The resistivity in the perpendicular direction was much larger than that in the lateral one for reasonably Cu-rich samples (ratios of $>10^9$). No less important is the fact that although the transport in the lateral direction was always ohmic, the transport in the vertical direction exhibited pronounced negative differential resistance, attributed to resonant tunneling from Cu into hole states in Cu_2O [45, 46].

3.1.6 Template-directed Electrodeposition

There are a number of studies on electrodeposition of semiconductors in the pores of various membranes forming nanowires of the semiconductor. Klein and associates electrodeposited CdSe and CdTe into the pores of Anopore membranes [47]. These membranes are 50-μm thick with closely spaced 200-nm-sized pores. One side of the membrane was sputter-coated with Au followed by Ni electrodeposition, which partially filled the pores, as a substrate. CdSe and CdTe were electrodeposited from acidic aqueous solutions containing $CdSO_4$ and SeO_2 or TeO_2. The semiconductors grow as compact wires that are many μm long and 200–300 nm in diameter. By electrodepositing first CdSe and then CdTe, nanowires were formed, which were compositionally graded over their length.

Chakarvarti and Vetter electrodeposited Se in the pores of nuclear track membranes [48]. These pores were relatively large (2.5 μm).

Using porous anodic aluminium oxide films, Routkevitchand associates electrodeposited very thin CdS nanowires [49]. The alumina membranes were typically 1–3-μm thick with pore sizes ranging from 9 to 35 nm. The CdS was electrodeposited, using ac electrodeposition, from a DMSO solution containing $CdCl_2$ and elemental S. The porous alumina, as anodized, was separated from the Al by a dense oxide forming a rectifying contact. AC deposition then resulted in formation of CdS only in the pores and not in cracks or defects in the film. The CdS wire dimensions closely followed those of the pores. For the smaller pores, the wires were apparently made up of single crystals joined in the axial direction, whereas for the larger pores, the coherence length, measured by XRD, was less than the pore diameter, suggesting that the wires were composed of several crystals also in the radial direction (or possibly, defected single crystals). It should be noted that even for films of CdS deposited by this technique the crystal size is very small (see preceding section). The CdS was hexagonal (wurtzite) with its c-axis aligned predominantly along the membrane thickness. HTEM studies of these nanowires confirmed the well-ordered and basal plane-textured CdS crystallinity [50]. Small quantum size effects were seen for the smallest wires, with band gaps varying from 2.36 to 2.42 eV [51].

The same technique to form CdS nanowires, only employing dc electrodeposition, has been used by Xu and associates [52]. In this case, the Al was etched away and a silver film evaporated on the membrane. The wires in this case were textured with the (10.1) plane parallel to the membrane surface [in contrast to the ac deposition where the (00.1) plane grew parallel to the surface]. This technique was extended to CdSe and CdTe nanowires [53].

This group also deposited CdS nanowires in porous alumina templates using an aqueous solution containing $CdCl_2$ and thioacetamide at pH = 4.6 [54]. This deposition is actually an electrochemically induced chemical solution deposition initiated by local pH changes caused by electrolysis [55]. In this case, the wires grew with the (00.1) plane parallel to the membrane surface.

3.1.7
Occlusion Electrodeposition of Composites

Apart from the layered Cu/Cu_2O composites described earlier, composites of nanocrystalline semiconductors with non-semiconductors (metals or polymers) have been electrodeposited by incorporation of the semiconductor phase from solution into the electrodepositing metal (also known as *occlusion*). WO_3 particles suspended in solution can be incorporated in electrodeposited polypyrrole [56] or polyaniline [57] films, and the resulting films exhibit electrochromism, both of the polymer and of the WO_3. Although this technique worked only for oxides with very low isoelectric points, the isoelectric point of oxides with relatively high points of zero charge could be increased by adsorption of anions, such as sulfate or iodide, from solution. In that manner TiO_2 was incorporated into electrodeposited polypyrrole films, and the resulting films showed a photoanodic photocurrent response because of the incorporated TiO_2 [58]. Very high concentrations of WO_3 – up to 53 wt% – were incorporated into electrodeposited polypyrrole [59]. High stirring rates and low current densities of plating favored high rates of incorporation. Incorporation of TiO_2 into polypyrrole was attributed to mechanical entrapment of the TiO_2 particles in the rough, soft polypyrrole. In addition, the electric field at the surface of the growing film was also ascribed a role [60].

In a similar manner, TiO_2 particles in suspension were incorporated into growing electrodeposited Ni films and the resulting films were photoelectrochemically active [61]. The morphology (and photoelectrochemical response) of TiO_2 was dependent on the electrodeposited metal matrix: deposition of Ni, Cu, Ag, and In matrices resulted in different dispersion geometries of the TiO_2 particles [62]. In the case of Cu, the photoelectrochemical response at a high pH showed both n-type and p-type polarities because of the formation of p-Cu_2O. The incorporation of alumina into electrodeposited Ni films was shown to be controlled by diffusion of the alumina particles to the Ni surface and their residence time at the surface [63]. CdS, as well as TiO_2 particles occluded into electrodeposited Ni, were shown to exhibit increased photoelectrochemical activity after potential cycling owing to passivation of the Ni matrix [64].

3.1.8
Sonoelectrochemical Formation of Nanocrystalline Semiconductors

Sonoelectrochemical synthesis has recently been used for the preparation of semiconductor nanocrystalline powders. In the sonoelectrochemical method, the ultrasound horn acts as both cathode and ultrasound emitter. This technique was used for preparing metal powders [65] and was extended to CdTe, although details of the CdTe particle size were not given [66]. CdSe nanocrystalline powders have been prepared by pulsed sonoelectrochemical reduction from an aqueous selenosulfate solution. The crystal size could be varied from X-ray amorphous up to 9 nm (sphalerite phase) by controlling the various electrodeposition and sonic parameters [67]. Crystal size was smaller for lower preparation temperatures, higher ultrasound intensity, and shorter current pulse width. These dependencies could be explained based on a pulse of electric current producing a high density of fine particles on the sonic tip, followed by a burst of ultrasonic energy, which removes the particles from the cathode into the solution and prevents them from growing. Similarly, PbSe nanocrystalline powders with a crystal size of 10 to 16 nm were prepared in the same manner from a similar solution using salts of Pb instead of Cd [68].

MoS_2 can be electrodeposited from a thiomolybdate solution [69, 70]. However, the deposit is apparently amorphous and requires annealing to crystallize. We reasoned that the high temperatures in and around ultrasonic cavitation bubbles might crystallize the electrodeposited MoS_2. Sonoelectrochemical formation of MoS_2 powders using the above solution indeed resulted in partial crystallization of the as-deposited MoS_2, but in the form of multiple closed shell fullerene-like structures [71]. The structures were polyhedra of typically several tens of nm in size comprising layers (typically about 10) of crystallized MoS_2 surrounding what was probably amorphous MoS_2. In some cases, nanotubes of MoS_2 were also obtained. Although the cause of formation of these structures is not known, it is likely to be related to the buildup and collapse of the cavitation bubbles.

References

1. J. Yahalom, *Surf. Coat. Technol.* **1998**, *105*, VII–VIII.
2. T. Cohen, J. Yahalom, W. D. Kaplan, *Rev. Anal. Chem.* **1999**, 18, 279.
3. C. R. Martin, D. T. Mitchell in *Electroanalytical Chemistry* (Eds.: A. J. Bard and I. Rubinstein), Marcel Dekker, Inc., New York, Basel 1999, p. 1, *Vol. 21.*
4. L. Sun, P. C. Searson, C. L. Chien, *Appl. Phys. Lett.* **1999**, *74*, 2803.
5. R. M. Penner, *Acc. Chem. Res.* **2000**, *33*, 78.
6. G. S. Hsiao, M. G. Anderson, S. Gorer et al., *J. Am. Chem. Soc.* **1997**, *119*, 1439.
7. S. Gorer, G. S. Hsiao, M. G. Anderson et al., *Electrochim. Acta* **1998**, *43*, 2799.
8. R. M. Nyffenegger, B. Craft, M. Shaaban et al., *Chem. Mat.* **1998**, *10*, 1120.
9. J. V. Zoval, R. M. Stiger, P. R. Biernacki et al., *J. Phys. Chem.* **1996**, *100*, 837.
10. J. V. Zoval, J. Lee, S. Gorer et al., *J. Phys. Chem. B* **1998**, *102*, 1166.
11. R. M. Stiger, S. Gorer, B. Craft et al., *Langmuir* **1999**, *15*, 790.
12. J. L. Fransaer, R. M. Penner, *J. Phys. Chem. B* **1999**, *103*, 7643.
13. H. Liu, R. M. Penner, *J. Phys. Chem. B* **2000**, *104*, 9131.
14. S. Gorer, R. M. Penner, *J. Phys. Chem. B* **1999**, *103*, 5750.
15. M. A. Anderson, S. Gorer, R. M. Penner, *J. Phys. Chem. B* **1997**, *101*, 5895.
16. S. Gorer, J. A. Ganske, J. C. Hemminger et al., *J. Am. Chem. Soc.* **1998**, *120*, 9584.
17. A. S. Baranski, W. R. Fawcett, *J. Electrochem. Soc.* **1980**, *127*, 766.
18. G. Hodes, A. Albu-Yaron, Optical, Structural and Photoelectrochemical Properties of Quantum Box Semiconductor Films, *Proc. Electrochem. Soc.* 1988, 88–14, Vol. 298.
19. G. Hodes, *Isr. J. Chem.* **1993**, *33*, 95.
20. D. Gal, G. Hodes, D. Hariskos et al., *Appl. Phys. Lett.* **1998**, *73*, 3135.
21. G. Hodes, T. Engelhard, A. Albu-Yaron et al., *Mat. Res. Soc. Symp.* **1990**, *164*, 81.
22. Y. Mastai, D. Gal, G. Hodes, *J. Electrochem. Soc.* **2000**, *147*, 1435.
23. A. Darkowski, M. Cocivera, *J. Electrochem. Soc.* **1985**, *132*, 2768.
24. J. von Windheim, M. Cocivera, *J. Electrochem. Soc.* **1987**, *134*, 440.
25. C. L. Colyer, M. Cocivera, *J. Electrochem. Soc.* **1992**, *139*, 406.
26. Y. Mastai, G. Hodes, *J. Phys. Chem. B* **1997**, *101*, 2685.
27. Y. Golan, L. Margulis, I. Rubinstein et al., *Langmuir* **1992**, *8*, 749.
28. Y. Golan, L. Margulis, G. Hodes et al., *Surf. Sci.* **1994**, *311*, L633.
29. Y. Golan, J. L. Hutchison, I. Rubinstein et al., *Adv. Mater.* **1996**, *8*, 631.
30. G. Hodes, Y. Golan, D. Behar et al., in *Nanoparticles and Nanostructured Films*, (Ed.: J. H. Fendler), Wiley-VCH, Germany 1998, 1.

31. D. Behar, I. Rubinstein, G. Hodes et al., *Superlattices Microstruct.* **1999**, *25*, 601.
32. Y. Golan, E. Ter-Ovanesyan, Y. Manassen et al., *Surf. Sci.* **1996**, *350*, 277.
33. Y. Golan, G. Hodes, I. Rubinstein, *J. Phys. Chem.* **1996**, *100*, 2220.
34. Y. Zhang, G. Hodes, I. Rubinstein et al., *Adv. Mater.* **1999**, *11*, 1437.
35. J. A. Switzer, M. J. Shane, R. J. Phillips, *Science* **1990**, *247*, 444.
36. J. A. Switzer, R. J. Phillips, T. D. Golden, *Appl. Phys. Lett.* **1995**, *66*, 819.
37. J. A. Switzer, R. P. Raffaelle, R. J. Phillips et al., *Science* **1992**, *258*, 1918.
38. T. D. Golden, R. P. Raffaelle, J. A. Switzer, *Appl. Phys. Lett.* **1993**, *63*, 1501.
39. R. J. Phillips, T. D. Golden, M. G. Shumsky et al., *Chem. Mat.* **1997**, *9*, 1670.
40. J. A. Switzer, C. J. Hung, B. E. Breyfogle et al., *Science* **1994**, *264*, 1573.
41. R. A. Vanleeuwen, C. J. Hung, D. R. Kammler et al., *J. Phys. Chem.* **1995**, *99*, 15 247.
42. J. A. Switzer, C. J. Hung, E. W. Bohannan et al., *Adv. Mat.* **1997**, *9*, 334.
43. J. A. Switzer, C. J. Hung, L. Y. Huang et al., *J. Mat. Res.* **1998**, *13*, 909.
44. E. W. Bohannan, L. Y. Huang, F. S. Miller et al., *Langmuir* **1999**, *15*, 813.
45. J. A. Switzer, C. J. Hung, L. Y. Huang et al., *J. Am. Chem. Soc.* **1998**, *120*, 3530.
46. J. A. Switzer, B. M. Maune, E. R. Raub et al., *J. Phys. Chem. B* **1999**, *103*, 395.
47. J. D. Klein, R. D. Herrick, D. Palmer et al., *Chem. Mat.* **1993**, *5*, 902.
48. S. K. Chakarvarti, J. Vetter, *J. Micromech. Microeng.* **1993**, *3*, 57.
49. D. Routkevitch, T. Bigioni, M. Moskovits et al., *J. Phys. Chem.* **1996**, *100*, 14 037.
50. J. L. Hutchison, D. Routkevitch, A. Albu-Yaron et al., *Microscopy of Semiconducting Materials 1997*, 1997, 389.
51. D. Routkevitch, T. L. Haslett, L. Ryan et al., *Chem. Phys.* **1996**, *210*, 343.
52. D. S. Xu, Y. J. Xu, D. P. Chen et al., *Chem. Phys. Lett.* **2000**, *325*, 340.
53. D. S. Xu, D. P. Chen, Y. J. Xu et al., *Pure Appl. Chem.* **2000**, *72*, 127.
54. D. S. Xu, Y. J. Xu, D. P. Chen et al., *Adv. Mater.* **2000**, *12*, 520.
55. K. Yamaguchi, T. Yoshida, T. Sugiura et al., *J. Phys. Chem. B* **1998**, *102*, 9677.
56. H. Yoneyama, Y. Shoji, *J. Electrochem. Soc.* **1990**, *137*, 3826.
57. H. Yoneyama, S. Hirao, S. Kuwabata, *J. Electrochem. Soc.* **1992**, *139*, 3141.
58. K. Kawai, N. Mihara, S. Kuwabata et al., *J. Electrochem. Soc.* **1990**, *137*, 1793.
59. F. Beck, M. Dahlhaus, *J. Appl. Electrochem.* **1993**, *23*, 781.
60. F. Beck, M. Dahlhaus, N. Zahedi, *Electrochim. Acta* **1992**, *37*, 1265.
61. M. Zhou, N. R. deTacconi, K. Rajeshwar, *J. Electroanal. Chem.* **1997**, *421*, 111.
62. N. R. de Tacconi, C. A. Boyles, K. Rajeshwar, *Langmuir* **2000**, *16*, 5665.
63. P. M. Vereecken, I. Shao, P. C. Searson, *J. Electrochem. Soc.* **2000**, *147*, 2572.
64. N. R. de Tacconi, H. Wenren, K. Rajeshwar, *J. Electrochem. Soc.* **1997**, *144*, 3159.
65. J. Reisse, H. Francois, J. Vandercammen et al., *Electrochim. Acta* **1994**, *39*, 37.
66. J. Reisse, T. Caulier, C. Deckerheer et al., *Ultrason. Sonochem.* **1996**, *3*, S147.
67. Y. Mastai, R. Polsky, Y. Koltypin et al., *J. Am. Chem. Soc.* **1999**, *121*, 10 047.
68. J. J. Zhu, S. T. Aruna, Y. Koltypin et al., *Chem. Mat.* **2000**, *12*, 143.
69. H. J. Byker, A. E. Austin, *J. Electrochem. Soc.* **1981**, *128*, 381 (CA no. 536).
70. E. A. Ponomarev, M. NeumannSpallart, G. Hodes et al., *Thin Solid Films* **1996**, *280*, 86.
71. Y. Mastai, M. Homyonfer, A. Gedanken et al., *Adv. Mat.* **1999**, *11*, 1010.

3.2 Macroporous Microstructures Including Silicon

C. Lévy-Clément
LCMTR, Thiais, France

3.2.1 Introduction

Porous materials are generally obtained by localized electrochemical corrosion under anodic polarization (anodization). Some metals such as aluminum or stainless steel can be made porous. Comparatively, a far larger number of semiconductors can be made porous, leading to important changes in the physical properties. The best and first known example is that of porous silicon (Si), which exhibits strong photoluminescence in the visible range instead of the weak luminescence of bulk Si. Since its discovery more than forty years ago [1–3] and recognition of its porous nature ten years later [4], it has attracted considerable research interest because of its morphological, electrical, and optical properties, which are different from those of bulk silicon [5]. Interest in porous Si during the 1970s and 1980s was because of the fact that it could easily be converted into thick oxide films and be used for dielectric (trench) isolation of active Si devices and in the full isolation by porous oxidized Si (FIPOS) process, resulting in less than 200 published papers, spanning a period of 35 years [5]. During that period of time, investigation of the localized corrosion of other semiconductors (groups IV, III–V, and II–VI semiconductors) raised limited interest, until the discovery in 1990 of the strong luminescence of porous silicon arising from quantum-confinement effects [6]. Since then, research work was mainly focussed on various aspects of porous Si, particularly on the rich variety of porous Si structures that can form in hydrofluoric acid (HF) – based solutions [5, 7–11], models evoked for explaining them [5, 12, 3], elucidation of the structure of luminescent material and the origin of its efficient luminescence [11, 16], and numerous technological applications of porous Si [17, 18]. Tremendous quantity of results (several thousand publications) have been analyzed all along the 1990s in various conference proceedings [19–26] and review articles [5, 8, 9, 11, 14–18, 27–30], as well as edited books [32–36]. Following the work on luminescent porous Si, a number of studies have been undertaken to render other semiconductors porous. This chapter will focus on the electrochemical aspects of porous semiconductors including porous Si.

3.2.2 Basic Principles

3.2.2.1 Thermodynamic Considerations

The general concept of the anodic dissolution process and reaction pathway was first explored with group-IV semiconductors (i.e. silicon and germanium) (1–3).

$$\mathrm{Si} + 6\mathrm{F}^- + (4-n)\mathrm{h}^+ \longrightarrow \mathrm{SiF_6}^{2-} + n\mathrm{e}^- \qquad (1)$$

h^+ and e^- are a valence band hole and a conduction band electron, respectively. Quadrivalent dissolution to Si(IV) is assumed and n is a number between 0 and 4. This work led Gerischer to develop a general model of semiconductor/electrolyte interfaces, particularly for III–V materials [37]. This model combines semiconductor band theory and electrochemistry. It assumes isoenergetic electron transfer across the semiconductor/solution interface, with the rate of reaction dependent on the overlap of the density of states.

When photoelectrochemical solar cells became popular in the 1970s, many reports appeared concerning the stability, dissolution, and flat-band potential of semiconductors in solutions. These papers investigated parameters such as the energy level of the band edges, which is critical for the thermodynamic stability of the semiconductor and how to determine the potential for the onset of the (photo)electrochemical etching [38–40]. The criterion for thermodynamic stability of a semiconductor electrode in an electrolyte solution is determined by the position of the Fermi level E_F with respect to the decomposition potential of the electrode with either the conduction band electrons ${}^{n}E_d$ or valence band holes ${}^{p}E_d$. Under illumination, the quasi-Fermi level replaces the Fermi level. The Fermi level is usually found within the band gap of the semiconductor and its position is not easily evaluated (especially the quasi-Fermi level of minority carriers). Therefore it was found more practical to use the conduction band minimum (E_c) and valence band maximum (E_v) as criteria for electrode corrosion. Thus, a semiconductor will be corroded in a certain electrolyte by the conduction band electrons if its decomposition potential is below the conduction band minimum ($E_c < {}^{n}E_d$) i.e. more positive than the conduction band. A similar statement can be given for the holes. The semiconductor electrode is unstable against anodic (photo)corrosion if its decomposition potential is higher (more negative) than the valence band maximum ($E_v > {}^{p}E_d$) of the semiconductor. In this case, the hole arriving at the interface carries out an oxidation reaction with the semiconductor or undergoes fast recombination with the electrons. The conditions of corrosion in an electrolyte are illustrated in Fig. 1.

Generally, the semiconductors are protected against cathodic (photo)corrosion in indifferent electrolytes because of the formation of a thin metallic film produced at the first instant of (photo)corrosion by the (photo)electrons, which for GaAs, for example, can be written as

$$GaAs + 3e^- \longrightarrow Ga^0 + As^{3-} \qquad (2)$$

The situation is different in the anodic regime. Gerischer [38] and Heller [42] pointed out that the majority of semiconductors (in the dark for p-type or under illumination for n-type) are unstable against corrosion with valence band (photo)holes, as it is illustrated in Figs. 1(c)

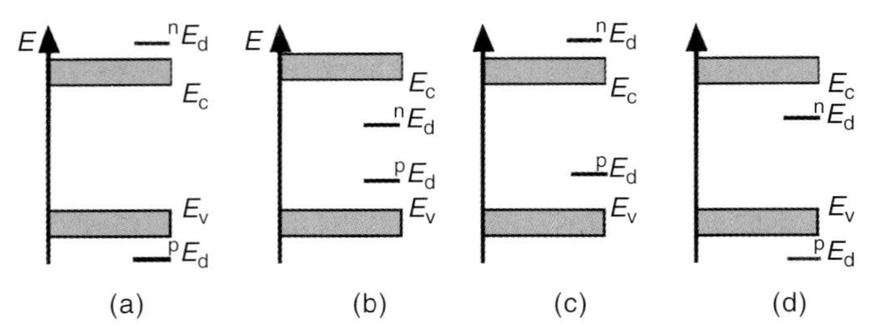

Fig. 1 Typical correlation between energy positions of band edges and decomposition potentials, controlling thermodynamic stability against photodecomposition. (a) stable, (b) unstable, (c) unstable against anodic decomposition, (d) unstable against cathodic decomposition (from Ref. 39).

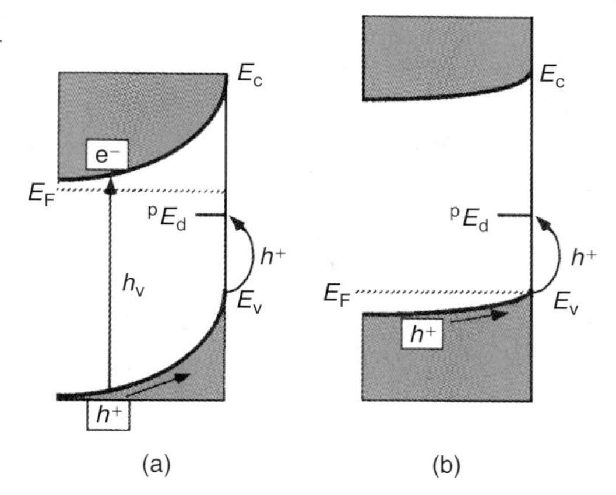

Fig. 2 Schematic drawings of the band diagram representing decomposition of semiconductor under anodization. (a) photogenerated holes in an *n*-type semiconductor, (b) bias-generated majority carriers in a *p*-type semiconductor (from Ref. 41).

and 2. The oxidation reaction induced by the holes in the valence band is the oxidation of the semiconductor itself, which for GaAs, for example, can be written as:

$$GaAs + 6h^{+} \longrightarrow Ga^{3+} + As^{3+} \quad (3)$$

Depending on the composition of the electrolyte, the ions from the semiconductor are either soluble in the electrolyte or are insoluble, and a new phase can form, generally, an oxide. By choosing an appropriate electrolyte, potential bias, and anodization time, the corrosion can be controlled.

3.2.2.2 Electrochemical Reactions

As for many other electrochemical reactions, two kinetic regimes are distinguishable in the anodic dissolution reaction [43]: (1) reaction kinetics is rate-limiting (kinetic control) and (2) the rate-limiting step is the diffusion of one of the reactants to (or products from) the electrode surface. The morphology and properties of the semiconductor surface are totally different in the two regimes. Generally, if the electrochemical corrosion reaction is carried out at sufficiently low current densities, the electrode kinetics is rate-limiting, in which case a rough or porous morphology is developed at the semiconductor surface. Under high current densities, a solid oxide is formed at the semiconductor surface because of mass transport of a limited supply of reactants or products. The reaction is diffusion-controlled and a smooth surface is consequently obtained. This is best demonstrated for Si in the presence of a solution of hydrofluoric acid (HF) [44, 45]. Current–voltage studies (Fig. 3) performed on Si-HF electrolyte junctions show that the main requirement for electrochemical porous Si formation is that anodic current densities below a critical value, i_{crit}, must be used. For a low-doped n-type semiconductor, light must be supplied.

The exact dissolution chemistry of silicon is still a matter of debate; however, ignoring the intermediate steps involved in the dissolution mechanism, the overall dissolution reaction for porous Si formation is as follows:

$$Si + 6HF + (2-\lambda)h^{+} \Longrightarrow SiF_6{}^{2-} + H_2 + 4H^{+} + \lambda e^{-}, \quad \text{with } \lambda \leq 1 \quad (4)$$

Porous Si formation is characterized by the fact that two charges are required to

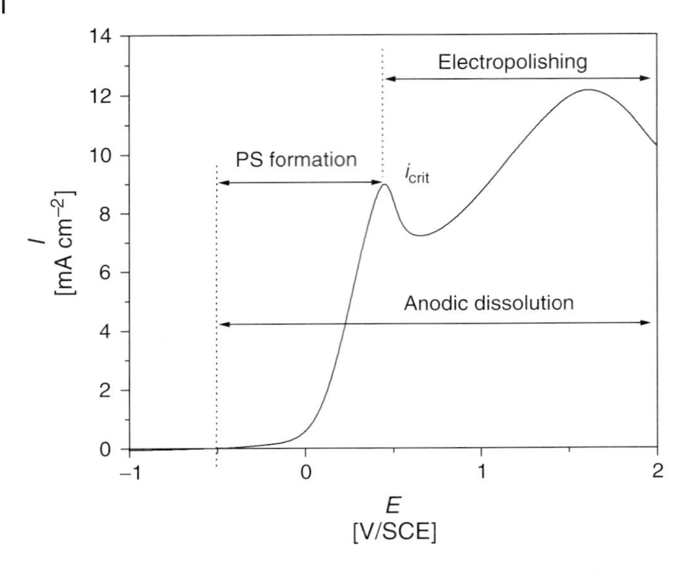

Fig. 3 Typical current–potential curve for *p*-type Si in the dark in HF.

dissolve one Si atom from the electrode, (i.e. the dissolution valence number equals 2) [1, 2, 46a,b] and H_2 is evolved. When current density is larger than i_{crit}, mass transfer to the solution limits the reaction and electropolishing occurs with a uniform dissolution of all the Si atoms at the surface of the wafer (dissolution valence number is 4). It is assumed that an anodic oxide is formed and is in turn chemically dissolved in HF. The overall dissolution reaction (Eq. (1) can then be better written as [46a,b])

$$Si + 2H_2O + 4h^+ \Longrightarrow SiO_2 + 4H^+ \quad (5a)$$

$$SiO_2 + 6HF \Longrightarrow SiF_6{}^{2-} + 3H_2O \quad (5b)$$

i_{crit} value depends on HF concentration. For low HF concentration (<5%) kinetic control of the electrochemical corrosion reaction is obtained at current densities no larger than 50–100 mA cm^{-2}.

3.2.2.3 **Influence of Crystallography**

Different crystallographic planes of a semiconductor electrode usually exhibit different reaction kinetics. It was found in III–V compounds in indifferent electrolytes that the (111)B face terminated with the anion plane (P, As) etched faster than the (111)A face containing the cations (Ga, In) [47]. The planes composed entirely of metal atoms react more slowly than any other crystal plane because of the stable metal oxide layer, which can be formed on such planes. Consequently on these planes termed "*etch stop*" planes, provision of reactants (diffusion control) is not rate-limiting. In Si, the (100) planes are known to etch faster than (111) planes in alkaline solutions. This property is at the origin of various applications, such as texturization of silicon surface [formation of pyramids on (100) planes], which allows reduction of reflectivity of the front surface of solar cells and Si micromachining [48]. The semiconductor surface may be shaped during the anodic dissolution as a result of the crystal reactivity at the different planes.

3.2.2.4 Terminology

Beside the diffusion-rate-limited electropolishing resulting in flat surfaces, two types of corrosion with kinetically controlled etching rate can be distinguished. When the corrosion slightly modifies the surface properties leading to the formation of surface microrelief structures, improvement of the surface chemistry and optoelectronic properties is observed. This corrosion process is named electrochemical etching (EC etching) or photoelectrochemical etching (PEC etching) when it is additionally light-controlled. On the other hand, the corrosion type that transforms the semiconductor into a porous material exhibiting different properties than those of the bulk material is referred to as porous etching or porous photoetching when it is light-assisted.

Following the IUPAC guidelines, porous media in general should be classified as the following: microporous, mesoporous, and macroporous, when the dominant pore sizes are <2 nm, in the 2–50-nm range and >50 nm, respectively [49]. In the porous Si literature, the term *microporous* has often been too loosely applied and bears no resemblance to the criteria accepted by the community studying porous carbons, glasses, and so on. Part of the problem originated from the fact that the optoelectronic properties of porous Si are largely determined by the skeleton size and not pore size. New terminology such as *nanoporous* Si has been frequently used to indicate a solid skeleton with dimensions small enough to constitute a nanostructure. The term nanoporous generally comprises those of microporous and mesoporous Si. To use consistent terminology in this paper, the word hole always refers to a defect electron, whereas a pore is understood as an etchpit whose depth exceeds its width.

Anodization under galvanostatic conditions is generally the preferred approach for reproducibility attaining a wide range of porosity or thickness, but studies under potentiostatic conditions have been performed. Porous Si films have also been generated by stain-etching without applying light or electrical bias. Strong oxidation agents added to the electrolyte are used to remove electrons from the conduction band and lead to chemical-etching by the resulting holes.

3.2.3 Porous Silicon

3.2.3.1 Anodic Dissolution

Several remarkable features characterize the anodic oxidation of Si in fluoride-containing solutions. First of all, the anodization is a multistep process that involves charge-transfer reaction both via the valence band (hole capture) and via the conduction band (electron injection) [8, 50–58]. The dissolution valence number $n = 2$. Hydrogen evolution is observed at potentials far more positive than the Nernst potential of the H/H_2 couple [5, 45, 55, 59]. SiF_6^{2-} is the final dissolution product and it is well established (from Fourier-transform infrared measurements) that during pore formation, the Si surface is mainly hydride-terminated (Si–H) [60, 61]. The two models proposed by Lehmann and Gösele [62] and Gerischer and coworkers [57, 58] take into account all the facts mentioned earlier. Both models include two electrochemical reactions (hole capture and electron injection) followed by a chemical oxidation reaction. It is generally accepted that the formation or decomposition of a crystal occurs preferentially at weakly coordinated sites, such as kink sites (present in steps on the surface or at the intersection of dislocations), where Si atoms have two or three "dangling" bonds exposed to the solution. Si atoms on kink site atoms have a binding energy, that is, on an average, half of that which they would have inside the crystal. In both models it is assumed that a valence band hole is captured in a Si−H surface species, resulting in an electron-deficient Si−H bond. The main difference between the two descriptions of the surface chemistry is the position of the positive charge in the electron-deficient Si−H bond. Lehmann and Gösele assume that the positive charge is located on the silicon side of the bond, resulting in a hydrogen radical [62]. Gerischer and coworkers assume that hole-trapping in the Si−H bond results in a silicon surface radical species and a proton [57]. In both cases the hole-capture step is followed by injection of an electron into the conduction band. The reaction steps at a kink site (in which the Si atom has two bonds linked to the lattice and two other bonds saturated by H-ligands) for the dissolution in aqueous HF are described in Fig. 4, at a molecular level with a two-dimensional representation, following the Gerischer and coworkers' model. The electrochemical reaction is initialized by a hole capture at a Si−H bond, which is thus weakened. The H atom is captured by an F^- ion as a proton H^+, leaving its electron at the Si atom ($\equiv Si^\bullet$ radical). By interaction with F^- ions, the unpaired electron can be excited up to the conduction band edge. The excited unpaired electron is injected into the conduction band, whereas the electrons of the F^- ion form a new very stable Si−F bond. The large electronegativity difference between Si and F is responsible for the polarization of the

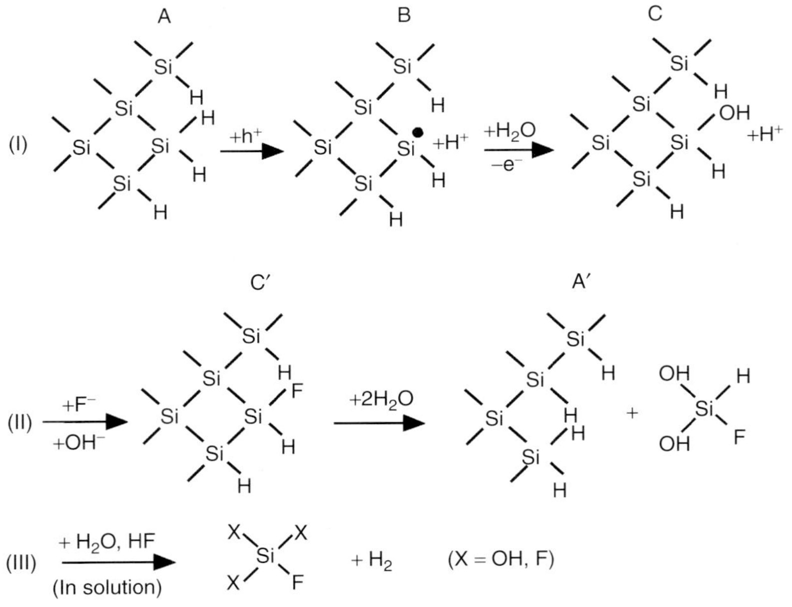

Fig. 4 Two-dimensional description of the reaction steps at a kink site on a silicon surface for the oxidation initiated by a hole (from Ref. 58).

two Si–Si back bonds underneath. This increases their chemical reactivity and favors the chemical disruption of one back bond. The splitting of the back bond may well occur by HF or by H_2O. Two new bonds are formed: Si–H at a back Si, and Si–F at the kink site (or Si–OH with further exchange of the OH ligand by F^-). The remaining Si–Si back bond is further weakened (and even more polarized) and will be quickly split by another HF (or H_2O) molecule. The Si atom at the kink site leaves the surface with a divalent oxidation state as a $HSiF_3$ molecule and the kink site situation is restituted at the next atom of the step (situation A′). The $HSiF_3$ molecule reacts chemically with water to give H_2 gas, as shown in the following reaction:

$$HSiF_3 + H_2O \Longrightarrow SiF_3OH + H_2 \quad (6a)$$

$$SiF_3OH + HF \Longrightarrow SiF_4 + H_2O \quad (6b)$$

However, attempts to detect the hypothetical intermediate $HSiF_3$ species during porous Si formation by in situ Fourier-transform electrochemically modulated infrared spectroscopy have been unsuccessful [61]. Calculations show that if such a species or a similar one exists, their lifetime must be shorter than 0.3 ms.

Undoubtedly, the most convincing evidence in support of the hypothesis of the electron-injection process is the observation of the photocurrent multiplication observed during anodization of n-type in HF solutions. [8, 53–56]. More than one charge carrier per absorbed photon contributes to the photocurrent measured in the external circuit. Photocurrent quantum efficiency, Q, (i.e. number of charge carriers measured in the external current by absorbed photon) is dependent on the incident photon flux. At low light intensity, Q is almost 4 but it drops to 2 as the photon flux is increased. An explanation for the current-quadrupling effect has been sought in terms of a model involving four consecutive electrochemical dissolution steps [53, 55, 56, 59]. In the initial step, a photogenerated hole is captured in a surface bond and a surface intermediate is formed. This intermediate is further oxidized to a Si(IV) species by three consecutive electrochemical steps, in which electrons are thermally excited to the conduction band. The variation of Q with the light intensity is explained by assuming that electron injection from surface intermediates and hole capture by surface intermediates are competing processes in two of the three charge-transfer steps.

3.2.3.2 **Influence of the Si Conduction Type on Current–Voltage Curves**

Although the chemical reactions at the silicon/HF interface are the same for n- and p-type silicon, there is a basic asymmetry in their electronic properties. For the dissolution of silicon under an "anodic bias", the p-type Si is forward-biased and the current is caused by thermally generated majority carriers. The n-type Si is reverse-biased and undergoes charge depletion. The current is characteristic of a minority carrier flow. For low-doped n-Si, anodic dissolution uses photogenerated minority carriers. The i-V curve for p-type Si has been discussed earlier and comparison with the i-V curves for n-Si are presented here. For n-Si under reverse bias condition, a dark current and/or a photocurrent are observed, depending on the doping density and illumination level.

In the dark: The i-V curve is characterized by a breakdown mechanism (Fig. 5, curve d). It occurs in Si for electrical field strength in excess of 3×10^5 V cm^{-1} [63].

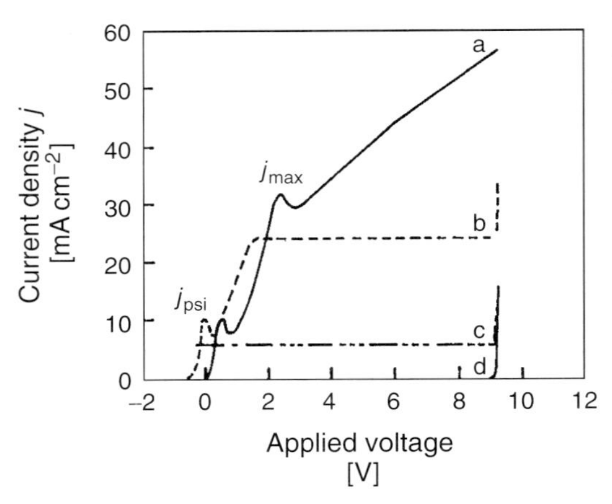

Fig. 5 Current–potential curves vs SCE for n-type lightly doped ($N_d = 10^{15}$ cm^{-3}) Si in 5% HF in the dark and under illumination. p-Type Si (curve a) is indicated for comparison.

When a potential that is large compared with kT is applied to the interface, a relatively large band-bending occurs. The thickness of the space charge layer, also termed *surface charge region* (SCR), becomes very small and the field becomes very strong. Electrons can tunnel directly from the surface Si atoms into the conduction band of the semiconductor, resulting in the dissolution of silicon [64–67]. The onset of the dark tunnelling current depends on the doping density. For flat surfaces, it occurs at high anodic overpotential ($V = 10$ to 15 V) for lightly doped n-type Si ($N_d = 10^{15}$ cm^{-3}) and at low voltage ($V = -0.2$ V vs SCE) for highly doped n-type Si ($N_d = 10^{18}$ cm^{-3}) [67].

Under illumination: The photocurrent induced by the light is not directly given by the number of photogenerated holes generated [68, 69] and as a result does not increase linearly with the light flux because of the electron-injection process. For example, as already mentioned, at very low illumination intensities (less than 5 μW cm^{-2}) for every one and up to three photogenerated holes, electrons may be injected into the conduction band by intermediate complexes, resulting in a doubling or quadrupling of the photocurrent, respectively [8, 53–56]. For lightly doped Si, the total current corresponds to the superposition of two currents: the light-generated hole current and the electron-injection current. The latter is not constant but depends on the absolute current density [68, 69]. For highly doped Si, the dark current caused by the electrical breakdown adds a third contribution to the total current (Fig. 6) [70].

In short, for high illumination intensities, the i-V curve of n-type Si is identical to that of p-type Si, with the exception of a cathodic shift of a few hundred millivolts [44, 45, 69, 71] (Fig. 5, curve b). The photocurrent varies with the light intensity up to 50 mW cm^{-2} [70]. At higher intensities, the current becomes almost independent of the light intensity, indicating that the rate-limiting process is no longer the supply of holes to the surface but rather the dissolution of the corrosion products. If the light flux is limited such that the photocurrent never reaches the i_{crit} value, the photocurrent levels off to a specific constant value independent of the applied voltage (Fig. 5, curve c) [44, 45, 69]. At potentials to the positive side of

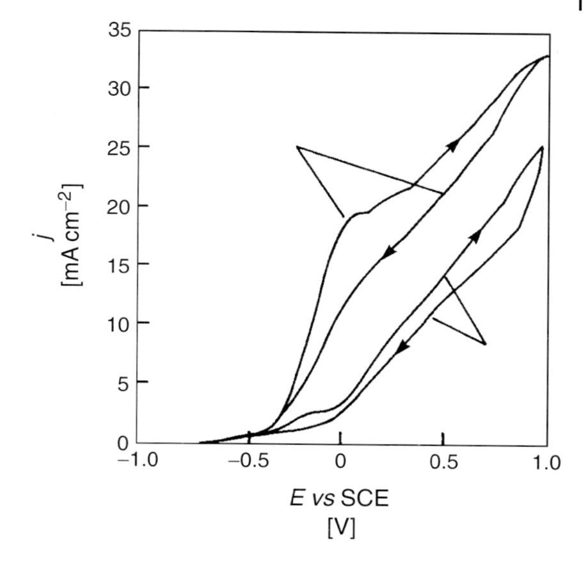

Fig. 6 Current–potential curves for n-type highly doped ($N_d = 10^{18}\ cm^{-3}$) Si in the dark and under illumination in 5% HF; scan rate = 100 mV s^{-1} (from Ref. 70).

i_{crit}, the electropolishing regime is never reached because of the lack of carriers. In this case it is possible to adjust the voltage and current independently by controlling the light flux. In particular, the depth of the SCR can be changed without changing the current density. In Sect. 3.5, a discussion on how this influences the morphology of the porous Si is presented [72–75].

3.2.3.3 Influence of the Solvent on Current–Voltage Curves

Experiments have been carried out in HF-contained acetonitrile (MeCN), dimethylformamide (DMF), and dimethylsulfoxide (DMSO) with tetrabutylammonium perchlorate (TBAP) as the supporting electrolyte [76–80]. The oxidation current for p-type (in the dark, $\rho = 100 - 0.2\ \Omega$ cm) and n-type Si (under illumination) increases linearly because of the combined solution and semiconductor resistance. Neither a critical current peak (i_{crit}) nor an electropolishing region is observed. There was no H_2 evolution when water-free HF was used.

3.2.3.4 Interfacial Potential Distribution and Pore-Formation Conditions

An anodically biased n-type semiconductor is divided into two regions: a space charge region (SCR), which is a depletion region and a field-free region, which extends after the SCR into the bulk region. The thickness of the SCR, W, at a planar surface under depletion conditions is given in Eq. (7). It is a function of the voltage drop in the SCR, V_{SCR}, and the doping level N_d:

$$W = \left[\frac{2\varepsilon_0\varepsilon_1 V_{SCR}}{eN_d}\right]^{0.5} \tag{7}$$

For example, W can vary more than one order of magnitude (from 0.5 to 0.02 μm for a 1 Volt polarization) between lightly ($N_d = 10^{15}\ cm^{-3}$ dopant) and highly doped Si ($N_d = 10^{18}\ cm^{-3}$), respectively.

On the electrolyte side, two regions can be distinguished: the Helmholtz layer, which is a region of specifically absorbed ions, water molecules, and acceptor ions, and the bulk region.

It is important to consider the relative magnitude of the potential drop in each region because the less rate-limiting reaction creates the greatest potential drop and therefore dictates the specific dissolution step. Impedance spectroscopy and capacity measurements of the silicon/HF system have helped to clarify part of the controversy in the potential distribution at the silicon–electrolyte interface [81–84]. The applied potential is partitioned across the electrochemical double-layer and the SCR in the silicon. The fraction of the potential dropped across the SCR is dependent on dopant concentration and etching conditions although *n*-type Si represents a limiting case wherein the applied potential is in the major part dropped across the SCR when no pores are present (flat surface). This means that the potential drop is equal to the difference between the applied potential, V, and the flat band potential, V_{fb} ($V_{SCR} = V - V_{fb}$). For *p*-type and degenerately doped Si, the process is more difficult to analyze quantitatively because of the partitioning of the potential and the nonuniform current distribution at the interface between the Si substrate and the porous layer. Under actual pore-formation conditions for all types of Si samples, one emerging opinion is that both the SCR and the Helmholtz regions control the anodic behavior of silicon. The Helmholtz region controls the charge-transfer reaction rate, whereas the semiconductor SCR controls the specific morphology [83, 84].

3.2.3.5 Many Different Morphologies

The formation of porous Si critically depends on the type of conductivity and doping level of the Si electrode, HF concentration in the electrolyte, nature of solvent in which HF is diluted, and applied current density (or voltage) [5, 7, 10, 11, 71, 85–87].

Porous Si exhibits distinctive appearances (Table 1). The size of the structures varies by three orders of magnitude, ranging from nanoporous silicon with pores and crystallites in the nanometer scale up to macroporous silicon with pore and pillar dimensions in the micrometer scale. Each of these variations has its origin in different physical effects. The main pores grow in ⟨100⟩ directions. Most pores extend from the surface down to the porous Si/bulk Si interface. Although side pore formations and branching may be extensive, pores are never seen to interconnect. Porous Si is macroscopically and roughly characterized by its porosity, f_{void}, which is the fraction of voids in the porous layer.

3.2.3.5.1 Nanoporous Silicon Including Micro and Mesoporous Si

It consists of a spongelike randomly distributed array of nanometer-sized silicon features. When the size of the features is less than 5 nm, quantum size effect caused by the confinement of the excitons is assumed, which gives rise to room temperature photoluminescence and electroluminescence in the visible range (1.6–1.8 eV). These properties have been intensively investigated for applications of light-emitting devices [6, 88], although porous Si is highly resistive in air ($\rho = 10^5 – 10^7\ \Omega$ cm) [89].

Microporous Si It originates from low-doped *p*-type silicon (1 Ω cm) [5–7, 85, 89, 90]. By employing modified transmission electron microscopy (TEM) specimen preparation such as direct specimen cleavage, it has been observed that the coral-like structure of anodized *p*-type Si is composed of highly interconnected micropores in the size range ≤2 nm (Fig. 7). High resolution transmission electron microscopy (HRTEM) images showed that strongly luminescent samples (>70% porosity) exhibit narrow undulating Si columns with

Tab. 1 Different varieties of porous Si.

Type	*Doping density*	*HF concentration*	*Current density/ voltage*	*Morphology*
p	10^{15} cm^{-3}	25% aqueous HF	50 mA cm^{-2}	Microporous [6]
p^+ and n^+	10^{17}–10^{19} cm^{-3}	25% aqueous HF	3–300 mA cm^{-2} breakdown tunneling	Mesoporous, [86, 89]
p	10^{14}–10^{15} cm^{-3}	30% aqueous HF	40 mA cm^{-2}	Macroporous [102]
	10^{16}–10^{17} cm^{-3}	8% HF in DMF or DMSO	10 mA cm^{-2}	Idem [80]
n	10^{15}–10^{18} cm^{-3}	5% aqueous	15 mA cm^{-2} front-side illumination	Duplex layer (nanoporous + macroporous) [70]
n, (100) oriented	2.10^{15} cm^{-3}	6% HF aqueous	10 mA cm^{-2} back-side illumination	Array of vertical macropores developed on prestructured nuclei [72]
n	2.10^{15} cm^{-3}	5% HF aqueous	10 V	20 μm deep channels [67]
	3.10^{18} cm^{-3}		2–15 mA cm^{-2}	Duplex layer (nanoporous + macroporous) [93]

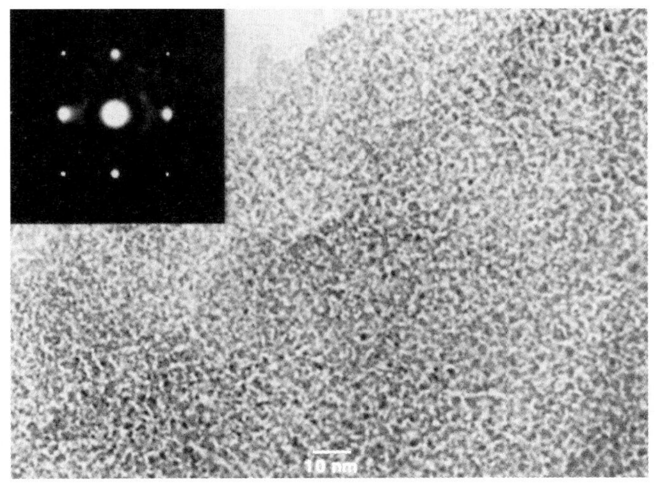

Fig. 7 TEM image and diffraction pattern from microporous Si formed on low-doped *p*-type Si (from Ref. 11).

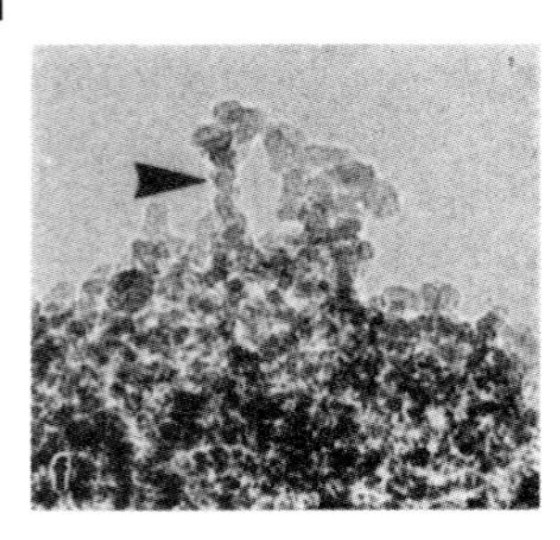

Fig. 8 TEM image (bright-field, under focus) of thin, high-porosity luminescent microporous. Columnar Si structures are arrowed (from Ref. 7).

diameter size down to <3 nm (Fig. 8). The porous material is completely crystalline. This material also presents a very large internal surface area (1000 m^2 cm^{-3}) and is therefore suitable for catalytic surface reactions. Freshly prepared microporous Si luminesces at room temperature. The position and intensity of the maximum of the luminescence peak depends on the porosity. It varies from 1000 nm to 880 nm for 64% and 77% porosity, respectively. Porosity of microporous Si, further subjected to a continuous chemical etching in 40% HF, increases from 70 to 80%. As a result, the photoluminescence output rises dramatically and the peak position is blue-shifted toward 780 nm (Fig. 9). Because the removal of Si constitutes an overall decrease in Si skeleton, this experience was considered as an indirect evidence that visible room temperature photoluminescence could be attributed to quantum size effect [6].

Mesoporous Silicon This kind of porous silicon forms on heavily (degenerated) doped silicon substrates, either n^+-type or p^+-type (0.01–0.001 Ω cm), under

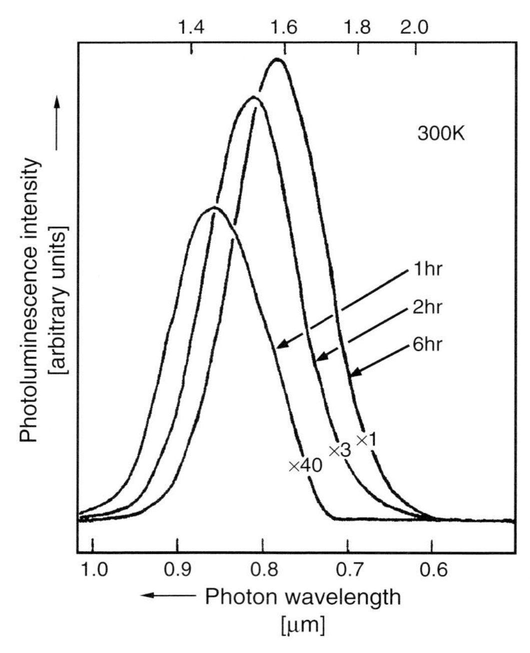

Fig. 9 Room temperature photoluminescence spectra of a freshly prepared microporous Si as a result of a partial chemical dissolution in 40% aqueous HF for the times indicated (from Ref. 6).

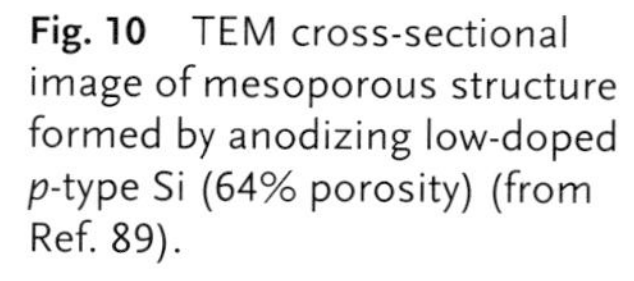

Fig. 10 TEM cross-sectional image of mesoporous structure formed by anodizing low-doped *p*-type Si (64% porosity) (from Ref. 89).

conditions that would produce micropores in low and moderately doped Si. The pore and pillar sizes are in the 10–50 nm range [5, 11, 85, 86, 89, 91, 92] (Fig. 10). The macroscopic surface is slightly darker-looking in comparison to bulk silicon, and of excellent smoothness. The internal surface area is important (100 $m^2\ cm^{-3}$). Mesoporous Si (like microporous Si) can be easily etched in low-concentration alkaline solutions. The mesoporous Si has been exploited for the fabrication of silicon on insulator (SOI) structures [94], using variations in porosity in buried layers with altered doping. It is also used as a sacrificial layer in microelectronic and thin film microcrystalline solar cells [95]. Mesoporous Si layers, depending on the porosity and layer width, present remarkable antireflective properties and as such can be used as an antireflective coating in crystalline Si solar cells [96–98]. Mesoporous Si exhibits a weaker luminescence than microporous Si. This result suggests that quantum size Si particles are present within the mesoporous structure.

3.2.3.5.2 **Macroporous Silicon** n-Type silicon (low, medium, and high) [67, 69–71, 99, 100] and very low-doped p-type (<5 Ω cm) are made macroporous in aqueous HF electrolyte [101, 103]. Low- and medium-doped p-type (5–0.2 Ω cm) is made macroporous in HF-containing organic electrolyte [79, 80, 03]. Macroporous Si consists of micrometer-sized pores and silicon pillars. The internal surface area is only 1 $m^2\ cm^{-3}$.

Macropore on *n* and n^+-Type Si Under Illumination Macropore formation on n-type in aqueous HF under illumination was studied in detail in the 1990s [70–72, 99, 100]. Macropore formation is anisotropic and occurs preferentially along the ⟨100⟩, ⟨010⟩, and ⟨001⟩ directions. Figure 11 shows that anisotropy should result in the formation of square-shaped tubular pores on a (100)-oriented substrate and triangular pores on a (111)-oriented substrate [71]. This has indeed been experimentally confirmed and macropores formed on (100) substrates are tubular with a square-sectional form and a round bottom (Fig. 12a) [70]. In the case of (111) substrates, they have an almost triangular section (Fig. 12b). In both cases the macropores delineate a regular network. The pore-to-pore distance equals twice the value of the SCR, which can be adjusted by doping level and applied voltage. The pore size can be controlled by doping, HF concentration, and applied current density.

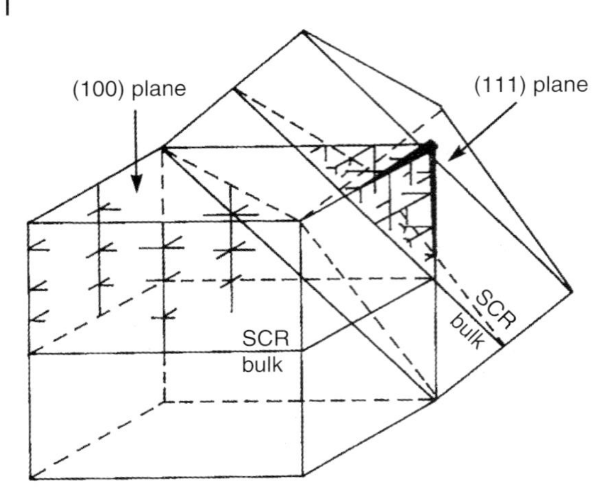

Fig. 11 Schema showing that the porous photoetching of Si occurs preferentially along the ⟨100⟩, ⟨010⟩, and ⟨001⟩ directions. Pores with a square shape and triangular shape are expected to be formed for (100)- and (111)-oriented substrates, respectively (from Ref. 71).

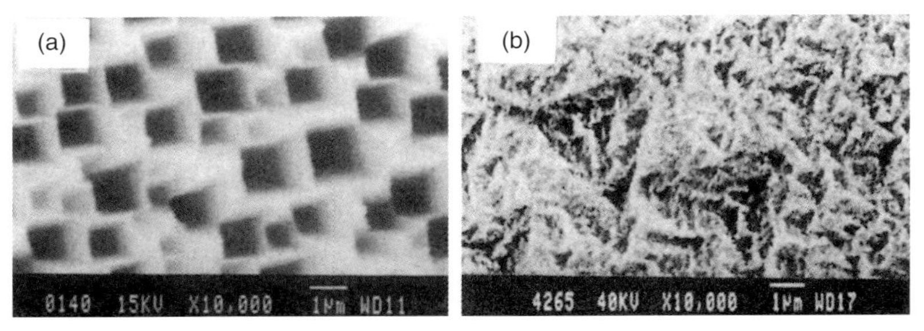

Fig. 12 Plan view SEM images of macroporous Si formed on highly doped ($N_d = 10^{18}$ cm^{-3}) *n*-type Si with a charge of 10 C cm^{-2} and after dissolution of the nanoporous layer. (a) (100)-oriented substrate; (b) (111)-oriented substrate (from Ref. 70).

The active state of the pore tips is due to the collection efficiency required for the holes initiating the dissolution process. The passivation of the pore walls is ascribed to a depletion of holes in the walls between the pores. This has been confirmed by spectroscopy impedance measurements on macroporous n^+-type Si [104].

When the silicon–electrolyte interface is front-side illuminated during the anodization, the macropore Si walls and the top of the Si pillars are covered by a nanoporous silicon layer (Fig. 13) [70, 91, 92, 99, 105, 106]. Cross-sections of electrochemically detached nanoporous layers well illustrate the morphology of the tubular and triangular macropores formed on (100)- and (111)-oriented substrates (Fig. 14). The nanoporous Si is characterized through a network of tangled filaments of nanometer width, some of them exhibiting quantum size effect (Fig. 14) [10, 91]. This nanoporous Si that appears fibrous was shown to be the source of the visible luminescence [92]. This is not the case when the interface is illuminated from the backside of the silicon substrate leading to macroporous Si formation only [72]. The possibility of controlling pore location with a standard photolithography technique and producing (with rear-light-controlled corrosion)

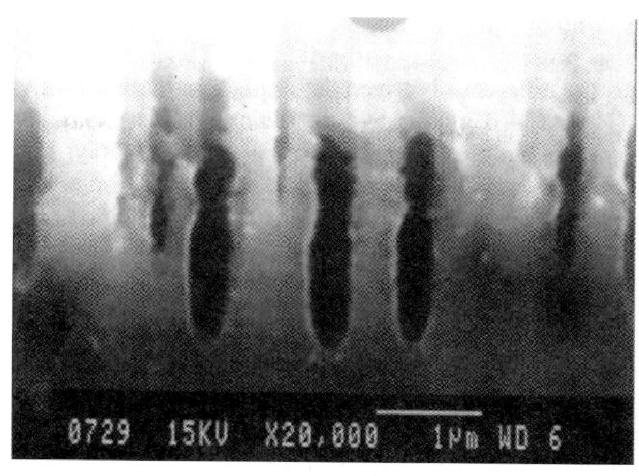

Fig. 13 Cross-sectional SEM image of porous Si formed by porous photoetching on (100) low-doped Si ($Q = 4$ C cm^{-2}), revealing the distribution of the nanoporous (top layer) and the macroporous layer (bottom layer) (from Ref. 70).

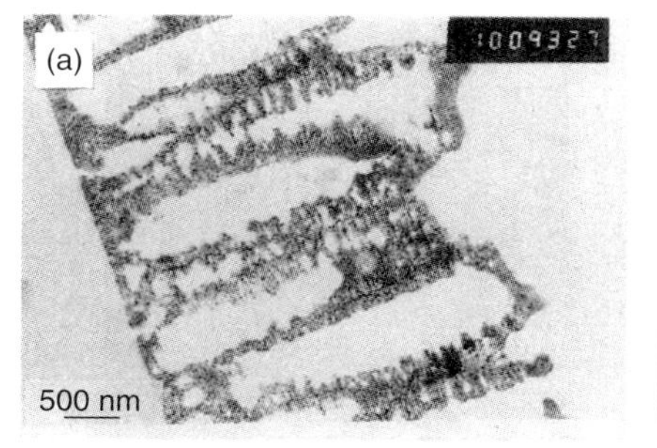

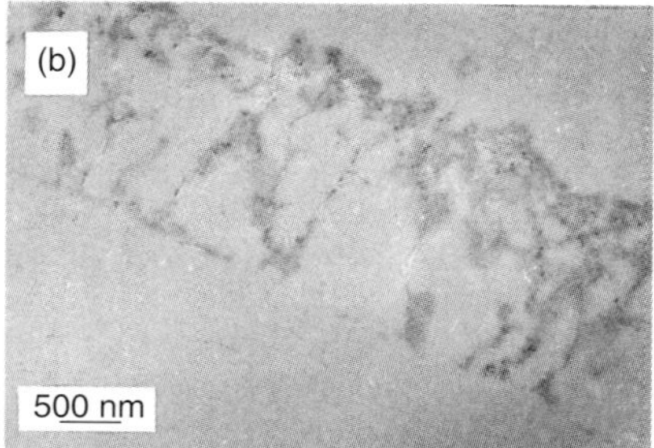

Fig. 14 TEM micrographs (cross-section) revealing the typical internal structure through the thickness of the nanoporous Si formed on highly doped n-type Si ($N_d = 10^{18}$ cm^{-3}) with a charge of $Q = 10$ C cm^{-2}: (a) (100) oriented; (b) (111) oriented (from Refs. 91).

well-defined macroporous structures in Si, with a pore aspect ratio of 250, has led to new technological applications such as two-dimensional photonic crystals with a band gap in the near and mid-infrared regions [75, 107–111], microcapacitors [73, 74], etc.

Macropore on n^+-Type Si in the Dark Macropore formation was first observed in 1972 by Theunissen who found formation of 20 μm deep channels that occurred on low-doped ($N_d = 2.10^{15}$ cm^{-3}) n-type Si in 5% HF under breakdown conditions (10 V bias) [67]. Recently, it has been shown that highly doped ($N_d > 2.10^{18}$ cm^{-3}) porous Si obtained under breakdown conditions (0 V/SCE) is constituted of a duplex layer of mesoporous and macroporous Si. Pore diameter (15–70 nm) and shape (round or square section) of mesopores depends on the amount of charge and current density used for its formation. Pore diameter of macroporous Si is constant and of the order of 0.3 μm [93]. The mesoporous Si layer can be dissolved in 0.1 M NaOH, whereas the macroporous Si was only dissolved in HF (48%): HNO_3 (69%) (1 : 9 ratio volume). This permits measuring the thickness of each layer by profilometry. Depending on the current density and amount of charge used during porous-etching, thicknesses up to 5 μm and 4 μm for nanoporous and macroporous layers respectively, have been measured [112]. The morphology resembles that shown in Fig. 13 for macroporous Si obtained under illumination, except that pore diameter and pore wall thicknesses are two times smaller (Fig. 15). When porous-etching is done in an organic electrolyte, the macropore shape is better defined and the diameter is larger (0.3–1 μm) [113].

Macropore on *p*-Type Si Macropore formation in *p*-type Si was recently evidenced [77–80, 101–103, 108]. Macroporous *p*-type Si was first observed on low-doped (>10 Ω cm) substrates. It can be formed in HF contained aqueous and organic solutions (Fig. 16a). Using specific HF/organic electrolytes

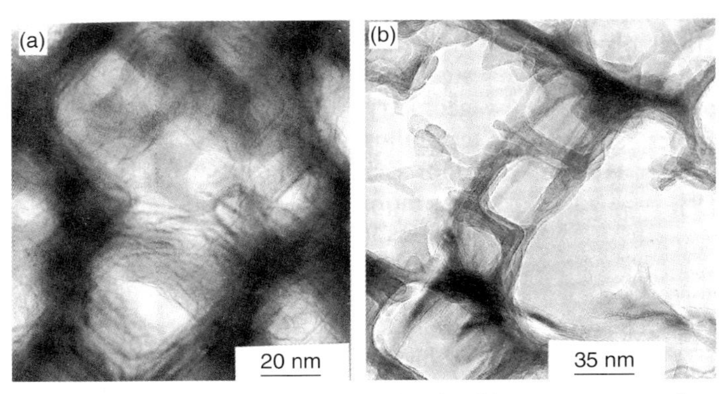

Fig. 15 HRTEM micrographs showing details of the nanometer-sized fibers (plan-view): (a) (100) oriented, $N_d = 10^{15}$ cm^{-3}; (b) (111) oriented, $N_d = 10^{18}$ cm^{-3} (from Refs. 91, 92).

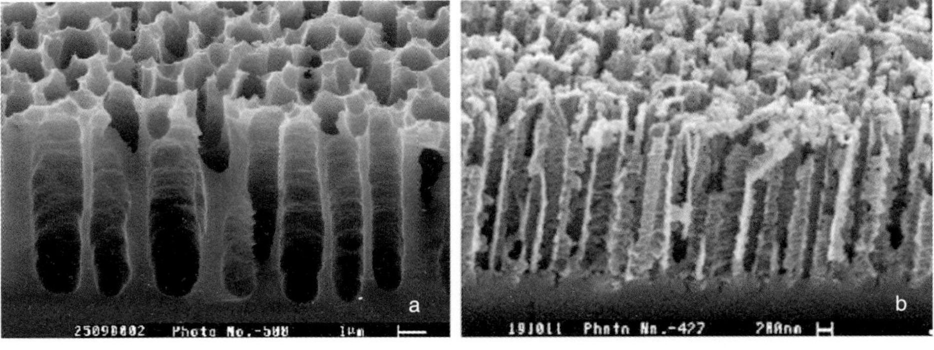

Fig. 16 Cross-sectional SEM images of macroporous Si formed on (100) *p*-type Si. (a) $N_a = 10^{15}$ cm^{-3} in 10% HF/dimethylsulfoxide; (b) $N_a = 10^{17}$ cm^{-3} in 8% HF/dimethylformamide.

[dimethylformamide (DMF) and dimethylsulfoxide (DMSO)] permitted macropore formation on (0.2 Ω cm) medium-doped Si (Fig. 16b) [80]. The macropores diameter and depth depend on the doping density. The diameter varies from a few microns to 0.2 μ. The macropore depth can reach several hundred microns when organic electrolytes are used [114].

3.2.3.6 **Models of Porous Si Formation**

Several qualitative models have been proposed to explain porous Si formation but none of them allow full explanation of the rich variety of morphology exhibited by porous Si and, in particular, the formation of the duplex layers (nano + macroporous). In addition, they possess very little predictive power. A majority of the models focussed on the pore propagation, whereas the mechanism of pore initiation received very little attention. A comprehensive review of the various models proposed to explain pore formation is found in excellent review articles by Smith and Collins [5], Parkhutik [12], and Chazalviel and coworkers [13]. Two main categories of models have been proposed. The first one is basically electrostatic in nature, based on the consideration that physical effects associated with the SCR play a major role in the pore-formation mechanism. The second category is based on computer simulations.

3.2.3.6.1 **Pore Initiation** The few proposed models are related to the surface chemistry of Si during pore initiation. In general, defects are invoked as active sites for pore initiation. However, scanning tunneling microscopy (STM) observations have clearly shown that in initial stages the dissolution is isotropic and uniform on the nanometer scale, meaning that it is independent of defects [115].

Kinetics of pore nucleation on (100) *p*-type (0.05 Ω cm) and *n*-type (1–4 Ω cm) was observed as small steps in fast-current and potential pulse transients. It was shown that the formation of porous Si could be better explained and correlated with experimental results if a silicon oxide intermediate was considered [116].

From the following observations that (1) Si dissolves with formation of molecular H_2 and protons, (2) only Si forms microporous layers with a single crystalline skeleton, and (3) hydrogen (Deuterium) species penetrate into the Si substrate upon porous Si formation, a model was proposed to explain pore initiation. It was assumed that hydrogen incorporation induces structural defects in Si, which may act as active sites for the localized Si dissolution. The model relates pore initiation to the selective dissolution of the hydrogen-induced structural defects at the surface of bulk silicon [117, 118].

Microetchpits One model developed to explain etchpit formation on II–VI semiconductors [119–121] has been extended to porous Si. This model attempts to address the initiation of the *n*-type macroporous morphology through nonuniformities in the photoetching current caused by nonuniform microfields existing around the dopant atoms nearest to the semiconductor–electrolyte interface [100], or alternatively near the semiconductor surface because of a trapped (localized) positive charge formed in the first instant of the photocorrosion at the interface. A calculation of the relevant parameters as a function of the doping density of the silicon wafer shows that up to $N_d = 10^{17}$ cm^{-3}, the Debye screening length is larger than the mean distance of the first donor atoms. This means that up to this doping value the donor charges are not shielded and that the donors are expected to have an influence on the etchpit pattern. The first layer of dopant (with a mean distance $a = 1/(2N_d^{1/3})$ from the semiconductor surface) contributes to a nonuniform photocurrent, which is high near the dopant atoms and can be ascribed to a photoinduced avalanche effect with a multiplication factor $M(E_a)$ where E_a is the electric field near the dopant atom [100, 122]. An important consequence of this model is that the surface layer becomes dopant-poor after PEC-etching. This assumption was supported by the reduced surface recombination velocity, as indicated by photovoltaic measurements, which show substantial gain in the short wavelength part of the photocurrent spectrum and increased fill factor, following PEC etching [123]. Once the initiation process sets in, other mechanisms, such as photocurrent focussing in the tip, may take place. A redistribution process starts. Some pores stop growing and terminate, whereas other pores continue to grow with increasing diameters. Lehmann and Föll [71, 72] showed that pore initiation at the surface is determined by the doping density of the Si substrate. From the initial 10^{10} cm^{-2} PEC-etched pits formed, only one in a hundred or one in a thousand survive and become pore tips. The density of the resulting pore tips is determined by the SCR width and therefore by the doping density.

3.2.3.6.2 Pore Propagation

Models Based on the Role of the Space Charge Region Two fundamental aspects of porous-etching are the electrostatics and electrochemical kinetics. Electrostatics is crucial in determining the sites wherein semiconductor valence band holes are available for the dissolution reactions (e.g. at the etchpit initiation sites or at the bottom of the pores). Electrochemical kinetics determines how fast holes react with the surface at sites when they are available. Specific surface chemistries can make the reactivity for holes strongly different from one semiconductor–electrolyte system to another. The disadvantages of the most popular models proposed to explain porous Si formation are that they do not take into account the chemical aspects that may influence the kinetics at the Si/electrolyte interface.

For pores to propagate, the pore walls have to be passivated and the pore tips to be active in the dissolution reaction. Consequently, a surface, which is depleted of holes, is passivated to porous-etching, which means that anodization is self-limiting.

One must emphasize the distinct role of the SCR for n- and p-type Si. In the case of p-Si, it acts as a barrier for the holes, and only near the pore tips is the field sufficiently large for opening gaps at the top of the barrier. The reaction may proceed entirely over the valence band ($n = 2$). In the case of n-type Si, the SCR rather acts as a sink channeling the holes to the pore tips. The porous Si may be formed on n-type Si under front and back illumination or under strong polarization (avalanche breakdown). On heavily doped n- and p-type Si, porous Si may be obtained by hole-tunneling. Eventual presence of intermediate Si oxide may favor the electron-tunneling process through the conduction band. In summary, the effects responsible for pore wall passivation (top row) and for passivation breakdown at the pore tip (middle row) as well as the resulting kind of porous Si structure together with substrate doping type (bottom row) are presented in Fig. 17 [86].

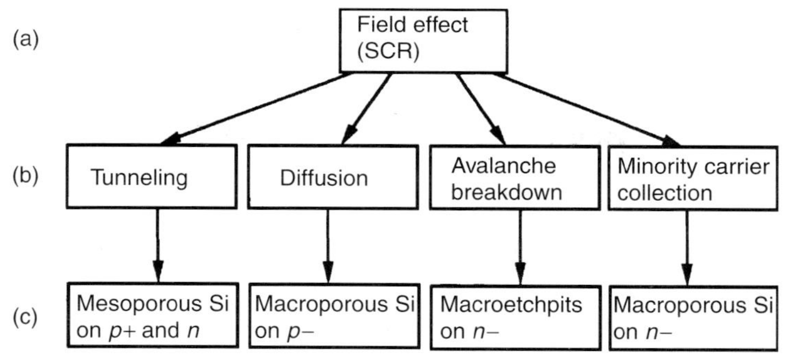

Fig. 17 Field effect is proposed to be responsible for pore wall passivation (a). Effects responsible for passivation breakdown at the pore tip (b) and the resulting porous silicon structure together with substrate doping type (c) (adapted from Ref. 86).

Models Relevant to Micropore Formation The mechanism of microporous Si formation from p-type is controversial. A puzzling question remains as to the unusually high stability of thin fibers or nanoparticles of Si to electrolytic attack or, in other words, what prevents nanometer-thick fibers/particles from being dissolved altogether as the etching front moves microns into the crystal bulk.

The first model, proposed by Beale and coworkers [85, 89], assumes that at potentials corresponding to pore growth the Fermi level of Si is pinned near the middle of the band gap due to of a high number of surface states. This results in a potential drop at the interface, which restricts the hole transfer from the substrate. Anodic currents may flow as a result of thermionic emission across the potential barrier. The local current flow is then determined by the potential height, which in turn is dependent on the local electric field. High electric field favors hole transfer. Hence, the electric field strength being greatest at the base pore, this results in dissolution only at the pore tips. Therefore, following pore initiation, the pores propagate as a result of the focussing effect of the electrical field. As a pore grows, the radius of curvature increases and the local field decreases and the pore growth decreases. To maintain the total current density, new pores start to grow. As soon as the fiber/particle diameter in microporous Si gets thin, its resistivity is sharply increased by the Fermi level pinning to surface states. Because the electrochemical dissolution is essentially a current-controlled process, further etching is impossible and a thin fiber/nanoparticle is stable. This model is often disregarded because Fermi level pinning by surface states must be assumed. Unfortunately electron paramagnetic resonance (EPR) measurements indicate that there is a low density of surface states in the porous layer [124, 125]. This has been confirmed by Searson and coworkers, using in situ impedance spectroscopy to measure interface states at [(111), $N_d = 10^{13}$ cm^{-3}] Si surfaces in fluoride solutions [126, 127]. They found, at pH between 3 and 6, a very low density of electrically active surface states equal to 2×10^{10} cm^{-2} corresponding to about one in every 10^5 surface states.

Nevertheless, the idea of barrier-lowering may remain relevant. For example, if part of the applied potential appears in the Helmholtz layer [83, 84], the barrier will be decreased by the corresponding amount. If the Helmholtz potential drop is proportional to the surface electric field, the barrier lowering will be inversely proportional to the SCR thickness.

A very often cited model is the so-called "quantum wire model" [62]. The aim of the model developed by Lehmann and Gosële was to explain the morphology observed for porous Si, which luminesces as a result of quantum confinement. As the dimensions of the particle decrease, the band gap increases (from 1.1 to 1.5 eV). Because of quantum confinement, the energy of valence band states is lowered in the walls between the micropores. The holes are transferred to the interface only at the pore bottoms. The quantum confinement model may be regarded as chemical passivation of pore wall dissolution, where the nanocrystalline silicon particles are the passivating species. However, this model seems to be contradictory to the well-established properties of isotype heterojunctions. A barrier height of a few hundred meV is known to be largely insufficient for blocking majority carrier transport in a heterojunction at room temperature for both the directions of electric current [128]. In addition, the "quantum wire model" has nothing to say about the formation of all other types of pores.

Models Relevant to Mesopore Formation
The Beale and coworkers' model has been extended to explain mesopore formation on degenerately doped p- and n-type Si (in the dark) for which hole-tunneling is the major charge-transfer mechanism [85, 89]. The probability of a hole-tunneling across the Si-electrolyte interface depends on both the potential barrier height and SCR width. The total electric field at the pore tip results in both a lowering of the potential barrier and a decrease of the SCR width. Hence, as mentioned earlier, the propagation of pores is favored. As the pores grow, the depletion layers of adjacent pores may overlap. This results in the probability of tunneling becoming negligible at the pore walls; that is, the reaction is passivated at the pore walls. Overlap of the depletion layers prevents branching and explains the observed columnar structure.

Zhang [66] and Searson [9] confirmed that model and evoked a "field-enhancement effect" at the tips of the pores wherein the current flow is controlled by a tunneling mechanism (Fig. 18). Their

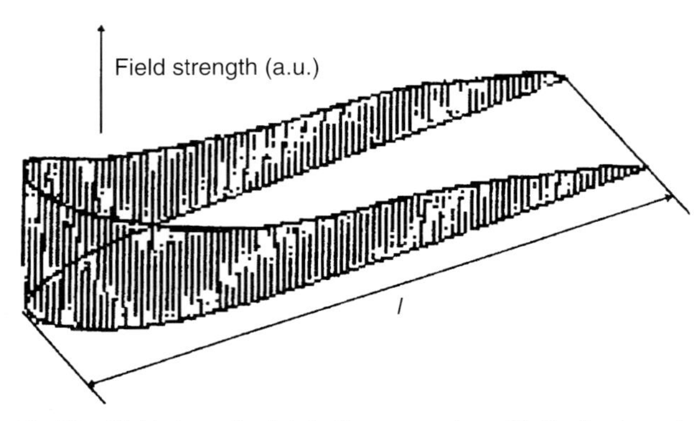

Fig. 18 Field strength distribution around an elliptically shaped pore tip (from Ref. 8).

model predicts the formation of highly oriented parallel pores for (100)-oriented substrates and gives a reasonable explanation for the regular pore distribution and spacing by considering the depletion layer width. The calculation of the current distribution around the pore front using a two-dimensional solution to Poisson's equation shows that the current is considerably greater at the pore tip as compared with the pore wall [9], accounting for the unidirectional growth of the pores. In the case of multiple adjacent pores, the equipotential lines in the region between the pore walls are shifted further away from the surface and into the bulk, effectively depleting the pore wall regions. The pore spacing is determined by the distance at which the region between the pore walls becomes completely depleted and is comparable to the calculated SCR width at a planar surface.

In 1972, Theunissen [67] observed formation of wide-etched channels for donor concentrations of less than $N_d = 2 \times 10^{17}$ cm^{-3} under high voltage (>10 V), in 2.5–5% HF, when anodization was performed in the dark. He proposed that at some pore tips the electrical field strength was sufficient to allow avalanche breakdown at the depletion layer, which then generated the necessary carriers for further pore growth. This was corroborated recently by Lehmann et al. [86] showing experimentally and by simulations based on the electrical field distribution present at the pore tip and pore walls that the mesopore formation is dominated by charge carrier tunneling, whereas avalanche breakdown is responsible for the formation of large etchpits.

Models Relevant to Macropore Formation
Macropores can be formed under illumination on n- and n^+-Si and in the dark on n^+-Si as well as on low- and medium-doped p-type Si.

1. n-type porous Si under illumination: There is a reasonable agreement on the formation of macroporous Si from n-type Si [71, 72]. The formation mechanism of n-type macroporous Si is ruled by the reversed-biased SCR at the solution interface.

Front-side illumination: The condition at the pore tip of an illuminated n-type electrode is different from that in the dark, because the presence of a breakdown field is not necessary to generate charge carriers. Every depression or pit in the surface of the n-type silicon anode bends the electric field in the SCR in a way that the concave surface regions become more efficient in collecting holes than the convex ones. Concave regions are etched preferentially and the pores start to grow, consequently enhancing the local current density [68].

After initiation of microetchpits, macropore formation occurs. The density of the resulting pore tips is determined by the SCR width and therefore by the doping density. The thickness of the remaining silicon walls between the pores is two times the SCR width and as a result the pore walls are charge carrier–depleted (Fig. 19). This finding has been confirmed by impedance spectroscopy studies [104].

Until now, no model has been proposed to explain the simultaneous formation of the duplex nanoporous–macroporous structure, when the n-Si and n^+-type Si are illuminated from the front side.

The striking property of the orientation dependence of the nanoporous and macroporous structures may be correlated with the observation that the critical current i_{crit} is greatest for the (100)-oriented substrate [72]. One possible explanation considers the process in which one Si–H

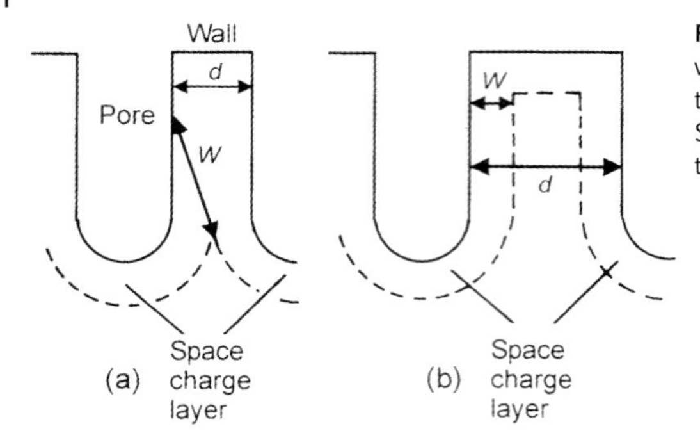

Fig. 19 Schema showing the effective width of the SCR. (a) when the wall thickness, *d*, is smaller than twice the SCR thickness, W, (b) when *d* is larger than twice W (from Ref. 66).

bond is substituted by a Si–F bond and may well be an indication that the surface chemistry is an extremely sensitive function of the silicon crystallographic orientation. For instance, the unreconstructed Si (111) surface is ideally terminated by one H atom per Si, forming a monohydride (≡SiH) [129], and the Si (100) surface by two H atoms per Si, forming a dihydride (=SiH_2) [130]. It has been proposed that because of the presence of the two H atoms per Si the dihydride-terminated (100) surface is sterically hindered [77, 78, 31]. The steric hindrance causes bond strain, enhances the chemical reactivity, and the dissolution will occur faster along the (100) planes [5, 72].

Rear illumination: When anodization is performed under illumination from the backside, holes move to the interface by diffusion until they are captured by the SCR, which accelerates them toward the pore bottoms. Lehmann [72] has clearly demonstrated the validity of the macropore propagation model by studying, in aqueous HF solution, the formation of a regular orthogonal pattern of cylindrical macropores on a rear-illuminated substrate. He initiated the pores by using standard lithography to produce a predetermined homogeneous pattern of pits and subsequently developed the pores by alkaline etching. Illumination with a suitable wavelength from the rear of the Si wafer favors the collection of the photogenerated holes in the bulk and their migration toward the tip of the initiated pits. Such an experiment is only possible for lightly doped *n*-type Si, because the diffusion length of the holes is comparable to the sample thickness. Stable macropore growth occurs when the current density is limited by hole generation and not by the applied bias. In addition, if the local current density at the pore tip, i_{tip}, exceeds the critical current density ($i_{\text{tip}} \geq i_{\text{crit}}$), no nanoporous silicon will form at the pore tip because of the presence of the electropolishing regime. For a given initiation pattern, the pore diameter, d_{p}, and wall thickness, W, are determined by the ratio between the applied current, i, and the critical current, i_{crit} [Eqs. (8) and (9)].

$$d_{\text{p}} = p\left(\frac{i}{i_{\text{crit}}}\right)^{0.5} \tag{8}$$

and

$$w = p\left[1 - \left(\frac{i}{i_{\text{crit}}}\right)^{0.5}\right] \tag{9}$$

The diameter of the macropores etched at different current densities, i, is a linear function of the square root of i/i_{crit} following Eq. (8). It was found that the wall thickness could be up to a factor 10 smaller or wider than the SCR width. The rate of pore growth is a function only of the critical current density i_{crit} (which is a function of the HF concentration and temperature).

The stable macropore formation obtained for an applied bias is sufficient to generate the critical current density (about 1 V). This understanding of the formation mechanism has allowed good control of the geometrical parameters of macropore arrays (Fig. 20). It has been shown recently that 100-μm wide pores with 2-μm wall thickness and 200-μm depth (Fig. 21) can be formed on highly resistive (2000–5000 Ω cm) n-type Si [132]. The demonstration that there is no restriction concerning the wall spacing opens the route to form vertical structure by (photo)electrochemistry.

Fig. 20 Cross-sectional SEM images and 45° bevel of n-type Si samples ($N_d = 10^{15}$ cm^{-3}) showing the predetermined patterns of macropores. (a) orthogonal array, (b) hexagonal array. Pore growth was induced by regular patterns of etchpits produced by standard lithography and subsequent alkaline etching (inset upper right) (from Ref. 72).

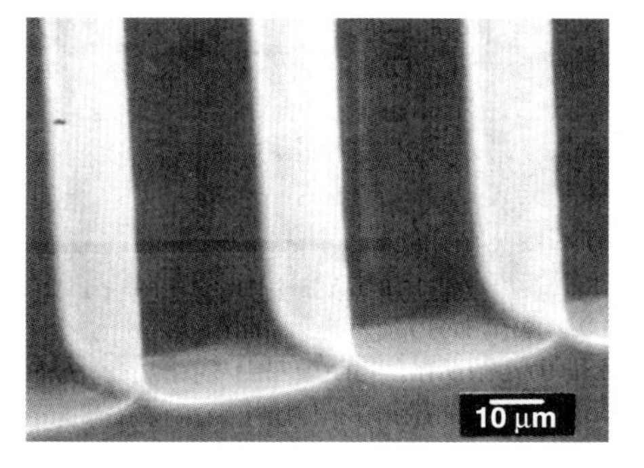

Fig. 21 Cross-sectional SEM image of macropores obtained from a prepatterned n-type Si (2000–5000 Ω cm) in 3% HF electrolyte with backside illumination (from Ref. 132).

A recent systematic study of macropore formation performed on various doped n-type Si substrates with rear illumination, by Föll and coworkers [106] showed that a strong influence of the SCR on the average macropore density is indeed observed in accordance with the Lehmann model [72] (i.e. an increased anodic bias decreases the density of pores), except for highly doped Si. It was observed that an increasing anodic bias increases the pore density, in contrast to the prediction. The pore growth seems to be dominated by the chemical-transfer rate and most likely calls for a chemical passivation mechanism of the macropore walls.

2. p-type Si: Formation of macroporous Si is surprising at first sight because in contrast to n-type Si, p-type electrodes are under forward bias conditions. An extended SCR is therefore not expected for a p-type electrode in the anodic regime. To explain macropore formation process on p-type Si, Lehmann and Rönnebeck [102] put forward the key role of diffusion across the space charge region. Especially, the field enhancement and the associated narrowing of the SCR at the pore tips are assumed to be responsible for the pore tip dissolution and pore wall passivation. This model predicts macropore formation up to $N_a = 10^{17}\ \text{cm}^{-3}$ doping level, but fails to explain why macroporous formation on moderately doped p-type Si (1–0.2 Ω cm) is only observed when HF-contained organic protophilic solvents (DMF and DMSO) are used [80].

Another model proposed by Kohl and coworkers [77, 78] is based upon the strain – induced preferential etching described earlier. The model accounts for the formation of macropores and highly branched micropores when the silicon is rendered porous in either nonaqueous or aqueous HF solutions, respectively. They suggest that the contrast between aqueous and nonaqueous etching can be attributed to two factors, the competition of OH^- with F^- for complexing Si, and the kinetically slow dissolution of oxide (or hydroxide) species formed in aqueous solutions.

Computer Simulations These have been attempted to obtain morphologies similar to those observed in porous Si. The models proposed to explain porous Si formation are similar in spirit to those previously used to understand the complex crystal-growth phenomena. Although porous Si formation is a dissolution process, similarity is found with the growth phenomenon. The models fall in two categories. The first type corresponds to the popular diffusion-limited aggregation (DLA) model, which is based on the diffusion of an electrostatic species such as hole (electron) to (from) the interface [133]. The second type is the Mullins-Sekerta instability model [134]. It consists of analyzing the linear instability of the Si/electrolyte interface by taking into account both the transport of holes in the semiconductor and ions in the electrolyte together with the surface tension of Si. The DLA model has first been used by Smith and coworkers to explain pore formation as a result of the diffusion of holes to the interface [135, 136]. Yan and Hu adopted a different approach by modeling the interfacial dynamics governing the formation of porous Si using a two-dimensional two-component resistor network model [137]. The ratio of the resistances of the two networks is used as a control parameter to simulate the advance of the interface according to stochastic dynamics, which involves local current. John and Singh developed a diffusion-induced nucleation model for the formation of porous Si based on two primary processes [138]. The diffusion of holes from the bulk to the surface is controlled mainly by (1) the SCR width (w) and (2) the drift-diffusion length (l) of holes inside the lattice. The theoretical models permit obtaining the porous Si morphology, which looks very similar to what is observed in practice but generally lack physical substantiation and serve as an illustrative facility rather than an analytical tool relevant for both scientific and practical applications.

The concept of the instability model used to explain porous Si formation was first introduced by Kang and Jorné [139, 140]. The model considers the pore nucleation at the Si surface as a mathematical problem of the instability of a planar interface toward small perturbations. The interface can be destabilized for an optimal deformation wavelength as a result of a competition between the destabilizing effect of hole diffusion and the stabilizing ones because of ion diffusion in the electrolyte and surface tension of Si. The optimal wavelength is expected to give an order of magnitude of the interpore spacing. The analytical and numerical stability analysis of the Si/electrolyte interface for PEC-etching of n-type Si was performed by Valance [141, 142]. This model allowed expression of the dissolution speed and derivation of the scaling laws for interpore spacing as a function of the doping level of Si and applied potential. Another important result obtained by Kang and Jorné was the relationship between the intensity of the rear illumination and the pore diameter: the higher the intensity, the larger the photocurrent per pore and the larger the pore diameter. Chazalviel and coworkers recently addressed the linear stability analysis of the interface to the case of p-type Si [13, 143]. It allowed an understanding of the observed changes in the distribution of structure sizes (from microporous to macroporous) as the layer thickened, and the dependence of the pore sizes on the resistivity of the starting material.

3.2.4 Porous Semiconductors: A Review

3.2.4.1 PEC and EC Etching (Table 2)

The first observation of intentional localized corrosion on semiconductors, other than Si, was purposely performed to corrugate the semiconductor surface, decrease the reflectivity, improve the optoelectronic properties, and consequently increase the performance of photoelectrochemical solar cells.

3.2.4.1.1 Group II–VI Semiconductors

Beside the porous Si, II–VI semiconductors are those for which the origin of deep microrelief (micro-etchpit) formation was studied first. Long ago, n-type CdS [144], CdSe [145, 146, 47], n- and p-type CdTe [119, 120, 148, 149], n-type $CdSe_{0.65}Te_{0.35}$ [150], ZnSe [121], and CdHgTe [151] have been found to undergo an extreme surface roughening under EC and PEC etching. The II–VI electrodes were exposed in aqua regia to a reverse bias of over 1 V/SCE accompanied by high-intensity illumination. The duration of the PEC-etching generally did not exceed 5 s.

The anodic dissolution valency is two, which implies the formation of elemental tellurium forming an insulating layer that hampers further transfer of photogenerated holes.

$$CdTe + 2h^{+} \longrightarrow {Cd_{(aq)}}^{2+} + Te^{0} \quad (10)$$

After removal of the insulating (photo)-corrosion products obtained by dissolution in a suitable medium, the surface contains

Tab. 2 PEC-etched semiconductors.

Semi-conductor	*Type*	*Band gap size (eV)*	*Conduction type*	*Electrolyte*	*Potential (V/SCE)/current density*	*Morphology*
CdS [144]	*d*	2.42	*n*, 2 Ω cm (4.5×10^{10} cm^{-2} donors)	HNO_3, HCl, H_2O (1/4/20)	+1 V, light, 3–4.5 s, 12 mA	Etchpits / λ laser (nm) 2×10^9 cm^{-2} / 457.9 2×10^8 cm^{-2} / 514
CdSe [146, 147]	*d*	1.72	*n*, 10 Ω cm	HNO_3, HCl, H_2O (1/4/20) 1 M Na_2SO_3 0.2 M $FeSO_4$ + 0.1 M H_2SO_4	+1 V, light, 3–4.5 s	1-μm etchpits
			10^{16}–10^{17} cm (1 Ω cm)			0.2 μm etchpits
CdTe (111) [119, 120, 148, 149]	*d*	1.44	*n*, 1000 Ω cm	HNO_3, HCl, H_2O (1/4/20)		1 μm triangular etchpits
			10^{16}–10^{17} cm^{-3} (5–0.05 Ω cm)			Idem 0.15 μm
ZnSe [121]	*d*	2.7	*n*, 4×10^{17} cm^{-3}	HNO_3, HCl, H_2O (1/4/20)	+1 V, light	Etchpits / λ laser (nm) 10^{10} cm^{-2} / 457.9 5×10^8 cm^{-2} / 472.7 0 / 488
InP (149)	*i*	1.35	*n*, 1–2×10^{18} cm^{-3}	1 M HCl	10 mA cm^{-2}, light, 1–120 min.	Elongated etchpits μm. Groove structure along (011) axis
WSe_2 [158]	*i* *d*	1.22 1.37	*n*	0.4–1 M $HClO_4$	+1 V, light	Microetchpits on van der Waals planes on initiated punctures

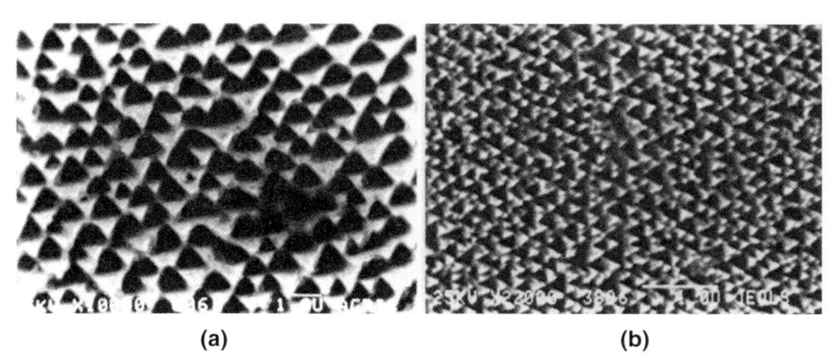

(a) (b)

Fig. 22 Plan view SEM images of photoetched (111) *n*-type CdSe. Influence of the doping concentration on the density of the micro-etchpits. (a) $N_d = 10^{16}$ cm^{-3}; (b) $N_d = 10^{17}$ cm^{-3} (from Refs. 145, 146).

a pattern of highly dense microetchpits ($>10^9$ cm^{-2}) whose density and shape vary with the doping density and crystallographic orientation of the material (Figs. 22a and 22b). The size of the microetchpits observed under scanning electron microscopy (SEM) is on the order of a few hundred nanometers. On (110) and (111) surfaces, the pits have parallelogram and triangular shapes, respectively. The PEC-etched (texturized) surfaces have not been described in terms of the formation of a porous layer, especially as the depth of the pores is not known. However, there is a striking similarity with some of the macroporous silicon morphologies. It has been suggested that the texturized surface arises from the preferential etching of surface defects in the vicinity of dopants [121]. The characteristic size (200 nm–1 μm) of the observed structures and its decrease with increasing sample doping were given as arguments in favor of this model.

3.2.4.1.2 **Group III–V Semiconductors** General trends observed with II–VI semiconductors are also observed on III–V semiconductors. Formation of etchpits was observed on *n*-InP in the presence of HCl solution, under illumination [47, 152–154]. The main difference with the II–VI compounds is that the corrosion products do not lead to a passivation layer and a long duration PEC etching renders the InP surface porous. This will be discussed in the next section.

3.2.4.1.3 **Layered Semiconductors** Layered semiconductors such as InSe [155], WX_2 (X = S, Se) [156–158] and SnS_2 (X = S, Se) [158] have been photoelectrochemically etched in a manner similar to the II–VI semiconductors. A review of this work is given in Ref. 158. The shape of the etchpits depended strongly of the crystallographic orientation of the etched surface. Once again, the purpose of this work was to improve the photoelectrochemical characteristics of electrodes. In the case of the van der Waals surface, known to be free of surface states (dangling bonds), no photoetchpits are formed upon PEC etching. On the contrary, the surface perpendicular to the van der Waals surface that has a significant number of dangling bonds is strongly corroded. Micrometer photo-etchpits can be initiated on WSe_2 van der Waals planes after creating defects by microscopic punctures at selected places [159] (Fig. 23). No porous layer has been observed.

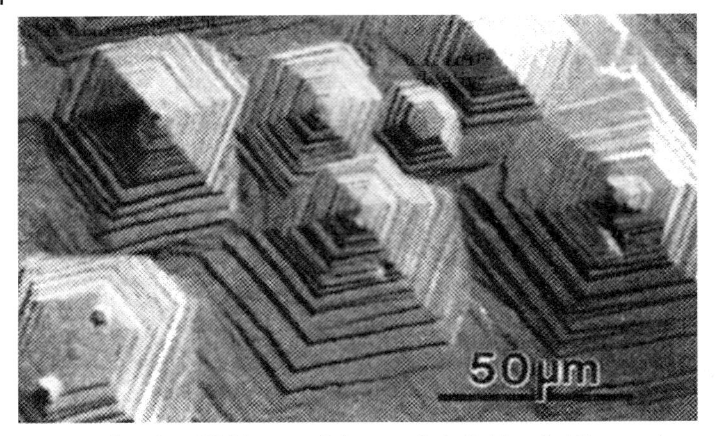

Fig. 23 Plan view SEM image of photo-etchpits initiated at the van der Waals surface of WS_2, after creating defects by microscopic puncture at selected places and upon PEC etching (adapted from Ref. 41).

3.2.4.2 **Porous-etching** (Table 3)

Following the work on luminescent porous Si, a number of studies have been undertaken to render other semiconductors nanoporous. The motivation of such studies was based on the fact that if they can exhibit tunable luminescence in a similar way to nanoporous Si, then common features or differences might reveal the mechanisms involved. Studies concern indirect band gap semiconductors such as those of column IV (Ge, $Si_{1-x}Ge_x$, SiC) and GaP, as well as direct band gap III–V alloys (GaAs and InP) and II–VI compounds (CdTe, $Cd_{0.95}Zn_{0.05}Te$, CdTe, and ZnTe).

3.2.4.2.1 **Porous Group-IV Semiconductors, Other Than Si** Various column-IV semiconductors have been made microporous using recipes inspired from those used in porous silicon: *p*-type Ge (0.66 eV), $Si_{x-1}Ge_x$ ($x = 0.04–0.4$) alloys, *n*- and *p*-type SiC (2.8–3.2 eV), *p*-type hydrogenated amorphous Si (a-SiH), and $Si_{1-x}C_x$ alloys have been made porous in ethanolic HF, either in the dark or under illumination. Depending on the experimental conditions (*n*- or *p*-type, in the dark or under illumination), the reported structure sizes generally range from 10 nm to 1 μm.

In the case of a-SiH and $Si_{1-x}C_x$, the thickness of the nanoporous layer is limited by the formation of macropores, an instability of the growth front attributed to the high resistivity of the starting material [160].

1. *p*-type Ge: Porous Ge [161–166] formed in 10% ethanolic HF on 6–10 Ω cm (100) *p*-type Ge under galvanostatic conditions (50 mA cm^{-2}) most probably has a similar nanostructure (~2 nm) to porous Si obtained on low-doped Si as suggested by extended X-ray absorption fine structure (EXAFS) measurements [164]. One main difference is that the porous Ge surface is much rougher than the porous Si surface. A 10-min. porous-etching produces a roughness greater than 5 μm, which is similar to that obtained from porous-etching Si for an hour under similar conditions. The

Tab. 3 Porous semiconductors other than Si

Semi-conductor	*Type*	*Band gap size (eV)*	*Conduction type*	*Electrolyte*	*Potential (V/SCE) or current density*	*Morphology*
Ge [161–166]	*i*	0.66	*p*, 6–10 Ω cm	(10%) HF	50 mA cm^{-2}, 10 min	Porous Ge, features less than 50 nm diameter
$Si_{0.8}Ge_{0.2}$ [166–175]	*i*		*p*, 10^{19} cm^{-3} 10^{17} cm^{-3}	HF (12.5–24%)/ C_2H_5OH/H_2O	10–40 mA cm^{-2}	Mesoporous pore diameter 10 nm, SiGe skeleton 5 nm diameter, porosity 70%
6H-SiC [176–195]	*i*	3	*p*, 2.2×10^{18} cm^{-3}	Diluted HF	50 mA cm^{-2}	Porous + roughened morphology and shallow depression (<150 nm).
			n, $1–3 \times 10^{18}$ cm^{-3}	2.5 M NH_4F	2.5 mA cm^{-2} 0.5–1 h + light	1–10 nm Interpore spacing, Crystallite morphology undetermined i.e. spongelike crystallite network, or wire-like

(continued overleaf)

Tab. 3 *(continued)*

Semi-conductor	*Type*	*Band gap size (eV)*	*Conduction type*	*Electrolyte*	*Potential (V/SCE) or current density*	*Morphology*
					20–25 V dark current mode	Porous 50 nm
					2.5 mA cm^{-2}, 0.5–1 h, light 365 nm or +1.5–2 V	Platelike 0.5–1 μm fiber thickness
GaAs [197–212]	*d*	1.424	*n*, 4×10^{16} cm^{-3}	0.1 M HCl	3V, 5 min 6 V, 5 min	Inhomogeneous: Pitting including GaAs needles sub μm, top covered by $AsCl_3$ crystallite layer and 20 nm fine features
			p, 9×10^{17} cm^{-3}	0.1 M HCl + $NH_4H_2PO_4$	1 V	Pitting
GaP [211–227]	*i*	2.26	*n*, 5×10^{17} cm^{-3}	(1/1) 40% HF/MeOH	20 mA cm^{-2}, 10 min	Porous layer 3 μm thick
				0.5 M H_2SO_4	−10 V, 2 h	100 nm, 50 μm thick, distance between pits 10–30 μm

InP [152–154, 228–235]	*i*	1.35	*n*, 2.5×10^{18} cm^{-3}	1 M HCl	10 mA cm^{-2}, light, 1–120 min.	Macropores 1–6 μm, Groove structure along ⟨011⟩ axis
				1 M HCl or 1 M HBr	2–5 V	
			n, (111) $2–3 \times 10^{18}$ cm^{-3}	15% HCl	15 mA cm^{-2} 2–5 min	50 nm size nanostructure
GaN [237]		3.36	5×10^{17} cm^{-3}	0.5 M H_2SO_4 +10% ethanol	Cycling between −1 and −2 V	40–120 nm size structure
CdTe $Cd_{0.95}Zn_{0.05}Te$ [239, 240]	*d*	1.44	*p*, 10^{15} cm^{-3}	0.5 M H_2SO_4	3 mA cm^{-2} 1.2 V	0.4–0.8 μm macropores
ZnTe [241–242]	*d*	2.23	*p*, 10^{14}–10^{15} cm^{-3}	$HNO_3 : HCl : H_2O$ (1 : 4 : 20)	300 mA cm^{-2} 5–25 sec.	(0.1–1 μm) macropores. Morphology depends on crystallographic orientation
					+1 V (100 mA cm^{-2}), 30 s	25 nm length columnar structures
TiO_2 (001) rutile [246–249]	*d*	3	*M*	0.5 M H_2SO_4	1–1.5 V + light (16 C)	Macropores 0.2 μm diameter
				1 M $HClO_4$	idem (45 C)	Macropores 0.5 μm diameter

room-temperature photoluminescence centered at 1.17 eV is very similar to the photoluminescence from porous Si, but weaker in magnitude by a factor of about 10 [163, 164]. Porous Ge made on 0.2–0.5 Ω cm p-type Ge does not exhibit visible luminescence [166].

2. p-type Si-Ge alloys [166–175]: Study of these compounds was undertaken because it was expected that the Si luminescence wavelength could be controlled by changing the Ge content (0.9 eV band gap) in the material.

   – Heavily doped ($N_a = 10^{17} - 10^{19}$ cm^{-3}): When porous Si–Ge was made by stain-etching, no significant spectral shift was observed in the photoluminescence spectrum of samples with the nominal $0.04 < x < 0.41$ composition, whereas the signal intensity became weaker with increased x [166]. When porous $Si_{0.95}Ge_{0.05}$ and $Si_{0.8}Ge_{0.2}$ were made electrochemically in 12.5–24% HF ethanolic solution and current density between 10 and 40 mA cm^{-2}, a column-like pore structure was obtained, closely analogous to that observed for mesoporous p^+-type porous Si of comparable 50% porosity [169]. Typical pore dimensions were 10 nm, and the remaining SiGe skeleton had a diameter of 5 nm. The existence of a local substructure with lower dimensions might be evoked to explain the visible photoluminescence within the quantum confinement model. The maximum photoluminescence band exhibited by porous Si-Ge is shifted by 0.2 eV to a lower energy compared to porous Si whereas the intensity of the photoluminescence was increased by a factor of 3 for porous $Si_{0.95}Ge_{0.05}$ and reduced by an order of magnitude for $Si_{0.8}Ge_{0.2}$ compared to porous Si [167, 169–174]. The photoluminescence was interpreted in terms of quantum size effects
   – Low-doped Porous $Si_{0.85}Ge_{0.15}$ (1–15 Ω Cm): The morphology of porous SiGe that originated from low-doped substrate could unfortunately not be analyzed by TEM because of the low density of the porous layers produced. Photoluminescence spectra were similar to those observed for highly doped porous samples [175].

3. n- and p-type SiC: Porous SiC, a promising candidate for blue-light emission devices, is, beside porous Si, the most-studied porous semiconductor. The band gap energy of SiC is considerably higher than that of Si and varies with the polytypes modification from 2.3, 3.0, and 3.2 eV for the 3C, 6H, and 4H main polytypes, respectively. The three polytypes can be made porous in HF and NH_4F electrolytes [176–195]. For n-type 6H-SiC ($N_d = 3 \times 10^{18}$ cm^{-3}) porous photoetching was done galvanostatically (2–40 mA cm^{-2}) under ultra violet (UV) photoassistance (365 nm, 0.1–0.2 W) [177–178, 181] or in the dark under tunnelling breakdown (20–25 V) [179]. For Al-doped p-type ($N_a = 2–5 \times 10^{18}$ cm^{-3}), it was performed galvanostatically (5–150 mA cm^{-2}) in the dark [178, 184]. p-Type porous SiC showed great similarity to p^+-porous Si sponge-like crystalline network morphology (Fig. 24) [178, 183]. The interpore spacing was found to be in the range of 1–10 nm with an average size of 6 nm. Low-doped

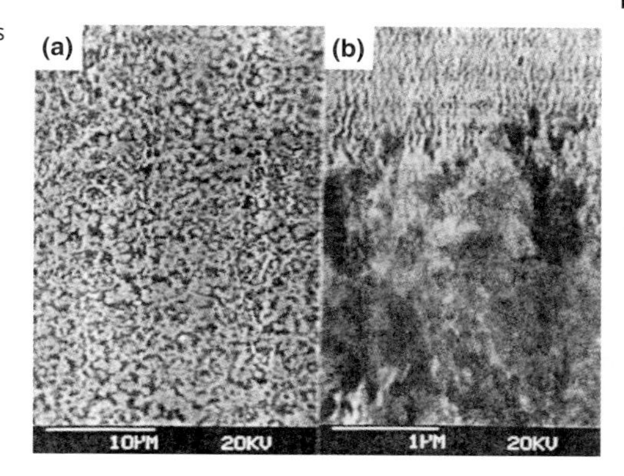

Fig. 24 SEM images of porous *n*-type 6H-SiC. (a) after porous-etching at +20–25 V (plan view); (b) after porous photoetching (cross-sectional view) (from Ref. 179).

porous *p*-type SiC (1.5 Ω cm) was very fragile and looked very dendritic. SiC structures down to 10 nm could be seen [186]. Dramatic difference in pore wall size was observed between the structure of porous *n*-type materials produced under light (10–30 nm pore size, 10–100 nm pore walls) and in the dark (0.1–1 μm pore walls) (Fig. 24) [184].

Pores in 4H-SiC first propagate nearly parallel to the substrate surface and gradually change direction toward the direction of the *c*-axis [195]. Similar anisotropic porous-etching was found in some 6H-porous SiC [189].

Intense blue–green luminescence (two order of magnitude stronger than that exhibited by crystalline SiC) were observed for porous *n*-type 6H-SiC produced by porous photoetching in HF-based electrolyte [176, 177, 181, 182]. The photoluminescence peaked at wavelengths (2.6 eV) lower than those corresponding to interband transition in crystalline SiC, thus showing the possibility of a related defect rather than quantum confinement mechanism of light emission from porous SiC [181, 182, 186, 190]. Similarity between photoluminescence properties of *n*- and *p*-type porous SiC layers with highly different morphology were observed [186]. The 3C, 6H, and 4H *n*-type porous SiC exhibited identical luminescent emission spectra with a broad peak around 2.6 eV (except for an additional broad shoulder at 1.9–2.1 eV for the 3C polytype) demonstrating a clear impedance of the band gap energy of a particular SiC polytype [182]. The photoluminescence of porous SiC with different microstructures was found to become enhanced as the microporous SiC changed from fibrous to porous dendritic. A weak blue electroluminescence was also observed on porous *n*-type 6H-SiC (0.1 Ω cm) [180, 183].

The electrical properties measured with the use of a metal-point probe contact showed for porous SiC obtained in the dark, current–voltage characteristics qualitatively similar to those of bulk SiC, whereas porous SiC obtained under light was insulating. This was interpreted in terms of Fermi level pinning at the surface of the 50 nm thick fibers corresponding to the SCR thickness for highly doped SiC. In the case of SiC, the reaction with HF is known to produce a carbon-enriched phase at the surface, which may provoke a high density of surface states. A model based on the Fermi level pinning was evoked to explain porous SiC formation [182].

3.2.4.2.2 **Porous III–V Semiconductors**

Porous III–V semiconductors are of great interest because of possible applications in optoelectronics and photonics. Study of the luminescence from porous direct band gap semiconductors such as GaAs and InP presents an additional interest because quantum confinement effects are well established in two-, one-, and zero dimensional electron systems prepared on crystalline III–V materials. Thus it would be possible to compare the light-emission properties between structures obtained by porous-etching and those produced in well-characterized lithographically defined structures directly. GaP was studied for comparison because it is an indirect band gap.

Variations in the conditions of formation of the III–V porous semiconductors and in the obtained morphologies have been observed [197–238]. The III–V semiconductors were rendered porous in HF and HCl solutions but other acidic solutions such as H_2SO_4 permitted obtaining porous GaAs and GaP. n- and p-types GaAs were also made porous in KOH and NH_4OH alkaline solutions. Generally the porous-etching was performed in the dark, under breakdown conditions (either by avalanche or tunneling) of the semiconductor-depletion layer. Porous photoetching was also used. The majority of these III–V substrates utilized were n-type heavily doped ($N_d = 5.10^{17} - 1.10^{18}$ cm^{-3}).

1. *Mesoporous:* Porous layers on GaAs and GaP surfaces with (100) orientation obtained under breakdown conditions after porous-etching in HCl and H_2SO_4 solutions show a network shaped like cellular structures. It is noteworthy that for GaP this additional feature is actually buried under the etched surface; the top layer still consisting of distinct and separate submicron pits whose number roughly scales with dislocation density of the original substrate. Pore formation can be triggered at surface defects created on purpose.

n-Type GaAs(10^{16} cm^{-3}) was made mesoporous in HCl- and Cl^--containing solutions [200, 201], when the electrode was polarized at a potential larger than 3 V, which is the critical potential [pore formation potential, (PFP)] at which breakdown of the Schottky barrier occurs. Pore formation occurs in two steps. In the first step, the crystallographic etching occurs at the (111) GaAs exposed planes. Then during a second phase, the electrochemical attack spreads out from the initial etchpits and leaves distinct patches behind. Within the patches a porous structure was found (Fig. 25). It has been suggested that the initial pits develop at the defects or dislocations and radial pore growth occurs through enhancement of the electric field at the space charge boundary. Polarization for a long time at a potential above PFP leads to a complete coverage of the surface by a black porous layer of 60–80% porosity. Typical structure size is about 100 nm, but generally few structural data are available to reveal the dimensionality of the porous skeleton. For p-type, porous GaAs is only obtained in phosphate buffer solution because a stable oxide film can be formed. The presence of Cl^- in the solution was also necessary [201]. The pore formation mechanism on n- and p-type GaAs illustrates the asymmetry that exists in the conduction properties of the semiconductor/electrolyte junction under anodic polarization. For n-type, the current-limiting force is the SCR and porous GaAs occurs in Cl^--contained electrolyte, whereas for the p-type, formation of a passivation oxide layer is necessary to establish the current limitation.

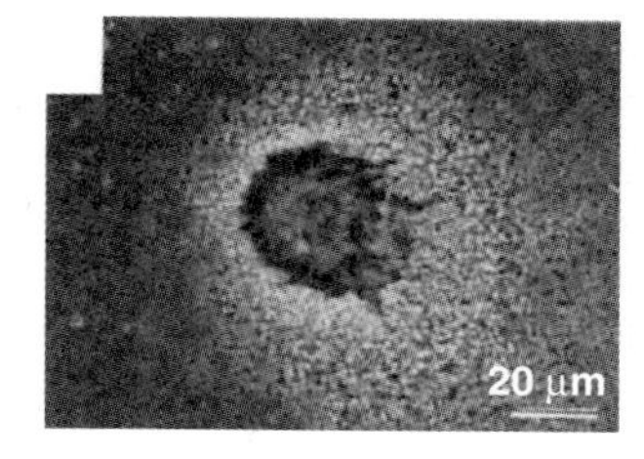

Fig. 25 Plan view SEM image of one of the etchpit formed on (100) *n*-type GaAs after porous-etching in 0.1 M HCl (from Ref. 200).

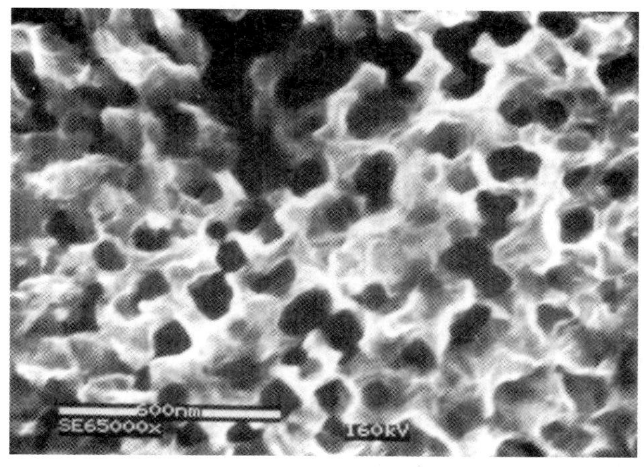

Fig. 26 Plan view SEM image of porous *n*-type GaP after porous-etching in the dark (+10 V vs SCE, 2 *h*, 0.5 M H_2SO_4) (from Ref. 218).

Porous-etching of (001)-oriented $N_d = 5.10^{17}$ cm^{-3} *n*-type GaP crystals was performed in 0.5 M H_2SO_4 at +10 V [214, 215]. After porous-etching, the surface looked flat though slightly spitted, and reflected light as well as a polished wafer. To reveal the porous structure below the flat surface, the electrode was dipped in an aqueous bromide solution to remove the top 0.5 μm of the GaP electrode. A pattern of porous domains was revealed (Fig. 26). The pore diameter is similar to the pore wall thickness and equals 90–120 nm. It was found that the nucleus (from which pores radiate to the thin walls of the domain) of each porous domain is located exactly below the site at which a pit emerges at the flat surface of the anodized electrode. The porosity of the layer was 25%. At an early stage of anodization, the porous domains are circular and have almost the same diameter. The domains expand until the pore fronts from neighboring domains meet, at which time the lateral etching stops [215, 220] (Fig. 27). When the applied voltage was temporally increased from 5 to 15 V during anodization, formation of annular traceries around each etchpit was observed (Fig. 28). This strongly suggests that the etching proceeds radially at the same rate in all directions away from the primary pores. Following a preliminary high-energy particle treatment of GaP wafer (Kr^+ preimplanted), prior to porous-etching, to create a large number of surface states, the observed

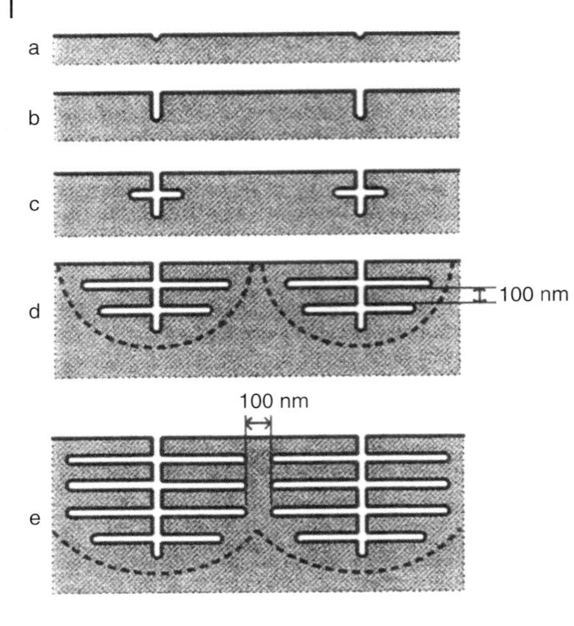

Fig. 27 Schematic illustration of various stages of GaP porous-etching: (a) initiation of pores at defect sites; (b) continued growth of the initial pores; (c) appearance of the first branches or secondary pores; (d) further branching, separate porous domains are formed defined by pore fronts which have a more or less hemispherical geometry. The pore walls are 100-nm thick; and (e) porous domains meet, creating 100-nm thick nanoporous walls. A porous layer now covers the entire sample.

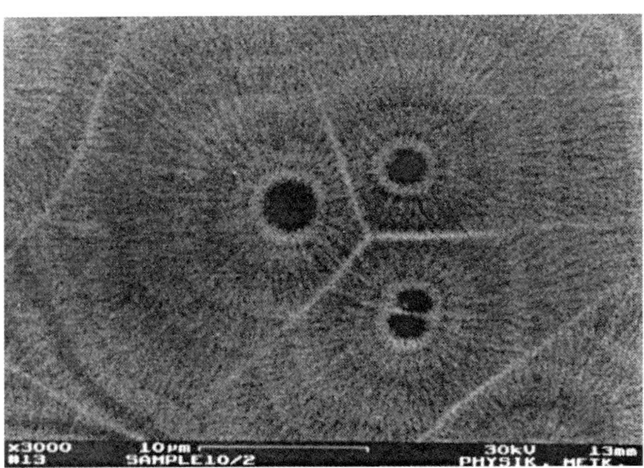

Fig. 28 SEM image of the top surface of GaP sample anodized under temporary variations of the applied voltage from +5 to +15 V (from Ref. 220).

porous morphology resembled that which was previously obtained after a long etching time and removal of the surface layer [220]. The high number of surface defects created by ion implantation provided a high density of etchpits, and thus a high density of primary pores with a wall thickness of 50–100 nm, sufficient to leave a GaP skeleton completely free of carriers, as a consequence of the SCR effect.

Porous-etching of InP done in the dark under tunneling breakdown conditions leads to the formation of small pores with a diameter around 50 nm on (111)-

and (100)-oriented heavily doped n-type substrates. The pore morphology depends strongly on the nature of the halogen acid present in the electrolyte. Blue, colorless, and gray layers were observed when the anodization was performed in 1 M HCl, HBr, and HF, respectively [233]. The pore diameter was larger in HBr than in HCl. In HF, microscopic inhomogeneity was present, with the presence of opaque regions suggesting a certain amount of occluded porosity. An enrichment of phosphorous within the porous structure obtained in HCl solution indicates a selective dissolution of indium. Porous InP was also obtained when the n-type electrode was continuously biased between 0.1 and 2 (3, 4) V. The pore diameter increased from 40 to 120 nm when the final bias varied from 2 to 4 V [233].

2. *Substrate Orientation:* Pore morphology depends on the doping density and crystallographic orientation of the surface. For example, it was shown for porous GaAs obtained in 25 M HF that the primary pores having triangular or hexagonal cross-section grow in the $\langle 111 \rangle$A direction (corresponding to the direction from Ga plane to As plane) [204, 205], which is in contrast to porous Si where the pores grow in the $\langle 100 \rangle$ direction. In both GaAs and Si, the pore tip is composed of the slowest reacting planes in the structure. This trend is general for all porous III–V semiconductors.

(111)-oriented surfaces: First, n-type GaAs was made porous in the dark in KOH and H_2SO_4 solutions. The porous layer contained "tunnels" of diameters ranging from 0.5 to 5 μm and a pore density of about 10^8 cm^{-2} [197, 198], which propagate only in the (111)A direction [199].

The first report on porous-etching of GaP was related to the formation of high-density networks of tunnels formed in chlorine-contained methanol. Tunnels were formed when the Ga (111)A face was in contact with the electrolyte and the electrode polarized at 120 V (20 mA cm^{-2}). When the P (111)B face was in contact with the electrolyte, complete dissolution of the surface occurred by chemical dissolution [213]. The results were compared with the channels observed by Theunissen on Si. Formation of the tunnels was interpreted in terms of a localized electrochemical attack resulting from the avalanche breakdown of the SCR. It was shown that in GaP, the tunnels were not related to dislocations.

Porous-etching in an HF/ethanol solution of a (111)A GaP surface leads to a structure consisting of triangular prism-shaped columns. In H_2SO_4, freestanding membranes of porous GaP (111) were obtained. The 2 μm thick membranes comprised arrays of isolated pores stretching perpendicular to the surface of the sample with average pore and skeleton thickness of about 50 nm (30 min, 50 mA cm^{-2} in the dark) [223].

Porous-etching of a (111)A InP surface in HCl/H_2O solution leads to a top layer with a pillar structure characterized by isolated columns stretching perpendicularly to the initial surface [228]. The shape and size of pores can be automatically fixed when porous-etching was combined with conventional electron-beam lithography with an appropriate mask pattern. Triangular pores with fairly good size and uniform and ordered InP triangular vertical pillar structures were obtained [229]. The average size of the pillar structures was reduced from 60–50 to 30–40 nm when the doping density of the substrate increased from $N_d = 10^{18}$ cm^{-3} to 10^{19} cm^{-3} (Fig. 29) [230]. Quantum-wire and quantum-box structures with low-size fluctuations were obtained.

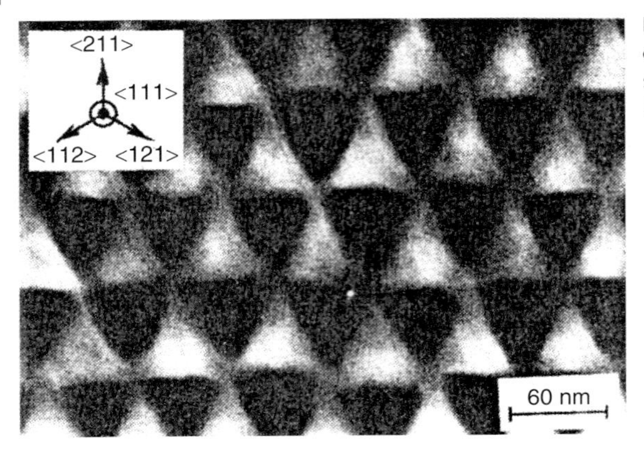

Fig. 29 Plan view SEM image of the porous (111)A n-type InP substrate with SiO_2 mask (from Ref. 230).

(001)-oriented surfaces: The direction of the pores depends strongly on the applied bias. On (001) InP (1 M HCl, $N_d = 10^{18}$ cm^{-3}), the pore growth changed the direction from ⟨111⟩ when the applied bias was below 3 V to ⟨100⟩ when it was over 4 V [231]. Below 3 V, the pores were aligned along the direction having a tilt angle of 63° with respect to the (100) plane resulting in a tree-like cross-sectional structure. At potentials above 4 V the branching disappeared and fairly regular straight pores basically running toward the (100) direction were observed. When the porous-etching is done at high over-potential, above 7 V, the surface morphology of porous InP looks similar to that of the porous n-type GaP shown in Fig. 26. (001)-oriented nanometer-sized straight pores, free of any branching with 250 nm pore diameter, 16 nm wall thickness, and 80 μm pore length were obtained on a 3 min treatment (Fig. 30) [231]. The formation of (001)-oriented pores was interpreted by the combination of an enhanced electric field at the pore tip and a short hole lifetime of 0.1 μs in InP. The pores exhibited square shapes defined by four crystalline (001) and (011) facets when a new solution (1 M HCl + HNO_3) was used [232].

3. *Macroporous obtained under illumination:* Ordered honeycomb hollow arrays are formed on n-type GaAs (001) substrates highly doped (2×10^{-3} Ω cm) in 29 wt% NH_4OH by porous photoetching

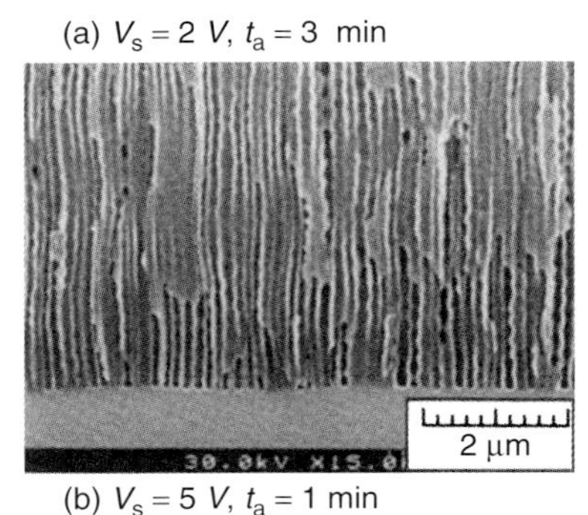

Fig. 30 Cross-sectional SEM image of porous n-type InP after porous-etching at +5 V during 1 min, in 1 M HCl (from Ref. 231).

with a bias voltage of 12 V. The hexagonal hollows exhibited 19 nm average diameter with a standard distribution of 200 nm. Self-organization formation of the hollows in porous GaAs was explained by repulsive interaction between the hollows [210].

The first work on porous InP, [(001)-oriented, $N_d = 10^{18}$ cm^{-3}], was accomplished with anodization in HCl electrolyte under illumination, for producing surface relief features in optoelectronic devices [152]. The pore location was controlled with standard photolithography, and well-defined macroporous structures with a pore aspect ratio up to 250 were produced by photoelectrochemistry in an analogous manner to the macroporous patterned n-Si. When the initial surface was mirror polished, the surface microstructure varied from 50 nm to several μm depending on the porous-etching duration and was strongly affected by the crystallography [47]. Using a homogeneous white light beam, Decker and coworkers [153] found that microetchpits develop preferentially along the ⟨011⟩ direction. Upon longer etching time, high depth/width ratio structures (grooves) of average period of 1 μm are developed (Fig. 31). The reflectance of such highly corrugated surfaces drops from 40 to 4% after photoetching at 20 mA cm^{-2} current density during 60 min. in 1.2 M HCl [154]. This effect can, in principle, be very useful in the development of new antireflective surfaces for solar cells.

Porous photoetching of n-type GaP in $H_2SO_4/H_2O/H_2O_2$ leads to the formation of macroporous GaP (random networks of single crystalline GaP with pore sizes of 150 nm and porosity of 35 to 50% for porous etched and porous photoetched, respectively), which appears to be the most strongly scattering material in visible light reported to date and localization effects are anticipated [224, 225].

4. *Photoluminescence:* Porous GaAs obtained in HCl exhibits visible region luminescence (around 540–570 and 850–870 nm) (Fig. 32). The green and infrared bands were interpreted by Schmuki and coworkers as being caused by a quantum confinement effect in GaAs microcrystallites and nanocrystallites, respectively [201, 207–209]; whereas

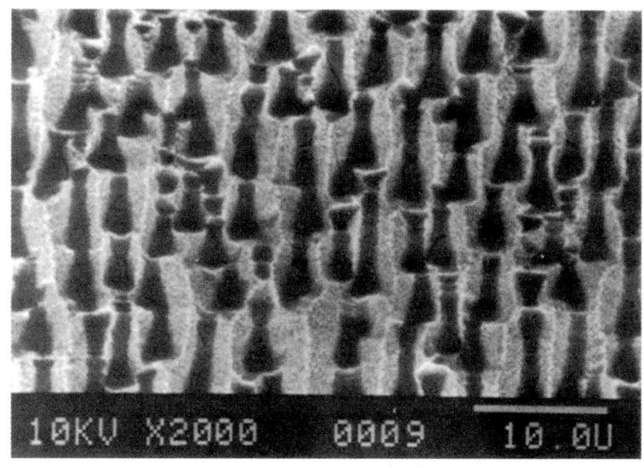

Fig. 31 SEM image of n-type InP after PEC-etching in 1.2 M HCl during 9.2 min (from Ref. 154).

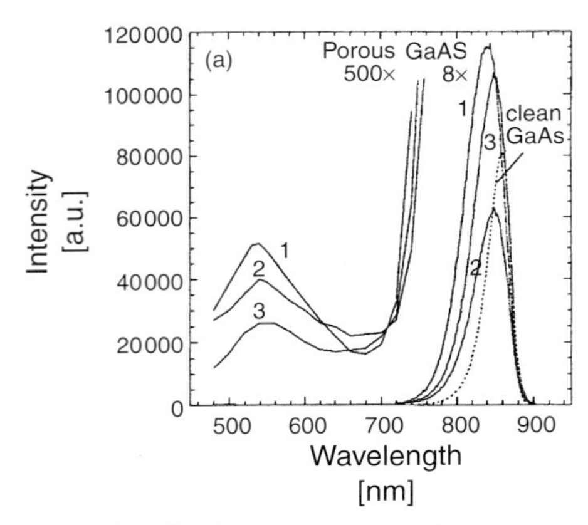

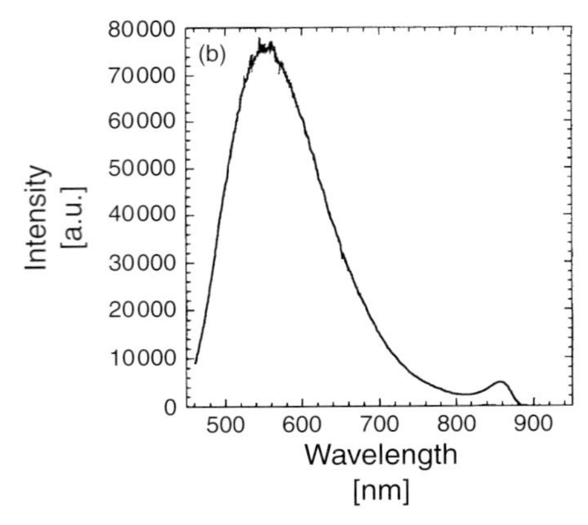

Fig. 32 Photoluminescence spectra (room temperature) of porous GaAs obtained in 0.1 M HCl: (a) for three different surface locations on porous GaAs formed by a potential sweep from −0.5 to 6 V (SCE); (b) highest "green" photoluminescence intensity observed from a specific location on a porous GaAs sample formed by a potential step (8 V/SCE, 5 min).

Finnie and coworkers identified the source of the green luminescence to be arsenic oxide microcrystals formed during porous-etching [211, 212] and that of the infrared band to the scattered excitation radiation, exciting luminescence from the relatively unperturbed outer regions of the etchpit [211].

Photoluminescence spectra of porous GaP obtained in HF solution exhibits three strong sharp emission lines in the orange spectral region (2.1–2.3 eV, 580 nm) and a weak emission between 300 and 700 nm with maxima in the blue (2.4 eV) and UV (3.2 eV) spectral regions [216]. The sharp lines are characteristic of bulk GaP. The blue and UV emission has been explained by Anedda and coworkers [216] on the basis of charge-carrier confinement in crystalline quantum wires of about 25-nm diameter. However, similar results observed by Kelly and coworkers when porous GaP is obtained in H_2SO_4 solution are interpreted, not in terms of quantum confinement effect, but as caused by gallium oxide (Ga_2O_3 or GaO_x) formed on the surface of porous GaP [217].

Contradictory results have been published on the photoluminescence of porous InP. Some authors reported blue-shifted photoluminescence peaks lying above the band gap energy interpreted in terms of quantum confinement [230]. Hamamatsu and coworkers reported that contrary to expectation of blue-shifted emission caused by quantum confinement at pore walls, porous (001) InP samples exhibited an intense red-shifted photoluminescence peak [231]. This was explained by the formation of a set of well-defined new surface state levels on anodized pore wall surfaces.

5. *GaN:* This semiconductor was made porous by continuous cycling of potential between −1 and 2 V in 0.5 M H_2SO_4+ 10% ethanol [237, 238].

3.2.4.2.3 **Porous II–VI Semiconductors:** Low-doped (10^{15} cm^{-3}) *p*-type CdTe [239], $Cd_{0.95}Zn_{0.05}Te$ [239, 240], and ZnTe [241–243] were made porous by anodization in the dark.

p-Type CdTe and $Cd_{0.95}Zn_{0.05}Te$: Porous-etching was performed in 0.5 M

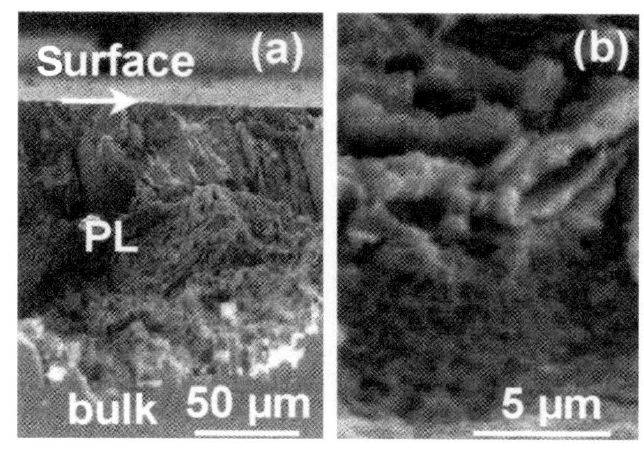

Fig. 33 Cross-sectional SEM images of p-type $Cd_{0.95}Zn_{0.05}Te$ after porous-etching ($Q = 100$ C cm^{-2}). (a) low magnification, (b) high magnification (from Ref. 240).

H_2SO_4 at $+1.2$ V (3 mA cm^{-2}) for several hours. A 100 µm thick layer consisting of highly porous network (0.5–0.8 µm macropore diameter) was observed beneath a polished surface, which was left intact [Fig. 33]. The reaction mechanism for the anodic dissolution of $Cd_{0.95}Zn_{0.05}Te$ is expected to be similar to that of n-CdTe [244, 245]. The reaction proceeds in two steps, the formation of elemental tellurium followed by its oxidation:

$$\mathrm{CdTe} + 2h^+ \longrightarrow \mathrm{Cd_{(aq)}}^{2+} + \mathrm{Te}^0 \quad (11)$$

$$\mathrm{Te}^0 + 4h^+ + 2\mathrm{H_2O} \longrightarrow \mathrm{HTeO_{2(aq)}}^+ + 3\mathrm{H_{(aq)}}^+ \quad (12)$$

Because elemental Te^0 is an intermediate species in the dissolution process, the surface of the porous layer is at least partially covered with Te^0. This was confirmed by XPS, a 14 nm thin Te porous layer lining the pore walls [239]. The results obtained under PEC etching with the formation of an insulating Te^0 layer are not necessary in conflict with the obtaining of porous CdTe and p-$Cd_{0.95}Zn_{0.05}Te$, which implies that no blocking insulating layer is formed. Indeed when anodic dissolution is slow, further oxidation of Te to $HTeO_{2(aq)}{}^+$ is not rate-limiting [Eq. (12)].

p-Type ZnTe: Porous-etching of p-type polycrystalline ZnTe has been undertaken to texturize the surface for photovoltaic applications [241, 243]. The goal was to increase the surface of ZnTe while decreasing the reflectivity. Low-doped ($N_a = 10^{14} - 10^{15}$ cm^{-3}) large grains polycrystalline ZnTe is rendered porous in acidic solution composed of HNO_3 : HCl : H_2O (1 : 4 : 20) either potentiostatically at 1 V vs SCE or galvanostatically at 300 mA cm^{-2} constant current density. After a few seconds and subsequent dissolution of the corrosion layer, a layer of macroporous ZnTe with a columnar structure is formed. Diameter of the pores (3–5 µm deep) varies from 0.1 to 1 µm upon increase of porous-etching duration from 5 to 25 secs. The pore wall thickness remains constant around 100 nm. The effect of grain orientation results in different pore shapes, for example, tubular or flat isosceles triangles or inverted pyramids with a rectangular base [243]. The grain boundaries remain largely undissolved. Enhanced photoresponse (30%) of a ZnTe/aqueous polysulfide photoelectrochemical cell is observed for the porous surface because of decrease in reflectivity [242]. Further porous-etching results in the formation of a needle-like structure with a triangular cross-section, the smallest structure reaching 25 nm typical size. Such porous layer exhibits blue-shifted luminescence compared to bulk ZnTe, which is attributed to quantum-confinement effect [241].

3.2.4.2.4 **Porous Oxides** Among the oxides, only titanium dioxide, TiO_2, has been made porous. TiO_2 is one of the few semiconducting materials highly stable against photocorrosion, with the conduction band-edge located so that photogenerated holes have sufficient energy to oxidize water or the reduced member of redox species involved. It plays a leading role in the photoenergy conversion and acts as an efficient photocatalyst to decompose harmful organic compounds as in the TiO_2 dye-sensitized photoelectrochemical cells. A highly developed surface morphology of TiO_2 is a crucial factor in these applications and porous-etching is one technique that allows surface enlargement. Porous photoetching of n-type TiO_2, under xenon lamp illumination, occurs in 0.05–0.5 M H_2SO_4 and 1 M $HClO_4$ (perchloric acid). Porous photoetching reaction was assumed to be a two-hole process following reaction (13) in H_2SO_4 [246]:

$$TiO_2 + SO_4^{2-} + 2h^+ \longrightarrow TiO{\cdot}SO_4 + \tfrac{1}{2}O_2 \qquad (13)$$

The TiO_2 crystals or sintered pellets were first treated in hydrogen at 650 °C for 30 min. to obtain n-type semiconductivity. Porous photoetching of 0.2–2 Ω cm (100) rutile type TiO_2 done potentiostatically (1–1.5 V vs SCE) renders the surface velvety black due to light confinement arising from the presence of square shaped macropores (0.3–1-μm diameter), whose morphology is similar to that of highly doped porous Si obtained by porous photoetching (Fig. 12a) [246, 247]. Oxygen gas is evolved at the TiO_2 surface. Increasing the SO_4^{2-} concentration and pH of the solution (from 0.45 to 1.65) results in the formation of spare macropores on a much finer scale (<0.1 μm). In concentrated 9 M H_2SO_4, corrosion of TiO_2 is inhibited. Addition of Co^{2+} in 0.5 M H_2SO_4 quenches the porous-etching, the photogenerated holes oxidizing preferentially SO_4^{2-} ions to $S_2O_8^{2-}$ ions, whereas addition of Co^{2+} in 1 M $HClO_4$ does not prevent porous-etching [248].

Porous photoetching was also done on sintered pellets of TiO_2. The influence of applied potential on corrosion mechanism is illustrated in Fig. 34. When anodization is done at 0.2 V vs SCE, the grains are selectively dissolved, whereas the grain boundaries are not dissolved creating a "skeleton structure". By contrast under 1 V bias, the grain boundaries of the pellet are selectively dissolved and a characteristic porous pattern appears on each grain surface [249].

3.2.5
Photoelectrochemical Properties of Porous Semiconductors

Porous-etching has been applied in the design of photoelectrochemical devices. The

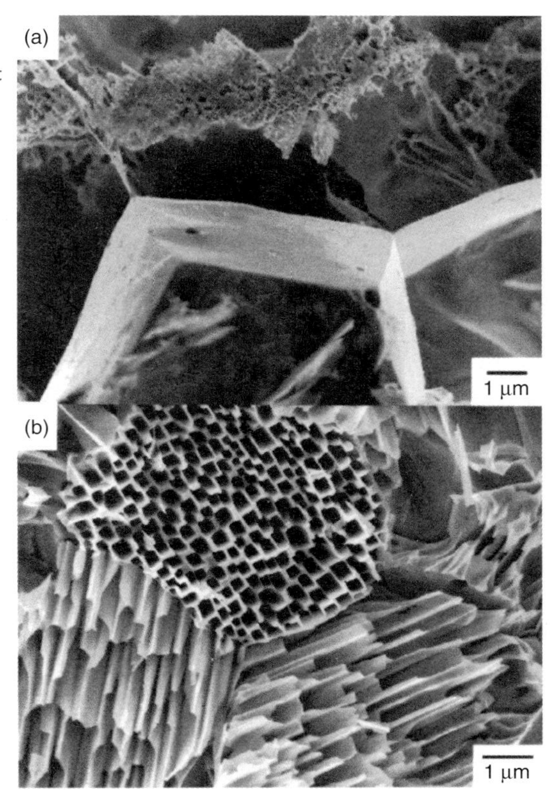

Fig. 34 SEM images of macroporous TiO_2 after porous photoetching of a sintered pellet of TiO_2. (a) +0.2 V bias, (b) +1 V bias (from Ref. 249).

porosity and pore wall size of porous semiconductors have different effects on the photocurrent depending on the band gap type (indirect or direct) of the crystalline material.

Photoelectrochemical cells with indirect band gap porous semiconductors such as *n*-type Si (obtained under illumination), SiC, and GaP (made porous in the reverse bias range) show spectacular results. Presence of a porous layer on lightly and highly doped Si strongly modifies the photocurrent potential curve of porous Si/HI (57%) junctions, whose rising part is shifted to more negative potentials by −0.25 V compared to a flat (nonporous) electrode (Fig. 35). In the case of lightly doped Si, an increase in the plateau photocurrent intensity close to 40% is also observed. In addition, the external quantum efficiency vs light wavelength of the porous electrode is greatly enhanced compared to the flat electrode, without any shift in the onset of the photocurrent [27]. Similar trends are observed for *n*-type porous GaP and porous SiC [215, 250]. Photocurrent increases by a factor 100 after porous-etching of GaP, whereas the internal quantum efficiency (corrected by the photon flux) of a 30 μm thick porous layer electrode is spectacularly enhanced with respect to a flat electrode for light absorbed in the indirect band gap and for sub–band gap light [214, 219] (Fig. 36). Porous *n*-type 6H-SiC electrode exhibits a similar behavior. The enhancement of the quantum

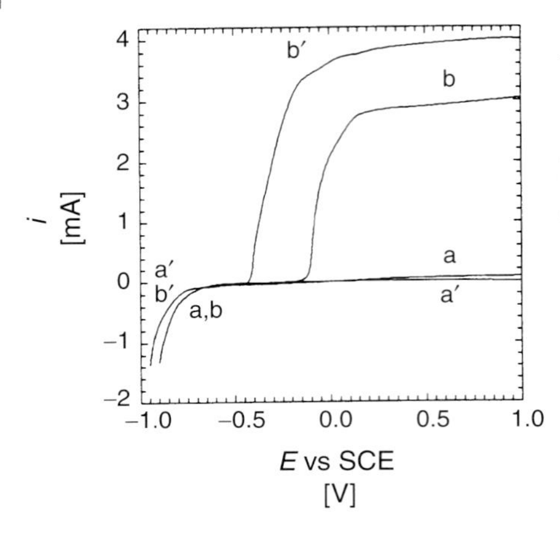

Fig. 35 (Photo)current potential curves for n-type Si ($N_d = 10^{15}$ cm^{-3}) in 57% HI, irradiance 50 mW cm^{-2} (tungsten halogen lamp), surface area = 0.21 cm^2. a) In the dark, b) under illumination, before porous photoetching, a′) in the dark, and b′) under illumination after porous photoetching (Q = 4 C cm^{-2}) (from Ref. 27).

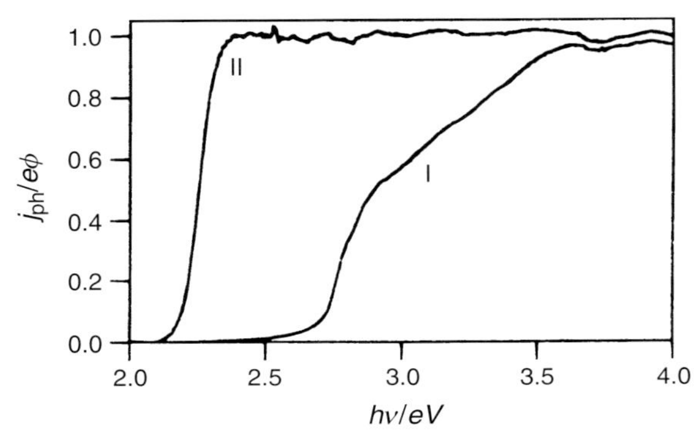

Fig. 36 Quantum efficiency ($j_{ph}/e\phi$) at 1 V (SCE) as function of photon energy for a polished n-type (100) GaP electrode (0 C cm^{-2}) and for the same electrode after porous-etching at +10 V (SCE) in the dark for the time needed to pass 16 C cm^{-2}. The indirect optical transition is at 2.24 eV and the direct transition at 2.76 eV (from Ref. 218).

efficiency can be attributed to two effects related to the nature of the porous layer. First, light scattering gives rise to a more effective absorption of photons in the mesoporous layer. Second, the size of the pore walls corresponds to the SCR width in the photocurrent range, where the band bending is sufficient for effective competition of charge separation with surface recombination. Because of the large internal surface area of porous semiconductors, the volume density of the interface states in the band gap may be very large. Such states can give rise to sub–band gap photocurrent as it has been observed for porous GaP and SiC. In addition, charge carriers in bulk band gap states, which normally do not contribute, can also give rise to photocurrent in a porous system.

For materials with a direct optical transition (short absorption length) and a low dopant density (wide SCR) or large diffusion for minority carriers, porous-etching is less interesting. This was demonstrated on II–VI porous semiconductors with direct band gap. Anodization of p-type CdTe and $Cd_{1-x}Zn_xTe$ contributes to changes in the photocurrent spectra. At short anodization time (EC etching) the photocurrent can increase, when surface roughening leads to a decrease in reflectivity. The decrease in photocurrent at longer anodization times (porous-etching) is due in part to the formation of the Te^0 layer produced during the porous-etching. But this is not the only effect [239]. Chemical removal of Te^0 using potassium cyanide (KCN) solution causes only partial recovery of the initial photocurrent before pore formation. The second effect is recognized as being caused by the porous morphology, which appears to have an effect of its own, leading to an increase in recombination. With p-type semiconductors rendered porous in the forward bias range, the size of the porous features corresponds to a SCR thickness and band bending that are markedly lower (0.2 V) than in the potential range, where surface recombination is negligible for a flat surface (nonporous electrode) (>1 eV). The weak band bending at the pore surfaces favors recombination, especially at sites where the pore walls are the thinnest, resulting in a decrease of the photocurrent.

3.2.6
Potential Applications of Porous Si and Other Porous Semiconductors

Although more work is needed to characterize and hence understand pore-nucleation processes, the fabrication methods, although not described in detail here are now well understood and controlled. Porous Si has been extensively studied using spectroscopic techniques, and our understanding of the interesting and unusual optical properties of this material is now well developed. Its recently established nanoscale structure and light-emitting properties have tremendously increased the interest of the scientific community, and new applications in different fields of technology, summarized in Table 4 together with the key properties of porous Si, have been investigated.

The efficient visible room temperature photoluminescence and electroluminescence of porous Si [6, 263] provide application possibilities within, for example,

Tab. 4 Porous Si as a multidisciplinary material (from Ref. 273)

Scientific area	*Key properties*	*Application example*	*References*
Optoelectronics	Light emission	Displays	251
Analytical chemistry	Porosity	Sensors	252
Optics	Tunable refractive index	Optical filter	253
Ultrasonics	Low thermal conductivity	Transducer	254
Surface science	High surface area	Catalyst	255
Microbiology	Biocompatibility	Bioreactor	256
Energy conversion	Low reflectivity	Solar cell	96–98
Astrophysics	Cosmic abundance	Dust chromophore	257
Microengineering	Lithographic patterning	Microsystems	258
Nuclear science	Conversion to oxide	Radiation-hard circuits	259
Electronics	Dielectric properties	Microwave circuits	260
Signal processing	Dynamic process	Process control	261
Medicine	Resorbability	Drug delivery	262

optoelectronics [251]. Porous Si is also an interesting candidate for application in the fields of gas and biosensors technologies [252, 266, 267]. This is due to the possibility of a wide range variation of characteristics such as pore size, volume porosity, specific surface area, and thickness through different choices of processing conditions like etching solution composition, current density, crystal orientation of the Si substrate, or temperature. Furthermore because porous-etching occurs at the pore bottoms, that is, at the interface between the etching solution and the silicon substrate, it is possible to fabricate multilayer structures, each layer having a different porosity (commonly called porous Si superlattices) that can be designed, for example, to function as Bragg reflectors [17, 268, 269], Fabry-Pérot filters [18, 270–272], or optical waveguides [253, 274]. The low reflectivity exhibited by porous Si has been used to design high-efficiency Si solar cells with a reduced number of steps in the processing of industrial cells [96–98]. Porous Si matrices are very efficient templates for implementation in medical therapeutics and biologic sensors [262].

Over the last ten years, porosity has emerged as a promising tool for controlling optical and photoelectrochemical properties in semiconductors other than Si, especially III–V compounds. In GaP, for instance, porosity leads to an enhanced photoresponse of solid/liquid junctions, whereas macroporous GaP appears to be the most strongly scattering material for visible light to date, and localization effects are anticipated [224, 225]. It has recently been shown that the nanostructured membrane of GaP produced by porous-etching are suitable for nonlinear optical applications [223]. In InP, porosity induces a sizeable birefringence at wavelength suitable for communication systems. In addition, Raman scattering spectroscopy has revealed the existence of surface related vibrational mode in both GaAs and GaP. These modes of the Frölich character exhibit a downward frequency shift with increasing porosity and/or dielectric constant of the surrounding medium that hints at new possibilities for optical phonon engineering in these materials [206, 221]. Porous GaN may be used as a buffer layer for growth of bulk and epitaxial GaN films with low stress [238].

3.2.7 Conclusion

A large variety of porous semiconductors can be produced chemically by stain-etching or electrochemically, whose structural, optical, and electrical properties are different from those of bulk semiconductors, opening the route to new potential technological applications. In view of its room temperature efficient visible light-emitting properties, porous Si was regarded as a prime candidate material for a new generation of Si chips, which might extend the functionality of Si technology from microelectronics into optoelectronics. Recent progress in realizing electroluminescence efficiencies comparable to the photoluminescence is promising. However, it is not yet clear if porous Si will finally enable optical and electronic technologies to be integrated on a single silicon chip. The major interests in porous Si might well be foreseen in other domains of applications described earlier, such as sensors, dielectric multilayer, Bragg reflectors, Microcavities, Waveguides, Photonic band gap systems, microcapacitors, photovoltaic, and biomedical applications.

Although the number of publications devoted to porous semiconductors is much less than that devoted to porous Si, some important experimental findings have recently been reported, in particular with porous III–V compounds. Electrochemistry offers an accessible and cost-effective approach for "drilling" vertical arrays of macropores in semiconductor materials. Similar to porous Si, quasi-uniformly distributed parallel pores can be introduced into III–V semiconductors, by photon-assisted or MeV-ion implantation-assisted anodic etching. Freestanding GaP membranes represent promising materials for advanced nonlinear optical applications or can make possible new applications such as gas-sensing devices.

Nevertheless, the field of porous semiconductors is still in its infancy and lot of work has to be done before foreseeing industrial devices using porous Si and other porous semiconductors, whose properties have so far been studied rather scantily.

Acknowledgments

My collaborators Stéphane Bastide and Christian Vard are warmly acknowledged for their help in preparing the manuscript. I am grateful to Prof. Leigh Canham for giving me an exhaustive list of references on porous semiconductors. This work was partially funded by CNRS-ECODEV and ADEME within the French Photovoltaic Program.

References

1. A. Uhlir, *Bell Syst. Tech. J.* **1956**, *35*, 333.
2. D. R. Turner, *J. Electrochem. Soc.* **1958**, *105*, 402–408.
3. D. R. Turner, *J. Electrochem. Soc.* **1960**, *107*, 810.
4. Y. Watanabe, Y Arita, T. Yokoyama et al., *J. Electrochem. Soc.* **1975**, *122*, 1351.
5. R. L. Smith, S. D. Collins, *J. Appl. Phys.* **1992**, *71*, R1–R22.
6. L. T. Canham, *Appl. Phys. Lett.* **1990**, *57*, 1046–1048.
7. A. G. Cullis, L. T. Canham, *Nature* **1991**, *353*, 335–338.
8. P. C. Searson, J. M. Macaulay, S. M. Prokes, *J. Electrochem. Soc.* **1992**, *139*, 3373–3378.
9. P. C. Searson, J. M. Macaulay, F. M. Ross, *J. Appl. Phys.* **1992**, *72*, 253.
10. C. Lévy-Clément, *Porous Silicon Science and Technology* (Eds.: J. C. Vial, J. Derrien) Les Editions de Physique, Paris, 1995, P. 329–344.
11. A. G. Cullis, L. T. Canham, P. D. J. Calcott, *J. Appl. Phys.* **1997**, *82*, 919–965.
12. V. Parkhutik, *Electrochem. Soc. Proc.* **1995**, *PV 95-25*, 105–114.
13. J.-N. Chazalviel, R. B. Werhspohn, F. Ozanam, *Mater. Sci. Eng., B, Solid State Mater. Advanced Technology* **2000**, *B69–70*, 1–10.
14. R. T. Collins, P. M. Fauchet, M. A. Tischler, *Phys. Today* **1997**, *50*, 244.
15. K. H. Jung, S. Shih, D. K. Kwong, *J. Electrochem. Soc.* **1993**, *140*, 3046–3064.
16. P. M. Fauchet, *J. Lumin.* **1996**, *70*, 294–309.
17. W. Theiss, *Surf. Sci. Rep.* **1997**, *29*, 91.
18. L. Pavesi, *La Rivista del Nuovo Cimento* **1997**, *20*, 1–126.
19. Special issue of *Phys. Status Solidi (a)* 2000, 182 No. 1.
20. (a) MRS Symposium Proceedings, Materials Research Society, Pittsburgh. Vol. 256, 1992, 20; (b) Vol. 283, 1993, 20; (c) Vol. 358, 1995, 20; (d) Vol. 452, 1997, 20; (e) Vol. 486, 1998.
21. *Advanced Luminescent Materials*, The Electrochemistry Society Proceedings Series, Pennington, NJ PV 95-25 1996.
22. The Electrochemistry Society Proceedings Series *Pits and Pores: Formation, Properties and Significance for Advanced Luminescent materials, Corrosion/Luminescence and Display Materials*, PV 97-7 1997.
23. Special issue of *J. Lumin.* *57*, **1993**; *80*, 1998.
24. (a) Special issue of *Thin Solid Films 255*, **1995**, 23; (b) *276*, **1996**, 23; (c) *297*, **1997**.
25. Special issue of *J. Porous Mater.* **2000**, *7*, 1–414.
26. Special issue of *Mater. Sci. Eng. B, Solid State Mater. Adv. Technol.* **2000**, *B69–70*, 1–215.
27. C. Lévy-Clément, A. Lagoubi, R. Tennee et al., *Electrochim. Acta.* **1992**, *37*, 877–888.

28. C. Lévy-Clément, *Semiconductor Microprocessing* (Eds.: H. J. Lewerenz, S. A. Campbell), Wiley and Sons, Chapter 5, 1998, pp. 177–215.
29. C. Lévy-Clément, S. Bastide, *Zeitschrift für Physikalische Chemie* **1999**, *212*, 123–143.
30. O. Bisi, *Surface Science Reports* **1999**, *38*, 1–126.
31. *Optical properties of Low dimensional Silicon Structures*, (Eds.: D. C. Benshael, L. T. Canham, S. Ossicini) NATO ASI Series, Vol. *244*, Kluwer Academic Publishers, Dordrecht, 1993.
32. *Porous Silicon Science and Technology* (Eds.: J. C. Vial, J. Derrien) Les Editions de Physique, Paris, 1995.
33. *Porous Silicon* (Eds.: Z. Chuan Feng, R. Tsu), World Scientific, New York, 1995.
34. *Properties of Porous Silicon* (Ed.: L. T. Canham) IEE INSPEEC, The Institution of Electrical Engineers, London, 1997.
35. *Structural and Optical Properties of Porous Silicon Nanostructures*, (Eds.: G. Amato, C. Delerue H. J. von Bardeleben), Gordon and Breach, Amsterdam, 1997.
36. *Light emission in Silicon: from Physics to Devices. Semiconductors and Semimetals*, Vol. *49*, (Ed.: D. J. Lockwood), Academic Press, London, 1998.
37. H. Gerischer, A. Mauerer, W. Mindt, *Surf. Sci.* **1966**, *4*, 431–439.
38. H. Gerischer, *Physical Chemistry: Vol. IXA/Electrochemistry* (Eds.: H. Eyring, D. Henderson, W. Jost) Academic Press, New York, 1970, pp. 463–542.
39. H. Gerischer, *J. Electroanal. Chem.* **1977**, *82*, 133–143.
40. A. J. Bard, M. S. Wrighton, *J. Electrochem. Soc.* **1977**, *124*, 1706–1710.
41. R. Tenne, in *Semiconductor Micromachining, Vol. 1: Fundamental Electrochemistry and Physics* (Eds.: S. A. Campbell, H. J. Lewerenz), J. Wiley and Sons Ltd, New York, 1998, pp. 139–175.
42. A. Heller, B. Miller, F. A. Thiel, *Appl. Phys. Lett.* **1981**, *38*, 282.
43. A. J. Bard, R. L. Faulkner, *Electrochemical Methods, Fundamendals and Applications*, Wiley and Sons, New York, 1980.
44. M. J. Eddowes, *J. Electroanal. Chem.* **1990**, *280*, 297–311.
45. M. J. Eddowes, *J. Electrochem. Soc.* **1990**, *137*, 3514–.
46. (a) R. Memming, G. Schwandt, *Surf. Sci.* **1966**, *4*, 109; (b) **1966**, *5*, 97–110.
47. P. A. Kohl, C. Wolowodiuk, F. W. Ostermayer Jr., *J. Electrochem. Soc.* **1983**, *130*, 2288–2293.
48. N. J. Hovel, *Semiconductors and Semimetals, Vol. 11, Solar Cells* (Eds.: R. K. Willardson, A. C. Beer), Academic Press, New York, 1975, pp. 225.
49. J. Rouquerol, D. Avnir, C. W. Fairbridge et al., *Pure Appl. Chem.* **1994**, *66*, 1739.
50. M. Matsumura, S. R. Morrison, *J. Electroanal. Chem.* **1983**, *144*, 113–120; (b) Ibid **1983**, *147*, 157–166.
51. H. Gerischer, M. Lübke, *Ber. Bunsenges. Phys. Chem.* **1987**, *91*, 394–398.
52. H. Gerischer, M. Lübke, *J. Electrochim. Soc.* **1988**, *135*, 2782.
53. H. J. Lewerenz, J. Stumper, L. M. Peter, *Phys. Rev. Lett.* **1988**, *61*, 1989–1992.
54. J. Stumper, H. J. Lewerenz, C. Pettenkofer, *Phys. Rev. B.* **1990**, *41*, 1592.
55. L. M. Peter, A. M. Borazio, H. J. Lewerenz, J. Stumper, *J. Electroanal Chem.* **1990**, *290*, 229–248.
56. J. Stumper, L. M. Peter, *J. Electroanal. Chem.* **1991**, *309*, 325.
57. H. Gerischer, P. Allongue, V. Costa Kieling, *Ber. Bunsenges. Phys. Chem.* **1993**, *97*, 753–757.
58. P. Allongue, V. Kieling, H. Gerischer, *Electrochim. Acta* **1995**, *40*, 1353–1360.
59. D. J. Blackwood, A. M. Borazio, R. Greef et al., *Electrochem. Acta* **1992**, *37*, 889.
60. L. M. Peter, D. J. Blackwood, S. Pons, *Phys. Rev. Lett.* **1989**, *62*, 308–311.
61. A. Venkateswara Rao, F. Ozanam, J.-N. Chazalviel, *J. Electrochem. Soc.* **1991**, *138*, 153–159.
62. V. Lehmann, U. Gosele, *Appl. Phys. Lett.* **1991**, *58*, 856.
63. S. M. Sze, *Physics of Semiconductor Devices*, 2nd ed., Wiley and Sons, New York, 1981, pp. 804.
64. R. Memming, *Comprehensive Treatise of Electrochemistry* (Eds.: B. E. Conway, J. O'M. Bockris, E. Yeager et al.,) Plenum, New York, 1983, Vol. *7*, p. 529.
65. H. Gerischer, *Electrode Process in Solid State Ionics* (Eds.: H. Kleitz, J. Dupuy), D. Reidel: Dordrecht, The Netherlands, 1976, p. 299.
66. X. G. Zhang, *J. Electrochem. Soc.* **1991**, *138*, 3750–3756.

67. M. J. J. Theunissen, *J. Electrochem. Soc.* **1972**, *119*, 351.
68. H. Föll, *Appl. Phys. A.* **1991**, *53*, 8–19.
69. V. Lehmann, H. Föll, *J. Electrochem. Soc.* **1988**, *135*, 2831–2835.
70. C. Lévy-Clément, A. Lagoubi, M. Tomkiewicz, *J. Electrochem. Soc.* **1994**, *141*, 958–967.
71. V. Lehmann, H. Föll, *J. Electrochem. Soc.* **1990**, *137*, 653–659.
72. V. Lehmann, *J. Electrochem. Soc.* **1993**, *140*, 2836–2483.
73. V. Lehmann, W. Hönlein, H. Reisinger et al., *Thin Solid Films* **1996**, *276*, 138–142.
74. S. Ottow, V. Lehmann, H. Föll, *J. Electrochem. Soc.* **1996**, *390*, 143, 385–390.
75. U. Grüning, V. Lehmann, C. M. Engelhardt, *Appl. Phys. Lett.* **1995**, *66*, 3254–3256.
76. E. Propst, P. A. Kohl, *J. Electrochem. Soc.* **1993**, *140*, L78–L80.
77. E. Propst, P. A. Kohl, *J. Electrochem. Soc.* **1994**, *141*, 1006–1013.
78. M. M. Rieger, P. A. Kohl, *J. Electrochem. Soc.* **1995**, *142*, 1490.
79. E. Ponomarev, C. Lévy-Clément, *Electrochem. Sol. State Lett.* **1998**, *1*, 42–45.
80. S. Lust, C. Lévy-Clément, *Phys. Status. Solidi. (A)* **2000**, *182*, 17–21.
81. P. C. Searson, X. G. Zhang, *J. Electrochem. Soc.* **1990**, *137*, 2539.
82. P. C. Searson, X. G. Zhang, *Electrochim. Acta* **1991**, *36*, 499.
83. F. Gaspard, A. Bsiesyl, M. Ligeon et al., *J. Electrochem. Soc.* **1989**, *136*, 3043–3046.
84. I. Ronga, A. Bsiesy, F.Gaspard et al., *J. Electrochem. Soc.* **1991**, *138*, 1403–1407.
85. M. I. J. Beale, *Appl. Phys. Lett.* **1985**, *46*, 86–88.
86. V. Lehmann, R. Stengl, A. Luigart, *Mater. Sci. Eng.* **2000**, *B69–70*, 11–22.
87. E. Ponomarev, C. Lévy-Clément, *J. Porous Mater.* **2000**, *7*, 51–56.
88. P. Steiner, F. Kozlowski, W. Lang, *Appl. Phys. Lett.* **1993**, *62*, 2700.
89. M. I. J. Beale, J. D. Benjamin, M. J. Uren et al., *J. Cryst. Growth* **1985**, *73*, 622–636.
90. J. P. Gonchond, A. Halimaoui, K. Ogura, *Microscopy of Semiconducting Materials*, (Eds.: A. G. Cullis, N. J. Long), IOP, Bristol, 1991, p. 235.
91. A. Albu-Yaron, S. Bastide, J. L. Maurice et al., *J. Lumin, "Light Emission From Silicon"*, (Eds.: J. C. Vial, L. T. Canham, W. Lang) **1993**, *57*, 67–71.
92. E. Galun, A. Lagoubi, R. Tenne, C. Lévy-Clément, *J. Lumin*, **1993**, *57*, 125–129.
93. P. Williams, C. Lévy-Clément, J. E. Peou et al., *Thin Solid Films* **1997**, *298*, 66–75.
94. G. Bomchil, A. Halimaoui, R. Herino, *Microelectron. Eng.* **1998**, *8*, 293–310.
95. T. Yonehara, K. Sakaguchi, N. Sato, *Appl. Phys. Lett.* **1994**, *64*, 2108–2110.
96. S. Strehlke, Q. N. Le, D. Sarti et al., *Thin Solid Films* **1997**, *297*, 291–295.
97. S. Strehlke, S. Bastide, C. Lévy-Clément, *Sol. Energy Mater. Sol. Cells* **1999**, *58*, 399–409.
98. S. Strehlke, S. Bastide, J. Guillet et al., *Mater. Sci. Eng.* **2000**, *B69–70*, 81–86.
99. C. Lévy-Clément, A. Lagoubi, D. Ballutaud et al., *Appl. Surf. Sci.* **1993**, *65/66*, 408–414.
100. E. Galun C. Reuben, S. Matlis et al., *J. Phys. Chem.* **1995**, *99*, 4132–4140.
101. R. B. Wehrspohn, J.-N. Chazalviel, F. Ozanam, *J. Electrochem. Soc.* **1999**, *146*, 3309–3314.
102. V. Lehmann, S. Rönnebeck, *J. Electrochem. Soc.* **1999**, *146*, 2968–2975.
103. E. Ponomarev, C. Lévy-Clément, *J. Porous Mater.* **2000**, *7*, 51–56.
104. W. M. Shen, M. Tomkiewicz, C. Lévy-Clément, *J. Appl. Phys.* **1994**, *76*, 3636–3639.
105. A. Albu-Yaron, S. Bastide, D. Bouchet et al., *J. Phys. I France* **1994**, *4*, 1181–1197.
106. M. H. Al Rafai, S. Ottow, M. Christophersen et al., *J. Electrochem. Soc.* **2000**, *147*, 627–635.
107. U. Grüning, V. Lehmann, *Thin Solid Films* **1996**, *276*, 151–154.
108. U. Grüning, V. Lehmann, S. Ottow et al., *Appl. Phys. Lett.* **1996**, *68*, 747–749.
109. A. Birner, S. Ottow, A. Schneider et al., *Phys. Status. Solidi. (A)* **1998**, *165*, 111–117.
110. F. Müller, A. Birner, U. Gösele et al., *J. Porous Mater.* **2000**, *7*, 201–204.
111. V. Lehmann, W. Hönlein, H. Reisinger et al., *Thin Solid Films* **1996**, *276*, 138–143.
112. S. Strehkle, *Thesis*, Paris, 1999.
113. M. Christophersen, J. Cartensen, H. Föll, *Phys. Status. Solidi (a)* **2000**, *182*, 45–50.
114. M. Christophersen, J. Cartensen, A. Feuerhake, M. Föll, *Mater. Sci. Eng.* **2000**, *B69–70*, 194–198.
115. P. Allongue, V. Costa-Kieling, H. Gerischer, *J. Electrochem. Soc.* **1993**, *140*, **1009**.
116. F. Ronkel, J. W. Schultze, *J. Porous Mater.* **2000**, *7*, 11–16.

117. P. Allongue, M. C. Bernard, J. E. Peou et al., *Thin Solid Films* **1997**, *297*, 1–4.
118. J. E. Peou, C. Henry de Villeneuve, P. Allongue et al., *Pits and Pores: Formation, Properties and Significance for Advanced Luminescent Materials, Corrosion Luminescence and Display Materials*, (Eds.: D. J. Lockwood et al.), The Electrochemical Society Proceedings Series, Pennington, NJ, 1997, Vol. 97-7, 83–91.
119. R. Tenne, G. Hodes, *Appl. Phys. Lett.* **1980**, *37*, 428–430.
120. R. Tenne, G. Hodes, *Surf. Sci.* **1983**, *135*, 453.
121. R. Tenne, H. Flaisher, R. Triboulet, *Phys. Rev. B* **1984**, *29*, 5799.
122. V. M. Dubin, *Surf. Sci.* **1992**, *274*,82.
123. C. Lévy-Clément, A. Lagoubi, M. Neumann-Spallart et al., *J. Electrochem. Soc.* **1991**, *138*, L69–70.
124. S. M. Prokes, W. E. Carlos, V. M. Bermudez, *Appl. Phys. Lett.* **1992**, *61*, 1447.
125. H. J. von Bardeleben, C. Ortega, A. Grosmann et al., *J. Lumin.* **1993**, *57*, 301.
126. G. Oskam, P. M. Hoffman, J. C. Schmidt et al., *J. Phys. Chem.* **1996**, *100*, 1801–1806.
127. P. M. Hoffman, I. E. Vermeir, P. Searson, *J. Electrochem. Soc.* **2000**, *147*, 2999–3002.
128. S. M. Sze, *Physics of Semiconductor Devices*, 2nd ed., Wiley and Sons, New York, 1981, p. 122.
129. V. A. Burrows, Y. J. Chabal, G. S. Higasi et al., *Appl. Phys. Lett.* **1988**, *53*, 998.
130. Y. J. Chabal, G. S. Higasi, K. Raghavachari et al., *J. Vac. Sci. Technol., A* **1988**, *7*, 2104.
131. J. J. Boland, *Surf. Sci.* **1992**, *261*, 17.
132. P. Kleimann, J. Linnros, S. Petersson, *Mat. Sci. Eng.* **2000**, *B69–70*, 29–33.
133. T. A. Witten, L. M. Sander, *Phys. Rev. Lett.* **1981**, *47*, 1400–1403.
134. W. W. Mullins, R. F. Sekerta, *J. Appl. Phys.* **1964**, *35*, 444.
135. R. L. Smith, S. D. Collins, *Phys. Rev. A* **1989**, *39*, 5409.
136. R. W. Bower, S. D. Collins, *Phys. Rev. A.* **1991**, *43*, 3165.
137. H. Yan, X. Hu, *J. Appl. Phys.* **1993**, *73*, 4324–4331.
138. G. C. John, V. A. Singh, *Phys. Rev. B.* **1995**, *52*, 11125–11131.
139. Y. Kang, J. Jorné, *J. Electrochem. Soc.* **1993**, *140*, 2258.
140. Y. Kang, J. Jorné, *Appl. Phys. Lett.* **1993**, *62*, 2224–2226.
141. A. Valence, *Phys. Rev. B* **1995**, *52*, 8323–8336.
142. A. Valence, *Phys. Rev. B* **1997**, *55*, 9706–9715.
143. R. B. Wehrspohn, F. Ozanam, J.-N. Chazalviel, *J. Electrochem. Soc.* **1999**, *146*, 3309–3314.
144. R. Tenne, V. Marcu, Y. Prior, *Appl. Phys. Lett.* **1985**, *A37*, 205–209.
145. R. Tenne, H. Mariette, C. Lévy-Clément et al., *Phys. Rev. B* **1987**, *15*, 1204–1207.
146. R. Tenne, C. Lévy-Clément, H. Mariette et al., *J. Cryst. Growth* **1987**, *86*, 826–833.
147. H. Homyonfer, R. Tenne, W. Giriat et al., *Phys. Rev. B* **1993**, *47*, 1244.
148. R. Tenne, *Appl. Phys. Lett.* **1983**, *43*, 201–203.
149. R. Tenne, V. Marcu, N. Yellin, *Appl. Phys. Lett.* **1984**, *45*, 1219–1221.
150. C. Lévy-Clément, J. Rioux, S. Litch et al., *J. Appl. Phys.* **1985**, *58*, 4703–4708.
151. C. Lévy-Clément, R. Triboulet, R. Tenne, *Sol. Energy Mater.* **1988**, *17*, 201–206.
152. P. A. Kohl, F. W. Ostermayer Jr., *Ann. Rev. Mater. Sci.* **1989**, *19*, 379–399.
153. F. Decker, D. A. Soltz, L. Cestato, *Electrochim. Acta* **1993**, *38*, 95–99.
154. N. G. Ferreira, D. A. Soltz, F. Decker et al., *J. Electrochem. Soc.* **1995**, *142*, 1348.
155. R. Tenne, B. Theys, J. Rioux et al., *J. Appl. Phys.* **1985**, *57*, 141–145.
156. R. Tenne, A. Wold, *Appl. Phys. Lett.* **1985**, *47*, 707–709.
157. D. Mahalu, A. Wold, R. Tenne et al., *J. Phys. Chem.* **1990**, *94*, 8012.
158. C. Lévy-Clément, R. Tenne, Photoelectrochemistry and Photovoltaics of Layered Semiconductors, Chapter IV, (Ed.: A. Aruchamy), Kluwer Academic Publishers, Norwell, MA, 1991, pp. 155–194.
159. D. Mahalu, L. Margulis, A. Wold, R. Tenne *Rapid Comm. Phys. Rev. B* **1992**, *45*, 1943.
160. F. Ozanam, R. B. Wehrspohn, J. N. Chazalviel, I. Solomon, *Phys. Stat. Sol. (a)* **1998**, *165*, 15.
161. K. H. Beckmann, *Surf. Sci.* **1966**, *5*, 187–196.
162. M. Sendova-vassileva, N. Tzenov, D. Dimova-malinskova et al., *Thin Solid Films* **1994**, *255*, 282–285.
163. S. Miyazawa, K. Shiba, K. Hirose et al., *Thin Solid Films* **1995**, *255*, 99.
164. S. Bayliss, Q. Zhang, P. Harris, *Appl. Surf. Sci.* **1996**, *102*, 390–394.

165. H. C. Choi, J. M. Buriak, *Chem. Commun.* **2000**, 1669–1670.
166. A. Ksendzov T. George, W. D. Pike et al., *Appl. Phys. Lett.* **1993**, *63*, 200–202.
167. S. Gardelis, P. Dawson, J. S. Rimmer et al., *Appl. Phys. Lett.* **1991**, *59*, 2118–2120.
168. H. Shi, Y. Zeng, Y. Wang et al., *Appl. Phys. A* **1993**, *57*, 573–575.
169. M. Schoisswohl, J. L. Cantin, M. Chamarro et al., *Phys. Rev. B* **1995**, *52*, 11898–11903.
170. K. Kolic, E. Borne, A. Sibai et al., *Thin Solid Films* **1995**, *255*, 279–281.
171. M. Schoisswohl, J. L. Cantin, S. Lebib et al., *J. Appl. Phys.* **1996**, *79*, 9301–9304.
172. D. Buttard, J. L. Cantin, M. Schoisswohl et al., *Thin Solid Films* **1997**, *297*, 233–236.
173. B. Ünal, P. Phillips, E. H. C. Parker et al., *Thin Solid Films* **1997**, *305*, 274–279.
174. S. Lebib, J. L. Fave, J. Cernogora et al., *J. Lumin.* **1999**, *80*, 153–157.
175. B. Ünal, M. Parkinson, S. C. Bayliss et al., *J. Porous Mater.* **2000**, *7*, 143–146.
176. S. Shor, B. Z. Weiss, A. D. Kurtz et al., *Appl. Phys. Lett.* **1993**, *62*, 2836–2838.
177. T. Matsumoto, T. Tamaki, T. Futagi et al., *Appl. Phys. Lett.* **1994**, *64*, 226–228.
178. J. S. Shor, L. Bernis, A. D. Kurtz et al., *J. Appl. Phys.* **1994**, *76*, 4045–4049.
179. A. O. Konstantinov, A. Henry, E. Janzën, *Appl. Phys. Lett.* **1994**, *65*, 2699–2701.
180. H. Mimura, T. Matsumoto, Y. Kanemitsu, *Appl. Phys. Lett.* **1994**, *65*, 3350–3352.
181. V. Petrova-Koch, O. Sreseli, G. Polisski et al., *Thin Solid Films* **1995**, *255*, 107–110.
182. A. O. Konstantinov, A. Henry, C. I. Harris et al., *Appl. Phys. Lett.* **1995**, *66*, 2250–2252.
183. H. Mimura, T. Matsumoto, Y. Kanemitsu, *Solid-State Electron.* **1996**, *40*, 501–504.
184. M. F. MacMillan, *J. Appl. Phys.* **1996**, *80*, 2412–2419.
185. V. P. Parkhutik, F. Namavar, E. Andrade, *Thin Solid Films* **1997**, *297*, 229–232.
186. O. Jessensky, F. Müller, V. Gösele, *Thin Solid Films* **1997**, *297*, 224–228.
187. H. W. Shim, K. C. Kim, Y. H. Seo et al., *Appl. Phys. Lett.* **1997**, *70*, 1757–1759.
188. V. P. Parkhutik, *J. Appl. Phys.* **1998**, *83*, 4647–4651.
189. A. M. Danishevskii, V. B. Shuman, A. A. Sitnikova et al., *Semicond. Sci. Technol.* **1998**, *13*, 1111–1116.
190. W. Shin, T. Hikosaka, H. S. Ahn et al., *J. Electrochem. Soc.* **1998**, *145*, 2456–2460.
191. J. N. Wang, Z. M. Chen, P. W. Woo et al., *Appl. Phys. Lett.* **1999**, *74*, 923–925.
192. J. E. Spanier, I. P. Herman, *J. Porous Mater.* **2000**, *7*, 139–142.
193. J. E. Spanier, I. P. Herman, *Phys. Rev. B* **2000**, *61*, 10437–10450.
194. J. E. Spanier, G. T. Dunne, L. B. Rowland et al., *Appl. Phys. Lett.* **2000**, *76*, 3879–3881.
195. S. Zangooie, H. Arwin, *Phys. Status Solidi. (A)* **2000**, *18*, 213–219.
196. T. Monguchi, K. Ono, Y. Baba et al., *J. Electrochem. Soc.* **2000**, *147*, 741–743.
197. J. P. Krumme, M. E. Straumanis, *Trans. Met. Soc. AIME* **1967**, *239*, 395.
198. M. M. Faktor, D. G. Fiddyment, M. R. Taylor, *J. Electrochem. Soc.* **1975**, *122*, 1566–1567.
199. D. H. Holt, *J. Mater. Sci.* **1988**, *23*, 1131.
200. P. Schmuki, J. Fraser, C. M. Vitrus et al., *J. Electrochem. Soc.* **1996**, *143*, 3316–3322.
201. P. Schmuki, D. J. Lockwood, B. F. Mason et al., *Appl. Phys. Lett.* **1996**, *69*, 1620–1622.
202. M. Hao, C. Shao, T. Soga et al., *J. Cryst. Growth* **1997**, *179*, 6661.
203. D. N. Goryachev, O. M. Sreseli, *Semiconductor* **1997**, *31*, 1192–.
204. G. Oskam, A. Natarajan, F. M. Ross et al., *Appl. Surf. Sci.* **1997**, *119*, 160–168.
205. F. M. Ross, G. Oskam, P. C. Pearson et al., *Phil. Mag. A* **1997**, *75*, 525–539.
206. I. M. Tiginyaru, G. Irmer, J. Monecke et al., *Semicond. Sci. Technol.* **1997**, *12*, 491–493.
207. P. Schmuki, L. E. Erickson, D. J. Lockwood et al., *Appl. Phys. Lett.* **1998**, *72*, 1039–1041.
208. P. Schmuki, L. E. Erickson, D. J. Lockwood et al., *J. Electrochem. Soc.* **1999**, *146*, 735–740.
209. D. J. Lockwood, P. Schmuki, H. J. Labbé et al., *Physica E* **1999**, *4*, 102–110.
210. Y. Morishita, S. Kawai, J. Sunagawa, *Jpn. J. Appl. Phys.* **1999**, *38*, L1156–L1158.
211. C. M. Finnie, X. Li, P. W. Bohn, *J. Appl. Phys.* **1999**, *86*, 4997–5003.
212. X. Li, P. W. Bohn, *J. Electrochem. Soc.* **2000**, *147*, 1740–1746.
213. B. D. Chase, D. B. Holt, *J. Electrochem. Soc.* **1972**, *119*, 314–317.
214. B. H. Erné, D. Vanmaekelbergh, J. J. Kelly, *Adv. Mater.* **1995**, *7*, 739–742.
215. D. Vanmaekelbergh et al., *Electrochim. Acta* **1995**, *10*, 689.
216. A. Anedda, A. Serpi, V. A. Karavanski et al., *Appl. Phys. Lett.* **1995**, *67*, 3316–3318.

217. A. Meijerink, A. A. Bol, J. J. Kelly, *Appl. Phys. Lett.* **1996**, *69*, 2807–2803.
218. B. H. Erné, D. Vanmaekelbergh, J. J. Kelly, *J. Electrochem. Soc.* **1996**, *143*, 305–314.
219. F. Iranzo Marin, M. A. Hamstra, D. Vanmaekelbergh, *J. Electrochem. Soc.* **1996**, *143*, 1137–1142.
220. I. M. Tiginyanu, C. Schwab, J.-J. Grob et al., *Appl. Phys. Lett.* **1997**, *71*, 3829–3831.
221. I. M. Tiginyanu, G. Irmer, J. Monecke et al., *Phys. Rev. B* **1997**, *55*, 6739–6742.
222. K. Kuriyama, K. Ushiyama, K. Ohbora et al., *Phys. Rev. B* **1998**, *58*, 1103–1105.
223. A. Sarua, I. M. Tiginyanu, V. V. Ursaki et al., *Solid. Stat. Comm.* **1999**, *112*, 581–585.
224. F. J. P. Schhurmans, D. Vanmaekelbergh, J. van de Lagemaat et al., *Science* **1999**, *284*, 141–143.
225. F. J. P. Schhurmans, M. Megens, D. Vanmaekelbergh et al., *Phys. Rev. Lett.* **1999**, *83*, 2183–2186.
226. W. H. Lubberhuizen, D. Vanmaekelberg, E. van Faassen, *J. Porous Mater.* **2000**, *7*, 147–152.
227. I. M. Tiginyanu, I. V. Kravetsky, J. Monecke et al., *Appl. Phys. Lett.* **2000**, *77*, 2415–2417.
228. T. Takizawa, S. Arai, M. Nakahara, *Jpn. J. Appl. Phys.* **1994**, *33*, L643–L645.
229. E. Kikuno, M. Amiotti, T. Takizawa et al., *Jpn. J. Appl. Phys.* **1995**, *34*, 177–178.
230. T. Takizawa, M. Nakahara, E. Kikuno et al., *J. Electron. Mater.* **1996**, *25*, 657–660.
231. A. Hamamatsu, C. Kaneshiro, H. Fujikura et al., *J. Electroanal. Chem.* **1999**, *473*, 223–229.
232. H. Fujikura, A. Liu, A. Hamamatsu et al., *Jpn. J. Appl. Phys.* **2000**, *39*, 4616–4620.
233. P. Schmuki, L. Santinacci, T. D. Djenizian et al., *Phys. Status Solidi (A)* **2000**, *182*, 51–61.
234. A. Sarua, G. Gärtner, G. Irmer et al., *Phys. Status Solidi (A)* **2000**, *182*, 207–211.
235. K. Chernoutsan, V. Dneprovskii, O. Shaligina et al., *Phys. Status. Solidi. (A)* **2000**, *182*, 347–352.
236. M. Mynbaeva, D. Tsvetkov, *Inst. Phys. Conf. Ser.* **1997**, *155*, 365.
237. J. van de Lagemaat, *Thesis*, Utrecht, 1998.
238. M. Mynbaeva, A. Titkov, A. Kryganovskii et al., *Appl. Phys. Lett.* **2000**, *76*, 1113–1115.
239. B. H. Erné, C. Mathieu, J. Vigneron et al., *J. Electrochem. Soc.* **2000**, *147*, 3759–3767.
240. B. H. Erné, A. Million, J. Vigneron et al., *Electrochem. Solid Stat. Lett.* **1999**, *2*, 619–621.
241. F. Zenia, R. Triboulet, C. Lévy-Clément et al., *Appl. Phys. Lett.* **1999**, *75*, 531–533.
242. F. Zenia, C. Lévy-Clément, R. Triboulet et al., *Electrochem. and Solid Stat. Lett.* **2000**, *3*, 73–76.
243. F. Zenia, C. Lévy-Clément, R. Triboulet et al., *Thin Solid Films* **2000**, *361–362*, 49–52.
244. D. Lincot, J. Vedel, *J. Cryst. Growth* **1985**, *72*, 426.
245. D. Lincot, J. Vedel, *J. Electroanal. Chem.* **1987**, *220*, 179.
246. L. A. Harris, R. H. Wilson, *J. Electrochem. Soc.* **1976**, *123*, 1010–1015.
247. Y. Nakato, H. Akanuma, J. I. Shimizu et al., *J. Electroanal. Chem.* **1995**, *396*, 35–39.
248. L. A. Harris, D. R. Cross, M. E. Gerstner, *J. Electrochem. Soc.* **1977**, *124*, 939–844.
249. T. Sugiura, T. Yoshida, H. Minoura, *Electrochem. Solid State Lett.* **1998**, *1*, 175–177.
250. J. van de Lagemaat, M. Plakman, D. Vanmaeckelbergh et al., *Appl. Phys. Lett.* **1996**, *69*, 2246–2248.
251. (Ed.: P. M. Fauchet) *IEEE J. Select. Topics Quantum Electron.* **1998**, *4*, 1020.
252. B. Lendt et al. *Analyt. Chem.* **1997**, *69*, 2877.
253. G. Lerondel et al. *J. Imaging Sci. Technol.* **1997**, *41*, 468.
254. H. Shinoda et al. *Nature* **1999**, *400*, 853.
255. J. H. Song, M. J. Sailor, *Comments Inorg. Chem.* **1999**, *21*, 69.
256. M. Wainwright et al. *Lett. Appl. Microbiol.* **1999**, *29*, 224.
257. V. G. Zubko et al. *Astrophys. J.* **1998**, 501.
258. T. E. Bell et al. *J. Micromech. Microeng.* **1996**, *6*, 361.
259. V. P. Bondarenko et al., *IEEE Trans Nucl. Sci.* **1997**, *44*, 1719.
260. C. Nam et al. *IEEE Microwave Guided Wave Lett.* **1997**, *7*, 236.
261. V. Parkhutik, E. Yu. Budnikiv, S. F. Timashev, *Mater. Sci. Eng. B* **2000**, *67/70*, 53–58.
262. L. T. Canham, *Adv. Mater.* **1995**, *7*, **1033**.
263. N. Koshida, H. Koyama, *Appl. Phys. Lett.* **1992**, *60*, 347.
264. S. Zangooie, R. Bjorklund, H. Arwin, *Sens. Actuators B* **1997**, *43*, 168.
265. T. Laurell, J. Drott, L. Rosengre et al., *Sens. Actuators B* **1996**, *31*, 161.
266. S. Setzu, G. Lérondel, R. Romestain, *J. Appl. Phys.* **1998**, *84*, 3129.

267. Z. Zangooie, R. Janson, H. Arvin, *J. Vac. Sci. Technol. A 1998, 16,* 2901.
268. S. Fronhoff, M. G. Berger, *Adv. Mater.* **1994**, *12,* 963.
269. M. G. Berger, M. Thönissen, R. Arhens-Fischer et al., *Thin Solid Films* **1995**, *255,* 313.
270. C. Mazzoleni, L. Pavesi, *Appl. Phys. Lett.* **1995**, *67,* 2893.
271. L. Pavesi, P. Dubost, *Semicond. Sci. Technol.* **1997**, *12,* 570.
272. A. Loni, L. T. Canham, M. G. Berger et al., *Thin Solid Films* **1996**, *276,* 143.
273. V. P. Parkhutik, L. T. Canham, *Phys. Status Solidi (A)* **2000**, *182,* 591–598.
274. V. P. Parkhutik, L. T. Canhamll, *Phy. Status Solidi. (A)* **2000**, *182,* 591–598.

3.3
Inorganic Nanoparticles with Fullerene-like Structure and Nanotubes; Some Electrochemical and Photoelectrochemical Aspects

Reshef Tenne
Weizmann Institute of Science, Rehovot, Israel

3.3.1
Introduction

The discovery of carbon fullerenes [1] and carbon nanotubes [2] has established a new paradigm in the chemistry of nanomaterials. Graphite, which is a quasi-two-dimensional (2D) (layered) compound, is the most stable polymorph of carbon in ambient conditions. Nonetheless, polyhedral carbon structures are the thermodynamically favorable form of carbon if the number of atoms in the particle is not allowed to grow beyond a few thousands. The driving force for the formation of closed-cage carbon nanostructures stems from the abundance of carbon atoms, which are only twofold bonded, in the periphery of graphite nanoparticles. By introducing pentagons into the otherwise hexagonal network, folding of the planar nanostructure takes place and is gradually closed onto itself when 12 such pentagons occur in the nanoparticle. However, it was hypothesized that this virtue should not be limited to carbon only. As shown in Fig. 1, the propensity of nanoparticles to form hollow closed structures is common to highly anisotropic layered materials such as MoS_2. Therefore, the formation of closed polyhedra and nanotubes is believed to be a generic property of materials with anisotropic (2D) layered structures [3, 4]. These new structures were denoted as inorganic fullerene-like structures (IF). In analogy to carbon fullerenes, other related structures, like multilayer polyhedra (onions) and nanotubes could be anticipated. Figure 2 shows a transmission electron microscope (TEM) image of multiwall fullerene-like particles (a) and a nanotube (b) of MoS_2. The validity of this concept has been confirmed over the last few years through numerous studies. These observations suggest that the phase diagram of elements, which form layered compounds such as Mo and S, include the new phase of hollow and closed nanomaterials (nanostructures) in the vicinity of the bulk 2D-phase. Therefore, if the otherwise planar crystallites are not allowed to grow beyond a certain size (less than say 0.2 microns), the closed-cage phase would be the thermodynamically preferred state. Globally however, the IF phase is less stable than the bulk lamellar structure. Both experiment and theory for numerous (2D) layered compounds have verified this concept.

Using a templated growth strategy, nanotubular structures from three-dimensional (3D) compounds, such as TiO_2 were also accomplished [5]. However, a clear distinction holds between nanotubular structures obtained from 3D and layered 2D compounds. Clearly, a pure 3D compound cannot form a perfectly ordered and flawless nanotubular or polyhedral structure because some of its bonds, particularly on the surface of the nanotube, remain unsatisfied. On the other hand, 2D (layered) compounds form perfectly crystalline closed-cage structures by introducing elements of lower symmetry, such as pentagons or rectangles (squares) into the otherwise hexagonal array of atoms in the crystallites' planes. For 3D compounds to form tubules, the bonds pointing outward of the folding plane must be passivated by a templating agent, such as alkyl amines, and so forth. Examples for this strategy are discussed later. Furthermore, discommensuration is inevitable for a multilayered

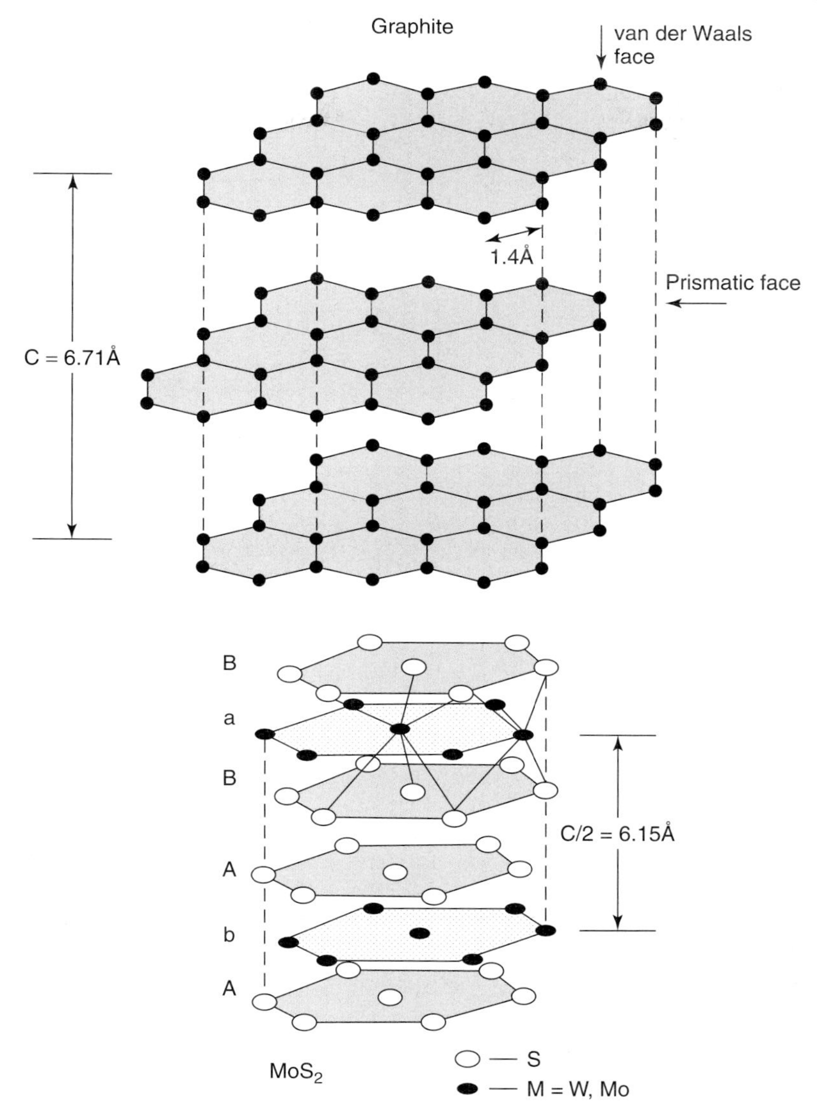

Fig. 1 Schematic drawings of graphite and MoS_2 nanoclusters. Note that in both cases the surface energy, which destabilizes the planar topology of the nanocluster, is concentrated in the prismatic edges parallel to the *c*-axis (||*c*) [3].

folded structure such as nanotubes. Discommensuration can be accommodated in multilayer nanotubes of 2D-layered compound because of the easy shear of the molecular layers. However, nanotubes of compounds with 3D structure can overcome the discommensuration by introducing dislocations, and hence the crystallinity of the nanotubes cannot be perfect in this case.

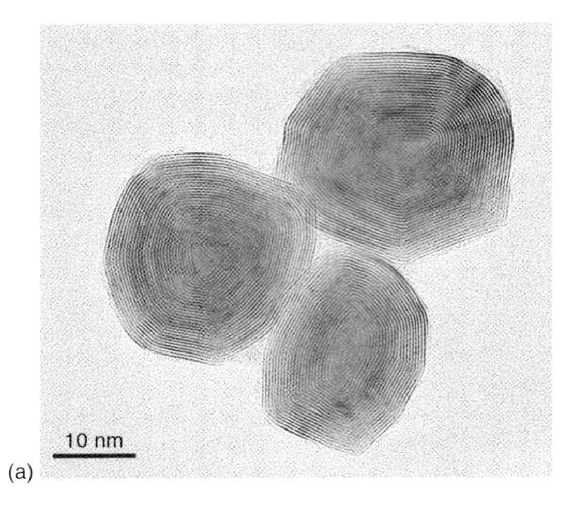

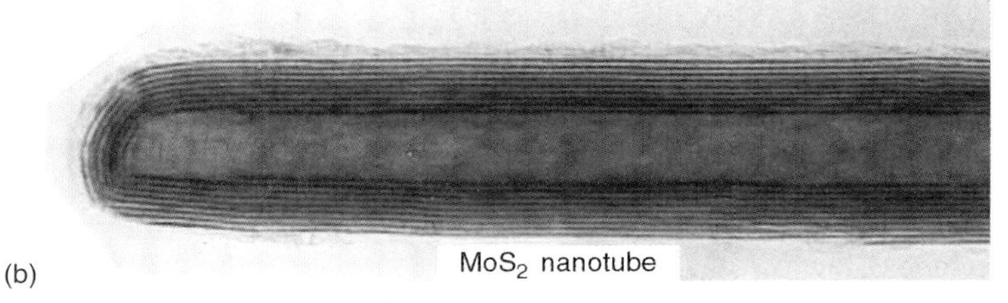

Fig. 2 TEM image of multiwall MoS_2 fullerene-like particles (a) and nanotube of MoS_2, (b) seven molecular layer thick [4]. The distance between each two layers is 0.61 nm. The *c*-axis is always normal to the surface of the nanotube.

Various synthetic strategies have been proposed, which yield large amounts of V_2O_5 [6], WS_2 [7–12], MoS_2 [13], and BN [14, 15] nanotubes and nested fullerene-like nanoparticles [16]. Generally, the synthesis of multiwall inorganic nanotubes or fullerene-like particles does not require a catalyst. Separation of a catalyst, which is necessary, for example, for the growth of single-wall carbon nanotubes, is generally a tedious and expensive procedure.

Structural aspects of nanoparticles with a closed-cage structure are also discussed in this review. Extensive mathematical research has been devoted to polyhedra made of a single layer [17]. Surprisingly however, geometrical analysis of polyhedra and nanotubular structures made of two and more interconnected layers has appeared only very recently as a result of the discovery of IF phases [18, 19]. Most inorganic layered materials consist of molecular layers with more than one kind of atom, with strong and oriented chemical bonds interconnecting between the different atoms within the molecular layer. Therefore, in this case, a mathematical analysis of interconnected multilayer polyhedra is called for.

It was found that different growth strategies lead to fullerene-like nanoparticles and nanotubes with quite distinct structures, for example, chiral versus achiral nanotubes, as discussed later. Size and shape control of MoS_2 nanotubes has been recently achieved [20], but has yet to be developed to other nanotubes and fullerene-like structures. Such control is required if one wishes to elucidate the properties of such polyhedra on an atomistic basis and study the structure–properties relationships. Also recently, the synthesis of nested nanoparticles of MoS_2 by the arc-discharge method has been demonstrated [21]. Preliminary work on the laser ablation of MoS_2 powder yielding MoS_2 octahedra with 578 Mo atoms in each nanoparticle, has been reported [22]. Synthetic methods, such as laser ablation and arc discharge, which have been used so successfully for the synthesis of size selective nanostructures of carbon and other materials will probably play an important role in future work.

So far, the properties of inorganic nanotubes have been studied rather scantily. Optical measurements in the UV-vis range and Raman scattering provided important clues regarding the electronic structure of these nanoparticles. Generically, semiconducting nanoparticles of 3D compounds exhibit a blueshift in the absorption and luminescence spectrum because of quantum size confinement of the electronic wave functions. In contrast, the band gap of semiconducting nanotubes, such as BN [23] and MoS_2 [19], shrinks with decreasing nanotube diameter. This effect is attributed to the strain in the folded structure and to the Brillouin zone folding stipulated by the new boundary conditions imposed on the energy bands in the cage structures.

Although the mechanical properties of inorganic nanotubes have not been investigated in detail so far, very good tips for scanning probe microscopy have been prepared from WS_2 nanotubes [24]. Their mechanical and chemical stability is attributed to their structural perfection and rigidity. Recent data indicates that BN nanotubes exhibit Young modulus in par with carbon nanotubes [25]. This is important, as long as BN nanotubes are insulators and carbon nanotubes are conductors, in general. Therefore, different range of applications may be called for the two kinds of ultrahigh moduli nanofibers.

The electrochemical and photoelectrochemical characterization of IF phases is starting to emerge as macroscopic amounts of the material become available. Some aspects of this preliminary work are discussed.

The potential applications of inorganic nanotubes as tips for scanning probe microscopy for the study of soft tissue rough surfaces and for nanolithography is further discussed in this review. Most importantly, these kinds of nanoparticles exhibit interesting tribological properties, which are discussed in some detail.

3.3.2 Classification of the Different Mechanisms for the Folding of Inorganic Compounds, which Yield Close-cage Structures and Nanotubes

Various mechanisms have been proposed for the formation of closed-cage structures or hollow whiskers from inorganic compounds. An early mechanism, proposed by Pauling [26] involves 2D compounds with a nonsymmetric unit cell along the c-axis, such as that of kaolinite. The structure of this compound consists of a stacking of molecular layers of SiO_2 tetrahedra connected to AlO_6 octahedra, the latter having a larger b parameter. To compensate for this geometric mismatch, hollow whiskers are formed, in which the AlO_6 octahedra occupy the outer perimeter, whereas the SiO_2 tetrahedra make the inner perimeter of the layer. In this geometry, all the chemical bonds are satisfied with relatively little strain. Consequently, the chemical and structural integrity of the compound is maintained. In analogy to the naturally occurring minerals, the so-called "misfit compounds" having a similar asymmetry along the *c*-axis were synthesized and investigated in various laboratories [27–30]. Some of the compounds belonging to this category exhibit lattice mismatch in the *b*-axis, and therefore their unit cell is inclined along the *c*-axis. Consequently, they crystallize in either a platelet form or as hollow whiskers.

Inorganic compounds with layered (2D) structure are known to have fully satisfied chemical bonds on their (0001) or van der Waals (basal) planes and consequently their (0001) surfaces are generally very inert. In contrast, the atoms on the prismatic $(10\bar{1}0)$ and $(11\bar{2}0)$ faces are not fully bonded and they are therefore chemically very reactive. When nanoclusters of a 2D compound are formed, the prismatic edges are decorated by atoms with dangling bonds, which store enough chemical energy to destabilize the planar structure. One way to saturate these dangling bonds is through a reaction with the environment, for example, with ambient water or oxygen molecules. However, in the absence of reactive chemical species, an alternative mechanism for the annihilation of the peripheral dangling bonds may be invoked. By introducing elements (rings) of lower symmetry, such as pentagons, squares, or triangles, hollow closed nanoclusters are formed [3]. For this process to take place, sufficient thermal energy is required to overcome the activation barrier associated with the bending of the layers (elastic strain energy). In this case, completely seamless and stable hollow nanoparticles are obtained in the form of either polyhedral structures or elongated nanotubes.

In several cases, nanostructures of 2D-layered compounds were shown to crystallize in scroll-like structures [31, 32]. Because of its open ends, scroll-like structures are, in general, less stable than the fully closed multiwall nanotubes. However, the stabilization of such nanostructures can be ascribed to the interlayer van der Waals interaction. It can be argued that the folded scroll-like structure is more stable than either a single layer (molecular) sheath or from a multilayer planar nanoparticle with the same number of atoms. In the first case, interlayer van der Waals interactions are absent, whereas in the second case, plenty of dangling bonds on the periphery of the nanoparticle destabilize the planar structure. The main obstacle for a full closure of the ends of a scroll-like structure can be attributed to kinetic barriers. Scroll-like structures were obtained by sonicating an aqueous solution of $GaCl_3$ [32]. Analysis of the product showed that it consists of the layered compound GaOOH. It is believed that a monolayer of this compound has been formed at the gas–liquid interface of the collapsing bubbles. Once the bubble has disappeared, the GaOOH monolayer becomes unstable in the liquid and it rapidly folds into a scroll-like structure. In V_2O_5-alkylamine nanotubes [31], the reaction conditions are too moderate to permit full closure of the cap and therefore, most of the nanotubes (nanoscrolls) are obtained either open-ended or with a partially closed cap.

However, annihilation of the dangling bonds in the periphery of the nanocluster is not the sole mechanism, which can lead to the formation of nanotubular or microtubular structures. In contrast to carbon, which forms structures derived from both sp^2 and sp^3 bonds, silicon is unable to form sp^2 related structures. Because one out of four sp^3 bonds of a given atom is pointing outward, the most stable fullerene-like structure in this case is a network of interconnected cages. This kind of network is realized in alkali-metal doped silicon clathrates [33], which were identified to have a connected fullerene-like structure [34]. In these compounds, Si polyhedra of twelve five-fold rings and two or four more six-fold rings share faces, and form a network of interconnected hollow cage structures, which can accommodate endohedral metal atoms. The clathrate compound $(Na,Ba)_x Si_{46}$ has been synthesized and demonstrated a transition into a superconductor at 4 K [35]. The electronic structure of these compounds is drastically different from that of sp^3 Si solid [36].

Another way for silicon to form cages is to establish a "core-shell" structure, in which the Si "core" atoms are arranged in a framework similar to the carbon atoms in C_{60}. The protruding bonds, which point out of the inner cage, are connected to a "shell" structure comprising Si atoms arranged in a distorted sp^3 bonding [37, 38]. However, such structures have eluded the experimenters and could not be synthesized, so far.

Closed-cage H,S,O compounds (hydridosilsesquioxanes) of the composition $Si_nO_{3n/2}H_n$ with $n = 4-36, 48, 60$ have recently been investigated theoretically [39]. Successful synthesis of some of these closed-cage structures has also been performed. The larger members in this homologous series contain 10 and 12 member rings, which consist of equal number of O and Si atoms, each. The smaller members contain rings of smaller number of atoms. The hydrogen atoms serve to passivate the extra dangling bond of Si, which protrude from the cage outward.

Foremost among the elements that pack in icosahedral (and other Archimedeans) structures is boron, its various hydrides (boranes), and related boron compounds [40]. This topic, which is covered by numerous reviews and books, is not discussed further in this review. Many other cage structures have been discussed in the literature; some of them are briefly discussed in Ref. 41.

Nanotubes and microtubes of a semi-crystalline nature can be formed by almost any compound using a templated growth mechanism. Amphiphilic molecules with a hydrophilic head group, such as carboxylate, amine or an $-OH$ group, and a hydrophobic carbon-based chain are known to form very complex phase diagrams, when these molecules are mixed together with water and an aprotic (nonaqueous) solvent [42]. Structures with tubular shape are typical for at least one of the phases in this diagram. This mode of packing can be exploited for the templated growth of inorganic nanotubes by chemically attaching a metal atom to the hydrophilic part of the molecule. Once the tubular phase has been established, the template for the tubular structure can be removed by calcination. In this way, stable metal-oxide nanotubes can be obtained from various oxide precursors [43]. Nonetheless, the crystallinity of these phases is far from being perfect, which is clearly reflected by their X-ray diffraction (XRD) and electron diffraction (ED) patterns. Thus, whereas a network of sharp diffraction spots is observed in the ED patterns of nanotubes from 2D (layered) compounds, the ED patterns of nanotubes from 3D (isotropic) materials, appears as a set of diffuse diffraction rings or streaks, which is indicative of their imperfect crystallinity. Also, the sharpness of the ED pattern in the latter case may vary from point to point, alluding to the difference in crystallinity of the different domains on the nanotube. Scroll-like nanostructures of PbS were obtained by the template growth [44]. In this case, reaction of H_2S

with tubular template molecules (Pb salts of *n*-alkanoic acids) induces crystallization of PbS nanoparticles, promoting the formation of scroll-like nanostructures. Here again, the crystallinity of the nanostructures is not uniform along the shells.

3.3.3
Synthesis of Inorganic Nanotubes and Fullerene-like Nanoparticles

Recently, numerous techniques for the synthesis of large amounts of WS_2 and MoS_2 multiwall inorganic nanoparticles with fullerene-like structures [16, 45, 46] and inorganic nanotubes of these compounds have been described [6–13, 47–51]. Each of these techniques is very different from the others and produces nanotube and fullerene-like material of somewhat different characteristics. This fact by itself indicates that IF materials (including nanotubes) of 2D metal-dichalcogenides are a genuine part of the phase diagram of the respective constituents. It also suggests that, with minor changes, these techniques can be used for the synthesis of IF materials from other inorganic layered compounds. Perhaps the most remarkable accomplishments among these are the synthesis of single-wall MoS_2 nanotubes [20] and MoS_2 octahedra [22].

One of the early synthetic methods for MS_2 nanotubes [10, 47] exploited the chemical vapor transport method, which is the standard growth technique for high-quality single crystals of layered metal-dichalcogenides (MX_2) compounds. According to this method, a powder of MX_2 (or M and X in the 1 : 2 ratio) is placed on the hot side of an evacuated quartz ampoule, together with a transport agent, such as bromine or iodine. A temperature gradient of 20–50 °C is maintained. After a few days, a single crystal of the same compound grows on the colder side of the ampoule. Almost accidentally, microtubes and nanotubes of MoS_2 were found in the colder end of the ampoule. These preliminary studies were extended to other compounds such as WS_2, and the method was optimized with respect to the production of nanotubes rather than to the bulk crystals. Whereas nanotubes produced by this method were found to consist of the 2H polytype [10], microtubes of diameters exceeding 1 mμ and length of a few hundreds micron were found to prefer the 3R polytype packing [49, 50].

Synthesis of MoS_2 nanotubes using a solid template has also been reported [51]. This synthesis is based on a generic deposition strategy, which has been proposed by Martin [52] and further perfected by Masuda and coworkers [53]. Nonuniform electrochemical corrosion of aluminum foil in an acidic solution produces a dense pattern of cylindrical pores, which serve as a solid template for the deposition of nanofilaments from a variety of materials. Thermal decomposition of a $(NH_4)_2MoS_4$ precursor, which was deposited from solution at 450 °C, and the subsequent dissolution of the alumina membrane in a KOH solution led to the isolation of large amounts of MoS_2 nanotubes. However, because of the limited thermal stability of the alumina membrane, the annealing temperatures of the nanotubes were relatively low and consequently their structure was imperfect. In fact, the MoS_2 nanotubes resembled more bamboo-shaped hollow fibers [51, 54].

The growth mechanism for the IF-MS_2 (M = Mo, W) materials by the sulfidization of the respective oxide nanoparticles has been elucidated in detail [16, 45, 46]. Here, the initial step is the sulfidization of the oxide nanoparticle surface at temperatures between 800–950 °C in an almost instantaneous reaction. Once the first sulfide layer enfolds the oxide nanoparticle, its surface is completely passivated, which prevents the sintering of the nanoparticles. In the next step, which may last a few minutes, a partial reduction of the oxide nanoparticle core by hydrogen takes place. In the third step, which is rather slow and may take a few hours, depending on the size of the nanoparticles and the temperature, a slow diffusion-controlled reaction of the reduced oxide core into sulfide takes place. It is important to note that the sulfidization of the suboxide core occurs along one single growth front in a highly ordered and concerted fashion. Although full commensuration is not possible between the two subsequent sulfide layers, the upper sulfide layer serves as a template in this case to the underlying growing sulfide layer. This growth habit is clearly observed in a high-resolution transmission electron microscope (HRTEM) [55]. The small differences in density between the oxide and the denser sulfide dictate a small but discernible hollow core for the fully sulfidized IF nanoparticle. Further study of the reaction suggested a unique mechanism for the sulfidization of the topmost oxide surface [46]. The first step in the reaction is an extraction of one oxygen atom from the oxide surface by reaction with hydrogen. In the subsequent step, extraction of another oxygen atom from the same site promotes a shear movement of the metal atoms. At the same time, reaction with sulfur atom leads to the replacement of the two oxygen atoms by a single sulfur atom on the surface. At intermediate temperatures (600–850 °C), this reaction proceeds in a highly ordered fashion, and therefore, a perfectly crystalline monolayer of sulfide covering the oxide nanoparticle is obtained. This mechanism entails a close proximity of hydrogen and sulfur atoms to the reaction site, which can be obtained by the surface-induced decomposition of H_2S gas. This mechanistic point may explain the higher yield of the IF-WS_2 (MoS_2) synthesis obtained with H_2S (+H_2), as compared to the reaction with sulfur and hydrogen, alone.

More recently, the growth mechanism of IF-MoS_2 nanoparticles was elucidated, and substantial control over the growth parameters has been achieved [16]. An overall schematic of the reactor is presented in Fig. 3(a), whereas Fig. 3(b) presents an exploded image of the inner reactor (a) and the reaction schematics. In this work, MoO_3 powder is placed in a bucket in the inner part of the reactor (a) and is heated to about 780 °C. Volatile $(MoO_3)_3$ clusters are formed and are carried down through the inner reactor (a) by the carrier gas (N_2). At some point they react with H_2 gas molecules, which diffuse through the nozzles (c) from the outer reactor (b). The mild reduction conditions yield reduced MoO_{3-x} clusters, which are less volatile and therefore coalesce and form MoO_{3-x} nanoparticles at the lower part of a. Once they grow to an adequate size (>5 nm), these suboxide nanoparticles react with H_2S gas, which also diffuses through the nozzles (c) into the inner reactor (a). In this reactor, the mass difference between the two gases, which determines their diffusivity through the nozzles c, is exploited to separate the reaction zones for the reduction and sulfidization. The sulfide coated oxide nanoparticles are carried by the carrier gas outside the reactor a. Because these nanoparticles are surface passivated they land on the ceramic filter (d) and the oxide to sulfide conversion reaction continues within the core without coalescence of the nanoparticles. This process yields a

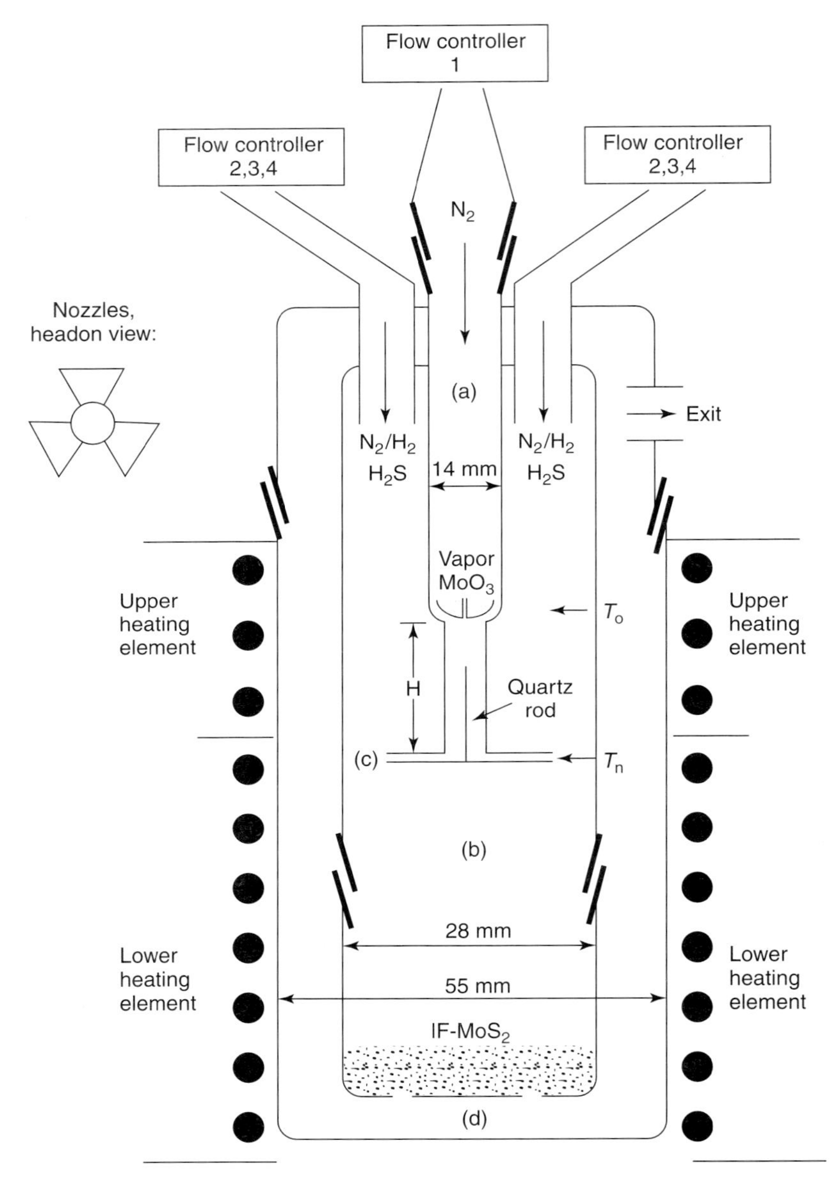

Fig. 3 (a) Schematic representation of the experimental vertical gas-phase reactor [16]: a-inner tube; b-middle tube; c-nozzles; d-external tube; (b) blow up of the inner reactor-a. The quartz rod is used to collect the various reaction products for analysis.

Scheme of the suboxide nanoclusters formation.

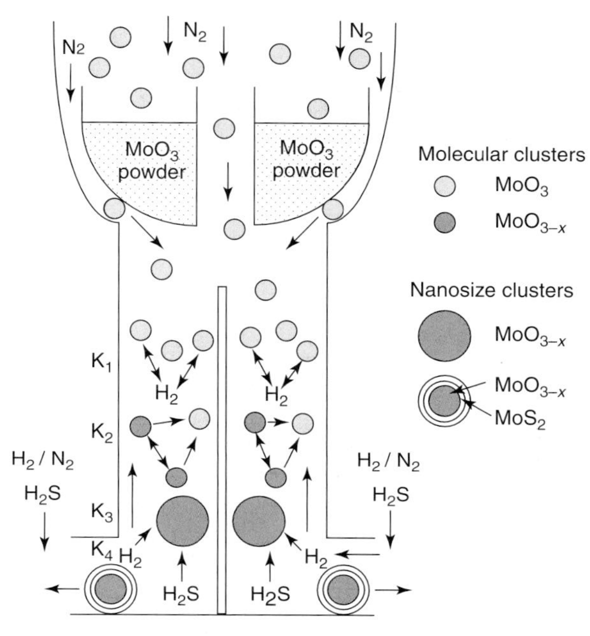

Fig. 3 *(Continued)*

pure IF-MoS_2 phase, and the control over the size and shape of the nanoparticles is quite good.

The synthesis of a pure phase of WS_2 nanotubes, 2–10-mμ long and with diameters in the range of 20–30 nm, has been recently reported [7]. Here, short tungsten oxide nanowhiskers (50–300-nm long) are prepared by heating a tungsten filament in the presence of water vapor. These short nanowhiskers are reacted with H_2S under mild reducing conditions. The length of the resulting nanotubes is determined by the interplay between three reactions: tip growth of the oxide nanowhisker, reduction, and sulfidization. In a strong reducing atmosphere (5% hydrogen in the gas mixture), relatively short nanotubes are obtained. On the other hand, long (up to 10 μm) and highly crystalline nanotubes are obtained by slowing down the rate of the reduction and sulfidization reactions. A typical assemblage of such nanotubes is shown at two magnifications in Fig. 4. One can visualize the growth process of the encapsulated nanowhisker as follows. The short oxide nanowhisker reacts with H_2S and forms a protective tungsten disulfide monomolecular layer, which covers the entire surface of the growing nanowhisker, except for its tip, which continues to grow uninterrupted. This WS_2 monomolecular skin prohibits coalescence of the nanoparticle with neighboring oxide nanoparticles, which therefore drastically slows their coarsening. Simultaneous condensation of $(WO_3)_n$ or $(WO_{3-x}\cdot H_2O)_n$ clusters on the uncovered (sulfur-free) nanowhisker tip, and their immediate reduction by hydrogen gas, lead to a reduced volatility of

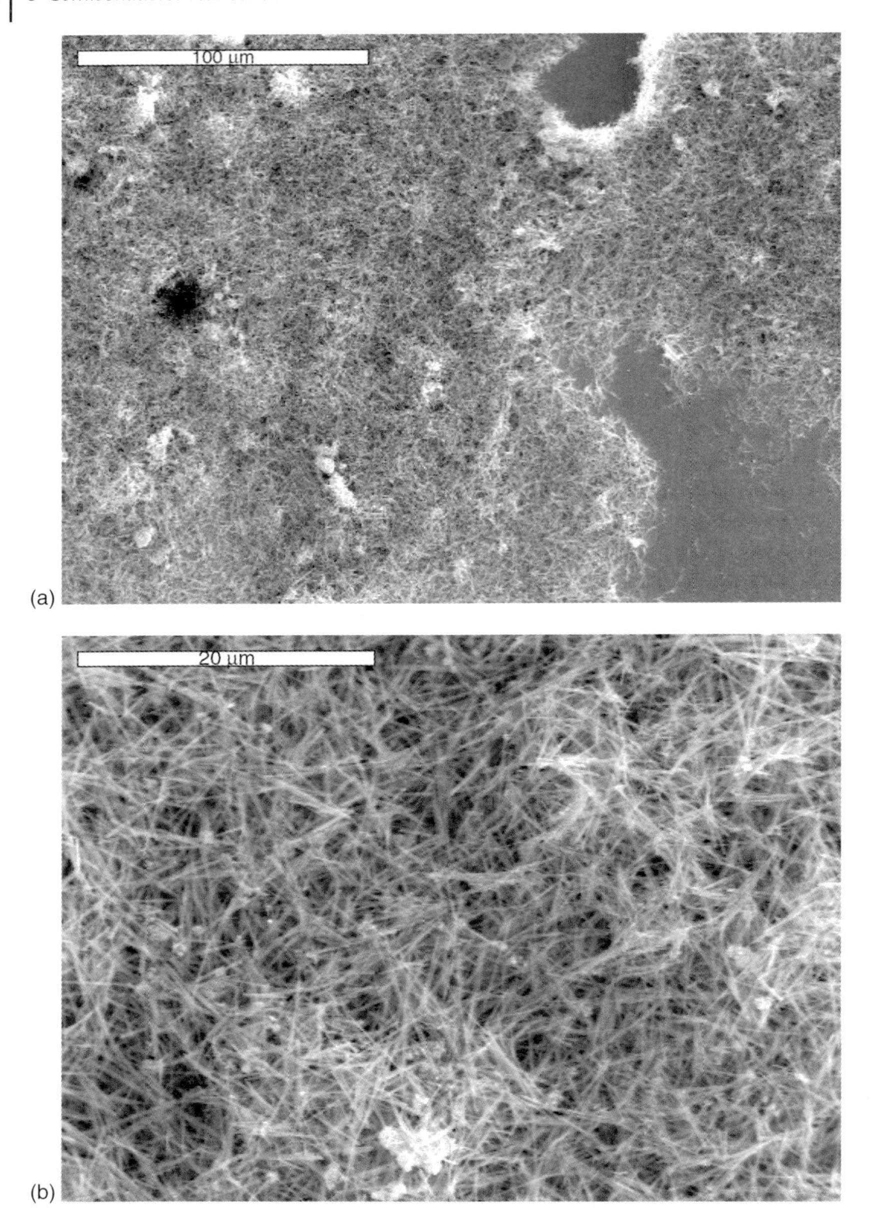

Fig. 4 SEM micrograph of a mat of long WS_2 nanotubes at two different magnifications [8].

these clusters and therefore to the tip growth. This concerted mechanism leads to a fast growth of the sulfide-coated oxide nanowhisker. Once the tungsten oxide source is depleted, the vapor pressure of the tungsten oxide in the gas phase decreases and the rate of the tip growth slowdown. It is believed that the source of the oxide clusters in the vapor phase stems from very small (<5 nm) WO_3 nanoparticles, which have higher vapor pressure than the bulk phase. Once they are depleted, the partial vapor pressure produced by the tungsten oxide clusters is lowered. This leads to the termination of the growth process because the rate of sulfidization of the oxide skin continues at the same pace and the exposed whisker tip becomes completely coated with the protective sulfide skin. This process leads to the formation of oxide nanowhiskers 2–10 μm long, coated with an atomic layer of tungsten sulfide. After the first layer of sulfide has been formed in an almost instantaneous process, a relatively fast reduction of the oxide into W_nO_{3n-1} phases takes place. These phases have been investigated in detail in the past. They consist of a WO_3 backbone with distorted ReO_3 structure, interrupted periodically by rows of edge sharing oxide octahedra-crystal shear planes (CS) [7, 55, 56]. The conversion of the reduced oxide core into tungsten sulfide is a rather slow diffusion-controlled process, which may last one to a few hours, depending on the experimental parameters. Note that during the gradual reduction of the oxide core, the CS planes in the oxide phase rearrange themselves and approach each other [55] until a stable reduced oxide phase, W_3O_8, is reached [56]. Another phase that is often found in tungsten oxide whiskers is the monoclinic $W_{18}O_{49}$ with pentagonal columns and hexagonal channels [57]. These phases provide a sufficiently open structure for the sulfidization to proceed until the entire oxide core is consumed and converted into the respective sulfide. Furthermore, the highly ordered nature of the reduced oxide serves as a kind of a template for a virtually dislocation-free sulfide layer growth. Further reduction of the oxide core would bring the sulfidization reaction to a halt [58]. It was shown that in the absence of the sulfide skin, the oxide nanowhisker is reduced rather swiftly to a pure tungsten nanorod. Therefore, the encapsulation of the oxide nanowhisker, which tames the reduction of the core, allows for the gradual conversion of this nanoparticle into a hollow WS_2 nanotube. The present model alludes to the highly synergistic nature between the reduction and sulfidization processes during the WS_2 nanotube growth and the conversion of the oxide core into multiwall tungsten sulfide nanotubes. It is important to emphasize that in this process, the diameter of the nanotube is determined by the precursor diameter. The number of WS_2 layers increases with the duration of the reaction until the entire oxide is consumed. This permits very good control of the number of WS_2 walls in the nanotube, but with the oxide core inside. The oxide whisker can be visualized as a template for the sulfide, which grows inward from outside. Unfortunately, this method does not lend itself to the synthesis of (hollow) single-wall nanotubes because long whiskers, a few nm thick, would not be stable.

The multiwall hollow WS_2 nanotubes obtained in this process are quite perfect in shape, which has a favorable effect on some of their physical and electronic properties. This strategy, that is, the preparation of nanowhiskers from an oxide precursor and their subsequent conversion into nanotubes, is also likely to become a versatile vehicle for the synthesis of pure nanotube phases from other 2D-layered compounds. A related method for the synthesis of WS_2 nanotubes is to first synthesize crystalline and long $W_{18}O_{49}$ nanowhiskers, and subsequently sulfidize these nanowhiskers [8, 11]. This method yields large amounts of very long (30 microns) WS_2 nanotubes, but often with open-ended caps.

Another strategy for the synthesis of inorganic nanotubes is through the sol-gel process, sometimes combined with hydrothermal or solvothermal methods [6, 31, 59, 60, 61]. Here, a metal organic compound (e.g. metal alkoxide) is dissolved together with a template-forming species in alcohol and a template structure for the growth of nanotubes is formed. The addition of small amounts of water leads to a slow hydrolysis of the organic metal compound, that is, the formation of metal-oxide sol, but the template structure is retained. On tuning of the pH, polycondensation takes place and the sol transforms into a gel, which consists of an −M−O− polymer with longer chains. In the case of the layered compound V_2O_5 [6, 31], a sol was first prepared by mixing vanadium (V) oxide triisopropoxide with hexadecylamine in ethanol and aging the solution while stirring, which resulted in the hydrolysis of the vanadium oxide. Subsequent hydrothermal treatment at 180 °C led to the formation of nanotubes with the formal composition $VO_{2.45}\cdot(C_{16}H_{33}NH_2)_{0.34}$. Figure 5 shows HRTEM images of such nanotubes (courtesy of Dr. F. Krumeich, ETH). Many of the nanotubes do not have a closed cap as shown in Fig. 5(a). A cross section of the nanotubes, shown in Fig. 5(b) indicates that the V_2O_5 nanotubes adopt a spiral growth mode. Whereas the dark image on the left (a) shows electronmicrograph of the cross sections of two nanotubes, the bright contrast on the right (b) shows the vanadium image obtained by an electron energy filter. The coincidence between these two kinds of images shows that the layers of the nanotubes consist of the vanadium atoms. These nanotubes are crystalline, which is also evident from their X-ray and ED patterns. Facile and (quite) reversible Li intercalation into such nanotubes has been demonstrated [61],

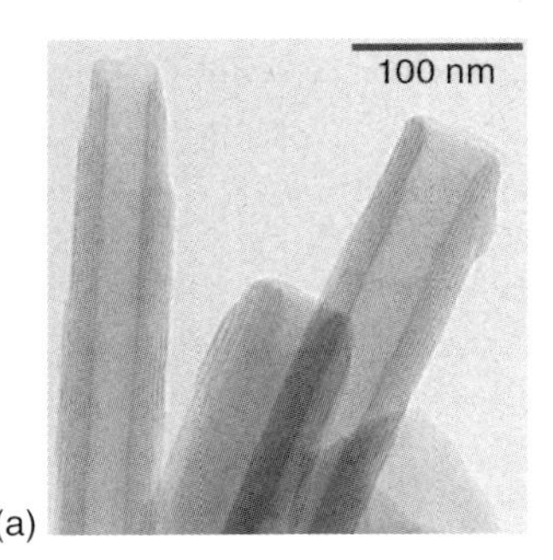

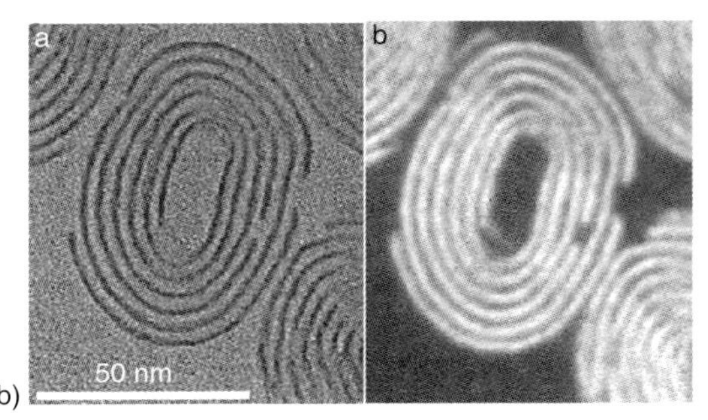

Fig. 5 HRTEM micrograph of $VO_{2.4}\cdot(C_{16}H_{33}NH_2)_{0.34}$ nanotubes (a rolled up superlattice of V_2O_5 layers separated by amphiphilic moieties with amine head group) [51]. (a) side-on view of three nanotubes with unclosed caps; (b) cross section of the nanotubes. On the left HRTEM image; on the right: electron energy filtered image of the vanadium.

which puts this material into the forefront of high-energy density battery research. A similar strategy has recently been used for the synthesis of nanotubes and nanorods from the layered compound $Mg(OH)_2$ [62].

Thus far, only metal oxide nanotubes have been synthesized by the hydrothermal templated growth. Whereas crystalline nanotubes were obtained from 2D (layered) oxides, various 3D oxide compounds resulted in semicrystalline or amorphous nanotubes only [63]. This kind of process could be extended to the synthesis of nanotubes from chalcogenide and halide compounds in the future.

In another procedure, carbon nanotubes were used as templates for the deposition of V_2O_5 nanotubes; the solid template was subsequently removed by burning the sample in air at 650 °C [64]. This strategy can easily be adopted for the synthesis of different oxide nanotubes, as shown in the following section.

The heating of an ammonium thiomolybdate compound with the formula $(NH_4)_2Mo_3S_{13}{\cdot}xH_2O$ was shown to lead to its decomposition around 673 K and to the formation of a nonhomogeneous MoS_2 phase, which also contains elongated MoS_2 particles (mackles) [65]. The mackles are ascribed to a topochemical conversion of the precursor particles, which form closed layers of MoS_2 on top of an amorphous core (possibly a-MoS_3). A slightly modified method also yielded nanotubes, which, however, do not exhibit a highly crystalline order [66]. One of the early demonstrations of a MoS_2 nanoparticles with fullerene-like structures was obtained by electron beam irradiation of MoS_2 crystallites [67]. The nanoparticles were quite spherical in shape, which is a reminiscent of the formation of carbon onions by a similar method [68]. Another strategy for the synthesis of nanomaterials with fullerene-like structures is through microwave irradiation of a suitable precursor [69, 70]. In this process, $Mo(CO)_6$ and $W(CO)_6$ powders were vaporized and subsequently mixed with a heated H_2S(1%)/argon atmosphere, under microwave irradiation. When the temperature of the reaction was raised to 580 °C, some nonperfect, fullerene-like structures could be observed.

Another method for the synthesis of MoS_2 nanoparticles with fullerene-like structure is by crystallization of amorphous MoS_3 (a-MoS_3) nanoparticles by application of microsecond long electrical pulses from the tip of a scanning tunneling microscope. These nanoparticles consist of IF-MoS_2 envelope and an *a*-MoS_3 core [71]. The MoS_2 shells were found to be quite perfect in shape and fully closed. This observation is indicative of the fast kinetics of crystallization of the fullerene-like structures. A self-propagating self-limiting mechanism has been proposed for this reaction [71]. According to this model, the process is maintained by the local heating because of the exothermic nature of the chemical reaction and the heat of crystallization. When the MoS_2 layers are completed and closed, the process self-extinguishes itself because of lack of the extra heat from the reaction.

A novel room temperature method for producing nested fullerene-like MoS_2 with an *a*-MoS_3 core using a sonoelectrochemical probe has been described [72]. MoS_2 nanotubes also occur occasionally in this product. Ultrasonically induced reactions are attributed to the effect of cavitation, whereby very high temperatures and pressures are obtained inside the imploding gas bubbles, which are formed in ultrasonically treated liquid solutions [73]. The combination of a sonochemical probe with electrochemical deposition has been investigated for some time now [74, 75]. Generally, the decomposition of the gas molecules in the bubble and the high-cooling rates lead to the production of amorphous nanomaterials. It appears, however, that in the case of layered compounds, crystalline nanomaterials with structures related to fullerenes, are obtained by sonochemical reactions [72]. In this case, the collapsing bubble serves as an isolated reactor and there is a strong thermodynamic driving force in favor of forming the seamless (fullerene-like) structure, rather than the amorphous or platelike nanoparticle [32, 72, 76]. In fact, a few similarities between the two foregoing methods for the preparation of IF-MoS_2, exist [71, 72, 76]. First, both processes consist of two steps: in the first step the a-MoS_3 nanoparticles are prepared and are subsequently crystallized by an electrical pulse [71] in one case, and by a sonochemical pulse in the other process [72]. A compelling factor in favor of the fast kinetics of fullerene-like nanoparticle formation is that, in both processes, the envelope is complete while the core of the nanoparticles (>20 nm) remains amorphous. Because the transformation of the amorphous core into a crystalline structure involves a slow outdiffusion of sulfur atoms; the core of the nanoparticles is unable to crystallize during the short (ns-μs) pulses. It is likely that the sonochemical formation of IF-MoS_2 nanoparticles can be attributed also to the self-propagating self-extinguishing process described earlier.

In a related study, scroll-like structures were prepared from the layered compound GaOOH by sonicating an aqueous solution of $GaCl_3$ [32]. This study again shows the preponderance of nanoparticles with a rolled-up structure from layered compounds. It has been pointed out [77] that in the case of nonvolatile compounds, the sonochemical reaction takes place on the interfacial layer between the liquid solution and the gas bubble. It is hard to envisage a gas bubble of such an asymmetric shape as the GaOOH scroll-like structure. Alternatively, it is proposed that a monomolecular layer of GaOOH is formed on the bubble's envelope, which rolls into a scroll-like shape once the bubble is collapsed. Rolled-up scroll-like structures were obtained by sonication of $InCl_3$, $TlCl_3$, and $AlCl_3$, which demonstrates the generality of this process [32]. Hydrolysis of this group of compounds (MCl_3) results in the formation of the layered compounds MOOH, which, on crystallization, prefer the fullerene-like structure. However, the hydrolysis of MCl_3 compounds may also lead to MOCl compounds, also having a layered structure, and its formation during the sonication of MCl_3 solutions has not been convincingly excluded. Figure 6(a) shows a nested structure of $Ni(OH)_2$ obtained by the sonochemical reaction of $NiCl_2$ solution (courtesy of Y. Rosenfeld Hacohen, Weizmann Institute). The magnified core of the same nanoparticle is shown in Fig. 6(b), which clearly reveals its nested fullerene-like structure with the hollow center. Nanoscale tubules with scroll-like structure have been obtained from potassium hexaniobate ($K_4Nb_6O_{17}$) by acid exchange and careful exfoliation in basic solution [78, 79]. The exfoliation process results in monomolecular layers, which are unstable against folding even at room temperature and consequently form the more stable scroll-like structures.

More recently, the formation of fullerene-like Tl_2O nanoparticles by the sonochemical reaction of $TlCl_3$ in aqueous solution was reported [80]. Tl_2O possesses the anti-$CdCl_2$ structure, with the oxygen layer sandwiched between two thallium layers. Currently, the yield of the IF-Tl_2O product is not very high (ca. 10%), but purification of this phase could be obtained by heating of the sample to 300 °C. The size and shape control of the fullerene-like particles is not easy in these types of reactions. Nonetheless, the fact that this is a room temperature process is rather promising, and future developments will hopefully permit better control of the reaction products. Perhaps, the most important aspect of this work is the fact that bulk Tl_2O is not stable in the ambient atmosphere, but the fullerene-like structure is found to be rather stable in these conditions. This fact stems from the closed (seamless) nature of the cage. In contrast to that, macroscopic platelets of this compound are unstable in the ambient because facile water and oxygen intercalation occurs through the edges into the van der Waals gap separating the molecular layers. This process disrupts the stacking of the molecular layers and leads to their exfoliation and rapid oxidation. This extra stability of IF nanostructures in ambient conditions is discussed in greater detail in the following section. It was discussed before in the context of the synthesis of IF-VS_2, which does not have a stable 2D macroscopic analog [81].

NbS_2 is a compound with a layered structure, exhibiting a transition to superconductivity at 6 K. Figure 7 shows an

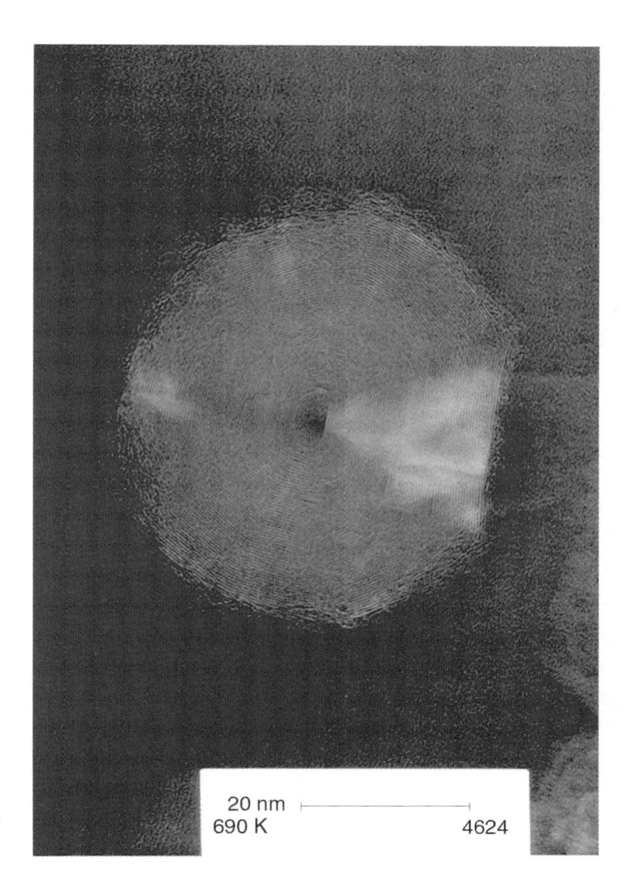

Fig. 6 TEM view of a nested fullerene-like $Ni(OH)_2$ nanoparticle. The layer to layer distance is 0.46 nm.

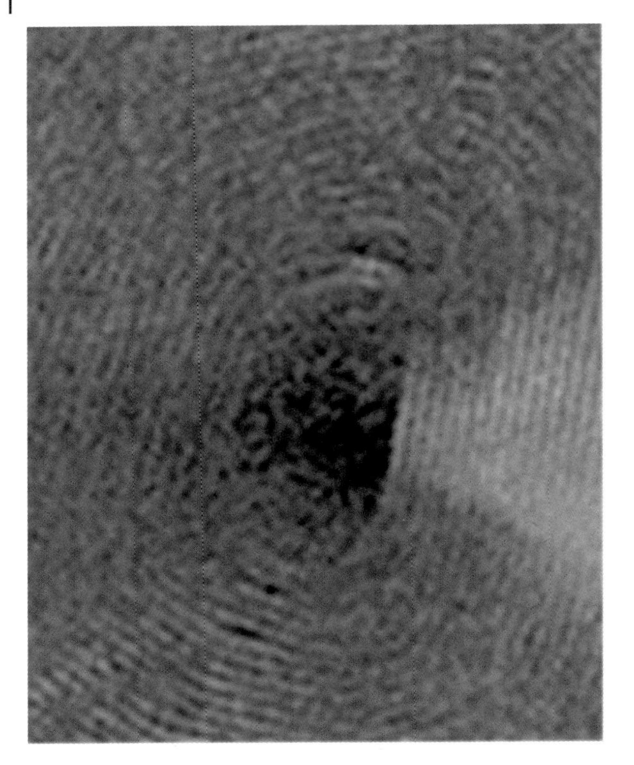

Fig. 6 *(Continued)*

"onion"-shaped NbS_2 nanoparticle, produced by the reaction of Nb_2O_5 with H_2S (courtesy of Dr. M. Homyonfer, Weizmann Institute). The size and curvature dependence of the superconducting transition temperature (and the critical fields) is of fundamental importance in this case, and serves as a stimulus for these studies. The electronic structure of NbS_2 nanotubes was calculated recently [82]. Optimized nanotube structures of these compounds are found to be stable and possess metallic behavior.

$NiCl_2$ is a layered compound with $CdCl_2$ structure, where the Ni layer is sandwiched between two chlorine layers and six Cl atoms surround each Ni atom in an octahedral arrangement. Strong ferromagnetic interactions occur between the Ni atoms, orienting the magnetic dipoles in the a-b plane of the layer ($\perp c$). Weak antiferromagnetic coupling between the Ni dipoles of adjacent layers lead to the antiferromagnetic coupling in this material with Néel temperature of 51 K. Spherical and polyhedral nanoparticles of $NiCl_2$ and nanotubes thereof have been reported [83]. Such nanostructures cannot be antiferromagnetic because there is no atomic layer with the same number of atoms in these structures. Furthermore, closed polyhedral structures with an odd number of layers (1 and 3) have been synthesized, which cannot be antiferromagnetic. Unfortunately, the synthesis of large amounts of these nanostructures proved to be rather difficult, mainly because of the hygroscopic nature of the compound.

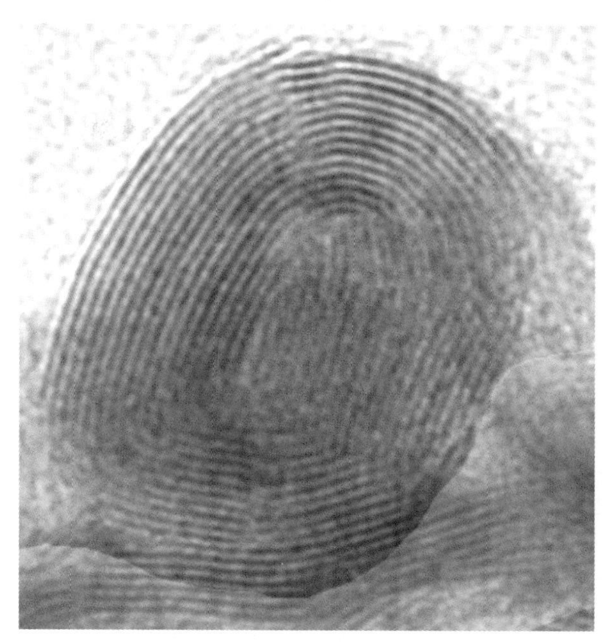

Fig. 7 TEM micrograph of NbS_2 nanoparticle with nested ("onion") fullerene-like structure (courtesy of Dr. M. Homyonfer, Weizmann Institute). The layer to layer distance is 0.59 nm.

$CdCl_2$ and its first hydrate $CdCl_2{\cdot}nH_2O$ have a layered structure. $CdCl_2$ has a hexagonal structure with two $CdCl_2$ layers in the unit cell, whereas the second is rhombohedral with three $CdCl_2{\cdot}nH_2O$ layers in the unit cell. Because of its hygroscopisity, the hydrate-free compound is not stable in ambient conditions. Careful drying of the powder produces the relatively stable $CdCl_2{\cdot}H_2O$ compound, although further hydration of the compound is inevitable in ambient conditions. Irradiation of this powder by the electron beam of a TEM leads to the loss of the water molecules and recrystallization to water-free $CdCl_2$ nanoparticles with closed-cage polyhedral structures [84]. Similarly, nested $CdCl_2$ structures were obtained by annealing $CdCl_2{\cdot}nH_2O$ in the oven at 750 °C under argon flow. Figure 8 shows a typical hexagonal closed structure of $CdCl_2$ obtained by e-beam irradiation of the $CdCl_2{\cdot}H_2O$ precursor powder. Admittedly, the detailed structure of $CdCl_2$ cages has not been unraveled as yet. Because bulk $CdCl_2$ is extremely hygroscopic, it is impossible to handle the bulk material in the ambient atmosphere. In contrast, the fullerene-like structures are perfectly stable in the ambient, which again is a manifestation of the kinetic stabilization of the closed-cage structure.

BN and $B_xC_yN_z$ nanotubes and fullerene-like structures have been synthesized by various laboratories in recent years. The most popular methods are the plasma arc and laser ablation techniques. The first report on the synthesis of BN nanotubes, using the arc-discharge technique, was by the Zettl group [85, 86]. Because

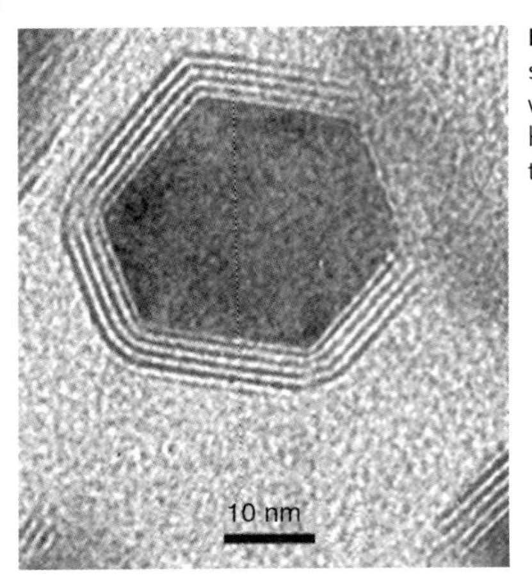

Fig. 8 TEM micrograph showing a $CdCl_2$ cage structure with four layers in the shell and hexagonal symmetry. The layer to layer distance is 0.58 nm [84].

BN is an insulator, a composite anode was prepared from a tungsten rod with an empty bore in the center, which was stuffed with a pressed hexagonal BN powder. For the cathode, water-cooled Cu rod was used. The collected gray soot contained a limited amount of multiwall BN nanotubes. It is possible that in this case the tungsten also serves as a catalyst. By perfecting this method, macroscopic amounts of double-wall BN nanotubes of a uniform diameter (2 nm) were obtained in large amounts [14]. An alternative route employed HfB_2 electrodes in nitrogen atmosphere [87]. This route led to the synthesis of BN nanotubes with varying number of walls, from a single-wall to multiple-wall nanotubes. The Hf was not incorporated into the tube and probably played the role of a catalyst. Using Ta instead of W as the metal anode, BN nanotubes with flat heads, alluding to the existence of three B_2N_2 squares in the cap have been observed [88]. In another synthetic approach, pyrolysis of CH_3CNBCl_3 complex in the presence of a Co catalyst provided $B_xC_yN_z$ nanotubes and nanofibers [89]. Clear evidence in support of the cap containing three B_2N_2 squares was obtained. This is to be contrasted with carbon nanotube caps, which contain six pentagons. Recently, long and quite perfect BN nanotubes were obtained by focusing a continuous CO_2 laser onto hexagonal boron nitride powder in N_2 gas atmosphere [90]. Furthermore, ropes consisting of hexagonal array of BN nanotubes have been observed in this study, which is indicative of the uniform diameter (2 nm) of the nanotubes. The synthesis of $B_xC_yN_z$ nanotubes and BN-cage structure has been intensively pursued by Bando's group (see for example Refs. 91, 92). Thus, a successful strategy for the synthesis of single- and multiple-wall $B_xC_yN_z$ nanotubes through chemical substitution of carbon nanotubes has been demonstrated [91]. Here, single-wall carbon nanotube bundles were thermally treated with BO_3 at 1523–1623 K under a nitrogen gas flow. The resulting nanotubes had diameters of 1.2–1.4 nm, which is similar to the precursor nanotubes. Electron beam irradiation of hexagonal boron nitride resulted in octahedral BN onions [92]. These structures are characterized by the presence of six B_2N_2 squares embedded in the hexagonal BN network.

Concentric multiwall BN/C nanotubes were prepared. Here, a spontaneous segregation of nanotubes of a different chemical composition was observed. Thus, concentric carbon nanotubes followed by BN and again carbon nanotubes were identified [93]. The segregation of the carbon nanotubes in the inner end outer surfaces of the concentric nanotubes structure is attributed to the lower surface energy of graphite as compared with hexagonal BN. Concentric BC_2N and carbon nanotubes have also been reported. Using laser ablation method, concentric CN and carbon nanotubes with SiO_2 core have been prepared [94]. Built-in semiconducting junctions can be fabricated using the composite BN−C nanotubes. Silver nanoparticles encapsulated within boron nitride nanocages were produced by mixing boric acid, urea, and silver nitrate and reduction at 700 °C under hydrogen atmosphere [95]. Generalization of this method to the encapsulation of other metallic nanoparticles was discussed.

As discussed earlier briefly, semicrystalline or amorphous nanotubes can be obtained from 3D compounds and metals by depositing a precursor on a nanotube template intermediately, and subsequently removing the template by calcination. If the template molecules are not removed and they are able to effectively passivate the dangling bonds of the compound, a perfectly crystalline nanotube composite can be obtained. However, after high-temperature calcination, the organic scaffold is removed and the inorganic oxide remains. Because a nanotube is the rolled-up structure of a 2D-molecular sheet, there is no way that all the chemical bonds of the 3D-inorganic compound will be fully satisfied on the nanotube inner and outer surfaces. Furthermore, the number of molecules increases with the diameter, and hence a full commensuration between the various molecular layers is not possible. Therefore, nanotubes of 3D compounds cannot form a perfectly crystalline structure, and the nanotube surface is not going to be inert. Nonetheless, there are certain applications such as in catalysis, where such a high-surface area pattern with reactive surface sites, that is unsaturated bonds, is highly desirable. The first report of SiO_2 nanotubes [96] was observed serendipitously during the synthesis of spherical silica particles by the hydrolysis of tetraethylorthosilicate in a mixture of water, ammonia, ethanol, and tartaric acid. More recently, nanotubes of SiO_2 [59, 60], TiO_2 [5, 97, 98], Al_2O_3, and ZrO_2 [5, 99], and so forth have been prepared by the self-assembly of molecular moieties on preprepared templates, which instigate uniaxial growth mode. One can distinguish between solid templates, such as carbon nanotubes, porous alumina, and soft templates, such as elongated micelles. In fact, there is almost no limitation on the type of inorganic compound that can be "molded" into this shape using this strategy. A related synthesis of tubular β-Ag_2Se crystals has been described [100]. Here, hydrothermal reaction between AgCl, Se, and NaOH lead to the formation of tubular structures with a hexagonal cross section.

We note in passing that although the production of carbon nanotubes does not lend itself to an easy scale-up, the tunability of the carbon nanotube radii and the perfection of its structure could be important for their use as a template for the growth of inorganic nanotubes with a controlled radius. This property can be rather important for the selective catalysis of certain reactions, where either the reaction precursor or the product must diffuse through the (inorganic) nanotube inner core.

The rational synthesis of peptide-based nanotubes by self-assembling of polypeptides into a supramolecular structures was demonstrated. This self-organization leads to peptide nanotubes, having channels of 0.8 nm in diameter and a few hundred nm long [101]. The connectivity of the proteins in these nanotubes is provided by weak bonds such as hydrogen bonds. These structures benefit from the relative flexibility of the protein backbone, which does not exist in nanotubes of covalently bonded inorganic compounds.

3.3.4 Thermodynamic, Structural, and Topological Considerations

The thermodynamic stability of the fullerene-like materials is rather intricate and far from being fully understood. IF structures are not expected to be globally stable, but they are probably the stable phase of a layered compound, when the nanoparticles are not allowed to grow beyond, say a fraction of a micron. Therefore, a narrow domain of conditions, where nanophases of this kind exist, is assumed. The existence zone of the IF phase on the binary-phase diagram must be very close to the existence zone of the layered compound itself. This idea is supported by a number of observations. For example, the W-S phase diagram provides a very convenient pathway for the synthesis of IF-WS_2. The compound WS_3, which is stable below 850 °C under excess of sulfur, is amorphous. This compound will therefore lose sulfur atoms and crystallize into the compound WS_2, which has a layered structure, on heating or when sulfur is denied from its environment. If isolated nanoparticles of WS_3 are prepared and they are allowed to crystallize under the condition that no crystallite can grow beyond 0.2 mμ, fullerene-like WS_2 (MoS_2) particles and nanotubes will become the favored phase. This principle serves as a principal guideline for the synthesis of bulk amounts of the IF-WS_2 phase [63], and WS_2 nanotubes in particular [7, 8]. Unfortunately, in most cases the situation is not as favorable and more work is needed to clarify the existence zone of the IF phase in the phase diagram (in the vicinity of the layered compound).

Another very important implication of the formation of nanoparticles with IF structures is that in several cases it has been shown that the IF nanoparticles are stable, but the bulk form of the layered compound is either very difficult to synthesize or is totally unstable. The reason for this surprising observation is probably related to the fact that the IF structure is always closed and hence it does not expose reactive edges and interacts only very weakly with the ambient, which in many cases is hostile to the layered compound. For example, Na intercalated MoS_2 is unstable in a moistured ambient because water is sucked between the layers and into the van der Waals gap of the platelet and exfoliates it. In contrast, Na intercalated IF-MoS_2 has been produced and was found to be stable in the ambient or even as alcoholic suspensions [81]. Coaxial nanotubes of MoS_2 and WS_2 intercalated with Ag and Au atoms were recently reported [102, 103]. The analogous phase in the bulk material has not been reported so far, which is another demonstration for this point. Chalcogenides of the first row of transition metals, such as $CrSe_2$ and VS_2, are not stable in the layered structure. However, Na intercalation endows extra stability to the layered structure because of the charge transfer of electrons from the metal into the partially empty valence band of the host [104]. Thus, for example, $NaCrSe_2$ and $LiVS_2$ form a superlattice in which the alkali metal layer and the transition metal layer alternate. The structure of this compound can be visualized akin to the layered structure of $CrSe_2$, in which the octahedral sites in the van der Waals gap between adjacent layers are fully occupied by the Na (Li) atoms. Notwithstanding, VS_2 nanoparticles with a fullerene-like structure, that is, consisting of layered VS_2 were found to be stable [81]. The unexpected extra stability of this structure emanates from the closed seamless structure of the IF, which does not expose the chemically reactive sites to the hostile environment. Similarly, γ-In_2S_3, which is unstable as a layered structure in the bulk (platelets), was also found to be stable in the IF form [81]. More recently, nanotubes of InS were obtained in a low temperature reaction between tributyl indium and H_2S in the presence of thiobenzene catalyst [105]. Until this work, the layered structure of InS was not known. This work emphasizes again on the relative stability of the nanotubular structure in the presence of moisture and oxygen compared to the instability of the macroscopic 2D-crystalline form. Many compounds such as GaN come in more than one crystalline structure, of which the layered structure may be one [106]. Although the layered polymorph is not stable in ambient conditions, this phase can nevertheless be synthesized under extreme conditions, and subsequently rapidly quenched to ambient conditions, where sluggish kinetics will slow its transformation into the stable phase (wurzite). On the same token, it is possible to assume that nanotubes and fullerene-like structures of GaN can be formed, for example by using similar strategies to the ones used for the BN synthesis. This idea opens new avenues for the synthesis of layered compounds, which could not be previously obtained or could not be exposed to the ambient, and therefore could only be studied to a limited extent. On the other hand, this concept provides a vehicle for the study of nanotubular structures with interesting properties, which could not be anticipated before.

Many layered compounds come in more than one stacking polytype [107]. For example, the two most abundant polytypes of MoS_2 are the 2H and 3R. The 2H polytype stands for a hexagonal structure with two S–Mo–S layers in the unit cell (AbA ··· BaB ··· AbA ··· BaB, and so forth). The 3R polytype has a rhombohedral unit cell of three repeating layers (AbA ··· BcB ··· CaC ··· AbA ··· BcB ··· CaC, and so forth). In the case of MoS_2, the most common polytype is the 2H form but the 3R polytype was found, for example, in thin MoS_2 films prepared by sputtering [108]. The nanotubes grown by the gas-phase reaction between MoO_3 and H_2S at 850 °C were found to belong to the 2H polytype [4, 107]. The same is true for WS_2 nanotubes obtained from WO_3 and H_2S [7, 8]. The appearance of the 3R polytype in such nanotubes can probably be associated with strain. For example, a "superlattice" of 2H and 3R polytypes was found to exist in MoS_2 nanotubes grown by chemical vapor transport [50]. Strain effects are invoked in explaining the preference of the rhombohedral polytype in both MoS_2 and WS_2 microtubes grown in the same way [49]. These observations indicate that the growth kinetics of the nanotubes and of thin films influence the strain-relief mechanism, and therefore different polytypes can be adopted by the nanotubes.

The trigonal prismatic structure of MoS_2 alludes to the possibility of forming stable point defects consisting of a triangle or a rhombus [58]. In the past, evidence in support of the existence of "bucky-tetrahedra" [110] and "bucky-cubes" [111], which have four triangles and six rhombi in their corners respectively, were found. However, the most compelling evidence in support of this idea was obtained in nanoparticles collected from the soot of laser ablated MoS_2 [22]. Detailed theoretical calculations indicate that rectangular and even octahedral elements are inherently stable in the nanotube tip [19]. Figure 9 shows the caps of MoS_2 nanotubes with zigzag (a) and armchair (b) structure [19]. These drawings show that only small distortions of the Mo−S bond and the S−Mo−S dihedral angles are necessary to close the cap by three rectangles or octahedron (and four rectangles). Sharp cusps and even a rectangular apex were noticed in WS_2 nanotubes [7, 8]. These features are probably a manifestation of the inherent stability of elements of symmetry lower than pentagons, such as triangles and squares, in the structure of MoS_2, and so forth. Point defects of this symmetry were generally not observed in carbon fullerenes [112], most likely because the sp^2 bonding of carbon atoms in graphite is not favorable for such topological elements. These examples and others illustrate the influence of the lattice structure of the layered compound on the detailed topology of the fullerene-like nanoparticle or of the nanotube cap obtained from such compounds.

All $B_xC_yN_z$ nanotubes are made of a hexagonal network of sp^2 bonded atoms, with three nearest neighbors to each atom [23, 85, 86, 113–116]. In the case of BC_2N nanotubes, two different arrangement of the sheet are possible, leading to two isomers with different structure and distinct in their electrical properties [114].

The chemical composition of the IF phase deviates only very slightly, if at all, from the composition of the bulk layered compound. Deviations from stoichiometry can only occur in the cap of the nanotube. In fact, even the most modern analytical techniques such as scanning probe techniques and high-resolution (spatial) electron energy loss spectroscopy are unable to resolve such tiny deviations from the stoichiometry, like the excess or absence of a single Mo (W) or S (Se) atom in the nanotube cap.

XRD studies have shown an expansion of 2–4% in the c-axis of multiwall IF structures (including inorganic nanotubes) [4, 21, 45, 46]. The shift to lower angles (larger

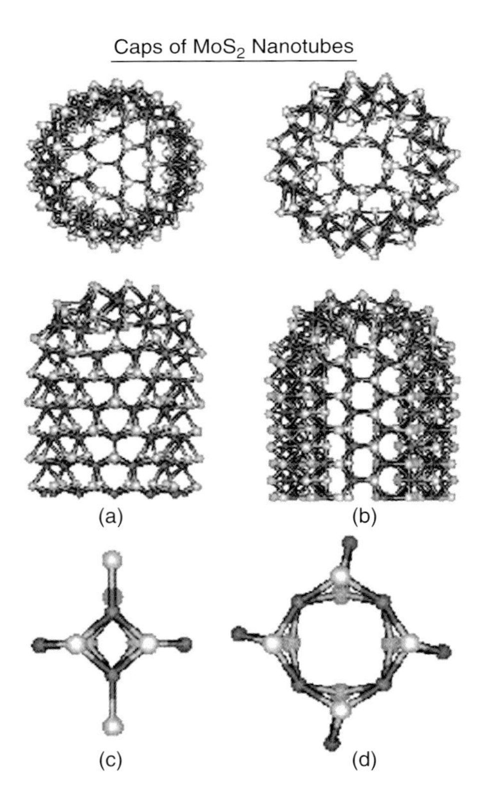

Fig. 9 Computer-generated pictures of (a) zigzag and (b) armchair nanotube caps [19]. Figure 9(c) and (d) are two arrangements of the MoS_2 molecules in the caps of zigzag and armchair nanotubes, respectively.

c-axis spacing) for the IF phase compared to the bulk material is a clear distinction of this phase and serves as a quality measure for the synthetic process. The average size of the nanoparticles can be calculated from the peak width. Note that full commensuration between the upper and lower layer of the multiwall IF nanoparticle or nanotubes is not possible because the number of atoms in the upper layer is always larger than in the underlying layer. However, the structural relationship between the different layers, which is typical for the different polytypes, such as 2H and 3R, is preserved.

3.3.5 Physical Properties

3.3.5.1 Band Structure Calculations

Earlier, a few groups used powerful theoretical tools to calculate the stability, bandstructure, and other physical properties of $B_xC_yN_z$ nanotubes and fullerene-like nanoparticles [23, 113–118]. A few striking conclusions emerged from these studies. First, it was found that B–B and N–N nearest neighbors do not lead to stable polyhedral structures. Instead, dissimilar B–N pair of atoms are thermodynamically preferred. This observation implies that B_2N_2 squares, rather than the five-member rings found in carbon fullerenes and nanotubes, are preferred in the case of BN polyhedra and nanotubes. Experimental verification for this hypothesis has been obtained in the work of a few groups [88, 119]. Secondly, in contrast to carbon nanotubes, which can be metallic or semiconducting depending on their chirality, all BN nanotubes were found to be semiconductors (insulators), independent of their chirality. Thirdly, whereas the smallest forbidden gap of zigzag $(n, 0)$ nanotubes was found to be a direct (Γ-Γ) one, an indirect band gap (Δ-Γ) is calculated for the armchair (n, n) nanotubes. Bulk BN material has an indirect band gap of 5.8 eV. This is to be contrasted with carbon nanotubes, which are either metallic or semiconducting, depending on their (n, m) values. The fourth point to be noted is that the strain in the nanotubes scales as $1/D^2$, where D is the nanotube diameter. The strain effect is predominant for nanotubes with small diameters and therefore overwhelmingly, the band gap of inorganic nanotubes was found to decrease with a decreasing diameter of the (inorganic) nanotubes. In contrast to that, the band gap of semiconducting carbon nanotubes increases with a shrinking diameter of the cage. It should be furthermore emphasized that generically, the band gap of semiconducting nanoparticles increases with a decrease in the particle diameter, which is attributed to the quantum size confinement of the electron wave function [120–122].

As mentioned in the previous section, BC_2N nanotubes have two isomers with distinctly different structure and properties. One of them with alternating carbon and B–N chains was predicted to be a semiconductor. In the armchair (n, n) configuration, the alternating conducting carbon and insulating B–N chains form a solenoid, which on proper doping can become a nanosize coil [113]. The electrical properties of BC_3 nanotubes are largely influenced by their packing arrangements. Theory shows [114] that concentric multiwall nanotubes of this kind are metallic, whereas isolated single wall nanotubes are a semiconductor.

Further work was carried out on nanotubes of the semiconducting layered compound GaSe [18]. In this compound, each atomic layer consists of a Ga–Ga dimer sandwiched between two outer selenium atoms in a hexagonal arrangement. This work indicated that some of the early observations made for BN and boro-carbonitride nanotubes are not unique to these layered compounds, and are valid for a much wider group of nanotubes. First, it was found that like the bulk material, GaSe nanotubes are semiconductors. Furthermore, the strain energy in the nanotube was shown to increase, and consequently the band gap was found to shrink as the nanotube diameter becomes smaller.

Recent theoretical work on (single wall) MoS_2 and WS_2 nanotubes [19, 123] confirmed these earlier observations. Although the lowest band gap of the armchair (n, n) nanotubes was found to be indirect, a direct transition was predicted for the zigzag $(n, 0)$ nanotubes (Fig. 10). Additionally, an $1/D^2$ dependence of the strain energy versus diameter was observed for these nanotubes. The strain is about one order of magnitude larger in these nanotubes as compared to carbon or BN nanotubes of the same diameter. The reason for this effect is the bulky nature of the triple S$-$M$-$S layer (vide infra) as compared to the one layer of carbon or BN in the respective nanotubes. In contrast to carbon nanotubes and with semiconductor nanoparticles in general, the band gap energy shrinks with a decreasing nanotube diameter. These findings, which can be attributed to both strain effects and Brillouin zone folding of the energy bands of the nanotubes suggest a new mechanism for optical tuning through strain effects in the hollow nanocrystalline structures of layered compounds. The existence of a direct gap in zigzag nanotubes is rather important because it suggests that such nanostructures may exhibit strong (electro) luminescence, which has not been observed for the bulk material.

The potential of synthesizing p-n or Schottky junctions, which are a built-in unit of the nanotubes, is probably a

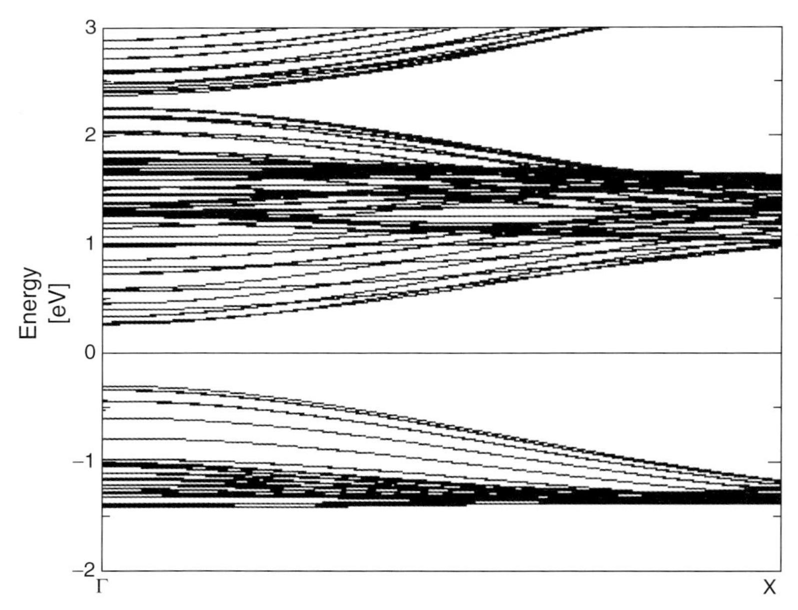

Fig. 10 Calculated band structure for WS_2 (22,0) nanotube [82]. (Courtesy of Prof. G. Seifert, Paderborn University.)

realistic proposition. There are various alternative ways to accomplish this task. One way is to convert the nanotube surface or tip into a metal. This can be achieved by various surface modifications (treatments), such as electrochemical deposition or doping. Another way is to attach a short organic moiety with a thiol head group and a metal at the rear, pointing outward. The self-assembly of the monolayer with loosely spaced metal atoms on the nanotube surface will endow a metallic conductivity to the surface film, which is in an intimate contact with the WS_2 nanotube, leading therefore to a Schottky junction. Another possible way to establish a junction is to synthesize the WS_2 nanotube on top of a metallic carbon nanotube. Similarly, synthesizing composite WS_2(semiconductor)/NbS_2(metal) nanotubes, is feasible. The electronic structure of NbS_2 nanotubes show that they possess a high density of states in the Fermi level, irrespective of their chirality, which is indicative of their high electrical conductivity [82]. Therefore, the synthesis of complex nanotubes with, for example, semiconducting WS_2(MoS_2) core and metallic NbS_2 shell, is believed to be doable and is also rather intriguing. MoS_2 monolayers were shown to exhibit a first-order phase transformation from 2H into 1T polytype, where each Mo atom is surrounded by six S atoms in an octahedral fashion [124–126]. The 1T polytype has half-filled Mo d level and is consequently a metal. There are already some indications that the MoS_2 octahedra produced by laser ablation [22] acquire the 1T (octahedral) arrangement of the S atoms around each Mo atom, rather than the trigonal prismatic structure of bulk MoS_2 [127]. Recent work also indicates that the tip of WS_2 and MoS_2 nanotubes consist of half octahedra, in which case it is likely that the nanotube body is a semiconductor, whereas the tip is essentially a metal. In this case, a built-in semiconductor (Schottky) junction could have been established inadvertently.

Black phosphorous (b-P) and arsine (As) crystallize in a layered structure, in which each atom is bound to three neighbors like atoms. In contrast to the flat hexagonal network of carbon atoms in graphite, the P-sheet (As) forms a puckered hexagonal honeycomb. To minimize the repulsive energy between the lone pairs of the two neighboring phosphorous atoms, they are arranged in opposite direction to each other in an alternating fashion. Using density functional tight-binding theory, the structure, electronic structure, and the mechanical properties of b-P nanotubes were derived [128–130]. The strain energy was found to scale as $1/D^2$, but is larger than that of carbon nanotubes of the same diameter D. The larger strain energy was attributed to the repulsion between the electron lone pairs of next nearest-neighbor atoms. Six five-member rings were found to establish the most stable apex for the nanotubes. Here too, the energy gap was found to shrink with a decreasing D.

The electronic structure of GaN nanotubes was also calculated [106] and was essentially in accordance with the band structure calculations of the other inorganic nanotubes. The band gap of nanotubes with diameter >2 nm is above 4 eV, and it shrinks with the nanotube diameter. Zigzag nanotubes are found to have a direct transition, which suggests that they could serve as an ultrasmall blue light-emitting source. The structure and stability of $CaSi_2$ nanotubes have been investigated but few details are currently available [129, 130].

The transport properties of inorganic nanotubes have not been reported so far.

A wealth of information exists for the transport properties of the bulk 2D-layered materials, which is summarized in a few review articles (see Refs. 107, 131).

3.3.5.2 Optical Studies in the UV and Visible

Measurements of the optical properties in this range of wavelengths can probe the fundamental electronic transitions in these nanostructures. Some of the aforementioned effects have in fact been experimentally revealed in this series of experiments [132]. As mentioned earlier, the IF nanoparticles in this study were prepared by a careful sulfidization of oxide nanoparticles. Briefly, the reaction starts on the surface of the oxide nanoparticle and proceeds inward, and hence the number of closed (fullerene-like) sulfide layers can be controlled quite accurately during the reaction. Also, the deeper is the sulfide layer in the nanoparticle, the smaller is its radius and the larger is the strain in the nanostructure. Once available in sufficient quantities, the absorption spectra of thin films of the fullerene-like particles and nanotubes were measured at various temperatures (4–300 K). The excitonic nature of the absorption of the nanoparticles was established, which is a manifestation of the semiconducting nature of the material (Fig. 11). In addition to the previous report [132], which included a detailed analysis of the A and B excitons, the present spectrum reveals also the C exciton. Therefore, in accordance with the theoretical analysis [19, 82], the present work shows the semiconductor behavior of the fullerene-like particles. This suggests that these new phases might be very suitable for photoelectrochemical and photocatalytic applications. The redshift in the exciton energy, which increased with the number of sulfide layers of the nanoparticles, was established [132]. The temperature dependence of the exciton energy was not very different from the behavior of the exciton in the bulk material. This observation indicates that the redshift in the exciton energy cannot be attributed to defects or dislocations in the IF material, but rather it is a genuine property of the inorganic fullerene-like and nanotube structures. In contrast to previous observations, IF phases with less than five layers of sulfide revealed a clear blueshift in the excitonic transition energy, which was associated with the quantum size effect. Figure 12 summarizes this series of experiments and the two effects. The redshift of the exciton peak in the absorption measurements, which is the result of the strain in the bent layer on one hand, and the blueshift for IF structures with very few layers and large diameter (minimum strain) on the other hand, can be discerned.

The WS_2 and MoS_2 nanotubes and the nested fullerene-like structures used for the experiments described in Figs. 11 and 12 had relatively large diameters (>20 nm). Therefore, the strain energy is not particularly large in the first few closed layers of the sulfide, but it nevertheless increases as the oxide core is progressively converted into sulfide, that is, closed sulfide layers of smaller and smaller diameter are formed. This unique experimental opportunity permitted a clear distinction to be made between the strain effect and the quantum size effect of the electronic wave function. In the early stages of the reaction, the strain is not very large and therefore the confinement of the exciton along the *c*-axis is evident from the blueshift in the exciton peak. The closed and therefore seamless nature of the MS_2 layer is analogous to an infinite crystal in the *a*-*b* plane and hence quantum size effects in this plane can be ruled out.

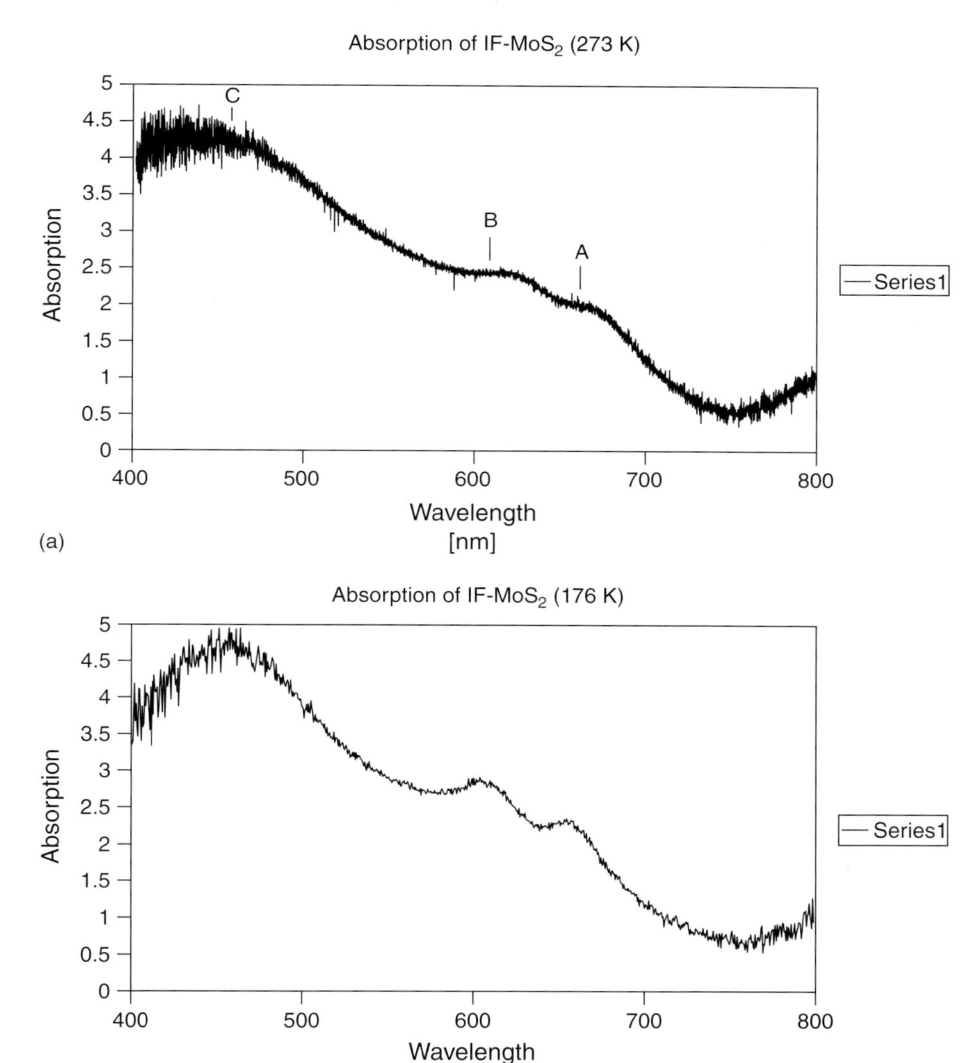

Fig. 11 Absorption spectra of IF-MoS_2 nanoparticles at 300 and 176 K. The position of the A,B, and C excitons of the bulk 2H-MoS_2 are marked with arrows.

However, there is a clear confinement effect observable perpendicular to the a-b plane, that is, in the c-direction. The quantum size effect in layered compounds was studied in the past [133, 134]. The energy shift because of this effect (ΔE_g) can be expressed as:

$$\Delta E_g = \frac{h^2}{4\,\mu_{\parallel}} {L_z}^2 \tag{1}$$

Here, $\mu_{\parallel}$ is the exciton effective mass parallel to the c-axis and L_z is the (average) thickness of the WS_2 nested structure ($L_z = n \times 0.6.2$ nm, where n

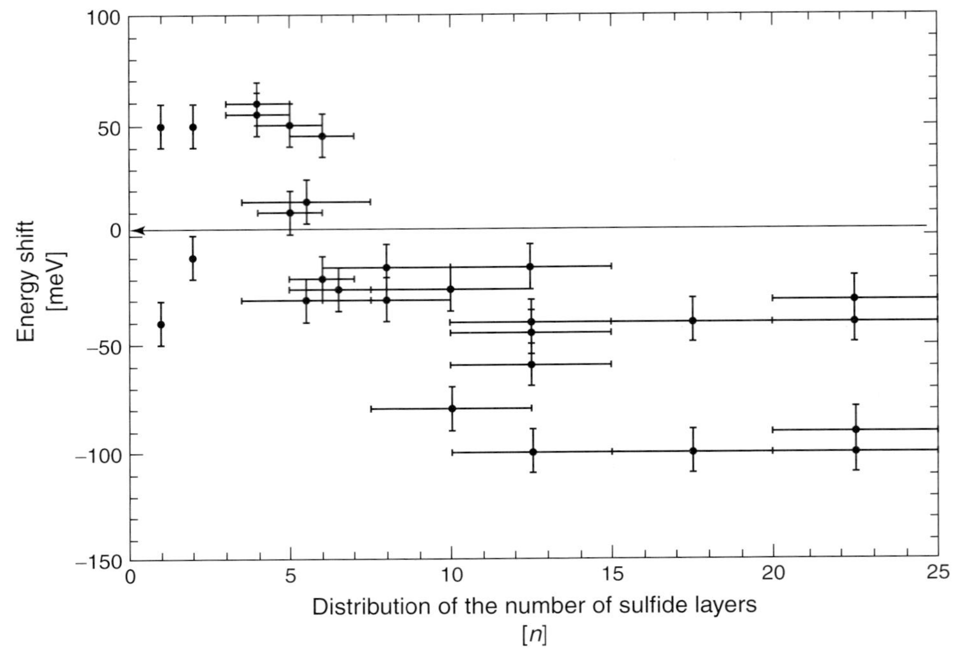

Fig. 12 The dependence of the A exciton shifts on the number of layers in the IF structure [132]. The error bar represents the distribution of the number of layers determined with TEM for each sample. The *y*-axis error bar is ± 10 meV.

is the number of WS_2 layers) in the nanoparticle. In a previous study of ultrathin films of 2H-WSe_2, ΔE_g of the A exciton was found to obey Eq. (1) over a limited thickness range. ΔE_g exhibited a linear dependence on $1/L_z{}^2$ for L_z in the range of 4–7 nm and became asymptotically constant for $L_z >$ 8 nm [133]. A similar trend is observed for IF-WS_2 and MoS_2 [132]. Therefore, the quantum size effect is observed for IF structures with a very small number of WS_2 layers ($n < 5$) and large diameter. Note that in the current measurements, IF films 150-nm thick were used, but because each IF structure is isolated and the exciton wave function cannot diffuse from one nanoparticle to the other, the quantum size effect can be distinguished in this case. Note also, that because of the (residual) strain effect, the energy for both the A and B excitons is smaller than for their bulk counterparts. The corresponding redshift in the absorption spectrum has also been found for MoS_2 nanotubes [51].

These studies suggest a new kind of optical tunability. Combined with the observation that achiral inorganic nanotubes are predicted to exhibit direct optical transitions [19, 23, 82, 106], new opportunities for optical device technology, based on GaN, or MoS_2 nanotubes as light-emitting diodes and lasers, could emerge in the future. The importance of strong light sources a few nm in size in nanotechnology can be appreciated from the need to miniaturize current submicron light sources for lithography.

3.3.5.3 Raman Spectroscopy

Raman and resonance Raman (RR) measurements of fullerene-like particles of MoS_2 have been carried out recently [135]. Using 488-nm excitation from Ar ion laser light source, the two strongest Raman features in the Raman spectrum of the crystalline particles, at 383 and 408 cm^{-1}, which correspond to the $E_{2g}{}^1$ and A_{1g} modes respectively (Table 1), were found to be dominant also in IF-MoS_2 and in MoS_2 platelets of a very small size. A distinct broadening of these two features could be discerned as the size of the nanoparticles was reduced. In analogy to the models describing quantum confinement in electronic transitions, it was assumed that quantum confinement leads to contributions of modes from the edge of the Brillouin-zone. Thus, phonon modes with a high density of states in the edge of the Brillouin-zone are expected to have significant contribution to the Raman spectra. Lineshape analysis of the peaks led to the conclusion that the phonons are confined in coherent domains of about 10 nm in size within the IF nanoparticles. Such domains could be associated with the faceting of the polyhedral IF structures.

RR spectra were obtained by using the 632.8-nm (1,96 eV) line of a He-Ne laser. Figure 13 shows the RR spectra of a few MoS_2 samples. Table 1 lists the peak positions and the assignments of the various peaks for the room temperature spectra. A few second-order Raman transitions were also identified. The intensity of the 226 cm^{-1} peak did not vary much by lowering the temperature, and therefore it cannot be assigned to a second-order transition. This peak was therefore attributed to a zone-boundary phonon, activated by the relaxation of the $q = 0$ selection rule in the nanoparticles. Lineshape analysis of the intense 460 cm^{-1} mode revealed that it is a superposition of two peaks at 456 and 465 cm^{-1}. The lower frequency peak is assigned to a second order 2LA(M) process, whereas the higher energy peak is associated with the A_{2u} mode, which is

Tab. 1 Raman peaks observed in MoS_2 nanoparticle spectra at room temperature and the corresponding assignments. All peak positions are in cm^{-1}

Bulk MoS_2[12]	*PL-MoS_2 5000 Å*	*PL-MoS_2 30 × 50 Å*	*IF-MoS_2 800 Å*	*IF-MoS_2 200 Å*	*Symmetry assignment*
177	179	180	180	179	$A_{1g}(M) - LA(M)$
		226	227	226	$LA(M)$
			248	248	
				283	$E_{1g}(\Gamma)$
382	384	381	378	378	$E_{2g}{}^1(\Gamma)$
407	409	408	407	406	$A_{1g}(\Gamma)$
421[10]	419		Weak	Weak	
465	460	455	452	452	$2 \times LA(M)$
		498	495	496	Edge phonon
526	529				$E_{1g}(M) + LA(M)$
		545	545	543	
572	572	~557	565	563	$2 \times E_{1g}(\Gamma)$
599	601	595	591	593	$E_{2g}{}^1(M) + LA(M)$
641	644	635	633	633	$A_{1g}(M) + LA(M)$

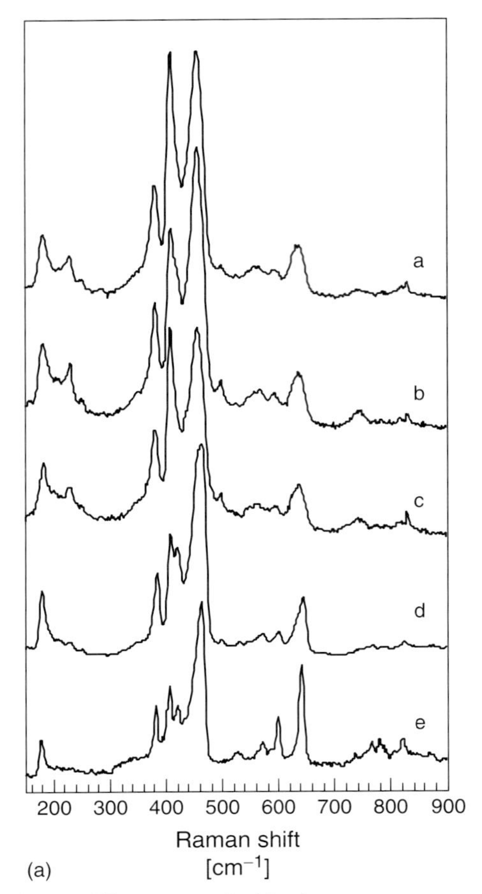

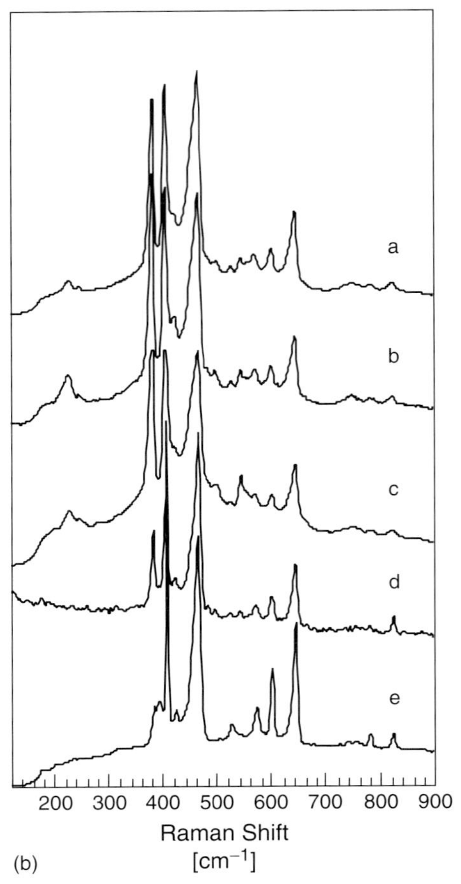

Fig. 13 RR spectra excited by the 632.8 nm (1.96 eV) laser line at room temperature (a) and 125 K (b), showing second order Raman bands for several MoS_2 nanoparticle samples [135, 136]: a. IF-MoS_2 (20 nm); b. IF-MoS_2 (80 nm); c. MoS_2 platelets (5×30 nm^2); 2H-MoS_2 (500 nm); e. 2H-MoS_2 bulk.

Raman inactive in crystalline MoS_2, but is activated by the strong resonance Raman effect in the nanoparticles.

3.3.5.4 Mechanical Properties

The mechanical properties of the inorganic nanotubes have only been investigated to a relatively small extent. The Young's modulus of multiwall BN nanotubes was measured within a TEM [25], and was found to be about 1.2 TPa, which is comparable to the values measured for carbon nanotubes. The Young's modulus of the b-P nanotubes was calculated [128]. The observed value, 300 Gpa, is 25% of the Young modulus of carbon nanotubes. The Poisson ratio of b-P nanotubes was calculated to be 0.25 in this work.

An elastic continuum model, which takes into account the energy of bending; the dislocation energy, and the surface energy, was used to describe the mechanical properties of multilayer cage structures to a first approximation [137].

A first-order phase transition from an evenly curved (quasi-spherical) structure into a polyhedral cage was predicted for nested fullerenes with shell thickness larger than about one-tenth of the nanotube radius. Indeed, such a transition was observed during the synthesis of IF-WS_2 particles [45, 46]. Initially the oxide nanoparticles enfolded with a few WS_2 molecular layers were found to be quasi-spherical. They transformed into a very faceted structure, when the thickness of the sulfide shell in the nanoparticles exceeded a few nm. Further theoretical analysis of the mechanical and elastic properties of IF-MS_2 onions has been undertaken recently [138]. First, by summing (integration) the interaction between the nanoparticles and the underlying substrate, the adhesion energy per unit area reads: $u = -A/12\pi d^2 (=100\ \mathrm{erg/cm^2})$ with A the Hamaker constant (of the order 10^{-12} erg), and d- an atomic cutoff for the van der Waals (vdW) interaction (0.165 nm). Considering a spherical fullerene-like particle of a radius R, the total adhesion energy of the nanoparticle with the substrate is: $E_A = -AR/6d$. Taking R as 60 nm yields $E_A = 6.10^{-11}$ erg = 1400 kT. The adhesion energy between two particles of this radius is smaller by a factor of two only. This calculation alludes to the appreciable adhesion of the nanoparticles to the underlying substrate surface or to each other. The adhesion between the two MS_2 layers in the "onion" can be calculated likewise and amounts to 10^6kT. Indeed, it was found that IF-MoS_2 nanoparticles form small clusters, which cannot be easily separated into isolated nanoparticles by ultrasonic treatment in various solvents. However, high resolution imaging of the nanoparticles by scanning probe microscopy techniques have eluded the experimenters. This fact was attributed to an easy tip induced sliding and rolling of the nanoparticles on the underlying substrates. This fact cannot be easily reconciled with the calculated high-adhesion energies of the nanoparticle to the underlying substrate, or to the observed tendency of the IF nanoparticles to form stable clusters. Obviously, the IF nanoparticles can be solvated by water from the ambient, which could lead to a significant reduction in the adhesion energy. This effect was not considered in the theoretical analysis [138]. This study has further indicated that deformation of the nanoparticles because of the adhesion or because of shearing forces of the fluid is small and consequently delamination of the nanoparticles is not likely. However, this study also showed that, whereas small pressure leads to reversible deformations of the nanoparticles, strong pressure brings about an (buckling) instability, which will eventually result in their delamination. Clearly, more work is needed to understand the physical properties of the IF nanoparticles better.

3.3.6 Electrochemistry and Photoelectrochemistry Using IF-Materials and Inorganic Nanotubes

The study of the electrochemical and photoelectrochemical behavior of the IF nanoparticles could not be done before sufficient amounts of the nanoparticles were available. Recently, a few groups reported the successful synthesis of macroscopic amounts of these nanoparticles [16, 45, 46, 31], which makes such studies feasible.

3.3.6.1 Electrodeposition of IF-MS_2 (M = W,Mo) Nanoparticles and their Photoelectrochemical Properties

Sodium intercalated IF-MS_2 powder was synthesized from sodium doped MO_3

oxide precursors [81]. Sonication of the products in alcoholic solutions led to the formation of stable suspensions. On the other hand, nonintercalated IF-MS_2 powders did not form stable suspensions even after prolonged sonication, and precipitated after a short while [81]. These results indicate that the intercalation of alkali metal atoms in the van der Waals gap of the IF particles led to a partial charge transfer from the alkali metal atom to the host lattice, which increased the polarizability of the nanostructures, enabling them to disperse in polar solvents. The transparency of the suspensions and their stability increased with the amount of alkali metal intercalated into the IF structures. Alcoholic suspensions prepared from IF powder (both fullerene-like particles and nanotubes), which contained large amount of intercalant ($>5\%$), were found to be virtually indefinitely stable. The optical absorption of the IF suspensions, measured in the solution, was found to be very similar to that of thin films of the same nanoparticles [132].

In the next step, thin films of IF nanoparticles were deposited onto a gold substrate by electophoretic deposition. Given the chemical affinity of sulfur to gold, it is not surprising that electrophoretic deposition led to relatively well adhering IF films. Furthermore, some selectivity with respect to the IF sizes and the number of MS_2 layers in the films was achieved by varying the potential of the electrode. The thickness of the film was controlled by varying the electrophoresis time. Because the nonintercalated IF particles do not form stable suspensions, films of such material were obtained by electrophoresis from vigorously sonicated dispersions of the IF powder. Alternatively, films of intercalated IF nanoparticles were prepared by evaporating the solvent from a dip-coated metal substrate; this method, however, resulted in poorly adhering films.

The optical absorption spectra of intercalated 2H-MoS_2 did not show appreciable changes on alkali metal intercalation up to concentrations of 30%, where a transition into a metallic phase at room temperature and a further transition into a superconductor at about 3–7 K was reported [139, 140]. Because the concentration of the intercalating metal atoms in the IF nanoparticles did not exceed 10%, no changes in the optical transmission spectra were anticipated nor were they found to occur. Also, the intercalation of alkali atoms in the IF particles induces n-type conductivity of the host.

The prevalence of dangling bonds on the prismatic faces of 2H-MS_2 crystallites leads to a rapid recombination of the photoexcited carriers. Consequently, the performance of thin film photovoltaic devices made of layered compounds has been disappointing. The absence of dangling bonds in IF material suggests that this problem could be alleviated here. Therefore, the photocurrent response of IF-WS_2 films in selenosulfate solutions (0.2 M Se and 0.4 M Na_2SO_3) was examined and compared to that of 2H-WS_2 films. The response was found to be very sensitive to the density of dislocations in the film. Figure 14 shows the quantum efficiency (number of collected charges and number of incident photons) of a typical IF-WS_2 film with a low density of dislocations, as a function of the excitation wavelength. On the other hand, films having nested fullerene-like particles with substantial amounts of dislocations exhibited a poor photoresponse and substantial losses at short wavelengths, which indicates that the dislocations impair the lifetime of excited carriers in the film [141]. Films of

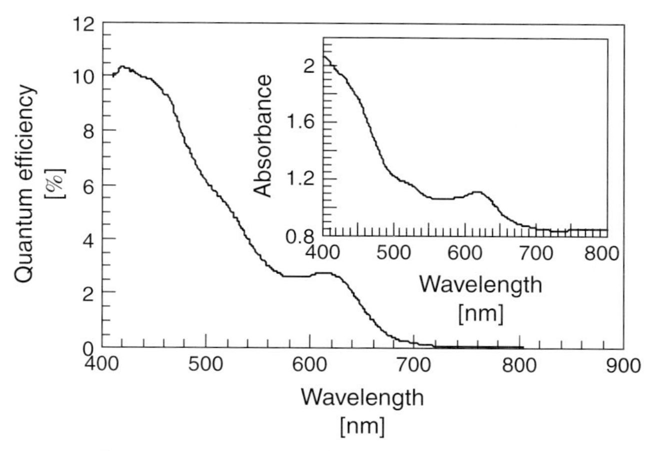

Fig. 14 Photoresponse spectra of thin films of IF-WS_2 deposited on a gold substrate. The three transitions caused by excitons A, B, and C are obvious in the figure. The absorption of a similar IF-WS_2 film deposited on a conductive (ITO) glass is shown in the inset [143].

nanotubes also showed substantial photoresponse. Finally, films made of 2H-WS_2 platelets (each about 1 μm in size), which are known to have many recombination centers on the prismatic $(10\bar{1}0)$ face, did not exhibit any measurable photoresponse under comparable conditions. The photocurrent decreased with negative bias, reaching zero at −1.0 V versus the Pt foil counterelectrode, thus affirming the n-type conductivity of the alkali-metal intercalated IF particles. The photoresponse of the IF films did not show any degradation after 48 hours of continuous illumination. Electrolyte electrotransmission (EET) spectra of the IF-WS_2 films were also recorded and were in accordance with previous studies of 2H-WS_2 crystals [142]. The inset of Fig. 14 shows the absorption spectrum of the films, which clearly reveals the direct excitonic transitions of the film at 2.02 (A exciton), 2.4 eV (B exciton), and 2.9 eV (C exciton), respectively [143].

Limited photoelectrochemical measurements were done also on films of SnS_2 nanoparticles with fullerene-like structure. Figure 15 shows the spectral response of films consisting of IF-SnS_2 nanoparticles with an oxide core, which were prepared by incomplete conversion of SnO_2 nanoparticles into the respective sulfide [143]. An indirect transition of approximately 2.03 eV was obtained by extrapolation of the photocurrent$^{1/2}$ versus photon energy. This band gap is found to be in accordance with published data for bulk SnS_2 [144]. The long wavelength tail extends to over 800 nm, beyond the bulk indirect band gap value, and is clearly a sub–band gap component. Sub–band gap photoresponse has been observed previously in semiconductor photoelectrodes, and was attributed to absorption in intra–band gap states.

The modest photoelectrochemical performance of the IF-based films can be ascribed to a number of factors. Perhaps the most critical issue is the charge transfer across the nanoparticles boundaries. The other issue is the charge transfer to the back contact itself. These issues will have to be studied in greater detail to identify the loss mechanisms and improve the

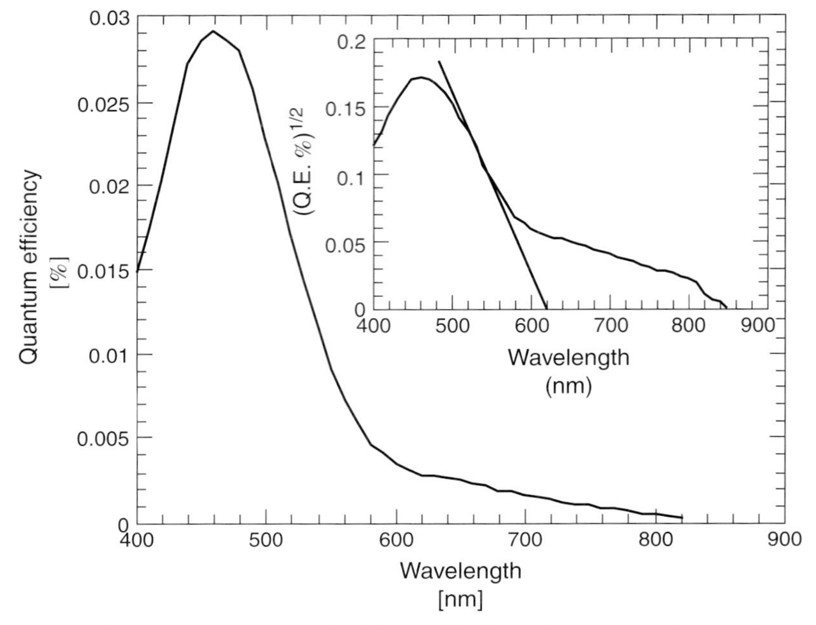

Fig. 15 Photocurrent spectrum of a fullerene-like SnS_2 film on a Au substrate. Inset: Plot of the (quantum efficiency)$^{1/2}$ versus wavelength. The extrapolation to zero signal yields a band gap of 2.03 eV [143].

light-induced charge collection efficiency of the films.

3.3.6.2 Electrochemical Studies with V_2O_5 and Metallic Nanotubes

The synthesis of $V_2O_{2.45}$(alkylamine) nanotubes, by a sol-gel reaction using vanadium oxide triisopropoxide in the presence of hexadecyl amine template and hydrothermal treatment was described [6, 31]. These nanotubes consist of concentric shells with alternating $V_2O_{2.45}$(alkylamine) layers. Most of them are obtained as scrolls. Further studies [61] showed that the nanotube material could be used as a Li insertion electrode. To obtain the electrode material, the nanotubes were refluxed, first, for 24 hours in ethanolic solution of NaCl, in order to remove the hexadecylamine template. Sodium ions were intercalated between the atomic layers of the nanotubes. The refluxed nanotubes were mixed (50%) with teflonized carbon black, and the mixture was pressed onto a titanium foil, which served as current collector. The electrochemical studies were carried out in a hermetically sealed three-electrode cell. The anode and cathode compartments were separated by a glass fiber separator and the electrolyte was 1 M $LiClO_4$ solution of propylene carbonate (PC).

The distance between the atomic layers in $V_2O_{2.45}$(TEMP), where TEMP = $C_{16}H_{33}NH_2$, was found to be about 3 nm before the reflux. It decreased to only 0.65 nm after the amine was soaked out. The surface area of the anode material was determined to be 35 $m^2\ g^{-1}$. Figure 16 shows the cyclic voltammogram of the electrodes with and without the template against Li^+/Li reference electrode. The cathodic current observed in potentials more negative than 2 V corresponds to Li insertion into the template-containing electrode, which can be described according to the following reaction:

$$\mathrm{XLi^+} + xe^- + \mathrm{VO_{2.45}(TEMP)_{0.34}} \longrightarrow \mathrm{Li}_x\mathrm{VO_{2.45}(TEMP)_{0.34}} \quad (2)$$

The oxidation step starting in potential more anodic than 3 V corresponds to the dissolution of the Li ions from the nanotube material. The large potential difference between the charge and discharge potentials is indicative of sluggish kinetics or insertion and dissolution reactions. This phenomena stems probably from the slowness of the diffusion of the Li ion into and out of the confined space between the nanotube walls. A specific charge and discharge of 120 $\mathrm{mAhg^{-1}}$ was obtained for the first five cycles. However, the specific charge decreased with the number of cycles, probably because of an irreversible reaction of the amine, and amorphization of the nanotubes. When the template was removed, the specific charge increased to 180 $\mathrm{mAhg^{-1}}$. Furthermore, the process was quite reversible with relatively a small kinetic barrier, as indicated by the small potential difference between the cathodic and anodic peaks. However, the specific charge decreased relatively fast in the first few cycles. The charge capacity leveled off after a few cycles and it reached the value of 100 $\mathrm{mAhg^{-1}}$ after 10 cycles. Further work is needed to elaborate the mechanism of the insertion and discharge processes and to further improve the electrode stability.

Metallic nanotubes can be synthesized using hard or soft template. Notwithstanding their incomplete crystallinity, their high surface area can be exploited for various electrochemical reactions. The large surface area of the nanotube material may lead to improved kinetics of the electrochemical reactions and to a decrease in the losses due to overvoltage of the reaction. A novel utilization of this principle is demonstrated in Ref. 145. Here, palladium nanotubes, a few microns long were

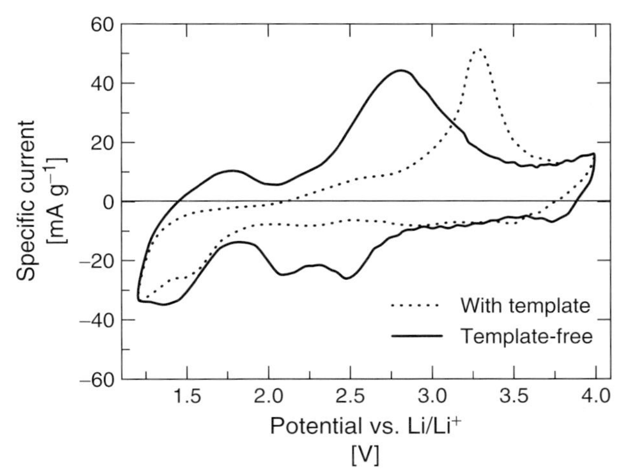

Fig. 16 Cyclic voltammograms of the $\mathrm{VO_{2.45}(TEMP)_{0.34}}$ and template-free nanotubes in 1 M $\mathrm{LiClO_4}$/polycarbonate electrolyte [61].

synthesized by first electrodepositing Cu nanotubes on a perforated polycarbonate template. Following this step, electroless deposition of the Pd, which displaced the Cu nanotubes into solution, and dissolution of the polycarbonate led to a suspension of Pd nanotubes. These nanotubes were subsequently mixed with Ni slurry and used as the negative electrode in nickel-metal hydride battery. The Pd nanotubes contributed only 1% to the weight of the negative electrode. Several charge–discharge cycles of the battery were carried out and compared to the reference battery, which did not contain the Pd nanotube in the cathode. The nanotube containing battery showed a substantially higher capacity compared to the reference battery. In another study [146], Co and Fe nanowires and nanotubes were directly electrodeposited inside the pores of a perforated polycarbonate template. The nanotubes were 5–6 micron long and their diameter varied between 10–80 nm. These studies indicate that metallic nanotubes may play an important role in future electrochemical devices of different kinds.

3.3.7 Applications

The spherical shape of the fullerene-like nanoparticles and their inert sulfur-terminated surface suggests that MoS_2 particles could be used as a solid-lubricant additive in lubrication fluids, greases, and even in solid matrices. Applications of a pure IF-MoS_2 powder could be envisioned in high-vacuum and microelectronic equipment, where organic residues with high vapor pressure can lead to severe contamination problems [147, 148]. Because the MoS_2 layers are held together by weak van der Waals forces; they can provide easy shear between two close metal surfaces, which slide past each other. At the same time, bulk MoS_2 particles, which come in the form of platelets, serve as spacers, eliminating contact between the two metal surfaces and minimizing the metal wear. Therefore, MoS_2 powder is used as a ubiquitous solid lubricant in various systems, especially under heavy loads, where fluid lubricants cannot support the load and are squeezed out of the contact region between the two metal surfaces. Unfortunately, MoS_2 platelets tend to adhere to the metal surfaces through their reactive prismatic $(10\bar{1}0)$ edges, in which configuration they tend to "glue" the two metal surfaces together, rather than serve as a solid lubricant. During the mechanical action of the engine parts, abrasion and burnishing of the solid lubricant produces smaller and smaller platelets, increasing their tendency to stick to the metal surfaces through their reactive prismatic edges. Furthermore, the exposed prismatic edges are reactive sites, which facilitate chemical oxidation of the platelets. These phenomena adversely affect the tribological benefits of the solid lubricant and lead to a relatively rapid disappearance of their beneficial effects. In contrast, the spherical IF-MS_2 nanoparticles are expected to behave like nano–ball bearings and under mechanical stress they would slowly exfoliate or mechanically deform to a rugby-shape ball as indicated in Ref. 138, but would not lose their tribological benefits until they are completely gone, or oxidize. To test this hypothesis, various mixtures of the solid powder and lubrication fluids were prepared and tested under standard conditions [149]. The beneficial effect of IF powder as a solid lubricant additive has been thus confirmed through a long series of experiments [149–152]; some of them are summarized in Table 2. It has to be emphasized that the IF-solid lubricant seem to be particularly effective under heavy loads. The use of a pure lubricating fluid without any solid lubricant, under these conditions, leads to a rapid wear and deterioration of the mating metal surfaces.

More recently, a number of studies have indicated that the IF material can serve as a dry solid lubricant [21, 153, 154]. In one study, IF-MoS_2 were produced by the arc-discharge technique and collected on a Ti foil forming a thin film of this material [21]. The IF-based film exhibited very low friction coefficients (>0.01), even under 45% humidity. Under similar conditions, a sputtered MoS_2 film exhibited a friction coefficient greater than 0.1 and rapid wear.

The mechanism of the action of the IF nanoparticles as additives in lubrication fluids is more complicated than was initially thought. First, it is clear that the more spherical the nanoparticles and the fewest structural defects they contain, the better is their performance as solid lubricant additives [152]. Three main mechanisms responsible for the onset of failure of the nanoparticles in tribological tests have been clearly identified. They include exfoliation of the nanoparticles; deformation into a rugby ball shape, and explosion. The partially damaged (deformed) nanoparticles are left with reactive edges and dislocations, which can undergo further oxidation, and a complete loss of their tribological action. Recent

Tab. 2 Wear (***W***) and friction (μ) coefficients of a steel block in contact with a steel disk for four kinds of solids lubricants mixed with mineral oil. Average particle size shown in parentheses at the head of each column

Exp.	*Velocity ($m\ s^{-1}$)*	*Load (N)*	*Conc. (wt%)*	*Co-eff.*	*Pure oil*	*2H-MoS_2 (4 µm)*	*2H-WS_2 (0.5 µm)*	*2H-WS_2 (4 µm)*	*IF-WS_2 (120 nm)*
	0.44	300	5	μ	0.07	0.07		0.05	0.03
a				***W***	$1.6 \cdot 10^{-8}$	$1.3 \cdot 10^{-8}$		$1.5 \cdot 10^{-8}$	$0.7 \cdot 10^{-8}$
				R_a	0.83	0.75		1.18	0.53
		1800	20	***W***		$7.9 \cdot 10^{-1}$			$5.1 \cdot 10^{-1}$
b	0.22		45	***W***		$3.9 \cdot 10^{-1}$		$2.3 \cdot 10^{-1}$	$1.3 \cdot 10^{-1}$
		600	60	μ			0.10	0.043	0.034
		300	60	μ			0.067	0.042	0.028
c	0.11 to	300 to	60	***W***			$4.6 \cdot 10^{-4}$	$2.8 \cdot 10^{-4}$	$1.9 \cdot 10^{-4}$
	0.44	3000		T °C			88		72

Note: (a) In this experiment the sliding track length was $1.27 \cdot 10^4$ m (8 hr). A commercial oil for transmission systems (Delcol) was used. ***W*** is given in $mm^3\ mm^{-1}\ N^{-1}$. Profiles of the wear track were measured by stylus profilometry. The results of the experiment with the powder of 2H-MoS_2 with 0.5-µm grain size was very similar to the results of the 4-µm 2H-MoS_2 powder and is therefore not reported in detail here. The average roughness (R_a) of the area of the wear-track was determined after the experiment and is also included. R_a is given in µm.
(b) In this experiment, the sliding track length was $2.38 \cdot 10^3$ m (6 hr).
(c) In this experiment, the load on the disk was increased from 300 N to 3000 N in steps of 300 N, each step lasting 300s. At each load, the velocity was increased from 0.11 to 0.44 m sec^{-1} in steps of 0.11 m sec^{-1} (3.5 hr per experiment). The temperature of the flat block was determined during the experiment by contacting a thermocouple to the block. Reported temperature is average after the initial run-in.

nanotribological experiments using the surface force apparatus with the IF-WS_2 lubricant mixed with tetradecane between two perpendicular mica surfaces revealed that material transfer from the IF nanoparticles onto the mica surface plays a major role in reducing the friction between two mica surfaces [150, 151]. No evidence in support of a rolling friction mechanism could be obtained in these studies. On the other hand, 2H-WS_2 platelets of a similar size exhibited poor tribological properties in similar experiments, and furthermore, no evidence for material transfer could be obtained in this case. It was argued [138] that the shear forces provided by the surface force apparatus are below the threshold necessary to onset the rolling of the nanoparticles. However, the shear rates of the tribological tests reported in Refs. 149, 152 are sufficient to induce rolling friction, according to the calculations. These experiments and many others carried out over the last few years, suggest an important application for these nanoparticles as an additive in lubrication fluids or greases.

Self-lubrication of mechanical parts can alleviate some of the technological complexities involved in the lubrication by fluids of mechanical systems, and the environmental impact of this technology. For example, the use of fluid lubricants in the automotive industry adds some 2–3% to the overall weight of the cars. Furthermore, the technological complexities of these intricate systems lessens their reliability. The used fluid lubricant must be processed or buried in special depots, which adds to the adverse effects of the automotive systems on the environment. Recently, self lubricating bronze-graphite sliding bearings impregnated with 3 to 5% IF-WS_2 nanoparticles, have been prepared and tested [153]. Figure 17 shows the effect of the IF impregnation on the loading capacity of these bearings. Whereas the unlubricated bearings could not withstand loads larger than about 35 kg, the IF impregnated bearings supported loads of over 90 kg before seizure. In addition, long-term tests showed that the lifetime of the self-lubricating bearing impregnated with the IF material can be extended by one to two orders of magnitudes. The remarkable effect of the IF material has been attributed to the slow release of the IF nanoparticles, which reside in the porous matrix of the self-lubricating bearings, as demonstrated in Fig. 18. 2H-WS_2 (MoS_2) platelets, which adhere to the metal surface, contribute very little to the self-lubrication. In contrast, the IF nanoparticles, which reside in the porous metal matrix, are slowly released to the surface and provide a very efficient means of self-lubrication mechanism. Similarly, iron-graphite self-lubricating bearings impregnated with IF-WS_2 nanoparticles were fabricated and tested [154]. These bearings are much harder and could carry much higher loads. Although the effect of the IF material was somewhat inferior in this case, substantial improvements in the loading capacity and the lifetime of the IF-impregnated bearings was observed. This series of studies suggest numerous applications for the IF materials in self-lubricating systems.

Another important field where inorganic nanotubes can be useful is as tips in scanning probe microscopy [24]. Here applications in the inspection of microelectronics circuitry have been demonstrated and potential applications in nanolithography are being contemplated. A comparison between a WS_2 nanotube tip and a microfabricated Si tip indicates that while the microfabricated conical-shaped Si tip is unable to probe the bottom of deep and narrow

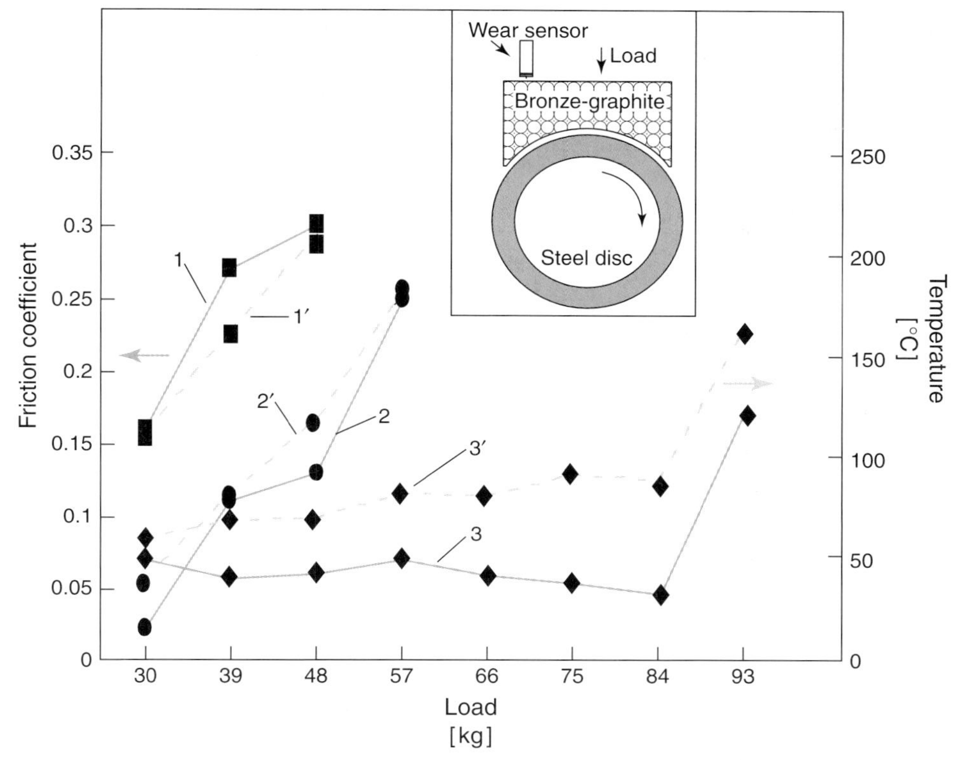

Fig. 17 Friction coefficient (1, 2, 3) and temperature (1′, 2′, 3′) versus load (in kg) of porous bronze-graphite block against hardened steel disk (HRC 52). In these experiments, after a run-in period of 10–30 hours, the samples were tested under a load of 30 kg and sliding velocity of 1 m sec^{-1} for 11 hours. Subsequently, the loads were increased from 30 kg with an increment of 9 kg and remained one hour under each load. (1, 1′) bronze-graphite sample without added solid-lubricant; (2, 2′) bronze graphite sample with 2H–WS_2 (6%); (3, 3′) the same sample with (5%) hollow (IF) WS_2 nanoparticles [153].

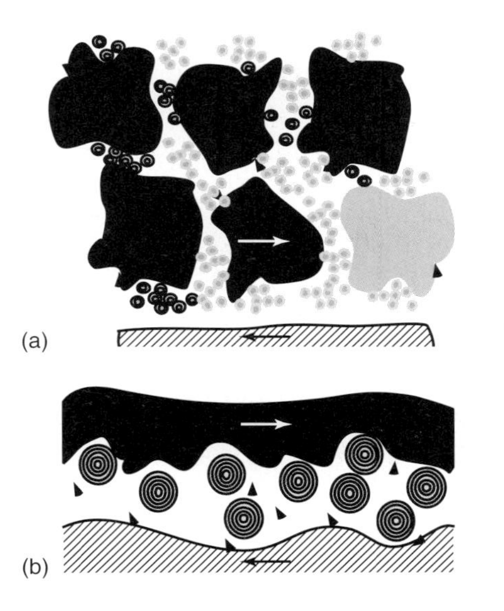

Fig. 18 Schematic illustration of the wear mechanism of porous metal matrices impregnated by solid lubricant particles. The shaded areas are representative of the metal grains; the concentric circles represent the fullerene-like nanoparticles.

grooves, the slender and inert nanotube tip can go down at least 1 μm deep and image the bottom of the groove faithfully [24]. This particular tip has been tested for a few months with no signs of deterioration, which is indicative of its resilience and passive surface. Other kinds of tips have been in use for high-resolution imaging using scanning probe microscopy in recent years. However, the present tips are rather stiff and inert and consequently they are likely to serve in high-resolution imaging of rough surfaces having features with large aspect ratio. Furthermore, inorganic nanotubes exhibit strong absorption of light in the visible part of the spectrum and their electrical conductivity can be varied over many orders of magnitude by doping and intercalation. This suggests numerous applications, in areas such as nanolithography, photocatalysis, and others.

3.3.8
Conclusion

Inorganic fullerene-like structures and inorganic nanotubes, in particular, are a generic structure of nanoparticles of inorganic layered (2D) compounds. Various synthetic approaches to produce these nanostructures are presented. In some cases, such as WS_2, MoS_2, BN, and V_2O_5,both fullerene-like nanoparticles and nanotubes are produced in gross amounts. However, size and shape control is still at its infancy. Study of these novel nanostructures has led to the observation of a few interesting properties and some potential applications in tribology, high-energy density batteries, and nanoelectronics.

Acknowledgment

I am indebted to Dr. Ronit Popovitz-Biro for the assistance with some of the TEM images and to Dr. Rita Rosentsveig for the synthesis of the IF-WS_2 nanoparticles. This work was supported in part by the following agencies: Israeli Ministry of Science (Tashtiot program); USA-Israel Binational Science Foundation; Israel science Foundation; Krupp von Bohlen and Halbach Stiftung (Germany); France-Israel R&D (AFIRST) Foundation; Israeli Academy of Sciences (First program).

References

1. H. W. Kroto, J. R. Heath, S. C. O'Brein et al., *Nature* **1985**, *318*, 162–163.
2. S. Iijima, *Nature* **1991**, *354*, 56–58.
3. R. Tenne, L. Margulis, M. Genut et al., *Nature* **1992**, *360*, 444–445.
4. Y. Feldman, E. Wasserman, D. J. Srolovitz et al., *Science* **1995**, *267*, 222–225.
5. B. C. Satishkumar, A. Govindaraj, E. M. Vogel et al., *J. Mater. Res.* **1997**, *12*, 604–606.
6. M. E. Spahr, P. Bitterli, R. Nesper et al., *Angew. Chem. Int. Ed.* **1998**, *37*, 1263–1265.
7. A. Rothschild, G. L. Frey, M. Homyonfer et al., *Mater.Res.Innov.* **1999**, *3*, 145–149.
8. A. Rothschild, J. Sloan, R. Tenne, *J. Am. Chem. Soc.* **2000**, *122*, 5169–5179.
9. A. Rothschild, R. Popovitz-Biro, R. Tenne, *J. Phys. Chem. B* **2000**, *104*, 8976–8981.
10. M. Remskar, Z. Skraba, M. Regula et al., *Adv. Mater.* **1998**, *10*, 246–249.
11. Y. Q. Zhu, W. K. Hsu, N. Grobert et al., *Chem. Mater.* **2000**, *12*, 1190–1194.
12. Y. Q. Zhy, W.K.Hasu, H.Terrones et al., *J. Mater. Chem.* **2000**, *10*, 2570–2577.
13. W. K. Hsu, B. H. Chang, Y. Q. Zhu et al., *J. Am. Chem. Soc.* **2000**, *122*, 10155–10158.
14. J. Cumings, A. Zettl, *Chem. Phys. Lett.* **2000**, *316*, 211–216.
15. J. Cumings, A. Zettl, *Chem. Phys. Lett.* **2000**, *318*, 497.
16. A. Zak, Y. Feldman, V. Alperovich et al., *J. Am. Chem. Soc.* **2000**, *122*, 11 108–11 116.
17. P. W. Fowler, D. E. Manolopoulos in *An Atlas of Fullerenes*, Oxford University Press, Cambridge, 1995.
18. M. Cote, M. L. Cohen, D. J. Chadi, *Phys. Rev. B* **1998**, *58*, R4277–R4280.

19. G. Seifert, H. Terrones, M. Terrones et al., *Phys. Rev. Lett.* **2000**, *85*, 146–149.
20. M. Remskar, private communication.
21. M. Chhowalla, G. A. J. Amaratunga, *Nature* **2000**, *407*, 164–167.
22. P. A. Parilla, A. C. Dillon, K. M. Jones et al., *Nature* **1999**, *397*, 114.
23. A. Rubio, J. L. Corkill, M. L. Cohen, *Phys. Rev. B* **1994**, *49*, 5081–5084.
24. A. Rothschild, S. R. Cohen, R. Tenne, *Appl. Phys. Lett.* **1999**, *75*, 4025–4027.
25. N. G. Chopra, A. Zettl, *Solid State Commun.* **1998**, *105*, 297–300.
26. L. Pauling, *Proc. Natl. Acad Sci.* **1930**, *16*, 578–582.
27. K. Kato, I. Kawada, T. Takahashi, *Acta Crystallogr.* **1977**, *B33*, 3437–3443.
28. L. Guemas, P. Rabu, A. Meerschut et al., *Mater. Res. Bull.* **1988**, *23*, 1061–1069.
29. G. A. Wiegers, A. Meerschut, *J. Alloys Compd.* **1992**, *178*, 351–368.
30. K. Suzuki, T. Enoki, K. Imaeda, *Solid State Commun.* **1991**, *78*, 73–77.
31. F. Krumeich, H. -J. Muhr, M. Niederberger et al., *J. Am. Chem. Soc.* **1999**, *121*, 8324–8331.
32. S. Avivi, Y. Mastai, G. Hodes et al., *J. Am. Chem. Soc.* **1999**, *121*, 4196–4199.
33. J. S. Kasper, P. Hagenmuller, M. Pouchard et al., *Science* **1965**, *150*, 1713.
34. C. Cros, M. Pouchard, E. P. Hagenmuller, *J. Solid State Chem.* **1970**, *2*, 570.
35. H. Kawaji, H. Horie, S. Yamanaka et al., *Phys. Rev. Lett.* **1995**, *74*, 1427–1429.
36. S. Saito, A. Oshiyama, *Phys. Rev. B* **1995**, *51*, 2628–2631.
37. E. Kaxiras, K. Jackson, *Phys. Rev. Lett.* **1993**, *71*, 727–730.
38. L. Zeger, E. Kaxiras, *Phys. Rev. Lett.* **1993**, *70*, 2920–2923.
39. D. Wichmann, K. Jug, *J. Phys. Chem. B* **1999**, *103*, 10087–10091.
40. N. N. Grennwood, A. Earnshaw in *Chemistry of the Elements*, Pergamon Press, Oxford, 1984.
41. R. Tenne, *Adv. Mater.* **1995**, *7*, 795–965.
42. *Micelles, Membranes, Microemulsions and Monolayers*, (Eds.: W. M. Gelbart, A. Ben-Shaul, D. Roux), Springer, New York, 1994.
43. W. Shenton, T. Douglas, M. Young et al., *Adv. Mater.* **1999**, *11*, 253–256.
44. S. W. Guo, L. Konopny, R. Popovitz-Biro et al., *Adv. Mater.* **2000**, *12*, 302–306.
45. Y. Feldman, G. L. Frey, M. Homyonfer et al., *J. Am. Chem. Soc.* **1996**, *118*, 5362–5367.
46. Y. Feldman, V. Lyakhovitskaya, R. Tenne, *J. Am. Chem. Soc.* **1998**, *120*, 4176–4183.
47. M. Remskar, Z. Skraba, F. Cléton et al., *Appl. Phys. Lett.* **1996**, *69*, 351–353.
48. M. Remskar, Z. Skraba, F. Cléton et al., *Surf. Rev. Lett.* **1998**, *5*, 423–426.
49. M. Remskar, Z. Skraba, C. Ballif et al., *Surf. Sci.* **1999**, *435*, 637–641.
50. M. Remskar, Z. Skraba, R. Sanjinés et al., *Appl. Phys. Lett.* **1999**, *74*, 3633–3635.
51. C. M. Zelenski, P. K. Dorhout, *J. Am. Chem. Soc.* **1998**, *120*, 734–742.
52. C. R. Martin, Acc. Chem. Res. **1995**, *28*, 61–68.
53. H. Masuda, K. Fukuda, *Science* **1995**, *268*, 1466–1468.
54. P. Santiago, D. Mendoza, E. Espinosa et al., submitted.
55. J. Sloan, J. L. Hutchison, R. Tenne et al., *J. Solid State Chem.* **1999**, *144*, 100–117.
56. E. Iguchi, *J. Solid State Chem.* **1978**, *23*, 231–239.
57. W. B. Hu, Y. Q. Zhu, W. K. Hsu et al., *Appl. Phys. A.* **2000**, *70*, 231–233.
58. L. Margulis, G. Salitra, R. Tenne et al., *Nature* **1993**, *365*, 113–114.
59. H. Nakamura, Y. Matsui, *J. Am. Chem. Soc.* **1995**, *117*, 2651–2652.
60. H. P. Lin, C. Y Mou, S. B Liu, *Adv. Mater.* **2000**, *12*, 103–106.
61. M. E. Spahr, P. Stoschitzki-Bitterli, R. Nesper et al., *J. Electrochem. Soc.* **1999**, *146*, 2780–2783.
62. Y. Li, M. Sui, Y. Ding et al., *Adv. Mater.* **2000**, *12*, 818–821.
63. M. Harada, M. Adachi, *Adv. Mater.* **2000**, *12*, 839–841.
64. P. M. Ajayan, O. Stephan, P. Redlich et al., *Nature* **1995**, *375*, 564–567.
65. A. Leist, S. Stauf, S. Löken et al., *J. Mater. Chem.* **1998**, *8*, 241–244.
66. P. Afanasiev, C. Geantet, C. Thomazeau et al., *Chem. Commun.* **2000**, 1001–1002.
67. M. José-Yacamán, H. Lopez, P. Santiago et al., *Appl. Phys. Lett.* **1996**, *69*, 1065–1067.
68. D. Ugarte, *Nature* **1992**, *359*, 707–709.
69. D. Vollath, D. V. Szabo, *Mater. Lett.* **1998**, *35*, 236–244.
70. D. Vollath, D. V. Szabó, *Acta Materialia* **2000**, *48*, 953–967.

71. M. Homyonfer, Y. Mastai, M. Hershfinkel et al., *J. Am. Chem. Soc.* **1996**, *118*, 7804–7808.
72. Y. Mastai, M. Homyonfer, A. Gedanken et al., *Adv. Mater.* **1999**, *11*, 1010–1013.
73. K. S. Suslick, S. -B. Choe, A. A. Cichovlas et al., *Nature* **1991**, *353*, 414–416.
74. T. J. Mason, J. P. Lorimer, D. J. Walton, *Ultrasonics* **1990**, *28*, 333–337.
75. A. Durant, J. L. Deplancke, R. Winand et al., *Tetrahedron Lett.* **1995**, *36*, 4257–4260.
76. R. Tenne, M. Homyonfer, Y. Feldman, *Chem. Mater.* **1998**, *10*, 3225–3238.
77. K. S. Suslick, D. A. Hammerton, R. E. Cline, *J. Am. Chem. Soc.* **1986**, *108*, 5641–5642.
78. G. B. Saupe, C. C. Waraksa, H -N Kim et al., *Chem. Mater.* **2000**, *12*, 1556–1562.
79. R. Abe, K. Shinohara, A. Tanaka et al., *Chem. Mater.* **1997**, *9*, 2179–2184.
80. S. Avivi, Y. Mastai, A. Gedanken, *J. Am. Chem. Soc.* **2000**, *122*, 4331–4334.
81. M. Homyonfer, B. Alperson, Yu. Rosenberg et al., *J. Am. Chem. Soc.* **1997**, *119*, 2693–2698.
82. G. Seifert, H. Terrones, M. Terrones et al., *Solid State Commun.* **2000**, *115*, 635–638.
83. Y. Rosenfeld Hacohen, E. Grunbaum, R. Tenne et al., *Nature* **1998**, *395*, 336.
84. R. Popovitz-Biro, A. Twerski, Y. Rosenfeld Hacohen et al., *Isr. J. Chem.*, in press.
85. N. G. Chopra, J. Luyken, K. Cherry et al., *Science* **1995**, *269*, 966–967.
86. Z. W. Sieh, K. Cherrey, N. G. Chopra et al., *Phys. Rev. B* **1995**, *51*, 11229–11232.
87. A. Loiseau, F. Willaime, N. Demoncy et al., *Phys. Rev. Lett.* **1996**, *76*, 4737–4740.
88. M. Terrones, W. K. Hsu, H. Terrones et al., *Chem. Phys. Lett.* **1996**, *259*, 568–573.
89. M. Terrones, A. M. Benito, C. Manteca-Diego et al., *Chem. Phys. Lett.* **1996**, *257*, 576–582.
90. T. Laude, A. Marraud, Y. Matsui et al., *Appl. Phys. Lett.* **2000**, *76*, 3239–3241.
91. D. Golberg, Y. Bando, W. Han et al., *Chem. Phys. Lett.* **1999**, *308*, 337–342.
92. D. Golberg, Y. Bando, O. Stéphan et al., *Appl. Phys. Lett.* **1998**, *73*, 2441–2443.
93. K. Suenaga, C. Colliex, N. Demoncy et al., *Science* **1997**, *278*, 653–655.
94. Y. Zhang, K. Suenaga, S. Iijima, *Science* **1998**, *281*, 973–975.
95. T. Oku, T. Kusunose, K. Niihara et al., *J. Mater. Chem.* **2000**, *10*, 255–257.
96. W. Stöber, A. Fink, E. Bohn, *J. Colloid Interface Sci.* **1968**, *26*, 62–69.
97. T. Kasuga, M. Hiramatsu, A. Hoson et al., *Langmuir* **1998**, *14*, 3160–3163.
98. T. Kasuga, M. Hiramatsu, A. Hoson, et al., *Adv. Mater.* **1999**, *11*, 1307–1311.
99. C. N. R. Rao, B. C. Satishkumar, A. Govindaraj, *Chem. Commun.* **1997**, 1581–1582.
100. J. Hu, B. Deng, Q. Lu et al., *Chem. Commun.* **2000**, 715–716.
101. M. R. Ghadiri, J. R. Granja, R. A. Milligan et al., *Nature* **1993**, *366*, 324–327.
102. M. Remskar, Z. Skraba, R. Sanjines et al., *Surf. Rev Lett.* **1999**, *6*, 1283–1287.
103. M. Remskar, Z. Skraba, P. Stadelmann et al., *Adv. Mater.* **2000**, *12*, 814–818.
104. J. M. Vandenberg-Voorhoeve, Physics and chemistry of materials with layered structures, *Optical and Electrical Properties*, (Eds.: P. A. Lee), D. Reidel Publishing Company, Dordecht-Holland, 1976, p. 447, Vol. 4, Ch. 8.
105. J. A. Hollingsworth, D. M. Poojary, A. Clearfield et al., *J. Am. Chem. Soc.* **2000**, *122*, 3562–3563.
106. S. M. Lee, Y. H. Lee, Y. G. Hwang et al., *Phys. Rev. B* **1999**, *60*, 7788–7791.
107. J. A. Wilson, A. D. Yoffe, *Adv. Phys.* **1969**, *18*, 193–335.
108. J. Moser, F. Lévy, F. Bussy, *J. Vac. Sci. Technol. A* **1994**, *12*, 494–500.
109. L. Margulis, P. Dluzewski, Y. Feldman et al., *J. Microscopy* **1996**, *181*, 68–71.
110. L. Margulis, R. Tenne, S. Iijima, *Microsc. Microanal. Microstruct.* **1996**, *7*, 87–89.
111. R. Tenne, *Adv. Mater.* **1995**, *7*, 965–995.
112. W. Qian, M. D. Bartberger, S. J. Pastor et al., *J. Am. Chem. Soc.* **2000**, *122*, 8333–8334.
113. Y. Miyamoto, A. Rubio, M. L. Cohen et al., *Phys. Rev. B* **1994**, *50*, 4976–4979.
114. Y. Miyamoto, A. Rubio, S. G. Louie et al., *Phys. Rev. B* **1994**, *50*, 18360–18366.
115. F. Jensen, H. Toftlund, *Chem. Phys. Lett.* **1993**, *201*, 95–98.
116. G. Seifert, P. W. Fowler, D. Mitchell et al., *Chem. Phys. Lett.* **1997**, *268*, 352–358.
117. K. Kobayashi, N. Kurita, *Phys. Rev. Lett.* **1993**, *70*, 3542–3544.
118. X. Xia, D. A. Jelski, J. R. Bowser et al., *J. Am. Chem. Soc.* **1992**, *114*, 6493–6496.
119. O. Stéphan, Y. Bando, A. Loiseau et al., *Appl. Phys. A* **1998**, *67*, 107–111.

120. A. D. Yoffe, K. J. Howlett, P. M. Williams, Cathodoluminescence studies in the SEM, *Scanning Electron Microscopy/1973 (Part II)*, 1973, 301–308.
121. D. A. B. Miller, D. S. Chemla, S. Schmitt-Rink, in *Optical Nonlinearities and Instabilities in Semiconductors* (Eds.: H. Haug), Academic Press, Orlando, FL, 1988, pp. 325.
122. M. L. Steigerwald, L. E. Brus, *Annu. Rev. Mater. Sci.* **1989**, *19*, 471–495.
123. G. Seifert, H. Terrones M. Terrones et al., *Solid State Commun.* **2000**, *114*, 245–248.
124. X. R. Qin, D. Yang, R. F. Frindt et al., *Phys. Rev. B* **1991**, *44*, 3490–3493.
125. S. J. Sandoval, D. Yang, R. F. Frindt et al., *Phys. Rev. B* **1991**, *44*, 3955–3962.
126. D. Yang, R. F. Frindt, *Mol. Cryst. Liq. Cryst. Sci. Technol., Sect. A* **1994**, *244*, 355–360.
127. M. J. Hebben, B.A. Parkinson, private communication.
128. G. Seifert, E. Hernández, *Chem. Phys. Lett.* **2000**, *318*, 355–360.
129. G. Seifert, T. Frauenheim, *J. Korean Phys. Soc.* **2000**, *37*, 89–92.
130. G. Seifert, E. Hernandez, to be published.
131. E. Bucher, Photovoltaic properties of solid state junctions of layered semiconductors, *Physics and Chemistry of Materials with Layered Structures*, (Eds.: A. Aruchamy), Kluwer Academic, Dordrecht, 1992, pp. 1–81, Vol. 14.
132. G. L. Frey, S. Elani, M. Homyonfer et al., *Phys. Rev. B* **1998**, *57*, 6666–6671.
133. F. Consadori, R. F. Frindt, *Phys. Rev. B* **1970**, *2*, 4893–4896.
134. M. W. Peterson, A. J. Nozik, Quantum size effects in layered semiconductor colloids, *Physics and Chemistry of Materials with Layered Structures*, (Eds.: A Aruchamy), Kluwer Academic, Dordrecht, 1992, 297–317, Vol. 14.
135. G. L. Frey, R. Tenne, M. J. Matthews et al., *J. Mater. Res.* **1998**, *13*, 2412–2417.
136. G. L. Frey, R. Tenne, M. J. Matthews et al., *Phys. Rev. B* **1999**, *60*, 2883–2893.
137. D. J. Srolovitz, S. A. Safran, M. Homyonfer et al., *Phys. Rev. Lett.* **1995**, *74*, 1779–1782.
138. U. S. Schwarz, S. Komura, S. A. Safran, *Europhys. Lett.* **2000**, *50*, 762–768.
139. L. T. Chadderton, D. Fink, Y. Gamaly et al., *Nucl. Instr. Methods Phys. Res. B* **1994**, *91*, 71–77.
140. R. B. Somoano, J. A. Woollam, *Intercalated Layered Materials*, (Eds.: F. Lévy), D. Reidel Publishing Company, Dordrecht, 1979, pp. 307–319.
141. H. J. Lewerenz, A. Heller, F. DiSalvo, *J. Am. Chem. Soc.* **1980**, *102*, 1877–1880.
142. J. Bordas, Optical and Electrical Properties, *Physics and Chemistry of Materials with layered Structures*, (Eds.: P. A. Lee), D. Reidel Publishing Company, Dordrecht, 1976, pp. 145–230.
143. B. Alperson, M. Homyonfer, R. Tenne, *J. Electroanal. Chem.* **1999**, *473*, 186–191.
144. G. Domingo, R. S. Itoga, C. R. Kannewurf, *Phys. Rev.* **1966**, 143, *143*.
145. V. Badri, A. M. Hermann, *Intl. J. Hydrogen Energy* **2000**, *25*, 249–253.
146. G. Tourillon, L. Pontonnier, J. P. Levy et al., *Electrochem. Solid-State Lett.* **2000**, *3*, 20–23.
147. I. L. Singer in *Fundamentals of Friction: Macroscopic and Microscopic Processes*, (Eds.: I. L. Singer H. M. Pollock), Kluwer, Dordrecht, 1992, p. 237.
148. F. P. Bowden, D. Tabor, in *Anchor*, Garden City, New York, 1973.
149. L. Rapoport, Yu. Bilik, Y. Feldman et al., *Nature* **1997**, *387*, 791–793.
150. Y. Golan, C. Drummond, M. Homyonfer et al., *Adv. Mater.* **1999**, *11*, 934–937.
151. Y. Golan, C. Drummond, J. Israelachvili et al., *Wear* **2000**, *245*, 190–195.
152. L. Rapoport, Y. Feldman, M. Homyonfer et al., *Wear* **1999**, *225-229*, 975–982.
153. L. Rapoport, M. Lvovsky, I. Lepsker et al. *Adv. Eng. Mater.* **2001**, *3*, 71–75.
154. V. Leshchinsky, E. Alyoshina, M. Lvovsky et al., submitted.

4
Solar Energy Conversion without Dye Sensitization

4.1
The Photoelectrochemistry of Semiconductor/Electrolyte Solar Cells

Maheshwar Sharon
Indian Institute of Technology, Bombay, India

4.1.1
Introduction

Statistical assessment suggests that 12 000 kg of coal per capita per year is being used by developed countries, whereas 150 kg is being used by developing countries. It is estimated that in the last century, man has consumed the energy equivalent to 4×10^{21} J and in the next century would need energy equivalent to 100×10^{21} J. The rate of use of fossil fuels (e.g. oil, gas, coal, etc.) has linearly increased with respect to utilization time. However, it cannot follow the same trend indefinitely. To avoid an undesirable situation arising due to the shortage of fossil fuels, scientists all over the world, are trying to take advantage of renewable energy sources such as solar, wind, ocean, and so on.

The sun is the cleanest and most abundant energy source. It is estimated that out of the total solar energy (3.8×10^{20} MW), earth receives about 1.7×10^{6} MW. Therefore, in recent years, attention has been paid to use solar energy for terrestrial applications. One of the most important aspects in using solar energy is its conversion from solar radiation into electrical energy. This is achieved by using semiconductor-based photovoltaic (PV) cells. Today's PV market is 151 MW per year corresponding to a value of about $0.7 - 1$ billion US$. This is a remarkable market but still far away from being a noticeable contribution to the world energy consumption. The major reason for the low penetration of PV today is the high cost. Broadly speaking, there are three types of PV cells (1) $n : p$ junction cell, (2) metal-semiconductor cell also known as Metal-Schottky junction, and (3) semiconductor-electrolyte junction. This section discusses the last type of cell.

In 1839, Becquerel [1] first discovered the PV phenomena in electrochemical systems. Brattain and Garret [2, 3] were pioneers explaining aspects of the properties of semiconductor–electrolyte interfaces. Fujishima and Honda [4] reported the first indication of a practical application of a photoelectrochemical (PEC) system in 1972. This paper sparked off a wave of investigations all over the world. It would be appropriate, however, to suggest that the interest in photoelectrochemistry of semiconductor blossomed only after the pioneering work of Gerischer [5] and Myamlin and Pleskov [6]. These studies led to the discovery of wet PV solar cell. This is popularly known as a PEC cell. It was believed that a PEC cell might be more economically viable as compared with the solid-state PV cell. Simplicity in fabricating the cell was the main reason for such an optimistic view. Moreover, as discussed in a later chapter, it has been possible to make a rechargeable battery with in situ storage capability by using a PEC cell [7–9]. The wet type PEC cell suffers from the instability of semiconductor in aqueous media. It has been realized that in spite of its simplicity and economical viability, the wet-type PEC cell cannot easily replace a silicon PV solar cell, unless we discover photoelectrochemically stable semiconductor materials possessing band gap approximately 1.4 eV. Till such time, future of (unsensitized) wet-type PEC solar cell appears gloomy. However, research utilizing a standard PEC cell configuration is not complex, facilitating testing of the photoactivity of new semiconductor materials. Current PEC cell research trends have also been directed toward sensitized semiconductors and applications in the field of decontamination of water from pathogenic bacteria [10], detoxification of water from toxic organic/inorganic materials, photography [11] or in any other field where instability of the semiconductor does not pose a real problem.

The present section is devoted to various aspects that are essential for the development of a PEC cell.

4.1.2 Description of a PEC Cell

A PEC cell consists of a photoactive semiconductor electrode (either *n*- or *p*-type) and a metal counter-electrode. Both these electrodes are immersed in a suitable redox electrolyte. The semiconductor is protected by insulation, so that only one of its surfaces is exposed to the redox electrolyte. In a regenerative PEC solar cell, the metal counter-electrode is expected to perform an electrochemical reaction that is the opposite of the process occurring at the semiconductor electrode. The counter-electrode should be electrochemically stable. It would be desirable that work function of counter-electrode is compatible with that of the Fermi level of the semiconductor. The matching of work function may not be logical with metallic electrodes such as platinum, but may be important with semiconducting electrode such as SnO_2.

On illumination of the semiconductor-electrolyte junction with a light having energy greater than the band gap of the semiconductor, photogenerated electron/holes are separated in the space charge region. The photogenerated minority carriers (holes for *n*-type or electron for *p*-type semiconductor) arrive at the interface of the semiconductor–electrolyte. Photogenerated majority carriers (i.e. electrons for *n*-type and holes for *p*-type semiconductor) accumulate at the backside of the semiconductor (i.e. the side that is not illuminated and not in contact with the redox electrolyte). With the help of a connecting wire, photogenerated majority carriers are transported to counter-electrode via a load (Fig. 1). These carriers at the counter-electrode electrochemically react with the redox electrolyte (i.e. reduction of redox electrolyte occurs with *n*-type or oxidation of redox electrolyte with *p*-type semiconductor).

Conversely, the photogenerated minority carriers generated at the interface of the semiconductor perform the opposite reaction, which occurred at the metal counter-electrode. Thus, photogenerated carriers are consumed. Because of these

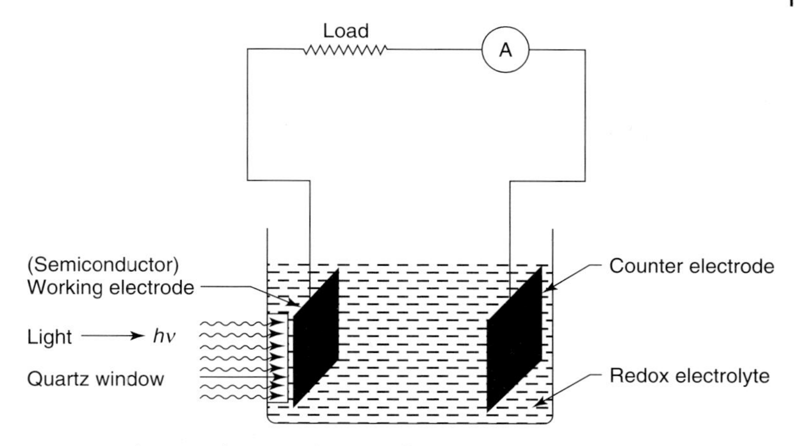

Fig. 1 A schematic diagram of a PEC cell.

two reactions, there is no net change in the electrolyte, and electrical power is produced in the load.

Details of the electrochemical reactions with respect to a PEC cell utilizing a *p*-type semiconductor dipped in $FeSO_4/Fe_2(SO_4)_3$ electrolyte are given as

At the *p*-semiconductor electrode

$$h\nu + p\text{-semiconductor} \longrightarrow e^-/h^+ \times \text{(photogenerated carriers)} \quad (1)$$

Separation of charge carrier (e^-/h^+) caused by junction formation

$$(e^-/h^+)_{\text{depletion region}} \longrightarrow e^-_{surf} + h^+_{bulk} \quad (2)$$

h^+_{bulk} is transported to the counter-electrode via a load and oxidizes redox electrolyte, that is,

$$h^+_{bulk} + Fe^{2+} \longrightarrow Fe^{3+} \quad (3)$$

The corresponding chemical reaction is

$$2FeSO_4 + 2h^+_{bulk} \longrightarrow Fe_2(SO_4)_3, \quad (4)$$

that is, near this electrode, one additional $SO_4^{=}$ ion is required to be transported from the counter-electrode or the bulk electrolyte to maintain the neutrality of the solution. At the counter electrode e^-_{surf} reduces the redox electrolyte, that is,

$$e^-_{surf} + Fe^{3+} \longrightarrow Fe^{2+} \quad (5)$$

Corresponding chemical reaction is

$$Fe_2(SO_4)_3 + 2e^-_{surf} \longrightarrow 2FeSO_4, \quad (6)$$

that is, at the counter-electrode, one additional $SO_4^{=}$ ion is formed that needs to be transported to the semiconductor or to the bulk solution to maintain its neutrality.

The reaction in this regenerative PEC is the conversion of photon energy into electrical energy without destroying chemical composition of the redox electrolyte or counter-electrode or semiconductor electrode, that is, $h\nu$ (photon energy) → electrical energy.

Since in such a PEC cell, both electrodes are immersed in the same electrolyte and specific reactions occur only at the semiconductor and the metal, there is no need for a separator. Hence, charge balance due to oxidation and reduction processes is maintained.

4.1.3
Types of PEC Cells

Figure 2 shows various possible PEC [12] cells. In a PEC cell (or electrochemical PV cell), optical energy is converted into electrical energy, with zero change in the free energy of redox electrolyte. In this cell, the electrochemical reaction occurring at the counter-electrode is opposite to the photoassisted reaction occurring at the semiconductor electrode (Fig. 3). Thus, the light energy is converted into electrical energy, with no change in the solution composition or electrode material. This is the case described earlier. In a photoelectrosynthetic cell, optical energy is converted into chemical energy with nonzero free energy change in the electrolyte. In this case, a net chemical change occurs on illumination. This type of cell can be further classified as photoelectrolytic cell ($\Delta G > 0$, Fig. 3) and photocatalytic cell ($\Delta G < 0$, Fig. 3), depending on relative location of the potentials of the two redox couples (e.g. O/R and O′/R′). In the latter type of cells, anodic and cathodic compartments need to be separated to prevent mixing of the two redox couples.

4.1.4
Aim of this Section

In this section, we shall be concerned with developing a regenerative PEC solar cell, where $\Delta G = 0$. From previous discussions, we understand how a PEC cell is fabricated. However, if we are to develop new materials for making an economically viable PEC cell, it is essential that we

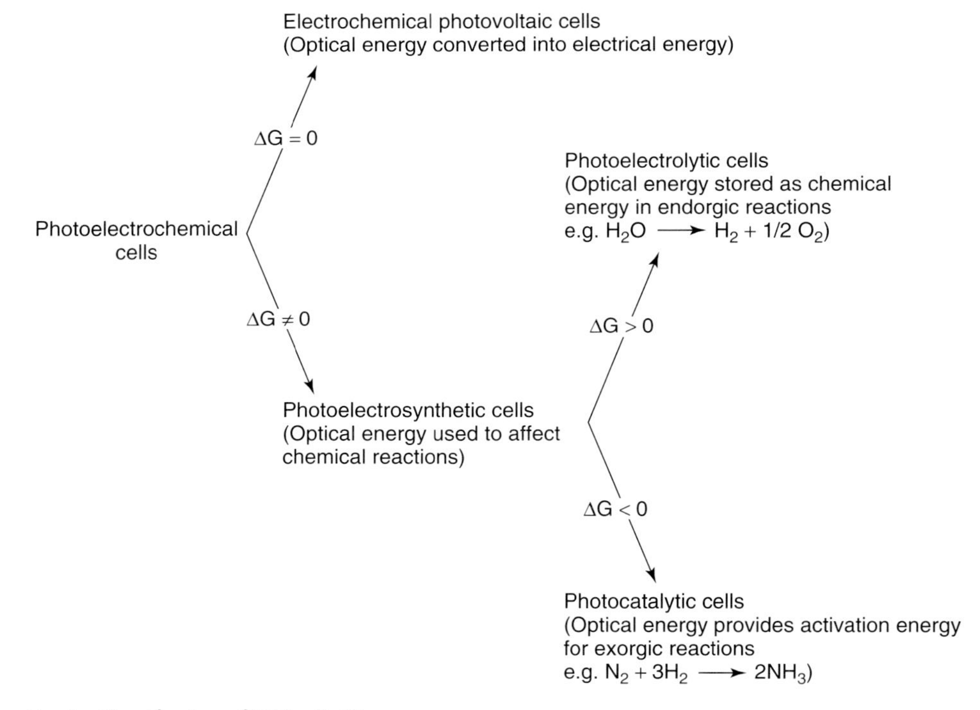

Fig. 2 Classification of PEC cells [8].

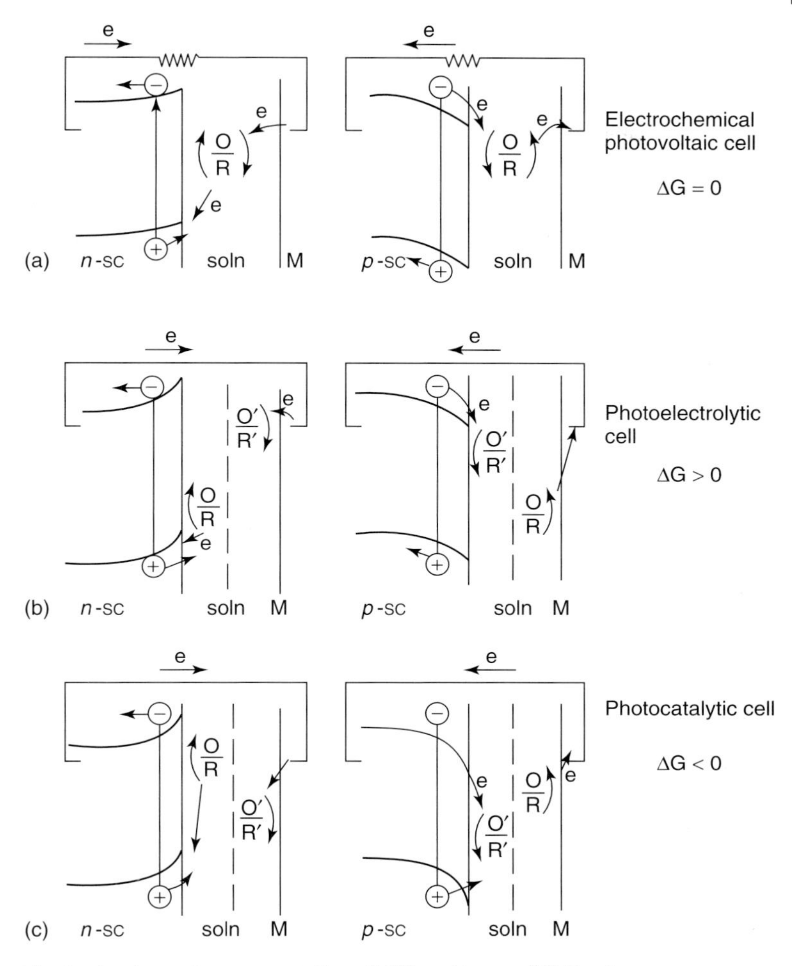

Fig. 3 A schematic representation of different types of PEC cells.

understand the principle, which controls the conversion of light energy into electrical energy. This knowledge will help us to design a desirable semiconductor. It is also necessary to realize that there are some constraints while employing a semiconductor and a redox electrolyte to make a PEC cell. Therefore, knowledge regarding potential energy of redox electrolyte and its relation with the band positions of semiconductor is beneficial. To accomplish this we also need to perceive a simple technique to procure information about the band position of the semiconductor. Finally, a simple experimental technique to maneuver the photocharacteristics of semiconductor-redox electrolyte is desirable. Unless, we have this information, it would be an extremely difficult task to launch production of new materials for developing an economically viable and photoelectrochemically stable PEC cell. In the forthcoming sections, efforts are made to provide this information as briefly as possible because discussions on each of these items are dealt extensively in other sections of this book.

4.1.5 Semiconductor-electrolyte Junction

When a semiconductor comes in contact with another material of Fermi level, different from the semiconductor, a junction is formed. This can be formed between n-type and p-type semiconductors, or a metal and a semiconductor, or a semiconductor and a redox electrolyte.

The semiconductor-electrolyte junction should not be confused with a metal-electrolyte junction. In the metal-electrolyte junction, the potential drop occurs entirely on the solution side and practically nil on the metal side, whereas for semiconductor–electrolyte interface, the potential drop occurs on the semiconductor side as well as the solution side.

For clarification of the type of junctions formed at the semiconductor-electrolyte, let us take an example of n-type semiconductor. In addition to possessing free electrons (referred to as the majority carrier), n-type semiconductor also possesses holes (referred to as the minority carrier). The concentration of holes is temperature-dependent and is equivalent to the intrinsic concentration of the carrier (which is related to the concentration of Frankel defects). It can be shown mathematically that the Fermi level of minority carrier lies at almost half the band gap position. On the other hand, the concentration of majority carriers as well as the Fermi level depends on doping concentration. Thus, the Fermi level of the majority carrier can lie anywhere between the conduction band edge and the intrinsic Fermi level that is situated at $\sim\frac{1}{2}E_g$.

Similarly, the redox electrolyte can also be viewed as a material possessing a specific Fermi level. For example, reduction of Fe^{3+} and Ce^{4+} requires 0.77 and 1.44 volts versus hydrogen electrode, respectively. Alternatively, we can say that the energy of electrons present in the reduced product of these materials must be equivalent to this potential. Hence, these potentials can be referred to as the work function (or Fermi level) of the redox electrolyte.

Let us imagine a situation where an n-type semiconductor is brought in contact with a redox electrolyte (Fig. 4a). Let us also assume that the electrochemical potential, that is, its redox potential, is almost equal to the intrinsic Fermi level of the semiconductor. This situation forces the electrons (majority carriers) to flow from the semiconductor to the electrolyte. This migration continues until the two Fermi levels achieve an equilibrium position (Fig. 4b). At this condition, no further migration of electron occurs and a dynamic equilibrium is established. In this situation, instead of electrochemically reacting with the redox electrolyte, these majority carriers accumulate at the interface of semiconductor–electrolyte to maintain neutrality of the material.

When electrons move toward the interface, lattice sites from where electrons are originated become positively charged (Fig. 4b). Creation of positively charged carriers occurs randomly in the semiconductor (shown by a curved vertical line in Fig. 4b), but are situated not very far from the interface. Any mathematical treatment for such random distribution or even approximated exponential distribution of charged ions becomes a complicated system to deal with.

Hence, a simplified mathematical model is visualized. It is assumed that all positively charged species instead of being scattered are situated in one hypothetical

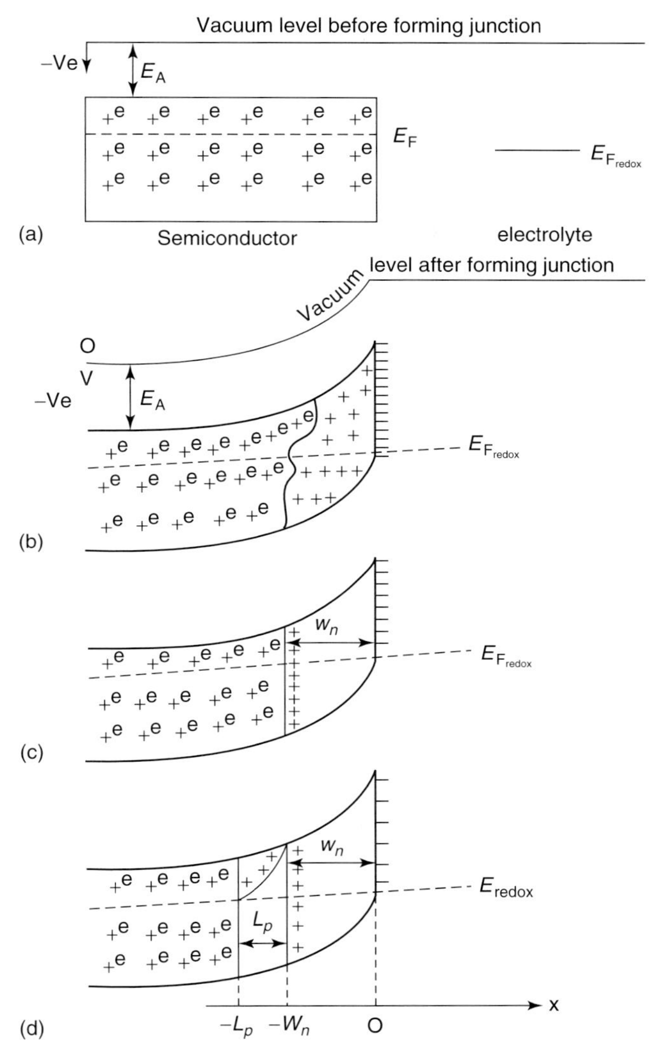

Fig. 4 Schematic diagrams showing the formation of band bending at a semiconductor–electrolyte interface. (a) A semiconductor possessing its charge homogeneously distributed. +/e refers to a situation in which lattice sites are neutral. A horizontal line of vacuum scale showing no variation in energy of electrons of semiconductor. (b) The contact of semiconductor with redox electrolyte disturbs the neutrality, and nonhomogeneous charge separation occurs. Vacuum scale takes the same shape as that of semiconductor to maintain its electron affinity. (c) A hypothetical model showing charges retained in two parallels planes separated by a distance w. (d) Creation of diffusion region as a result of accumulation of minority carriers at plane situated at distance $x = -w$.

plane at a distance "w" from the interface (Fig. 4c). We can get a better idea of this hypothetical plane by considering various positive charges $q_1, q_2, q_3, \ldots q_n$ situated at $x_1, x_2, x_3, \ldots x_n$ distances from the interface of the semiconductor. The total electrical field created between the negative charges situated at the interface (i.e. at $x = 0$) and positive charges situated at various distances would be sum of the product of charges divided by their respective distances.

That is, total electrical field created

$$(\mathrm{F}) = \frac{q_1 e^-}{x_1} + \frac{q_2 e^-}{x_2} + \cdots + \frac{q_n e^-}{x_n} \approx \frac{Q^+_{\text{total}} Q^-_{\text{total}}}{w} \quad (7)$$

where Q^+_{total} is total positive charge of minority carriers, equivalent to the total charge of majority carriers accumulated at the interface (Q^-_{total}) and "w" is the distance of the hypothetical plane from the interface ($x = 0$), which contains Q^+_{total} total charge, such that $(Q^+_{\text{total}}\ Q^-_{\text{total}})/w =$ F. It is worth noting that "w" is not equal to the sum of all distances (i.e. $w \neq x_1 + x_2 + x_3 + \cdots + x_n$).

This hypothetical visualization of charge distribution near the interface (i.e. its presence in one plane) reveals the possibility of formation of charges in two parallel planes. One plane appears at the interface (which is populated with the majority carriers at $x = 0$) and another at a depth w (which is populated with charges equivalent to total charges developed at $x = -w$). The depth between these two planes, (Fig. 4c), is called the *space charge width* designated as w_n (for n-type) and w_p (for p-type). It should be understood that concentration of minority carriers at $x = -w_n$ would be much greater than the intrinsic concentration of hole at $x \approx -\alpha$. In this mathematical model, it is assumed that all charges present within the distance of $x = 0$ and $x = -w$ are regular and fully ionized. Moreover, within the space charge region, there are no free carriers. In other words, entire lattice sites falling within the space charge region are fully ionized. It is also assumed that electrostatic field becomes zero beyond $x = -w_n$.

Should the changes in distribution of electrons (majority carriers) in n-type be abrupt at the interface as compared to the bulk or should it change linearly or exponentially? The most reasonable assumption would be that the change in concentration of majority carrier between the bulk and interface is exponential. This situation can be illustrated by an exponential bending in the conduction band (Fig. 4). Similarly, the valence band must also bend in a similar fashion to maintain the difference between the conduction band and valence band, which is equal to the band gap of the semiconductor. Since difference between the Fermi level and conduction band is related to the magnitude of dopant's concentration, this difference must be maintained even after the formation of band bending. However, it is difficult to present it pictorially. Hence, we compromise by depicting it as a horizontal broken line.

The position of the conduction band on the vacuum scale also reflects the magnitude of electron affinity of the semiconductor. This value is not altered because the semiconductor has come in contact with a redox electrolyte. Assuming the vacuum level to follow a similar bend as the conduction band solves this problem (Fig. 4). Considering all these factors, the entire energy diagram of the semiconductor after making its contact with redox electrolyte may be expressed as shown in Fig. 5. The variation in charges and potential within the space charge region are shown in Fig. 6.

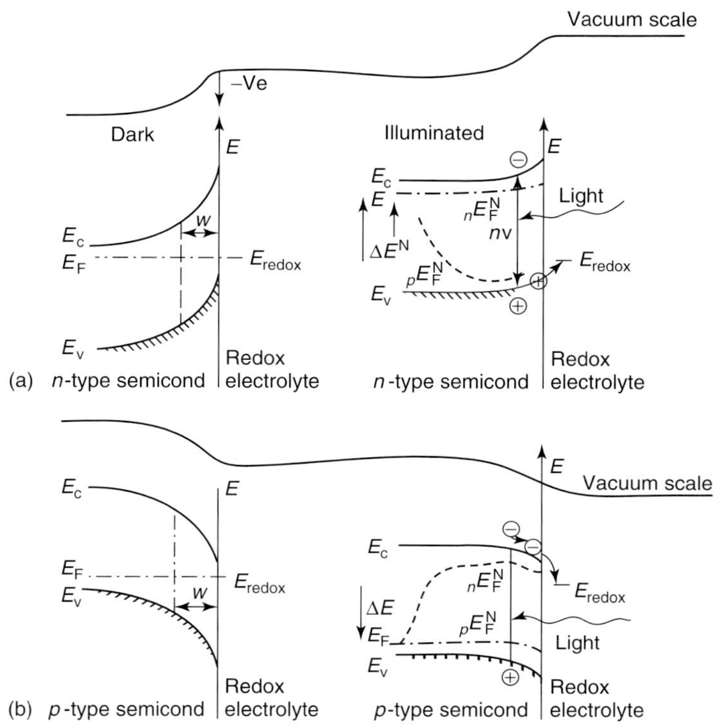

Fig. 5 Nature of band bending in the dark and under illumination with (a) an *n*-type semiconductor-electrolyte and (b) a *p*-type semiconductor–electrolyte interface. Vacuum scale changes its shape to match the shape of band bending. Variation in Fermi level for minority carriers is shown by broken line (for illuminated semiconductor).

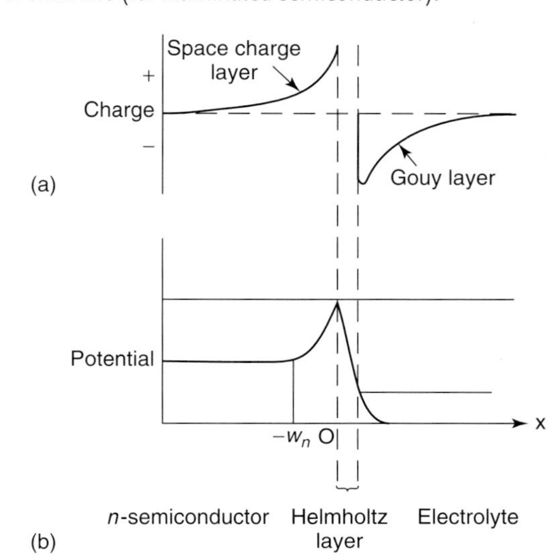

Fig. 6 Schematic diagrams showing charge and potential distribution in an *n*-semiconductor–electrolyte interface. (a) Variation in charge distribution within the space charge region and in Gouy layer. (b) Variation in potential within the space charge region, and electrolyte. In Helmholtz layer it changes linearly, whereas in Gouy layer it decreases exponentially.

4.1.6 What Kind of Force Exists in this Region?

As per this model, minority carrier concentration at $x = -w_n$ would be larger than the intrinsic carrier concentration. Moreover, it is also reasonable to assume that intrinsic carriers are homogeneously distributed in the bulk region. Therefore, the total magnitude of charge present at $x = -w_n$ must also decrease exponentially until its concentration becomes equal to the intrinsic equilibrium concentration (Fig. 4d). Let us assume that at $x = -L_p$ we establish such a position, then the depth between $x = -w_n$ and $x = -L_p$ is called the *diffusion region*. In the diffusion region, thus we have an exponential concentration gradient established (Fig. 4d).

What are possible reasons that prevent majority carrier to increase to an infinite value at $x = 0$? There must be an opposing force operating to establish an equilibrium condition. What is that force? When a semiconductor comes in contact with redox electrolyte, the majority carrier tends to move toward the interface. When these carriers move toward the interface, they create similar number of oppositely charged carriers (which can be termed as minority carriers). These minority carriers oppose the flow of majority carriers toward the interface. Since no mobile carriers can be stationed in the space charge region, the concentration of minority carriers has to be established in the diffusion region only. Thus, we see that the force that allows establishment of an equilibrium condition (or the force that opposes the flow of majority carrier toward the interface) is the force that allows the flow of freshly generated minority carriers toward $x = -\alpha$. The flow of minority carrier is also referred to as *drift flow*. When these two confronting forces become equal in magnitude, an equilibrium condition is established. This condition limits the concentration of majority carrier at the interface.

4.1.7 Magnitude of Potential Developed at the Interface

What would be the magnitude of potential developed between the charges accumulated at the interface ($x = 0$) and at plane $x = -w_n$ (or for p-type semiconductor at $x = -w_p$)? The driving force for these carriers depends on the magnitude of the potential difference between the Fermi level of the semiconductor and the redox potential of the electrolyte (i.e. $E_F - E_{F(redox)}$). This potential is known as the *contact potential* (θ). Can we fabricate a PEC cell, which gives a contact potential equal to the band gap value of the semiconductor? In other words, can we form a PEC cell with a semiconductor (whose $E_c \approx E_F$) and redox electrolyte (whose $E_{redox} \approx E_v$), such that the contact potential $(\theta) = E_g$? The approximate Fermi level of the semiconductor (i.e. the intrinsic semiconductor) is approximately equal to half the band gap of the semiconductor (i.e. $\frac{1}{2}E_g$). Therefore, redox electrolyte cannot lower the Fermi level of n-type semiconductor beyond $\frac{1}{2}E_g$. This condition puts a restriction to the maximum achievable contact potential (θ), and is equal to $\frac{1}{2}E_g$ value. This also suggests that for a given semiconductor, the most suitable electrolyte would be the one that has a redox potential that is almost equal to the intrinsic Fermi level of the semiconductor.

In conclusion, we see that when a semiconductor comes in contact with a redox electrolyte, two types of regions are formed: a space charge region of width "w" and diffusion region of width "L," in addition to its bulk region. In the bulk region, there is no accumulation of charge.

4.1.8
Effect of Junction Illumination

We shall now examine the effect of illumination of these three regions separately. When photon of energy greater than band gap (known as band gap light) falls in the space charge region, electrons from the valence band get excited to the conduction band, leaving behind a positive charge (i.e. hole) in the valence band. These carriers experience an electrostatic potential in the space charge region. The magnitude of the electrical field is very high in this region. For example, with a contact potential of 1 V and space charge width of 1000 Å, an electrostatic field of around 10^5 V/cm is formed. This field is strong enough to prevent the recombination of photogenerated electron/holes pairs in the space charge region. As a result, photogenerated holes migrate toward the interface and electrons migrate toward the diffusion region (Fig. 5). When illumination is conducted in the diffusion region, electron/hole pairs also may be formed. In this region, there is only equilibrium drift current (established due to flow of minority carrier toward the bulk of semiconductor) and no electrostatic field. Therefore, photogenerated electrons/holes of diffusion region experience only the drift current as a driving force to prevent their recombination. To maintain the equilibrium condition, electrons are directed to move toward the bulk and holes toward the interface. Drift current being not very strong, the majority photogenerated carriers in this region undergo recombination. Finally, if bulk region is illuminated, photogenerated electrons/holes would undergo 100% recombination, as this region has no force of any kind to prevent them from recombining. Thus, the net result of these illuminations is that photogenerated minority carriers accumulate at the interface and photogenerated majority carriers accumulate at the bulk of the semiconductor.

While photogenerated holes at the interface perform oxidation of the redox electrolyte, photogenerated electrons move via the load toward the counter-electrode to perform reduction of the redox electrolyte (Fig. 1). Thus, electrical energy is created from light energy.

4.1.9
Efficiency of Conversion of Light Energy into Electrical Energy

After constructing a PEC cell, it is necessary to find the efficiency of conversion of light energy into electrical energy. For this purpose, a PEC cell is connected with a resistance (as a load), variable D.C. power source and an ammeter in series (Fig. 7a). A current-voltage characteristic is measured by applying different potentials to the PEC cell and measuring the corresponding current. This is called *dark current* (Fig. 7b). The PEC cell is then illuminated with a constant source of light. The current-voltage characteristic is again measured as before. This current is the photocurrent (Fig. 7b). It is observed that the dark current is constant in the reverse bias condition. This current is known as the *saturation current* (I^{o}). The magnitude of the saturation current ($I^{\mathrm{o}}_{\mathrm{ph}}$) increases when the PEC cell is illuminated. The difference between the two saturation currents is the photocurrent (I_{ph}) that is produced from the PEC cell. However, the efficiency of power conversion by the PEC cell is calculated from the current-voltage characteristics, plotted in the fourth quadrant. It is observed that

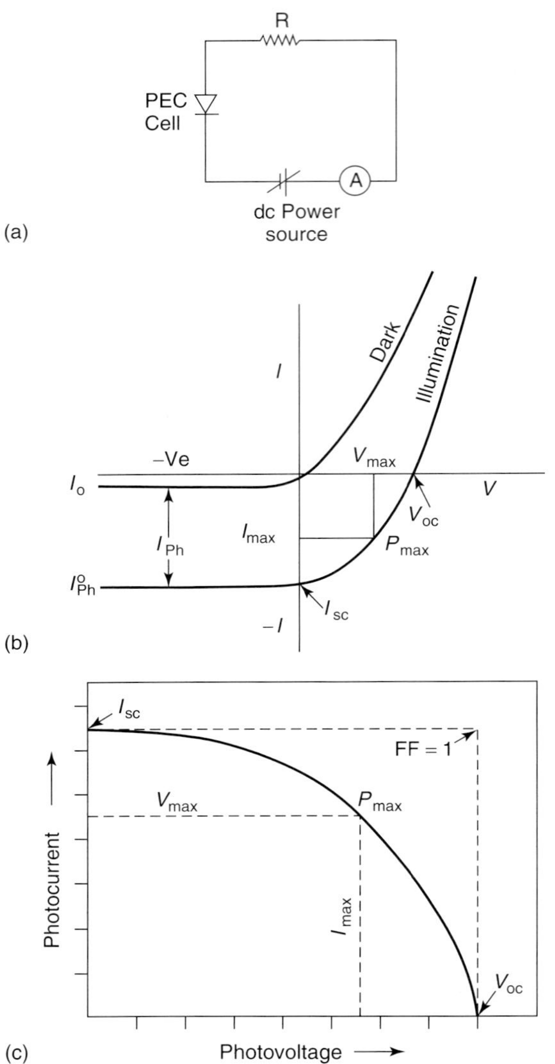

Fig. 7 Schematic diagram for (a) measuring current-voltage characteristics of a PEC connected in series with a load (R) and a variable dc power source and an ammeter (A). (b) current-voltage characteristics of a PEC cell under dark and illuminated condition (c) current-voltage characteristics (Fig. 7b) of a PEC cell replotted assuming its photocurrent to be a positive quantity.

the maximum photopotential is obtained when photocurrent is zero. This potential is known as the *open-circuit potential* (V_{oc}). Similarly, a maximum photocurrent is observed when photopotential is zero. This photocurrent is known as *short-circuit current* (I_{sc}). One would get zero power by working at these values. To find the maximum power, which can be drawn from the PEC cell, a square or rectangle with maximum area is drawn (Fig. 7b,c). The intercept of the square at the current-voltage curve is the magnitude of maximum power (P_{max}) that can be drawn from the PEC cell. Similarly, the intercept on the potential axis is the magnitude of the maximum photopotential (V_{max}) and intercept on current axis is the value of maximum photocurrent (I_{max}) that can be drawn from the PEC cell. These characteristics are obtained when the PEC cell is connected in series with a load of resistance (R). To determine the maximum power load that can be connected to the PEC cell, the current-voltage curve is drawn for various resistances, and the corresponding P_{max} is determined. The resistance that gives the maximum P_{max} is taken as the *maximum load resistance* (R_{max}) that can be connected in series with a PEC to drive the maximum power. The power efficiency of the PEC cell is calculated from the ratio of P_{max} to the power input (P_{in}) to the cell. The nature of the junction formed between the semiconductor and the electrolyte can also be obtained from the current-voltage characteristic (Fig. 7b,c). This is obtained from a factor known as *fill factor* (FF), which is given by:

$$\mathrm{FF} = \frac{I_{max} V_{max}}{I_{sc} V_{oc}} \tag{8}$$

FF decides how well the curve approximates a rectangle. In an ideal condition FF = 1. Generally, current-voltage curve instead of plotting in fourth quadrant is plotted in the first quadrant (Fig. 7c). Note, that the photocurrent, although shown as a positive current in figure, depends whether an n-type (i.e. negative current) or p-type (i.e. positive current) semiconductor is utilized.

4.1.10 Factors Considered in Selecting a Semiconductor

From the previous discussions, it is clear that material for a PEC cell must be able to absorb band gap light in the space charge region. In other words, the reciprocal of the *absorption coefficient* (i.e. $1/\alpha$) of band gap light preferably should be equal to the space charge width (w). While the material should have large space charge width ($\approx 1/\alpha$) to ensure entire absorption of band gap light in the space charge region, the diffusion region should also be large. This is because the magnitude of photocurrent flowing through the semiconductor depends on the width of the diffusion region. It will be seen in the forthcoming section that concentration of dopant as well as the redox electrolyte forming the junction can control the width of the space charge region. Moreover, the FF should also approach to unity. Both the space charge width and the FF depend mainly on quality of semiconductor-electrolyte junction. However, width of the diffusion region is the specific property of the material. Furthermore, the contribution of redox electrolyte in giving a better efficiency of a PEC cell is significant and is later discussed.

4.1.11 Energy Levels in Redox Couples

4.1.11.1 How Do We Compare Redox Potential with Fermi Level of a Semiconductor?

At the gas-solid interface, the vacuum level is taken as the reference energy level; the energy of an electron in vacuum is taken as zero (i.e. $E_{vaccum} = 0$). The energy of a bound electron in the solid is referred as a negative value in this reference scale. In electrochemistry, the standard zero level of the energy is the potential of hydrogen ions (at unit activity), which is in equilibrium with hydrogen gas at one atmosphere pressure. The electrode potential of any other redox electrolyte is referenced to this potential. With this hydrogen scale, the value of redox potential can be either positive or negative. Conventionally, positive redox potentials are shown below zero, whereas

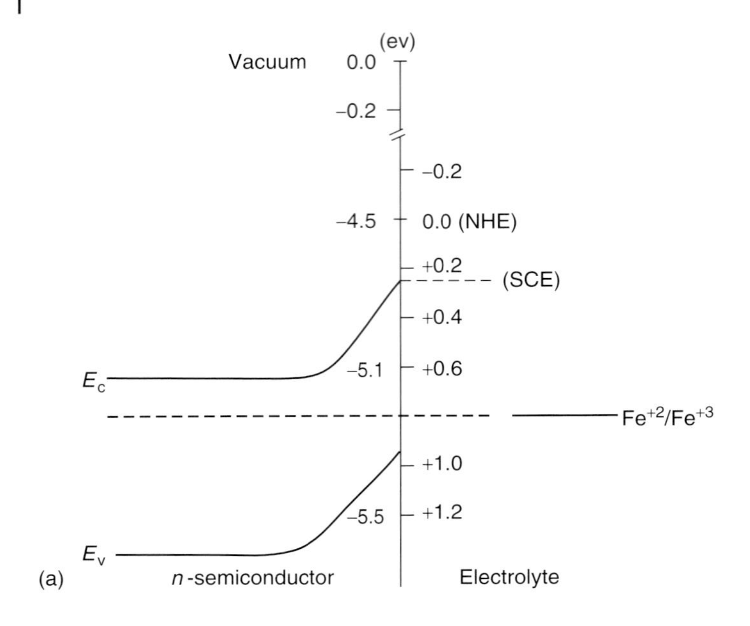

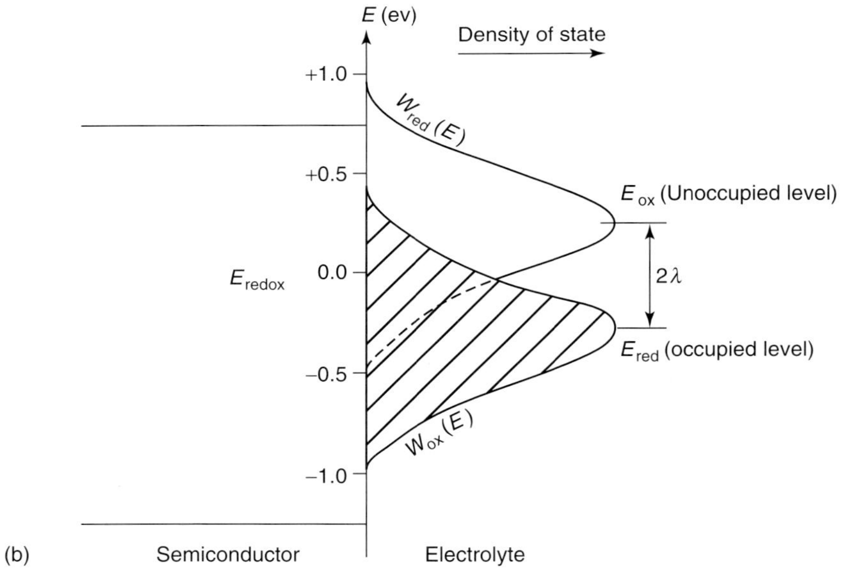

Fig. 8 Comparison of energy levels of redox electrolyte with that of semiconductor. (a) Band diagram of a semiconductor (in vacuum scale) and redox potential of electrolyte (in normal hydrogen scale (NHE)). Redox potential is shown to possess one value for its oxidized/reduced species. (b) Energy levels of redox electrolyte showing its distribution and variation in energy for oxidized and reduced levels.

negative redox potentials are shown above the zero (Fig. 8a).

When dealing with semiconductor-electrolyte junction, we have to consider both of these energy scales. Therefore, there is a need to find out the relationship between them, so that one can draw useful conclusion from the band diagram of the semiconductor–electrolyte interface (Fig. 8a).

Considering the free energy changes associated with the reduction of silver, one can derive the relationship between E_{vacuum} and the NHE. The energy involved with various reactions can be presented by considering the following four steps [8]:

$$\begin{array}{lll} Ag^{+}_{(g)} + e^{-} & = Ag_{(g)} & -7.64\ eV \\ Ag_{(g)} & = Ag_{(s)} & -2.60\ eV \\ Ag^{+}_{(aq)} & = Ag^{+}_{(g)} & +4.96\ eV \\ Ag_{(s)} + H^{+}_{(aq)} & = Ag^{+}_{(aq)} + \frac{1}{2}H_2 & +0.80\ eV \\ \hline H^{+} + e^{-} & = \frac{1}{2}H_2 & -4.48\ eV \end{array} \quad (9)$$

In this analysis, "g" indicates the gas phase, "s" the solid, and "aq" the aqueous solution phase. Thus, the relation between the two scales, namely, E_{vacuum} and NHE can be represented as:

$$E_{F(redox)}\ (eV) = E^{o}_{redox(NHE)}\ (volt) - 4.48\ (eV) \quad (10)$$

$E^{o}_{redox(NHE)}$ and $E_{F(redox)}$ are the potentials of a redox couple on the electrochemical NHE scale and vacuum scale, respectively.

4.1.11.2 Do the Energy Levels for Reduction and Oxidation Process Possess the Same Value?

The energy level of the redox electrolyte shown in Fig. 8(a) is an oversimplified model because it indicates that the energy levels for the reduced and oxidized species of the redox couple, $Fe^{2+/3+}$, are the same. This also means that if electrons were to be transferred from electrolyte to the semiconductor and back, it would come back to the same energy level of the electrolyte. Since the charge density on ions is different, the extent of hydration differs. Therefore, it has been proposed that the energy states of reduced and oxidized species should not be discrete but are broadened into thermal distribution of states caused by thermal fluctuation, interaction with solvents, and interaction with directly bonded ligands. The position of the localized energy states in a polar solvent depends greatly on whether the state is occupied or not because of the strong interaction with the solvent. Fluctuation in the structure of hydrated ions thus extends the energy of the electronic quantum states in the redox couple around two most probable values. The energy change caused by the variation in solvation shell can be represented in terms of harmonic oscillators. On the basis of the concepts proposed by Levich [13] and the distribution function proposed by Gerischer [14–16] a model for the energy levels in solution is discussed in this section. In this model, the ground state vibration level of only the first electronic state is considered relevant to electron transfer processes at the electrode. The energy change caused by vibrations in the solvent shell is represented in terms of harmonic oscillations. The potential energy is considered to vary around the ground state to an equal degree. The resulting distribution functions for oxidized and reduced species are given as:

$$W_{ox}(E) = \frac{[\exp\{-(E - E^{o}_{ox})^2/4\pi kT\lambda\}]}{(4\pi kT\lambda)^{1/2}} \quad (11)$$

$$W_{red}(E) = \frac{[\exp\{-(E - E^{o}_{red})^2/4\pi kT\lambda\}]}{(4\pi kT\lambda)^{1/2}} \quad (12)$$

where E^o_{ox} and E^o_{red} are the most probable energies for the unoccupied and occupied quantum states in solution, respectively. The parameter λ represents the activation energy for the process of transferring the solvation shell structure from equilibrium condition of one species to the most probable structure of the other. This is also called *reorientation energy*. The density of states for reduced and oxidized species at a particular energy are given as

$$D_{red}(E) = C_{red} \times W_{red}(E) \quad (13)$$

$$D_{ox}(E) = C_{ox} \times W_{ox}(E) \quad (14)$$

where C_{red} and C_{ox} are the concentrations of the reduced and oxidized species, respectively. The energy distribution is a Gaussian bell shaped curve (Fig. 8b). In the harmonic oscillator approximation, the reorientation energy of reduced and oxidized species are equal and are given by

$$E^o_{redox} = E^o_{ox} - \lambda = E^o_{red} + \lambda \quad \cdots \quad (15)$$

or,

$$E^o_{ox} - E^o_{red} = 2\lambda \quad \cdots \quad (16)$$

Bockris and Khan [17], however, have suggested that energy is not symmetrically distributed around the central point (i.e. the ground state) but in a Maxwell type distribution.

The most interesting concept derived from these views is that like a semiconductor, the oxidized and reduced species of redox electrolyte are linked with a conduction band (unoccupied electronic state) and a valence band (occupied electronic state). The energy necessary to transfer an electron from reduced to oxidized state is 2λ, analogous to the band gap of a semiconductor E_g.

For an effective charge transfer to occur, the energy level of the semiconductor must match with energy level of the redox electrolyte. For example, an electron transferred from the valence band to redox electrolyte would be facilitated if the level of the former matches the oxidized level of the later. Alternatively, if electrons are to be transferred to the conduction band from the redox electrolyte, the conduction band must match the reduced level of the redox electrolyte. Thus, knowledge of such distribution functions of the redox electrolyte helps in understanding the I–V characteristics of a semiconductor, especially when the λ value of the electrolyte approaches the material's band gap. Under such a condition, the rectifying character of semiconductor-electrolyte junction does not control the nature of anodic and cathodic current, that is, an *n*-type semiconductor would give a cathodic current in addition to the anodic current, and the reverse would be true for a *p*-type semiconductor. With this type of condition, one either has to select an electrolyte whose redox potential (i.e. its Fermi level) is such that only one of the two levels of the redox electrolyte (E^o_{ox} or E^o_{red} as the requirement may be) matches the concerned band edge level of the semiconductor. Alternatively, bias would have to be applied to the electrode to stop the unwanted current.

4.1.11.3 **Effect of Charge Distribution in the Electrolyte**

What is the role of the constituents of the redox electrolyte present at the interface of the semiconductor-electrolyte junction? Choice and optimization of the electrolyte has a substantial effect on PEC solar cells [18–19]. Before a semiconductor is immersed in an electrolyte, anions and cations freely and randomly move in the solution. As a result of this movement, no specific spatial accumulation of ions occurs in the solution. This situation alters as soon as we insert a semiconductor into the electrolyte. Because of the difference between the redox potential of the electrolyte and the Fermi level of the semiconductor, accumulations of the majority carriers occur at the interface of the semiconductor. These majority carriers attract (hydrated) cations toward the interface. Since they may be hydrated, instead of directly coming in contact with the semiconductor, they may approach the semiconductor spaced by water molecule. This situation creates a double layer at the interface: a negative charge at the interface of an n-type semiconductor and a positive charge (i.e. cation) separated by a water molecule, in the solution. Interface of the semiconductor that contains the majority carrier can be assumed as one plane possessing these charges. At a water molecule distance from the interface, we have another plane carrying the positive charge of cations. This plane is termed as *Helmholtz plane*. This double layer (known as *Helmholtz double layer*) thus develops an electrostatic potential. Because of formation of the double layer, as compared to the bulk solution, a large concentration of cations thus gets accumulated at the interface. In a dilute solution, however, it is observed that the concentration of cations (i.e. after the double layer) decreases exponentially to a point where there is no accumulation of charge and the solution behaves almost like the bulk solution. The region between the Helmhotlz plane and the plane where there is no accumulation of charge is called the *Guoy-Chapman diffusion region* (Fig. 6a). The variation in potential in Helmholtz region is expected to be linear, but it is exponential in Guoy-Chapman region (Fig. 6b). Thus, as with a semiconductor, we can visualize the formation of three charge regions in the solution: the Helmholtz region, Gouy-Chapman region, and the bulk region, (where there is no accumulation of charge).

4.1.12
Capacitance of the Space Charge Layer

What is the role of the various regions formed on the property of the semiconductor-electrolyte junction? Can we take some advantages from them to procure some useful parameters of the semiconductor-electrolyte junction? We shall examine this briefly.

It has been seen earlier that knowledge of band edge positions of the semiconductor and distribution of energy in the redox system are helpful in selecting a suitable redox electrolyte to get an efficient charge transfer across the semiconductor–electrolyte interface. The band edge positions of a semiconductor can be determined if we know the value of the *flat band potential* (i.e. Fermi level position) and the band gap of the semiconductor. The flat band position is equivalent to the magnitude of the biasing potential applied to the semiconductor so as to make the space charge region (i.e. depletion region) width (w_n for n-type semiconductor) zero. This potential is equivalent to Fermi level of the semiconductor. One of the easiest methods to determine the flat band potential is by studying a differential capacitance of the semiconductor–electrolyte interface.

We have seen that a semiconductor-electrolyte junction possesses three regions in the semiconductor and three in the electrolyte. These layers act as parallel-plate capacitors in series, so that the resultant reciprocal capacitance is the sum of the reciprocal of the capacitance for all the capacitors.

$$1/C_{\text{(total)}} = 1/C_{\text{SC}} + 1/C_{\text{diff}} + 1/C_{\text{H}} + 1/C_{\text{G}} \quad (17)$$

Capacitance due to the bulk region in an electrolyte and a semiconductor are overlooked because there is no charge accumulation in these two regions. It has also been experimentally observed that magnitude of the capacitance of the diffusion layer in semiconductor (C_{diff}), Guoy-Chapman layer (C_G), and the Helmhotlz layer (C_H) are very high compared to that of the space charge layer (C_{SC}). Under this condition, the reciprocal of capacitance of the space charge layer becomes equivalent to reciprocal of total capacitance of the semiconductor-electrolyte junction (i.e. $1/C_{(total)} = 1/C_{SC}$). Hence, if we can measure the total capacitance formed with the semiconductor-electrolyte junction, understanding of the capacitance of space charge layer can be obtained.

Measurement of the potential drop across the space charge layer yields information about the capacitance formed with the semiconductor-electrolyte junction; this capacitance and its measurement are described in a separate chapter.

4.1.13 Semiconductor-electrolyte Junction Under Illumination

In Sect. 4.4, we examined the effect of the illumination of semiconductor-electrolyte junctions, using a light source of photon energy greater than the band gap of semiconductor. We shall now try to develop a model to get a quantitative estimation of the photocurrent from the magnitude of the concentration of photogenerated carriers.

4.1.13.1 Gärtner's Model

Wolfgang Gärtner [20] derived a relation for photocurrent generated in space charge region and diffusion region after illuminating a semiconductor-metal junction. Since, semiconductor–metal interface behaves similar to semiconductor-electrolyte junction, Gärtner's model has been successfully applied to PEC cell, as well [21–22]. The photocurrent (J_{scl}) due to carriers generated in the space charge region is given as

$$J_{scl} = qI^o(1 - e^{-\alpha w}), \quad \cdots \quad \cdots \quad (18)$$

where q = charge of the electron; I^o = incident photon flux (number of photons $s^{-1}\,cm^{-2}$); α = absorption coefficient for photon of energy $h\nu$ (cm^{-1}); w = space charge width (cm). This is designated as w_n or w_p for n-type and p-type semiconductor, respectively; J_{scl} = photocurrent density ($A\,cm^{-2}$).

Similarly, an equation for photocurrent originated from the diffusion region is given as:

$$J_{diff} = -qI^o\left\{\frac{L_{min}}{1 + L_{min}}\right\}e^{-\alpha w} - qI_{min}\left(\frac{D_{min}}{L_{min}}\right) \quad \cdots \quad (19)$$

where L_{min} = width of diffusion layer (cm). In the n-type semiconductor, it will be due to holes (designated as L_p) where as in p-type due to electron (designated as L_n). I_{min} = equilibrium concentration of minority carriers (cm^{-3}), which is designated as p_o and n_o in n-type and p-type semiconductors, respectively. D_{min} = Diffusion coefficient for minority carrier ($cm^2 s^{-1}$). For holes in n-type it will be designated as D_p and for electrons in p-type semiconductor it will be designated as D_n. J_{diff} = photocurrent due to photogenerated carriers in the diffusion region.

Total photocurrent passing through the junction would be sum of these two components.

$$J_{\text{total}} = \{qI^{\circ}(1 - e^{-\alpha w})\} + \left[-qI^{\circ}\left\{\frac{\alpha L_{\text{min}}}{1+\alpha L_{\text{min}}}\right\} \times e^{-\alpha w} - qI_{\text{min}}\left(\frac{D_{\text{min}}}{L_{\text{min}}}\right)\right] \quad (20)$$

For a large band gap semiconductor, I_{min}, D_{min}, and L_{min} would be very small; hence, their contribution can be neglected. Thus, the Eq. (20) simplifies to

$$J_{\text{total}} = -qI^{\circ}\left\{\frac{1 - e^{-\alpha w}}{(1+\alpha L_{\text{min}})}\right\} \quad \cdots \quad (21)$$

It is important to remember that the polarity of photocurrent is negative, as one would get experimentally (Fig. 7b).

4.1.13.1.1 **Application of Gärtner Model** Equation (21) can be slightly modified to include a term "*quantum yield* (ϕ)" to get information about some useful parameters of semiconductors, for example, band gap, diffusion length, space charge width, and so on. Quantum yield is defined as the ratio of photocurrent (J_{photo}) to the total flux of light (I°) used to illuminate the semiconductor: that is, quantum yield $(\phi) = J_{\text{photo}}/I^{\circ}$.

Equation (21) can thus be rewritten as

$$(1 - \phi) = \frac{e^{-\alpha w}}{1 + \alpha L_{\text{min}}}$$

Taking the logarithm of this equation, we have

$$-\ln(1-\phi) = \alpha w + \ln(1 + \alpha L_{\text{min}}) \quad (22)$$

In Eq. (22), there are two variables, α (absorption coefficient of the semiconductor), which depends on the wavelength of light, and w, (the space charge width), which is related to the dopant's concentration and magnitude of the bias. Therefore, it is useful to examine Eq. (22) by keeping one of these variables constant.

4.1.13.1.2 **When α is Constant during Photocurrent Measurement** This condition is achieved when the junction is illuminated with a monochromatic light and the photocurrent is measured as a function of applied potential (V), using a potential close to the flat band potential. Substituting the value $w = [(2\varepsilon\varepsilon_o/eN_A)(V - V_{\text{fb}})]^{1/2}$ to Eq. (22) we have

$$-\ln(1-\phi) = \alpha\left(\frac{2\varepsilon_o\varepsilon}{qN_A}\right)^{1/2}(V - V_{\text{fb}})^{1/2} + \ln(1+\alpha L_{\text{min}}) \quad (23)$$

Equation (23) suggests that if $-\ln(1-\phi)$ is plotted against $(V)^{1/2}$, a linear graph is obtained. Intercept of the linear plot on the potential axis, when $-\ln(1-\phi) = 0$, would give the value of the flat band potential (V_{fb}) because none of the terms of $\alpha(2\varepsilon_o\varepsilon/qN_A)^{1/2}(V - V_{\text{fb}})^{1/2}$ except $(V - V_{\text{fb}})^{1/2}$ can be zero. A word of caution is necessary in this type of calculation. This equation is valid only under the condition of band bending. Therefore, it is always better to extrapolate the linear plot from the region where $(V - V_{\text{fb}}) > 0$, that is, from the region where forward bias is much greater than the V_{fb} values.

Slope of the linear region where $(V - V_{\text{fb}}) > 0$ gives the value of $\alpha(2\varepsilon_o\varepsilon/qN_A)^{1/2}$ for the wavelength of light used during illumination. If N_A is known from Hall measurements or from capacitance measurement, α can be obtained.

If the experiment is repeated at different wavelengths of light, one can get the value of α for different wavelengths also. This information is extremely useful because $1/\alpha$ is almost equal to the penetration depth of the corresponding wavelength of light. It is desirable to know the magnitude of $(1/\alpha)$ for band gap light, so that efforts could be made to select a suitable redox electrolyte such that $(1/\alpha) \sim w$. In addition, we can also calculate L_n from magnitude of the intercept (i.e. $\ln(1 + \alpha L_\mathrm{n})$) of the linear plot on y-axis. From L_n the *lifetime of the carrier* (τ) can be calculated by using the following equation

$$L_\mathrm{n} = \left(\frac{\mu_\mathrm{n} k T \tau}{\mathrm{q}}\right)^{1/2} \quad (24)$$

μ_n the mobility of the carrier, can be obtained from Hall measurement.

From these measurements, we can thus evaluate the flat band potential (which is equal to E_F for heavily doped semiconductor), absorption coefficient (α) for band gap light, diffusion length (L) and lifetime of the carrier (τ).

4.1.13.1.3 **When '*w*' is Constant During Photocurrent Measurement** Using $\alpha = \{A(h\nu - E_\mathrm{g})^{n/2}\}/h\nu$ and keeping w constant, Eq. (22) becomes

$$-\ln(1 - \phi) = \left\{\frac{A(h\nu - E_\mathrm{g})^{n/2}}{h\nu}\right\} \times w + \ln(1 + \alpha L_\mathrm{min}) \quad (25)$$

When $(\alpha L_\mathrm{min}) \ll 1$, Eq. (25) can be rearranged to get

$$-[\{\ln(1 - \phi)\}h\nu]^{2/n} = Aw(h\nu - E_\mathrm{g}) \quad (26)$$

In order to utilize Eq. (26), a semiconductor-electrolyte junction is illuminated by different wavelengths of light, keeping the applied potential to a PEC cell constant (i.e. under a constant forward potential). A plot of $[\{\ln(1 - \phi)\}h\nu]^{2/n}$ versus $h\nu$ would thus result in a linear graph. Let us assume that the value of the intercept on the x-axis at $[\{\ln(1 - \phi)\}h\nu]^{2/n} = 0$ is 's'. At this condition, none of the terms of $Aw(h\nu - E_\mathrm{g})$ can be made zero except $(h\nu - E_\mathrm{g})$. This would mean that the value of $(h\nu)$ (i.e. the intercept on $(h\nu)$-axis "s") should be equal to E_g. If we get the linear plot with $n = 1$, the band gap is a direct type and if $n = 4$, it is an indirect one. Some semiconductors may give linear plot for both values of n, indicating simultaneous presence of both types of band gap.

Thus, application of the Gärtner model to data obtained from photocurrent versus potential measurement yields several important parameters of a semiconductor-electrolyte junctions (e.g. E_g, V_fb, L_min, τ, and w).

4.1.14 Effect of Counter-Electrode

Besides the considerations on the parameters of the semiconductor and the redox electrolyte, performance of the counter-electrode also matters for designing an efficient PEC cell. The counter-electrode in a PEC cell is a charge carrier collecting material, and catalyzes the electron transfer, but is not chemically involved in the electrochemical reaction.

The counter-electrode material should be electrically conducting, economical, and chemically stable in an electrolyte. It should minimize polarization losses during exchange of electrons at the interface. In other words, polarizability of the counter-electrode should be small in a given redox electrolyte, indicating that the electrode reaction occurring at the counter-electrode is reversible and fast. This maximizes the potential obtained from the semiconductor–electrolyte interface.

For many redox reactions, noble metals such as Pt, Au have been preferred as counter-electrode by electrochemists. However, those are expensive for terrestrial application, and materials such as carbon (graphite, glassy carbon), and degenerated transparent [23] semiconductors (SnO_2 or ITO) have attracted the attention of scientists for their use as counter-electrodes. For aqueous polysulfide redox reactions, Hodes and coworkers [24] have shown that the use cobalt sulfide as an electrocatalyst gives much better performance than high surface area platinum or carbon electrodes.

When a counter-electrode other than inert platinum electrode is used, it is necessary to confirm whether the electrochemical reaction occurring at the counter-electrode is as reversible as a platinum counter-electrode [25]. Sharon and coworkers [26] has studied the efficiency of three types of counter-electrodes (ITO, Pt and Platinum-coated ITO) in a PEC cell of configuration: α-PbO/Fe $(CN)_6^{4-}$ (0.1M), $Fe(CN)_6^{3-}$ (0.01M) pH 9.2/(counter-electrode). They observed that while platinum counter-electrode gives the best performance (Fig. 9a), surface modification of ITO counter-electrode with platinum (Fig. 9b) gives almost similar performance to that obtained with platinum. However, untreated highly conducting ITO electrode gives considerably low power characteristics (Fig. 9c). Transparent counter-electrode, in addition, avoids the problem of light absorption by the redox electrolyte [25]. With a transparent counter-electrode, it is possible to place the semiconductor very close to the counter-electrode, which also serves as a window for the PEC cell. This backwall configuration is especially helpful when very concentrated electrolytes are used [27].

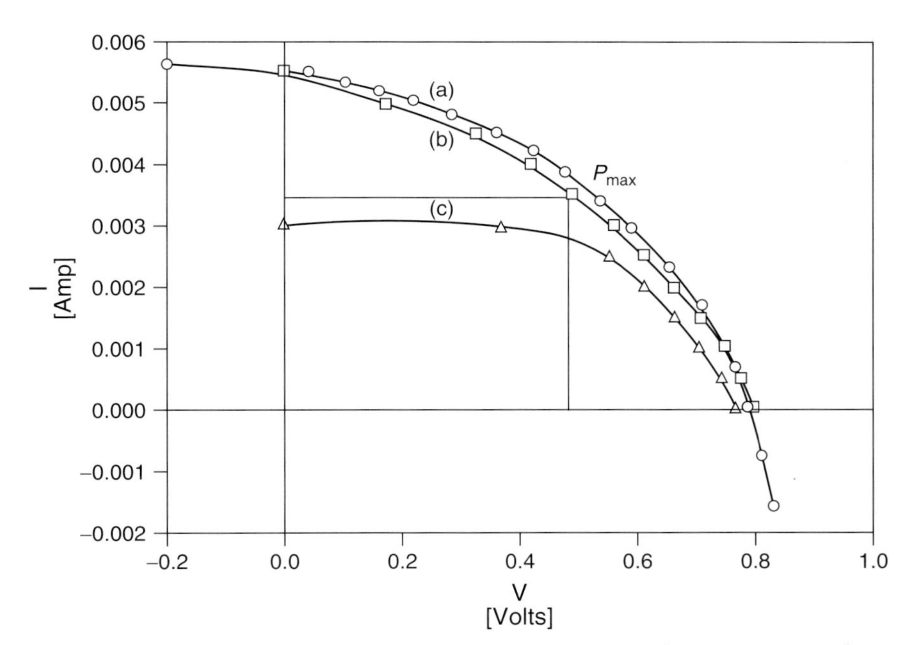

Fig. 9 Current-voltage characteristic of a PEC cell (α-PbO/Fe $(CN)_6^{4-}$ (0.1M), $Fe(CN)_6^{3-}$ (0.01M) pH 9.2/(counter-electrode). Three types of counter-electrodes were used: (a) conducting ITO electrode, (b) surface modified ITO electrode using platinum and (c) platinum electrode.

4.1.14.1 Effect of Surface States

It is important to appreciate that even if a material possess desirable characteristics, as discussed in the previous sections, it may not exhibit the desired efficiency of light energy conversion into electrical energy. For example, in spite of semiconductor and redox electrolyte possessing desired Fermi levels experimentally determined contact potential might be much less than the anticipated value. Presence of surface state is one of the factors responsible to show these deficiencies. Hence, efforts are required to minimize the adverse effects of surface states.

Surface states mainly originate from the lack of balance in the valence states of atoms/ions present at the surface as compared to atoms/ions present beneath the surface of thin film or single crystal. Consequently, atoms/ions present at the surface possess either more or less number of electrons as compared with those present in the bulk material. This situation creates surface states of various types. All of them behave like a donor or an acceptor sites on the surface. Thus, irrespective of the type of surface states, they tend to trap photo-generated carriers at the surface of the semiconductor. These trapped carriers are released at a slow rate. Moreover, these surface states can also cause pinning of the Fermi level. Pinning of the Fermi level takes away the freedom of Fermi level to adjust its position while attaining the equilibrium condition. Thus, pinning of the Fermi level decreases the magnitude of anticipated contact potential. Hence, for an efficient PEC cell, the semiconductor should be free from surface states. It is impossible to fully eliminate surface states. However, this effect can be minimized. Since these surface states behave either as a site possessing positive or negative charge, it is possible to neutralize their charges by controlling pH of the solution.

Sharon and coworkers [28] have developed a simple technique to find the pH at which the effect of surface states is minimum. Their model also gives information regarding the magnitude of other parameters such as flat band potential and band bending potential. The experiment is based on the fact that concentration of a majority carrier at the surface of the semiconductor (in dark condition) is controlled by the extent of band bending formed in the space charge region. Similarly, freedom of majority carrier to move freely over the surface also depends on the nature and magnitude of the surface states present over the surface. Therefore, the magnitude of surface states present on the surface influences the surface conductivity. The activation energy for the surface conductivity should thus also be related to the nature and magnitude of the surface states.

For surface conductivity measurement, two ends of a thin film of semiconductor are connected with platinum wires. The entire surface except the top surface of the film is covered with an insulating paint. This film is then immersed in a solution of known pH whose temperature is controlled very accurately. Surface resistance is measured versus temperature (normally in the range of $-10\,°C$ to $+25\,°C$). This experiment is repeated at different pH values. Log of the surface conductivity is plotted versus reciprocal of the absolute temperature to get the activation energy. This calculation is carried out for each pH. Finally, a plot of activation energy versus pH is plotted, which shows either a minimum with n-type semiconductor or a maximum with p-type semiconductor. These pH values correspond to point of zero charge (pzc) of the material.

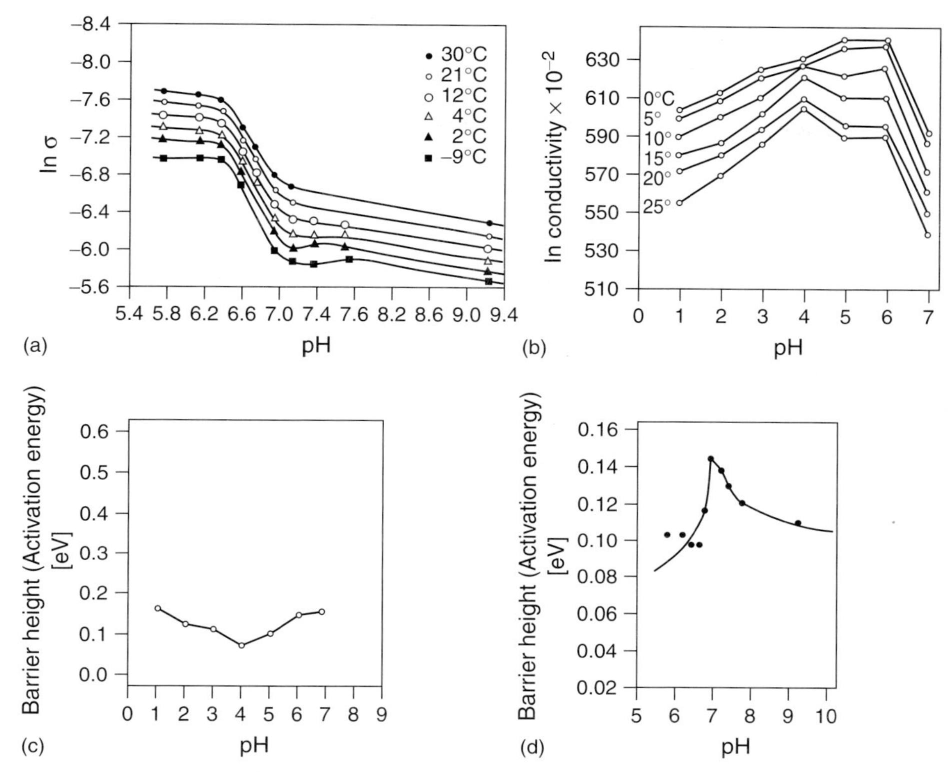

Fig. 10 Surface conductivity measurements of thin films of α-Fe_2O_3 and *p*-SnSe[21]. Log of surface conductivity versus pH at different temperatures for (a) *n*-α-Fe_2O_3 (b) *p*-SnSe. Activation energy of surface conductivity versus pH for (c) *n*-α-Fe_2O_3 and (d) *p*-SnSe.

In Fig. 10, typical results of thin films *n*-α-Fe_2O_3 (a,c) and *p*-SnSe (b,d) are shown. *n*-α-Fe_2O_3 gives a minimum at pH 4.1, which is equivalent to its *pzc*. A reverse behavior is obtained with *p*-SnS showing a maximum at pH 7.5, which is equivalent to its *pzc*. This is the only nondestructive method available to get information about pH that corresponds to *pzc* for large area thin-film semiconductor material.

A PEC cell operating at the pH equivalent to the *pzc* will experience the least hindrance caused by the surface states, and may be able to show its anticipated PEC behavior. PEC operating at *pzc* would be able to show its predicted contact potential because the surface of the semiconductor would have no excess charge for pinning the Fermi level. Moreover, photogenerated carriers will also not be trapped by the surface states as their charges are neutralized by pH of the solution.

4.1.15
Photodecomposition of Semiconductor

In the dark many semiconductors are stable in contact with a given electrolyte, but start decomposing on illumination of the junction. This and energy conversion efficiency are the most serious problems faced in PEC devices. If the photo-generated minority carriers are not rapidly used in oxidation or reduction by the electrolyte immediately, they have a tendency to react with the semiconductor material. The semiconductor electrode therefore starts corroding. PEC etching is being dealt in a greater depth in a separate chapter. However, the easiest way to establish PEC stability of the semiconductor is by measuring the photocurrent of a PEC cell over a longer period under continuous illumination. If semiconductor is stable, the photocurrent will show a constant value over a period of illumination.

Unfortunately, most of the low band gap materials studied so far have been found to undergo such corrosions. Therefore, there is a need to develop newer materials for developing a photoelectrochemically stable semiconductor.

4.1.15.1 Materials for PEC Cell

Because of the limitation of space, it is difficult to discuss details of materials studied for various types of PEC cell. Some of the materials and the redox electrolytes used for studying PEC cells are tabulated in Tables 1 and 2. Nevertheless, Sharon

Tab. 1 Some of the chalcogenide materials studied for developing PEC cell in various redox electrolytes

Material	*Electrolyte system*	*Reference*
n-CdS	NaOH,S,Na_2S; NaOH,Na_2Te; NaOH,$Na_2(SCH_2COO)S$; KCl, $KFe(CN)_6^{3-/4-}$; NaI, I_2 in CH_3CN	31–33, 35, 39
n-CdSe	NaOH, Na_2S, S; NaOH, Na_2Se; NaOH, Na_2Te; NaOH, Na_2S,S, Se; NaOH,$Na_2(SCH_2COO)S$, Se; KOH,Na_2S,S_n; $Fe(CN)_6^{3-/4-}$ pH $>$ 8, Cesium polysulphides, S,$CuSO_4$	31–33, 36–38, 40–42
n-CdTe/*p*-CdTe	NaOH-Na_2Se, NaOH-Na_2Te, K_2Se-Se-KOH	34, 35, 40, 43
n-$CdInS_2$, $CuInSe_2$, $CuInS_2$	NaOH-Na_2S-S	44–45,100
n-$Re_6Se_8Cl_2$	I_3^--I^--H_2SO_4	46
n-$CdIn_2Se_4$	KOH-S-Na_2S	47
n-$CuIn_5S_8$	Sulfide, Polysulfide,	48
n-WS_2	Halide solution,	49
p-WS_2	$CH_3CN/(n\text{-}Bu_4N)ClO_4$,	50
n-MoS_2	Et_2N-halide ion in CH_3CN, NaI-I_2(aq), Fe^{2+}/Fe^{3+},	51, 55, 56
n-$MoSe_2$	I/I_2, Fe^{2+}/Fe^{3+}, $(Et_4N)X/2$ in CH_3CN; X = halogen, HBr/Br_2 pH = 1	52–53, 58
n-WSe_2	$Fe(CN)^{3-/4-}$, NaI/I_3^--$Na_2SO_4^-$; H_2SO_4,pH = 0	52, 80
p-WSe_2	$Fe^{2+/3+}$-H_2SO_4	59
n-SnS	Ce^{4+}/Ce^{3+}, 0.5 M; H_2SO_4	70
n-Bi_2S_3	NaOH-Na_2S-S	60–61

Tab. 2 Some of the phosphide/oxide and other materials studied for developing PEC cell in various redox electrolytes

Compound	*Electrolyte*	*Reference*
n-InP	$NaOH/Fe(CN)_6{}^{3-/4-}$; $NaOH$-Na_2Te, VCl_2-VCl_5-HCl	62
p-InP	$NaOH/Fe(CN)_6{}^{3-/4-}$; $NaOH$-Na_2Te, VCl_2-VCl_5-HCl	63
p-GaP	VCl_2-$VCl5$-Na_2SO_4; $pH = 2.5$	64
ZnP_2, Zn_3P_2	$NaOH/Fe(CN)_6{}^{3-/4-}$	65
Fe_2O_3	$NaOH/Fe(CN)_6{}^{3-/4-}$	66
$YFeo_3$	$Fe(CN)_6{}^{3-/4-}$, $pH = 7$ & 9	67
Pb_3O_4	$IO_3{}^-/I^-$	68
BaTiO3	Ce^{3+}/Ce^{4+}, Fe^{2+}/Fe^{3+}; $pH = 2$	69
$PbO_{(2-x)}$	$Fe(CN)_6{}^{3-/4-}$, $pH = 9$	71
n-GaAs	$Fe^{2+/3+}$; $I^-/I_3{}^-$; S^{2-}/S_n^{2-}; $Fe(CN)_6{}^{3-/4-}$	43, 74–76, 79–80
p-GaAs	$I^-/I_3{}^-$ $pH = 0$; KCl-VCl_2/VCl; Eu^{3+}	57, 78, 81–83
p-Si	HCl-VCl_2/VCl_3	84–87
n-Si	$FeCp_2$; *n*-Bu_4N-ClO_4 in ethanol; $FeCp_2$ = ferrocene; $I^-/I_3{}^-$-$FeCl_2/FeCl_3$	54, 87, 89–93
$GaAs_{0.72}P_{0.28}$	Nonaqueous electrolyte	78, 81

Tab. 3 Chronological improvements in solar to electrical conversion efficiencies of some regenerative PEC solar cells

Year	*Solid/dissolved salts*	*% Efficiency*	*Reference*
1998	$AlGaAs/Si/V^{2+/3+}$	~19	96
1991	Nano TiO_2 (dye sensitized)/$I_3{}^-$	~7	97
1990	*n*-$CdSe/KFe(CN)_6{}^{3-}$	~16	98
1987	*n*-$GaAs/Se^{2-}$ (Os^{3+})	~15	99
1985	*n*-Cd(Se,Te)/S^{2-}	~13	100
1982	*n*-$CuInSe_2/I^-$	~9	101
1982	*n*-WSe_2/I^-	~10	102
1981	*n*-$MoSe_2/I^-$	~9	103
1981	*p*-InP/$V3^+$	~11	104
1978	*n*-$GaAs/Se_2$-(Ru^{2+})	~12	105

and coworkers [29, 30] have extensively reviewed these materials.

Examples of chronological improvements in high solar to electrical conversion efficiency PEC's are summarized in Table 3. Highest reported conversion efficiencies include more than 9% for dye sensitized thin film of *n*-TiO_2 (nanostructure) in polyiodide electrolyte [97] and more than 16% for single-crystal *n*-CdSe in cyanide modified $K_3Fe(CN)_6{}^{3-}$ electrolyte [98]. The synthesis of single-crystal semiconductor material is not cost effective, being both time intensive and limited to formation of smaller surface area of materials. A variety of techniques have been developed to provide polycrystalline or amorphous thin-film semiconductors. A particular advantage of electrodeposition (plating) techniques is a one-step chemical synthesis of the semiconductor. PV, as well as PEC solar cells requires the use of thin-film semiconductors to be cost competitive. Each of these techniques consumes relatively small amounts of material and can be applied to develop large surface area electrodes. Modes of film preparation include slurry techniques, chemical vapor deposition, wackercast, rf sputtering, vacuum coevaporation, sol gel, chemical bath deposition, hot pressing, and electroplating. These techniques have been used to prepare thin films of *n*-CdS, *n*-CdSe, *n*-CdTe, *n*-GaAs, *p*-InP, *n*-Si and *p*-Si, *n*-TiO_2 and used in PEC solar cells [97, 105].

Advantageously, thin film PEC's have exhibited photocurrent lifetimes significantly greater than single-crystal PEC's. This phenomenon has been attributed to the high microscopic surface area, and resulting lower effective microscopic photocurrent density, passing through the thin-film devices. The longest lifetime PEC device to date, slurry deposited thin-film *n*-Cd (Se, Te) electrodes immersed in modified polysulfide solutions, has demonstrated outdoor operation for approximately one year. In this system, photoelectrode and electrolyte stability was achieved by optimization of the solution alkalinity, using alternative cations, and modifying the ratio of dissolved sulfur to sulfide in solution [106]. The recent advance in dye-sensitized PEC devices requires future efforts to extend the lifetimes of the requisite dyes [97].

But unfortunately none of these materials show a long-term PEC stability to warrant making of a commercially viable PEC cell. Interestingly, it has been hinted that phosphides of Co, Fe, Mo, Ni, V, and W (band gap in vicinity of 1 eV) may be photoelectrochemically stable. Among the oxide materials, PbO_2 ($E_g = 1.6$ eV) is worth exploring because it is highly stable in acidic medium. Unfortunately, due to high oxygen vapor pressure, PbO_2 exits as a degenerated semiconductor (i.e. it behaves like a metal). If suitable doping could open the band gap, then PbO_2 can become one of the best materials for making commercially viable PEC cell. Semiconducting carbon is another class of material that needs to be extensively studied for this purpose. Earlier, diamond was the only known semiconductor of carbon. But in the recent past, it has been possible to synthesize semiconducting carbon with a band gap in the vicinity of 1.0 to 2.8 eV. Sharon and coworkers [107, 108] has recently reported a PEC cell from a semiconducting carbon made from the pyrolysis of camphor. In addition, efforts should be made to reduce the band gap of stable oxides such as TiO_2 into a low band gap material by manipulating to obtain their mixed oxides. Mixed oxides of Mo & W also need similar attention. Sharon and coworkers [109, 110] have tried to develop a theoretical model to predict the variation in band gap by varying the composition of components in mixed oxides. In addition to this, there is a need to develop a theoretical model, to predict PEC stability of materials. Such models will go a long way to minimize the number of experiments needed in developing materials for making a commercially viable PEC cell. It is necessary to emphasize that in spite of the simplicity in fabrication, the future of economically viable PEC depends entirely on the development of suitable low band gap, photoelectrochemically stable materials.

4.1.16 Laser Scanning Technique for a Large Area Electrode

There are materials, which can form varieties of nonstoichiometric compounds. For example, PbO_2 can coexist as PbO_x where x can have values from 0 to 2. Among these oxides, most photoactive oxide is $PbO_{0.8}$. If $PbO_{0.8}$ oxide is to be prepared by anodic oxidation of lead, it is possible that surface of the lead film might contain other forms of oxides as well. Thus, a PEC cell prepared by anodization of lead may give a low photoresponse, not because $PbO_{0.8}$ is an inappropriate material but because anodized surface may be contaminated with other forms of oxides as well. There seems to have no nondestructive technique available to confirm the uniformity of surface with photoactive species, especially of a large area electrode. Sharon and coworkers [111] has recently developed a laser scanning system to overcome this problem. A laser beam is scanned over the entire surface of the film, and photocurrent is measured at each illuminated point. A 3D plot of photocurrent versus (x, y) distances is made. If the composition of materials formed over the large surface area electrode contains photoactive oxide, a uniform distribution of photocurrent would be observed. Failing to get uniform photocurrent would be an indication of nonuniformity of the surface with the desired material. Though this technique is not specific about the composition, it is a very useful tool to give a guideline for improving the preparation technique especially for a large area electrode. The resolution of this experiment, however, depends on the scanning size of the laser beam. For example, the 3D laser scanning map of an anodized lead sheet (2.5 cm^2) shows that the photocurrent is not as uniform (Fig. 11a) as obtained with film anodized by condition shown in Fig. 11(b).

4.1.17 Summary

In this section, efforts are made to explain the basic principles of a regenerative PEC solar cell. Greater emphasis is made on

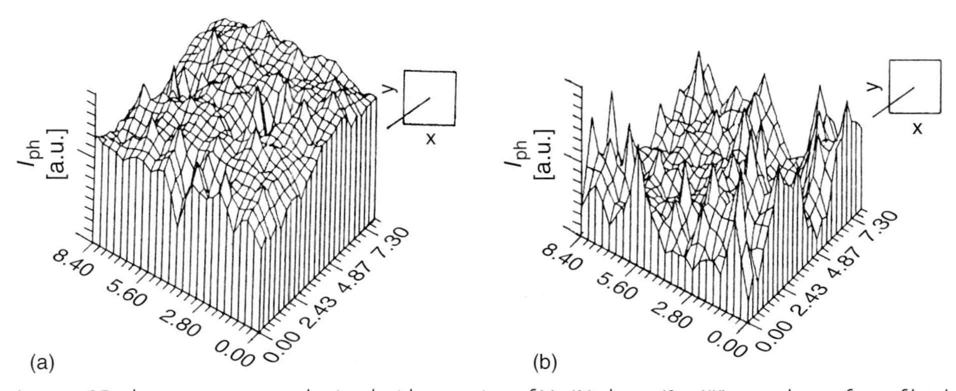

Fig. 11 3D photocurrent map obtained with scanning of He/Ne laser (3 mW) over the surface of lead sheet anodized in the potential range (a) −0.32 to +0.08 V and (b) −0.22 to +0.08 V versus SCE [26].

criteria for selection of a semiconductor and a redox electrolyte to make a PEC cell operating at its maximum performance. Effect of light on illuminating a junction formed with semiconductor–electrolyte interface is discussed with a view to characterize various parameters influencing the efficiency of a PEC cell. The role of counter-electrode and minimizing of the surface states to achieve a desirable performance of a PEC cell are also discussed. Whenever, a large surface area electrode of semiconductor of A_xB_y type is prepared especially by an electrochemical method, uniform deposition of material over the entire surface with similar composition is always a problem to establish. A laser scanning technique has been discussed for this purpose. Although PEC cells with efficiency as high as 16% have been developed, because of its instability problem, it has not been possible to commercialize. A section is devoted to discuss the material aspect of the PEC and a future direction of research needed to fabricate a commercially viable PEC. A chronological improvement made in PEC are also shown in a tabular form.

Acknowledgment

I am extremely grateful to all my students who have contributed a lot via their Ph.D. thesis and research work, which enabled me to perceive the complexity of PEC cell. Their interactions enriched me to evolve a simpler way to comprehend the intricacy of a PEC cell. This has immensely helped me in penning this section. I am also grateful to Dr. Michael Neumann-Spallart, CNRS, Meudon, France, for going through the manuscript and for his valuable suggestions, which helped me in formulating this chapter.

References

1. A. E. Becquerel, *C.R. Acad. Sci.* **1839**, *9*, 561.
2. W. H. Brattain, C. G. B. Garret, *Semiconductor surface phenomena in Semiconducting materials* Proc. Conf. Univ. Reading, England, 1951, 37.
3. W. H. Brattain, C. G. B. Garrett, *Phys. Rev.* **1955**, *99*, 376.
4. A. Fujishima, K. Honda, *Nature* **1972**, *238*, 37.
5. H. Gerischer in *Semiconductor electrochemistry: in Physical Chemistry: An advanced treatise* (Eds.: H. Eyring, D. Henderson, W. Jost), 9A, Academic Press, New York, 1970, p. 463.
6. V. A. Myamlin, Yu. V. Pleskov, *Electrochemistry of Semiconductors*, Plenum Press, New York, 1967.
7. M. Sharon, P. Veluchamy, C. Natrajan, C. D. Kumar, *Electrochim. Acta* **1991**, *36*(7), 1107.
8. S. Licht, G. Hodes, R. Tenne et al., *Nature* **1987**, *326*, 863.
9. S. Licht, B. Wang, T. Soga et al., *Appl. Phys. Lett.* **1999**, *74*, 4055.
10. B. Pal, M. Sharon, D. Kamat, *Chem. Environ. Res.* **1996**, *5*(1–4), 51.
11. H. Yoneyama, N. Takahashi, K. Susumu et al., *J. Chem. Soc. Chem. Commun.* **1992**, *9*, 716.
12. A. J. Nozik in *Photoelectrochemical devices for solar energy conversion*, NATO advance study institute series, Series B; Physics, Plenum press, New York, 1981, p. 280, Vol .69.
13. V. G. Levich in *Physical Chemistry – An Advanced Treatise*, (Eds.: H. Eyring, D. Henderson, W. Jost), Academic Press, New York, 1970, Vol. 9B, Chapter 12.
14. H. Gerischer, *Surf. Sci.* **1969**, *13*, 265.
15. H. Gerischer, in *Special Topics in Electrochemistry*, (Ed.: P. G. Rock), Elsevier, Amsterdam, 1977.
16. H. Gerischer, *Surf. Sci.* **1969**, *18*, 97.
17. J.Ó.M Bockris, S. U. M. Khan, *J. Electrochem. Soc.* **1985**, *132*, 2648.
18. S. Licht, *Nature* **1987**, *330*, 148.
19. S. Licht, *Sol. Energy Mater. Sol. Cells* **1995**, *38*, 353.
20. W. W. Gärtner, *Phys. Rev.* **1959**, *116*, 84.
21. I. Mukhopadhyay, M. Sharon, *Electrochim. Acta.* **1997**, *42*(1), 67–72.
22. I. Mukhopadhyay, M. Sharon, *Sol. Energy Mater. Sol. Cells* **1997**, *45*, 141.

23. J. F. Gibbons, G. W. Cogan, C. M. Gronet et al., *Appl. Phys. Lett.* **1984**, *44*, 539.
24. G. Hodes, J. Manassen, D. Cahen, *J. Electrochem. Soc.* **1980**, *127*, 544.
25. R. Tenne, M. Peisach, C. A. Rabe et al., *J. Electroanal. Chem.* **1989**, *269*, 389.
26. M. Sharon, S. Ghosh, *J. Solid State Electrochem.* **1999**, *4*, 52.
27. S. Licht, F. Forouzan, *J. Electrochem. Soc.* **1995**, *142*, 1539.
28. M. Sharon, B. M. Prasad, K. Basavaswaran, *Indian J. Chem.* **1989**, *28A*, 935.
29. M. Sharon, K. Basavaswaran, N. P. Sathe, *J. Sci. Ind. Res.* **1985**, *44*, 593.
30. M. Sharon, G. Tamizhmani, *J. Mater. Sci.* **1986**, *21*, 2193.
31. R. N. Noufi, P. A. Kohl, A. J. Bard, *J. Electrochem. Soc.* **1978**, *125*, 375.
32. M. S. Wrighton, A. B. Ellis, S. W. Kaiser, *J. Am. Chem. Soc.* **1976**, *98*, 1635.
33. S. Chandra, S. K. Pandey, R. C. Agrawal, *Sol. Cell 1980, 1*, 36.
34. W. J. Danaher, L. E. Lyons, *Nature* **1978**, *271*, 139.
35. J. M. Bolts, A. B. Ellis, K. D. Legg et al., *J. Am. Chem. Soc.* **1977**, *99*, 4826.
36. R. Noufi, D. Tench, L. F. Warren, *J. Electrochem. Soc.* **1980**, *127*, 2709.
37. A. Aruchamy, A. Venkatarathnam, G. V. Subba Rao et al., *Electrochem. Acta.* **1982**, *27*, 701.
38. H. Gerischer, *J. Electroanal. Chem.* **1975**, *58*, 263.
39. M. Tsuiki, H. Minoura, T. Nakamura et al., *J. Appl. Electrochem.* **1978**, *8*, 523.
40. A. B. Ellis, S. W. Kaiser, M. S. Wrighton et al., *J. Am. Chem. Soc.* **1977**, *99*, 2839.
41. R. Tenne, G. Hodes, *Appl. Phys. Lett.* **1980**, *37*, 428.
42. K. Rajeshwar, *J. Appl. Electrochem.* **1985**, *15*, 1.
43. A. Heller, K. C. Chang, B. Miller, *J.Am. Chem. Soc.* **1978**, *100*, 684.
44. L. Fornarini, M. Lazzari, L. P. Bicelli et al., *J. Power Source* **1981**, *6*, 371.
45. Y. Mirovsky, D. Cahen, *Appl. Phys. Lett.* **1982**, *40*, 727.
46. N. LeNagard, A. Perrin, M. Sergent et al., *Mater. Res. Bull.* **1985**, *20*, 835.
47. R. Tenne, Y. Mirovsky, Y. Greenstein et al., *J. Electrochem. Soc.* **1982**, *129*, 1506.
48. B. Scrosati, L. Fornarini, G. Razzini et al., *J. Electrochem. Soc.* **1985**, *132*, 593.
49. J. A. Baglio, G. S. Calabrese, E. Kamieniecki et al., *J. Electrochem. Soc.* **1982**, *129*, 1461.
50. J. A. Baglio, G. S. Calabrese, D. J. Harrison et al., *J. Am. Chem. Soc.* **1983**, *105*, 2246.
51. H. Tributsch, *Ber. Bunsen-Ges Phys. Chem.* **1977**, *81*, 361.
52. H. Tributsch, H. Gerischer, C. Clemen et al., *Ber. Bunsen-Ges Phys. Chem.* **1979**, *83*, 655.
53. L. F. Schneemeyer, M. S. Wrighton, *J. Am. Chem. Soc.* **1980**, *102*, 6964.
54. T. Skotheim, L. G. Peterson, O. Inganäs et al., *J. Electrochem. Soc.* **1982**, *129*, 1737.
55. G. Djemal, N. Muller, U. Lachish et al., *Sol. Energy Mater.* **1981**, *5*, 403.
56. R. Audas, J. C. Irwin, *J. Appl. Phys.* **1981**, *52*, 6954.
57. A. J. Bard, F.-R.-F. Fan, *J. Am. Chem. Soc.* **1980**, *102*, 3677.
58. H. Tributsch, *Ber. Bunsen-Ges Phys. Chem.* **1978**, *82*, 169.
59. H. Gerischer, W. Kautek, *Ber. Bunsen-Ges Phys. Chem.* **1980**, *84*, 645.
60. B. Miller, A. Heller, *Nature* **1976**, *262*, 680.
61. P. K. Mahapatra, C. B. Roy, *Sol. Cells* **1982–1983**, *7*, 225.
62. M. S. Wrighton, A. B. Ellis, J. M. Bolts, *J. Electrochem. Soc.* **1977**, *124*, 1603.
63. Y. Ramprakash, S. Basu, D. N. Bose, *J. Indian Chem. Soc.* **1981**, *LVIII*, 153.
64. M. Tomkiewicz, J. M. Woodall, *Science* **1977**, *196*, 990.
65. H. von Känel, L. Gantert, R. Hauger et al., *Hydrogen Energy Prog.* **1984**, *5*, 969.
66. M. Sharon, B. M. Prasad, *Sol. Energy Mater.* **1983**, *8*, 457.
67. M. Sharon, B. M. Prasad, *Electrochim. Acta* **1985**, *30*(3), 331.
68. M. Sharon, S. Kumar, S. R. Jawalekar, *Bull. Mater. Sci.* **1986**, *8*(3), 415.
69. M. Sharon, A. Sinha, *Sol. Energy Mater.* **1984**, *9*, 391.
70. M. Sharon, K. Basavaswaran, *Sol. Cells* **1988**, *25*, 97.
71. P. Veluchamy, M. Sharon, M. Shimizu et al., *J. Electroanal. Chem.* **1994**, *371*, 205.
72. A. Heller, B. Miller, S. S. Chu et al., *J. Am. Chem. Soc.* **1979**, *101*, 7633.
73. A. J. Bard, F.-R.-F. Fan, B. Reichman, *J. Am. Chem. Soc.* **1979**, *101*, 7633.
74. K. Rajeshwar, M. Kaneko, A. Yamado, *J. Electrochem. Soc.* **1983**, *130*, 38.
75. K. Rajeshwar, P. Singh, J. Du Bow, *Trans. ASME J. Sol. Energy Eng.* **1982**, *104*, 146.

76. A. Barrasse, H. Cachat, G. Horowitz et al., *Rev. Phys. Appl.* **1982**, *17*, 801.
77. D. S. Ginley, R. M. Bie Feld, B. A. Parkinson et al., *J. Electrochem. Soc.* **1982**, *129*, 145.
78. W. S. Hobson, A. B. Ellis, *Appl. Phys. Lett.* **1982**, *41*, 891.
79. S. Gourgaud, D. Elliott, *J. Electrochem. Soc.* **1977**, *124*, 102.
80. A. Heller, B. Miller, B. A. Parkinson, *J. Electrochem. Soc.* **1979**, *126*, 954.
81. M. G. Chris, S. L. Nathan, *Nature* **1982**, *300*, 733.
82. R. Memming, *J. Electrochem. Soc.* **1978**, *125*, 117.
83. A. J. Bard, R. E. Malpas, K. Itaya, *J. Am. Chem. Soc.* **1979**, *101*, 2535.
84. H. J. Lawerenz, M. Lubke, *Appl. Phys. Lett.* **1981**, *39*, 798.
85. W. M. Ayers, *J. Electrochem. Soc.* **1982**, *129*, 1644.
86. A. J. Bard, F.-R.-F. Fan, G. A. Hope, *J. Electrochem. Soc.* **1982**, *129*, 1647.
87. T. S. Jayadevaiah, *Appl. Phys. Lett.* **1974**, *25*, 399
88. A. J. Bard, F.-R.-F. Fan, H. S. White et al., *J. Am. Chem. Soc.* **1980**, *102*, 5142.
89. M. S. Wrighton, A. B. Bocarsly, E. G. Walton et al., *J. Electroanal. Chem.* **1979**, *100*, 283.
90. M. S. Wrighton, K. D. Legg, A. B. Ellis et al., *Proc. Natln. Acad. Sci. USA* **1977**, *74*, 4116.
91. M. S. Wrighton, R. G. Austin, A. B. Bocarsly et al., *J. Am. Chem. Soc.* **1978**, *100*, 1602.
92. M. S. Wrighton, J. M. Bolts, A. B. Bocarsly et al., *J. Am. Chem. Soc.* **1979**, *101*, 1378.
93. R. M. Candea, M. Kastner, R. Goodman et al., *J. Appl. Phys.* **1976**, *47*, 2724.
94. A. J. Nozik, *Appl. Phys. Lett.* **1976**, *29*, 150.
95. A. Heller, M. Robbins, K. J. Bachmann et al., *J. Electrochem. Soc.*, **1978**, *125*, 831.
96. S. Licht, O. Khaselev, T. Soga et al., *Electrochem. Soc. State Lett.* **1998**, *1*, 20.
97. B. O'Regan, M. Grätzel, *Nature* **1991**, *353*, 737.
98. S. Licht, D. Peramunage, *Nature* **1990**, *345*, 330.
99. S. Licht, D. Peramunage, *Nature* **1991**, *354*, 440.
100. B. J. Tufts, I. L. Abrahams, P. G. Santangelo et al., *Nature* **1987**, *326*, 861.
101. S. Licht, R. Tenne, G. Dagan et al., *Appl. Phys. Lett.* **1985**, *46*, 608.
102. S. Menezes, H. J. Lewerenz, K. J. Bachmann, *Nature* **1983**, *305*, 615.
103. G. Kline, K. Kam Keung, R. Ziegler et al., *Sol. Energy. Mater.* **1982**, *6*, 337.
104. G. Kline, K. Kam, D. Canfield et al., *Sol. Energy. Mater.* **1981**, *4*, 301.
105. A. Heller, B. Miller, *Appl. Phys. Lett.* **1981**, *38*, 282.
106. B. A. Parkinson, A. Heller, B. Miller, *Appl. Phys. Lett.* **1978**, *33*, 521.
107. S. Licht, *J. Phys. Chem.* **1986**, *90*, 1096.
108. K. Mukhopadhyay, I. Mukhopadhyay, M. Sharon et al., *Carbon*, **1997**, *35*, 863.
109. M. Sharon, I. Mukhopadhyay, K. Mukhopadhyay, *Sol. Energy Mater. Sol. Cells* **1997**, *45*(1), 35.
110. K. Murali Krishna, M. Sharon, M. K. Mishra, *J. Phys. Solids* **1996**, *57*(5), 615.
111. K. Murali Krishna, M. Sharon, M. K. Mishra et al., *Electrochem. Acta* **1996**, *41*(13), 1999.
112. M. Sharon, I. Mukhopadhyay, S. Ghosh, *J. Solid State Electrochem.* **1999**, *3*, 141.

4.2 Photoelectrochemical Solar Energy Storage Cells

Stuart Licht
Technion – Israel Institute of Technology, Haifa, Israel

4.2.1 Introduction

Although society's electrical needs are largely continuous, clouds and darkness dictate that photovoltaic solar cells have an intermittent output. A photoelectrochemical solar cell (PEC) can generate not only electrical but also electrochemical energy and provide the basis for a system with an energy-storage component (PECS). Sufficiently energetic insolation incident on semiconductors can drive electrochemical oxidation/reduction and generate chemical, electrical, or electrochemical energy. Aspects include efficient dye-sensitized or direct solar to electrical energy conversion, photoetching, photoelectrochemical water-splitting, environmental cleanup, and solar energy storage cells. This chapter focuses on photoenergy storage concepts based on photoelectrochemical processes, but includes a necessary comparison to other methods proposed for the conversion and storage of solar energy. The PEC uses light to carry out a chemical reaction, converting light to chemical energy. This fundamental difference of the photovolatic solar cell's (PV) solid–solid interface, and the PEC's solid–liquid interface has several ramifications in cell function and application. Energetic constraints imposed by single band gap semiconductors have limited the demonstrated values of photoelectrochemical solar to electrical energy conversion efficiency to 16%, and multiple band gap cells can lead to significantly higher conversion efficiencies [1, 2a,b]. Photoelectrochemical systems may facilitate not only solar to electrical energy conversion but have also led to investigations in photoelectrochemical synthesis, photoelectrochemical production of fuels, and photoelectrochemical detoxification of pollutants as discussed in other chapters in this volume.

4.2.1.1 Regenerative Photoelectrochemical Conversion

In illuminated semiconductor systems, the absorption of photons generates excited electronic states. These excited states have lifetimes of limited duration. Without a mechanism of charge separation their intrinsic energy would be lost through relaxation (recombination). Several distinct mechanisms of charge separation have been considered in designing efficient photoelectrochemical systems. At illuminated semiconductor–liquid interfaces, an electric field (the space charge layer) occurs concurrent with charge–ion redistribution at the interface. On photogeneration of electron-hole pairs, this electric field impedes recombinative processes by oppositely accelerating and separating these charges, resulting in minority carrier injection into the electrolytic redox couple.

This concept of carrier generation is illustrated in Fig. 1(a) (for an n-type PEC) and has been the theoretical basis for several efficient semiconductor-redox couple PEC cells. Illumination of the electrode surface with light, whose photon energy is greater than the band gap, promotes electrons into the conduction band leaving holes in the valance band. In the case of a photoanode, band-bending in the depletion region drives any electron that is promoted into the conduction band into the interior of the semiconductor and holes

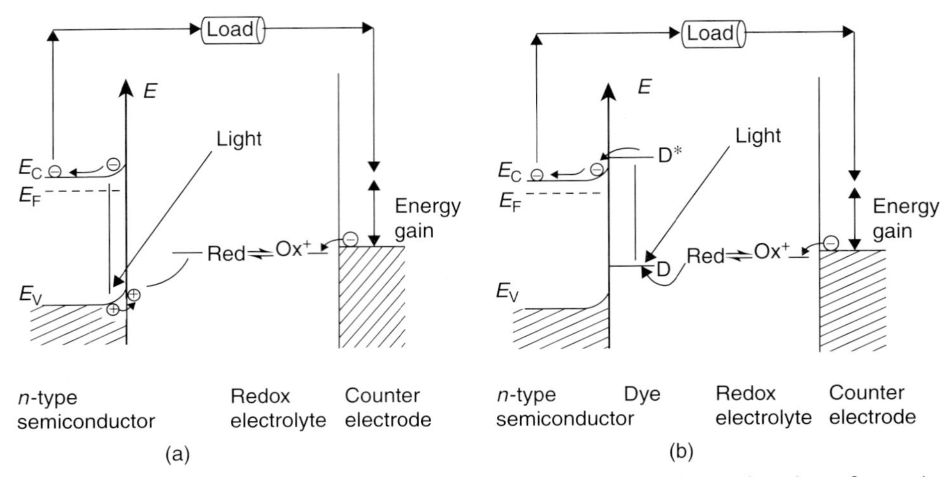

Fig. 1 Carrier generation under illumination arising at (a) the semiconductor–liquid interface and (b) the semiconductor–dye sensitizer–liquid interface.

in the valance band toward the electrolyte, where they participate in an oxidation reaction. Electrons through the bulk drive an external load before they reach the counter electrode or storage electrode, where they participate in a reduction process. Under illumination and open circuit, a negative potential is created in a photoanode, and as a result the fermi level for the photoanode shifts in the negative direction, thus reducing the band-bending. Under illumination with increasing intensity, the semiconductor fermi level shifts continually toward negative potentials until the band-bending effectively reduces to zero, which corresponds to the flat band condition. At this point, a photoanode exhibits its maximum photovoltage, which is equal to the barrier height.

Excitation can also occur in molecules directly adsorbed and acting as a mediator at the semiconductor interface. In this dye sensitization mode, the function of light absorption is separated from charge carrier transport. Photoexcitation occurs at the dye and photogenerated charge is then injected into a wide band gap semiconductor. This alternative carrier generation mode can also lead to effective charge separation as illustrated in Fig. 1(b). The first high solar to electric conversion efficiency example of such a device was demonstrated in 1991 [3] through the use of a novel high surface area (nanostructured thin film) n-TiO_2, coated with a well-matched trimeric ruthenium complex dye immersed in an aqueous polyiodide electrolyte. The unusually high surface area of the transparent semiconductor coupled to the well-matched spectral characteristics of the dye leads to a device that harvests a high proportion of insolation.

4.2.1.2 Photoelectrochemical Storage

PECs can generate not only electrical but also electrochemical energy. Figure 2 presents one configuration of a PEC combining in situ electrochemical storage and solar-conversion capabilities; providing continuous output insensitive to daily variations in illumination. A high solar to

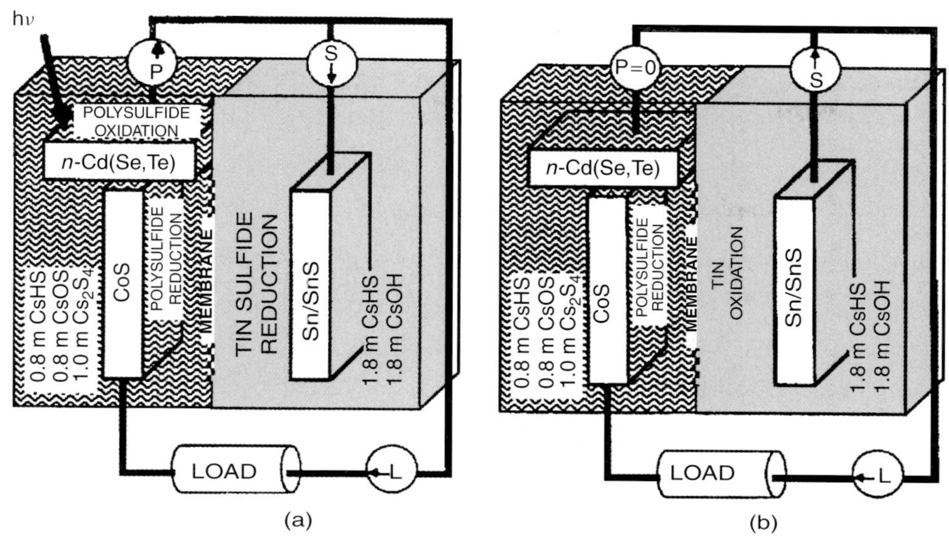

Fig. 2 Schematic of a photoelectrochemical solar cell combining both solar conversion and storage capabilities. (a) Under illumination; (b) in the dark.

electric conversion efficiency cell configuration of this type was demonstrated in 1987 and used a Cd(Se,Te)/Sx conversion half cell and a Sn/SnS storage system, resulting in a solar cell with a continuous output [4]. Under illumination, as seen in Fig. 2(a), the photocurrent drives an external load. Simultaneously, a portion of the photocurrent is used in the direct electrochemical reduction of metal cations in the device storage half-cell. In darkness or below a certain level of light, the storage compartment spontaneously delivers power by metal oxidation, as seen in Fig. 2(b).

4.2.2 Comparative Solar-Storage Processes

4.2.2.1 Thermal Conversion and Storage

Solar insolation can be used to directly activate a variety of thermal processes; the enthalpy is stored physically or chemically and then either directly utilized or released upon reversal of the storage process. To date, the predominant nonbiologic utilization of solar energy is to heat a working fluid that is maintained in an insulated enclosure, storing a portion of the incident solar radiation for future use. Limitations of this approach include the low energy available per unit mass of the storage medium, and low efficiencies of thermal to mechanical and thermal to electrical energy conversion. A variety of passive and dynamic optical concentrators have been studied to compensate for these limitations.

The high temperatures generated by concentrated solar power have been utilized to drive highly endothermic reactions. The reverse reaction releases the chemically stored energy as thermal energy. Various systems have been investigated such as in 1985 [5]:

$$SO_3 \longrightarrow SO_2 + \tfrac{1}{2}O_2 \quad \Delta H = 98.94 \tag{1}$$

4.2.2.2 Photochemical Storage

In photochemical processes, photon absorption creates a molecular excited state or stimulates an interband electronic transition in a semiconductor that induces a molecular change. Comprehensive reviews, including those by Gratzel [6], Kalyanasundaram [7], and Harriman [8], have discussed various aspects of photochemical energy conversion. A photoactivated molecular excited state can drive either (1) photodissociation (2) photoisomerisation or (3) photoredox reactions. Processes based on semiconductors may involve photovoltaic or photoelectrochemical systems.

A substantial photochemical effort research has centered on the photoredox storage of solar energy. Molecular photoredox processes use electron transfer from photoinduced excited states:

$$A + h\nu \longrightarrow A^* \qquad (2)$$

The electron transfer may be either direct

$$A^* \longrightarrow A^+ + e^- \qquad (3)$$

or a variety of indirect processes as exemplified by either

$$A^* \longrightarrow B;$$

followed by

$$B \longrightarrow B^+ + e^- \qquad (4)$$

or

$$A^* + B \longrightarrow AB;$$

followed by

$$AB \longrightarrow AB^+ + e^- \qquad (5)$$

Solar-activated photodissociation processes generally involve cleavage of a simple molecule into several energetic products. Limitations of this approach include the limited absorption of solar energy by the molecule, low quantum yield, rapid back reaction, and difficulties separating the product species. An example of storage by such photodissociation processes is exemplified by

$$2NOCl \longrightarrow 2NO + Cl_2 \qquad (6)$$

In photoisomerization, an absorbed photon activates molecular rearrangement and conversion of organic molecules into strained isomers. The products are stored. Despite attractive features including a high heat storage capacity and good thermal stability, most systems tested have poor efficiencies. These systems necessitate transformation to strained conformations at high energies; energies consistent with wavelengths below 450 nm. This excludes much of the energy inherent in the Air Mass 1 (AM1) solar spectra. The stored thermal energy is released on catalytic-induced reversion to the starting components. An example has been presented by Kutal for the photocatalytic transformation of norbornadiene to quadricyclane [9].

Photochemical redox reactions can generate fuel formation, including H_2, CH_3, and CH_3OH. Because of its availability, the splitting of water to produce H_2 has been the focus of particular attention. H_2O is transparent to near UV or visible radiation, and therefore sensitization is required to drive the water-splitting process. In early attempts on photoredox-splitting of water by Heidt and McMillian, the process was sensitized by using solution redox species such as $Ce^{3+/4+}$ [10].

$$Ce^{3+}(aq) + H^+ \longrightarrow Ce^{4+}(aq) + \tfrac{1}{2}H_2 \qquad (7)$$

$$2Ce^{4+}(aq) + H_2O \longrightarrow 2Ce^{3+}(aq) + 2H^+ + \tfrac{1}{2}O_2 \qquad (8)$$

These processes have displayed poor quantum yields. As in photoisomerization

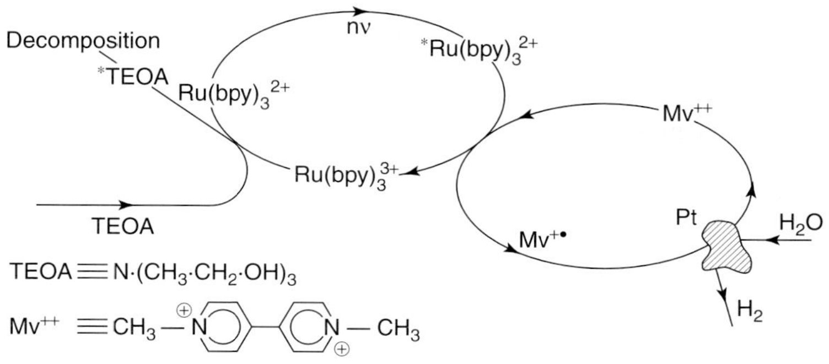

$TEOA \equiv N\cdot(CH_3\cdot CH_2\cdot OH)_3$

Scheme 1

processes, these reactions are also generally driven only by high-energy radiation (short wavelength) and cannot efficiently convert incident AM1 solar radiation.

H_2 or O_2 generation from water is a multielectron process. Optimization of photoredox-splitting of water necessitates the presence of a catalyst to mediate this complex multielectron transfer. In one such process, a sacrificial reagent triethanolamine (TEOA) is consumed irreversibly in the process, as denoted in Sch. 1 [9]:

Direct multielectron processes are rare and instead incorporate one or more radical intermediate steps. These reactive intermediates are susceptible to unfavorable side reactions, resulting in substantial losses in the energy-conversion process. Kinetically favored back reactions further reduce the overall conversion efficiency. The engineering of these complex molecular organizations provides a substantial scientific challenge and have generally resulted in systems with low conversion efficiencies.

4.2.2.3 Semiconductor Photoredox Storage

Semiconductor surfaces have been used as sensitizers to drive photochemical conversion and storage of solar energy. In principle, this should lead to a higher level of photon absorption and more effective charge separation. Both effects can substantially increase solar photochemical-conversion efficiency, but these systems have not yet displayed high efficiencies of fuel generation or long-term stabilities.

Photoredox processes at semiconductor electrodes generating fuels or products other than hydrogen, including methanol and ammonia, have been attempted with low overall yields. The photoelectrolysis of HI into H_2 and I_3^- at *p*-InP electrodes has been described [11]. These H_2 and I_3^- photogenerated products are prime candidates for a fuel cell. Analogous advanced systems, in which the photoelectrochemically generated fuels have been successfully recombined to generate electrical energy, are discussed later in this section.

A system exemplifying photoelectrochemical synthesis to generate hydrogen is water photoelectrolysis. An early demonstration of water photoelectrolysis used TiO_2 (band gap 3.0 eV) and was capable of photoelectrolysis at ~0.1% solar to chemical energy–conversion efficiency [12]. The semiconductor $SrTiO_3$ was demonstrated to successfully split water in a direct photon-driven process by Bolts and Wrighton (1976), albeit at low solar energy–conversion efficiencies [13].

The high $SrTiO_3$ band gap, E_g, of 3.2 eV creates sufficient energetic charge to drive the photoredox process. This excludes the longer wavelength photons and corresponds to only a small fraction of incident solar radiation. To improve the solar response, E_g has to be lowered; in a single band gap system, an optimum efficiency can be expected around 1.4 eV.

In photoelectrochemical water-splitting systems, corrosion of the semiconductor photoelectrodes can pose a significant problem. Most surface-stabilizing redox reactions compete with oxygen and hydrogen generation and must be excluded from these systems. To enhance the solar response of high band gap materials, techniques such as dye sensitization and impurity sensitization have been attempted, although with little improvement [14]. Semiconductor surfaces have been modified to protect low band gap materials against photocorrosion [15, 16]. A self-driven photoelectrochemical cell consisting of Pt-coated *p*-InP and Mn-oxide-coated *n*-GaAs has been demonstrated to operate at 8.2% maximum efficiency to generate H_2 and O_2 under simulated sunlight [17], and more recently a two band gap cell in a tandem arrangement has been used to split water at 12% efficiency [18]. A multijunction GaAs, Si cell has been recently used to drive water-splitting at over 18% solar to electrical-conversion efficiency [19].

Colloids and suspensions of semiconductors have been used for the photoredox-splitting of water. The principle advantage of a fine suspension is the large active surface area available. Reaction rates of H_2 and O_2 generation have been enhanced by loading the particles with small deposit of precious metals, and although significant progress was made in this direction, a practical system is yet to be demonstrated [11].

4.2.3 Modes of Photoelectrochemical Storage

Conversion of a regenerative PEC to a photoelectrochemical storage solar cell (PECS) can incorporate several increasingly sophisticated solar energy conversion and storage configurations.

4.2.3.1 Two-Electrode Configurations

A variety of two-electrode configurations have been investigated as PECS systems. Important variations of these photoelectrochemical conversion and storage configurations are summarized in Table 1. In each case, and as summarized in Fig. 3 for the simplest configurations, exposure to light drives separate redox couples and a current through the external load. There is a net chemical change in the system, with an overall increase in free energy. In the absence of illumination, the generated chemical change drives a spontaneous discharge reaction. The electrochemical discharge induces a reverse current. In each case in Table 1, exposure to light drives separate redox couples and current through the external load.

Consistent with Fig. 1, in a regenerative PEC, illumination drives work through an external load without inducing a net change in the chemical composition of the system. This compares with the two-electrode PECS configurations shown in Fig. 3(a) and (b). Unlike a regenerative system, there is a net chemical change in the system, with an overall increase in free energy. In the absence of illumination, the generated chemical change drives a spontaneous discharge reaction. The electrochemical discharge induces a reverse current. Utilizing two quasi-reversible chemical processes, changes taking place in the system during illumination can be reversed in the dark. Similar to

Tab. 1 Important two-electrode photoelectrochemical conversion and storage configurations

Scheme	*Electrode 1*	*Electrolyte(s)*	*Electrode 2*
I	SPE	| Redox A Redox B |	CE
II	SPE	| Redox A-membrane-Redox B |	CE
III	SPE	| Redox A |	Redox B_{CE}-CE
IV	SPE-Redox A_{SPE}	| Redox B |	CE
V	SPE	| Redox A-membrane-Redox B |	SPE

Note: Components of these systems include a semiconductor photoelectrode (SPE) and a counter electrode (CE). At the electrode–electrolyte interface, redox couples "A" or "B" are either in solution (| Redox |), counter electrode–confined (| Redox B_{CE}-CE) or confined to the semiconductor photoelectrode (SPE-Redox A_{SPE}|).

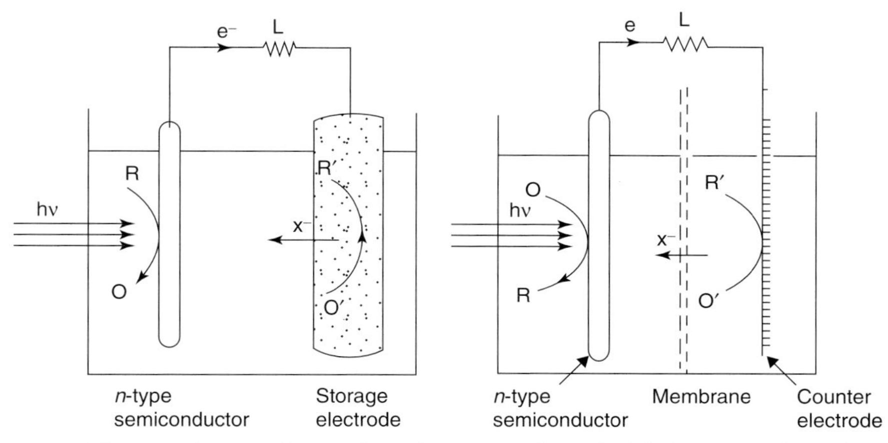

Fig. 3 Schematic diagram of a two-electrode storage cell. On the left the storage electrode contains an insoluble redox couple, and on the right a soluble redox couple, with storage represented as

$$R + O' \underset{}{\overset{h\nu}{\rightleftharpoons}} O + R'$$

a secondary battery, the system discharges producing an electric flow in the opposite direction and the system gradually returns to the same original chemical state.

Each of the cells shown in Fig. 3 has some disadvantages. For both bound (Fig. 3left) and soluble (Fig. 3right) redox couples, the redox species may chemically react with and impair the active materials of the photoelectrode. Furthermore, during the discharge process, the photoelectrode is kinetically unsuited to perform as a counter electrode. In the absence of illumination, the photoelectrode P, in this case a photoanode, now assumes the role of a counter electrode by supporting a reduction process. For the photoanode to perform efficiently during illumination (charging), this very same reduction process should be inhibited to minimize photooxidation back reaction losses. Hence, the same photoelectrode cannot efficiently fulfill the duel role of being kinetically sluggish to reduction during illumination and yet being kinetically facile to the same reduction during dark discharge. The configuration represented in Fig. 3 has another disadvantage, the disparity between the small surface area needed to minimize photocurrent dark current losses and the large surface area necessary to minimize storage polarization losses to maximize storage capacity [20].

4.2.3.2 Three-Electrode Configurations

Several of the two-electrode configuration disadvantages can be overcome by considering a three-electrode storage cell configuration as shown in Fig. 4. In Fig. 4, the switches E and F are generally alternated during charge and discharge. During the charging, only switch E may be closed, facilitating the storage process, and during discharge, E is kept open while F is closed. In this case, chemical changes that took place during the storage phase are reversed, and a current flow is maintained from the storage electrode to a third (counter) electrode that is kept in the first compartment. To minimize polarization losses during the discharge, this third electrode should be kinetically fast to the redox couple used in the first compartment. Still an improved situation would be to

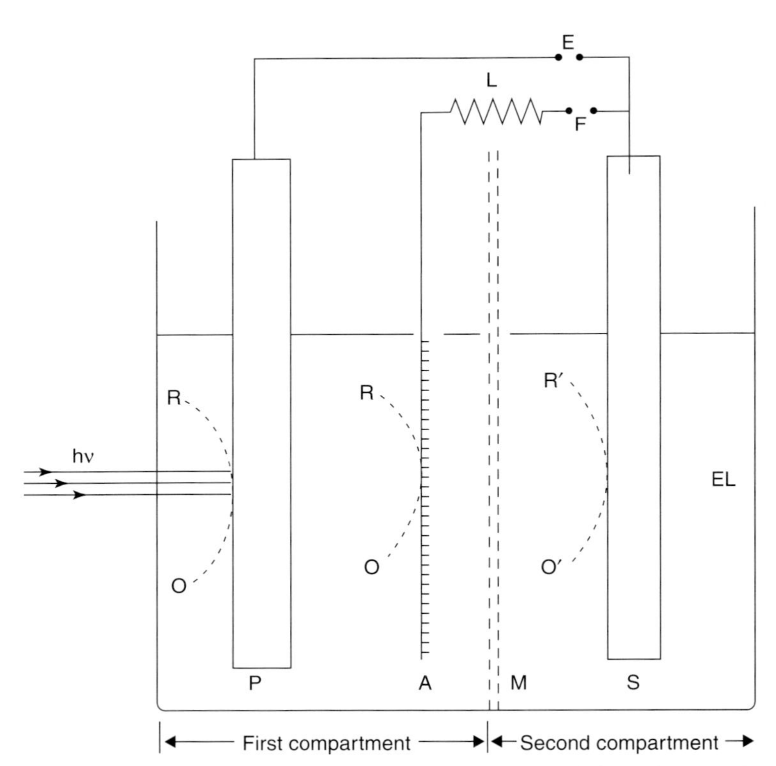

Fig. 4 Schematic diagram for a storage system with a third electrode (counter electrode) in the photoelectrode compartment. P = Photoelectrode, A = Counter electrode, M = Membrane, S = Storage electrode, EL = Electrolyte, E, F = Electrical switch, and L = Load.

have both switches closed all the time. In this case, electric current flows from the photoelectrode to both counter and storage electrodes. The system is energetically tuned such that when insolation is available, a significant fraction of the converted energy flows to the storage electrode. In the dark or diminished insolation, the storage electrode begins to discharge, driving continued current through the load. In this system, a proper balance should be maintained between the potential of the solar energy–conversion process and the electrochemical potential of the storage process. There may be residual electric flow through the photoelectrode during dark cell discharge, as the photoelectrode is sluggish, but not entirely passive, to a reduction process. This can be corrected by inserting a diode between the photoelectrode and the outer circuit.

4.2.4 Optimization of Photoelectrochemical Storage

The power obtained is the product of voltage and current, and consideration of the photocurrent is as important as the photovoltage. If the band-bending is sufficiently large, then the minority carrier redox reaction, which is essential to maintain the photocurrent, can compete effectively with the recombination of photogenerated electron hole pairs. This recombination represents a loss of absorbed photo energy. Therefore, an objective is to maintain a high band-bending and at the same time a significant photovoltage. A photoanode creates a negative photovoltage under illumination, which results in reducing the band-bending. In principal, one way to accomplish high band-bending is to choose a very positive redox couple in the electrolyte. The converse is true for a photoanode. Improvements relating to its stability and conversion efficiency are of paramount importance.

4.2.4.1 Improvements of the Photoresponse of a Photoelectrode

To improve the solar response of a photoelectrode, a proper match between the solar spectrum and the band gap of the semiconductor should be maintained. When a single band gap semiconductor is used, a band gap in the vicinity of 1.4 eV is most desirable from the standpoint of optimum solar-conversion efficiency. An important criterion is that the minority carrier that is driven toward the semiconductor–electrolyte interface should not participate in a photocorrosion reaction that is detrimental to the long-term stability of the photoelectrode. Photocorrosion can be viewed in terms of either kinetic or thermodynamic considerations and the real cause may be a mixture of both. From thermodynamic perspective, a photoanode is susceptible to corrosion if the fermi level for holes is at a positive potential with respective to the semiconductor corrosion potential [21]. The corrosion can be prevented or at least inhibited by choosing a redox couple that has its E_{redox} more negative than that for the corrosion process [22, 23]. The kinetic approach has been to allow another desired redox process to occur at a much faster rate than the photocorrosion reaction [13]. Other attempts to minimize the photocorrosion has been to coat the photoelectrode surface with layers such as Se [24] and protective conductive polymer films [25], and to search for alternate low band gap semiconductors [26]. Extensive reviews on the performance and stability of cadmium chalcogenides include those by Cahen and coworkers, 1980 [27] and Hodes, 1983 [28]. Etching of photoelectrode surface has been recognized and widely used as an important treatment to achieve high-conversion efficiency [29]. This effect is mostly attributed to removal of surface states that may act as trapping centers for photogenerated carriers. A related procedure called *photoetching*, initially developed for CdS and then applied to a wide variety of semiconductors, improves the photoelectrode performance and preferentially removes the surface defects acting as recombination centers [30].

In addition to the variety of etching procedures, several other surface treatments have been used to improve photoelectrode performance. Examples include a Ga^{3+} ion dip on CdSe [31], $ZnCl_2$ dip on thin film CdSe [32], Ru on GaAs [33], Ru on InP [34], and Cu on CdSe [34]. Reasons explaining the effectiveness of these dips range from a decrease of dark current to electrocatalysis by surface-deposited metal atoms. Solution phase chemistry of the electrolyte is an important parameter that has been shown to dramatically influence photoeffects. The equilibrium position of the redox couple will affect equilibrium band-bending. A photoanodic system with a solution containing a more positive potential redox couple causes a greater band-bending, which in turn leads to a higher photovoltage and efficient carrier separation under normal experimental conditions.

4.2.4.2 Effect of the Electrolyte

Semiconductor photoeffects in a complex redox electrolyte are largely affected by the solution properties such as solution redox level, interfacial kinetics (adsorption), conductivity, viscosity, overall ionic activity, solution stability, and transparency within a crucial wavelength region. Redox electrolytes are known to inhibit unfavorable phenomena such as surface recombination and trapping [35]. In addition, the solution redox couple may induce a favorable influence on the PEC system by improved charge-transfer kinetics leading to improved stability of the photoelectrode [36]. Additives incorporated in redox electrolytes are known to enhance the performance of PEC systems. Addition of small concentrations of Se in polysulfide electrolyte is known to improve the stability of CdSe (single crystal)/polysulfide system [22]. In this case, Se improves the PEC performance by reducing S/Se exchange and by increasing the dissolution of the photooxidized product S, which is the rate-determining step in the oxidation of sulfide at the anode. Addition of Cu^{2+} into the I^-/I_3^- electrolyte is known to improve the stability of $CuInX_2$ photoelectrode considerably [37], and in the same electrolyte, tungsten and molybdenum dichalcogenide photoelectrochemistry can be substantially improved by addition of Ag^+, or other metal cations, and shift of the $I^-/I_3^- E_{redox}$ [38].

In the case of CdSe/polysulfide system, solution activity, conductivity, efficiency of the photoanode (fill factor), charge-transfer kinetics at the interface, and the stability of the photoelectrode are known to exhibit improvements in the trend $Li > Na > K > Cs >$ for alkali polysulfide electrolyte. This trend is explained in terms of the secondary cation effect on electrochemical anion oxidation in concentrated aqueous polysulfide electrolytes [39]. In the case of Cd(Se,Te)/polysulfide system, the efficiency of light energy conversion is improved by using a polysulfide electrolyte without added hydroxide because of the combined effect of increasing the solution transparency, relative increase of S_4^{2-}, and decrease in S_3^{2-} in solution. For the same photoelectrode–electrolyte system, an optimum photoeffect was observed for a solution containing a sulfur–sulfide ratio of 1.5 : 2.1 with 1 : 2 molal potassium sulfide concentrations because of the combined effect of optimized solution viscosity, transparency, activity, and shift in solution redox level [40]. Stability of the polysulfide redox electrolyte, which is another parameter that determines the long-term performance of a PEC cell, has been shown to increase with sulfur and alkali metal sulfide concentration and to decrease with either increasing $-OH^-$ concentration or at high ratio of added sulfur to alkali metal sulfide [39, 41]. The combined polysulfide electrolyte optimization can substantially enhance cadmium chalcogenide photoelectrochemical conversion.

Chemical composition of the electrolyte is a particularly important parameter in PEC systems based on complex electrolytes, such as polysulfide or ferro/ferricyanide. In the latter redox couple, replacement of a single hexacyano ligand strongly changes the photoelectrochemical response of illuminated *n*-CdSe [42], and addition of the KCN to the electrolyte can increase *n*-CdSe and *n*-CdTe photovoltage by 200 mV [43].

4.2.4.3 Effect of the Counter Electrode

In a photoanodic system, even at moderate current densities, the occurrence of sluggish counter electrode kinetics for the cathodic process will cause significant polarization losses and diminish the photovoltage. Minimization of these kinetic limitations necessitates a counter electrode with good catalytic properties. For example, as shown by [34], CoS on stainless steel or brass electrodes exhibits electrocatalytic properties toward polysulfide reduction and overpotentials as low as 1 mV cm^2 mA^{-1} has been realized. Composition of a particular redox electrolyte may have a bearing on the extent of counter electrode polarization [32, 39].

In PEC systems, a compromise is maintained to simultaneously optimize the photoelectrode efficiency, stability, and electrolytic properties of the electrolyte. Practical PEC systems often require large working and counter electrodes and their geometric configuration within the PEC system will effect mass transport and effective cell current. In some cases, advantageous use has been made of selective sluggish counter electrode kinetics toward certain cathodic processes. For example, carbon is a poor cathode for H_2 evolution compared to Pt, and the direct hydrogenation of anthraquinone at a PEC cathode has been avoided by using a carbon anode [44]. In this case, such hydrogenation represents an undesirable side reaction.

4.2.4.4 Combined Optimization of Storage and Photoconversion

An efficient photoelectrochemical conversion and storage system requires not only an efficient functional performance of the separate cell components but also a system compatibility. In the combined photoelectrochemical storage system, simultaneous parameters to be optimized include

1. minimization of light losses reaching the photoelectrode,
2. high photoelectrode–conversion efficiency of solar energy,
3. close potential match between the photopotential and the required storage-charting potential,
4. high current and potential efficiency of the redox storage process,
5. high energy capacity of the redox storage,
6. reversibility (large number of charge–discharge cycles of the redox storage),
7. stability of the photoelectrode,
8. stability of the electrolyte,
9. stability of the counter electrode,

10. economy and cost effectiveness, and
11. reduced toxicity and utilization of environmentally benign materials.

A photoelectrochemical solar cell implicitly contains an electrolytic medium. In the majority of laboratory PEC configurations, incident light travels through the electrolytic medium before illuminating the photoelectrode. The resultant light absorbance by the electrolyte is a significant loss, which is avoided by use of a back cell configuration. For example, the substantial absorptivity of dissolved polysenide species has been avoided in a *n*-GaAs photoelectrochemistry through the use of the back wall cell configuration presented in Fig. 5 [45].

The photoelectrochemical system shown in Fig. 4 is a combination of a photoelectrode, electrolyte, membrane, storage, and a counter electrode. As an example of challenges that may arise in the combined photoconversion and storage system, consider an *n*−CdSe/polysulfide/tin sulfide version of Fig. 4 and consisting of Cell 1.

Cell: 1. CdSe | HS^-, OH^-, $S_x{}^{2-}$| Membrane | HS^-, OH^-| SnS | Sn

With illumination, the cell exhibits simultaneous photoelectrode, counter electrode and storage reactions, and equilibria including

Photoanode:

$$HS^- + OH^- \longrightarrow S + 2e^- + H_2O \tag{9}$$

Photocompartment equilibria:

$$S + S_x^{2-} \longrightarrow S_(x + 1)^{2-} \tag{10}$$

Counter electrode:

$$S + 2e^- \longrightarrow HS^- + OH^- \tag{11}$$

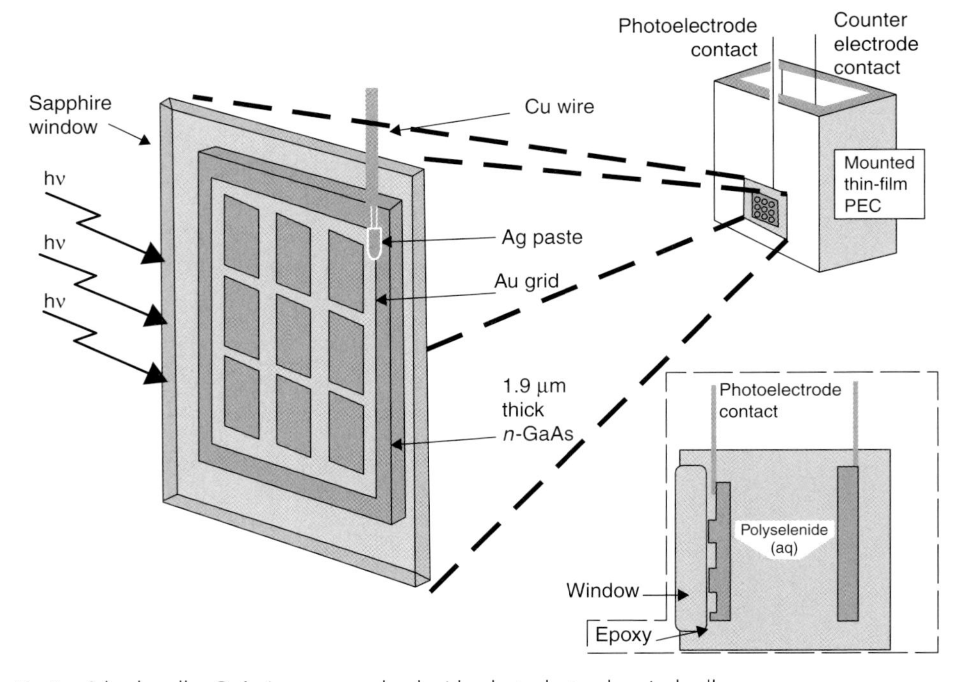

Fig. 5 A back wall *n*-GaAs/aqueous polyselenide photoelectrochemical cell.

Storage electrode:

$$SnS + H_2O + 2e^- \longrightarrow Sn + HS^- + OH^- \quad (12)$$

Unlike the case of the analogous regenerative PEC system, in the preceding equations, sulfur formed at the photoanode (and dissolved as polysulfide species, $S_{(x+1)}{}^{2-}$ for x = 1 to 4) is not balanced by the reduction that is taking place at the counter electrode because of the simultaneous reduction process taking place at the storage electrode. As a result, sulfur is accumulated in the photoelectrode compartment and is removed only in the subsequent discharge process. This dynamic variation in electrolyte composition may have a profound influence on the stability of the photoelectrode and electrolyte and on cell potential. Hence, to minimize these effects, either excess polysulfide must be included in the photocompartment or a limit must be set to the maximum depth of cell charge and discharge.

Another important consideration is the energy compatibility between the photoconversion and the storage processes. This compatibility is referred to as the voltage optimization. Figure 6 presents the combined IV characteristics for an ideal photoelectrode and current–voltage curves for two alternative redox processes; process A and process B. V_{ph} is the maximum

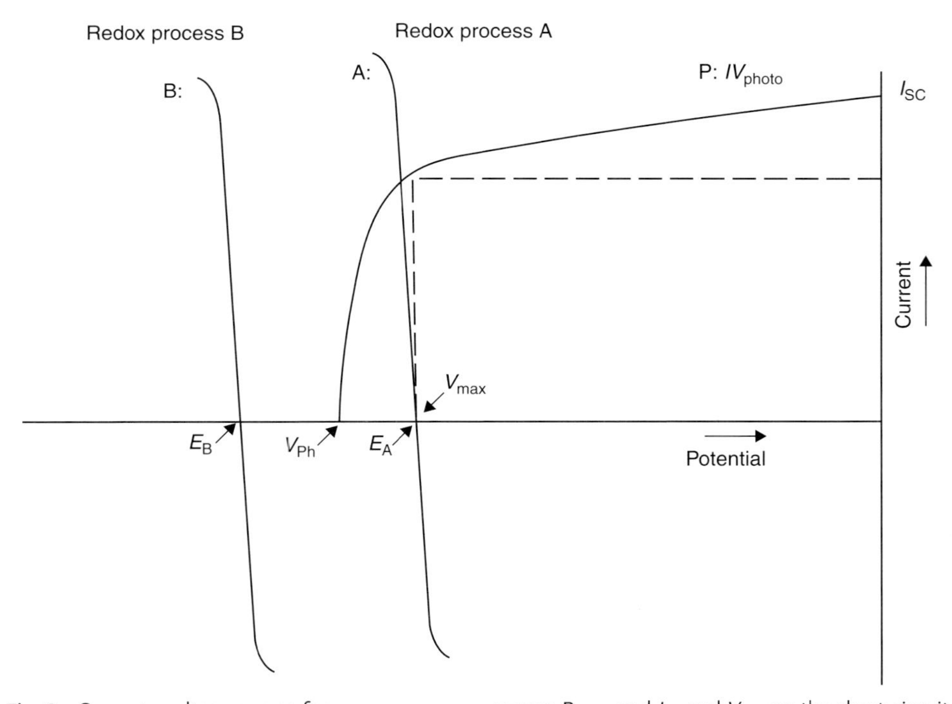

Fig. 6 Current–voltage curves for electrochemical storage processes, A or B. Process A may be charged by the photodriven current–voltage curve P, whereas process B may not. In the photodriven *IV* curve P, V_{max} is the voltage corresponding to the point of maximum power, P_{max}, and I_{sc} and V_{ph} are the short-circuit current and open-circuit photopotentials, respectively. E_A and E_B refer to the redox potentials for redox processes, A and B, respectively.

photovoltage that can be generated. I_{sc} is the short-circuit photocurrent corresponding to maximum band-bending. In Fig. 6, consider the electrochemical process represented by curve B. This process is located outside the region of potentials generated by the photoelectrode; it does not represent a potential storage system to be driven by a single photoelectrode. In such a case, a serial combination of more than one photoelectrode would be necessary. For a redox process to be a potential candidate for a redox storage system, the storage and photodriven current–voltage curves should intersect. Whereas V_{ph} and I_{sc} correspond to zero power, the point P_{max} shown in Fig. 6 corresponds to the maximum power point. Solar energy conversion is accomplished at its maximum efficiency only during operation in the potential vicinity of P_{max}.

By adjusting the electrical load L, shown in Fig. 4, the system can be constrained to operate near its maximum power efficiency. In this case, if the counter electrode is not polarized, the potential difference between photoelectrode and the counter electrode will be close to V_{max}. If one chooses a facile redox process for the storage electrode, as indicated by the sharply rising IV curve for process A in Fig. 6, with E_{redox} in the vicinity of V_{max}, then the potential during charge and discharge of the storage process will remain near V_{max}. As a result, the potential will be a highly invariant current variation through the load L, regardless of insolation intensity. This situation represents an ideal match between solar energy conversion and storage processes within a PECS. Nonideality occurs with poor voltage-matching or kinetic limitations and polarization losses associated with the counter, storage, or photoelectrodes.

Ideally, the membrane used to separate the two-cell compartments, as indicated in Fig. 4, must be permeable only to ions that will transport charge, but that will not chemically react or otherwise impair any electrode. The permeability of membranes is generally less than ideal. Different membranes permit other ions and water to permeate to a varying degree [46, 47]. Gross mixing of active materials across the membrane causes them to combine chemically and in the process lose energy. Favorable qualities that a membrane should exhibit are low permeability toward chemically reactive ions, low resistivity, mechanical integrity, and cost effectiveness.

4.2.5 High Efficiency Solar Cells with Storage

4.2.5.1 Multiple Band Gap Cells with Storage

A limited fraction of incident solar photons have sufficient (greater than band gap) energy to initiate charge excitation within a semiconductor. Because of the low fraction of short wavelength solar light, wide band gap solar cells generate a high photovoltage but have low photocurrent. Smaller band gap cells can use a larger fraction of the incident photons, but generate lower photovoltage. Multiple band gap devices can overcome these limitations. In stacked multijunction systems, the topmost cell absorbs (and converts) energetic photons, but it is transparent to lower energy photons. Subsequent layer(s) absorb the lower energy photons. Conversion efficiencies can be enhanced, and calculations predict that a 1.64-eV and 0.96-eV two–band gap system has an ideal efficiency of 38% and 50%, light of 1 and 1000 suns intensity, respectively. The ideal efficiency increases to a limit of 72% for a 36–band gap solar cell [48].

Recently, high solar conversion and storage efficiencies have been attained with a system that combines efficient multiple band gap semiconductors, with a simultaneous high capacity electrochemical storage [49, 50]. The energy diagram for one of several multiple band gap cells is presented in Fig. 7, and several other configurations are also feasible [1, 2a, b]. In the figure, storage occurs at a potential of $E_{redox} = E_{A+/A} - E_{B/B+}$. On illumination, two photons generate each electron, a fraction of which drives a load, whereas the remainder ($1/xe^-$) charges the storage redox couple. Without light – the potential falls below E_{redox} – the storage couple spontaneously discharges. This dark discharge is directed through the load rather than through the multijunction semiconductor's high dark resistance.

Cell: 2. In Fig. 8, an operational form of the solar conversion is presented and a storage cell described by the Fig. 7 energy diagram. The single cell contains both multiple band gap and electrochemical storage, which unlike conventional photovoltaics, provides a nearly constant energetic output in illuminated or dark conditions. The cell combines bipolar AlGaAs ($E_g = 1.6$ eV) and Si ($E_g = 1.0$ eV) and AB_5 metal hydride/NiOOH storage. Appropriate lattice-matching between AlGaAs and Si is critical to minimize dark current, provide ohmic contact without absorption loss, and maximize cell efficiency. The NiOOH/MH metal hydride storage process is near ideal for the AlGaAs/Si because of the excellent match of the storage and photocharging potentials. The electrochemical storage processes utilizes MH oxidation and nickel oxyhydroxide reduction:

$$MH + OH^- \longrightarrow M + H_2O + e^-;$$
$$E_{M/MH} = -0.8 \text{ V vs SHE} \qquad (13)$$

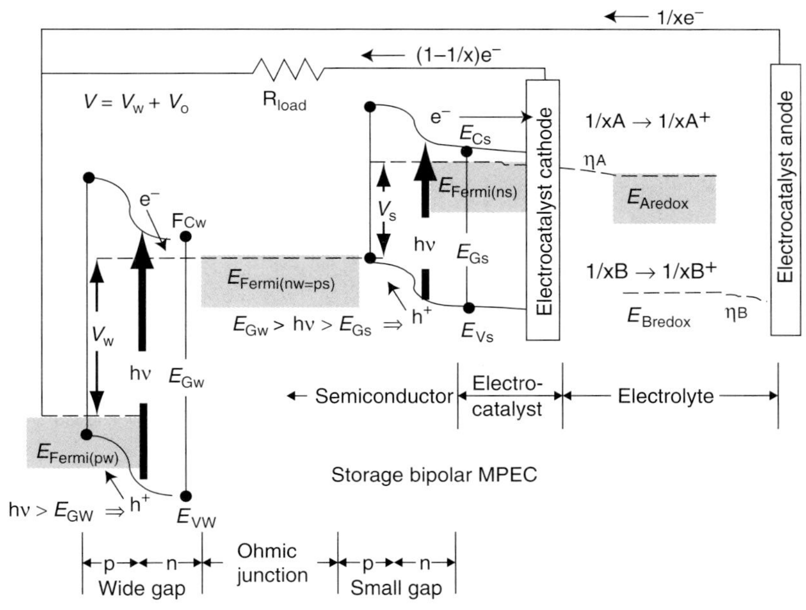

Fig. 7 Energy diagram for a bipolar band gap indirect ohmic storage multiple band gap photoelectrochemical solar cell (MBPEC).

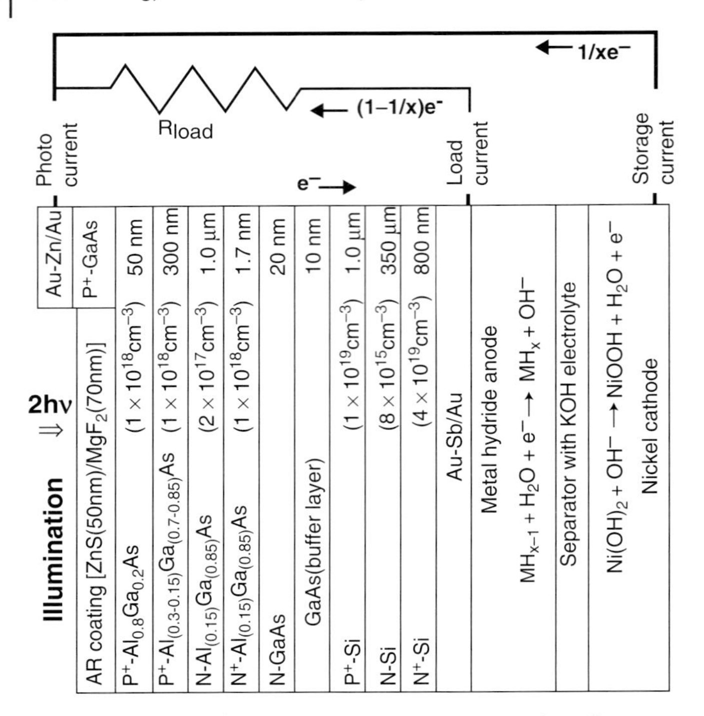

Fig. 8 The bipolar AlGaAs/Si/MH/NiOOH MBPEC solar cell.

$$\mathrm{NiOOH} + \mathrm{H_2O} + \mathrm{e^-} \longrightarrow \mathrm{Ni(OH)_2} + \mathrm{OH^-};$$

$$E_{\mathrm{NiOOH/Ni(OH)_2}} = 0.4\ \mathrm{V\ vs\ SHE} \quad (14)$$

As reported [49] and as shown in Fig. 9, the cell generates a light variation insensitive potential of 1.2–1.3 V at total (including storage losses) solar–electrical energy conversion efficiency of over 18%.

A long-term indoor cycling experiment was conducted to probe the stability of the AlGaAs/Si metal hydride storage solar cell [50]. Unlike the variable insolation of Fig. 9, in each 24-hour cycle, a constant simulated AM0 (135.3 mW $\mathrm{cm^{-2}}$) illumination was applied for 12 hours, followed by 12 hours of darkness, and the cell potential, and storage (charge and discharge) currents monitored as a function of time over approximately an eight-month period. Figure 10 presents representative results for two-day periods occurring 0, 40, 140, and 240 days into the experiment. As summarized in the lower curves of the figure, the load potential is again nearly constant, despite a 100% variation in illumination (AM0/dark) conditions. Over a 24-hour period, the load potential increases by ~2% as the cell charges with illumination, followed by a similar decrease in potential as stored energy is spontaneously released in the dark. The cycles exhibited in Fig. 10 are representative, and as observed exhibit little variation on the order of weeks, and exhibit a variation of ~1% over a period of months. In this figure, photopower is determined as the product of the measured cell potential and measured photocurrent. Power over load is determined as the

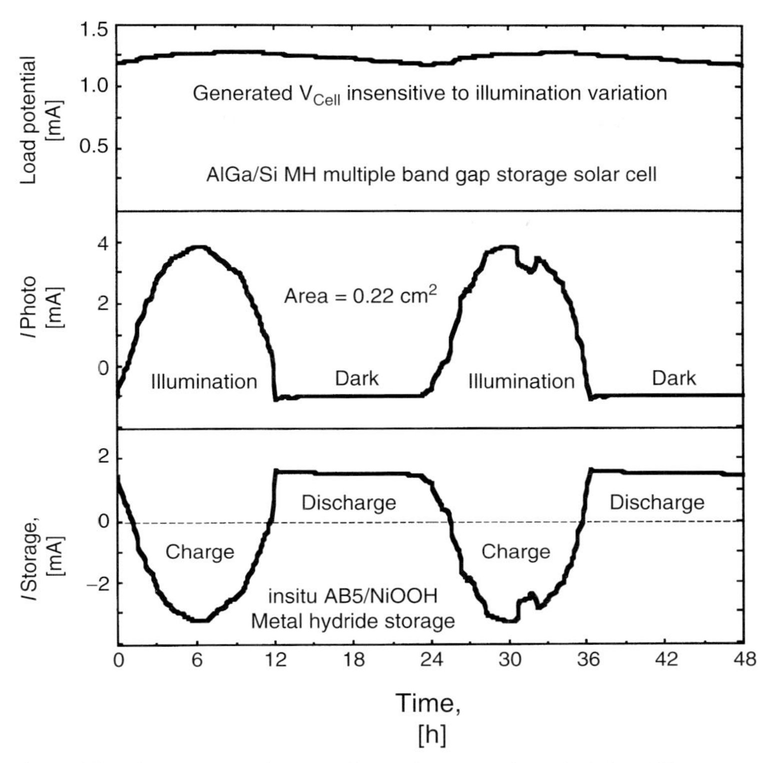

Fig. 9 Two days measured conversion and storage characteristics of the AlGaAs/Si/MH/NiOOH MBPEC solar cell.

product of the measured cell potential and measured load current.

Under constant 12-hour (AM0) illumination, the long-term indoor cycling cell generated a nearly constant photocurrent density of 21.2 (constant to within 1% or ± 0.2 mA cm^{-2}), and as seen in the top curves of Fig. 10, a photopower that varied by $\pm 3\%$. The cell's storage component exhibits the expected increase in charging potential with cumulative charging, which moves the system to a higher photopotential. The observed increase in photopower during 12 hours of illumination is because of this increase in photopotential with cumulative charging. A majority of the photogenerated power drives the redox cell, and the remainder consists of the power over load during illumination, as illustrated in Fig. 10. In the dark, inclusive of storage losses, the stored energy is spontaneously released and this power over load during both 12-hour illumination and 12-hour dark periods is also summarized. The cell is a single physical–chemical device generating load current without any external switching.

4.2.5.2 **PECS Driving an External Fuel Cell**

In the early 1980s, Texas Instruments, Inc. developed an innovative program based on a hybrid photovoltaic storage that used imbedded multilayer photoanode and photocathode silicon spheres and was

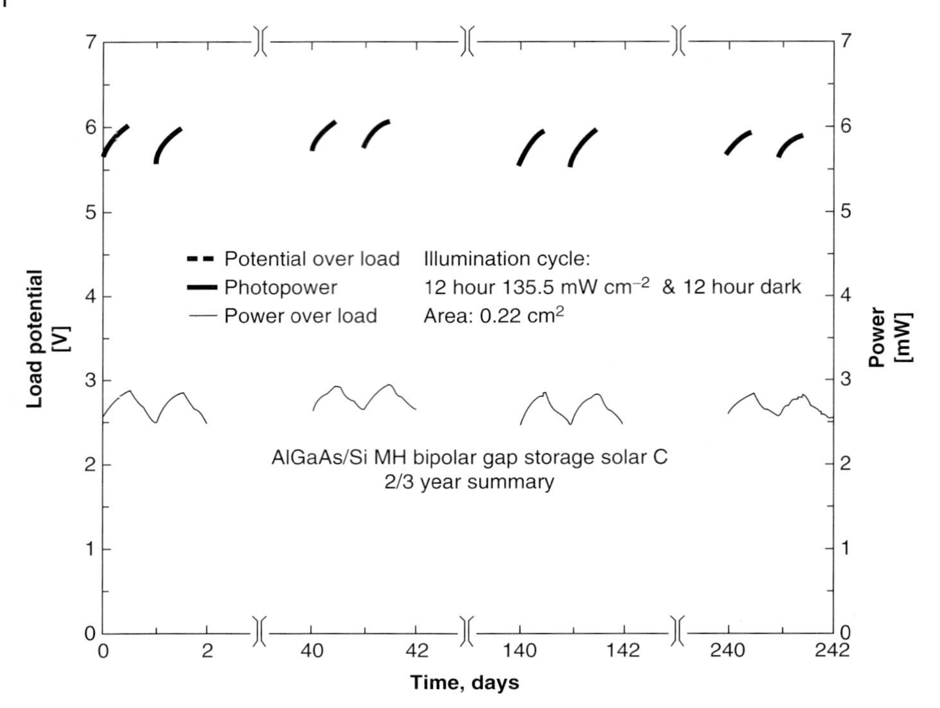

Fig. 10 Eight-months photopotential and power characteristics of the AlGaAs/Si/MH/NiOOH solar cell under fully charged AM0 conditions. Each day, the cell is illuminated for 12 hours and is in the dark for 12 hours.

designed to provide a close match between their maximum power point voltage and a solution phase bromine oxidation process in acidic solution. The program was discontinued, but the system has several attractive features.

Cell: 3 [51]. In the bromine-imbedded Si sphere system, energy stored as bromine is recovered in an external hydrogen bromine fuel cell. The conversion and storage reaction and cell configuration are summarized by

$$2HBr + H_2O + h\nu \longrightarrow H_2 + Br_2 + H_2O \qquad (15)$$

Photo anode:

Contact metal | Ohmic contact | n-Si | p-Si | Surface metal | Solution

Photo cathode:

Contact metal | Ohmic contact | p-Si | n-Si | Surface metal | Solution

A description of the Texas Instrument cell action is provided in Fig. 11.

4.2.6 Other Examples of Photoelectrochemical Storage Cells

Photoelectrochemical storage cells of different configurations have been suggested, designed, and tested for their performance, under sunlight or artificial illumination. Although none of these configurations has attained the high solar to electrical conversion and storage efficiency of the system in the earlier sections, they

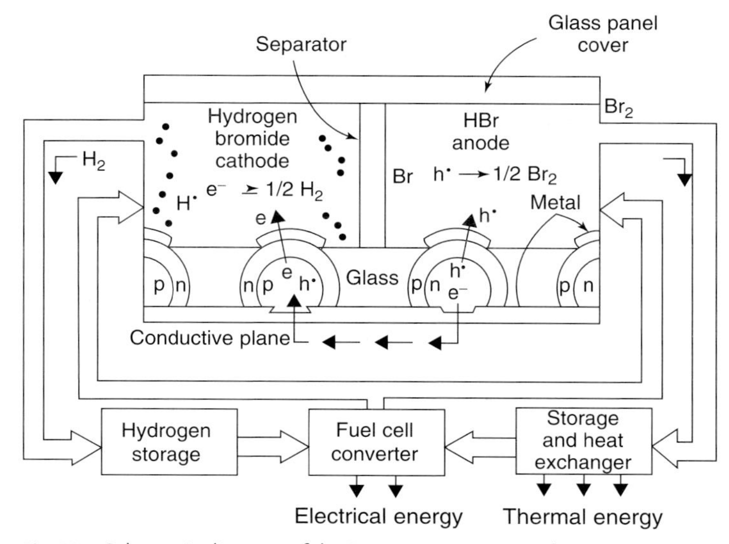

Fig. 11 Schematic diagram of the Texas Instruments Solar Energy System. Illumination occurs through the electrolyte to produce hydrogen gas that can be stored as a metal hydride and bromine that can be stored as aqueous tribromide. A hydrogen-bromine fuel cell is used to convert chemical to electrical energy and regenerate the hydrogen bromide electrolyte in a closed loop cyclic system. Thermal energy can be extracted through a heat exchanger.

are of significant scientific interest and form a solid basis for further development toward future systems. The following three sections, covering PECS with either a solution, solid, or intercalation storage redox processes, and provides a brief summary of many of these investigations, with a particular emphasis on their performance.

4.2.6.1 PECS Cells with Solution Phase Storage

Cell: 4 [52]. This is an example of the use of photoexcitable absorbers to promote a redox process using the following reaction sequence:

$$TH^+ + 2Fe^{2+} + 3H^+ \longrightarrow TH_4{}^{2+} + 2Fe^{3+}$$
$$(TH^+ = \text{Thionene acetate}) \quad (16)$$

The back reaction between Leucothionene ($TH_4{}^{2+}$) and Fe^{3+} is slow. Leucothionene is oxidized at the SnO_2 electrode.

$$TH_4{}^{2+} \longrightarrow TH^+ + 2e^- + 3H^+ \quad (17)$$

Ferric cation is reduced at a Pt electrode in a second compartment.

As a result, the concentration ratio of Fe^{3+}/Fe^{2+} is increased in the first compartment and decreases in the second compartment, which is equivalent to a difference in chemical potential. The system returns to its original uncharged state by discharging in the dark. Only 60 mV of potential difference is equivalent to a decade change of the Fe^{3+}/Fe^{2+} concentration ratio. Hence, this concentration cell does not generate a significant potential and the power density is low. The cell has the configuration

SnO_2 | 0.0001 M Thionene Acetate, 0.01 M $FeSO_4$, 0.001 M $Fe_2(SO_4)_3$ H_2SO_4, pH = 1.7–2 | Ion-exchange membrane | 0.01 M $FeSO_4$ 0.001 M $Fe_2(SO_4)_3$ | Pt

Studied cell characteristics: Polycrystalline photoelectrode, Pt counter electrode, E_{redox} of $Fe^{3+}/Fe^{2+} = 0.77$ V vs SHE, standard hydrogen electrode illumination 40–50 mW cm^{-2} by a tungsten lamp, initial current during discharge = 9.1 μA, and initial voltage = 10.9 mV.

Cell: 5 [53]. The photoelectrode used in this investigation was a powder-pressed sintered pellet. The high band gap of the semiconductor used (n-Pb_3O_4; E_g = 2.1 eV) and electron hole recombinations at grain boundaries have contributed to an observed low-conversion efficiency, which drives the overall cell reaction:

$$I^- + 6OH^- + 6Fe^{3+} \underset{\text{Light}}{\overset{\text{Dark}}{\rightleftarrows}} IO_3^- + 3H_2O + 6Fe^{2+} \qquad (18)$$

A saturated salt bridge is used between the cell compartments to minimize membrane IR loss, but it allows the active redox species to mix and chemically combine across the junction. The cell has the form

n-Pb_3O_4 | 0.1 M Fe^{3+}, Saturated Fe^{2+} | Salt bridge | 0.1 M IO_3^-, 0.1 M I^- | Pt

Studied cell characteristics: Polycrystalline photoelectrode, Pt counter electrodes, three-electrode configuration as in Fig. 4., E_{redox} of $Fe^{3+}/Fe^{2+} = 0.77$ V vs SHE, E_{redox} of $IO_3^-/I^- = -0 - 26$ V vs SHE, Illumination 60 mW cm^{-2} quartz halogen lamp, conversion efficiency = 0.09%, FF = 0.38, photopotential = 172 mV, V_{max} = 172 mV, charge efficiency of the battery = 74%, potential difference between Pt electrodes before charging = 720 mV, and after charging = 840 mV.

Cell: 6 [54]. In this cell, a wide gap semiconductor (E_g = 3.3), $BaTiO_3$ is used, capable of absorbing only near UV radiation and comprising less than 5% of available solar energy. This limits its practical use for solar energy conversion. The following storage couple is used:

$$Ce^{3+} + Fe^{3+} \underset{\text{Dark}}{\overset{\text{Light}}{\rightleftarrows}} Ce^{4+} + Fe^{2+} \qquad (19)$$

The storage system operates well away from the maximum power point of the semiconductor device, and therefore storage and discharge efficiency is poor. The cell uses a salt bridge between the two compartments and is of the form

Single crystal | n-$BaTiO_3$ | 0.1 M $Ce_2(SO_4)_3$, 0.005 M $Ce(SO_4)_2$ | Salt bridge | 0.1 M $Fe_2(SO_4)_3$, 0.005 M $FeSO_4$ | Pt

Studied cell characteristics: Pt counter electrodes, three-electrode configuration as in Fig. 4, E_{redox} of Ce^{4+}/Ce^{3+} = 1.45 V vs SHE, E_{redox} of Fe^{3+}/Fe^{2+} = 0.77 V vs SHE, illumination sunlight, conversion efficiency = 0.01%, FF = 0.26, photopotential = 730 mV, V_{max} = 0.33 V, charge efficiency 15%, potential across two Pt terminals of the charged cell = 0.60, and short-circuit current = 0.12 mA.

Cell: 7 [55]. In this study, attempts have been made to improve the behavior of a MoS_2 electrode in HBr electrolyte. Equilibration between the electrode and HBr has been improved by subjecting the electrode to a dark anodic potential. A ratio of 10 between the areas of counter and working electrode is another favorable feature in this study to minimize the polarization resistance at the counter electrode. Nafion-315 membrane contributes only a moderate resistance of 20 ohm cm^2. The storage reaction and cell configuration are

$$2Br^- + 3I_2 \underset{\text{Dark}}{\overset{\text{Light}}{\rightleftharpoons}} Br_2 + 2I_3^- \quad (20)$$

n-$MoSe_2$ | 0.1 M HBr, 0.01 M Br_2 | Nafion 315 | 1 M KI, 0.18 M I_2 | Pt

Studied cell characteristics: Single crystal photoelectrode, Pt counter electrodes, cell configuration as in Fig. 4, E_{redox} of Br/Br^- = 1.087 vs SHE, E_{redox} of I_3^-/I^- = 0.534 V vs SHE. Illumination 200 mW cm^{-2} Xe lamp, conversion efficiency = 6.2%, potential across two Pt terminals of the charged cell = 0.49, and short-circuit current = 0.5 mA.

Cell: 8 [55]. The next cell also uses a Nafion membrane, but makes use of n-CdSe to drive a polysulfide–polyselenide storage couple. Low-output power density is the biggest drawback in this cell. The storage reaction and cell configuration are

$$S^0 + Se^{2-} \underset{\text{Light}}{\overset{\text{Dark}}{\rightleftharpoons}} S^{2-} + Se^0 \quad (21)$$

n-CdSe | 1 M in Na_2S, S, NaOH | Nafion 315 | 1 M in M Na_2Se, Se, NaOH | Pt

Studied cell characteristics: Polycrystalline photoelectrode, Pt counter electrodes, cell configuration as in Fig. 4, E_{redox} of polysulfide electrolyte = −700 mV vs SCE, saturated calomel electrode, E_{redox} of $Se_2{}^{2-}/Se^{2-}$ = −800 mV vs SCE, illumination 100 mW cm^{-2} Xenon lamp, conversion efficiency 4%, FF = 0.45, photovoltage = −400 mV, charged cell has an open-circuit voltage of 60 mV, and initial current across a 100 ohm across Pt electrode = 0.5 mA.

Cell: 9 [44]. This study uses organic redox species for energy-storage purposes. Stability of the n-WSe_2 photoanode in iodine electrolyte and the stability of anthraquinone redox couple have been demonstrated in this study. Any H_2 evolution would carry out direct hydrogenation of AQ and associated side reactions, and therefore a carbon electrode has been selected because of the H_2 over potential on this electrode. The cell underwent several deep charge and discharge cycles with reproducible performance. The storage reaction and cell form are

$$2I^- + AQ + 2H^+ \underset{\text{Dark}}{\overset{\text{Light}}{\rightleftharpoons}} AQH_2 + I_2 \quad (22)$$

n-WSe_2 | 1 M KI, 0.1 M Na_2SO_4, 0.5 M H_2SO_4 | Saturated KCl bridge | 5×10^{-2} M AQ, 0.5 M H_2SO_4 | C

Studied cell characteristics: Single-crystal photoelectrode, C counter electrode during charging, Pt during discharging. The cell configuration is similar to that in Fig. 4. E_{redox} of I^{3-}/I^- = 0.534 V vs SHE, E_{redox} of AQ/AQH_2, illumination 150 mW cm^{-2} He–Ne laser (632.8 nm), conversion efficiency = 9%, discharge across a 10 ohm load produces a current of 1 mA cm^{-2}, and open-circuit voltage 200 mV.

Cell: 10 [44]. This study uses a p-WSe_2 photocathode rather than n-WSe_2. During the cell discharge, oxidation of AQH_2 at the surface of p-WSe_2 indicates that the electrode has the duel role of being a cathode during the charging and being the anode during the discharge. As discussed earlier, this limits the activity and low-current densities were observed. The storage reaction and cell configuration are

$$AQ + 2H^+ + 2I \underset{\text{Dark}}{\overset{\text{Light}}{\rightleftharpoons}} AQH_2 + I_2 \quad (23)$$

p-WSe_2 | 5×10^{-2} M AQ | Saturated | 1 M KI, 0.5 M H_2SO_4 | Pt
Single crystal | 0.5 M H_2SO_4 | Salt bridge | 0.5 M Na_2SO_4 |

Cell: 11 [55]. The theoretical band gap of WSe_2 provides a near ideal single band gap match for the solar spectrum. But the following cell has some disadvantages. These include the low solubility of the storage redox couple employed, MV^{2+} and $MV^{+\cdot}$ and the possibility of undesirable side reactions of the radical ion $MV^{+\cdot}$. Using dual (*n*-type and *p*-type) photoelectrodes expands the potential regime one can access for the redox-storage couple. The storage reaction and cell configuration are

$$2I^- + 2MV^{2+} \underset{\text{Dark}}{\overset{\text{Light}}{\rightleftharpoons}} 2MV^{+\cdot} + I_2 \quad (24)$$

$n\text{-}WSe_2 \mid I^- \mid MV^{2+} \mid p\text{-}WSe_2$

4.2.6.2 PECS Cells Including a Solid Phase–Storage Couple

The earlier experimental investigations, Cells 2–9, use only solution phase redox couples. However, as indicated in the following examples, a solid phase–storage couple may also be employed, which in principle tends to increase the cell's storage capacity.

Cell: 12 [57]. Having at least one component in insoluble form may add compactness into the cell configuration, although low conductivity of the insoluble active component may cause significant polarization losses associated with the storage electrode, as exemplified by the low conductivity of silver (chloride) in one of the next cells. The next four cells use a TiO_2 polycrystalline photoelectrode. In the first cell, the storage reaction and cell configuration are

$$2H_2O + 4Ag^+ \underset{\text{Dark}}{\overset{\text{Light}}{\rightleftharpoons}} 4H^+ + 4Ag + O_2 \quad (25)$$

TiO_2 | 1 M HNO_3, 1 M KNO_3 | Anion Specific Membrane | 1 M $AgNO_3$, 1 M KNO_3 | Ag

Studied cell characteristics: Polycrystalline photoelectrode, Pt counter electrodes, cell configuration is similar to Fig. 4, E_{redox} of O_2, H^+/H_2O couple = 1.23 V vs NHE at pH = 1, E_{redox} of $Ag/Ag^+ = 0.80$ V vs NHE, normal hydrogen electrode illumination 500 W Hg lamp, conversion efficiency = 1%, photopotential = 0.28 V vs NHE, open-circuit voltage of the charged cell = 0.28 V, and short-circuit current = 0.3 mA cm^{-2}.

Cell: 13 [57]. Of the four TiO_2 Cells 10–13, the following cell exhibited the highest short-circuit discharge current and voltage. However, during the charging process, a stationary concentration of Ce^{4+} was observed in the photoanode compartment. This suggest the existence of competing process that consumes the oxidized species Ce^{4+}. The later is known to participate in photochemical reactions under illumination [10]. Considering the low concentration of the reduced form of active materials used with the photoanode, there is a possibility that the water oxidation becomes the dominant process during charging. In this study, it was observed that with a passage of a charge of 10 coulomb during charging, Ce^{4+} present in the photoanode compartment accounted for only 22% of the charge. In this second TiO_2 photoelectrode cell, the storage reaction and cell configuration are

$$Ce^{3+} + Ag^+ \underset{\text{Dark}}{\overset{\text{Light}}{\rightleftharpoons}} Ce^{4+} + Ag \quad (26)$$

TiO_2 | 1 M HNO_3, 0.05 M $Ce_2(SO_4)_3$, 0.1 M $Ce(SO_4)_2$ | Anion Specific Membrane | 1 M $AgNO_3$, 1 M KNO_3 | Ag

Studied cell characteristics: Polycrystalline photoelectrode, Pt counter electrode, cell configuration illumination etc. are similar to the earlier cell, E_{redox}

of Ce^{4+}/Ce^{3+} vs SHE, charge efficiency of the cell without stirring = 18%, open-circuit voltage of the charged cell = 0.76, and initial short-circuit current 1.3 mA cm^{-2}.

Cell: 14 [57]. In this third TiO_2 photoelectrode cell, the storage reaction and cell configuration are

$$2Fe^{2+} + Cu^{2+} \underset{\text{Dark}}{\overset{\text{Light}}{\rightleftharpoons}} 2Fe^{3+} + Cu \quad (27)$$

TiO_2 | 1 M KNO_3, 0.01 M $FeSO_4$ | Anion Specific Membrane | 0.025 M $CuSO_4$, 1 M KNO_3 | Cu

Studied cell characteristics: Polycrystalline photoelectrode, E_{redox} of Cu^{2+}/Cu = 0.34 V vs NHE, E_{redox} of Fe^{3+}/Fe^{2+} = 0.77 vs NHE, open-circuit voltage of the charged cell = 0.3, and short-circuit current 1.5 mA cm^{-2}.

Cell: 15 [57]. The wide band gap of TiO_2 is not an appropriate match to the solar spectrum. In this fourth TiO_2 photoelectrode cell, the storage reaction and cell configuration are

$$Fe^{2+} + AgCl \underset{\text{Dark}}{\overset{\text{Light}}{\rightleftharpoons}} Fe^{3+} + Ag + Cl^- \quad (28)$$

TiO_2 | 0.2 M KCl, 0.01 M $FeCl_2$ |Anion Specific Membrane | 0.2 M KCl | AgCl | Pt

Studied cell characteristics: Polycrystalline photoelectrode, cell configuration and illumination are the same as in the previous cell, E_{redox} Fe^{3+}/Fe^{2+} = 0.77 V vs NHE, E_{redox} of $AgCl/Ag^+$ = 0.22 V vs NHE, open-circuit voltage of the charged cell = 0.39 V, and short-circuit current = 0.4 mA cm^{-2}.

Cell: 16 [58]. In this next cell, Ni is deposited during charge at 80% charge efficiency. Losses may be because of the competing reaction of H_2 evolution. Cell voltage of the charged cell is higher than the photovoltage available, which indicates the possible influence of another redox couple $Ni(OH)_2/NiOH^-$ occurring at a higher redox potential. Only about 55% of the charge stored can be recovered during discharge. The possibility of self-discharge reactions because of imperfect permeability of the membrane has been cited as a possible cause, and is further complicated by the complex ferro/ferricyanide equilibria that is known to occur (Licht, 1995). In this cell, the storage reaction and cell configuration are

$$2Fe(CN)_6{}^{4-} + Ni^{2+} \underset{\text{Dark}}{\overset{\text{Light}}{\rightleftharpoons}} Fe(CN)_6{}^{3-} + Ni \quad (29)$$

n-GaP | 0.2 M K_2SO_4, pH = 6.7, 0.05 M $K_3Fe(CN)_6$, 0.05 M $K_4Fe(CN)_6$ | Anion Specific Membrane | 0.05 M K_2SO_4, 0.2 M $NiSO_4$, 0.06 M $NiCl_2$ | Pt

Studied cell characteristics: Single crystal photoelectrode, Pt counter electrodes, cell configuration is similar to Fig. 4. E_{redox} of $Fe(CN)_6{}^{3-}/Fe(CN)_6{}^{4+}$ is 0.36 V vs NHE, E_{redox} of Ni^{2+}/Ni = −0.25 V vs NHE, illumination 500 W Hg lamp, conversion efficiency 13% for 450–540-nm region, photovoltage = 0.63 V, open-circuit voltage of the charged cell = 0.75 V, short-circuit current = 4.3 mA cm^{-2}, and charge efficiency = 55%.

Cell: 17 [55]. The conversion efficiency data in the following cell reflect the poor quality of the GaAs material that was used, although in other studies, there has been higher efficiency GaAs PEC (without storage). In this study, significant polarization was observed and performance data of the storage cell was not reported. The storage reaction and cell configuration are

$$\mathrm{Cd + Se_2{}^{2-} + 2OH^- \underset{Light}{\overset{Dark}{\rightleftharpoons}} Cd(OH)_2 + 2Se^{2-}} \quad (30)$$

n-GaAs | 0.1 M Na_2Se, 0.1 M Se, 1 M NaOH | Nafion | 2 M NaOH | Cd

Studied cell characteristics: Single crystal photoelectrode, Pt counter electrode, cell configuration is similar to Fig. 4, E_{redox} of $Se^{2-}/Se_2{}^{2-}$ = −800 mV vs SCE, E_{redox} of $Cd/Cd(OH)_2$ = −1050 mV vs SCE. Illumination 100 mW cm^{-2} Xe lamp, conversion efficiency = 4%, FF = 0.53, photopotential = −500 mV, and short-circuit discharge current of the storage cell in the dark using Pt electrodes = 14.6 mA cm^{-2}.

Cell: 18 [55]. The next two cells use a polycrystalline n-CdSe photoanode. The following cell exhibited steady current-time and voltage-time curves during the photoelectrochemical charging and dark discharging. The flat discharge curve prevailed until the capacity of the sulfide electrolyte is exhausted. The storage reaction and cell configuration are

$$\mathrm{Cd + S + 2OH^- \underset{Light}{\overset{Dark}{\rightleftharpoons}} Cd(OH)_2 + S^{2-}} \quad (31)$$

n-CdSe | 0.1 M in NaOH, Na_2S, 1 M in S, Na_2Se, Se | Nafion | 2 M NaOH | Cd

Studied cell characteristics: Polycrystalline photoelectrode, Pt counter electrodes, cell configuration is similar to Fig. 4. E_{redox} of $S_x{}^2/S^{2-}$ = −700 mV vs SCE, E_{redox} of $Cd(OH)_2/Cd$ = −1050 mV vs SCE, illumination 100 mV cm^{-2} Xe light, conversion efficiency = 4%, FF = 0.45, photovoltage = −400 mV during discharge through Pt and Cd electrodes with a 100-ohm load, and a current of 8.3 mA cm^{-2} flowed at cell voltage close to 175 mV.

Cell: 19 [59]. In this study, the possibility of using organic semiconductors to drive storage processes is demonstrated. The process is in principle similar to a concentration cell. During photocharging, Prussian Blue (PB, $Fe_4[Fe^{II}(CN)_6]_3$) is reduced at the photocathode and PB is oxidized at the anode. In the dark, the redox process involving PB is reversed producing an electron flow. Process ability, stability, and lack of photocorrosion make these low band gap organic materials very attractive for photoelectrochemical applications. However, they are defect-based systems, and the very low conversion efficiencies and self-discharge appear to outweigh these benefits. The storage reaction and cell configuration are

$$\text{bilayer electrode: } \mathrm{Fe_4^{II}[Fe^{II}(CN)_6]_3{}^{4-} + 4h^+ \underset{Dark}{\overset{Light}{\rightleftharpoons}} Fe_4^{III}[Fe^{II}(CN)_6]_3} \quad (32)$$

$$\text{counter electrode: } \mathrm{Fe_4^{III}[Fe^{II}(CN)_6]_3 + 4e^- \underset{Dark}{\overset{Light}{\rightleftharpoons}} Fe_4^{II}[Fe^{II}(CN)_6]_3{}^{4-}} \quad (33)$$

ITO | P3MT | PB | 0.2 M KCl, 0.1 M HCl | PB | ITO

Studied cell characteristics: Illumination 500 W Xenon lamp, the ITO/P3MT electrode has open-circuit voltage = 0.44 V, short-circuit photocurrent 0.09 μA cm^{-2}, and charge efficiency of the storage cell = 40%.

Cell: 20 [60]. Metal ions introduced into a solid β-alumina lattice behave like ions in solution. This study illustrates a compact solid-state storage cell that can be charged using solar energy. During charging, Fe and Ti change their oxidation state and the charge balance is maintained by the migration of Na^+ ions from one phase to the other. In the actual cell design, an *n*-type semiconductor is connected to the alumina phase containing Ti and *p*-material is connected to the phase containing Fe. Limitations are the comparatively slow diffusion of ions in the solid electrolyte and resistance to ionic movement at various phase boundaries, and lower the energy output during discharge. In this device, back wall illumination demands the use of very thin semiconductor layers to minimize absorption losses and has the general form

n-semiconductor | $Na_2O.11(AlFeO_3)$ | $Na_2O.11(Al_2O_3)$ | $Na_2O.11(AlTiO_3)$ | *p*-semiconductor

Cell: 21 [47]. In this detailed study, selection of a Nafion-315 membrane was done on the basis of (1) stability in high alkaline sulfide solutions, (2) low IR drop, and (3) low permeability to sulfide. Maintaining an area ratio of 1 : 8 between photo and storage electrodes has minimized polarization at the storage electrode. The storage system was driven by three semiconductor PEC devices connected in series. Charging was done up to 90% of the capacity followed by complete discharge. Overall observed charge efficiency was 83%. Although the system was not fully optimized with respect to photoelectrode, electrolyte, and storage, voltage efficiency of 75% was obtained during discharge. Discharge curves were flat until the stored active material was fully consumed. The storage reaction and cell configuration are

$$S^{2-} + Zn(OH)_4{}^{2-} \underset{\text{Dark}}{\overset{\text{Light}}{\rightleftharpoons}} S^0 + Zn + 4OH^- \quad (34)$$

n – CdSe | 1 M in NaOH, Na_2S, S Nafion-315 | 0.1 M ZnO, 1 M NaOH | C

Studied cell characteristics: Polycrystalline photoelectrode, Ni counter electrode, basic cell configuration is based on Fig. 4, E_{redox} of $S_x{}^2/S^{2-} = 0.500$ V vs SHE, E_{redox} of $Zn/Zn(OH)_4{}^{2-} = -1.25$ V vs SHE, artificial illumination, conversion efficiency = 3%, photovoltage = −0.50 V, during discharge through 75-ohm load between C and Ni discharge current = 10 mA, and voltage = 0.6 V.

Cell: 22 [61]. This cell takes advantage of photocorrosion to drive a storage cell. Under illumination, *n*-CdSe is decomposed and *p*-CdSe is electroplated, and the reverse occurs during cell discharge. However, photoactivity depends on an optimized semiconductor surface, and in an environment where the surface is changed constantly, the surface optimization is lost. This and the poor kinetics of the *p*-type photoreduction result in a continual deterioration of the photoactivity and cause low photoefficiency and low-discharge power density. The storage reaction and cell configuration are

$$CdSe + 2h^+ \underset{\text{Dark}}{\overset{\text{Light}}{\rightleftharpoons}} Se^0 + Cd^{2+} \quad (35)$$

and the other electrode in photoelectroplated by Cd

$$CdTe + 2e^- \underset{\text{Dark}}{\overset{\text{Light}}{\rightleftharpoons}} Cd^0 + Te^{2-} \quad (36)$$

n-CdSe | 0.1 M $CdSO_4$ | *p*-CdTe

Cell: 23 [46]. This is a detailed study of a thin film cell with moderately high outdoor solar efficiency, high storage efficiency, and an output that is highly invariant despite changing illumination. This study provides extensive details of the choice of photoelectrode, membrane, and electrochemistry of the tin–tin sulfide redox storage. Cd(Se,Te) electrodes, compared to CdSe, improve the band gap match and increase solar-conversion efficiency. Two photoelectrodes in series were used to provide a voltage match to the storage redox couple in a cell of the form of Fig. 12.

The conversion and storage reactions and cell configuration are presented as

$$SnS + 2e \underset{\text{Dark}}{\overset{\text{Light}}{\rightleftharpoons}} S^{2-} + Sn \text{ Storage} \quad (37)$$

$$S^{2-} \underset{\text{Dark}}{\overset{\text{Light}}{\rightleftharpoons}} S + 2e \text{ Photo electrode} \quad (38)$$

n-Cd(Se,Te) | 2 M in NaOH, Na_2S, S | Redcad Membrane | 2 M in NaOH, Na_2S | SnS | Sn

Studied cell characteristics: Bipolar series polycrystalline photoelectrode, CoS counter electrodes, cell configuration is as shown in Fig. 4 without the need of switches E or F. E_{redox} of $S/S^{2-} = -0.48$ V vs NHE, E_{redox} of $SnS/Sn,S^{2-} = -0.94$ V vs NHE. Illumination sunlight, 500 mWhr cm^{-2} per day, conversion efficiency 6–7%, photovoltage = −600 mV, and storage efficiency >90%. After two weeks of continuous operation the overall solar to electrical efficiency (including conversion and storage losses) is 2–7%.

Cell: 24 [4]. The earlier cell is improved by a series of solution phase optimizations (cesium electrolyte with low hydroxide and optimized polysulfide), to provide a higher photopotential and improved stability and also the use of a single crystal, rather than thin film, Cd(Se,Te) to also improve photopotential and cell efficiency, as described earlier in Fig. 12. Because of the higher photopotential, only a single photoelectrode is required to match the storage potential and high overall efficiencies are observed. The cell has the design as shown in beginning of the chapter (Fig. 2) and uses conversion and storage reactions described in the earlier cell and a configuration

n-Cd(Se,Te) | 0.8 M Cs_2S, 1 M Cs_2S_4 | Redcad Membrane | 1.8 M Cs_2S | SnS | Sn

Studied cell characteristics: The PEC had a power conversion efficiency of 12.7% under 96.5 mW cm^{-2} insolation and voltage at maximum power point was of −1.1 V vs SHE, sufficient to drive the SnS/Sn storage system. Under direct illumination, the 0.08 cm^2 single crystal photoelectrode generated more than 1.5 mA through the 3 cm^3 SnS electrode driving SnS reduction while supporting 0.33 mA through a 1500 load simultaneously at a photogenerated 0.495 V. In the dark spontaneous oxidation drive, the load with storage efficiency over 95%. The total conversion efficiency, including conversion and storage losses, was 11.8%.

4.2.6.3 PECS Cells Incorporating Intercalation

In photointercalation, illumination drives insertion storage into layer type compounds [62]. The photointercalation process can be characterized as

$$TX_2 + e^-(h) + p^+(h) + M_{sol}{}^+ \longrightarrow TM_{IN}X_2 + p^+ \quad (39)$$

where TX_2 is generally a nonintercalated transition metal dichalcogenide. For this process to occur without the assistance of an external power source, a counter electrode is driven at an electrode potential negative to that of the layer type intercalating electrode. The process is generally restricted to p-type materials. The development of this concept has been slow because of dearth of materials that are stable semiconductors and at the same time behave as

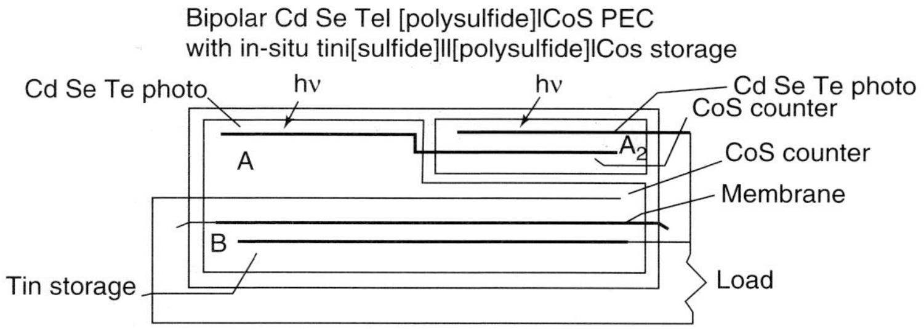

Fig. 12 A bipolar thin film photoelectrochemical solar cell with in situ storage. Compartments A and A_2 contain alkali polysulfide solution and compartment B contains alkali sulfide solution.

intercalating compounds that are able to exchange guest ions and molecules with an electrolyte in a reversible manner, and yet that is not disruptive to photon absorption.

Cell: 25 [63]. In this cell, E_{redox} of copper thiophosphate is variable depending on the degree of intercalation. A limitation of this system is poor-discharge kinetics and low-energy density of the discharge. The cell configuration is given by

Cu_3PS_4 | 0.02 M CuCl | CH_3CN | Cu_2S

Studied cell characteristics: E_{redox} of $Cu^+/Cu^0 = -0.344$ vs NHE, illumination 117 mW cm^{-2} Xe lamp, photopotential = 100 mV, charging current < 50 μA cm^{-2}, and discharge current < 10 μA cm^{-2}.

Cell: 26 [64]. This cell illustrates another all solid state design for a thin storage cell. *p*-Cu_xS changes its electrode potential with changes in its composition. During charging, Cu is oxidized at *n*-CdS surface while it is reduced at the Cu electrode. Between the two electrodes Cu^+ ion transport process takes place in the solid state electrolyte. The cell configuration is given by

Cu | *n*-CdS | *p*-Cu_2S | $RuCl_4I_5Cl_{3.5}$ | Cu

Cell: 27 [64]. As with the earlier cell, this final cell requires a very thin design because to reach the junction, light has to travel several layers. The cell functions in the same manner as the earlier cell, and the configuration is given by

Conductive Glass | Cu | Cu^+ Conducting solid electrolyte | *p*-Cu_2Te | *n*-CdTe | Mo

4.2.7 Summary

Conversion and storage of solar energy is of growing importance as fossil fuel energy sources are depleted and stricter environmental legislation is implemented. Although society's electrical needs are largely continuous, clouds and darkness dictate that photovoltaic solar cells have an intermittent output. Photoelectrochemical systems have the potential to not only convert but also store incident solar energy. Design component and system considerations and a number of photoelectrochemical solar cells with storage have been reviewed in this chapter.

Acknowledgment

S. Licht is grateful to Dharmasena Peramunage and for support by the BMBF *Israel–German* Cooperation.

References

1. S. Licht, O. Khaselev, T. Soga et al., *Electrochem. Solid State Lett.* **1998**, *1*, 20.
2. (a) S. Licht, O. Khaselev, T. Soga et al., *J. Phys. Chem. 102*, 2536; (b) *ibid*, 2546.
3. B. O'Regan, M. Grätzel, *Nature* **1991**, *353*, 737.
4. S. Licht, G. Hodes, R. Tenne et al., *Nature* **1987**, *326*, 863.
5. G. De Maria, L. D'Alesso, E. Coffari et al., C. A. Tiberio, *Solar Energy* **1985**, *35*, 409.
6. M. Grätzel, *Ber. Bunsenges. Phys. Chem.*, **1980**, *84*, 981.
7. K. Kalyanasundaram, *Coord. Chem. Rev.* **1982**, *46*, 159.
8. A. Harriman, *Photochemistry* **1986–1987**, *19*, 509.
9. K. Kutal, *Chem. Educ.* **1984**, *60*, 882.
10. L. J. Heidt, A. F. McMillan, *Science* **1953**, *117*, 712.
11. R. Memming, *Top. Curr. Chem.* **1988**, *143*, 79.
12. A. Fujishima, K. Honda, *Nature* **1972**, *238*, 37.
13. J. M. Bolts, A. B. Bocarsly, M. C. Palazzotto et al., *J. Am. Chem. Soc.* **1979**, *101*, 1378.
14. H. P. Maruska, K. A. Ghosh, *Solar Energy* **1978**, *20*, 443.
15. G. Hodes, L. Thompson, J. Dubow et al., *J. Am. Chem. Soc.* **1983**, *105*, 324.
16. R. C. Kaintala, B. Zelenay, J. OM. Bockris, *J. Electrochem. Soc.* **1986**, *133*, 248.
17. R. C. Kaintala, J. OM. Bockris, *Int. J. Hydrogen Energy* **1988**, *13*, 3712.
18. O. Khaselev, J. Turner, *Science* **1998**, *280*, 4212.
19. S. Licht, B. Wang, T. Soga et al., H. Tributsch, *J. Phys. Chem. B.* **2000**, *104*, 8920.
19. S. Licht, B. Wang, T. Soga et al., H. Tributsch, *J. Phys. Chem. B.* **2000**, *104*, 8920.
20. J. Manassen, G. Hodes, D. Cahen, *CHEMTECH* **1981**, 112.
21. A. J. Bard, M. S. Wrighton, *J. Electrochem. Soc.* **1977**, *124*, 1706.
22. A. Heller, G. P. Schawartz, R. G. Vadimisky et al., *J. Electrochem. Soc.* **1978**, *25*, 1156.
23. G. Hodes, J. Manassen, D. Cahen, *Nature* **1976**, *261*, 403.
24. K. W. Frese, *Appl. Phys. Lett.* **1982**, *40*, 2712.
25. F.-R. F. Fan, B. Wheeler, A. J. Bard et al., *J. Electrochem. Soc.* **1981**, *128*, 2042.
26. B. Miller, S. Licht, M. E. Orazem et al., Photoelectrochemical Systems, *Crit. Rev. Surf. Chem.* **1994**, *3*, 29.
27. D. Cahen, G. Hodes, J. Manassen et al., *ACS. Symp. Ser. No.* **1980**, *146*, 369.
28. G. Hodes, in *Energy Resources Through Photochemistry and Catalysis*, (Ed.: M. Gratzel), Academic Press, New York, 1983, p. 242.
29. A. Heller, K. C. Chang, B. Miller, *J. Electrochem. Soc.* **1977**, *124*, 697.
30. R. Tenne, G. Hodes, *Appl. Phys. Lett.* **1980**, *37*, 428.
31. M. Tomkiewicz, I. Ling, W. S. Parsons, *J. Electrochem. Soc.* **1982**, *129*, 2016.
32. J. Reichman, M. A. Russak, *J. Appl. Phys.* **1982**, *52*, 708.
33. B. A. Parkinson, A. Heller, B. Miller, *J. Electrochem. Soc.* **1979**, *126*, 954.
34. G. Hodes, J. Manassen, D. Cahen, *J. Electrochem. Soc.* **1980**, *127*, 544.
35. A. J. McEroy, M. Etman, R. Memming, *J. Electroanal. Chem.* **1985**, *190*, 2212.
36. J. Reichman, M. A. Russak, *J. Electrochem. Soc.* **1984**, *131*, 796.
37. D. Cohen, Y. W. Chen, *Appl. Phys. Lett.* **1984**, *45*, 746.
38. S. Licht, N. Myung, R. Tenne et al., *J. Electrochem. Soc.* **1995**, *142*, 840; S. Licht, N. Myung, ibid, L129; ibid, 8412.
39. S. Licht et al., *J. Electrochem. Soc.* **1985**, *132*, 1076; ibid, **1986**, *133*, 52; ibid, 272; ibid, 277; ibid, 269.
40. S. Licht, *Solar Energy Mat. Solar Cells* **1995**, *38*, 3012.
41. S. Licht, *Nature* **1987**, *330*, 148.
42. S. Licht, D. Peramunage, *Solar Energy* **1994**, *52*, 197.
43. S. Licht, D. Peramunage, *Nature* **1990**, *345*, 330.
44. B. Keita, L. Nadjao, *J. Electroanal. Chem.* **1984**, *163*, 171.
45. S. Licht, F. Forouzan, *J. Electrochem. Soc.* **1995**, *142*, 1539; ibid, 1546.
46. S. Licht, J. Manassen, *J. Electrochem. Soc.* **1987**, *134*, 1064.
47. P. Bratin, M. Tomkiewicz, *J. Electrochem. Soc.* **1982**, *129*, 2649.
48. C. H. Henry, *J. Appl. Phys.* **1980**, *51*, 4494.
49. S. Licht, B. Wang, T. Soga et al., *Appl. Phys. Lett.* **1999**, *74*, 4055.
50. B. Wang, S. Licht, T. Soga et al., *Solar Energy Mat. Solar Cells.* **2000**, *64*, 311.
51. J. White, F-R. Fan, A. J. Bard, *J. Electrochem. Soc.* **1985**, *132*, 544.

52. G. W. Murphy, *Solar Energy* **1978**, *21*, 403.
53. M. Sharon, S. Kumar, N. P. Sathe et al., *Solar Cells* **1984**, *12*, 353.
54. M. Sharon, A. Singha, *Int. J. Hydrogen Energy* **1982**, *7*, 557.
55. P. G. Ang, P. A. F. Sammells, Faraday Discussions of the Chemical Society, General Discussions 1980, No. 70, Photoelectrochemistry, St. Catherine's College, Oxford, Sept 8–10.
56. F. -R. F. Fan, H. S. White, B. L. Wheeler et al., *J. Am. Chem. Soc.* **1980**, *102*, 5142.
57. H. Hada, K. Takaoka, M. Saikawa, Y. Yonezawa, *Bull. Chem. Soc. Jpn.* **1981**, *54*, 1640.
58. Y. Yonezawa, M. Okai, M. Ishino et al., *Bull. Chem. Soc. Jpn* **1983**, *56*, 2873.
59. M. Kaneko, K. Takagashi, E. Tsuchida, *J. Electroanal. Chem.* **1987**, *227*, 2512.
60. A. Sammells, A. Ang, *Patent* 1980, US 4,235,9512.
61. H. J. Gerritsen, W. Ruppel, P. Wurfel, *J. Electrochem. Soc.* **1984**, *131*, 2037.
62. H. Tributsh, *Appl. Phys.* **1986**, *23*, 61.
63. G. Betz, S. Fiechter, H. Tributsch, *J. Appl. Phys.* **1987**, *62*(11), 4597.
64. T. Tonomura, K. Teratoshi, 1986, JP Patent 62,249,366; ibid., JP Patent 62,249, 3612.

4.3 Solar Photoelectrochemical Generation of Hydrogen Fuel

Maheshwar Sharon
Indian Institute of Technology, Bombay, India

Stuart Licht
Israel Institute of Technology, Technion, Israel

4.3.1 Introduction

Photoelectrolysis is a vast field and it is difficult to cover all its aspects in this chapter. Therefore, this chapter confines its discussions to photoelectrolysis of water to obtain hydrogen gas. Hydrogen (H_2) is an important renewable source of energy, as water is the main precursor for hydrogen. Water (H_2O), on electrolysis produces hydrogen and oxygen (O_2) in the ratio of 2 : 1. Moreover, when hydrogen is utilized as a fuel and oxidized to release its heat chemically (burnt) or electrochemically (in a hydrogen or air fuel cell), it produces water by consuming similar ratio of hydrogen and oxygen. Unlike other sources of energy such as coal, gas, and oil, hydrogen is a very clean source of energy. Hence, the hydrogen ⇔ water cycle seems to provide the most appropriate path toward a renewable source of energy.

4.3.2 Theoretical Consideration for Water Electrolysis

Solar energy–driven water splitting combines several attractive features for energy utilization. Both the energy source (sun) and the reactive media (H_2O) are readily available and are renewable, and the resultant fuel (generated hydrogen) and the emission with fuel consumption (H_2O) are each environmentally benign. Insolation (solar radiation) on semiconductors can generate significant electrical, electrochemical, or chemical energy [1–3]. Efficient solar-driven water splitting requires a critical balance of the energetics of the solar conversion and the solution phase redox processes. The UV and visible energy-rich portion of the solar spectrum is transmitted through H_2O (Fig. 1 [4] and Fig. 2 [5]). Therefore, sensitization, such as via semiconductors, is required to derive the water-splitting process. In a solar photoelectrolysis system, the redox-active interfaces can be in indirect or direct contact with the photosensitizer and can comprise either an ohmic or a Schottky junction. Independent of this interface composition, the various parameters in models predicting solar water-splitting conversion efficiency may be combined into general parameters: (1) related to losses in optical energy conversion, η_{photo} or (2) related to losses in redox conversion of H_2 and O_2, $\eta_{electrolysis}$. A combination of these parameters yields an overall solar to electrolysis efficiency (excluding storage and utilization losses) as

$$\eta_{photoelectrolysis} = \eta_{photo} \times \eta_{electrolysis} \quad (1)$$

Early photoelectrolytic attempts used solution redox species such as $Ce^{3+/4+}$ and displayed poor quantum yield [6]. Further studies utilized semiconductors, such as TiO_2 (band gap, $E_g = 3.0$ eV) [7] or $SrTiO_3$ ($E_g = 3.2$ eV) [8]. The wide E_g excludes longer wavelength insolation, leading to poor efficiency. H_2 and O_2 evolutions have been enhanced using large surface area or catalyst addition, but energy conversion efficiency remains low [9]. Early photoelectrolysis systems also combined p-type and n-type photoelectrodes [10, 11]. A two or more band gap configuration can provide efficient matching of the solar spectra.

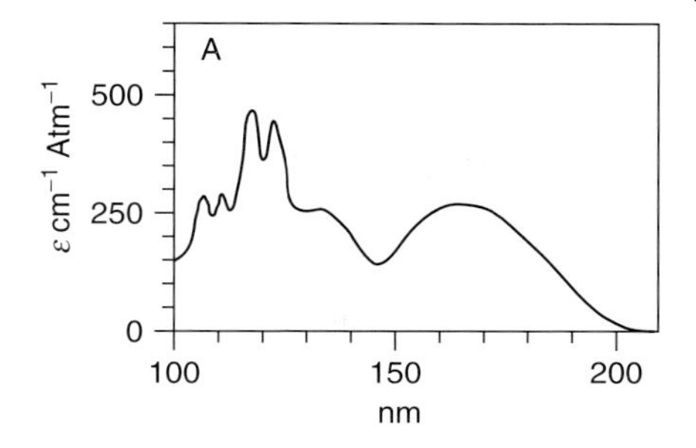

Fig. 1 Absorption spectrum of water (after K. W. Atanabe and M. Zelikoff, *J. Opt. Soc. Am.* **1953**, *43*, 753).

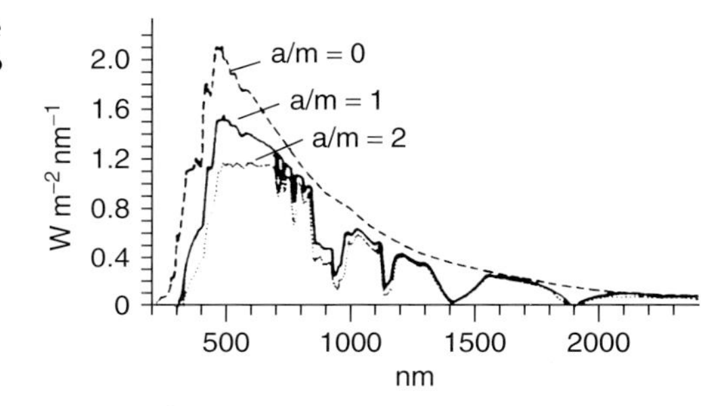

Fig. 2 Solar spectral irradiance (w m^{-2} nm^{-1}) at air-mass ratio (a/m) of 0, 1, and 2 (after Nikola Getoff, *Int. J. Hydrogen Energy* **1990**, *15*(6), 407).

p-InP and n-GaAs were demonstrated at 8.2% efficiency to generate H_2 and O_2 [12]. A $GaInP_2$/GaAs photoelectrolysis cell was demonstrated at 12.4% efficiency [13] and a AlGaAs/Si photoelectrolysis cell was demonstrated at 18.3% efficiency [14].

A limited fraction of incident solar photons has sufficient energy greater than E_g to initiate charge excitation within a semiconductor. A semiconductor configuration can drive water electrolysis under the conditions in which the generated photovoltage, V_{photo}, is greater than the electrolysis potential, V_{H_2O}. A complex energetic challenge exists to choose a semiconductor system with band gap at a combined maximum power point voltage tuned to V_{H_2O}, in a system also providing high η_{photo} (high fill factor, photopotential, and photocurrent). Incident light of sufficiently energetic photons are absorbed; less energetic photons are transmitted. Absorbed photons, $h\nu > E_{Gw}$, can stimulate e^-/h^+ pair excitation. A pn-type or a Schottky junction electrochemical or a solid-state junction can inhibit charge recombination, driving charge at photopotential, V_{photo}. In these junctions, V_{photo} is constrained by the saturation current, j_o, through to yield a maximum power photopotential, V_{pmax}, less than 60% of the band gap in efficient devices [15].

$$V_{photo}(j_{photo}) = V_w + V_s = \left(\frac{nkT}{e}\right) \ln\left[1 + \frac{j_{photo}}{j_{o-w}}\right] \times \left[1 + \frac{j_{photo}}{j_{o-s}}\right] \quad (2)$$

$$V_{pmax} < 0.6(E_{Gw} + E_{gs}) \quad (3)$$

To initiate electrolysis V_{photo} must be greater than V_{H_2O}. The thermodynamic potential, $E^o_{H_2O}$, for the water-splitting reaction is given by

$$H_2O \longrightarrow H_2 + \tfrac{1}{2}O_2; \quad E^o_{H_2O} = E^o_{O_2} - E^o_{H_2}; \quad E^o_{H_2O}(25\,^\circ C) = 1.229\ V \quad (4)$$

The subsequent challenge is to optimize sustained water electrolysis, without considerable additional energy losses. Effective water electrolysis must occur at a potential, V_{H_2O}, near the photocell point of maximum power. V_{H_2O} is greater than $E^o_{H_2O}(= -\Delta G_{H_2O}/nF)$ as a result of overpotential losses, ξ, in driving an electrolysis current density, j, through both the O_2 and the H_2 electrodes:

$$V_{H_2O}(j) = E_{O_2}(j) - E_{H_2}(j)$$
$$= [E^o_{O_2} + \xi_{O_2}(j)] - [E^o_{H_2} + \xi H_2(j)] \quad (5)$$

Planar platinum and Pt_{black} are examples of effective H_2 electrocatalysts. Minimization of ξ_{O_2} is a greater challenge. In the absence of competing redox couples, the faradic efficiency of H_2 and O_2 evolution approaches 100%, and the $\eta_{electrolysis}$ is determined by the current limited $V_{H_2O}(j)$:

$$\eta_{electrolysis} = \frac{E^o_{H_2O}}{V_{H_2O}(j)};$$
$$\eta_{electrolysis}(25\,^\circ C) = \frac{1.229\text{ V}}{V_{H_2O}(j)} \quad (6)$$

Thermodynamic heat considerations can be applied to photoconversion. Some amount of heat, Q, is always be lost due to charge carrier relaxation and vibrations. This heat difference involves an entropy turnover $Q/T = \Delta S$. Photoenergy conversion efficiencies can be optimized by minimizing ΔH_{ph}. For band gap excitation, ΔH_{ph} corresponds to E_g. If the absorbed photon energy is much higher than E_g, then ΔH_{ph} is correspondingly higher. If E_g is much larger than the required electrolysis energy, ΔG_{H_2O}, then the predicted efficiency is low. The efficiency is higher if the photovoltaic system is matched to the chemical one. Consequently, there is a thermodynamic photoefficiency describing the generation of electrochemical free energy for electrolysis:

$$\frac{\Delta G_{H_2O}}{\Delta H_{ph}} \quad (7)$$

In reality, how small can ΔH_{ph} be made, as compared with ΔG_{H_2O}?

For photosynthesis, it has been estimated that of the 1.8-eV excitation energy of chlorophyll, at least 0.6–0.8 eV are lost before the energy can be stored in stable chemical products. Photosynthesis, however, requires many subsequent electron transfer steps, all of which contribute to efficiency losses. The molecules involved are also complicated and quite unstable compounds maintained through self-organization. Photoelectrolysis systems with many fewer components, and with fewer degrees of freedom in terms of chemical reactivity, can better approach the ideal energy conversion efficiency. Contrarily, when fuel is consumed to reduce power (resulting in a fuel cell with efficiency $\Delta G/\Delta H$) the optimal efficiency for electrolysis must include entropy considerations, and this strongly depends on the temperature enthalpy change ($\Delta H_{H_2O}(T)$):

$$\eta_{electrolysis-opt} = \frac{\Delta H_{H_2O}}{\Delta G_{H_2O}} \quad (8)$$

Practical electrolysis has to include losses so that the efficiency becomes

$$\eta_{electrolysis-opt} = \frac{\Delta H_{H_2O}}{\Delta G_{H_2O} + \text{losses}} \quad (9)$$

A thermoneutral potential, $E_{tn}(E_{tn} = \Delta H_{H_2O}/zF)$, is defined in which no heat turnover is observed and $E_{tn} = 1.48$ V for water electrolysis [16]. If, because of effective catalysis, the total electrolysis cell voltage is close to 1.48 V, then $\eta_{electrolysis} \approx 1$.

If we return to Eq. (1) and substitute Eqs. (7) and (9), we obtain

$$\eta_{\text{photoelectrolysis}} = \eta_{\text{photo}} \times \eta_{\text{electrolysis}} = \frac{\Delta G_{H_2O}}{\Delta H_{\text{ph}}}\left(\frac{1 + \text{losses}}{\Delta G_{H_2O}}\right) \quad (10)$$

Two conclusions arise from Eq. (10). First, it is seen that efficiency can be maximized if electrochemical "losses" can be made small as compared with the Gibbs free energy change for water electrolysis, ΔG_{H_2O}. Second, it is seen that ΔH_{H_2O}, the enthalpy of water electrolysis (with little temperature dependence), is involved in determining the overall efficiency as well as ΔH_{ph}, the enthalpy of photogenerated charge carriers. If these two can be properly matched, a maximum overall efficiency may be accomplished.

4.3.3 Photoelectrochemical Cell for Photoelectrolysis

Conversion of solar energy into electrical energy can be achieved by using a photoelectrochemical (PEC) solar cell configuration similar to that described in the earlier chapter. In brief, a PEC cell consists of two electrodes separated by a suitable redox electrolyte. Both electrodes could either be *n*-type and *p*-type semiconductors or one electrode could be a semiconductor (either *n*-type or *p*-type) and the other electrode could be a noncorrosive metal. If both electrodes are made of semiconductor, then the anodic electrode should be made from *n*-type semiconductor and the cathodic electrode should be made from *p*-type semiconductor. This last configuration adds the additional requirement of photocurrent matching through the two photoelectrodes. Regarding *n*-type semiconductor (Fig. 3), during its illumination photogenerated carriers are generated, which assist in the oxidation of electrolyte at the interface of the semiconductor and electrolyte. A reverse phenomenon takes place if a *p*-type semiconductor is used, that is, photogenerated electrons are generated at the semiconductor–electrolyte interface to

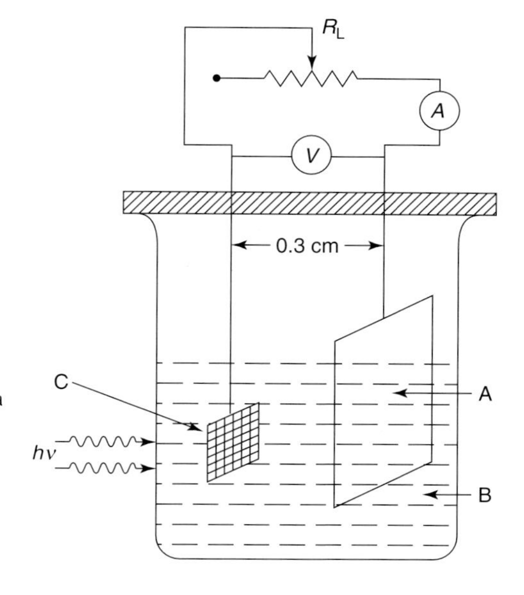

Fig. 3 A schematic representation of a photoelectrochemical cell. A, B, and C are counter electrode, electrolyte, and semiconductor electrode, respectively. R_L is used to vary the potential (V) to measure photocurrent (A) flowing across two electrodes.

initiate reduction of the electrolyte. Thus, in a PEC cell (Fig. 3), if water is used instead of a redox electrolyte, oxidation of water can take place at the illuminated *n*-type semiconductor (C) and reduction can take place at counter electrode (A). However, for photodecomposition of water rather than a single redox couple, two different redox couples are used at the two immersed electrodes, with a provision to collect hydrogen and oxygen separately.

4.3.4 Photoelectrolysis of Water

Fujishima and Honda [7, 17] were the first to show the possibility of decomposing water through a PEC cell (Fig. 4a). They used a photoelectrochemical cell with an anode made up of a semiconductor electrode of *n*-TiO_2 connected to a platinum black counter electrode through an external load. Because photopotential developed by the TiO_2 electrode was insufficient, they had to apply an external bias (E_b) to facilitate the photoelectrolysis of water. The energy level of semiconductor and other photoelectrochemical reactions are shown in Fig. 4(b). The semiconductor anode was illuminated with UV radiation. On illumination of TiO_2 electrode, the photogenerated holes oxidize water to produce oxygen and the photogenerated electrons are transferred to counter electrode to perform the reduction of water at the platinum electrode.

4.3.4.1 Energetics of Photodecomposition of Water

To accomplish photodecomposition of water, so that the PEC cell supplies the entire potential, energetic of the cell should meet the requirements as shown in Fig. 5. It is assumed that Fermi energy of *n*-type semiconductor (or its flat band potential) is equivalent to water reduction potential. Alternatively, instead of using metal as a counter electrode, one could use a *p*-type semiconductor (Fig. 6). Because

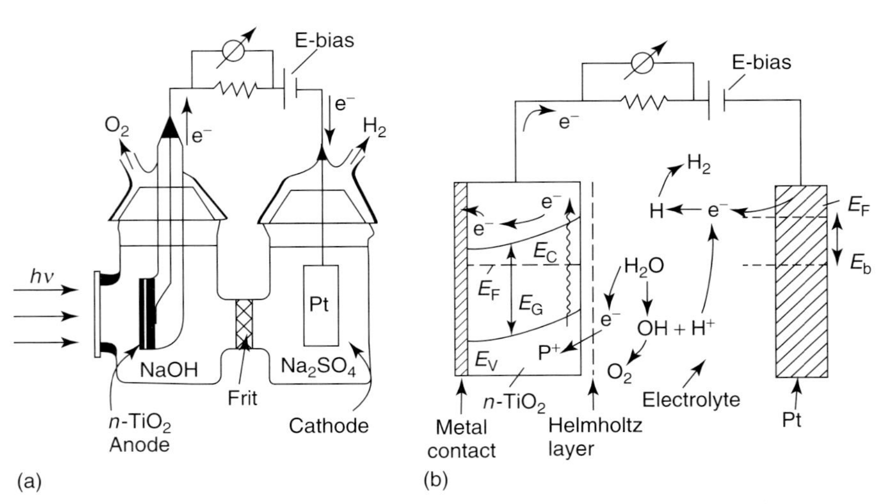

Fig. 4 (a) Fujishima-Honda cell with *n*-TiO_2 photoanode and Pt-cathode. (b) Schematic energy level diagram of the cell. E_V-valence band, E_C-conduction band, E_F-Fermi level, E_G-energy gap (for *n*-TiO_2; $E_G = 3.0$ eV), E_b-bias voltage, p^+-hole (after Nikola Getoff, *Int. J. Hydrogen Energy* **1990**, *15*(6), 407).

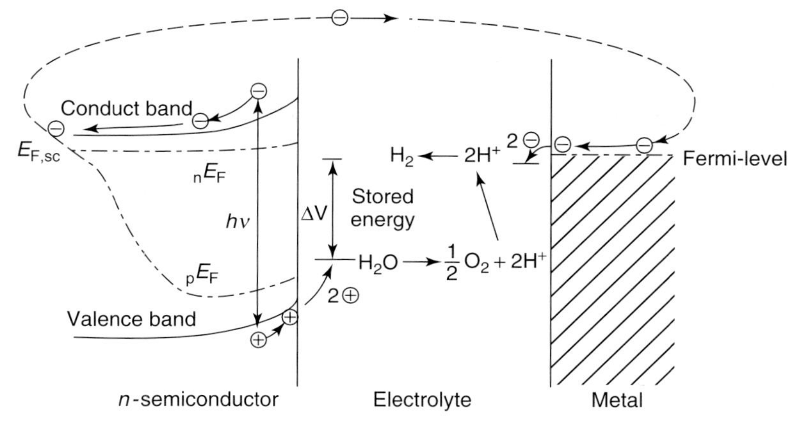

Fig. 5 Energy scheme of a cell with one *n*-type semiconductor electrode for photoelectrolysis of water. ΔV is stored energy for electrolysis. ${}_pE_F$ is Fermi level of photogenerated holes known as quasi-Fermi level, ${}_nE_F$ is Fermi level of electron. (after Heinz Gerischer, *Pure Appl. Chem.* **1980**, *52*, 2649).

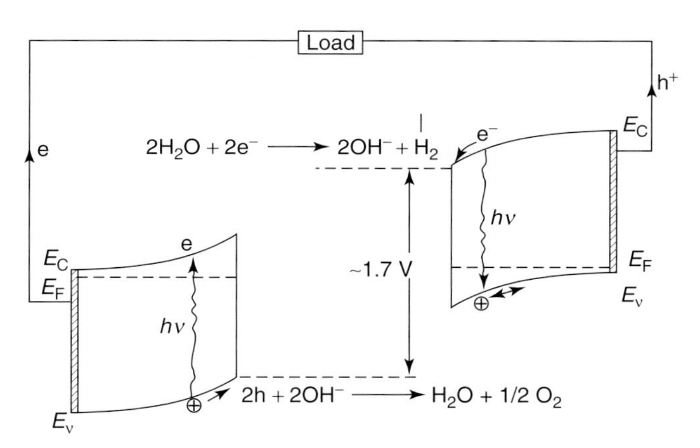

Fig. 6 Energy scheme of a cell utilizing *n*-type semiconductor and *p*-type semiconductor. The difference between the energy levels of valence band edge of *n*-type and conduction band edge of *p*-type should be approximately equal to potential needed to electrolyze water (V_{H_2O}).

the objective is to electrolyze water with a freely available energy, light source for the illumination has to be solar energy. Considering factors responsible to produce photocurrent or photopotential generated by semiconductors of various band gaps utilizing solar energy of various wavelengths and intensities, one can calculate theoretical achievable efficiency for various band gaps. This relationship gives a parabolic curve with a maximum theoretical efficiency of about 30% for a band gap of 1.4 eV (Fig. 7). But, a semiconductor with a band gap of 1.4 eV would give a maximum photopotential of ≅700 mV (assuming its flat band potential

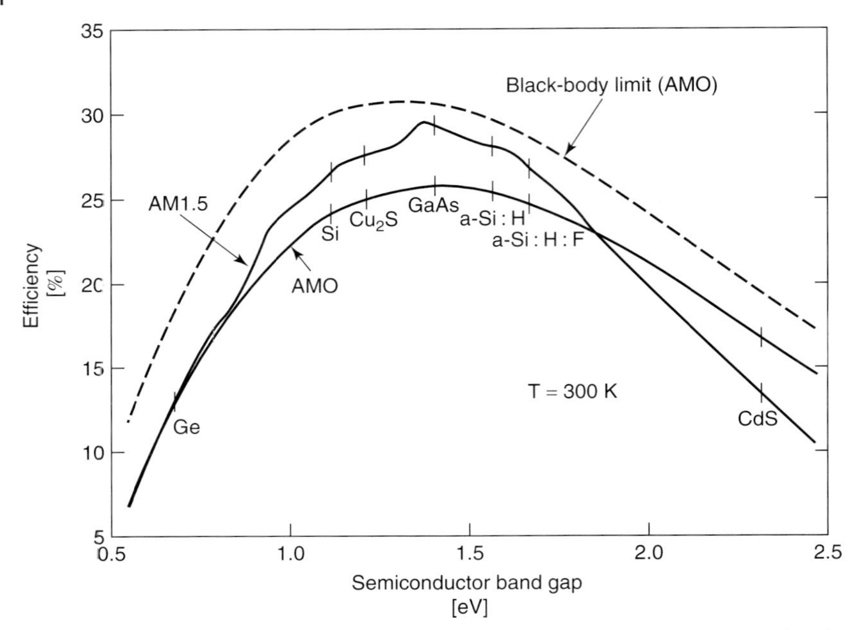

Fig. 7 Theoretically calculated conversion efficiency of solar cell materials versus band gap for single junction cells (after Adolf Goetzberger, Christopher Hebling, *Sol. Energy Mater. Sol. cells* **2000**, *62*, 1).

matches with the reduction potential of water), which is not sufficient to electrolyze water. On the contrary, for obtaining the necessary electrolysis photopotential using a single semiconductor, a semiconductor of much larger band gap is needed (Fig. 5). With such material, biasing is needed, as is case with n-TiO_2 (Fig. 4). Moreover, as a result of high band gap, n-TiO_2 absorbs only about 2% of the solar spectra (Fig. 2). Therefore, a semiconductor of large band gap would correspondingly give very low photocurrent (hence, lowphotoconversion efficiency), making the photoelectrolysis processes uneconomical.

4.3.4.2 Multiple-type PEC Cells

The calculations from the earlier section suggest that there is a need to introduce some improvements in the PEC cell to attain a sufficient solar energy conversion with photopotential larger than V_{H_2O}. Sharon and Rao [19] developed a photoelectrochemical cell with a semiconductor separating two types of electrolytes: one electrolyte forming an ohmic contact and the other forming a Schottky-type contact. It is postulated that like the formation of a Schottky-type barrier, if the magnitude of the Fermi level of the semiconductor and the redox electrolyte is same, then the contact between them should be ohmic. This provided an opportunity to visualize a PEC cell in which the semiconductor acts like a separator between two types of redox electrolytes. The electrolyte is selected such that the front side of the semiconductor (i.e. the side to be illuminated) forms a Schottky junction. The backside of the semiconductor is kept in contact with another redox electrolyte, which gives an ohmic contact. The front side of the semiconductor acts as a normal PEC cell and its backside acts as a counter electrode. This type of cell has been classified as a Sharon-Schottky-type cell [20]. To simplify this type of cell, this concept was extended to make a multiple-type PEC cell. It is assumed that a metal forms an ohmic contact with its own oxide. In this cell, three electrodes made of n-type semiconductor (n-Pb_3O_4) deposited over a metal (lead) are arranged in an array (Fig. 8a). The front side of each electrode contains n-Pb_3O_4, which is dipped in a suitable redox electrolyte to form a Schottky-type barrier. On illumination of n-Pb_3O_4, electrode photogenerated holes perform oxidation of redox electrolyte. Photogenerated electrons migrate in opposite directions to arrive at the metal electrode (lead). Because this metal gives an ohmic contact, photogenerated electrons are easily transferred to the redox electrolyte to perform reduction of the electrolyte. Thus, while the front side (i.e. the illuminated side) performs oxidation of the redox electrolyte, the backside performs reduction of the electrolyte (and also acts like a counter electrode). When more than one such electrodes are used (Fig. 8a), a vectorial migration of holes takes place from left to right side while vectorial migration of electrons takes place from right to left side. These carriers (electrons and holes) are transferred to the load by the two outer electrodes, that is,

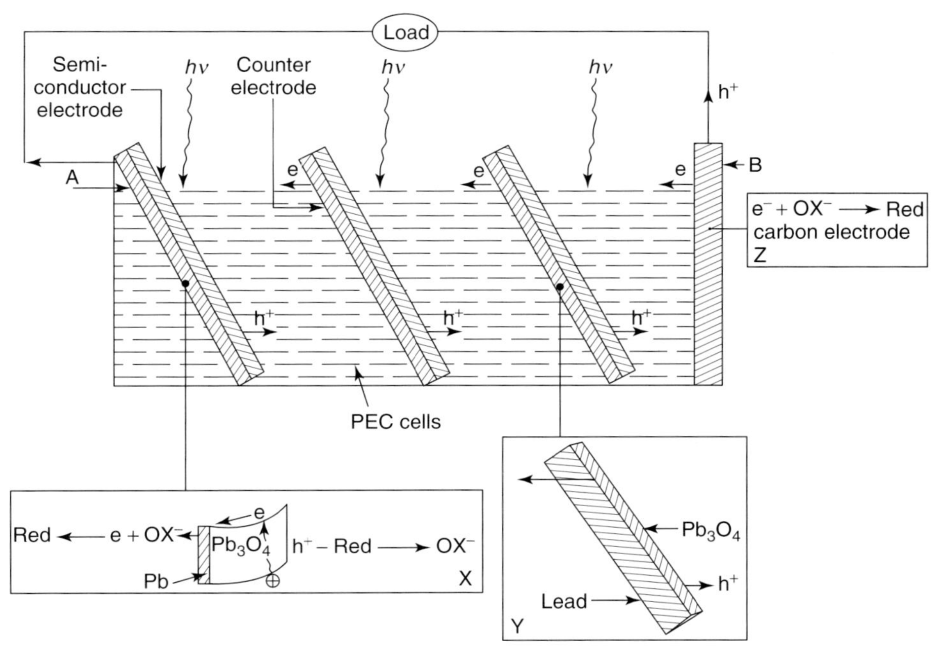

Fig. 8 Schematic representation of multiple PEC cells connected by the help of redox electrolyte. Inset X shows the energy level of semiconductor Pb_3O_4 electrode deposited over metal lead. Photoelectrochemical reactions occurring at two ends of the semiconductor is shown. Inset Y shows the ohmic contact between the metal and the semiconductor and the direction of flow of photogenerated electron and hole. Electron is extracted to the load via metal A and the hole is extracted via inert metal B. Inset Z describes reduction process occurring at inert carbon electrode-B.

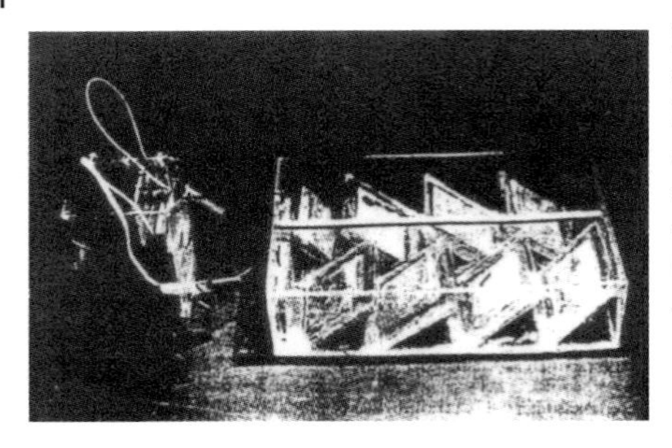

Fig. 9 A photograph of a multiple PEC cells (7 in number) connected in series by the help of redox electrolyte as described in Fig. 7. The output of this cell can be given to a normal electrolytic cell for electrolysis of water (after M. Sharon et al., *Electrochim. Acta* **1991**, *36*(7), 1107–1126).

by the backside of the last left side electrode (A) and metallic electrode on the extreme right side of the cell (B). The total photopotential of such system is equal to the photopotential of one PEC cell times the number of semiconductor electrodes used [20]. This arrangement gives an opportunity to get the required electrolysis photopotential, even with the semiconductor of band gap $\approx$1.4 eV. In addition, the requirement for matching the conduction band edge (or flat band potential) with the electrode potential for hydrogen evolution and the valence band edge with electrode potential of oxygen evolution (Figs. 5 and 6) becomes redundant. Photocurrent, however, would depend on the area of the individual semiconductor exposed to solar radiation (assuming each semiconductor to be of the same area). Employing this cell as a source of electrical power, it also can be used for electrolyzing water (Fig. 9).

4.3.4.3 Bipolar Cell

The bipolar cell of Sharon-Schottky cell has been further developed. Bard and coworkers [21, 22] developed a bipolar semiconductor photoelectrode array (Fig. 10) and studied its application to light-driven water splitting and electrical power generation. They used five n-TiO_2 electrodes in series. In the subsequent developments [23], they devised a similar bipolar cell (Fig. 11) with CoS/CdSe. In this cell, also a salt bridge is used to complete the electrical circuit. CoS is used to make ohmic contact with

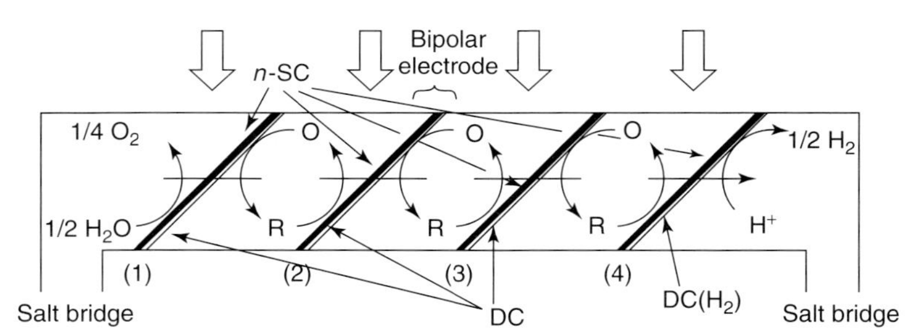

Fig. 10 A bipolar cell to show the vectorial transfer of electrons in one direction. On the extreme left hand side oxygen is evolved and on the extreme right hand side hydrogen gas is released. All other cells behave similar to a PEC cell. Salt bridge is used to complete electrical contact. n-SC – n-TiO_2, dark electrocatalyst reaction $(O \rightarrow R)_j$, light-initiated electrocatalyst $(R \rightarrow O)_j$, and dark electrocatalyst for H_2 evolution DC(H_2) (after A. J. Bard et al., *J. Phys. Chem.* **1986**, *90*, 4606 and A. J. Bard et al., *J. Electrochem. Soc.* **1988**, *135*, 567).

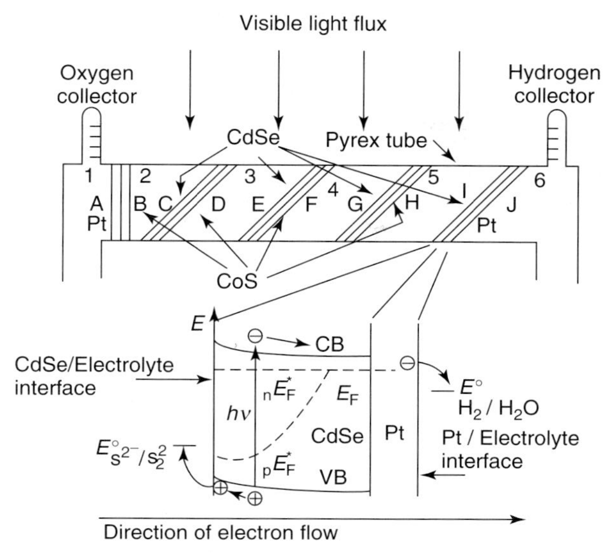

Fig. 11 (a) Schematic representation of water photoelectrolysis cell. A, J: Pt; B, D, F, H: CoS; C, E, G, I: CdSe. Solutions 1,6: KOH (1M); 2–5 Na_2S (1M), S (1M), KOH (1M). For H_2 and O_2 generation solutions 1 and 6 are connected with KOH bridge. (b) Expansion shows energetics of bipolar panel (after A. J. Bard et al., *J. Phys. Chem.* **1987**, *91*, 6).

n-CdSe as well as with the electrolyte. Four CoS/CdSe electrodes are used. Polysulfide is used as an electrolyte in these PEC cells. In this arrangement, the PEC cell is formed among (B, C), (D, E), (F, G), and (H, I). Sides C, E, G, and I form Schottky-type barriers while sides B, D, F, and H act like a counter electrode. The first and the last electrodes are made of platinum. The energy level diagram to show the flow of electron is shown in Fig. 11.

Several oxides (e.g. TiO_2 [7], $SrTiO_3$ [8], *n*-SiC [24], *p*-GaP [25], Fe_2O_3 [26], etc.) and chalcogenides such as CdSe [23] have been used as anodes for water photodecomposition. *n*-TiO_2, even today seems to be one of the most important materials, because it has been possible to extend its spectral response into visible portion of the solar spectrum through sensitization with organic dyes. A separate chapter is devoted for sensitization of semiconductor electrode. Hence, this will not be discussed here any further.

4.3.5
Recent Developments

Principal solar water-splitting models had predicted similar dual band gap photoelectrolysis efficiencies of only 16% and 10–18% [3, 28], respectively, whereas recently dual band gap systems were calculated to be capable of attaining over 30% solar photoelectrolysis conversion efficiency [14]. The physics of the earlier models were superb, but their analysis was influenced by dated technology, and underestimated the experimental η_{photo} attained by contemporary devices or underestimated the high experimental values of $\eta_{electrolysis}$ which can be attained. For example, Ref. 3 estimates low values of η_{photo} (less than 20% conversion) because of the assumed cumulative relative secondary losses that include 10% reflection loss, 10% quantum-yield loss, and 20% absorption loss. Experimentally, a cell containing illuminated AlGaAs/Si RuO_2/Pt_{black} was demonstrated to evolve H_2 and O_2 at record solar–driven water electrolysis efficiency. Under illumination, bipolar configured $Al_{0.15}Ga_{0.85}As$ (E_g = 1.6 eV) and Si (E_g = 1.1 eV) semiconductors generate open circuit and maximum power photopotentials of 1.30 and 1.57 V, respectively, well suited to the water electrolysis thermodynamic potential:

$$H_2O \longrightarrow H_2 + \tfrac{1}{2}O_2;$$
$$E^o_{H_2O} = E_{O_2} - E_{H_2};$$
$$E^o_{H_2O}(25\,^\circ C) = 1.229\ V \qquad (11)$$

The $E^o_{H_2O}$/photopotential-matched semiconductors are combined with effective water electrolysis O_2 or H_2 electrocatalysts, RuO_2, or Pt_{black}. The resultant solar photoelectrolysis cell drives sustained water-splitting at 18.3% conversion efficiencies [14]. These recent developments in hydrogen generation at high solar energy conversion efficiency are detailed in this volume in the chapter titled "Optimizing Photoelectrochemical Solar Energy Conversion: Multiple Band Gap and Solution Phase Phenomena."

4.3.6 Conclusion

In this section, efforts are made to discuss the thermodynamics of photoelectrolysis of water using a PEC cell. To facilitate the generation of required potential for the photoelectrolysis of water, discussions are made on some modified PEC cells popularly known as *bipolar cell*. Potential developed in bipolar cells, amounts to connecting large number of PEC cells in series. The advantage of bipolar cells is that it can operate with low band gap semiconductor, yet provide desired photopotential and high solar to electrical efficiency. Although photopotential of such device is equal to the photopotential of one-photoelectrode times the number of PEC cells connected in series, photocurrent of system depends on the intensity of solar radiation falling on an individual photoelectrode. But it is important to realize that none of semiconductors so far developed can be used for prolonged photoelectrolysis of water, because of their inherent instability towards photocorrosion. Success of making bipolar cells as a commercial viable system entirely depends on the development of photoelectrochemically stable low band gap semiconductor.

Acknowledgment

S. Licht is grateful to Helmut Tributsch for his review and suggestions on theoretical sections of this chapter and thankful for the support by the BMBF Israel-German Cooperation and the Berman-Shein Solar Fund. Maheshwar Sharon is thankful to his students and in special to G. Ranga Rao who contributed toward development and modification of PEC cell.

References

1. M. A. Green, K. Emery, K. Bucher et al., *Propgr. Photovolt.* **1999**, *11*, 31.
2. S. Licht, B. Wang, T. Soga et al., *Appl. Phys. Lett.* **1999**, *74*, 4055.
3. J. R. Bolton, S. J. Strickler, J. S. Connolly, *Nature* **1985**, *316*, 495.
4. K. W. Atanabe, M. Zelikoff, *J. Opt. Soc. Am.* **1953**, *43*, 753.
5. Nikola Getoff, *Int. J. Hydrogen Energy* **1990**, *15*(6), 407.

6. L. J. Heidt, A. F. McMillan, *Science* **1953**, *117*, 75.
7. A. Fujishima, K. Honda, *Nature* **1972**, *238*, 37.
8. J. M. Bolts, M. S. Wrighton, *J. Phys. Chem.* **1976**, *80*, 2641.
9. R. Memming, *Top. Curr. Chem.* **1988**, *143*, 79.
10. A. Nozik, *Appl. Phys. Lett.* **1976**, *29*, 150.
11. J. White, F. -R. Fan, A. J. Bard, *J. Electrochem. Soc.* **1985**, *132*, 544.
12. R. C. Kaintala, J. Ö. M. Bockris, *J. Int. Hydrogen Energy* **1988**, *13*, 375.
13. O. Khaselev, K. Turner, *Science* **1998**, *280*, 425.
14. S. Licht, B. Wang, S. Mukerji et al., *J. Phys. Chem. B* **2000**, *104*, 8920.
15. R. Memming in *Photochemical Conversion and Storage of Solar Energy*, (Eds.: E. Pelizzetti, M. Schiavello), Kluwer Academic Publishers, Netherlands, 1991, pp. 193–212.
16. F. Gutman, O. J. Murphy in *Modern Aspects of Electrochemistry* (Eds.: White, Bockris, Conway) 1983, P5.
17. A. Fujishima, K. Honda, *Bull. Chem. Soc. Jpn.* **1971**, *44*, 1148.
18. Heinz Gerischer, *Pure Appl. Chem.* **1980**, *52*, 2649.
19. Maheshwar Sharon, G. Ranga Rao, *Indian J. Chem.* **1986**, *25A*, 170–172.
20. M. Sharon et al., *Electrochim. Acta* **1991**, *36*(7), 1107–1126.
21. A. J. Bard et al., *J. Phys. Chem.* **1986**, *90*, 4606.
22. A. J. Bard et al., *J. Electrochem. Soc.* **1988**, *135*, 567.
23. Tooru Inoue, Toshihiro Yamase, *Chem. Soc. Japan, Chem. Lett.* **1985**, 869.
24. H. Honeyama, H. Sakamoto, H. Tamura, *Electrochim. Acta* **1979**, *277*, 637.
25. Lynn C. Schumacher, Suzanne Mamiche-Afara, Michael F. Weber et al., *J. Electrochem. Soc.* **1985**, *132*(12), 2945.
26. M. F. Weber, M. J. Digman, *Int. J. Hydrogen Energy* **1986**, *11*, 225.
27. Ibid, *J. Electrochem. Soc.* **1984**, *131*, 1258.
28. Adolf Goetzberger, Christopher Hebling, *Sol. Energy Mater. Sol. cells* **2000**, *62*, 1.

4.4 Optimizing Photoelectrochemical Solar Energy Conversion: Multiple Bandgap and Solution Phase Phenomena

Stuart Licht
Technion – Israel Institute of Technology, Haifa, Israel

4.4.1 Introduction

This chapter focuses on two concerted efforts to achieve stable, high solar energy conversion efficiency using a variety of photoelectrochemical systems. The first effort explores the use of multiple band gap semiconductor systems. Limiting constraints of multiple band gap photoelectrochemical energy conversion as well as practical configurations for efficient solar to electrical energy conversion have been probed [1–6]. Such systems are capable of better matching and utilization of incident solar radiation (insolation). Efficient solar cells, solar storage cells, and solar hydrogen generation systems are discussed and demonstrated.

The principles of photoelectrochemical phenomena systems can differ substantially from conventional solid-state physics. Photoelectrochemical systems can be characterized not only by semiconductor but also often by electrolytic limitations. The second half of the chapter focuses on the substantial improvements to photoelectrochemical energy conversion attained by understanding and optimizing of such solution phase phenomena [7, 8].

4.4.2 Multiple Band Gap Photoelectrochemistry

4.4.2.1 Theory of Multiple Band Gap Solar Cell Configurations

Radiation incident on semiconductors can drive electrochemical oxidation or reduction and generate chemical, electrical, or electrochemical energy. Energetic constraints imposed by single band gap semiconductors had limited values of photoelectrochemical solar to electrical energy conversion efficiency to date to 12–16% [9, 10]. Multiple band gap devices can provide efficient matching of the solar spectra [11–15]. A two or more band gap configuration will lead, per unit surface area, to more efficient solar energy conversion, and in the solid-state, multiple band gap solar cells have achieved more than 30% conversion efficiency of solar energy [14]. The fundamental benefits of multiple band gap photoelectrochemistry have been recognized [16]. In this section an overview of the energetics of distinct multiple band gap photoelectrochemical solar cell (MPEC) configurations is introduced. The MPEC configurations can lead to higher conversion efficiency than previously observed for single band gap solar cells.

A limited fraction of incident solar photons have sufficient (greater than band gap) energy to initiate charge excitation within a semiconductor. Because of the low fraction of short wavelength solar light, wide band gap solar cells generate a high photovoltage but have low photocurrent. Smaller band gap cells can use a larger fraction of the incident photons but generate lower photovoltage. As shown in Fig. 1, schematic, multiple band gap devices can overcome these limitations. In stacked multijunction systems, the topmost cell absorbs (and converts) energetic photons but is transparent to lower energy photons. Subsequent layer(s) absorb the lower energy photons. Conversion efficiencies can be enhanced, and calculations predict that a 1.64-eV and 0.96-eV two-band gap system has an ideal efficiency of 38% and 50% for 1 and 1000 suns concentration, respectively.

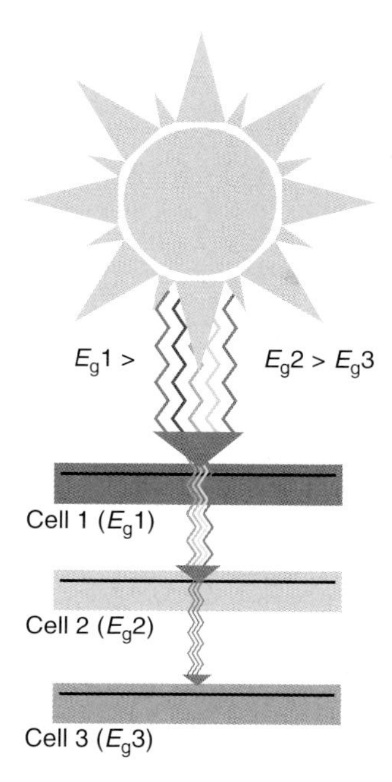

Fig. 1 In stacked multijunction solar cells, the top cell converts higher energy photons and transmits the remainder onto layers, each of smaller band gap cell than the layer above, for more effective utilization of the solar spectrum.

The ideal efficiency increases to a limit of 72% for a 36-band gap solar cell [11].

Present solid-state (photovoltaic) multijunction photovoltaics exist in one of two modes, either splitting (in which the solar spectrum is optically separated before incidence on the cells) or stacked devices. The latter mode has been developed as either monolithic (integrated) or mechanically (discrete cells layered with transparent adhesive) stacked cells. Most monolithic stacked multijunction photovoltaics under development use III–V semiconductors in at least one of the layers and take advantage of the variations in band gap and close lattice match achieved with other related III–V alloys. For example, GaInP has been used as a wide gap top cell, or GaSb and GaInAsP as small gap lower cells, for III–V top-layer multijunction cells; other cells use silicon or CIS cells as a lower layer [12–15].

The photopotential of a single pn junction is:

$$V_{\mathrm{pn}} = \left(\frac{nkT}{e}\right) \ln\left[\frac{1 + j_{\mathrm{ph,pn}}}{j_{\mathrm{o,pn}}}\right] \quad (1)$$

where V_{pn} is constrained by the photocurrent density through the junction, $j_{\mathrm{ph,pn}}$. $j_{\mathrm{o,pn}}$ is the saturation current (of a reversely polarized diode) and is described by the Shockley equation [17].

The photopotential of a single liquid Schottky junction is given by:

$$V_{\mathrm{Sch}} = \left(\frac{nkT}{e}\right) \ln\left[\frac{1 + j_{\mathrm{ph,Sch}}}{j_{\mathrm{o,Sch}}}\right] \quad (2)$$

Depending on the relative rates of charge transfer, $j_{\mathrm{o,Sch}}$ may be constrained by either solid-state or electrochemical limitations, and is respectively termed the *saturation current* or the *equilibrium exchange current* [17]. A single representation of either a pn or Schottky junction is schematized in the upper center of Fig. 2, by a junction generating a voltage V.

A variety of distinct MPEC configurations are possible, each with advantages and disadvantages [1]. The simplest MPEC configurations contain two adjacent band gaps. Adjacent band gap layers can be aligned in the cell in either a bipolar or a less conventional inverted manner. In either the bipolar or inverted cell configuration, the PEC solid–electrode interface can consist of either an ohmic or a Schottky interface. The ohmic interface can consist of either direct (semiconductor–electrolyte)

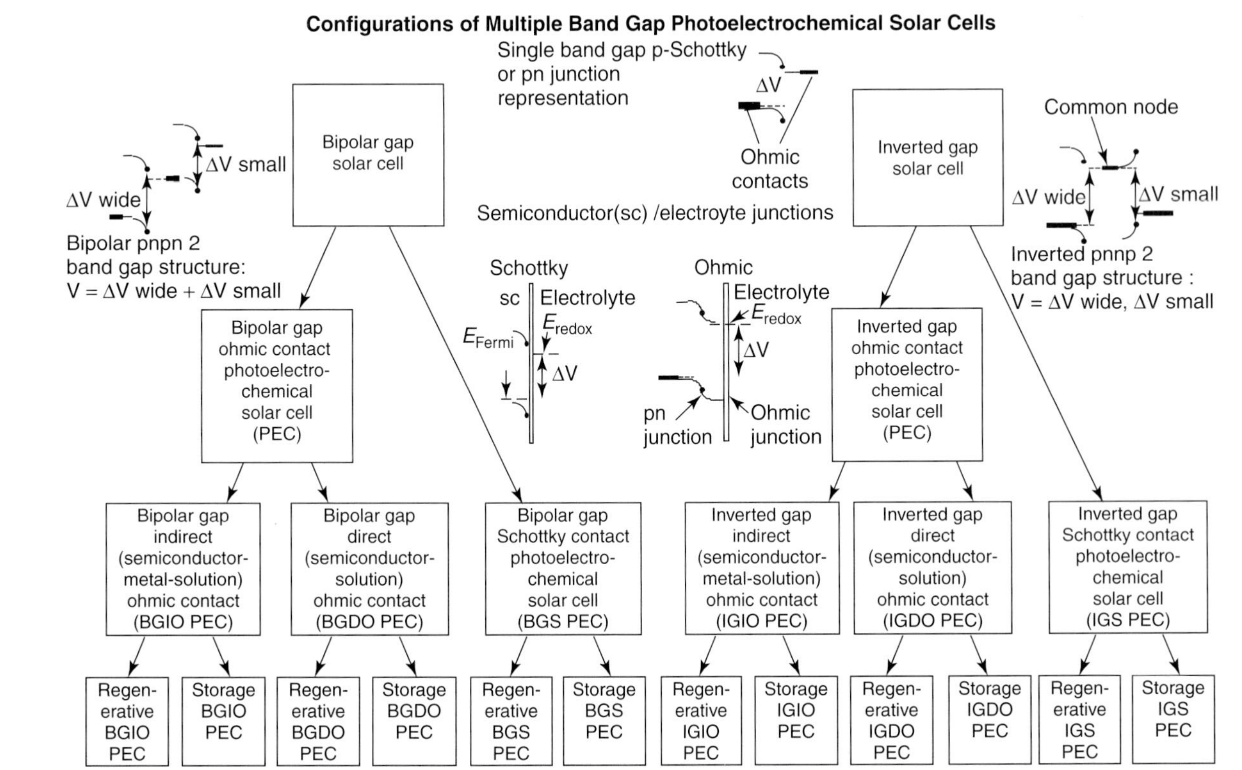

Fig. 2 Relation of the twelve representative MPEC configurations comprising regenerative ohmic cells: (1) bipolar gap direct ohmic regenerative, (2) bipolar gap indirect ohmic regenerative, (3) inverted gap direct ohmic regenerative, (4) inverted gap indirect ohmic regenerative; storage ohmic cells: (5) bipolar gap direct ohmic storage, (6) bipolar gap indirect ohmic storage, (7) inverted gap direct ohmic storage, (8) inverted gap indirect ohmic storage; regenerative Schottky cells: (9) bipolar gap Schottky regenerative, (10) inverted gap Schottky regenerative; and storage Schottky cells: (11) bipolar gap Schottky storage, and the (12) inverted gap Schottky storage configuration.

or indirect (semiconductor–metal and/or electrocatalyst–electrolyte) interfaces. Finally, each of these modes can be configured as either regenerative or storage solar cells. As shown in Fig. 2, these modes provide twelve alternate configurations available for MPECs.

Compared with the inverted case, the bipolar arrangement provides a conceptually simpler PEC but generates a large open-circuit photopotential, V_{oc}. As shown for a regenerative MPEC in either scheme on the left side of Fig. 2, the generated bipolar photovoltage, V_{photo}, is the sum of the potentials of the individual band gap layers minus cathodic and anodic polarization losses in driving a regenerative redox couple:

$$V_{photo} = V_W + V_S - (\eta_{cathode} + \eta_{anode}) \tag{3}$$

The energy diagram of a bipolar band gap photocathodic electrochemical Schottky configuration is presented in Fig. 3(a). The scheme comprises a two-photon/one-electron photoelectrochemical process ($2h\nu \rightarrow e^-$), which may be generalized for an n band gap configuration, to an n photon process ($nh\nu \rightarrow e^-$). Light shown incident from the left side of the configuration first enters the wide band gap layer(s) in which more energetic photons are absorbed; less energetic photons are transmitted through this upper layer and are absorbed by the small band gap layer. The resultant combined potential of the photodriven charge sustains reduction at the photocathode interface, and drives extractable work through the external load, R_{load}. The wide "w" and small "s" band gap layers are denoted with respective valence and conduction bands, E_V and E_C, and band gap:

$$E_{GW} = E_{CW} - E_{VW}; \quad E_{GS} = E_{CS} - E_{VS} \tag{4}$$

Wide band gap layer charge separation occurs across a pn junction space charge field gradient, while charge separation in the small band gap is maintained with a field formed by the Schottky semiconductor–electrolyte interface. In the bipolar Schottky MPEC configuration, generated charge flows through all layers of the cell, providing the additional constraint:

$$j_{ph,pn} = j_{ph,Sch} \tag{5}$$

In an alternate bipolar regenerative configuration, the bipolar (or multiple) band gap configurations may contain consecutive space charge field gradients generated only via solid-state phenomena. This is presented in Fig. 3(b) for the case of two consecutive bipolar pn junctions. The lowest semiconductor layer (the small band gap n-type layer in the figure) may remain in direct contact with the electrolyte, but the contact is ohmic and is not the source of the small band gap space charge field. A similar indirect ohmic contact bipolar regenerative configuration may also be derived from this figure. In this case, the lowest semiconductor layer is restricted to electronic and not ionic contact with the electrolyte through use of an intermediate (bridging) ohmic electrocatalytic surface layer. This can facilitate charge transfer to the solution phase redox couple, and prevent any chemical attack of the semiconductor. In the bipolar cases (including Schottky, direct or indirect ohmic configurations), the photopower generated by a bipolar regenerative MPEC is given by the product [17]:

$$P_{bipolar\ regenerative} = j_{ph}[V_W + V_S - (\eta_{cathode} + \eta_{anode})] \tag{6}$$

Adjacent band gap layers in a multiple band gap configuration can also be aligned

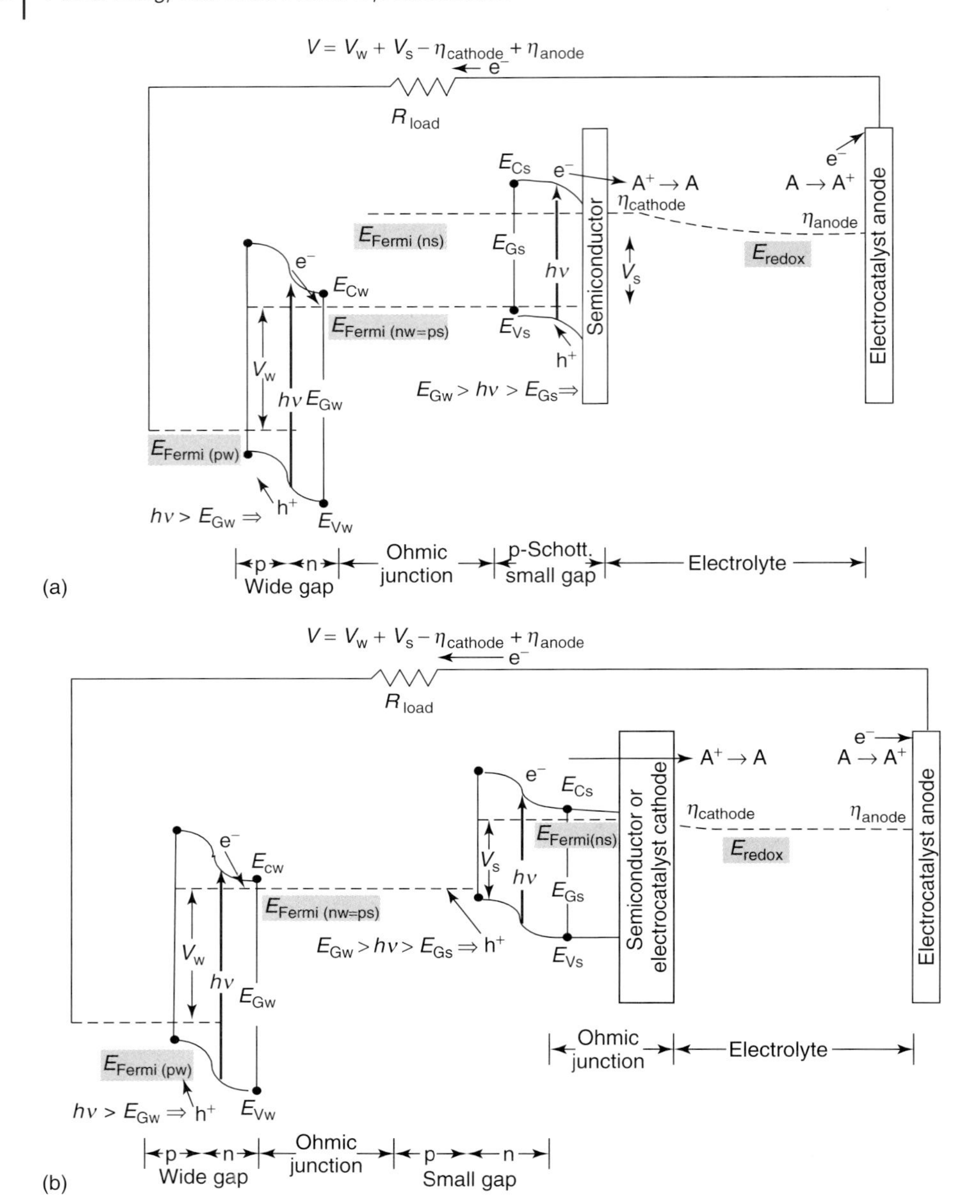

Fig. 3 Energy diagrams for multiple band gap photoelectrochemistry. Elements of bipolar or inverted band gap, Schottky, ohmic, regenerative, and storage configurations are illustrated. (a): Bipolar band gap Schottky regenerative MPEC. (b): Bipolar band gap (direct or indirect) ohmic regenerative MPEC.(c): Inverted band gap indirect ohmic regenerative MPEC.(d): Inverted gap (direct or indirect) ohmic storage MPEC.

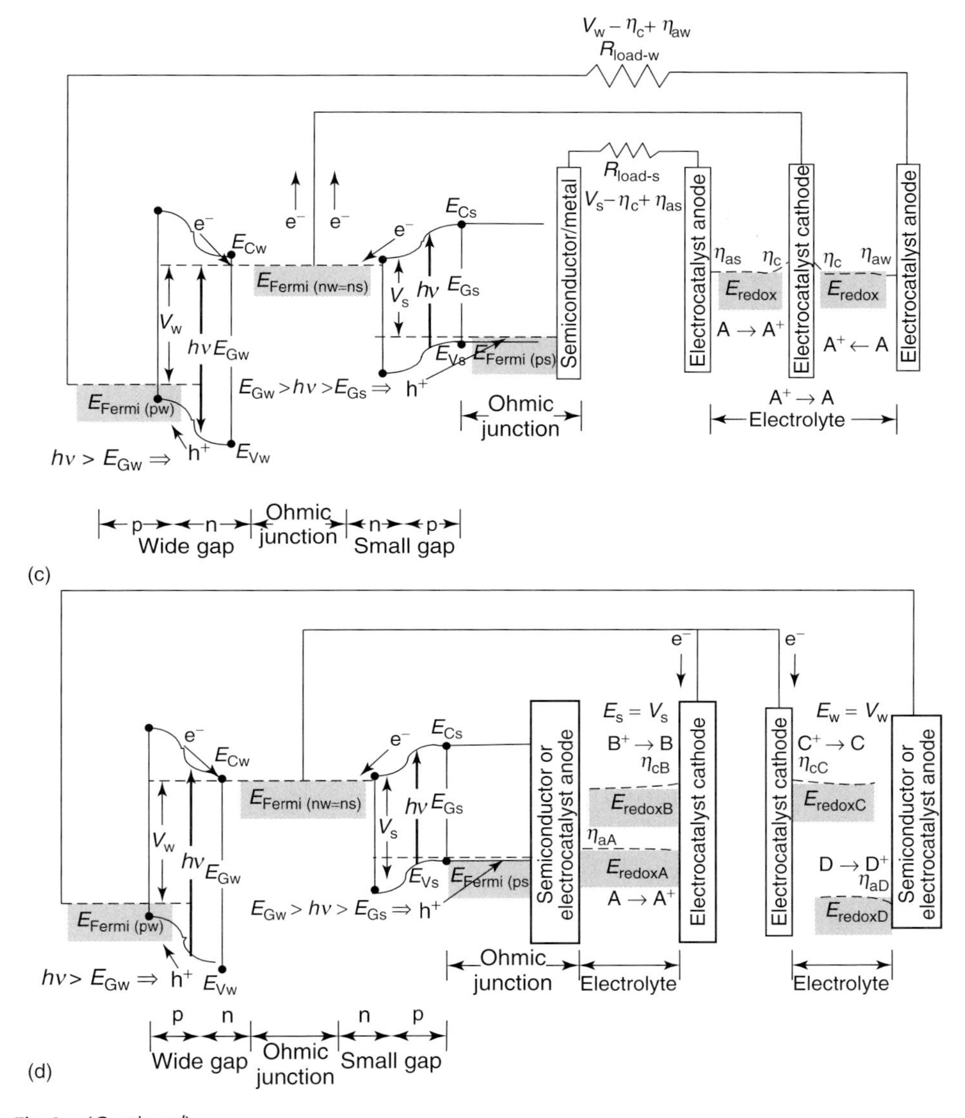

Fig. 3 (*Continued*)

in a less conventional inverted manner. As shown in Fig. 3(c), a dual gap inverted cell generates two smaller photopotentials, which can be separately applied to the same PEC, minimizing electrolysis losses. This figure presents an inverted pn/np (wide band gap or small band gap) regenerative ohmic configuration, compared to the pn/pn in Fig. 3(b). As seen for the inverted pn/np case, a quasi fermi level is shared by the conduction bands of the wide band gap *n*-type and the small band gap *n*-type layers. Light incident from the figure left side generates charge through a common intermediate node, comprising a two photon/two electron photoelectrochemical process ($2h\nu \rightarrow 2e^-$):

$$2h^+e^- + h\nu(h\nu > E_{Gw}) + h\nu(h\nu > E_{Gs}) \longrightarrow 2h^+ + 2e_w{}^{-*} + 2e_s{}^{-*} \quad (7)$$

Photoinduced holes, generated in accord with Eq. (7), drive solution phase oxidation at separate electrocatalytic anodes. A challenge to bipolar MPEC use is that all generated charge must flow through subsequent cell layers, as seen in the diagrams in the left side of Fig. 3. This imposes a current matching constraint on each of the individual junctions of any monolithic bipolar device. This constraint is removed with inverted band gaps, in which photocurrent is independent for each band gap, flowing through the common node illustrated in the figure. Unlike the bipolar cases, for the inverted regenerative (Schottky, direct or indirect ohmic) configurations, the photopower generated is a separate sum of the different layers, where for the two band gap case [17]:

$$P_{\text{inverted regenerative}} = j_{\text{ph-w}} V_w + j_{\text{ph-s}} \times V_s - (\eta_{\text{cathode}} + \eta_{\text{anode}}) \quad (8)$$

Electrochemical energy storage configurations provide an energy reservoir that may compensate for the intermittent nature of terrestrial insolation. When different anodic and cathodic redox processes are driven, electrochemical energy storage can be accomplished. Storage can consist of both in situ or photoelectrolysis cells. Until recently, photoelectrolysis cells, in which the electrolyte solvent is oxidized and/or reduced to provide a chemical fuel, had been inefficient [6]. In situ secondary redox couples, added to the electrolyte, have been simpler to optimize, and efficient single band gap in situ PEC storage cells have been demonstrated [18]. Secondary redox storage must be accomplished at sufficiently low potentials to prevent losses due to simultaneous (undesired) solvent electrolysis. However, bipolar band gap photoelectrochemistry imposes large photopotentials. These are avoided through the inverted band gap configuration, as exemplified in Fig. 3(d), in which the photopotentials generated in the respective small and wide band gap portions of the tandem cell, V_w and V_s drive two separate electrochemical storage processes:

$$\begin{aligned} &D + C^+ \longrightarrow D^+ + C; \\ &E_{C^+/C} - E_{D^+/D} < V_w \text{ and} \\ &A + B^+ \longrightarrow A^+ + B; \\ &E_{B^+/B} - E_{A^+/A} < V_s \end{aligned} \quad (9)$$

4.4.2.2 **Bipolar Band Gap PECs**

A bipolar gap direct ohmic photoelectrochemical system comprises either a bipolar band gap pnpn/electrolyte ohmic photoelectrochemical cell, with reduction occurring at the photoelectrode–electrolyte interface and regenerative oxidation occurring at the electrolyte–counter electrode (anode) interface or alternately: a npnp/electrolyte bipolar band gap with oxidation occurring at the semiconductor–electrolyte interface and regenerative reduction occurring at the electrolyte–counter electrode interface. In the bipolar gap direct ohmic photoelectrochemical system, direct refers to the direct contact between semiconductor and solution, and ohmic indicates this interface is an ohmic rather than a Schottky junction. This facilitates study of several characteristics of bipolar multiple band gap systems, without the added complication of simultaneous parameterization of a direct Schottky barrier at the electrolyte interface.

The examples presented here utilize combine multijunction solid-state layers consisting of a bipolar AlGaAs ($E_{Gw} = 1.6$ eV) wide band gap, overlaid on a Si ($E_{Gs} = 1.0$ eV) small band gap, and used in an electrolytic cell [2]. Light absorption by the electrolyte can interfere with the cell and should be avoided. Figure 4 overlays the optical characteristics of the solid and solution phase of the AlGaAs/Si solid-state and $V^{3+/2+}$ electrolyte optimized components within a bipolar gap photoelectrochemical cell. Solution transmission is measured through a pathlength that is typical (1 mm) of many experimental front wall photoelectrochemical cells. As is evident, light-transmission interference will occur for the top AlGaAs layers, and bottom Si layers through this or substantially shorter electrolyte pathlengths. The solid-state component includes a graded band emitter, varying from $Al_{(0.3-0.15)}Ga_{(0.7-0.85)}As$ with overlayers of p^+-$Al_xGa_{x-1}As$ on n-$Al_xGa_{x-1}As$. The growth sequence and graded band emitter layer improve collection efficiency [2]. The Si bottom cell consists of a p^+-Si, n-Si, and n^+-Si multijunction. The band edges observed in the figure at approximately 800 nm and 1100 nm are consistent with the respective AlGaAs and Si band gaps.

For efficient electron/hole pair charge generation, incident photons need to be localized within the multiple band gap

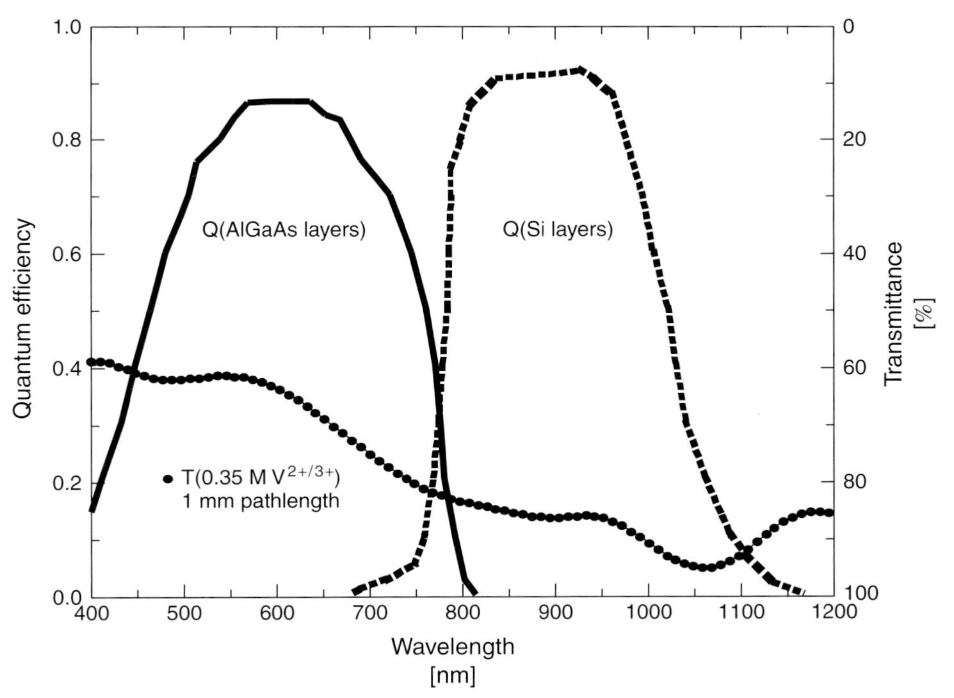

Fig. 4 Overlay of the optical characteristics of the solid and solution phase of the AlGaAs/Si solid-state and $V^{3+/2+}$ electrolyte constituents within a bipolar gap photoelectrochemical cell. Transmission of the $V^{3+/2+}$ electrolyte is measured through a pathlength of 1 mm. As described in the text, the Si bottom cell consists of a p^+-Si, n-Si and n^+-Si multijunction. The $Al_{(0.3-0.15)}Ga_{(0.7-0.85)}As$ top cell utilizes a graded band emitter.

semiconductor small and wide band gap regions, rather than lost through competitive electrolyte light absorption. As seen in Fig. 4, the vanadium electrolyte can significantly block light, over a wide range of visible and near infrared wavelengths, from entering the wide and small band gap layers of the multiple band gap photoelectrochemical cell. This deleterious effect is prevented by use of the back wall multiple band gap photoelectrochemical cell presented in Fig. 5. Light does not pass through the solution. As shown, illumination enters directly through antireflection films of 50 nm ZnS situated on 70 nm MgF_2. An evaporated Au-Zn/Au grid provides electrical contact to the wide gap AlGaAs layers through a bridging $p^+ =$ GaAs layer. Internally, a bridging GaAs buffer layer provides an ohmic contact between the wide band gap AlGaAs junctions and the lower Si layers. An intermediate contact layer indicated as "Au" is used only for probing separated characteristics of the wide and small band gap junctions, and is not utilized in the complete cell. Photo generated charge at the indicated silicon electrolyte interface induces solution phase vanadium reduction, and a carbon counter electrode provides an effective (low polarization) electrocatalytic surface for the reverse process in a regenerative cell, in accord with:

$$\begin{aligned} V^{3+}(+h\nu) &\longrightarrow V^{2+} + h^+; \\ V^{2+} &\longrightarrow V^{3+} + e^- \end{aligned} \qquad (10)$$

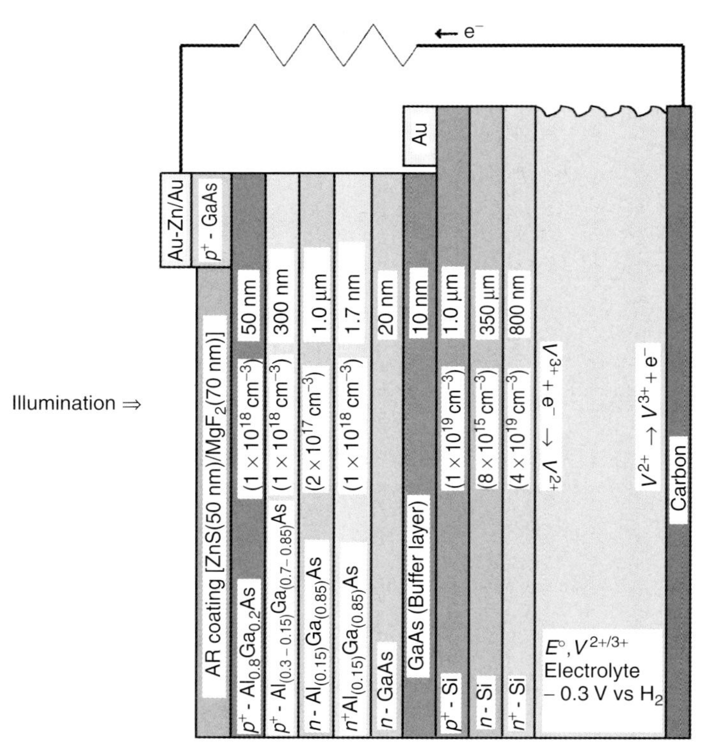

Fig. 5 Schematic description of the components in the bipolar gap direct ohmic AlGaAs/Si-$V^{3+/2+}$ PECs.

Heller, Miller and coworkers have previously shown that the p-Si surface is capable of sustaining minority carrier injection into the solution and stabilizes reduction of the $V^{3+/2+}$ redox couple [19]. However, at n-Si this does not appear to be the case for reduction of the $V^{3+/2+}$ redox couple by majority carrier injection into the solution. Thermodynamically silicon can be spontaneously oxidized in aqueous solution. This process may occur despite the photoinduced generation of reductive charge at the semiconductor–electrolyte interface. This is observed as the time-dependent decrease of photocurrent under constant illumination, during photoinduced reduction of V^{2+} (by majority carrier injection at n-Si) indicated in the inset Fig. 6. This is consistent with the onset of surface passivation forming a layer that is passive to charge transfer and diminishes the photocurrent. The addition of a low concentration of HF to the electrolyte can remove this passivating layer and permit sustained photocurrent. This is illustrated in the inset of Fig. 6, in which the addition of 0.02 and 0.2 M HF improves the photocurrent stability of this bipolar direct ohmic MPEC. Also as shown, no further improvement in the photocurrent stability is observed when the HF concentration is further increased from 0.2 to 0.5 M HF. As presented in the main portion of Fig. 6, the 0.2 M HF modified electrolyte stabilizes the photocurrent at the n-Si interface for the measured period of several hours [2]. As will be subsequently shown, photocurrent stability is further improved through use of a electrode catalyst bridging the silicon–electrolyte interface.

As can be seen in Eq. (6), maximization of the photopower necessitates minimization of the anodic and cathodic polarization losses, η_{anode} and

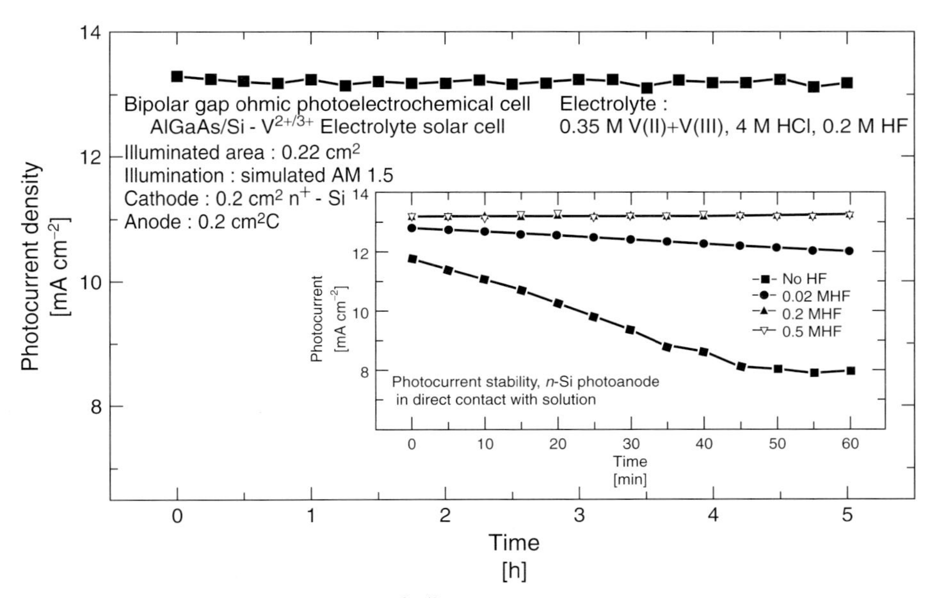

Fig. 6 Photocurrent stability in several $V^{3+/2+}$ aqueous electrolytes of the bipolar band gap direct ohmic AlGaAs/Si-$V^{3+/2+}$ photoelectrochemical cell (measured indoors using a tungsten halogen lamp to simulate outdoor AM 1.5 insolation).

$\eta_{cathode}$, during charge transfer through the photoelectrode and counter electrode interfaces. In the current domain investigated, polarization losses are highly linear for both anodic and cathodic processes at 2.5 to 3.5 mV cm^{2}mA^{-1}, and can create small but significant losses on the order of 10–100 millivolts in the MPEC.

Figure 7 presents the outdoor characteristics of the bipolar direct ohmic AlGaAs/Si/V$^{2+/3+}$ photoelectrochemical cell under solar illumination. The system comprises the individual components illustrated in Fig. 5 and uses a HF containing aqueous vanadium electrolyte to improve photocurrent stability (0.35 M V(II)+V(III), 4 M HCl, 0.2 M HF). The photoelectrochemical characteristics of the cell were determined under 75-mW cm^{-2} insolation. As shown under illumination, the AlGaAs/Si/V$^{2+/3+}$ electrolyte photoelectrochemical solar cell exhibits an open-circuit potential, V_{oc} = 1.4V, a short-circuit photocurrent, J_{sc} = 12.7 mA cm^{-2}, a fill factor, FF = 0.81, determined from the fraction of the maximum power, P_{max}, compared to the product of the open-circuit potential and short-circuit current.

The multiple band gap solar to electrical conversion efficiency of 19.2% compares favorably to the maximum 15 to 16% solar to electrical energy conversion efficiency previously reported for single band gap PECs [9, 10]. Small photoelectrochemical efficiency losses can be attributed to polarization losses accumulating at the solution interfaces. Under illumination, a photocurrent density of 13 mA cm^{-2} seen in Fig. 7 is consistent with polarization

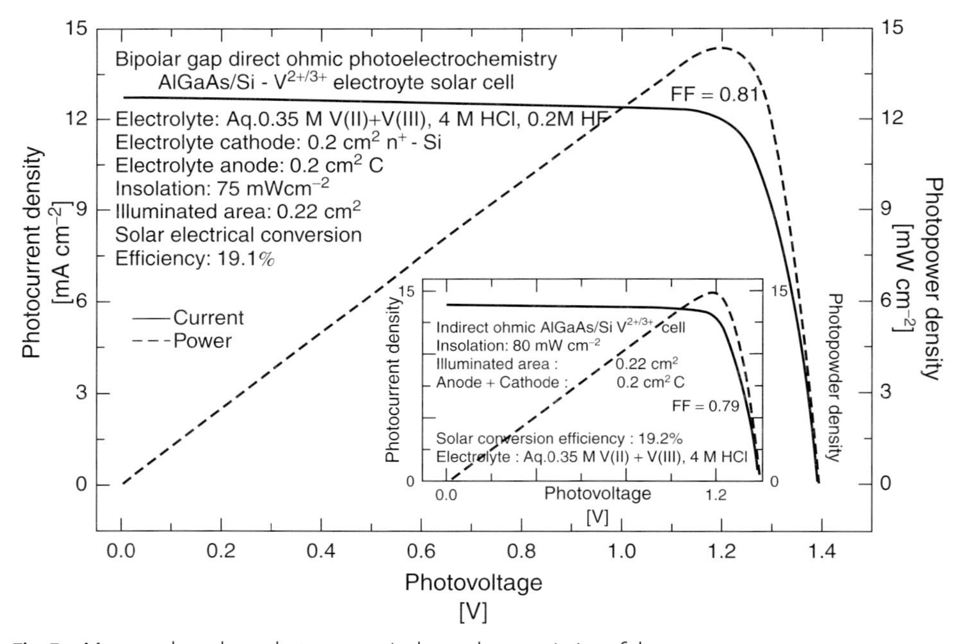

Fig. 7 Measured outdoor photocurrent/voltage characteristics of the bipolar gap direct ohmic AlGaAs/Si-V$^{3+/2+}$ photoelectrochemical solar cell. Inset: Measured outdoor photocurrent/voltage characteristics of the bipolar gap indirect ohmic AlGaAs/Si-V$^{3+/2+}$ photoelectrochemical solar cell.

losses of approximately 0.04 V, accumulating both at the anode and the cathode. At the observed maximum power point photovoltage, in excess of 1.2 V in Fig. 7, this combined 0.08 V loss is consistent with an effective loss in efficiency of 0.3 to 0.4% compared to an analogous solid-state photovoltaic cell configurations.

A common disadvantage of photoelectrochemical systems is photoinduced corrosion of the semiconductor that originates at the semiconductor–solution interface. The corroded surface inhibits charge transfer that diminishes photocurrent. A stable solid–solution interface, which both facilitates charge transfer and impedes semiconductor photocorrosion, is provided by an electrocatalyst placed between the semiconductor and the electrolyte. The multiple band gap photoelectrochemical cell can utilize this electrocatalyst interface, as well as a bipolar series arrangement of wide and small band gap semiconductors to enhance energy conversion. The photoelectrochemical characterization presented in the inset of Fig. 7 summarizes a modified GaAs/Si-$V^{3+/2+}$ MPEC. In this indirect ohmic photoelectrochemistry, electrolyte induced photocorrosion of the silicon is entirely inhibited by utilization of an electrocatalyst (a second carbon electrode) bridging charge transfer between the semiconductor and the electrolyte.

Bipolar gap indirect ohmic photoelectrochemistry comprises either: a bipolar band gap pnpn/electrolyte ohmic photoelectrochemical cell, with reduction occurring at the semiconductor–electrocatalyst–electrolyte interface and regenerative oxidation occurring at the electrolyte–counter electrode (anode) interface or alternately: an npnp/electrolyte cell, with oxidation occurring at the semiconductor–electrocatalyst–electrolyte interface and regenerative reduction occurring at the electrolyte–counter electrode interface. In these systems, indirect refers to the catalyst interface that bridges the semiconductor and solution, and ohmic indicates this interface is an ohmic rather than a Schottky junction. The photocathodic bipolar direct ohmic MPEC photoelectrochemistry comprises a two-photon/one-electron photoelectrochemical process ($2h\nu \rightarrow e^-$) injecting charge into an electrocatalyst and then into solution. Energy conversion can occur either through Fig. 8's photocathodic driven redox process:

$$h\nu\ \text{pn(wide gap)|pn (small gap)| electrocatalyst|redox couple| electrocatalyst anode} \quad (11)$$

or a photoanodic driven redox couple:

$$h\nu\ \text{np(wide gap)|np (small gap)| electrocatalyst|redox couple| electrocatalyst cathode} \quad (12)$$

Two highly stable aqueous phase redox couples, iodide and polysulfide, are used in these studies to further this bipolar indirect multiple band gap photoelectrochemistry. Platinum and cobalt sulfide, respectively, provide an effective (low overpotential) electrocatalyst for a wide range of iodide and polysulfide electrolytes [2]. These electrolytes have also been utilized in bipolar band gap solar cells. Table 1 summarizes the results of a variety of bipolar two band gap solar cells [2, 3]. In this table, comparison of the solid-state and direct photoelectrochemical cells shows that energy conversion efficiency of the photoelectrochemical cell approach that of the solid-state device. In the bipolar direct MPEC cell, the majority of the photopower loss in these photoelectrochemical cells was attributed to polarization

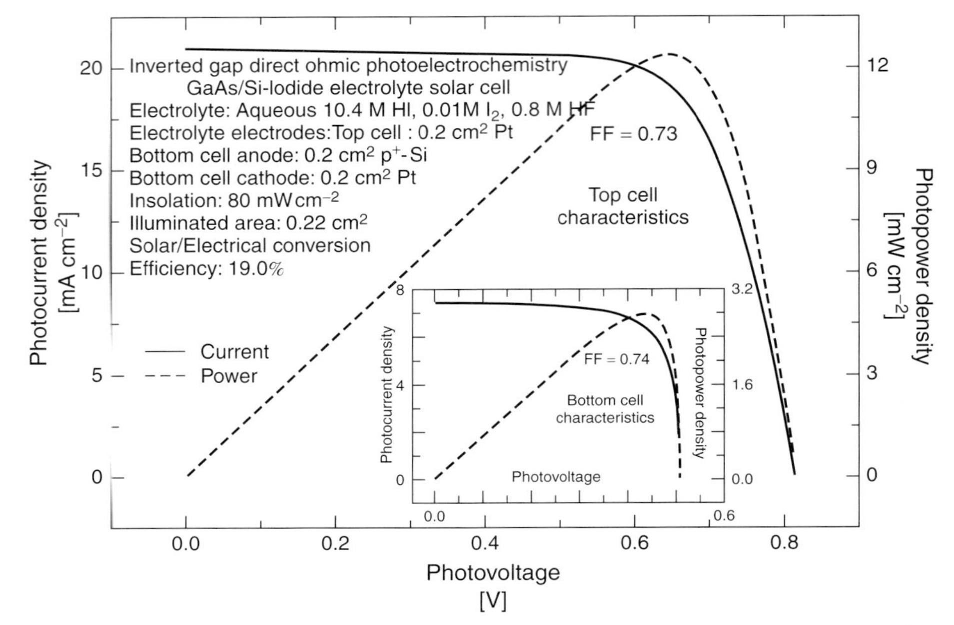

Fig. 8 Measured outdoor characteristics of the inverted band gap direct ohmic GaAs/Si/I_3^-, 3/2I^-/Pt MPEC. The top cell consists of the GaAs/Pti/I_3^-, 3/2I^-/Pt portion of the cell. The lower cell consists of the Si/I_3^-, 3/2I^-/Pt portion of the cell. Main figure: top layer cell photocurrent/voltage characteristics. Inset: bottom layer cell photocurrent/voltage characteristics.

Tab. 1 Comparison of bipolar solid-state or bipolar regenerative ohmic PECs under solar illumination

Cell configuration	V_{oc} *[V]*	I_{sc} *[mA cm^{-2}]*	P_{max} *[mW cm^{-2}]*	*FF*	*Insolation [mW cm^{-2}]*	*Conversion [Efficiency %]*
Bipolar solid state						
AlGaAs/Si	1.409	15.6	17.7	0.81	90.2	19.6
Bipolar direct $V^{2+/3+}$						
AlGaAs/Si/V^{2+}/V^{3+}/C	1.393	12.7	14.4	0.81	75.0	19.1
Bipolar indirect $V^{2+/3+}$						
AlGaAs/Si/C/V^{2+}/V^{3+}/C	1.383	14.0	15.3	0.79	80.1	19.2
Bipolar indirect sulfide						
AlGaAs/Si/CoS/S_2^{2-}/S_4^{2-}/CoS	1.456	15.8	18.5	0.80	94.1	19.7
Bipolar indirect iodide						
AlGaAs/Si/Pt/I_3^-/I^-/Pt	1.409	15.9	18.1	0.81	94.1	19.2

Note: Cells utilize multijunction wide band gap AlGaAs layers over smaller band gap Si layers and one of the indicated pairs of electrolyte/electrocatalyst electrodes indicated as Direct $V^{2+/3+}$: 0.35 M V(II)+V(III), 4 M HCl, 0.2 M HF at carbon; Indirect $V^{2+/3+}$: 0.35 M V(II)+V(III), 4 M HCl at carbon; Indirect Sulfide 1 M K_2S_2, 1 M KOH at CoS; Indirect Iodide: 10.4 M HI, 0.01, 4 M I_2 at Pt.
PEC, photoelectrochemical solar cells.

losses arising at the solid–solution interfaces. In certain indirect MPEC cases, as seen in Table 1, the photoelectrochemical energy conversion efficiency is comparable to the solid-state cell process. In this case, combined electrolytic polarization losses are less than or equal to the potential drop from resistance losses in the back contact of the solid-state device.

4.4.2.3 Inverted Band Gap PECs

Inverted photoelectrochemistry permits efficient energy, and maintained smaller photopotentials than in comparable bipolar MPECs. To probe inverted semiconductor direct ohmic regenerative electrochemistry, an electrolyte such as iodide was driven by a GaAs ($E_{GW} = 1.4$ eV) wide band gap, and an Si ($E_{GS} = 1.0$ eV) small band gap were utilized [3]. In the inverted configuration, bottom and top cells are utilized simultaneously. Simultaneous to the photopower generated by the bottom portion (Si) of the MPEC, the top (GaAs) portion also generates photopower. A platinum counter anode, provides an effective (stable, low polarization) electrocatalytic surface for the reverse process in both the top (wide band gap) and bottom (small band gap) driven regenerative cells. At the n-Si interface, photocurrent stability can be improved with the addition of HF to an acidic iodide electrolyte, and the electrolyte used was 10.4 M HI, 0.01 M I_2, 0.8 M HF. Previous studies have shown that other electrolytes and surface modifications will also enhance photocurrent stability through the (single band gap) silicon–electrolyte interface [19–21]. The inset of Fig. 8 presents the inverted direct ohmic outdoor characteristics of the ($Si/I_3^-/3/2I^-$) bottom cell portion, and the main portion of the figure presents the characteristics of the GaAs driven $I_3^-/3/2I^-$ portion of the cell under solar illumination.

Photoelectrochemical characteristics were determined under 80 mW cm^{-2} insolation and the generated photocurrent is highly stable. As shown under illumination, the Si bottom cell exhibits an open-circuit potential, $V_{oc} = 0.51$V, a short-circuit photocurrent, $J_{sc} = 7.4$ mA cm^{-2} a fill factor, FF $= 0.73$, and a maximum power, $P_{max} = 2.8$ mW cm^{-2}. Simultaneous to the photopower generated by the bottom portion of the MPEC, the top (GaAs) portion also generates photopower. The main portion of Fig. 8 presents the outdoor characteristics of the top cell ($GaAs/Pt/I_3^-$, $3/2I^-/Pt$) portion during solar illumination of the complete inverted direct ohmic $GaAs/Si/I_3^-, 3/2I^-/Pt$ MPEC. In this portion of the cell, GaAs is isolated from the iodide electrolyte via a Pt electrocatalytic anode. The Pt electrode is stable in this electrolyte, and hence the top cell portion of the photocurrent appears to be fully stable. Under the same 80 mW cm^{-2} insolation, photopower is generated simultaneous to the photopower presented in Fig. 8. As shown under illumination, the GaAs cell exhibits an open-circuit potential, $V_{oc} = 0.81$ V, a short-circuit photocurrent, $J_{sc} = 21.0$ mA cm^{-2}, a fill factor, and a maximum power, $P_{max} = 12.4$ mW cm^{-2} at FF $= 0.74$. The total power generated is the sum of the simultaneous extractable power generated by each component.

Table 2 compares inverted solid-state photovoltaics with direct and indirect regenerative ohmic PECs, each containing a GaAs wide band gap and an inverted aligned Si small band gap, and a variety of redox couples. As in the case of the bipolar systems summarized in Table 1, the solid-state device exhibits a marginally enhanced energy conversion efficiency compared to

Tab. 2 Comparison of inverted solid-state or inverted regenerative ohmic PECs under solar illumination

Cell configuration	V_{oc} *[V]*	I_{sc} *[mA cm^{-2}]*	P_{max} *[mW cm^{-2}]*	*FF*	*Insolation [mW cm^{-2}]*	*Conversion [Efficiency %]*
Inverted solid-state						
bottom: Si	0.552	22.8	3.6	0.74		
top: GaAs	0.855	8.7	14.2	0.74		
GaAs/Si			17.7		90.2	19.6
Inverted direct iodide						
Bottom: Si/I_3^-/I^-/Pt	0.510	7.4	2.8	0.74		
Top: GaAs/Pt/I_3^-/I^-/Pt	0.813	21.0	12.4	0.73		
GaAs/Si/I_3^-/I^-/Pt			15.2		80.1	19.0
Inverted indirect iodide						
Bottom: Si/Pt/I_3^-/I^-/Pt	0.541	9.2	3.7	0.74		
Top: GaAs/Pt/I_3^-/I^-/Pt	0.866	24.5	14.9	0.70		
GaAs/Si/Pt/I_3^-/I^-/Pt			18.5		94.1	19.7
Inverted indirect sulfide						
Bottom: Si/CoS/S_2^{2-}/S_4^{2-}/CoS	0.556	8.9	3.8	0.76		
Top: GaAs/CoS/S_2^{2-}/S_4^{2-}/CoS	0.855	24.0	14.9	0.73		
GaAs/Si/CoS/S_2^{2-}/S_4^{2-}/CoS			18.7		94.1	19.8
Inverted indirect $V^{2+/3+}$						
Bottom: Si/C/V^{2+}/V^{3+}/C	0.523	7.7	3.0	0.73		
Top: GaAs/C/V^{2+}/V^{3+}/C	0.823	20.8	12.3	0.72		
GaAs/Si/C/V^{2+}/V^{3+}/C			15.3		80.1	19.2

Note: Cells utilize multijunction wide band gap GaAs layers over smaller band gap Si layers and one of the indicated pairs of electrolyte or electrocatalyst electrodes, indicated as Direct Iodide: 10.4 M HI, 0.01, 4 M I_2, 0.8 M HF at Pt; Indirect Iodide: 10.4 M HI, 0.01, 4 M I_2 at Pt; Indirect Sulfide 1 M K_2S_2, 1 M KOH at CoS; Indirect $V^{2+/3+}$: 0.35 M V(II)+V(III), 4 M HCl at carbon.

the analogous direct regenerative MPEC. This difference may be attributed to diminished charge transfer through the semiconductor–electrolyte interface. However these differences are small and are only slightly larger than those caused by the uncertainty of ± 1 mW cm^{-2} in the measured insolation level. Each of the sulfide or iodide indirect regenerative MPECs probed in this study exhibit conversion efficiency comparable to, or better than, the solid-state device in Table 2. This indicates that electrolytic polarization losses in these cells are comparable to resistance losses of the semiconductor back contact in the solid-state MPEC. The marginally lower conversion efficiency of the $V^{2+/3+}$ indirect regenerative MPEC in this table is consistent with the higher polarization losses of 3 mV cm^2 mA^{-1} that we have measured for $V^{2+/3+}$ at carbon, compared to those of less than 2 mV cm^2 mA^{-1} previously measured at Pt and CoS, respectively, for the iodide and sulfide MPECs [2, 3].

4.4.2.4 Bipolar Band Gap Solar Storage Cells

One limitation to solar cells is that, while most of our electrical needs are continuous, clouds and darkness dictate that solar energy is intermittent in nature. PECs can generate not only electrical but also electrochemical energy, and provide a single device to solve the problem of the intermittent nature of solar energy. In 1987 we presented a highly efficient single band gap photoelectrochemical cell combining in situ electrochemical storage and solar conversion capabilities, providing continuous output insensitive to daily variations in illumination. The 1987 configuration was demonstrated with a n-Cd(Se,Te)/polysulfide electrolyte conversion half cell and a Sn/SnS storage system, resulting in a single cell operating continuously at an overall efficiency of 11% [18]. Under illumination, photocurrent droves an external load. Simultaneously, a portion of the photocurrent was used in the direct electrochemical reduction to a metal (Sn) in the device storage half cell. In darkness or below a certain level of light, the storage compartment delivers power by metal oxidation.

Recently, high solar conversion and storage efficiencies have been attained with a system that combines efficient multiple band gap semiconductors, with a simultaneous high-capacity electrochemical storage [4, 5]. The energy diagram for one of several multiple band gap cells is presented in Fig. 3(d), and as described in Fig. 2, several other configurations are also feasible. In the figure, storage occurs at a potential of $E_{redox} = E_{A+/A} - E_{B/B+}$. On illumination, two photons generate each electron, a fraction of which drives a load, while the remainder ($1/xe^-$) charges the storage redox couple. Without light, the potential falls below E_{redox} and the storage couple spontaneously discharges. This dark discharge is directed through the load, rather than through the multijunction semiconductor's high dark resistance.

In Fig. 9 is presented an operational form of the solar conversion and storage cell described by the Fig. 2 energy diagram. The single cell contains both multiple band gap and electrochemical storage, which unlike conventional photovoltaics, provides a nearly constant energetic output in illuminated or dark conditions. The cell combines bipolar AlGaAs ($E_g = 1.6$ eV) and Si ($E_g = 1.0$ eV) and AB_5 metal hydride/NiOOH storage. The NiOOH/MH metal hydride storage process is near ideal for the AlGaAs/Si because of the excellent match of the storage and photocharging potentials. The electrochemical storage processes utilizes MH oxidation and nickel oxyhydroxide reduction [22]:

$$MH + OH^- \longrightarrow M + H_2O + e^-; \quad E_{M/MH} = -0.8 \text{ V versus SHE} \quad (13)$$

$$NiOOH + H_2O + e^- \longrightarrow Ni(OH)_2 + OH^-; \quad E_{NiOOH/Ni(OH)_2} = 0.4 \text{ V vs SHE} \quad (14)$$

As shown in Fig. 10, the cell generates a light variation insensitive potential of 1.2–1.3 V at total (including storage losses) solar/electrical energy conversion efficiency of 18%. Over an eight-month period of daily cycles, under constant 12-hour (AM0) illumination, the long-term indoor cycling cell generated a nearly constant photocurrent density of 21.2 (constant to within one percent or ± 0.2 mA cm^{-2}), and a photopower that varied by ±3% [5]. The cell is a single physical/chemical device that generates load current without any external switching.

4.4.2.5 Bipolar Band Gap Solar Hydrolysis (hydrogen generation) Cells

Solar energy–driven water splitting combines several attractive features for energy utilization. Both the energy source (sun) and the reactive media (water) are readily available and are renewable, and the resultant fuel (generated hydrogen) and the emission with fuel consumption (water)

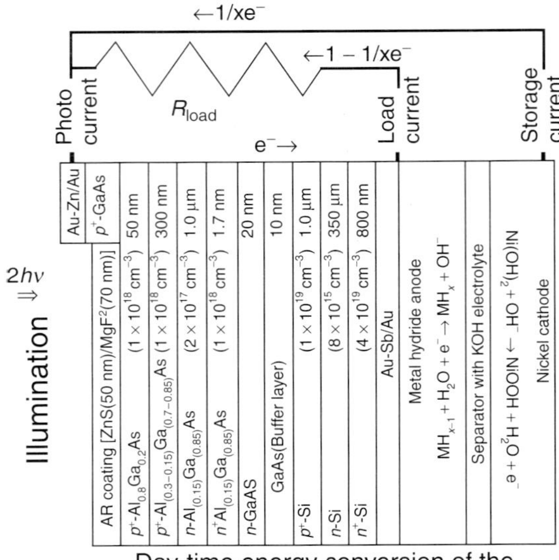

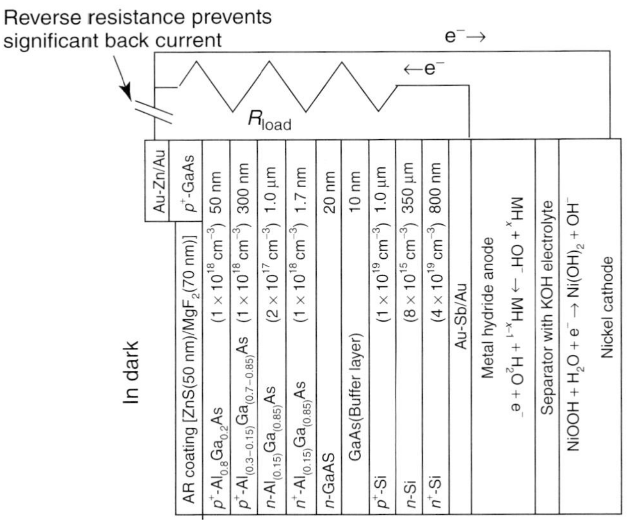

Fig. 9 The bipolar AlGaAs/Si/MH/NiOOH MBPEC solar cell.

are each environmentally benign. The UV and visible energy rich portion of the solar spectrum is transmitted through H_2O. Therefore sensitization, such as via semiconductors, is required to drive the water-splitting process. In a solar photoelectrolysis system, the redox active interfaces can be in indirect or direct

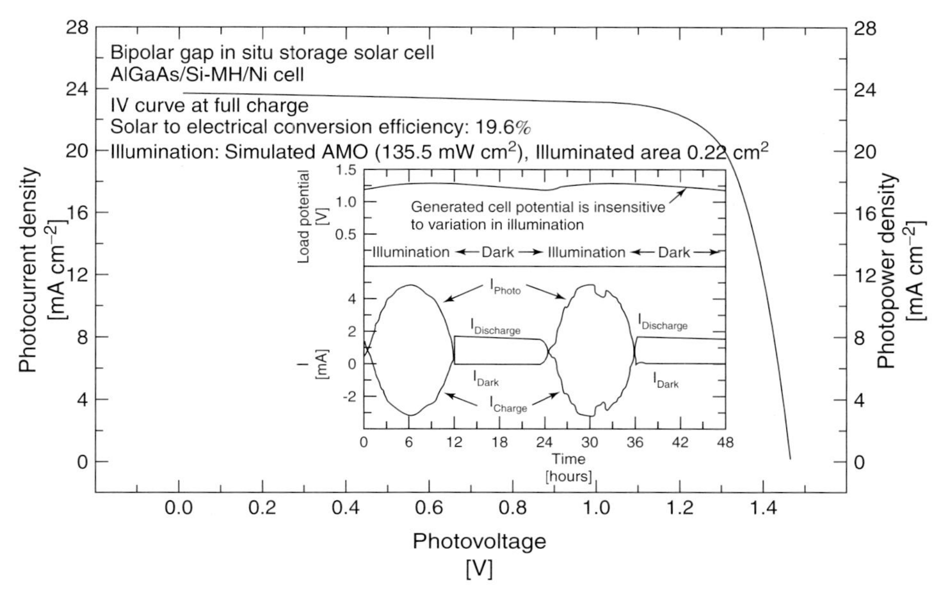

Fig. 10 Photoconversion current and power characteristics of the AlGaAs/Si/MH/NiOOH solar cell under fully charged AM0 conditions. Inset: Two days conversion and storage characteristics of the AlGaAs/Si/MH/NiOOH solar cell under simulated AM0 insolation. In the first 12 hours of each of the two 24-hour cycles, the AM0 illumination is varied from 0 to AM0 back to 0, using a graded diffuse filter; the second 12 hours are without illumination.

contact with the photosensitizer, and comprise either an ohmic or Schottky junction. Independent of this interface composition, the various parameters in models predicting solar water–splitting conversion efficiency, may be combined into two general parameters: (1) related to losses in optical energy conversion, η_{photo}, or (2) related to losses in redox conversion of water to H_2 and O_2, $\eta_{electrolysis}$. Combined, these yield an overall solar electrolysis efficiency (excluding storage and utilization losses) as:

$$\eta_{photoelectrolysis} = \eta_{photo} \times \eta_{electrolysis} \quad (15)$$

The thermodynamic potential, $E^o{}_{H2O}$, for the water splitting reaction is given by:

$$H_2O \longrightarrow H_2 + 1/2O_2;$$
$$E^o_{H_2O} = E^o_{O_2} - E^\circ_{H_2};$$
$$E^o_{H_2O}(25\,^\circ C) = 1.229\ V \quad (16)$$

The subsequent challenge is to optimize sustained water electrolysis, without considerable additional energy losses. Effective water electrolysis must occur at a potential, V_{H_2O}, near the photocell point of maximum power. V_{H_2O} is greater than $E^o_{H_2O}$ $(=-\Delta G_{H_2O}/nF)$ as a result of overpotential losses, ζ, in driving an electrolysis current density, j, through both the O_2 and the H_2 electrodes:

$$V_{H_2O}(j) = E_{O_2}(j) - E_{H_2}(j)$$
$$= \left(E^o_{O_2} + \zeta_{O_2}(j)\right) - (E^o_{H_2} + \zeta_{H_2}(j)) \quad (17)$$

A cell containing illuminated AlGaAs/Si RuO_2/Pt_{black} is demonstrated to evolve H_2 and O_2 at record solar-driven water electrolysis efficiency [6]. Under illumination, bipolar configured $Al_{0.15}Ga_{0.85}As$ (E_g = 1.6 eV) and Si (E_g = 1.1 eV) semiconductors generate open-circuit and maximum power photopotentials of 1.30 and 1.57 V, well suited to the water electrolysis thermodynamic potential of $E^o_{H_2O}(25\,^\circ C) = 1.229$ V. The $E^o_{H_2O}$/photopotential-matched semiconductors are combined with effective water electrolysis O_2 or H_2 electrocatalysts, RuO_2 or Pt_{black}. The resultant solar photoelectrolysis cell drives sustained water splitting at 18.3% conversion efficiencies. Alternate dual band gap systems are calculated to be capable of attaining more than 30% solar photoelectrolysis conversion efficiency.

Figure 11 presents the measured current/voltage characteristics for the AlGaAs/Si photocell at AM0 illumination. Further details and characterization of the cell are available [6]. Under illumination, this bipolar cell generates an open-circuit potential of 1.57 V, which is considerably larger the thermodynamic potential for the water-splitting reaction. Also included are the photopotential dependence of solar to electrical conversion efficiency. A portion of the high η_{photo} domain lies at a potential above $E^o_{H_2O}$, and in principle is sufficiently energetic to drive efficient photoelectrolysis.

Planar platinum and Pt_{black} are effective H_2 electrocatalysts. At low current densities (0.5 mA cm^{-2}), the observed Pt overpotential is low ($\zeta_{H_2} = -17$ mV), and even smaller ($\zeta_{H_2} = -13$ mV) with the Pt_{black} electrocatalyst, and smaller yet ($\zeta_{H_2} =$

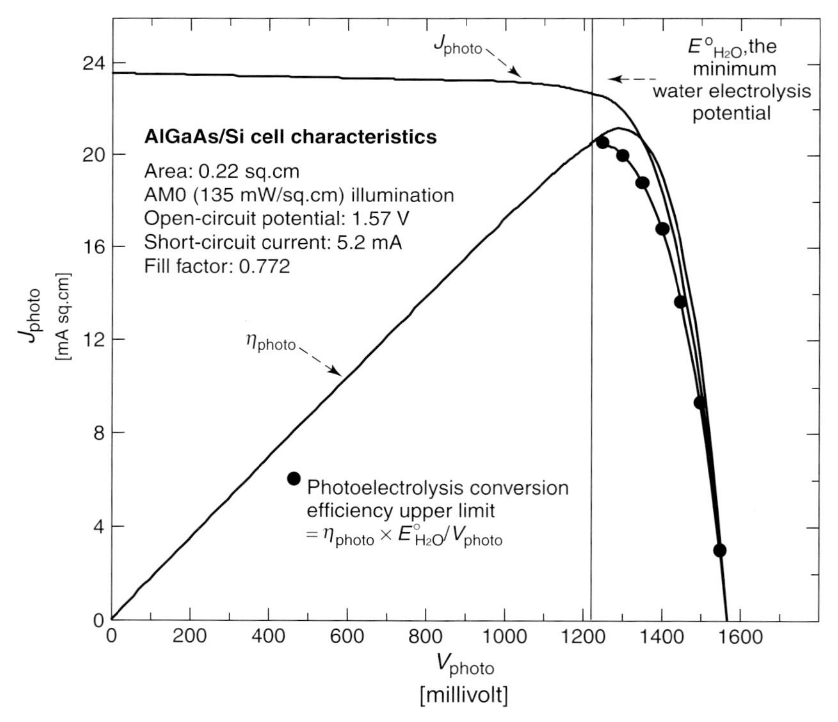

Fig. 11 AlGaAs/Si I–V characteristics at AM0 light intensity (135 mW cm^{-2}). $\eta_{photo} = 100\% \times (J_{photo} * V_{photo})/P_{illumination}$. Using the indicated, measured η_{photo} and V_{photo}, the upper limits of photoelectrolysis efficiency are calculated as $\eta_{photo} \times 1.229\ \text{V}/V_{photo}$.

−7 mV) with a low level of convection (stirring) to improve mass transport and prevent observed gas buildup on the electrode surface. Minimization of ζ_{O_2} is a greater challenge. In this case Pt is a poor electrocatalyst. ζ_{O_2} may be decreased ~500 mV, utilizing an RuO_2 electrode. The kinetics of this effective oxygen electrocatalyst [23] have been attributed to catalysis by the intervening RuO_4/RuO_2 redox couple [24, 25]. In the absence of competing redox couples, the faradaic efficiency of H_2 and O_2 evolution approaches 100%, and $\eta_{electrolysis}$ is determined by the current limited $V_{H_2O}(j)$:

$$\eta_{electrolysis} = E^o_{H_2O}/V_{H_2O}(j);$$
$$\eta_{electrolysis}(25\,^\circ C) = 1.229V/V_{H_2O}(j) \qquad (18)$$

The Fig. 12 inset contains the Eq. (18)–determined $\eta_{electrolysis}$. The limiting maximum $\eta_{photoelectrolysis}$ can be readily determined from the solar to electrical conversion efficiency. Expanding Eq. (15) with Eq. (18), $\eta_{photoelectrolysis}$ is diminished from η_{photo} by the potential of the stored energy compared to the potential at which the water electrolysis occurred:

$$\eta_{photoelectrolysis}(T) = \frac{\eta_{photo}(T) \times E_{H_2O}(T)}{E_{O_2}(T) - E_{H_2}(T)};$$
$$\frac{\eta_{photoelectrolysis}(25\,^\circ C) = 1.229V \times \eta_{photo}}{V_{H_2O}} \qquad (19)$$

Figure 12 includes these limiting $\eta_{photoelectrolysis}$ values because the AlGaAs/Si system is determined from measured values of η_{photo} at various values of V_{photo}. AlGaAs/Si solar photoelectrolysis at conversion efficiencies exceeding 20% (Fig. 11) are in principle possible at potentials approaching the E_{H_2O} limit. The main portion of Fig. 12 presents the water electrolysis potential, V_{H_2O}, determined under stirred or quiescent conditions, and measured in 1 M $HClO_4$ using equal areas of the optimized RuO_2 and Pt_{black} electrodes. These measurements of the electrolysis current, as a function of potential, enable us to predict the AlGaAs/Si photocurrent at which photoelectrolysis will occur.

Our photoelectrolysis cell consists of illuminated AlGaAs/Si RuO_2/Pt_{black} electrolysis. With these active electrocatalysts, high values of $\eta_{electrolysis}$ are insured by using large surface areas of the electrolysis electrodes, compared to the illuminated area. This is accomplished without increasing the illuminated electrode area, as schematically represented in the lower portion of Fig. 13, by utilizing a large vertical depth of electrolysis electrodes compared to the cross section of illumination. Specifically, the 10 cm^2 Pt_{black} and RuO_2 electrodes utilized are large compared to the 0.22 cm^2 illuminated area. In the Fig. 13(a) is the H_2 and O_2 electrolysis current generated by this cell as a function of time. The average photocurrent of 4.42 mA (generating a current density of 20.1 mA cm^{-2} at the illuminated electrode area, and 0.44 mA cm^{-2} at the electrolysis electrodes) corresponds to a photopotential of 1.36 V in Fig. 11 comparable to the equivalent electrolysis potential at 0.5 mA cm^{-2} in Fig. 12. The overall efficiency is determined by the 1.229 V energy stored as H_2 and O_2 and the incident photopower (135 mW cm^{-2}):

$$\eta_{photoelectrolysis} = 100\% \times 20.1\ \text{mA cm}^{-2} \times 1.229\ \text{V}/135\ \text{mW cm}^{-2} = 18.3\% \qquad (20)$$

4.4.2.6 **Higher Solar Production Rates of Hydrogen Fuel are Attainable**

In Eq. (15), the significance of the electrolysis compared to photo components

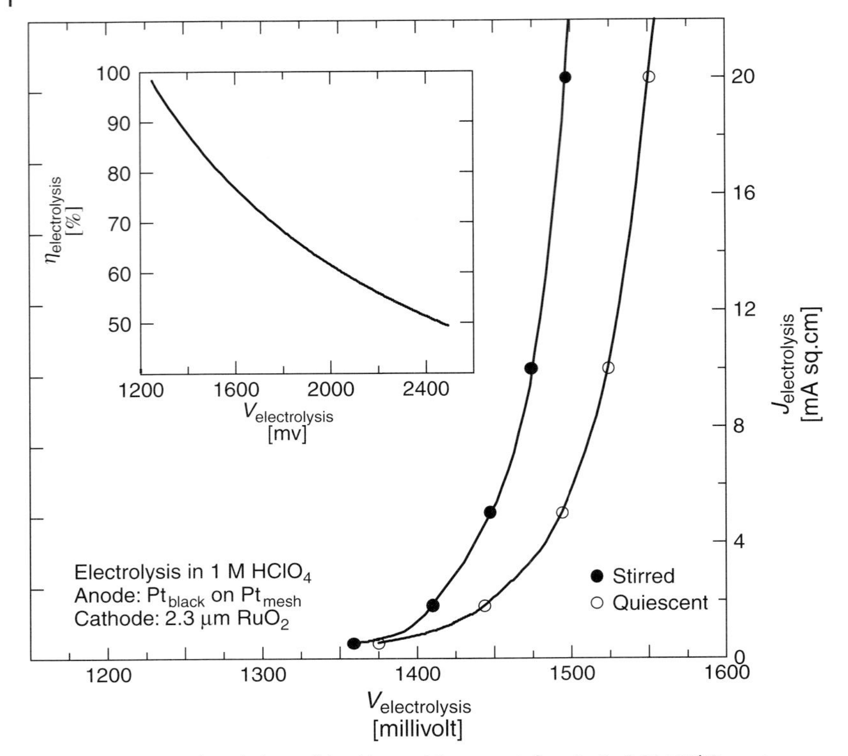

Fig. 12 Measured variation of the V_{H_2O} with current density in 1 M $HClO_4$, using equal area 2.3 μm RuO_2 and Pt_{black}. Inset: Calculated η_{photo}, in a 100% faradaic-efficient process, as a function of V_{H_2O}.

of the conversion efficiency is evident on analysis of a report of a 12.4% photoelectrolysis cell. The cell contained a Pt coated $GaInP_2$ hydrogen electrode in bipolar contact with a GaAs junction driving an oxygen electrode [26]. At their specified 11 sun–illuminated electrolysis, current density of 120 mA cm^{-2}, ζ_{H_2} will exceed 300 mV at Pt. Oxygen losses will be even larger, and the cell will have a total overpotential loss >700 mV. Two such cells will have losses $>E^o_{H_2O}$; unnecessary losses if the cell was illuminated at 1 sun or employed larger surface area or more effective electrocatalysts. Under these conditions, two $GaInP_2$/GaAs cells placed in series should drive three water electrolysis cells in series, effectively increasing the relative photoelectrolysis efficiency by 50%.

Previous principal solar water splitting models predict similar dual band gap photoelectrolysis efficiencies of only 16%, and 10–18%, respectively [27, 28]. Each are lower than our observed water splitting efficiency discussed below. The physics of these models were superb, but their analysis was influenced by dated technology and underestimated the experimental η_{photo} attained by contemporary devices or underestimated the high experimental values of $\eta_{electrolysis}$, which can be attained. For example, Boltonand coworkers, estimates low values of η_{photo} (less than 20% conversion) due to assumed cumulative relative

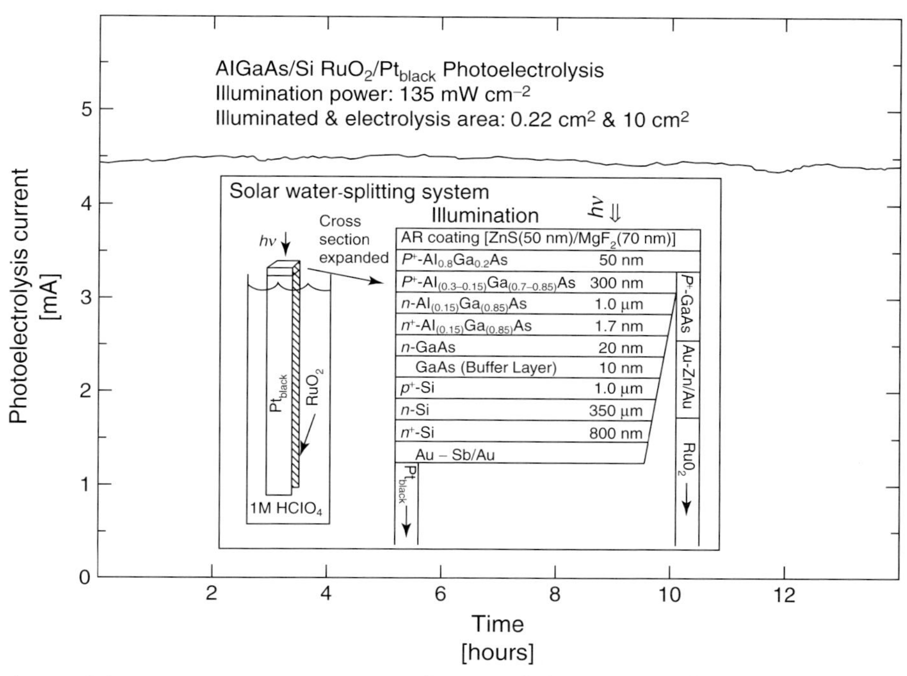

Fig. 13 Schematic representation (inset) and measured characteristics of the illuminated AlGaAs/Si RuO_2/Pt_{black} photoelectrolysis cell. Further details of the layered AlGaAs/Si structure are given in Ref. 6.

secondary losses that include: 10% reflection loss, 10% quantum-yield loss, 20% absorption loss. As summarized in Table 3 demonstrated η_{photo} are substantially higher than 20%. Many η_{photo} that have been carefully reviewed or monitored are in excess of 30% [29].

Each of the cells in the Table 3 exhibits an open-circuit photopotential significantly greater than the minimum potential necessary to split water. The majority of these cells can generate a photopotential in excess of 2 volts. An unnecessary limit of one multiple band gap photoexcitation per electrolysis would under-utilize V_{photo}, diminishing $\eta_{electrolysis}$. For example, a GaInP/GaAs cell has a maximum power photopotential of 2.0 to 2.1 V and an open-circuit potential of 2.3 V [12]. Two such cells in series will efficiently drive three 1.3 to 1.4 V water electrolysis cells in series, and as discussed here at a water electrolysis at 1.36 V (Fig. 12 inset) yielded electrolysis efficiencies of more than 90%. It is reasonable that with larger surface area, or more effective electrocatalysis, these efficiencies will approach 95%. Using this range of $\eta_{electrolysis} = 90-95\%$:

$$\eta_{photoelectrolysis}(\text{predicted maximum}) = \eta_{photo} \times 90\text{–}95\% \quad (21)$$

Table 3 includes predicted maximum $\eta_{photoelectrolysis}$ using observed η_{photo} of various dual band gap sensitizers. It is seen that solar water splitting efficiencies may be viable at up to double the amount of that previously predicted. Efficient, three or more multiple band gap photoelectrolysis

Tab. 3 Predicted and measured photoelectrolysis efficiencies. Calculated $\eta_{photoelectrolysis}$ are from Eq. (21)

Photovoltaic	Light Level	η_{photo} Measured	$\eta_{photoelectrolysis}$ Predicted Maximum	Measured
GaInP/GaAs	1 sun	30.3% (Ref. 30)	27–29% (Ref. 6)	
GaInP/GaAs	180 sun	30.2% (Ref. 30)	27–29% (Ref. 6)	
GaAs/Si	350 sun	29.6% (Ref. 30)	27–28% (Ref. 6)	
GaAs/GaSb	100 sun	32.6% (Ref. 30)	29–31% (Ref. 6)	
InP/GaInAs	50 sun	31.8% (Ref. 30)	29–30% (Ref. 6)	
GaAs/GaInAsP	40 sun	30.2% (Ref. 30)	27–29% (Ref. 6)	
AlGaAs/Si	1 sun	21.2% (Ref. 15)	19–20% (Ref. 6)	18.3% (Ref. 6)
$GaInP_2$/GaAs	11 sun			12.4% (Ref. 27)
InP/GaAs	1 sun			8.2% (Ref. 30)

will be expected to be capable of attaining even higher efficiencies.

4.4.3 Solution Phase Phenomena

4.4.3.1 Solution Phase Chemistry Optimization

In PECs, the manifold conditions in which solution phase chemistry constrains the kinetics and thermodynamics of photoelectrochemical charge transfer have been probed. Modification of the solution phase chemistry, can substantially impact on photoelectrochemical properties by (1) enhancing facile charge transfer, (2) suppressing competing reactions and suppressing both (3) electrode and (4) electrolyte decomposition products, as well as (5) substantially affecting the open-circuit photovoltage. These limitations have been studied in a variety of redox couples and electrolytes. A chemical mechanism was presented [7, 8] for photoelectrochemical solar to electrical energy conversion which emphasized understanding of the distribution of species in photoelectrochemical electrolytes. The approach systematically modifies observed photoelectrochemical kinetic and thermodynamic phenomena. This systematically probes the approach in studies on polysulfide, ferrocyanide, polyselenide, and polyiodide PECs.

4.4.3.2 *n*-Cd Chalcogenide/Aqueous Polysulfide Photoelectrochemistry

In 1976, the first regenerative PECs with substantial and sustained solar to electrical conversion efficiency were demonstrated. These PECs are based on *n*-type cadmium chalcogenide (S, Se or Te) electrodes immersed in aqueous polychalcogenide electrolytes. The cells were introduced by Hodes, Cahen, and Manassen [31], Wrighton and coworkers [32], and Heller and Miller [33] and were capable of converting up to 7% of insolation to electrical energy. Most investigations of these systems focused on solid-state and interfacial aspects of these PECs and photodriven oxidation of polysulfide at the photoelectrode was represented:

$$S^{=} + 2hn \xrightarrow{n\text{-cd(Se,Te)}} S + 2e^{-} \qquad (22)$$

However as represented in Fig. 14(c), the number and type of species in polysulfide photoelectrolytes is considerably more complex than represented by Eq. (22). The

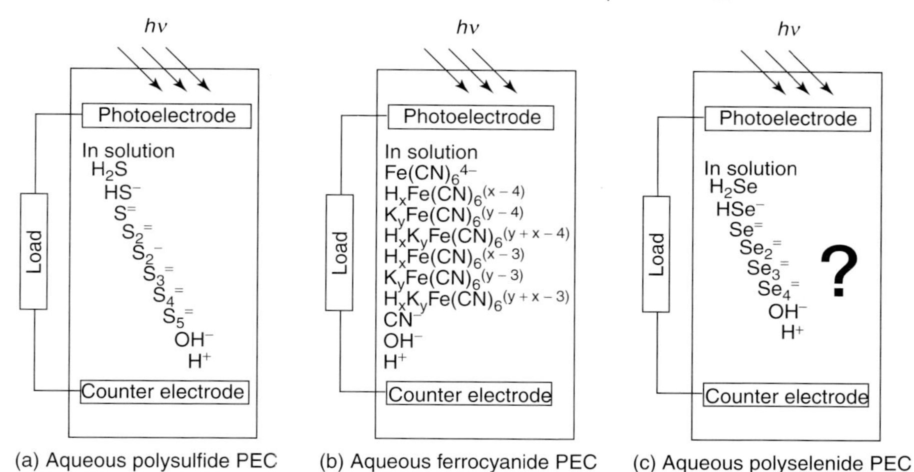

Fig. 14 Chemical species coexisting in polysulfide (a), ferrocyanide (b) and polyselenide (c) aqueous electrolytes used in regenerative PECs.

addition to water of a simple soluble sulfide salt and sulfur to water gives rise to a wide distribution of species in solution. The equilibria constraining this distribution has been investigated by Licht and coworkers [34–39], Gigenbach [40], and Teder [41].

It has been shown that modification of the distribution of species in aqueous polysulfide solution can be correlated to variations in transparency, conductivity, activity, and cadmium chalcogenide photoelectrochemistry. Specifically, the separate effects of hydroxide and pH modification [42], sulfur [43] and sulfide [44], and cation [45, 46] were investigated in terms of speciation and photoelectrochemical phenomena. Previously cadmium chalcogenide polysulfide PECs had generally employed electrolytes composed 1 molar in sodium sulfide, sulfur and sodium hydroxide and resulted in solar conversion efficiencies of approximately 7% [31–33]. However, added hydroxide is to be minimized in these cells, cesium is the preferred cation, and a sulfur to sulfide ratio of approximately 1.5 to 1 resulted in a near doubling of the conversion efficiency [47]. Figure 15 presents *n*-Cd(Se,Te) photocurrents measured at various applied potentials in a traditional and modified electrolyte. As seen in the figure, the cumulative effect of polysulfide electrolyte modifications on photoelectrochemical solar to electrical energy conversion by *n*-Cd(Se,Te)/aqueous polysulfide PECs can be considerable.

A greater percentage of photogenerated holes utilized in constructive oxidation of polysulfide results in enhanced photocurrents. This has the additional benefit of fewer oxidizing holes available for attack on the semiconductor (photocorrosion). As seen in Fig. 16, this results in enhanced photocurrent stability of the PEC. This study showed that with solution optimization, not only the photocurrent but also the polysulfide electrolyte exhibits enhanced lifetime, both approaching one-year operation outdoors [48].

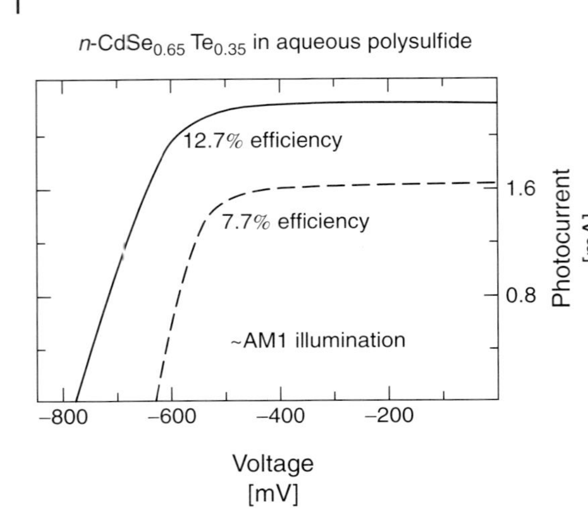

Fig. 15 Potentiostatic photocurrent voltage characteristics for an illuminated n-$CdSe_{0.65}Te_{0.35}$ single crystal immersed in either of two types of aqueous polysulfide electrolyte: Top curve in 1.8 M Cs_2S and 3 M sulfur:; bottom curve is in 1 M NaOH, 1 M Na_2S, 1 M sulfur. The photocurrent voltage curves were obtained outdoors, and solar to electrical conversion efficiencies are indicated.

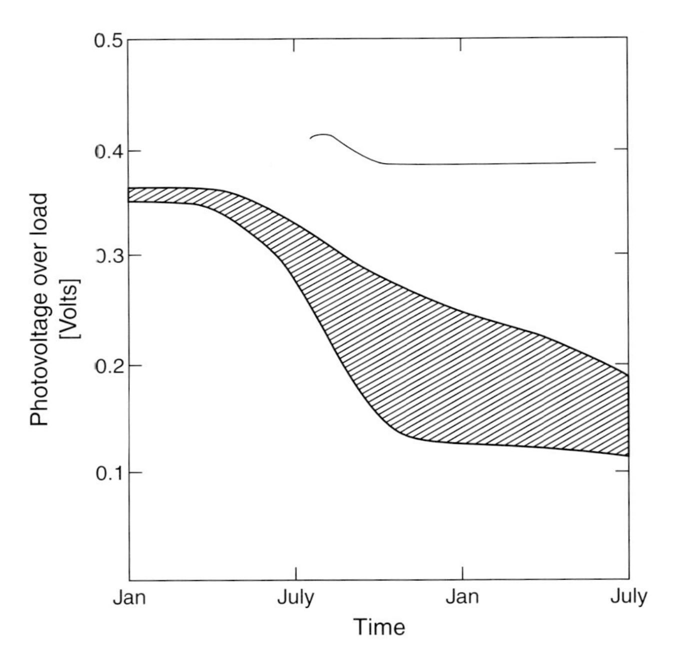

Fig. 16 Long-term outdoor stability curves of thin film $CdSe_{0.75}Te_{0.35}$ in electrolytes of either 1.8 M S (solid line) or 2.0 M KOH, 1.4 M Na_2S, and 2.6 M S (shaded region). The sodium/potassium electrolyte cell results are average results over separate cells, with initial conversion efficiencies from 3.5 to 4.5%. The cesium electrolyte cell had 4.6% conversion efficiency potential measured under a load chosen for initial maximum power.

4.4.3.3 *n*-Cd Chalcogenide/Aqueous Ferrocyanide Photoelectrochemistry

Reichman and Russak [49] and Freeze [50] had reported high (12 to 14%) conversion efficiencies for n-CdSe/($Fe(CN)_6^{3-/4-}$) solar cells in electrolytes containing a 1 : 1 ratio of aqueous $Fe(CN)_6^{4-}$ to $Fe(CN)_6^{3-}$ salts in a highly alkaline environment. However, Rubin and coworkers [51] had shown that these n-CdSe/$Fe(CN)_6^{3-/4-}$ solar cells can have a limited stability and photocurrents decrease within the order of hours. Their attempts to enhance the photoconversion stability of these systems have included cationic surface modification of the n-CdSe photoelectrode and isolation of the wavelength dependence of the surface instability [51, 52]. In these studies, photo-oxidative processes in n-CdSe/$Fe(CN)_6^{3-/4-}$ PECs were typically represented:

$$Fe(CN)_6^{4-} + hn \xrightarrow{n\text{-CdSe}} Fe(CN)_6^{3-} + e^- \quad (23)$$

The role of the various redox active constituents in constraining n-CdSe/$Fe(CN)_6^{3-/4-}$ photo-oxidative energy conversion had not been previously investigated, and ferricyanide and ferrocyanide salts exhibit a complex aqueous equilibria. However, in a manner analogous to the polysulfide electrolytes, and as illustrated in Fig. 14(b), these salts exhibit a complex aqueous equilibria, resulting in a variety of Fe(II) and Fe(III) species. As with the polysulfide electrolytes, we have probed the equilibria constraining the distribution of species in these solutions. This speciation can be controlled and substantially effects n-CdSe photo-oxidative energy conversion. The photoelectrochemical effect and solution stability effects of pH, cation modification and ratio of ferrocyanide to ferricyanide, have been studied by Licht and Peramunage [9, 53–55], and also the systematic variation of the primary photo-oxidized species. As seen in Fig. 17, substitution of a single ligand in the

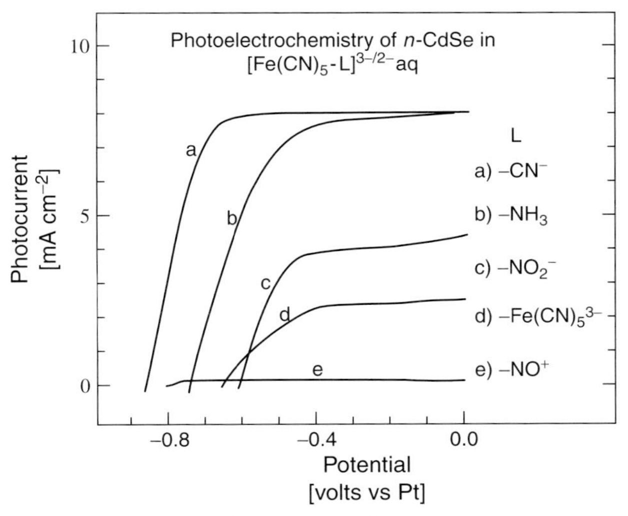

Fig. 17 Potentiostatic photocurrent voltage curves for an illuminated n-CdSe single crystal in several aqueous modified ferro/ferricyanide electrolytes in which 1 of the hexacyano ($Fe(CN)_6$) has been replaced by the indicated ligands.

hexacyanoferrate species has a substantial effect on photovoltage and photocurrent, Licht and Peramunage [56].

As seen in Fig. 18, addition of potassium cyanide to an alkaline ferrocyanide electrolyte substantially enhances both photocurrent stability (figure inset) and photovoltage (figure outset), Licht and Peramunage [9]. It was proposed by Bocarlsy and coworkers [57] that in addition to $Fe(CN)_6{}^{4-}$, CN^- was also photooxidized (to CNO^-) by the photoelectrode. Our subsequent investigations indicated that although cyanide may be chemically oxidized to cyanate, no cyanide was photoelectrochemically oxidized by the semiconductor Licht and Peramunage [53–56].

4.4.3.4 *n*-GaAs/Aqueous Polyselenide Photoelectrochemistry

n-GaAs/aqueous polyselenide PECs have shown stable efficient solar to electrical conversion. Photodriven oxidation of polyselenide can be written as

$$Se^{=} + 2hn \xrightarrow{n\text{-GaAs}} Se + 2e^- \quad (24)$$

or:

$$2Se^{=} + 2hn \xrightarrow{n\text{-GaAs}} Se_2{}^{=} + 2e^- \quad (25)$$

Parkinson, Heller, and Miller [58] and Lewis and coworkers [59] have shown that metal ion (Ru^{3+}, Os^{3+}) treatment of the *n*-GaAs surface leads to high solar to electrical conversion efficiencies in these cells

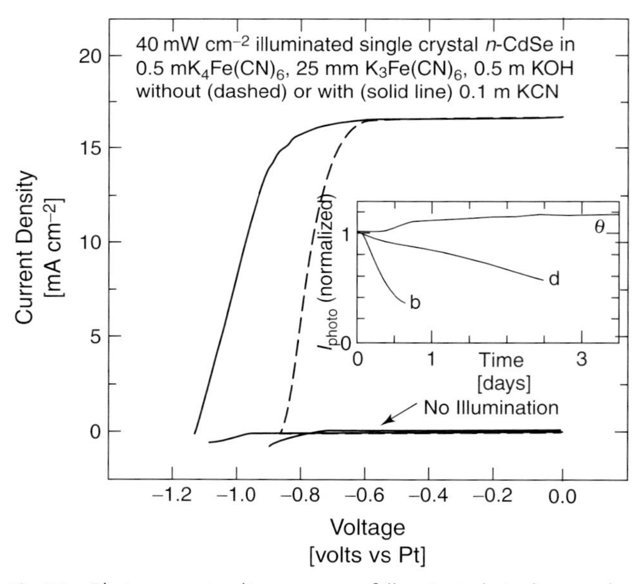

Fig. 18 Photocurrent-voltage curves of illuminated single crystal *n*-CdSe immersed in alkaline potassium ferrocyanide electrolytes with and without added cyanide. Inset: Photocurrent stability of in several electrolytes. "e" is the only electrolyte with cyanide. Electrolytes "d" and "e" contain low ferricyanide. Electrolyte "b" contains high ferricyanide. Specifically, e: 0.25 m $K_4Fe(CN)_6$, 0.01 m $K_3Fe(CN)_6$, 0.5 m KOH 0.5 m KOH, 0.1 m KCN; d: 0.25 m $K_4Fe(CN)_6$, 0.01 m $K_3Fe(CN)_6$, 0.5 m KOH; b: 0.25 m $K_4Fe(CN)_6$, 0.25 m $K_3Fe(CN)_6$, 0.5 m KOH. In an electrolyte comparable to "b" but with pH = 3.8, photocurrent diminished within seconds.

(with up to 15% conversion efficiencies reported). Lewis and coworkers [60] have further investigated 0.1 M to 1.0 M concentration K_2Se electrolytes. The first and second acid dissociation constant of hydrogen selenide of $pK_1 = 3.9$ and $pK_2 = 13.0$ have been well characterized by several investigators including Myers and coworkers [61]. However, as represented in the right hand portion of Fig. 14, the fundamental equilibria constraining the distribution of polyselenide species was not characterized. Measurements of these equilibria are a first step in determining the regenerative reactive species in polyselenide medium and in probing how they effect photoelectrochemical energy conversion.

The equilibria constraining the formation of diselenide, triselenide, and tetraselenide may be written:

$$Se_3^{=} + Se^{=} \longleftrightarrow 2Se_2^{=} \quad (26)$$

$$2Se_4^{=} + Se^{=} \longleftrightarrow 3Se_3^{=} \quad (27)$$

with equilibrium constants for the equilibrium given by K_{23} and K_{34}, respectively. These equilibria were probed by study of rest potential variation with solution composition and by isolation of the near UV absorption peaks of the various polyselenide species, and yield equilibrium constants of $pK_{23} = -0.65$ and $pK_{34} = -4.2$. When combined with K_1 and K_2 these provide a description of polyselenide speciation [62, 63].

Using the understanding of polyselenide speciation in solution, the rationale electrolyte modification of aqueous polyselenide photoelectrochemical solar cell gas been investigated. *n*-GaAs photocurrent, photovoltage, and photopower are affected by the distribution of hydroselenide,

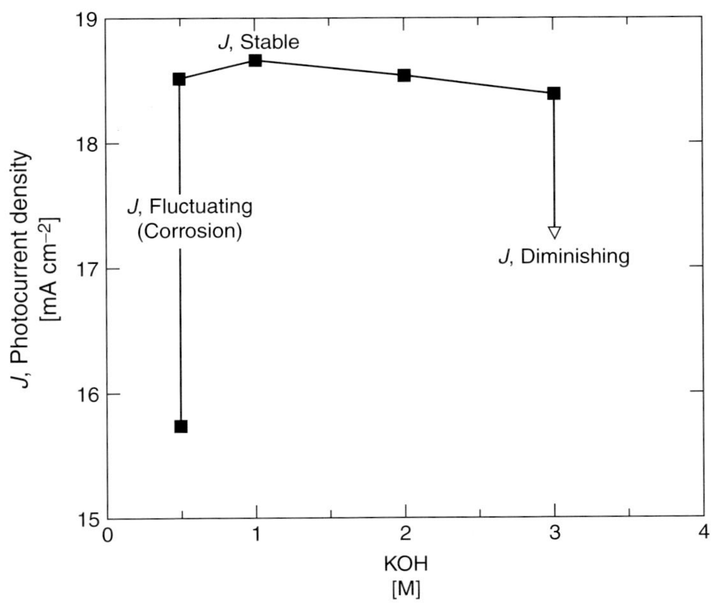

Fig. 19 Short-circuit photocurrent density, J_{SC}, for a 75 mW cm^{-2} tungsten-halogen illuminated single crystal *n*-GaAs immersed 1 M K_2Se, 0.01 M Se and in varying KOH concentration.

selenide, and polyselenide in solution. Polyselenide electrolytes containing 1 to 2 m hydroxide and 1 m selenide (as K_2Se) enhance n-GaAs photocurrent density stability (Reported as "J" in Fig. 19), whereas photocurrent is unstable at either 0.5 m KOH or 3.0 m KOH. The presence of polyselenide species in low concentrations (as 0.01 m dissolved selenium) are important for desorption of selenium from the n-GaAs. Furthermore, electrolytes enriched in dissolved selenium, (as up to 0.2 m dissolved selenium) enhance n-GaAs photovoltage. As seen in Fig. 20, the photopower (the product of the photocurrent and the applied potential) as well as the photopower improves with an increase in selenium. However, these electrolytes diminish electrolyte transmittance necessitating use of an alternate back wall cell configuration discussed in Licht and Forouzan [62, 63].

4.4.3.5
Aqueous Polyiodide Photoelectrochemistry

The aqueous polyiodide (I^-/I_3^-) is the only redox couple shown to be compatible with efficient oxidative photoelectrochemistry at n-type transition metal dichalcogenide including WSe_2, MoS_2, WS_2, and $MoSe_2$. Following the pioneering work of Tributsch and coworkers [64], several groups including Parkinson, Heller, and Miller [65] and Tenne and Wold [66] have reported n-WSe_2 or n-$MoSe_2$/aqueous polyiodide solar to electrical energy conversion of more than 10% and/or photocurrent stability in excess of 10^5 coulombs cm^{-2}. The photodriven oxidation of iodine at tungsten diselenide

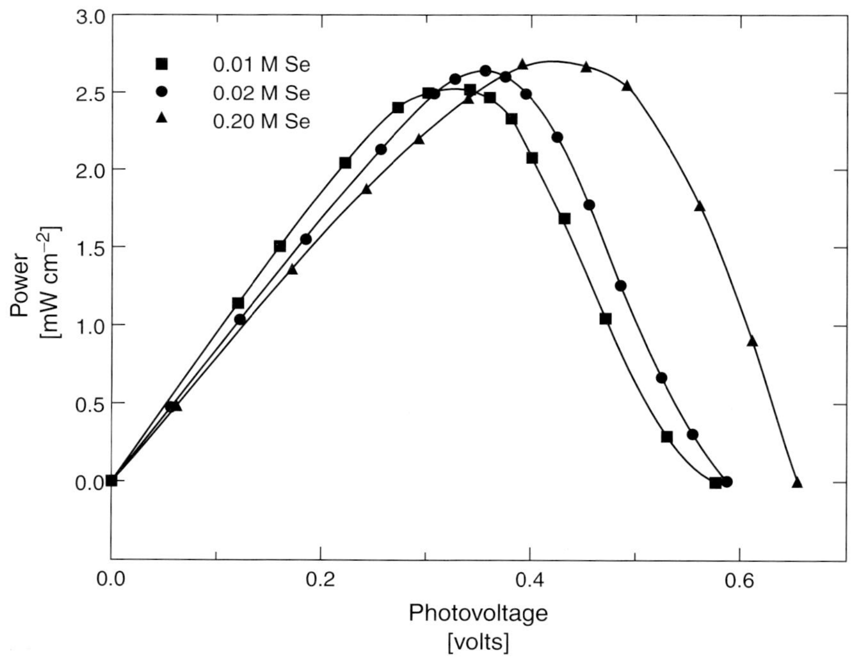

Fig. 20 Photopower variation of a 92 mW cm^{-2} tungsten-halogen illuminated backwall n-GaAS/polyselenide PEC as a function of increasing selenium concentration.

has been described in one of two manners:

$$I^- + hn \xrightarrow{n\text{-}WSe_2} I^\circ + e^- \quad (28)$$

or:

$$3I^- + 2hn \xrightarrow{n\text{-}WSe_2} I_3^- + 2e^- \quad (29)$$

Significant species in aqueous polyiodide solution can include: I^-, I_3^-, I_4^{2-}, I^-, OH^-, and H^+. Constrained by pH and the equilibria [67, 68]:

$$I^- + I_2 \longleftrightarrow I_3^- \quad K_3 = 723 \quad (30)$$

$$I_3^- + I^- \longleftrightarrow I_4^{2-} \quad K_4 = 0.20 \quad (31)$$

To a lesser extent the solution phase species IO_3^-, HIO, $H_2O\ I^+$, I_5^-, and I_6^{2-} can also occur. Figure 21 presents the relative variation of polyiodide speciation in solutions containing 0.01 molal iodine and 1 to 12 molal iodide and which may be compatible with n-WSe_2 regenerative photoelectrochemistry. As emphasized in the figure inset, in these electrolytes the concentration of added iodide dominates iodine in these solutions. The distribution of species in these solutions is calculated in accordance with Eqs. (30) and (31). As can be seen in the figure inset, little (less than 0.3%) of the added iodine exists as solution

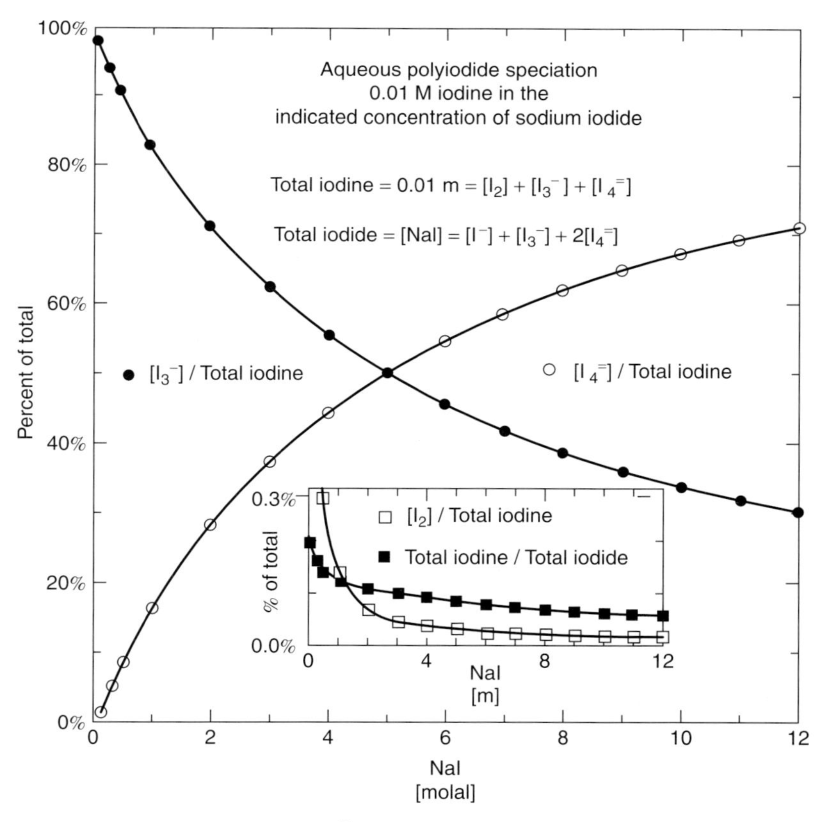

Fig. 21 The fractions of I_3^- and $I_4^=$ (compared to total iodine) in aqueous polyiodide solutions as a function of added iodide concentration. Inset: The fractions of solution phase iodine (compared to total iodine) and total iodine (compared to total iodide).

phase iodine, and the maximum $[I_2]$ is $\ll 1 \times 10^{-4}$ m. As seen in the figure outset, the bulk (greater than 99.7%) of the added iodine resides as the polyiodide species I_3^- and I_4^{2-}. In iodide concentrations less than 5 m, I_3^- is the predominant polyiodide, whereas at higher iodide concentrations, I_4^{2-} predominates. Longer chain species, I_5^- and I_6^{2-}, if they exist, have upper limit concentrations of 1×10^{-6} m and 1×10^{-8} m, respectively. Hence, in these aqueous polyiodide electrolytes, I_2, I_5^- and I_6^{2-} are not at substantial concentrations in these photoelectrolytes, and a primary oxidizable ion is I^- and a primary reducible ion is either I_3^- or I_4^{2-}.

Choice of cation in aqueous polyiodide solution can effect *n*-tungsten dichalcogenide photoelectrochemistry. The alkali cations do not substantially interact with solution phase iodide. However, other metal cations including Ag^+, or Zn^{2+} or Cd^{2+}, will create a series of complexes with dissolved iodide including: AgI_2^-, AgI_3^{2-}, ZnI_3^-, ZnI_4^{2-}, CdI_4^2 and so on. These complexes can shift rest potentials or either enhance or diminish charge transfer from the semiconductor surface. Because of the limitations on silver iodide solubility ($K_{sp}(AgI) = 10^{-16}$) high (molal) level concentrations of silver will dissolve only in solutions that facilitate complex formation as in concentrated NaI or KI solutions. In particular, we have shown that the presence of high concentrations of dissolved silver is advantageous to *n*-tungsten diselenide photoelectrochemistry. As seen in Fig. 22, silver dissolved as $AgNO_3$ enhances the voltage at maximum point. The extent of the improvement varies with the initial condition of the individual *n*-WSe_2 crystal, and general improvements are a 15 to 45 mV increase in the voltage of maximum power, V_{max}, and a 5 to 20% relative increase in power. These improvements are sustained in the silver-bearing electrolytes. On returning the electrode to the polyiodide electrolyte without silver, the PEC gradually (on the order of hours) returns to the silver free PEC behavior. This improvement appears to provide a mechanism for long-lasting suppression of *n*-WSe_2 exposed edges and recombination sites as further discussed by Licht and Myung [67, 68].

4.4.4 Concluding Remarks

In the 1970s through 1990s health concerns (air pollution) and political concerns (localized shortages) were cited in the need for development of alternative energy sources. Today, in addition to these concerns, growing awareness of carbon dioxide emissions as a green house gas, as well as the economic realization that oil and coal are better allocated as raw materials for pharmaceuticals and plastics, than for electrical energy generation, provide impetus to the technological development of renewable energy sources. Solar energy remains the principal symbol of a clean, abundant renewable energy source. Society's energy needs are continuous. Unlike purely solid-state, photovoltaic processes, the advantage of combined solar/electrochemical processes is that energy is not only converted but also may be stored for future use when sunlight is not so frequently available.

Limiting constraints of multiple band gap photoelectrochemical energy conversion, as well as practical configurations for efficient solar to electrical energy conversion have been probed [1–6]. Such systems are capable of better matching and utilizing of incident solar radiation (insolation). Efficient solar cells, solar storage cells,

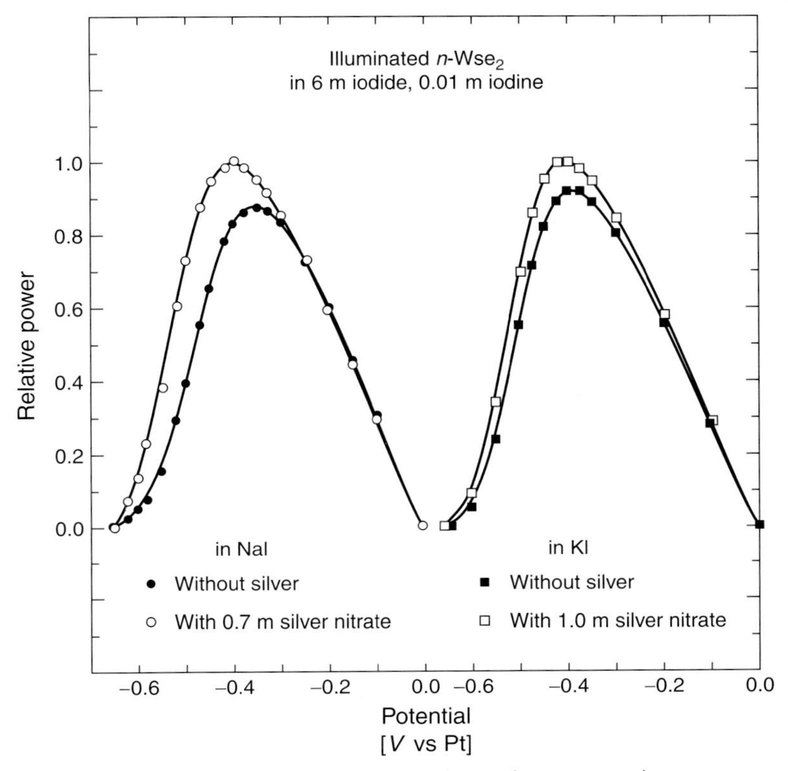

Fig. 22 The effect of dissolved silver on the relative photopower-voltage characteristics for illuminated single crystals of n-WSe_2. As indicated on the figure, four polyiodide electrolytes are used consisting of either: (6 m KI, 0.01 m I_2) or (6 m KI, 0.01 m I_2, 1.0 m $AgNO_3$) or (6 m NaI, 0.01 m I_2) or (6 m NaI, 0.01 m I_2, 0.7 m $AgNO_3$) as indicated. Relative power is determined by comparison to the maximum photopower measured in the silver bearing electrolyte.

and solar hydrogen generation systems are discussed and demonstrated.

Along with conventional parameters constraining photovoltaic devices, modification of the electrolyte (solution phase) chemistry is a key to understanding the mechanism of energy conversion and device characteristics of PECs. The fundamental importance of modification of the electrolyte in terms of the distribution of species in photoelectrolytes and the pragmatic importance in terms of enhanced solar to electrical conversion efficiency is reiterated by studies in polysulfide, ferrocyanide, polyselenide, and polyiodide electrolyte regenerative PECs.

The advantage of multiple band gap semiconductor processes is in the ability to improve the energy match to the solar spectra, and hence improved utilization/conversion of the incident solar radiation (insolation). Limiting constraints of multiple band gap semiconductor/electrolyte energy conversion, as well as practical configurations for solar to electrical energy conversion have been probed. Such combined systems are capable of highly efficient utilization of solar energy. Efficient solar cells, solar storage cells, and solar hydrogen generation systems are discussed and demonstrated. Alternate processes discussed include both an inverted (1 photon per e^-) or bipolar ($n \geq 2$ photons per e^-) arrangement of successive band gaps, which can contain either a Schottky or an ohmic photoelectrochemical solution interface, and can drive either regenerative single, or multiple different, redox reactions. The latter case permits solar energy storage or solar water splitting evolving H_2, and provides an energy reservoir which may compensate for the intermittent nature of solar energy. Generated H_2 is attractive as a clean, renewable fuel.

Experimental configurations examined include GaAs/Si or AlGaAs/Si photodriven redox couples, each with a solar to electrical conversion efficiency of 19 to 20%. A related solar cell is configured with electrochemical storage, and which provides a nearly constant energetic output in illuminated or dark conditions. Similarly, multiple band gap semiconductors can also be utilized to generate hydrogen fuel by solar-driven water splitting. A cell containing illuminated AlGaAs/Si RuO_2/Pt_{black} evolves H_2 and O_2 at record (18.3%) efficiency. Contemporary models underestimate the attainable efficiency of solar energy conversion to water splitting, and future multiple band gap systems are calculated as capable of attaining more than 30% solar efficiency.

Acknowledgment

S. Licht is grateful for support by the BMBF Israel-German Cooperation.

References

1. S. Licht, O. Khaselev, T. Soga et al., *Electrochem. Solid State Lett.* **1998**, *1*, 20.
2. S. Licht, O. Khaselev, P. A. Ramakrishnan et al., *J. Phys. Chem. B* **1998**, *102*, 2536.
3. S. Licht, O. Khaselev, P. A. Ramakrishnan et al., *J. Phys. Chem. B* **1998**, *102*, 2546.
4. S. Licht, B. Wang, T. Soga et al., *Appl. Phys. Lett.* **1999**, *74*, 4055.
5. B. Wang, S. Licht, T. Soga et al., *Sol. Energy Mater. Sol. Cells* **2000**, *64*, 311.
6. S. Licht, H. Tributsch, B. Wang et al., *J. Phys. Chem. B* **2000**, *104*, 8920.
7. S. Licht, *Sol. Energy Mater. Sol. Cells* **1995**, *38*, 305.
8. S. Licht, *Nature* **1987**, *330*, 148.
9. S. Licht, D. Peramunage, *Nature* **1990**, *345*, 330.
10. B. Miller, S. Licht, M. E. Orazem et al., Photoelectrochemical systems, *Crit. Rev. Surface Chem.* **1994**, *3*, 29.
11. C. H. Henry, *J. Appl. Phys.* **1980**, *51*, 4494.
12. D. J. Friedman, S. R. Kurtz, K. Bertness et al., *Prog. Photovolt.* **1995**, *3*, 47.
13. J. P. Benner, J. M. Olson, T. J. Coutts, in *Advances in Solar Energy*, Chap. 4 (Eds.: K. W. Boer), American Solar Energy Society, Inc., 1992, 125, Vol. 7.
14. M. A. Green, K. Emery, K. Bucher et al., *Progr. Photovolt.* **1996**, *4*, 321.
15. T. Soga, T. Kato, M. Yang et al., *J. Appl. Phys.* **1995**, *78*, 4196.
16. J. White, F-R. Fan, A. J. Bard, *J. Electrochem. Soc.* **1985**, *132*, 544.
17. R. Memming, in *Photochemical Conversion and Storage of Solar Energy*, (Eds.: E. Pelizzetti, M. Schiavello), Kluwer Acad. Pub., Netherlands, 1991, pp. 193.
18. S. Licht, G. Hodes, R. Tenne et al., *Nature* **1987**, *330*, 148.
19. A. Heller, H. J. Lewerenz, B. Miller, *J. Am. Chem. Soc.* **1981**, *103*, 200.
20. C. Lévy Clément, A. Lagoubi, M. Neumann-Spallart et al., *J. Electrochem. Soc.* **1991**, *138*, L69.
21. H. Kobayashi, T. Kubota, N. Toshikawa et al., *J. Electroanal. Chem.* **1995**, *398*, 165.
22. S. R. Ovshinsky, M. A. Fetcenko, J. Ross, *Science* **1993**, *260*, 176.
23. G. Lodi, E. Sivieri, A. Debattisti et al., *J. Appl. Electrochem.* **1978**, *8*, 135.
24. N. Alonso-Vante, H. Colell, U. Stimming et al., *J. Phys. Chem.* **1993**, *97*, 7381.
25. P. Salvador, N. Alonso-Vante, H. Tributsch, *J. Electrochem. Soc.* **1998**, *145*, 216.
26. O. Khaselev, J. Turner, *Science* **1998**, *280*, 425.

27. J. R. Bolton, S. J. Strickler, J. S. Connolly, *Nature* **1985**, *316*, 495.
28. M. F. Weber, M. J. Digman, *Int. J. Hydrogen Energy* **1986**, *11*, 225.
28a. M. F. Weber, *J. Electrochem. Soc.* **1984**, *131*, 1258.
29. M. A. Green, K. Emery, K. Bucher et al., *Prog. Photovolt.* **1999**, *11*, 31.
30. R. C. Kainthla, B. Zelenay, JO'M. Bockris, *J. Electrochem. Soc* **1986**, *133*, 248.
31. G. Hodes, J. Manassen, D. Cahen, *Nature* **1976**, *261*, 403.
32. A. B. Ellis, S. W. Kaiser, M. S. Wrighton, *J. Am. Chem. Soc.* **1976**, *98*, 1635.
33. B. Miller, A. Heller, *Nature* **1976**, *262*, 680.
34. S. Licht, G. Hodes, J. Manassen, *Inorg. Chem.* **1986**, *25*, 2486.
35. S. Licht, J. Manassen, *J. Electrochem. Soc.* **1987**, *134*, 918.
36. S. Licht, *J. Electrochem. Soc.* **1988**, *135*, 2971.
37. F. Forouzan, K. Longo, D. Peramunage et al., *Anal. Chem.* **1990**, *62*, 1356.
38. K. Longo, F. Farouzan, D. Peramunage et al., *J. Electroanal. Chem. Interfacial Electrochem.* **1991**, *318*, 111.
39. D. Peramunage, F. Forouzan, S. Licht, *Anal. Chem.* **1994**, *66*, 378.
40. W. Giggenbach, *Inorg. Chem.* **1971**, *10*, 1333.
41. A. Teder, *Sr. Papperstodn* **1969**, *72*, 245.
42. S. Licht, Joost Manassen, *J. Electrochem. Soc.* **1985**, *132*, 1077.
43. S. Licht, G. Hodes, Joost Manassen, *J. Electrochem. Soc.* **1986**, *133*, 272.
44. S. Licht, J. Manassen, *J. Electrochem. Soc.* **1986**, *133*, 277.
45. S. Licht, R. Tenne, H. Flaisher et al., *J. Electrochem. Soc.* **1986**, *133*, 52.
46. S. Licht, R. Tenne, H. Flaisher et al., *J. Electrochem. Soc.* **1984**, *131*, 5920.
47. S. Licht, R. Tenne, G. Dagan et al., *Appl. Phys. Lett.* **1985**, *46*, 608.
48. S. Licht, *J. Phys. Chem.* **1986**, *90*, 1096.
49. J. Reichman, M. Russak, *J. Electrochem. Soc.* **1984**, *131*, 796.
50. K. W. Freeze Jr., *Appl. Phys. Lett.* **1982**, *40*, 275.
51. H. D. Rubin, B. D. Humphrey, A. B. Bocarsley, *Nature* **1984**, *308*, 339.
52. H. D. Rubin, D. J. Arent, A. B. Bocarsley, *J. Electrochem. Soc.* **1985**, *132*, 523.
53. S. Licht, D. Peramunage, *Nature* **1991**, *354*, 440.
54. S. Licht, D. Peramunage, *J. Electrochem. Soc.* **1992**, *139*, L23.
55. S. Licht, D. Peramunage, *J. Electrochem. Soc.* **1992**, *139*, 1791.
56. S. Licht, D. Peramunage, *Sol. Energy* **1994**, *52*, 197.
57. G. Seshardi, J. K. M. Chun, A. B. Bocarsly, *Nature* **1991**, *352*, 508.
58. B. A. Parkinson, A. Heller, B. Miller, *J. Electrochem. Soc.* **1979**, *126*, 954.
59. B. J. Tufts, I. L. Abrahams, P. G. Santengelo et al., *Nature* **1987**, *326*, 861.
60. L. G. Casasgrand, B. J. Tufts, N. S. Lewis, *J. Phys. Chem.* **1991**, *95*, 1373.
61. D. E. Levy, R. J. Myers, *J. Phys. Chem.* **1990**, *94*, 7842.
62. S. Licht, F. Forouzan, *J. Electrochem. Soc.* **1995**, *142*, 1539.
63. S. Licht, F. Forouzan, *J. Electrochem. Soc.* **1995**, *142*, 1546.
64. W. Kautek, H. Gerischer, H. Tributsch, *J. Electrochem. Soc.* **1981**, *128*, 2471.
65. B. A. Parkinson, A. Heller, B. Miller, *Appl. Phys. Lett.* **1978**, *33*, 521.
66. R. Tenne, A. Wold, *Appl. Phys. Lett.* **1985**, *47*, 707.
67. S. Licht, N. Myung, *J. Electrochem. Soc.* **1985**, *142*, 845.
68. S. Licht, N. Myung, *J. Electrochem. Soc.* **1995**, *142*, L129.

5
Dye-Sensitized Photoelectrochemistry

5.1
Dye-Sensitized Regenerative Solar Cells

Augustin J. McEvoy and Michael Grätzel
Ecole Polytechnique Fédérale de Lausanne, Lausanne, Switzerland

5.1.1
Overview

At present, dye-sensitized photosystems provide the only technically and economically credible alternative to solid-state photovoltaic devices. The concept reconciles the electrochemical stability of wide band gap photoelectrodes with the efficiency of photovoltaic devices having a more extended absorption spectrum in the visible range. Optical absorption and charge separation take place on distinct sites within these photovoltaic cells. Hence, oppositely charged species are restricted to separate phases, so conventional recombination losses are inhibited. In consequence, device photoconversion efficiency is better maintained at low light levels than with conventional semiconductor solid-state junction devices. Various configurations of the dye-sensitized concept are under investigation, including electrochemical and heterojunction variants.

5.1.2
Device Concepts

It should be remembered that the first observation of the photovoltaic effect by A. -E. Becquerel [1] in 1839 was at a solid–liquid interface, with a semiconductor photoelectrode contacting an electrolyte, and was technically a photoelectrochemical cell. Nevertheless, the modern photovoltaic industry is established exclusively on solid-state devices, with semiconductors of appropriate band gap as optical absorbers and junctions to solids with different conduction mechanisms as the sites of charge separation and photovoltage generation. In commerce are found homojunction devices, in which the semiconductors are chemically identical, differing only in impurity content and therefore in free carrier polarity; heterojunctions, in which the semiconductors of opposite carrier polarity are chemically different, for example, cadmium sulfide or copper indium-gallium selenide contacts; and finally Schottky junctions in which the contacting phase is a metal. However, all these are solid-state devices in which conduction throughout is by an electronic mechanism. None is a photoelectrochemical device that has an electrolyte as the contacting phase or that uses an ionic conduction mechanism. Awareness of photoelectrochemistry in the photovoltaic world therefore remains marginal, although this awareness is rapidly growing. Intensive research over the past two decades has led to an inescapable conclusion that those semiconductors, whose band gaps are sufficiently wide to give a long-term stability under illumination, in contact with an electrolyte and a technically useful photovoltage are insensitive to visible light, requiring higher-energy ultraviolet photons to excite charge carrier pairs. A narrower band gap, compatible with the photoconversion of visible light, is indicative of weaker chemical bonding of the semiconductor and hence liability to photocorrosion and incompatibility with a stable extended lifetime as an energy-conversion device. The resolution of this dilemma lies in the separation of light absorption and charge-separation functions by sub–band gap sensitization of the semiconductor with an electroactive dye.

The realization in practice of this option is simply presented. Analogous to the band gap of the semiconductor is the HOMO-LUMO (highest occupied molecular orbital ⇒ valence band/lowest unoccupied molecular orbital ⇒ conduction band) gap of a molecule. When the latter lies above the conduction band edge of a semiconductor substrate, an electron entering the LUMO on photoexcitation of the dye by a photon within its absorption spectrum energy range can be injected into the semiconductor. The positively charged oxidized dye molecule can then be neutralized and its ground state restored by reaction and charge exchange with the electrolyte, which in turn receives an electron from a metallized counterelectrode. With a closed external circuit and under illumination, the device then constitutes a photovoltaic energy-conversion system, which is regenerative and stable, and is functionally comparable to its solid-state

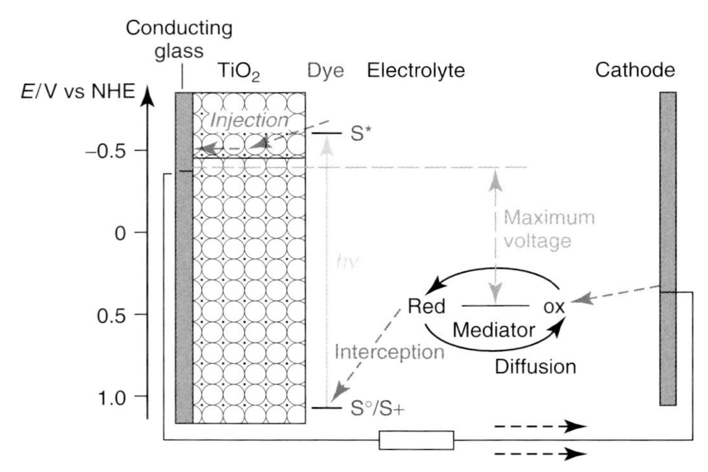

Fig. 1 Schematic of the structure and function of the dye-sensitized electrochemical photovoltaic cell. The nanoporous sensitized semiconductor photoanode receives electrons from the photoexcited dye, which is thereby oxidized and which in turn oxidizes the mediator, a redox species dissolved in the electrolyte. The mediator is regenerated by reduction at the metallic cathode, while the electrons circulate in the external circuit.

junction analog. A typical wide band gap semiconductor for photoelectrochemistry with stability against photocorrosion is titanium dioxide. Titanium dioxide has a band gap of 3.1 eV and is therefore insensitive to visible light. The selected dye must have a HOMO-LUMO gap corresponding to a photon energy in the visible or infrared parts of the spectrum, with an optimum near the solar spectra maximum of 1.4 eV. This mechanism is illustrated schematically in Fig. 1. However, before presenting the relevant characteristics of a semiconductor electrode that is apt for chemisorption of a dye and therefore photovoltaic application, a background of the history and development of the sensitization of semiconductors is necessary.

5.1.3 Photography and Photoelectrochemistry

Although artisanal "virtual reality" by painting is as old as human culture, synthetic "virtual reality" by technical reproduction of images is relatively recent, with the first photography by Daguerre in 1837 and the introduction of the silver halide process by Fox-Talbot in 1839. Thereafter, the art advanced so rapidly that the nineteenth century events – the wars, the explorations, the early factory age and the introduction of steam power – are vividly documented by photographs. Photography remained an art for a century, becoming a science only with the theoretical analysis of the process by Gurney and Mott [2] in 1938. Nevertheless, an empirical skill in the formulation of photographic emulsions had evolved in the interval, even in the absence of a fundamental understanding. Early emulsions, referred to as "orthochromic," had a distinct insensitivity to midspectral and red light, so that their gray scale reproduction was not realistic of the visual impression of the same scene. So many early photographs therefore give an impression of an unnatural "cleanliness" of the scene imaged that considerable efforts were expended to find emulsions capable of more realistic response. The problem is now recognized to be due to the semiconductor nature of the halide grains responsible for optical absorption in photography, which are activated essentially by the photoelectric effect. They have band gaps typically in the range 2.7 to 3.2 eV, and the photoresponse is effectively negligible for wavelengths longer than 460 nm. It was however noted that the use of gelatine as a support medium for the alkali halide grains in the photographic emulsion could significantly modify the film spectral response. Only in the last century was it finally demonstrated that an organo-sulfur compound present in calf-skin gelatine was responsible for this extension of sensitivity [3]. The mechanism now recognized is the induction of a nanostructure of silver sulphide on the halide surface of each grain, giving the first application of the sensitization of a semiconductor heterojunction! Even more significant in the history of sensitization is the work of Vogel, Professor of photochemistry, spectroscopy and photography at the then Königliche Technische Hochschule of Berlin, who in the 1870s investigated the effect of dyes on emulsions and found that the halide spectral response extended through the visible and even into the infrared region [4]. Not only did this facilitate his spectroscopic research and contribute to astronomy but also made possible the commercial "panchromatic" broad-spectrum black-and-white film with realistic gray scale reproduction, and later with spectrally selective dyes – the modern era of color photography.

Meanwhile, the photoelectric effect was a focus of scientific interest, with a history synchronized with that of photography. The concept of dye enhancement of the photoeffect was carried over from photography to the sensitization of an electrode already in 1887 by Moser [5] (Fig. 2) using the dye erythrosin on silver halide electrodes and confirmed by Rigollot in 1893 [6]. In the retrospectively quaint report, written substantially before the Einstein theory of the photoelectric effect, Moser records his observations on dye-induced enhancement as an increased photopotential (*V*) rather than the more fundamental current (*A*), despite the title – "Strengthened Photoelectric Current Through Optical Sensitization."

This parallel evolution of sensitization in photography and in photoelectrochemistry still seems to surprise each generation of photochemists. That the same dyes were effective for both processes was recognized among others by Namba and Hishiki [7] at the 1964 International Conference on Photosensitization in Solids, which can now be seen as a seminal event in the history of dyes in photochemistry. It was also recognized then that the dye should be adsorbed on the semiconductor surface as a closely packed monolayer for maximum sensitization effectiveness [8]. The conference also provided the venue for the identification of the sensitization mechanism through charge transfer from a dye excited state; until then, the matter was in dispute, with an energy-coupling process being

Notiz über Verstärkung photoelektrischer Ströme durch optische Sensibilisirung.[1]

Von Dr. James Moser.

(Aus dem physikalisch-chemischen Laboratorium der Wiener Universität).

(Vorgelegt in der Sitzung am 23. Juni 1887.)

Ich erlaube mir mitzutheilen, dass ich die von Herrn E. Becquerel entdeckten photoelektrischen Ströme erheblich dadurch verstärken konnte, dass ich die beiden chlorirten, jodirten oder bromirten Silberplatten in einer Farbstofflösung, z. B. Erythrosin, badete.

Beispielsweise war zwischen zwei chlorirten Silberplatten die elektromotorische Kraft im Sonnenlicht 0·02, zwischen zwei anderen in gleicher Weise behandelten, aber gebadeten Platten 0·04 Volt.

Bisher sind nur an jodirten Platten von Herrn Egoroff elektromotorische Kräfte beobachtet, und zwar bis $^1/_{15}$ Volt. Ich konnte bei jodirten und bromirten Platten durch Baden in Erythrosin $^1/_4$ Volt erreichen.

Ich halte es für meine Pflicht, schon an dieser Stelle Herrn Max Reiner, der mir bei diesen Versuchen assistirt, meinen verbindlichsten Dank auszusprechen.

---

[1] Akadem. Anzeiger Nr. XVI.

Fig. 2 A copy of the first paper on sensitization published by James Moser in 1887.

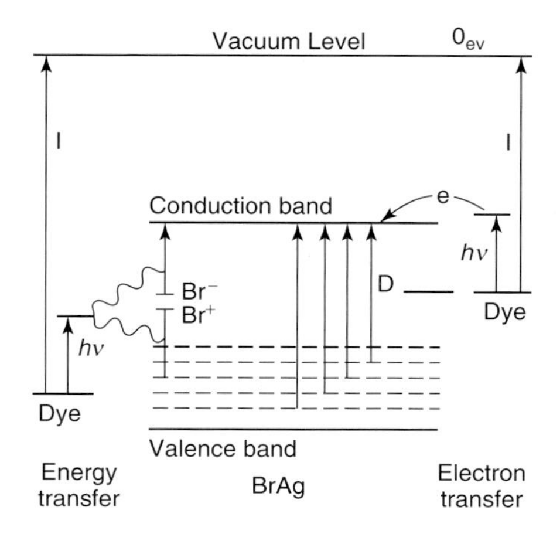

Fig. 3 The dilemma of 1968 – energy resonance or electron injection?

proposed, as illustrated in Fig. 3 taken from the conference proceedings [9]. The understanding became definitive with the work of Gerischer and Tributsch on sensitization of zinc oxide [10]. However, the fact remained that sensitization efficiency remained low, as research systematically investigated dyes in solution or those adsorbed on single-crystal substrates. In retrospect, it appears that there was an excessive influence of solid-state engineering practice in that the disordered surfaces were expected to promote surface recombination and therefore loss of excited charge carriers, whereas single-crystal surfaces, being better defined physically and chemically, make the effects of sensitization more evident. The results of this presumption were slow to emerge.

5.1.4
Sensitization of Powders and Rough Surfaces

With the ongoing interest in sensitization during the 1970s, and the idea that photocatalysts could possibly carry out electrolytic reactions such as the evolution of hydrogen from water, a phase of experimental development evolved with the sensitization of semiconductor powders [11]. With the ecological awareness of time, photochemical production particularly of energy-related chemicals was seen as a form of "artificial photosynthesis." The prototype energy-converting dye provided by nature is indeed chlorophyll, a molecule consisting of a central magnesium atom surrounded by a nitrogen-containing porphyrin ring. Variations are due to minor modifications of certain side groups. Chlorophyll is in turn similar in structure to hemoglobin, the oxygen-carrying iron-based pigment found in blood. Given that the development of the dye-sensitized cell arose out of an interest in artificial photosynthesis, the adoption of porphyrin-like organometallic dyes as sensitizers was logical. However, although nature confines itself to magnesium and iron for its principal pigments, the synthetic chemist can access the whole range of metallic elements. The use of ruthenium pyridyl complexes now has more than twenty years of development history. Clark and Sutin had already used a trisbipyridyl ruthenium complex in 1977 to sensitize titanium dioxide to sub–band gap illumination [12], but in solution only. Charge transfer could only occur after diffusion of the ion to the semiconductor, so the efficiency of the sensitization was very low. By 1980, the idea of chemisorption of the dye through an acid carboxylate group bonding to the metal oxide surface had also been established [11] so that the sensitizer was immobilized and it formed the required monomolecular film on the semiconductor substrate which facilitated charge transfer by electron injection. The carboxylated trisbipyridyl dye

("RuL_3") therefore became the prototype sensitizer in photochemistry. The subsequent development of sensitizers through to the panchromatic "black dye" that constitutes the current state of the art is dealt with in detail in a separate chapter.

As a substrate for the chemisorption of sensitizing dyes, titania, TiO_2, particularly in the anatase form, gradually became dominant [13]. It has many advantages for sensitized photochemistry and photoelectrochemistry, being a low-cost, widely available, nontoxic and biocompatible material, and as such is even used in health care products and domestic applications such as paint pigmentation. The objective had also evolved to concentrate on photovoltaic devices rather than on photosynthesis. Progress thereafter was incremental until the announcement in 1991 [14] of the sensitized electrochemical solar cell, with the remarkable conversion efficiency at that time of 7.1% under solar illumination, a synergy of structure, substrate roughness, dye photochemistry, counterelectrode kinetics and electrolyte redox chemistry. That evolution has continued progressively since then, with certified efficiency now more than 10% [15]. There has been a continuous development of the electrolyte solvent, aqueous systems being eliminated in favor of organics and molten salts. However, finding substitutes for the original iodine-iodide redox system has proved to be difficult although recent work in our laboratory shows that alternative redox couples are finally at hand.

5.1.5
Substrate Development and Fabrication

If molecular design and engineering has underpinned the evolution of efficient, stable sensitizer dyes, the materials science of nanoporous ceramic films is the basis for the understanding of semiconductors. The nanoporous structure permits the specific surface concentration of the sensitizing dye to be sufficiently high for the total absorption of the incident light that is necessary for efficient solar energy conversion, since the area of the monomolecular distribution of adsorbate is 2 to 3 orders of magnitude higher than the geometric area of the substrate. Moreover, in the electrochemical system, this high degree of roughness does not promote charge carrier loss by recombination, since the electron and the positive charge find themselves within picoseconds on opposite sides of the liquid–solid interface. In addition, the photoexcited carrier in the n-type TiO_2 is an electron, a majority carrier, and therefore not subject to depletion by carriers of the opposite polarity. Solid-state junction cells, on the other hand, are minority carrier devices. The carrier loss mechanisms are comparatively slow [16], and although conventionally referred to as recombination by analogy with the solid-state process, the loss of a photoexcited electron from the semiconductor should rather be regarded as a recapture of the electron by an oxidized dye species, or a redox capture, when the electron reacts directly with the iodine in the electrolyte. Either can occur on a millisecond timescale, but much slower than electron injection or transport to the rear contact layer.

The original substrate structure used for our early photosensitization experiments was a fractal derived by hydrolysis of an organo-titanium compound but it has since been replaced with a nanostructure deposited from colloidal suspension. This evidently provides a much more reproducible and controlled porous high surface area nanotexture. Further, since it is compatible with screen-printing technology, it anticipates future production requirements.

While commercially available titania powders, produced by a pyrolysis route from a chloride precursor, have been successfully employed, the present optimized material is the result of a procedure described by Brooks and coworkers [17]. A specific advantage of the hydrothermal technique is the ease of control of the particle size and hence of the nanostructure and porosity of the resultant semiconductor substrate. The relevant preparation flow diagram is given and the product is illustrated by the accompanying micrograph (Fig. 4; Table 1). Figure 5 presents data on the control of substrate porosity by the powder-preparation parameters.

5.1.6 Dye Sensitization in Heterojunctions

Because the sensitizing dye itself does not provide a conducting functionality but is distributed at an interface in the form of immobilized molecular species, it is evident that for charge transfer, each molecule must be in intimate contact with both conducting phases. It is clear that this applies to the porous wide band gap semiconductor substrate into which the photoexcited chemisorbed molecules inject electrons. It is also evident that in the photoelectrochemical form of the sensitized cell, the liquid electrolyte penetrates into the porosity, thereby permitting the intimate contact with the charged dye molecule that is necessary for charge neutralization after the electron loss by exchange with the redox system in solution. In the analogous nonliquid electrolyte case, it is not immediately evident that an interpenetrating network of two conducting solids can be so easily established that an immobilized molecule at their interface can exchange charge carriers with both. However, initial results [18,19] are promising. In both cases, the charge transport materials are deposited by spin coating from the liquid phase in order to achieve the necessary intimate contact. In the latter case, a mixture of polymers was used, which phase-separates spontaneously on removal of a solvent, whereas the Lausanne laboratory introduces a solution of the conducting compound into a previously sensitized nanostructure. The charge transfer material currently used is a spirobifluorene, proprietary to Hoechst [20] as shown in Fig. 6. As a matter of technical precision, if these materials function in the cell as a hole conductor, the device is a sensitized nanostructured heterojunction in which the charge transport is entirely of electronic nature (Fig. 7). However, if the molecules accept positive charge to become cations, for which there is initial spectroscopic evidence [18], the charge transfer mechanism within this organic phase can be considered a redox equilibration, and the device

Fig. 4 SEM image of the surface of a mesoporous film prepared from the hydrothermal TiO_2 colloid.

Tab. 1 Flow diagram for the preparation of TiO_2 colloids and mesoporous films [17]

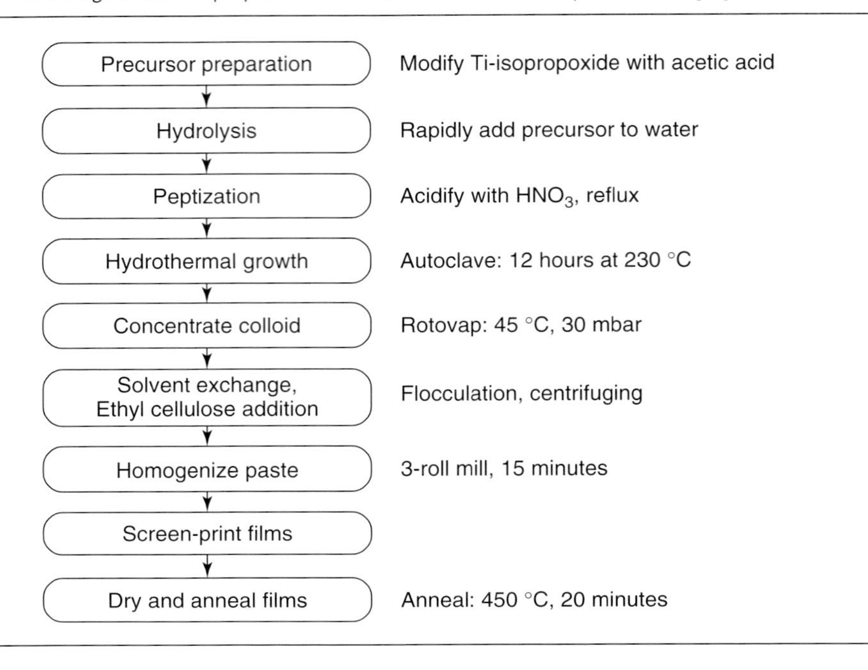

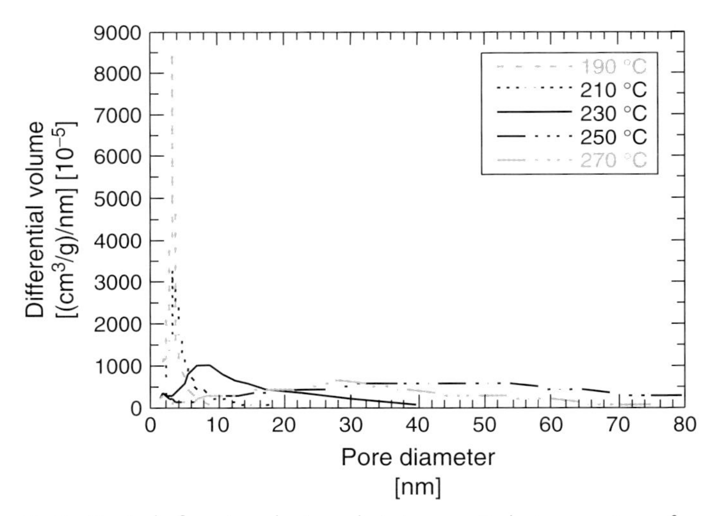

Fig. 5 Control of semiconductor substrate porosity by temperature of hydrothermal processing.

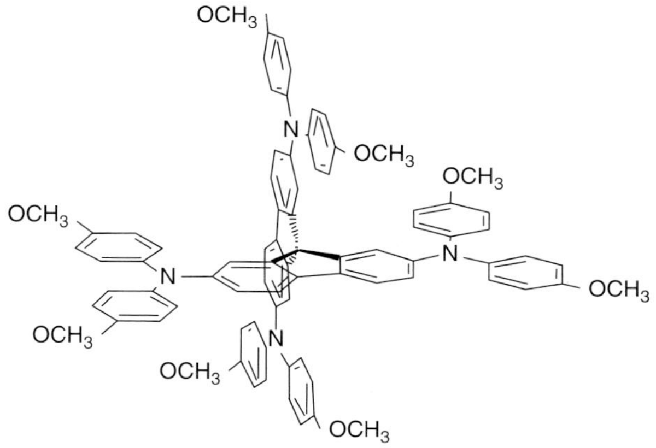

Fig. 6 Structure of spirobifluorene conducting material. The bisfluorene structures are perpendicular, conjoined through a carbon site common to both.

is photoelectrochemical. This distinction may have practical implications.

5.1.7
Commercial Prospects

The status of the dye-sensitized device as the only verified alternative to solid-state junction devices has already been discussed. It must be recognized that the solid-state devices, particularly the silicon *p-n* junction cells benefit from forty years of industrial and development experience, the technology transfer from the silicon-based electronics industry, and even the widespread availability of high-quality silicon at low cost resulting from the expansion of that industry. The procedures for high-yield fabrication of silicon devices, both crystalline and amorphous, are well understood, with costing well established on the basis of decades of solid industrial experience. For dye-sensitized cells, in contrast, fabrication procedures require development and costing based on estimates of the requirements of chemical processes rather than the silicon metallurgy with elevated temperatures and vacuum technology required for conventional cells. Equally, it is well known that the substitution of an established technology by an upcoming alternative requires that the new concept has definite advantages and no clear disadvantages. It is therefore noted with some satisfaction that several companies in Europe, Japan, and Australia have taken up the challenge and are currently engaged under license

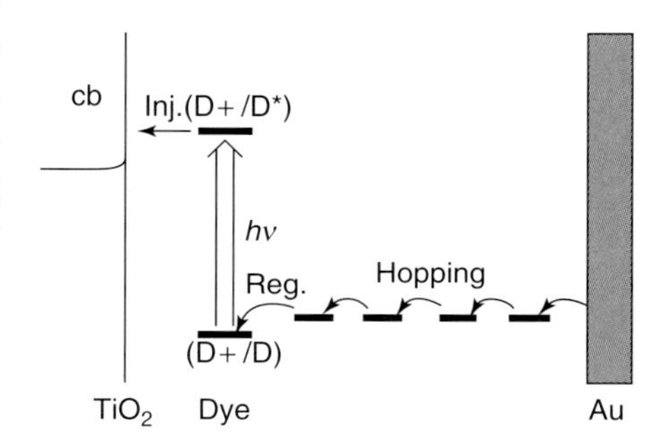

Fig. 7 Charge transport by hopping in the dye-sensitized solid-state photovoltaic cell.

in the venture of industrialization and commercialization of dye-sensitized photovoltaic cells.

5.1.8 Conclusion

The development of reproducible and stable photovoltaic devices adapted for manufacturing processes has proceeded in evolutionary steps, with each component optimized and verified for compatibility with system requirements. The dye-sensitized nanocrystalline solar cell, either the electrochemical device or the closely related sensitized heterojunction [18], provides a credible alternative to the conventional semiconductor junction solid-state cell. Time and the market will tell if it can compete successfully.

Acknowledgments

The team at EPFL greatly appreciates the vote of confidence represented by the licenses taken up by the industry for this type of solar cell. In the present work, we do acknowledge the initiative and innovative spirit of our coworkers, past and present, in the areas of dye synthesis, catalysis and electrochemistry, and semiconductor thin film fabrication.

References

1. A.-E. Becquerel, *C. R. Acad. Sci.* **1839**, *9*, 561.
2. R. W. Gurney, N. F. Mott, *Proc. Roy. Soc. A* **1938**, *164*, 151.
3. T. H. James, in *The Theory of the Photographic Process*, 4th. Ed., MacMillan, New York, 1977.
4. W. West, *Proc. Vogel Centenary Symp., Photogr. Sci. Eng.* **1974**, *18*, 35.
5. J. Moser, *Monatsh. Chem.* **1887**, *8*, 373.
6. H. Rigollot, *C. R. Acad. Sci.* **1893**, *116*, 873.
7. S. Namba, Y. Hishiki, *J. Phys. Chem.* **1965**, *69*, 774.
8. R. C. Nelson, *J. Phys. Chem.* **1965**, *69*, 714.
9. J. Bourdon, *J. Phys. Chem.* **1965**, *69*, 705.
10. H. Gerischer, H. Tributsch, *Ber. Bunsen-Ges. Phys. Chem.* **1968**, *72*, 437.
11. M. P. Dare-Edwards, J. B. Goodenough, A. Hamnet et al., *Faraday Discuss. Chem. Soc.* **1980**, *70*, 285.
12. W. D. K. Clark, N. Sutin, *J. Am. Chem. Soc.* **1977**, *99*, 4676.
13. J. DeSilvestro, M. Grätzel, L. Kavan et al., *J. Am. Chem. Soc.* **1985**, *107*, 2988.
14. B. O'Regan, M. Grätzel, *Nature* **1991**, *335*, 737.
15. M. K. Nazeeruddin, P. Pechy, Th. Renouard et al., *J. Am. Chem Soc.* **2001**, *123*, 1613.
16. S. A. Hague, Y. Tachibana, R. Willis, J. E. Moser, M. Grätzel, D. R. Klug, J. R. Durrant, *J. Phys. Chem. B* **2000**, *104*, 538.
17. K. G. Brooks, S. D. Burnside, V. Shklover et al., *Ceramic Transactions 109: Processing and Characterisation of Electrochemical Materials and Devices*, (Eds.: P. N. Kumta), American Ceramic Society, 2000, p. 115.
18. U. Bach, D. Lupo, P. Comte et al., *Nature* **1998**, *395*, 583.
19. J. J. M. Halls, C. A. Walsh, N. C. Greenham et al., *Nature* **1995**, *376*, 498.
20. D. Lupo, J. Salbeck, Intern. patent PCT/EP96/03944.

5.2 Dyes for Semiconductor Sensitization

Md. Khaja Nazeeruddin and Michael Grätzel
Swiss Federal Institute of Technology, Lausanne, Switzerland

5.2.1 Introduction

5.2.1.1 General Background

The extensive use of conventional energy supplies is causing many serious problems to the global environment. Moreover, sooner or later the fossil fuels are going to be exhausted and if nuclear power is not to be depended on solely, alternative methods for harnessing solar power, which is clean, nonhazardous, enormous, and ever lasting have to be developed. There are numerous ways to convert the solar radiation directly into electrical power or chemical fuel [1]. However, the capital cost of such devices is not attractive for large-scale applications.

Dye-sensitized solar cell technology is an interesting, inexpensive, and promising alternative to those of traditional solid-state photovoltaics. The process of extending the sensitivity of transparent materials such as titanium dioxide (TiO_2) to the visible spectra is called *spectral sensitization*. The topic of sensitization is very old [2–5] and without going into the history of the sensitization, the progress made during last 10 years on dye-sensitized solar cells is reviewed. The intention here is to restrict the discussion exclusively to inorganic dyes (transition metal complexes) based largely on our own work.

The photoelectrode of a dye-sensitized solar cell consists of 8–20 μm film of mesoporous nanocrystalline TiO_2 particles containing a monolayer of anchored dye molecules, which absorb light, freeing electrons to carry electric current. The low production costs, stability, and the efficiency of these cells has generated renewed interest in many groups around the world [6–15]. The added advantages of these cells are availability and nontoxicity of the main component that is TiO_2, which is even used in paints, cosmetics, and health care products like hip joints and other orthopedic implants.

Efficient sensitization of large band gap semiconductors to the visible and the near infrared (IR) solar spectrum can be achieved by engineering dyes at a molecular level, which is the topic of this chapter. A dye-derivatized mesoporous titanium film is one of the key components in dye-sensitized solar cells. The electrochemical and photophysical properties of the ground and the excited state of the dye play an important role in the charge-transfer (CT) dynamics at the semiconductor interface (Chapter 5.3 in Volume 6).

The photophysical and photochemical properties of group VIII metal complexes using polypyridyl ligands have been thoroughly investigated during the last three decades [16–20]. The main thrust behind these studies is to understand the energy and electron-transfer processes in the excited state and to apply this knowledge to potential practical applications such as dye-sensitized solar cells and light-driven information processing [21].

Particularly, ruthenium (Ru) (II) complexes have been used extensively as CT sensitizers on nanocrystalline TiO_2 films [22]. The choice of ruthenium metal is of special interest for a number of reasons: (1) because of its octahedral geometry, one can introduce specific ligands in a controlled manner; (2) the photophysical, photochemical, and the electrochemical properties of these complexes can be tuned in a predictable way; (3) it possesses stable and accessible oxidation states from I to IV; (4) it forms very inert bonds with imine nitrogen centers [23, 24].

Iron (Fe) that is in the first row transition metal of the periodic table, and the same group as ruthenium, is an inexpensive and abundant metal. However, the photophysical and the electrochemical properties of its coordination complexes are difficult to tune in an expected fashion [25]. The other notable disadvantage of this metal is weaker ligand field splitting compared to ruthenium and osmium (Os). On the other hand, osmium being in the third row transition metal of the periodic table has a

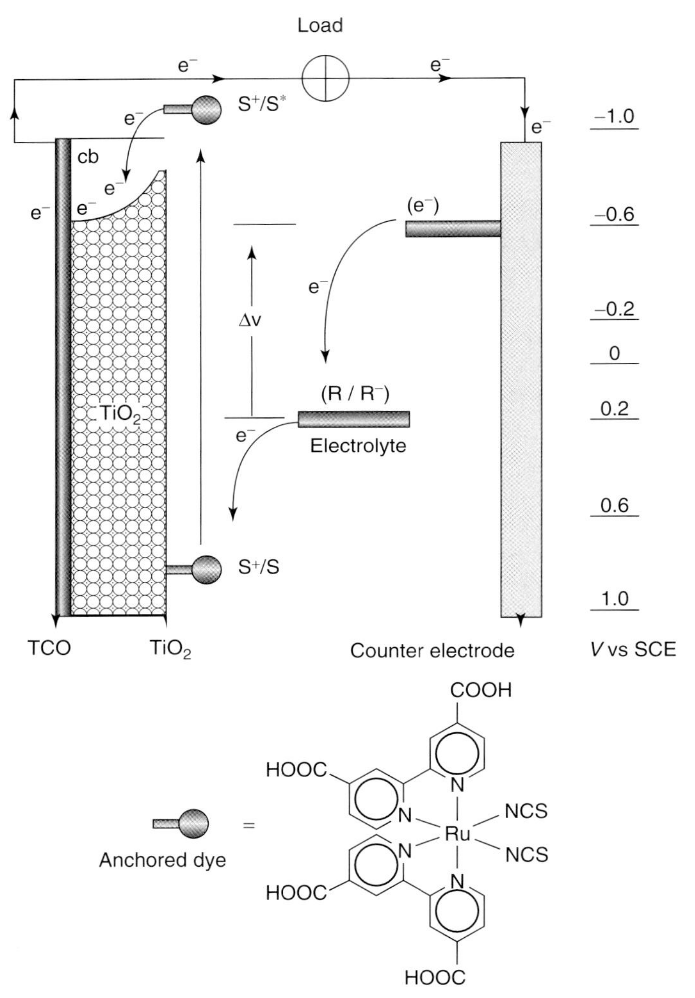

Fig. 1 Operating principles and energy level diagram of dye-sensitized solar cell. S/S^+ = Sensitizer in the ground and oxidized state; S^*/S^+ = Sensitizer in the excited state R/R^- = Redox mediator.

stronger ligand field splitting compared to ruthenium. Moreover, the spin-orbit coupling in osmium complexes leads to enhanced spectral response in the red region [26]. Nevertheless, the low abundance of this metal restricts its use for large-scale applications.

5.2.1.2 Operating Principles of the Dye-Sensitized Solar Cell

The details of the operating principles of the dye-sensitized solar cell are given in Fig. 1. The photo-excitation of the metal-to-ligand charge transfer (MLCT) of the adsorbed dye leads to injection of electrons into the conduction band of the oxide. The original state of the dye is subsequently restored by electron donation from an electrolyte, containing the redox system (iodide/triiodide). The injected electron flows through the semiconductor network to arrive at the back contact and then through the external load to the counter electrode. At the counter electrode, reduction of triiodide in turn regenerates iodide, which completes the circuit.

5.2.1.3 Incident Photon to Current Efficiency and Open-Circuit Photovoltage

The incident monochromatic photon-to-current conversion efficiency (IPCE), defined as the number of electrons generated by light in the external circuit divided by the number of incident photons as a function of excitation wavelength is expressed in the Eq. (1) [27]. The open-circuit photovoltage is determined by the energy difference between the fermi level of the solid under illumination and the Nernst potential of the redox couple in the electrolyte (Fig. 1). However, the experimentally observed open-circuit potential using various transition metal complexes is smaller than the difference between the conduction band and the redox couple because of differences in recombination rates [28].

$$\text{IPCE} = \frac{(1.25 \times 10^3) \times \underset{(\text{mA cm}^{-2})}{\text{photocurrent density}}}{\text{wavelength (nm)} \times \underset{(\text{W m}^{-2})}{\text{photon flux}}} \quad (1)$$

5.2.2 Molecular Sensitizers

5.2.2.1 Formation of Complexes

One can view the formation of a complex between metal ion and ligands as an electrostatic attraction between positively charged metal and negatively charged ions or the negative ends of the dipoles of neutral ligands. A metal ion or atom can act as a discrete center about which a set of ligands is arranged in a definite way. The general concept of a ligand (unidentate) is donating of an electron pair to the central metal atom. The number of ligands per metal center is generally either four or six, with others being rare. The approach of the negative, or the neutral, ligands toward a charged metal repels electrons residing in d-orbitals and raises their energies both with respect to those residing on the ligands and the metal.

In an octahedral ligand field, the five degenerate d-orbitals split into degenerate t_{2g} (d_{xy}, d_{xz}, d_{yz}), and e_g ($d_{x^2-y^2}$, d_{z^2}) sets of orbitals. The splitting value increases by about 40% as one moves from the first row (3d) to the second (4d) and third (5d) row transition metal ions [29]. A schematic representation for splitting of d-orbital and the position of ligand orbitals in an octahedral complex containing three bidentate 2,2′-bipyridyl ligands is shown in Fig. 2.

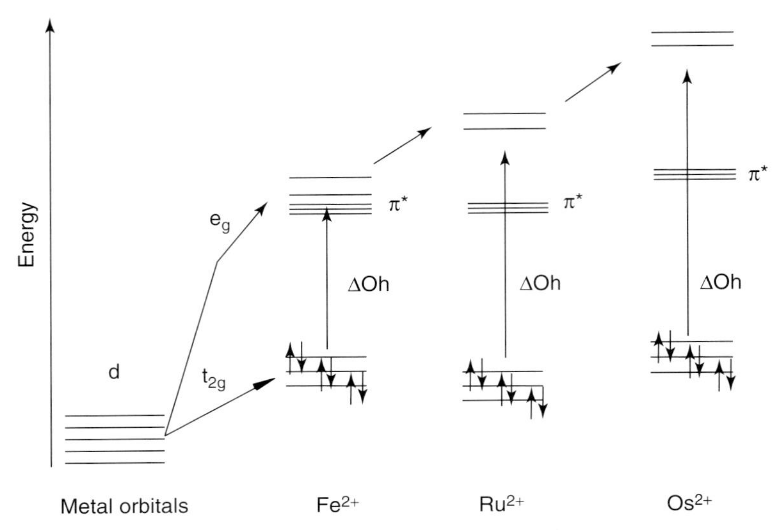

Fig. 2 A d-orbital splitting pattern for Fe^{2+}, Ru^{2+} and Os^{2+} in an octahedral field. The relative e_g levels with respect to t_{2g} and ligand π^*-orbitals are shown arbitrarily.

5.2.2.2 **Photophysical Properties**

The photophysics and photochemistry of polypyridyl complexes of ruthenium can be understood with the aid of the energy diagram shown in Fig. 3. In these complexes there are three possible excited states: (1) metal centered (MC), which are due to promotion of an electron from t_{2g} to e_g orbitals; (2) ligand centered (LC) that are $\pi-\pi^*$ transitions; and (3) CT excited states, which are either MLCT or ligand-to-metal (LMCT). An electronic transition from metal t_{2g} orbitals to empty ligand orbitals without spin-change allowed, is called a singlet-singlet optical transition. The allowed transitions are identified by large extinction coefficients. The transitions with spin change are termed singlet-triplet optical transitions, which are forbidden and are usually associated with a small extinction coefficient. However, the excited singlet state may also undergo a spin flip, resulting in an excited triplet state. This process is called intersystem crossing (ISC). The possible deexcitation processes are radiative and nonradiative. The radiative decay of a singlet and triplet excited states are termed fluorescence and phosphorescence, respectively.

The intense colors in 2,2′ bipyridyl complexes of iron, ruthenium, and osmium are due to excitation of an electron from metal t_{2g} orbitals to the empty π^*–orbitals of the conjugated 2,2′ bipyridyl. The photoexcitation of this MLCT excited state can lead to emission. However, not all complexes are luminescent because of the different competing deactivation pathways. This aspect is beyond the scope of this chapter; the interested reader can refer to a number of publications on this subject [16–20]. The other potential deactivation pathways for the excited dye are donation of an electron (called oxidative quenching, Eq. 2) or the capture of an electron (reductive quenching, Eq. 3) or transfer of its energy to other molecules or

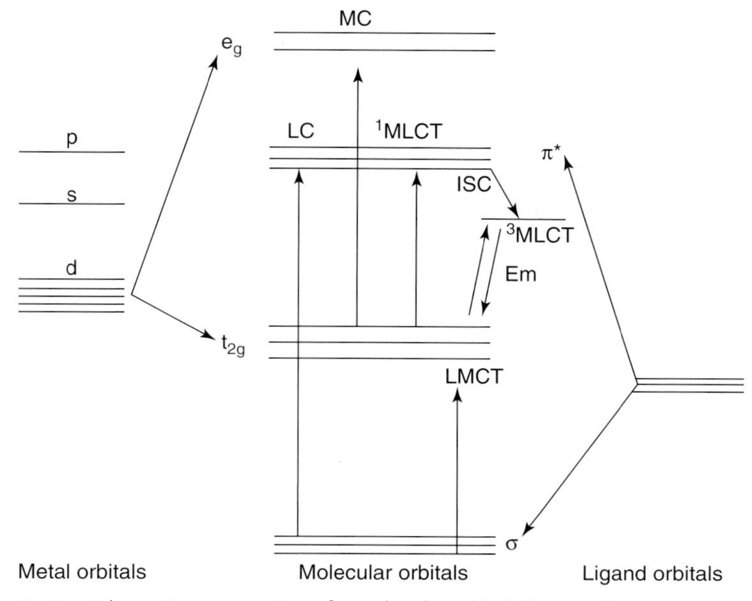

Fig. 3 Schematic presentation of a molecular orbital diagram for an octahedral d^6 metal complex involving 2,2′-bipyridyl type ligands in which various possible transitions are indicated.

solvent (Eq. 4).

$$D^* + Q \longrightarrow D^+ + Q^- \quad (2)$$

$$D^* + Q \longrightarrow D^- + Q^+ \quad (3)$$

$$D^* + Q \longrightarrow D + Q^* \quad (4)$$

5.2.2.3 **Ground and Excited State Redox Potentials**

The ground state oxidation and reduction potentials of a complex can be obtained by cyclic voltammetric studies. An approximate value of the excited state redox potential can be extracted from the potentials of the ground state couples and the zero-zero excitation energy (E_{0-0}) according to Eqs. (5) and (6). The zero-zero energy can be obtained from 77 K emission spectrum of the sensitizer [30]. The excited state redox potential of a sensitizer plays an important role in the sensitization process.

$$E(S^*/S^-) = E(S/S^-) + E_{0-0} \quad (5)$$

$$E(S^+/S^*) = E(S^+/S) - E_{0-0} \quad (6)$$

5.2.2.4 **Requirements of the Sensitizers**

The ideal sensitizer for a single junction photovoltaic cell should absorb all the solar photons below a wavelength of about 920 nm [31]. In addition, it should fulfill several demanding requirements: (1) It should possess suitable excited state redox properties; (2) It must firmly be grafted to the semiconductor oxide surface and inject electrons into the conduction band with a quantum yield of unity; (3) Its ground state redox potential should be sufficiently high that it can be regenerated rapidly via electron donation from the electrolyte or a hole conductor, (4) It should have a long-term photo and thermal stability to sustain at least 10^8 redox turnovers under illumination corresponding to about 20 years of exposure to natural sunlight at Air Mass (AM) 1.5.

On the basis of extensive screening of hundreds of ruthenium complexes, it was discovered that the sensitizer excited state oxidation potential should be at least – 0.9 V versus Saturated calomel reference electrode (SCE), to inject electrons efficiently onto TiO_2 conduction band. The ground state oxidation potential should be about 0.5 V versus SCE, to be regenerated rapidly via electron donation from the electrolyte (iodide/triiodide redox system) or a hole conductor. A significant decrease in electron injection efficiencies will occur if the excited and ground state redox potentials are lower than these values [32].

5.2.2.5 Absorption Spectral Properties of Metal Complexes

5.2.2.5.1 Tuning of MLCT Transitions

To absorb the light below 920 nm, the sensitizer's redox level needs to be tuned. The MLCT transitions and the redox properties of ruthenium polypyridyl complexes can be tuned in two ways. First, by introducing a ligand with a low-lying π^*-molecular orbital and second, by the destabilization of the metal t_{2g} orbitals with a strong donor ligand. The limiting conditions for an efficient sensitizer makes these two extreme situations incompatible. Because, in the former case, the excited state is not sufficiently energetic to enable electron transfer into the conduction band of the TiO_2 semiconductor, and in the latter case, the easily oxidizable complex cannot be reduced back by a suitable electron relay like I^-.

To illustrate the tuning aspects of the MLCT transitions in ruthenium polypyridyl complexes, the well known $[RuL_3]$ (L = 4,4′dicarboxylic acid-2,2′-bipyridine) type of complex can be considered. This complex shows strong visible band at 466 nm, because of CT transition from metal t_{2g} highest occupied molecular orbitals (HOMO) to π^*-lowest unoccupied molecular orbitals (LUMO) of the ligand (Fig. 3). The Ru(II)/(III) oxidation potential is at 1.3 V, and the ligand based reduction potential is at −1.5 V

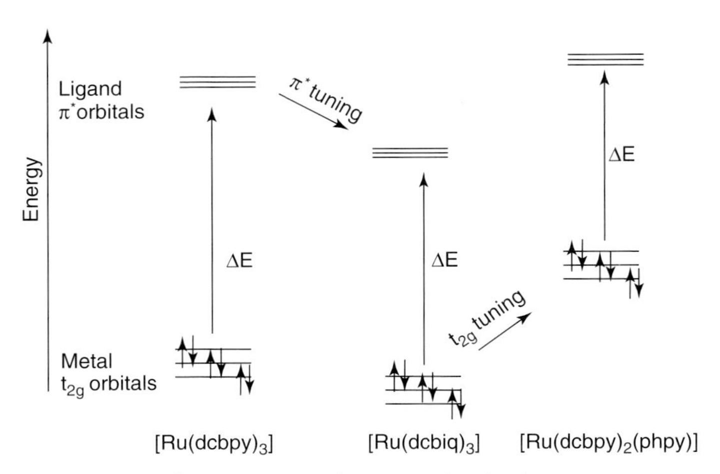

Fig. 4 Tuning of HOMO (t_{2g}) and LUMO (π^*) orbital energy in various ruthenium polypyridyl complexes.

versus SCE [33]. From spectro chemical and electrochemical studies of polypyridyl complexes of ruthenium, it has been concluded that the oxidation and reduction potentials are the best indicators of the energy levels of the HOMO and LUMO [34]. The energy between the metal t_{2g} orbitals and π^*-orbitals can be reduced either by raising the energy of the t_{2g} or by decreasing the energy of the π^*-orbitals with donor acceptor ligands, respectively (Fig. 4). In the following sections, ways to tune HOMO and LUMO energy levels by introducing various ligands will be discussed.

5.2.2.5.2 **Spectral Tuning in "Push-Pull" Type of Complexes** The lowest energy MLCT transition of ruthenium polypyridyl complexes of the type [RuL_3], (L = 4,4′dicarboxylic acid-2,2′-bipyridine), can be lowered so that it absorbs more in the red region of the visible spectrum by substituting one of the bidentate ligand by a donor bidentate ligand (pushing ligand). The remaining two electron-withdrawing bidentate ligands act as pull ligands. A list of complexes is included in Table 1, where the absorption maxima of the complexes are tuned from 466 nm to 560 nm by introducing different donor ligands. The oxidation potential in these complexes were tuned from 1.3 V to 0.3 V versus SCE, going from [RuL_3] to [RuL_2L'] L′ = phenylpyridine). Thus, the energy of the HOMO is varied to an extent of 1 V in this series of complexes [35]. Meyer and coworkers have used the same concept to synthesize black absorbers by judiciously selecting push and pull ligands [36].

It is interesting to note the magnitude of the spectral shift for the lowest energy CT transition is ≈0.5 eV, and the shift in the oxidation potential ≈1 eV. This clearly shows that not all the HOMO tuning is translated into the spectral shift of the complex. The apparent 0.5 eV, difference is involved in raising the energy of π^*-orbitals of the pulling ligands caused by the pushing ligands [35].

5.2.2.5.3 **Spectral Tuning in Complexes Containing Hybrid Donor Ligands** The absorption and luminescence spectral properties of complexes of the type [Ru(dmbip)(dcbpy)(X)] and [Ru(dmbip)(dcbiq)(X)], where dmbip = 2,6-bis(1-methylbenzimidazol-2′-yl)pyridine, dcbpy = 4,4′-$(COOH)_2$-

Tab. 1 Absorption, emission and redox properties of ruthenium complexes containing "push-pull" ligands

Complex	*Abs. max. (nm)*	*Emission max. (nm)*	*V vs SCE*	*Reference*
$[Ru(dcbpy)_3]^{4-}$	466	640	1.3	33
$[Ru(dcbpy)_2(dmg)]^{2-}$	470	665	1.2	35
$[Ru(dcbpy)_2(Me_2bpy)]^{2-}$	474	655	1.2	35
$[Ru(dcbpy)_2(2COOpy)]^{3-}$	490	712	0.80	35
$[Ru(dcbpy)_2(bpe)]^{2-}$	500	716	0.78	35
$[Ru(dcbpy)_2(dea\text{-}bpy)]^{2-}$	506	732	0.60	35
$[Ru(dcbpy)_2(dh\text{-}phen)]^{2-}$	516	760	–	35
$[Ru(dcbpy)_2(NCS)_2]^{4-}$	518	750	0.60	22
$[Ru(dcbpy)_2(phpy)_2]^{3-}$	560	865	0.3	35

Tab. 2 Absorption, emission, and redox properties of ruthenium complexes containing "hybrid donor" ligands

Complex	*Abs. max. (nm)*	*Emission max. (nm)*	*V vs SCE*	*Reference*
[Ru(dmbip)(dcbpyH)(Cl)]	516	–	0.67	32
[Ru(dmbip)(dcbpy)(H_2O)]	508	780	0.71	32
[Ru(dmbip)(dcbpy)(OH)]	528	800	–	32
[Ru(dmbip)(dcbpy)(NCS)]	502	770	0.80	32
[Ru(dmbip)(dcbpy)(CN)]	486	750	0.92	32
[Ru(dmbip)(dcbpy)($N(CN)_2$)]	492	760	0.60	38
[Ru(dmbip)(dcbpy)(Cl_3-pcyd)]	502	–	0.74	38
[Ru(dmbip)(dcbpy)(NCN)]	528	–	–	39
[Ru(dmbip)(5,5dcbpy)(NCN)]	542	–	–	38
[Ru(dmbip)($dcbiqH_2$)(NCS)]	580	900	1.15	32
[Ru(dmbip)($dcbiqH_2$)(Cl)] Cl	608	>950	0.83	32

2,2′-bipyridine, dcbiq=4,4′-$(COOH)_2$-2,2′-biquinoline and X = Cl^-, NCS^-, CN^-, $N(CN)_2^-$, NCN^-, and H_2O are summarized in Table 2. The 2,6-bis(1-methylbenzimidazol-2′-yl)pyridine ligand is a hybrid ligand that contains donor groups (benzimidazol-2′-yl) and an acceptor group (pyridine). The complexes containing these ligands show intense ultraviolet (UV) bands at 362 and 346 due to intraligand $\pi-\pi^*$ transition of 2,6-bis (1-methylbenzimidazol-2′-yl)pyridine. The bands at 308 and 244 nm are because of the $\pi-\pi^*$ transition of dcbpy ligand. On acidification, the 308 nm band shifts to 314 nm as a result of the protonation of the carboxylate groups, whereas the 362 and 342 nm bands remain unchanged in acidic and basic pH.

The MLCT bands of these complexes are broad and red shifted by approximately 140 nm, compared to $[Ru(dcbpy)_3]$. The lowest-energy MLCT transitions within this series were shifted from 486 to 608 nm, and the HOMO level varied over an extent of 0.45 V versus SCE. The energy of the MLCT transition in these complexes decreases with the decrease in the π-acceptor strength of the ancillary ligand, CN^-, NCS^-, H_2O, NCN^- Cl^-. The red shift of the absorption maxima in complexes containing dcbiq compared to dcbpy acceptor ligand is because of the low π^*-orbitals of biquinoline. The resonance Raman spectra of these complexes for excitation at 568 nm show bands predominantly associated with the dcbpy and dcbiq ligands indicating that the lowest excited state is a metal to dcbpy or dcbiq ligand CT state [32, 37, 38].

5.2.2.5.4 **Influence of Nonchromophoric Ligands on MLCT Transitions** This section compares spectral response of the three complexes of the type [Ru(P-terpy)(Me_2 bpy)(NCS)] **1**, (P-terpy = 4-phosphonato-2, 2′ : 6′,2″-terpyridine, Me_2bpy] = 4,4′-dimethyl-2,2′-bipyridine), $[Ru(H_2dcbpy)_2(NCS)_2]$ **2**, (H_2dcbpy = 4,4′-dicarboxy-2, 2′-bipyridine) and $(TBA)_2[RuL(NCS)_3]$ **3**, (L′ = 4,4′,4″-tricarboxy-2,2′ : 6′,2″-terpyridine) where the number of nonchromophoric ligands such as NCS^- are varied from one to three. The absorption and luminescence spectral properties of complex **1**–**3** are summarized in Table 3.

Tab. 3 Absorption, emission, and redox properties of ruthenium complexes containing NCS ligands

Complex	*Abs. max. (nm)*	*Emission max. (nm)*	*V vs SCE*	*Reference*
$[Ru(pterpy)(Me_2bpy)(NCS)]$	500	780	0.86	39
$[Ru(tcterpy)(Me_2bpy)(NCS)]$	510	800	0.94	39
$[Ru(H_2dcbpy)_2(NCS)_2]$	535	830	0.85	22
$[Ru(dcbpy)(Me_2bpy)_2(NCS)_2]$	527	760	0.78	39
$[Ru(terpy)(NCS)_3]$	580	820	0.54	40
$[Ru(pterpy)(NCS)_3]$	580	820	0.52	43
$[Ru(tcterpy)(NCS)_3]$	620	950	0.60	40

The absorption maxima are listed for the intense lowest-energy MLCT bands in ethanol, although the spectra possess additional absorption features at lower and higher energies.

The interesting feature of these complexes is a broad MLCT absorption bands in the visible region. The most intense MLCT transition maxima of complexes **1**, **2**, and **3** are at 500, 535, and 620 nm, respectively. The 120 nm red shift of **3** compared to **1** is the result of an increase in the energy of the metal t_{2g} orbitals caused by introducing nonchromophoric ligands and a slight decrease in the LUMO of the 4,4′,4″-tricarboxy-2,2′ : 6′,2″-terpyridine. The electrochemical data are consistent with the above assignments [22, 39, 40].

5.2.2.5.5 **Influence of the Position of Carboxyl Groups on MLCT Transitions** Ruthenium polypyridyl compounds, of the type $[RuL_2(X)_2]$, (where L = 4,4′-dicarboxylic acid-2,2′-bipyridine, 5,5′-dicarboxylic acid-2,2′-bipyridine, and 6,6′-dicarboxylic acid-2,2′-bipyridine; X = Cl^-, CN^- and NCS^-) have been reported [41–43]. The lowest MLCT absorption maxima of $[RuL_2(NCS)_2]$ complex (L = 2,2′-bipyridine) is seen at 510 nm in ethanol. By substituting two carboxylic acid groups at the 4,4″-position of 2,2-bipyridine ligand the MLCT maxima red shifted to 535 nm. However, substitution at the 5,5′-position of 2,2′-bipyridine shifted the maxima further into the red, 580 nm. On the contrary, the MLCT maxima blue shifted (500 nm) by substituting at the 6,6′ positions of 2,2′-bipyridine.

The enhanced red response of complexes containing the 5,5′-dicarboxylic acid-2,2′-bipyridine is because of a decrease in the energy of the π^*-orbitals, which makes them attractive as sensitizers for nanocrystalline TiO_2 films. Bignozzi and coworkers found that the IPCE of complexes having the 5,5′-dicarboxylic acid-2,2′-bipyridine ligands were lower than the analogous complexes that contain 4,4′-dicarboxylic acid-2,2′-bipyridine [43]. They rationalized the low efficiency of the sensitizers containing 5,5′-dicarboxylic acid-2,2′-bipyridine ligands in terms of low excited state redox potentials.

5.2.2.5.6 **MLCT Transitions in Geometric Isomers** Isomerization is another exciting approach toward tuning the spectral properties of metal complexes (Fig. 5) [44–46]. The electronic properties of *cis*- and *trans*-isomers of $[RuL_2(X)_2]$, (L = 4,4′-dicarboxylic acid-2,2′-bipyridine; X = Cl^-,

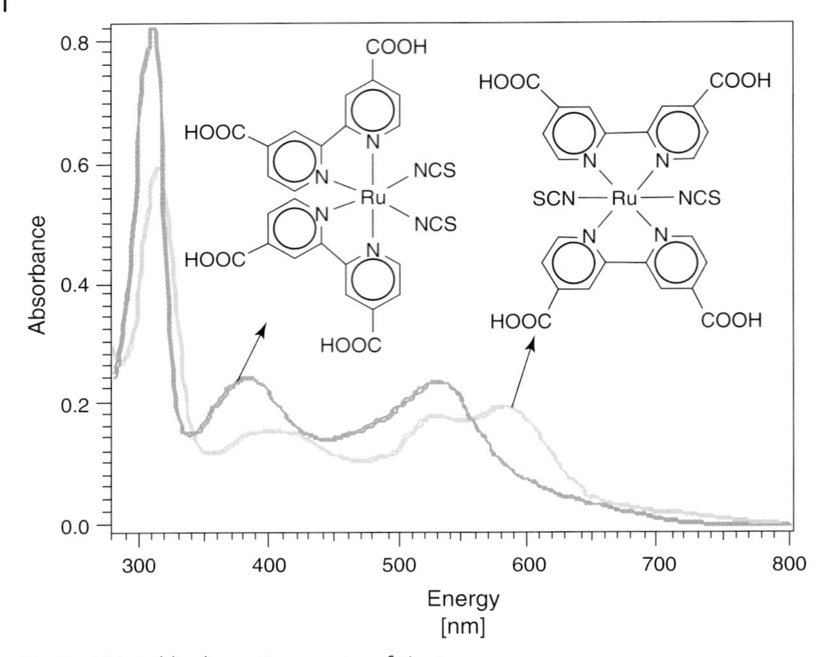

Fig. 5 UV-visible absorption spectra of *cis*-, *trans*-bis(2,2′-bipyridine-4,4′-dicarboxylic acid) ruthenium (II) $(NCS)_2$ complexes.

Tab. 4 Electronic Spectral Data of the *cis*- and *trans*-bis(4,4′-dicarboxylic acid-2,2′-bipyridine)Ruthenium (X_2) Complexes ($X = Cl^-$, H_2O, NCS^-) in DMF

Complex	*Abs. max. (nm)*	*Emission max. (nm)*	*Reference*
cis-$[Ru(H_2dcbpy)_2(Cl)_2]$	590	847	47
cis-$[Ru(H_2dcbpy)_2(H_2O)_2]$[a]	500	750	44
cis-$[Ru(H_2dcbpy)_2(NCS)_2]$	535	830	47
trans-$[Ru(H_2dcbpy)_2(Cl)_2]$	690	>900[b]	47
trans-$[Ru(H_2dcbpy)_2(H_2O)_2]$	530	800[b]	47
trans-$[Ru(H_2dcbpy)_2(NCS)_2]$	581	870	47
cis-$[Ru(bpy)_2(H_2O)_2]$	480	660	45
trans-$[Ru(bpy)_2(H_2O)_2]$	495	700	46

DMF, dimethylformamide.
[a] in water at pH 4.5.
[b] Weak and broad emission maximum.

H_2O and NCS^-) are collected in Table 4. The UV-visible absorption spectrum of the *trans*-dichloro complex in DMF solution shows at least three MLCT absorption bands in the visible region at 690, 592, and 440 nm. On the other hand, the *cis*-dichloro complex in DMF solution shows only two distinct broadbands in the visible region at 590 and 434 nm, which was assigned to a MLCT transitions.

The lowest energy MLCT band in the *trans*-dichloro complex is significantly red shifted compared to the *cis*-dichloro complex. This red shift is due to stabilization of the LUMO of the dcbpy ligand in the *trans*-species relative to the *cis*-species [47].

The red shift (108 nm) of the lowest energy MLCT absorption in the spectrum of the *trans*-dichloro complex compared to the spectrum of the *trans*-dithiocyanato is due to the strong σ donor property of the Cl^- compared to the NCS^- ligand. The chloride ligands cause destabilization of the metal t_{2g} orbitals, and raising them in energy relatively closer to the ligand π^*-orbitals result in a lower energy MLCT transition. According to Lever's electrochemical parameterization, the chloride ligand has a parameter value of −0.24 and the thiocyanate ligand has a value of −0.06. The electrochemical and the absorption data are in good agreement with reported parameterization scale for ligands [48].

5.2.2.5.7 Spectral Tuning in Heteroleptic Sensitizers

The synthesis and characterization of heteroleptic complexes of ruthenium $[Ru(L^1)(L^2)(L^3)]$ starting from a polymeric complex $Ru(CO)_2Cl_2$ have been reported [49–51]. The synthetic methodology of such heteroleptic complexes involves several steps, resulting in very low yields. Recently, a novel synthetic route has been developed to introduce different ligands on a ruthenium (II) precursor, namely, the $RuCl_2(DMSO)_4$ complex, which is a versatile and well-characterized starting material [52]. A range of mixed ligand sensitizers of the type $[Ru(L^1)(L^2)(L^3)]$ have been prepared and their spectroscopic, luminescent, and redox properties are reported [53, 54]. The absorption spectral data of these complexes show a red shift with increasing donor strength of the L^3 ligands. In these complexes, L^3 being the dithiocarbamate ligand, shifts the Ru(II) t_{2g} level upward without significantly affecting the π^*-level of the dcbpy ligand thereby increasing the MLCT absorption in the red portion of the visible region.

5.2.3 Hydrophobic Sensitizers

The other important aspect in dye-sensitized solar cells is water-induced desorption of the sensitizer from the surface. Extensive efforts have been made in our laboratory to overcome this problem by introducing hydrophobic properties in the ligand. The heteroleptic complexes containing hydrophobic ligands of the type $[Ru(dcbpy)(mhdbpy)(NCS)_2]$ **1**, $[Ru(dcbpy)(dtdbpy)(NCS)_2]$ **2** $[Ru(dcbpy)(mddbpy)(NCS)_2]$ **3** (dcbpy = 4,4′-dicarboxy-2,2′-bipyridine, mhdbpy = 4-methyl-4′-hexadecyl-2,2′-bipyridine and dtdbpy = 4,4′-ditridecyl-2,2′-bipyridine, mddbpy = 4-methyl-4′-didodecyl-2,2′bipyridine) have been synthesized (Fig. 6). The photocurrent action spectra of these complexes show broad features covering a large part of visible spectrum and displays a maxima around 550 nm, where the incident monochromatic IPCE exceeds 80%. The performance of these hydrophobic complexes as CT photosensitizers in nanocrystalline TiO_2-based solar cell shows excellent stability toward water-induced desorption [55].

The rate of electron transport in dye-sensitized solar cell is a major element of the overall efficiency of the cells. The electrons injected into the conduction band from optically excited dye, can traverse the TiO_2 network and can be collected at the transparent conducting glass or can react

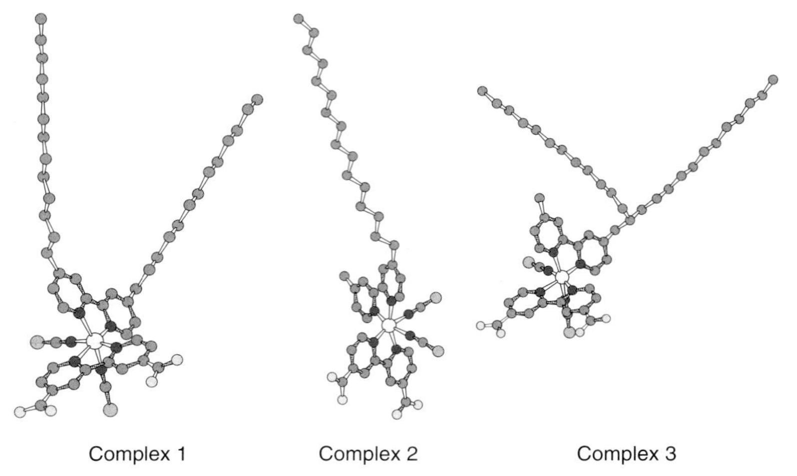

Fig. 6 Structure of complexes [Ru(dcbpy)(dtdbpy)$(NCS)_2$] **1**, [Ru(dcbpy)(mhdbpy)$(NCS)_2$] **2**, and [Ru(dcbpy)(mddbpy)$(NCS)_2$] **3**.

with a redox mediator. The reaction of injected electrons into the conduction band with a redox mediator gives undesirable dark currents, significantly reducing the charge-collection efficiency, and thereby decreasing the total efficiency of the cell. This problem is drastically suppressed by using hydrophobic sensitizers of the type shown in Fig. 6. The most likely explanation for the dark-current suppression is that the long chains of the sensitizer interacts laterally to form an aliphatic network, thereby preventing triiodide from reaching the TiO_2 surface.

5.2.4 Near IR Sensitizers

5.2.4.1 Phthalocyanines

Phthalocyanines possess intense absorption bands in the near IR region and are known for their excellent stability rendering them attractive for photovoltaic applications [56]. They have been repeatedly tested in the past as sensitizers of wide band gap oxide semiconductors and obtained poor incident photon to electric current conversion yields remaining under 1%, which is insufficient for solar cell applications [57–60]. One of the reasons for such low efficiencies is aggregation of the dye on the TiO_2 surface. This association often leads to undesirable photophysical properties such as self-quenching and excited state annihilation. However, the advantage of this class of complexes is the near IR response which is very strong, having extinction coefficients of close to 50 000 M^{-1} cm^{-1} at 650 nm compared to the polypyridyl complexes, which have small extinction coefficients at this wavelength.

5.2.4.2 Ruthenium Phthalocyanines

The bis(3,4-dicarboxypyridine)Ru(II)1,4,8,11,15,18,22,25-octamethyl-phthalocyanine (JM3306) shows a visible absorption band at 650 nm (ε 49 000 M^{-1} cm^{-1}) and a phosphorescence band located at 895 nm. The triplet state lifetime is 474 ns under anaerobic conditions. The emission is entirely quenched when JM3306 is adsorbed onto a nanocrystalline TiO_2 film.

The very efficient quenching of the emission of JM3306 was found to be due to the electron injection from the excited singlet/triplet state of the phthalocyanine into the conduction band of the TiO_2 [61]. The photocurrent action spectrum is shown in Fig. 7, where the incident IPCE is plotted as a function of wavelength. The feature is extending well into the near IR region displaying a maximum at around 660 nm where the IPCE exceeds 60%. These are by far the highest conversion efficiencies obtained with the phthalocyanine type sensitizers.

It is fascinating to note that this class of dyes inject efficiently into the conduction band of TiO_2, despite the fact that the pyridyl orbitals do not participate in the $\pi-\pi^*$-excitation which is responsible for the 650 nm absorption band. This phenomenon shows that the electronic coupling of the excited state of the dye to the Ti (3d) conduction band manifold is strong enough through this mode of attachment to render charge injection very efficient. These results establish a new pathway for grafting dyes to oxide surfaces through axially attached pyridine ligands.

5.2.4.3 Phthalocyanines Containing 3d Metals

Several zinc (II) and aluminum (III) phthalocyanines substituted by carboxylic acid and sulfonic acid groups were anchored to nanocrystalline TiO_2 films and tested for their photovoltaic

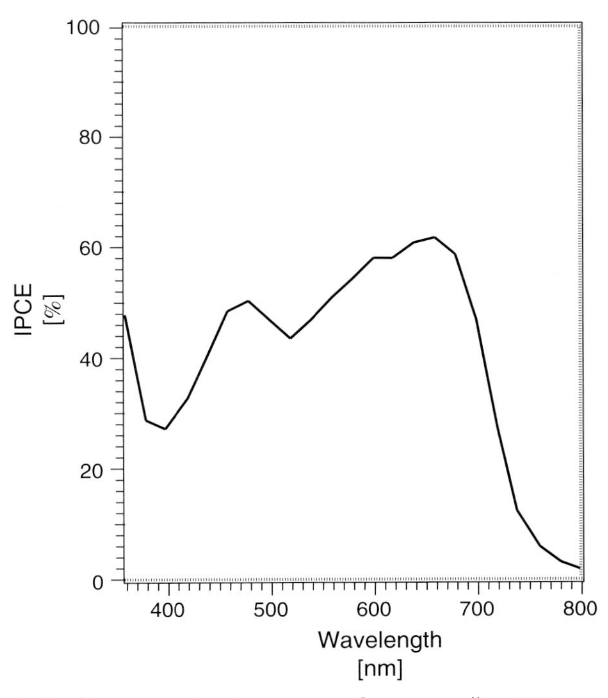

Fig. 7 Photocurrent action spectra of nanocrystalline TiO_2 films sensitized by bis(3,4-dicarboxypyridine)Ru(II) 1,4,8,11,15,18,22,25-octamethyl-phthalocyanin. The incident IPCE is plotted as a function of wavelength.

behavior using LiI/LiI_3/propylene carbonate electrolyte [62]. Interestingly, zinc (II) 2,9,16,23-tetracarboxyphthalocyanine exhibited 45% monochromatic current conversion efficiency at 700 nm. It is shown that electron injection to TiO_2 occurs from the excited singlet state of the phthalocyanine derivatives. The inherent problem of phthalocyanine aggregation is reduced considerably in 3d metal phthalocyanines by introducing 4-*tert*-butylpyridine and $3\alpha,7\alpha$-dihydroxy-5β-cholic acid (cheno) into the dye solution. The added 4-*tert*-butylpyridine in all likelihood coordinates to the metal in the axial position and thereby prevents aggregation of the phthalocyanines.

This type of sensitizers opens up new avenues for improving the near IR response of nanocrystalline injection solar cell. In addition, important applications can be foreseen for the development of photovoltaic windows transmitting part of the visible light. Such devices would remain transparent to the eye, while absorbing enough solar energy photons in the near IR to render efficiencies acceptable for practical applications.

5.2.5 Mononuclear and Polynuclear Metal Complexes of Group VIII

5.2.5.1 Iron Complexes

The photoconversion efficiency for iron complexes [FeL_3] and [$FeL_2(X)_2$], ($L =$ 4,4′-dicarboxylic acid-2,2′-bipyridine and 5,5′-dicarboxylic acid-2,2′-bipyridine) has been reported [63]. The [$FeL_2(X)_2$] type complexes show two broad MLCT bands in the visible region. In both cases, the photoconversion efficiency of the high-energy band was reported in the range of 40%. Nonetheless, there is virtually no photocurrent contribution from the lower energy MLCT band of the complexes containing 5,5′-dicarboxylic acid-2,2′-bipyridine ligand because of lack of driving force for electron injection.

5.2.5.2 Osmium Complexes

Although there are a large number of osmium polypyridyl complexes, very few have been used as sensitizers in dye-sensitized solar cells. The osmium complexes have several advantages compared to their ruthenium analogues: (1) osmium has a stronger ligand field splitting compared to ruthenium, (2) the spin-orbit coupling leads to excellent response in the red region because of enhanced singlet-triplet CT transitions [26]. Heimer and coworkers have synthesized [$OsL_2(CN)_2$] complex ($L = 4$, 4′-dicarboxylic acid-2,2′-bipyridine) and found that the complex is extremely stable under irradiation in a homogeneous aqueous solution compared to the analogous ruthenium complex [64]. The greater photo-stability for osmium is consistent with a stronger crystal field splitting of metal d-orbital that inhibits efficient population of d-d states. However, a factor of 10 lower injection efficiency for the osmium complex when anchored on TiO_2 electrode was reported compared to the ruthenium analog and rationalized in terms of slow regeneration rates for the oxidized dye [65].

Lewis and coworkers have synthesized and characterized two osmium complexes [$OsL_2(CN)_2$] and [OsL_3] ($L = 4$,4′-dicarboxylic acid-2,2′-bipyridine). In their studies they found nearly 80% incident monochromatic IPCE [66].

5.2.5.3 Polynuclear Complexes

Polynuclear systems in which more than one complex is connected by a bridging ligand have been studied in several laboratories [67–72]. Here studies have been restricted exclusively to the cyano-bridged (CN-bridged) trinuclear ruthenium complexes of the type shown in Fig. 8. Scandola and coworkers have synthesized and characterized several of these complexes with respect to their photophysical and electrochemical properties [73]. In such complexes, one can invoke possible antenna effects in the light harvesting process by introducing various fragments in which the photophysical properties are tuned to have directionality in the excited state. The presence of numerous metal centers should allow for different tailored fragments containing various ligands [72, 74].

Potential application of these complexes as photosensitizers for light energy harvesting relies heavily on demanding synthetic schemes. The CN-bridged trinuclear complex **1** (Fig. 8) has been found to be an excellent photosensitizer for spectral sensitization of the wide band gap semiconductor TiO_2 [75]. Among the possible MLCT excited states associated with the three metal centers, the one with lowest energy is based on the central unit carrying the ligand 4,4′-dicarboxy-2,2′-bipyridine (dcbpy). When coated on a TiO_2-electrode, the complex anchors through the carboxyl unit of this central unit. Photons absorbed by the higher energy peripheral units efficiently transfer excitation energy to this central unit, accounting for its excellent performance.

To understand the principles of operation of polynuclear complexes and possibly improve their performance, the "tunability" of photophysical and redox properties has been examined in a number of such complexes. These complexes are composed of polypyridyl units of Ru and Os as chromophores and cyanide as the bridging ligand [76]. By appropriate synthetic procedures, it is possible to control the mode of coordination of the bridging cyanide to a given chromophoric unit as a nitrile (C-bonded) or isonitrile (N-bonded) [74–77]. By varying the nature of

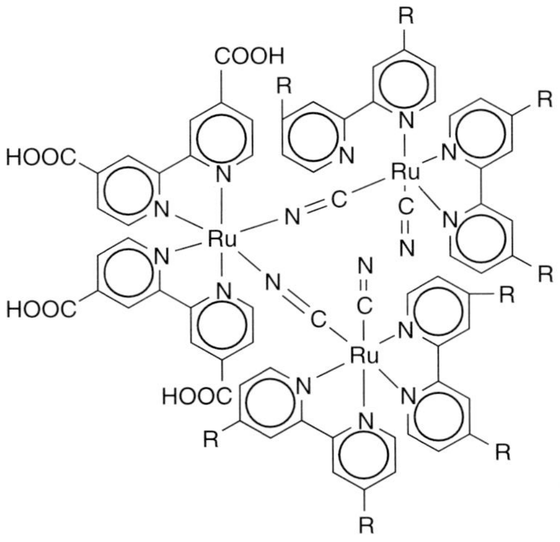

1 : R = H ; **2** : R = COOH ; **3** : R = CH_3 ; **4** : R = C_6H_5

5 = [(CN)(bby)$_2$ Os-CN-Ru(dcbpy)$_2$-NC-Os(bpy)$_2$(CN)]

6 = [(H_2O)(bby)$_2$ Ru-CN-Ru(dcbpy)$_2$-NC-Ru(bpy)$_2$(H_2O)]

Fig. 8 Structure of CN-bridge trinuclear ruthenium complex .

the metal center, the mode of coordination of the bridging ligand, and the spectator ligands, it is possible to vary the location and the direction of excited state electron transfer process.

5.2.6
Surface Chelation of Polypyridyl Complexes onto the TiO_2 Surface

5.2.6.1 Anchoring Groups

The functional groups serve as grafting agents for the oxide surface of the TiO_2 films. The grafting of polypyridyl complexes onto the oxide surface, which allows for electronic communication between the complex and the substrate, is an important target in dye-sensitized solar cells. Several ruthenium complexes containing carboxylic acid and phosphonated polypyridyl ligands are described [78–83]. The carboxylic and phosphonate functionality serves as an anchoring group to immobilize the complex on the nanocrystalline TiO_2 films. The immobilized sensitizer absorbs a photon to produce an excited state, which efficiently transfers its electron onto the TiO_2 conduction band. To achieve high quantum yields of the excited state electron-transfer process the dye ideally needs to be in intimate contact with the semiconductor

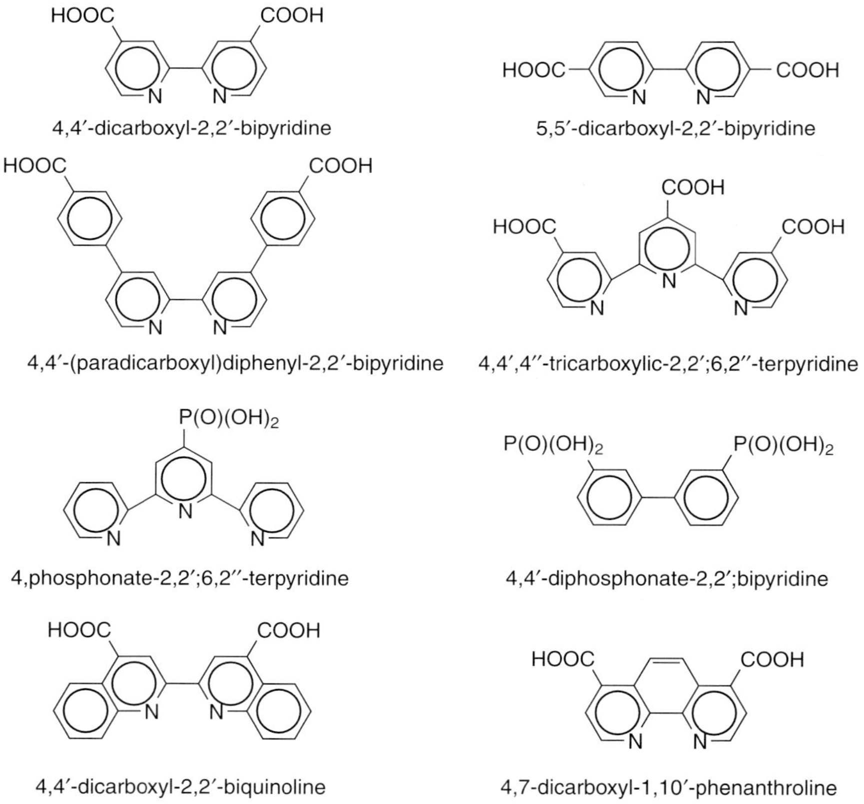

4,4′-dicarboxyl-2,2′-bipyridine

5,5′-dicarboxyl-2,2′-bipyridine

4,4′-(paradicarboxyl)diphenyl-2,2′-bipyridine

4,4′,4″-tricarboxylic-2,2′;6,2″-terpyridine

4,phosphonate-2,2′;6,2″-terpyridine

4,4′-diphosphonate-2,2′;bipyridine

4,4′-dicarboxyl-2,2′-biquinoline

4,7-dicarboxyl-1,10′-phenanthroline

Fig. 9 Structure of the ligands containing anchoring groups.

surface. The ruthenium complexes that have carboxylic acid and phosphonic acid groups show electron-transfer processes in the range of 80 to 90%. The near quantitative electron injection efficiency indicate a close overlap of the ligand π^*-orbitals and the titanium 3d orbitals [84].

Figure 9 shows the polypyridine ligands with different anchoring groups that have been employed for surface derivation of ruthenium complexes. Goodenough and coworkers have derivatized a mixed ligand complex [Ru(bpy)$_2$(dcbpy)] on TiO_2 and SnO_2 films in the presence of dicyclohexylcarbodimide [84].

The interaction between the adsorbed sensitizer and the semiconductor surface has been addressed using resonance Raman and Fourier transform infrared (FT-IR) spectroscopy. The carboxylic acid functional group could adsorb on the surface in a unidentate, bidentate, or bridging fashion. Yanagida and coworkers concluded that the sensitizer *cis*-dithiocyanato bis(2,2′-bipyridine-4,4′-dicarboxylate) ruthenium (II) complex (N3) binds to the surface using ester-like and chelating linkages [85]. Woolfrey and coworkers have reported that the N3 complex anchor on the surface of TiO_2 as a bidentate or bridging mode using two carboxylate groups per dye [86]. However, Fillinger and Parkinson studied the adsorption behaviour of the N3 sensitizer and found that the initial binding involves one carboxylate, with subsequent binding of two or more carboxylate groups on the surface [87].

It is noteworthy that, Bignozzi and coworkers achieved sensitization of TiO_2 with a cyanobridged dimer of Ru-Re polypyridyl complex where the coupling of the MLCT exited state to the conduction band manifold was through space and did not involve the anchoring group [88].

5.2.6.2 Acid-Base Equilibria of the Anchoring Groups

The grafting properties of complexes containing the anchoring group onto the TiO_2 surface and the factors responsible for differences in the attachment of various anchoring groups are essential for the design of novel sensitizers [89]. The difference in binding properties of the complexes containing carboxyl and phosphonate anchoring groups may stem from the differences in the pK_a values of the ligands. Knowledge of the pK_a values of ionizable groups may aid in indicating the possible nature of the attachment. The acid base properties of transition metal complexes containing ligands with protonatable functional groups can provide important information about the nature of binding properties on the oxide surface.

5.2.6.3 Acid-Base Equilibria of the 4,4-Dicarboxy-2,2′-Bipyridine and its Complexes

The ground state pK_a values of [Ru(dcbpy)$_2$(NCS)$_2$] (N3) were determined from the relationship between the change in the optical density or the peak maximum with the pH for a given wavelength [90]. When acid is added to an alkaline solution of N3, changes in the electronic spectrum occur as shown in Fig. 10. The plot of λ-max change versus pH for N3 shows the expected sigmoidal shape, with the pH at the inflection point giving two ground state pK_a values at 3 and 1.5 ± 0.1.

In a related complex of the type [Ru(bpy)$_2$(dcbpy)], two inflection points at pH 2.7 and <0.5 were found suggesting that the dissociation of the carboxylic acid groups of 4,4-dicarboxy-2,2′-bipyridine ligand is a sequential process. The pK_{a1} of the free 4,4-dicarboxy-2,2′-bipyridine ligand is 4.2 and the second one is below 2. The difference between the free ligand and

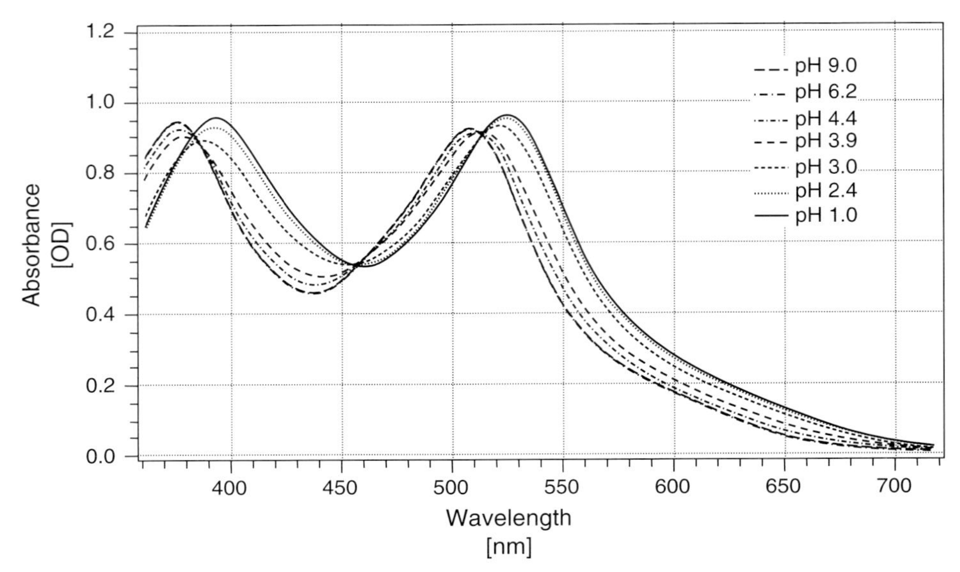

Fig. 10 Absorption spectral changes of *cis*-dithiocyanatobis(2,2′-bipyridine-4,4′-dicarboxylic acid) ruthenium (II) complex as a function of pH.

the complex pK_{a1} can be considered as a measure of donor strength of the ligand. The difference in the pK_a values of $[Ru(dcbpy)_2(NCS)_2]$ and $[Ru(bpy)_2(dcbpy)]$ is rationalized in terms of the donor or acceptor properties of nonprotonatable ligands [91].

5.2.6.4 Acid-Base Equilibria of the Phosphonato Group

The 4-phosphonato-2,2′ : 6′,2″-terpyridine is protonated in four steps in the pH region from 1.5 to 7 as shown in Sch. 1. Out of four protons, two are involved in protonation of the pyridyl nitrogens of terpyridyl, and the remaining two protons are on the phosphonated group. Since the adjacent nitrogens of 2,2′-bipyridyl ligand accept only one proton down to pH 1.6, it can be safely assumed that 4-phosphonato-2,2′ : 6′,2″-terpyridine accepts protons in nonadjacent positions as shown in Sch. 1. The central nitrogen of the 4-phosphonato-2,2′ : 6′,2″-terpyridine is sterically restricted and less basic because of the doubly protonated form.

The dissociation constants of 4-phosphonato-2,2′ : 6′,2″-terpyridine ligand were measured by pH-nuclear magnetic resonance (NMR) and potentiometric titration methods. The 4-phosphonato-2,2′ : 6′,2″-terpyridine ligand shows three measurable pK_as at 6.0, 4.65, and 3.55. The fourth pK_a was estimated to be below 2 [92]. Spectrophotometric titration of the ruthenium complex of the type [Ru(P-terpy)(Me_2bpy)(NCS)] (P-terpy = 4-phosphonato-2,2′ : 6′,2″-terpyridine and Me_2bpy = 4,4′-dimethyl-2,2′-bipyridine) exhibits two ground state pK_a values at 6 and 4, which can be assigned to pK_{a1} and pK_{a2}, respectively.

The emission maximum of [Ru(P-terpy)(Me_2bpy)(NCS)] shows a blue shift with decreasing pH, and apparent pH dependent excited state lifetimes.

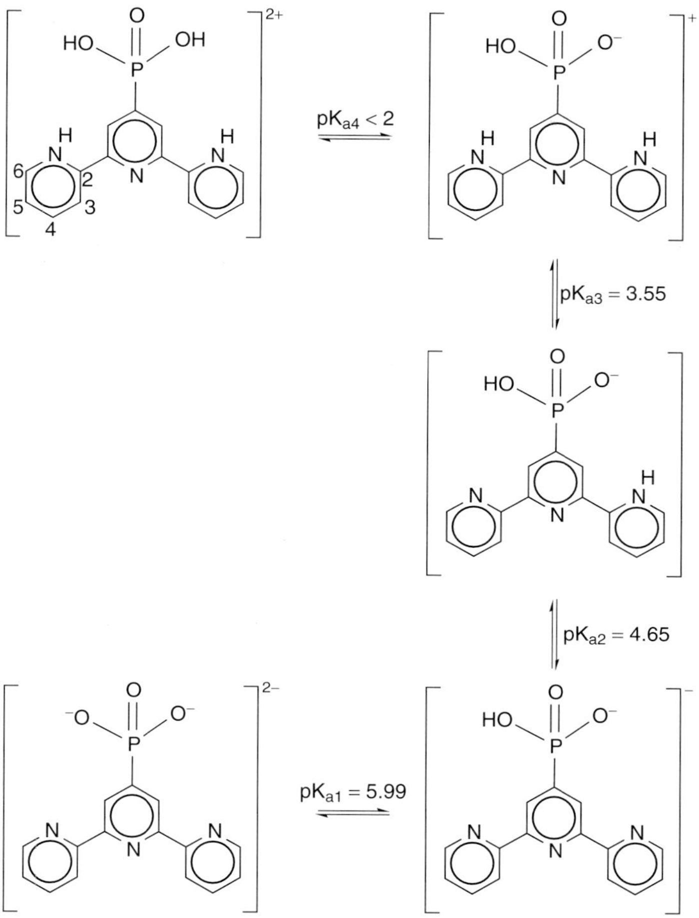

Scheme 1 Acid-base equilibria of 4-phosphonato-2,2′ : 6,2″-terpyridine.

The complex [Ru(P-terpy)(Me_2bpy)(NCS)] shows two excited state pK_a^*s at 6.5 (pK_{a1}^*) and 4.5 (pK_{a2}^*). Comparison of the ground and excited state pK_as of complex [Ru(P-terpy)(Me_2bpy)(NCS)] demonstrate that the excited state has a metal to 4-phosphonato-2,2′ : 6′,2″-terpyridine ligand CT character. The resonance Raman data of [Ru(P-terpy)(Me_2bpy)(NCS)] show that the energy levels of the two ligands are very close and in the excited state mostly 4-phosphonato-2,2′ : 6,2″-terpyridine peaks are enhanced compared to the Me_2bpy [39].

Schmehl and coworkers has measured ground and excited state pK_a of several ruthenium polypyridyl complexes containing phosphonate groups. They reported two ground state pK_a values approximately 2 and 6.3 for [Ru(bpy)$_2$ (bpppH$_2$)]. The excited pK_{a1}^* of [Ru(bpy)$_2$ (bpppH$_2$)] is more basic than the ground state pK_{a1} suggesting that the MLCT transition is localized on the bppH2 ligand [93].

5.2.7
Stability and Performance of the Dyes

5.2.7.1 Stability of the Inorganic Dyes

The thermal stability of ruthenium (II) complexes of the type $[Ru(H_2dcbpy)_2(NCS)_2]$ **1**, $(Bu_4N)_2[Ru(Hdcbpy)_2(NCS)_2]$ **2**, $(Bu_4N)_4[Ru(dcbpy)_2(NCS)_2]$ **3**, and $(Im)_4[Ru(dcbpy)_2(NCS)_2]$ **4**, where dcbpy = 2, 2′-bipyridyl-4,4′dicarboxylate, Bu_4N = tetrabutylammonium, and Im = dimethylethylimidazolium, has been studied using thermoanalytical techniques, IR, UV-visible, and ^1H NMR spectroscopic methods [94]. These complexes show remarkable stability in both nitrogen and air atmospheres at high temperatures, ranging from 180 °C to 250 °C for 1. The only process that is observed at lower temperatures is the dehydration that occurs between 40 to 110 °C. Higher temperature processes, including deamination of the counterion as well as decarboxylation and finally decomposition of the complex, occur between 200 to 400 °C with different characteristics observed in air and in nitrogen (Fig. 11).

The decarboxylation reaction is an endothermic process in a nitrogen atmosphere and overlaps with decomposition of the complexes. In air, on the other hand, it is an exothermic process distinctively separated from decomposition. Higher thermal stability is observed for **1** and **2** when anchored onto nanocrystalline TiO_2 films. The activation energy of decarboxylation is estimated for **1** in the free form (≈115 kJ mol^{-1}) and on TiO_2 (≈126 kJ mol^{-1}). In addition, the deamination temperature increases by about 35 °C when Bu_4N are replaced by Im cations. A surprising observation is the two-step deamination in dyes with four organic cations, which is due to the

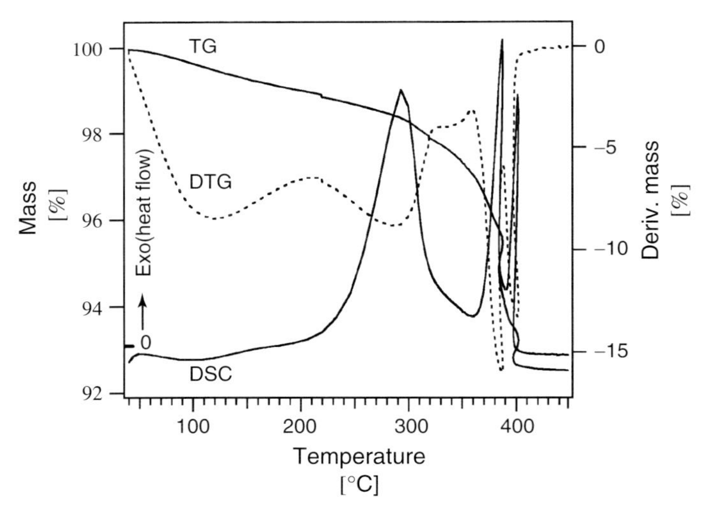

Fig. 11 The combined thermogravimetric (TG), derivative thermogravimetric (DTG), and differential scanning calorimetry (DSC) curves of complex $(Bu_4N)_4[Ru(dcbpy)_2(NCS)_2]$ adsorbed on TiO_2 in dynamic air atmosphere. Loss of water, exothermic decomposition of amine (between 200–330 °C) and decarboxylation starting at 360, ending at 390 °C.

difference in the extent of interaction between these cations and the carboxylate groups.

5.2.7.2 Effect of Protons Carried by the Sensitizer on the Performance

To obtain high conversion efficiencies, optimization of the short-circuit photocurrent (i_{sc}) and open-circuit potential (V_{oc}) of the solar cell is essential. The conduction band of the TiO_2 is known to have a Nernstian dependence on pH [80]. The fully protonated sensitizer $[Ru(H_2dcbpy)_2(NCS)_2]$ **1**, on adsorption transfers most of its protons to the TiO_2 surface charging it positively. The electric field associated with the surface dipole generated in this fashion enhances the adsorption of the anionic ruthenium complex and assists electron injection from the excited state of the sensitizer in the titanium conduction band, favoring high photocurrents. However, the open-circuit potential is lower as a result of the positive shift of the conduction band edge induced by the surface protonation.

On the other hand, the sensitizer $(Bu_4N)_4[Ru(dcbpy)_2(NCS)_2]$ **3**, that carries no protons shows high open-circuit potential compared to the complex **1**, because of the relative negative shift of the conduction band edge induced by the adsorption of the anionic complex, although consequently, the short-circuit photocurrent is lower. Thus, there should be an optimum degree of protonation of the sensitizer (for example $(Bu_4N)_2[Ru(Hdcbpy)_2(NCS)_2]$ **2**, which contains two protons) to obtain optimum short-circuit photocurrent and open-circuit potential, which determines the power conversion efficiency of the cell.

The performance of the three sensitizers **1**, **2**, and **3** that contain different degrees of protonation were studied on nanocrystalline TiO_2 electrodes [90]. The photocurrent action spectra obtained with a monolayer of these complexes coated on TiO_2 films are shown in Fig. 12. The redox electrolyte consisted of a solution of 0.6 mol dimethylpropyl imidazolium iodide, and 0.05 mol I_2, in methoxyacetonitrile.

The incident monochromatic IPCE is plotted as a function of excitation wavelength. The IPCE value in the plateau region is 80% for complex **1**, whereas for complex **3** it is only about 66%. In the red region, the difference is even more pronounced. Thus, at 700 nm the IPCE value is twice as high for the fully protonated complex **1** as compared to the deprotonated complex **3**. Consequently, the short-circuit photocurrent

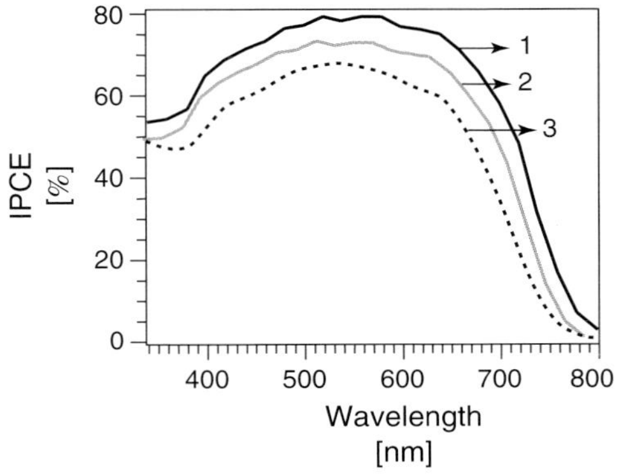

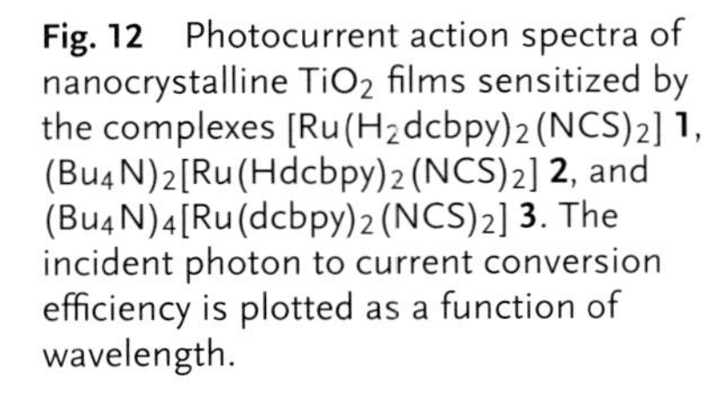
Fig. 12 Photocurrent action spectra of nanocrystalline TiO_2 films sensitized by the complexes $[Ru(H_2dcbpy)_2(NCS)_2]$ **1**, $(Bu_4N)_2[Ru(Hdcbpy)_2(NCS)_2]$ **2**, and $(Bu_4N)_4[Ru(dcbpy)_2(NCS)_2]$ **3**. The incident photon to current conversion efficiency is plotted as a function of wavelength.

is 17.5–18.5 mA cm^{-2} for complex **1**, whereas it is only about 12–13 mA cm^{-2} for complex **3**. However, there is a trade-off inasmuch as the photovoltage is about 200 mV higher for the latter as compared to the former sensitizer. Nevertheless, this is insufficient to compensate for the current loss. Hence, the photovoltaic performance of complex **2** carrying two protons is superior to that of compounds **1** and **3** that contains four and no protons, respectively. The doubly protonated form of the complex is therefore preferred over the other two sensitizers for sensitization of nanocrystalline TiO_2 films.

5.2.7.3 Comparison of IPCE Obtained with Various Sensitizers

Figure 8 in chapter 5.3, shows the photocurrent action spectrum of a cell containing various sensitizers. The broad feature covers the entire visible spectrum and extending into the near IR region up to 920 nm, for $(Bu_4N)_3[RuL(NCS)_3]$ ($L' =$ 4,4′,4″-tricarboxy-2,2′ : 6′,2″-terpyridine) complex. The incident IPCE value in the plateau region being about 80% for $(Bu_4N)_3[RuL(NCS)_3]$ complex. Taking the light losses in the conducting glass into account the efficiency of electric current generation is practically 95% over a broad wavelength range extending from 400 to 700 nm. The overlap integral of the absorption spectrum with the standard global AM 1.5 solar emission spectrum for $(Bu_4N)_3[RuL(NCS)_3]$ complex yields a photocurrent density of 20.5 mA cm^2. The open circuit potential is 720 mV and the fill factor is 0.7. These results were confirmed at the National Renewable Energy Laboratory (NREL), Golden, Colorado, USA (see Chapter 5.3, Fig. 10).

The complexes $[RuL_3]$, $[RuL_2(RuL''_2(CN)_2)_2]$, and $[RuL_2(NCS)_2]$ ($L = 4,4'$-dicarboxy-2,2′-bipyridine; $L'' =$ 2,2′-bipyridine) under similar conditions, show an IPCE value between 70–80% in the plateau region. Although the IPCE values are comparable with that of $(Bu_4N)_3[RuL(NCS)_3]$, the total integrated current decreased significantly because of increasing blue shift of the spectral response of the later complexes. This shows that the ideal sensitizer should absorb most of the visible light in order to achieve high currents.

5.2.8 Synthesis and Characterization

5.2.8.1 Synthetic Strategies for Ruthenium Complexes

Several groups have synthesized various ruthenium complexes using different starting materials. Kelly, Meyer, Vos, and Decon's groups have used a polymeric starting complex $[Ru(CO)_2Cl_2]$ for stepwise synthesis of complexes containing different ligands [49, 51, 95]. Recently, the synthesis of heteroleptic complexes starting from the $[Ru(Cl)_2(dmso)_4]$ complex [54] have been reported. For homoleptic complexes Dwyer and coworkers used $K_4[Ru_2Cl_{10}O]$ in weakly acidic aqueous solution containing the appropriate ligand [96]. The same procedure has been employed with $RuCl_3 . 3H_2O$ as source material. Wilkinson's group used ruthenium blue solutions in the synthesis of variety of complexes [97]. Other groups have used $RuCl_3 . 3H_2O$ as starting material for the synthesis of complexes of the type $[RuL_3]$, $[RuL_2L'']$ and $[RuL_2(X)_2]$ in DMF [98]. The advantage of the solvent DMF is that it provides high refluxing temperature and at the same time acts as a reducing agent for the ruthenium (III).

5.2.8.2 Purification

Purity is an indispensable requirement of any sensitizer for better efficiency of the dye-sensitized solar cell. Although well worked-out procedures exist for the efficient purification of metal complexes, it was found that the isolation of complexes at its isoelectric point and followed by column purification using Sephadex LH-20 gel resulted in analytically pure samples.

5.2.8.3 Characterization

The most commonly used methods for characterization of ruthenium sensitizers are NMR, IR, Raman, UV-vis, emission, cyclic voltammetry, high-performance liquid chromatography (HPLC) and X-ray crystallography.

5.2.9 Conclusion

By appropriate synthetic methodology, it has been possible to prepare a wide range of complexes and to examine the concept of tunability of their photophysical and redox properties. Different synthetic methods to introduce diverse ligands, which have been judiciously selected to reconcile the tasks of the sensitizer, to afford vectorial electron injection into the semiconductor and efficient solar light harvesting, were discussed. Analysis of the acid-base properties of transition metal complexes containing ligands with protonatable functional groups provide important information about the nature of binding properties on the oxide surface. Comparison of the ground state and excited state pK_as provide useful data for the identification of the ligand involved in the lowest energy MLCT transition, which is essential in the design of photosensitizers for nanocrystalline solar cells. The effects exerted by the proton content of the sensitizer on both the short-circuit photocurrent and open-circuit photovoltage of dye-sensitized nanocrystalline solar cells were also shown. The molecular engineering aspect of the sensitizers for enhanced spectral response into the visible and near IR region is clearly illustrated.

The development of efficient dye-sensitized solar cells would expand our ability to better harvest the sun's energy and decrease our dependency on fossil fuels and nuclear power.

Acknowledgment

We acknowledge financial support of this work by the Swiss Federal Office for Energy (OFEN) and the Institute for Applied Photovoltaics (INAP), Gelsenkirchen, Germany. We thank Drs. R. Humphry-Baker, S. M. Zakeeruddinn, F. P. Rotzinger and K. Kalyanasundaram for their valuable discussions. Authors are indebted to Dr. P. Infelta for his assistance in handling various applications during the preparation of this chapter.

References

1. R. D. McConnell, *Future Generation Photovoltaic Technologies, AIP Conference Proceedings 404*, Denver, 1997.
2. H. W. Vogel, *Ber. Dtsch. Chem. Ges.* **1873**, *6*, 1302.
3. R. Gleria, R. Memming, *Z. Phys. Chem.* **1976**, *98*, 303.
4. W. D. K. Clark, N. Sutin, *J. Am. Chem. Soc.* **1977**, *99*, 4676.
5. A. Hamnett, M. P. Dare-Edwards, R. D. Wright et al., *J. Phys. Chem.* **1979**, *83*, 3280.
6. M. Grätzel, *Prog. Photovoltaics* **2000**, *8*, 171.
7. S. Y. Huang, G. Schlichthörl, A. J. Nozik et al., *J. Phys. Chem. B* **1997**, *101*, 2576.
8. J. van de Lagemaat, N.-G. Park, A. J. Frank, *J. Phys. Chem.* **2000**, *104*, 2044.
9. R. Argazzi, C. A. Bignozzi, G. M. Hasselmann et al., *Inorg. Chem.* **1998**, *37*, 4533.
10. K. Murakoshi, G. Kano, Y. Wada et al., *J. Electroanal. Chem.* **1995**, *396*, 27.
11. A. Zaban, S. Ferrere, J. Sprague et al., *J. Phys. Chem. B* **1997**, *101*, 55.
12. H. Sughihara, L. P. Sing, K. Sayama et al., *Chem. Lett.* **1998**, 1005.

13. C. Nasr, S. Hotchandani, P. V. Kamat, *J. Phys. Chem. B* **1998**, *102*, 4944.
14. A. Solbrand, A. Henningsson, S. Södergren et al., *J. Phys. Chem. B* **1999**, *103*, 1078.
15. K. Tennakone, G. R. R. A. Kumara, I. R. M. Kottegoda et al., *Chem. Commun.* **1999**, 15.
16. G. A. Crosby, *Acc. Chem. Res.* **1975**, *8*, 231.
17. T. J. Meyer, *Acc. Chem. Res.* **1978**, *11*, 94.
18. A. Juris, V. Balzani, F. Barigelletti et al., *Coord. Chem. Rev.* **1988**, *84*, 85.
19. K. Kalyanasundaram, *Coord. Chem. Rev.* **1982**, *46*, 159.
20. R. Krause in *Structure and Bonding*, Vol. 67, Springer-Verlag, Berlin, Heidelberg, 1987, pp. 1–52.
21. K. Kalyanasundaram, M. Grätzel, *Coord. Chem. Rev.* **1998**, *77*, 347.
22. M. K. Nazeeruddin, A. Kay, I. Rodicio et al., *J. Am. Chem. Soc.* **1993**, *115*, 6382.
23. J. J. Rack, H. B. Gray, *Inorg. Chem.* **1999**, *38*, 2.
24. H. Taube, *Angew. Chem., Int. Ed. Engl.* **1984**, *23*, 329.
25. P. S. Braterman, J. I. Song, R. D. Peacock, *Inorg. Chem.* **1992**, *31*, 555.
26. E. M. Kober, T. J. Meyer, *Inorg. Chem.* **1983**, *22*, 1614.
27. A. Hagfeldt, M. Grätzel, *Chem. Rev.* **1995**, *95*, 49.
28. D. Cahen, G. Hodes, M. Grätzel et al., *J. Phys. Chem.* **2000**, *104*, 2053.
29. D. M. Roundhill in *Photochemistry and Photophysics of Metal Complexes*, Plenum Press, New York, 1994.
30. S. Ernst, W. Kaim, *Inorg. Chem.* **1989**, *28*, 1520.
31. M. X. Tan, P. E. Laibinis, S. T. Nguyen et al., *Prog. Inorg. Chem.* **1994**, *44*, 21.
32. M. K. Nazeeruddin, E. Muller, R. Humphry-Baker et al., *J. Chem. Soc., Dalton Trans.* **1997**, 4571.
33. M. K. Nazeeruddin, K. Kalyanasundaram, M. Grätzel, *Inorg. Synth.* **1997**, *32*, 181.
34. V. Balzani, A. Juris, M. Venturi et al., *Chem. Rev.* **1996**, *96*, 759–833.
35. K. Kalyanasundaram, M. K. Nazeeruddin, *Chem. Phys. Lett.* **1992**, *193*, 292.
36. P. A. Anderson, G. F. Strouse, J. A. Treadway et al., *Inorg. Chem.* **1994**, *33*, 3863.
37. S. Ruile, O. Kohle, P. Péchy et al., *Inorganica Chimica Acta* **1997**, *261*, 129.
38. O. Kohle, S. Ruile, M. Grätzel, *Inorg. Chem.* **1996**, *35*, 4779.
39. S. M. Zakeeruddin, M. K. Nazeeruddin, P. Pechy et al., *Inorg. Chem.* **1997**, *36*, 5937.
40. M. K. Nazeeruddin, P. Pechy, M. Grätzel, *Chem. Commun.* **1997**, 1705.
41. M. K. Nazeeruddin, R. Humphry-Baker, P. Pechy et al., *10th International Conference on Photochemical Conversion and Storage of Solar Energy*, (Interlaken, Switzerland), 1994, pp. 201.
42. P. -H. Xie, Y. J. Hou, B. W. Zhang et al., *J. Chem. Soc., Dalton Trans.* **1999**, 4217.
43. R. Argazzi, C. A. Bignozzi, T. A. Heimer et al., *Inorg. Chem.* **1994**, *33*, 5741.
44. M. K. Nazeeruddin, S. M. Zakeeruddin, R. Humphry-Baker et al., *Inorg. Chim. Acta* **1999**, *296*, 250.
45. M. A. Masood, B. P. Sullivan, D. J. Hodges, *Inorg. Chem.* **1994**, *33*, 5360.
46. B. Durham, S. R. Wilson, D. J. Hodges et al., *J. Am. Chem. Soc.* **1980**, *102*, 600.
47. M. K. Nazeeruddin, S. M. Zakeeruddin, R. Humphry-Baker et al., *Coord. Chem. Rev.* **2000**, *208*, 213.
48. A. B. P. Lever, *Inorg. Chem.* **1990**, *29*, 1271.
49. J. M. Clear, J. M. Kelly, C. M. O'Connell et al., *J. Chem. Soc., Chem. Commun.* **1980**, 750.
50. D. J. Cole-Hamilton, *J. Chem. Soc., Chem. Commun.* **1980**, 1213.
51. D. S. C. Black, G. B. Deacon, N. C. Thomas, *Inorg. Chim. Acta* **1982**, *65*, L75.
52. I. P. Evans, A. Spencer, G. Wilkinson, *J. Chem. Soc., Dalton Trans.* **1973**, 204.
53. K. A. Maxwell, M. Sykora, J. M. DeSimone et al., *Inorg. Chem.* **2000**, *39*, 71.
54. S. M. Zakeeruddin, M. K. Nazeeruddin, R. Humphry-Baker et al., *Inorg. Chem.* **1998**, *37*, 5251.
55. S. M. Zakeeruddin, M. K. Nazeeruddin, P. Péchy et al., *Chem. Mater.*, communicated.
56. D. Wöhrle, D. Meissner, *Adv. Mater.* **1991**, *3*, 129.
57. C. D. Jaeger, F. F. Fan, A. J. Bard, *J. Am. Chem. Soc.* **1980**, *102*, 2592.
58. A. Giraudeau, F. Ren, F. Fan et al., *J. Am. Chem. Soc.* **1980**, *102*, 5137.
59. A. Giraudeau, F. F. Fan, A. J. Bard, *J. Am. Chem. Soc.* **1980**, *102*, 5137.
60. H. Yanagi, S. Chen, P. A. Lee et al., *J. Phys. Chem.* **1996**, *100*, 5447.
61. M. K. Nazeeruddin, R. Humphry-Baker, B. A. Murrer et al., *J. Chem. Soc., Chem. Commun.* **1998**, 719.

62. M. K. Nazeeruddin, R. Humphry-Baker, M. Graetzel et al., *J. Porphyrins and Phthalocyanines* **1999**, *3*, 230.
63. S. Ferrere, *Chem. Mater.* **2000**, *12*, 1083.
64. T. A. Heimer, C. A. Bignozzi, G. J. Meyer, *J. Phys. Chem.* **1993**, *97*, 11987.
65. M. Alebbi, C. A. Bignozzi, T. D. Heimer et al., *J. Phys. Chem.* **1998**, *102*, 7577.
66. G. Sauvé, M. E. Cass, S. J. Doig et al., *J. Phys. Chem.* **2000**, *104*, 3488.
67. C. A. Bignozzi, M. T. Indelli, F. Scandola, *J. Am. Chem. Soc.* **1989**, *111*, 5192.
68. Y. Lei, T. Buranda, J. F. Endicott, *J. Am. Chem. Soc.* **1990**, *112*, 8820.
69. C. A. Bignozzi, R. Argazzi, C. Chiorboli et al., *Coord. Chem. Rev.* **1991**, *111*, 261.
70. G. Denti, S. Serroni, S. Campagna et al., in *Perspectives in Coordination Chemistry*, (Eds.: A. F. Williams, C. Floriani, A. E. Merbach), Helvetica Chimica Acta Verlag and VCH, Basel, Switzerland and Weinheim, Germany, 1992, p. 153.
71. K. Kalyanasundaram, M. K. Nazeeruddin, *Inorg. Chim. Acta* **1994**, *226*, 213.
72. K. Matsui, M. K. Nazeeruddin, R. Humphry-Baker et al., *J. Phys. Chem.* **1992**, *96*, 10 587.
73. F. Scandola, M. T. Indelli, C. Chiorboli et al., *Top. Curr. Chem.* **1990**, *158*, 73.
74. M. K. Nazeeruddin, M. Grätzel, K. Kalyanasundaram et al., *J. Chem. Soc., Dalton Trans.* **1993**, 323.
75. M. K. Nazeeruddin, P. Liska, J. Moser et al., *Helv. Chim. Acta* **1990**, *73*, 1788.
76. M. K. Nazeeruddin, R. Humphry-Baker, M. Grätzel, *J. Phys. Chem.*, communicated.
77. K. Kalyanasundaram, M. Grätzel, M. K. Nazeeruddin, *Inorg. Chem.* **1992**, *31*, 5243.
78. C. A. Bignozzi, R. Aragazzi, C. J. Kleverlaan, *Chem. Soc. Rev.* **2000**, *29*, 87.
79. S. A. Trammell, J. A. Moss, J. C. Yang et al., *Inorg. Chem.* **1999**, *38*, 3665.
80. S. G. Yan, J. T. Hupp, *J. Phys. Chem.* **1996**, *100*, 6867.
81. G. Will, G. Boschloo, S. Nagaraja Rao et al., *J. Phys. Chem. B* **1999**, *103*, 8067.
82. B. Lemon, J. T. Hupp, *J. Phys. Chem. B* **1999**, *103*, 3797.
83. B. Jing, H. Zhang, M. Zhang et al., *J. Mater. Chem.* **1998**, *8*, 2055.
84. S. Anderson, E. C. Constable, M. P. Dare-Edwards et al., *Nature* **1979**, *280*, 571.
85. K. Murakoshi, G. Kano, Y. Wada et al., *J. Electroanal. Chem.* **1995**, *396*, 27.
86. K. S. Finnie, J. R. Bartlett, J. L. Woolfrey, *Langmuir* **1998**, *14*, 2744.
87. A. Fillinger, B. A. Parkinson, *J. Electrochem. Soc.* **1999**, *146*, 4559.
88. R. Argazzi, C. A. Bignozzi, T. A. Heimer et al., *Inorg. Chem.* **1997**, *36*, 2.
89. P. Péchy, F. P. Rotzinger, M. K. Nazeeruddin et al., *J. Chem. Soc. Chem. Commun.* **1995**, 65.
90. M. K. Nazeeruddin, S. M. Zakeeruddin, R. Humphry-Baker et al., *Inorg. Chem.* **1999**, *38*, 6298.
91. K. Kalyanasundaram, M. K. Nazeeruddin, *Inorg. Chem.* **1990**, *29*, 1888.
92. M. K. Nazeeruddin, S. M. Zakeeruddin, R. Humphry-Baker et al., *Inorg. Chem.* **2000**, *39*, 4542.
93. M. Montalti, S. Wadhwa, W. Y. Kim et al., *Inorg. Chem.* **2000**, *39*, 76.
94. M. Amirnasr, M. K. Nazeeruddin, M. Grätzel, *Thermochim. Acta* **2000**, *348*, 105.
95. P. A. Anderson, G. F. Strouse, J. A. Treadway et al., *Inorg. Chem.* **1994**, *33*, 3863.
96. F. P. Dwyer, J. E. Humpoletz, R. S. Nyholm, *J. Proc. Roy. Soc. N. S. Wales.* **1946**, *80*, 212.
97. D. Rose, G. Wilkinson, *J. Chem. Soc. A* **1970**, 1791.
98. R. Krause in *Structure and Bonding*, Vol. 67, Springer-Verlag, Berin, Germany, 1987, pp. 1–52.

5.3 Charge Transport in Dye-sensitized Systems

Jenny Nelson
Imperial College of Science, Technology and Medicine, London, United Kingdom

5.3.1 Introduction

5.3.1.1 Dye-sensitized Systems for Photovoltaics

In this chapter, "Dye-sensitized System" is used to refer to the electrochemical junction, or heterojunction between a porous, wide band gap semiconducting film, which has been sensitized for visible light by an adsorbed dye species and a hole conducting medium. By far the most important application is the dye-sensitized solar cell (DSSC), which has been responsible for much of the research aimed at understanding and optimizing these systems. [1–3].

Charge transport in DSSCs is particularly intriguing because of the long lifetimes and long diffusion lengths for charge-separated species in these systems. In other photovoltaic devices, based on organic materials, diffusion lengths for charge-separated species are typically tens or hundreds of nm, severely limiting the quantum efficiency. In the DSSC, photocarrier diffusion lengths are of the order of microns, in spite of the proximity of positive and negative charge carriers. The improvement is in part due to the function of the sensitizer, which separates the process of photoinduced charge separation from the main recombination process. These features give the DSSC its remarkably high light to electric current efficiency first reported in 1988 [4], and has enabled the development of 10% efficient solar cells.

Dye-sensitized systems are not suited to interpretation by conventional device physics methods on account of two features: firstly, the mesoscopic phase separation of electron and hole conductors, which makes the porous material unable to sustain large electric fields; and secondly, the separation, through the use of sensitizers, of optical absorption from charge transport in either material. Efforts to understand the photovoltaic action of the DSSC are leading to a reassessment of basic principles and the possibilities of novel photovoltaic designs.

5.3.1.2 Dye-sensitized Systems for Other Applications

Other applications for dye-sensitized systems are emerging. A particular example is the dye-sensitized photoelectrochromic window, where the absorption of light in a dye-sensitized electrode leads to charging of a counter electrode, resulting in a change in its optical density [5, 6]. Novel, indirectly related applications are emerging in biotechnology, where the sensitization of semiconducting films with proteins, leads to an electronic signal in response to the biochemical action of the protein [7]. Charge transport in mesoporous systems is relevant to non-dye-sensitized applications, such as lithium intercalation batteries [8] and electrochromics [9].

Another aspect is the use of sensitized systems to study the properties of either the sensitizer or the film. For example, the use of transparent porous TiO_2 films to study the optoelectronic properties of PbS quantum dots (QD) [10] and the use of sensitizers to study electron dynamics in the nanocrystalline film [11].

This chapter is set out as follows: in Sect. 5.3.2 we summarize the structure and function of a dye-sensitized junction; Sect. 5.3.3 reviews the material properties of the electron and hole conducting components; Sect. 5.3.4 reviews some basic theoretical principles and attempts to develop a "device physics" for dye-sensitized devices; Sect. 5.3.5 reviews the experimental technique for studying charge transport and discusses the main results and the models that have been proposed; finally some points of consensus and outstanding questions are summarized in Sect. 5.3.6.

5.3.2 Structure and Function of Dye-sensitized Junctions: An Overview

5.3.2.1 General Structure of Dye-sensitized Electrochemical Junction

Dye-sensitized devices are based on an electrochemical junction or heterojunction, containing three media: electron conductor, sensitizer, and hole conductor. Light absorbed in the sensitizer causes the injection of an electron into the electron conductor and a positive charge into the hole conductor. Electron and hole conductor are separately connected through selective contacts to the terminals of the device. Charges successfully reaching the external circuit as a result of illumination constitute the photocurrent. In addition to light-induced charge transfers, we need to consider bias-induced, dark charge-transfer reactions. In the case of a solar cell at operating point, light forward biases the junction to stimulate a reverse current, which opposes the photocurrent. In a dye-sensitized device, unusually, this reverse current contains both light and dark components. The key charge transfer reactions are illustrated in Fig. 1.

Key Processes

Light absorption:

$$h\nu + \mathrm{D} \longrightarrow \mathrm{D}^* \tag{1}$$

Light-induced forward electron transfers:

$$\mathrm{D}^* \longrightarrow \mathrm{e(sc)} + \mathrm{D}^+ \tag{2}$$

$$\mathrm{HC_{Red}} + \mathrm{D}^+ \longrightarrow \mathrm{HC_{Ox}} \tag{3}$$

$$\text{transport of e(sc)} \tag{4}$$

$$\text{transport of } \mathrm{HC_{Red}} \text{ and } \mathrm{HC_{Ox}} \tag{5}$$

$$\mathrm{e(sc)} \longrightarrow \mathrm{e(n)} \tag{6}$$

$$\mathrm{e(p)} + \mathrm{HC_{Ox}} \longrightarrow \mathrm{HC_{Red}} \tag{7}$$

Dark reverse electron transfers:

$$\mathrm{e(sc)} + \mathrm{HC_{Ox}} \longrightarrow \mathrm{HC_{Red}} \tag{8}$$

$$\mathrm{e(n)} \longrightarrow \mathrm{e(sc)} \tag{6*}$$

$$\mathrm{HC_{Red}} \longrightarrow \mathrm{HC_{Ox}} + \mathrm{e(p)} \tag{7*}$$

Light induced reverse electron transfers:

$$\mathrm{e(sc)} + \mathrm{D}^+ \longrightarrow \mathrm{D} \tag{9}$$

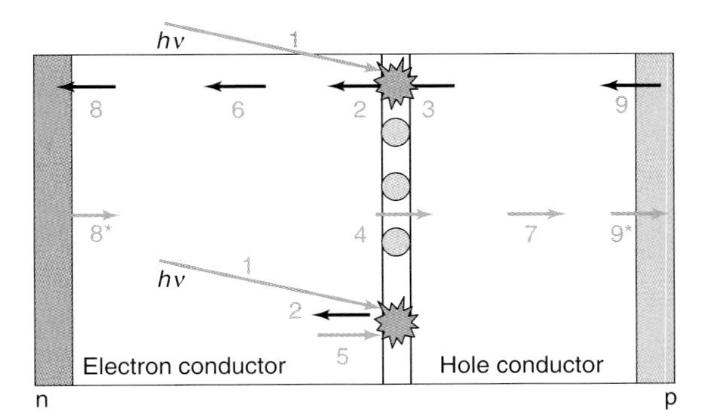

Fig. 1 Key electron transfer reactions in a dye-sensitized device.

where D, D^*, and D^+ refer to the dye, the excited state of the dye, and the dye cation; e(sc), e(n), and e(p) refer to the electron in the semiconductor (electron conductor), in the n-contact and in the p-contact material; and HC_{Red} and HC_{Ox} refer to the reduced and oxidized state of the hole conductor. Following charge transfer, electrons and holes are transported through the respective mediums by electron conduction (6) and hole conduction (7). We have assumed that hole transport is more appropriately described as a redox process and electron transport as a simple electron conduction. In discussing the net transport of electrons, we need to include processes (4), (5), and (6), whereas net hole transport involves (3), (4), and (7).

In addition, there is the energy transfer process because of relaxation of the excited state of the dye

$$D^* \longrightarrow D + h\nu \text{ or heat} \qquad (1^*)$$

which does not result in a forward electron transfer and may be counted with the reverse electron transfers as a loss process.

We define forward as the direction of charge transfer under photovoltaic operating conditions, and will use "forward" bias to mean the direction of potential difference in those conditions (p terminal positive with respect to n). Note that this condition corresponds to negative applied potentials in electrochemical terms.

5.3.2.2 Energetic Structure

In a functioning dye-sensitized device, the net current output is the sum of all the forward electron transfer processes less all of the reverse processes. For a net forward current under (steady state) illumination, processes (2) and (3) must be kinetically more favorable than (4), and (3) must be preferred to (5). These conditions enable the continual photoionization and regeneration of the sensitizer under steady state illumination. In energetic terms, the Gibbs free energy of the excited sensitizer, D^*, plus the vacant electron acceptor state in the semiconductor, should exceed that of the dye cation plus the occupied acceptor state, $D^+ + e(TiO_2)$. This is usually represented on a simplified potential energy diagram as a sensitizer excited state of potential energy higher than the conduction band of the semiconductor. Similarly, the redox potential μ_{RedOx} of the redox couple in the hole conductor should be sufficient to reduce the dye cation. This is usually represented as a dye ground state of potential energy which is lower than the redox potential of the electrolyte. These relationships are illustrated in Fig. 2 together with the energetically preferred directions for charge transfer. Such potential energy diagrams do not normally distinguish the entropic contribution to the driving force, which arises from the much greater density of states per unit surface area in the semiconductor than in the dye. This effect is estimated to contribute up to 0.1 eV to the driving force [12]. Contact materials should allow facile transfer of electrons from electron conductor and positive charge from electrolyte.

5.3.2.3 Physical Structure and Materials

In the most commonly studied configuration of the DSSC, the electron conductor is a wide band gap, nanocrystalline, metal oxide film. Colloidal titanium dioxide is most often used although other wide band gap metal oxides are candidates. The sensitizer is a transition metal-based (usually ruthenium), organic dye, with excited state free energy sufficient to reduce the semiconductor, and containing ligands, such as carboxylates or phosphonates, which facilitate bonding to the semiconductor

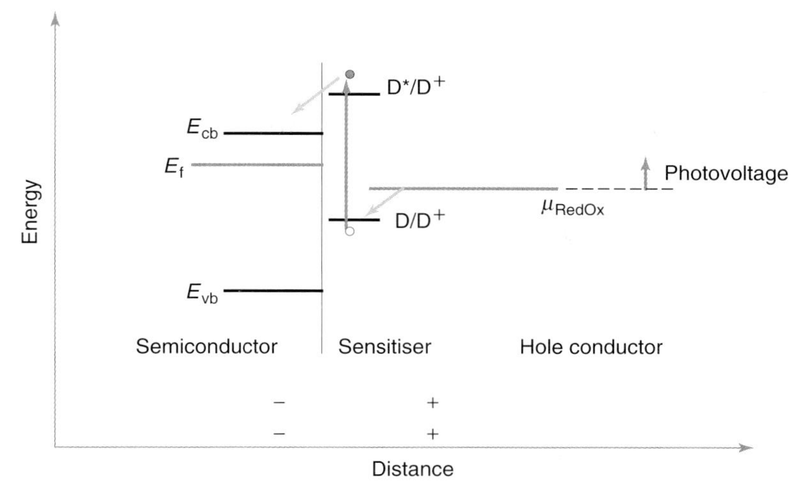

Fig. 2 Potential energy diagram of DSSC. For net forward electron transfer the oxidation potential of the dye-excited state, D^*/D^+ should be higher than the semiconductor conduction band, and the redox potential of the hole conductor should be higher than the dye-ground state potential D/D^+. Absorption of light by the dye causes charge separation across the interface, resulting in a splitting of the electrochemical potential and hence a photovoltage.

surface. The hole conductor is traditionally a polar liquid electrolyte, such as propylene carbonate or acetonitrile, containing the iodide–triiodide redox couple, although latterly solid-state ionic and hole conductors have been substituted. In many cases, the electrolyte contains a lithium salt, which facilitates electron accumulation in the TiO_2 and apparently improves charge separation [13]. In the TiO_2–liquid electrolyte system, ohmic contact is made to the oxide through a fluorine doped SnO_2 conducting glass substrate, and to the electrolyte through a platinum coated counter electrode, which catalyzes the regeneration of the redox couple. Material properties and requirements are described here and further details are given in Refs. 2, 14.

A unique feature of DSSCs is that the electron conductor (the oxide film) is both porous and electronically connected. The morphology of a typical TiO_2 film is illustrated in Fig. 3. Crystallites of 5–20 nm in size are sintered together into a random porous network, leaving a network of voids of similar volume fraction and dimensions. In the DSSC, the dye is adsorbed on to the entire film surface, whereas the voids are filled with the hole conductor. The enormous microscopic surface area, which is typically 1000 times the geometrical surface area for a 10-μm-thick film, leads to a much higher optical depth than for a planar interface, whereas the pores allow redox ions to travel between all points on the surface and the counter electrode [2]. Figure 4 illustrates the typical device structure of the DSSC.

5.3.2.4 Charge Carriers and Charge Transfer Processes

The charge transfer processes are summarized here. In the dark, at equilibrium, there is no net charge transfer. When a forward bias (negative potential) is applied

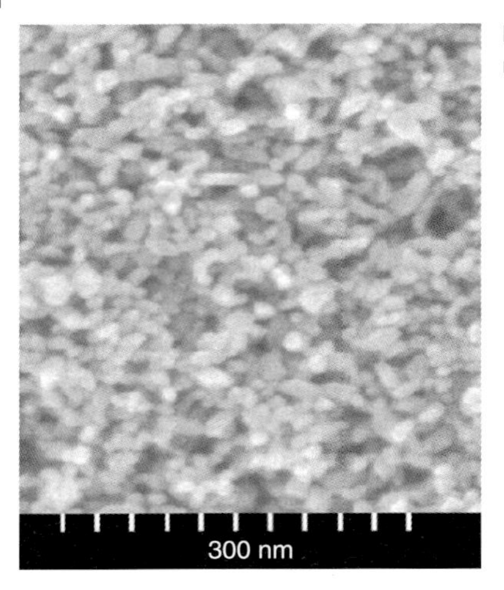

Fig. 3 Scanning electron micrograph of film surface (from Ref. 15).

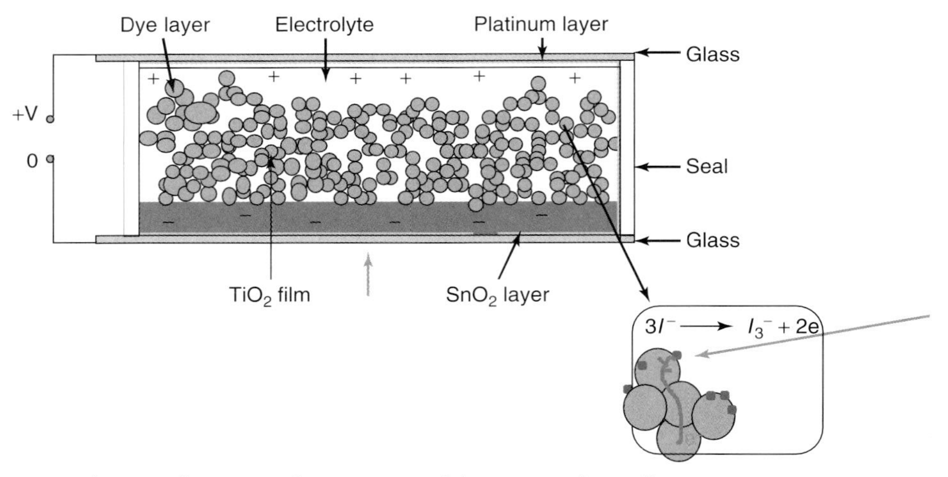

Fig. 4 Schematic illustration of arrangement of dye-sensitized PV cell. Electron conduction to the conducting glass contact proceeds through the TiO_2, whereas hole conduction by the liquid electrolyte proceeds through the pores of the TiO_2 film.

in the dark, electrons flow from substrate into the film and oxidized species from counter electrode toward the interface. Electrons and oxidized species combine at the interface through process (4), which in the iodide–triiodide redox couple can be summarized as

$$e_{TiO_2} + \frac{1}{2}I_3^- \longrightarrow \frac{3}{2}I^-. \qquad (10)$$

Probable intermediate reaction schemes are detailed in Ref. 16 and the solution phase redox couple has been probed in depth [17]. In the steady state, ion populations are maintained by the oxidation of I_3^- at the counter electrode, which delivers the electron to the external circuit. This recombination process is responsible for the dark current from the junction.

Under illumination, an absorbed photon results in the injection of an electron into the oxide film (2). The electron is transported through the film to the substrate (6). The reduced form of the hole conductor (I^-) restores the electron to the dye cation (3), summarized by the reaction

$$\frac{3}{2}I^- + D^+ \longrightarrow D + \frac{1}{2}I_3{}^-. \qquad (11)$$

The reaction may be detailed by the sequence:

$$D^+ + I^- \longrightarrow D + I\bullet$$
$$I\bullet + I^- \longrightarrow I_2\bullet^-$$
$$2I_2\bullet^- \longrightarrow I_3{}^- + I^-$$

The ($I_3{}^-$) is transported to the counter electrode (7) and is there reduced to regenerate I^- through the donation of electrons from the external circuit. This charge transfer constitutes the photocurrent. Some electrons are lost through recombination with the oxidized form of the hole conductor (4) or the dye cation (5) before being collected. Both recombination processes amount to the light-induced recycling of charge around the interface. Both are accelerated by the light driven forward biasing of the junction, leading to kinetic competition for the dye cations ((3) versus (5)) and, effectively, for the photoinjected electrons ((4) and (5) versus transport, (6)). Note that, for the iodide–triiodide redox couple, the reaction schemes for (3) and (4) are not necessarily first order and cannot be described by a rate constant. The rates of these processes depend on the concentration of dye cations (i.e. illumination) and electrons (i.e. applied bias), leading to illumination and bias dependent reaction kinetics.

The primary *species* involved in charge transport are thus electrons in the oxide film and the reduced and oxidized forms of the hole conductor. The extremely high surface area means that interfacial electron transfer reactions and electrostatic interactions between species in the two conducting media are particularly important. In addition, the following secondary species may influence charge transport:

- *Cations in the hole conductor.* These influence the electrostatics of the system by reorganizing in response to motion of electrons in the oxide film (which may improve electron mobility), and by screening redox species from electrons.
- *Intercalated or surface adsorbed species in the electron conductor.* Protons and Li ions, if present in the hole conductor, may be capable of intercalating into the oxide film, affecting both the potential energy for electron stabilization and the ease of electronic conduction. Surface absorbed cations influence electronic states near the surface. Adsorption of OH^- groups influences electronic states and proton transport. Water molecules, which are expected to be present around the oxide surface even in dry solvent, may be dissociatively adsorbed leaving adsorbed OH species and mobile protons, which influence electronic states and conductivity.
- *Charge hopping between dye cations.* To a first approximation, dyes are normally assumed to be immobilized on the surface and function only to inject charge to other media. However, there is limited evidence for hole migration between dyes, as discussed later.
- *Holes.* In the case of band gap illumination, holes will be generated in the electron conductor. These may act

as recombination centers for electrons (a loss process), or may be scavenged by the hole conductor (a photovoltaic effect). However, the wide band gap means that holes are seldom generated, and except in the case of ultraviolet (UV) irradiation, can be ignored.

5.3.2.5 Potential Distribution

To sustain the forward electron transfer described earlier, the hole-conducting material should have a lower effective work function than the electron conductor. As in any electronic junction, the electrochemical potential or Fermi level in the two materials aligns at equilibrium, leaving a potential difference equal to the difference in work functions to be accommodated in the two materials. In a conventional electrochemical junction, the charge density in the layer of electrolyte close to the interface (the Helmholtz layer) is much greater than the fixed charge density in the semiconductor, and the potential difference is accommodated mainly in the semiconductor by band bending. In a lightly doped semiconductor, this band bending extends over a depletion region some microns in width, beyond which the semiconductor is neutral.

In the porous geometry of the DSSC, the situation is different. The high charge density of the electrolyte means that the electrolyte tends to maintain the outer surface of the metal oxide film at a constant potential, and individual nanoparticles making up the film are too small to sustain the junction potential difference. Calculations using the Albery–Bartlett model indicate that only some 50 mV of potential can be dropped between the surface and center of a 10-nm radius particle of (lightly *n*-type) TiO_2 [18–20]. Therefore, if the film surface is an equipotential, large electric fields cannot be sustained within the porous material. This leaves two regions where the potential can be dropped: at the interface between the TiO_2 film and the $SnO_2:F$ (fluorine doped SnO_2) rear conducting contact; and across the interface layer between film and electrolyte, which includes outer atomic layers of film and Helmholtz layer. It is most likely that the potential drop is shared between the two interfaces [19, 20]. Bisquert has shown by solving Poisson's equation for a columnar model of the film that the largest electric fields within the film are established at the interface with the SnO_2 substrate [19]. Large fields cannot extend beyond a few nanoparticle radii from the rear substrate (Fig. 5). The same conclusion was reached earlier with more qualitative arguments in Refs. 21, 22. The proportion of the potential difference that can be taken up by the film and rear substrate is determined by morphology and materials properties; the remainder must be taken up by the Helmholtz layer [23, 12], leading to large electric fields at the interface, extending only a few nm into the electrolyte [24], and a film which is largely depleted of electrons in equilibrium. The size of the potential drop at the Helmholtz layer is believed to be influenced by dye adsorption, which polarizes the surface layer by introducing protons, and by the redox potential of the electrolyte. Under applied reverse bias, the potential drop increases and an increasing proportion is taken up in the Helmholtz layer. When the junction is forward biased either electrically or by illumination, the concentrations of both electrons in the film and oxidized species ("holes") in the electrolyte increase. The applied bias will be taken up, in general, by both the TiO_2–TCO (transparent conducting oxide) and the TiO_2–hole conductor interface layers, reducing the potential difference and modifying the electric field at those interfaces. However, according to Cahen and coworkers, neither potential drop is sufficient on its own to account for the open circuit voltage of the cell, and this is instead controlled by the difference between the redox potential of the electrolyte and the conduction band of the TiO_2 [12]. This is an important distinction from conventional photovoltaic devices, where the difference in work functions controls V_{oc}. It is supported by studies by Pichot and Gregg of the photovoltage of cells deposited on different work function substrates, which indicate that the photovoltage is not influenced by the potential drop at the TiO_2–TCO interface [25].

A key consequence of the potential distribution in the cell is that once separated, the electrons and holes in their respective media do not experience large electrostatic forces because of the junction, and are not driven primarily by built-in electric fields.

An additional effect of applied bias is the reorganization of solvent ions to minimize the energy of the system. A more negative charge on the TiO_2 electrode will attract cations toward the film surface, where they act to screen and stabilize electrons in the film. Very small cations such as Li^+, which are able to intercalate into the interstices of the crystal structure, may be drawn in. These processes clearly modify the electrostatic potential distribution. In the steady state, they appear as changes in the effective transport parameters; in transient studies, currents due to ion movement may be resolved [26].

5.3.2.6 Relevance to Device Performance

Key requirements for good performance in any photovoltaic device are high light absorption, efficient charge separation, and efficient charge transport. In the DSSC, light absorption is ensured by the large optical depth of the sensitized porous film. Efficient charge separation in DSSC is provided by the much faster rate of injection (2) relative to excited state decay, and is discussed elsewhere in this Volume. Efficient charge transport is therefore the key. Having identified the species and

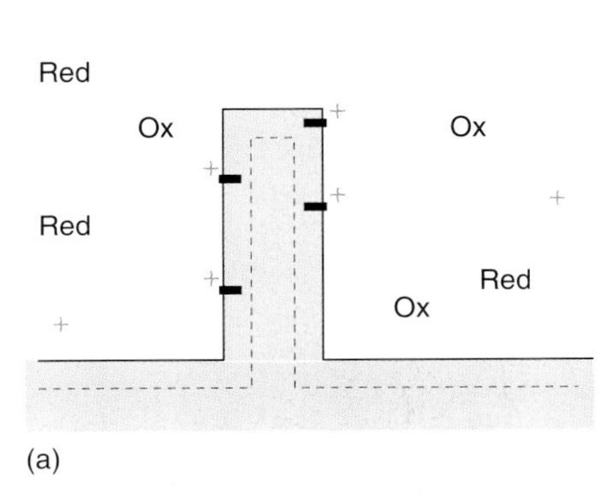

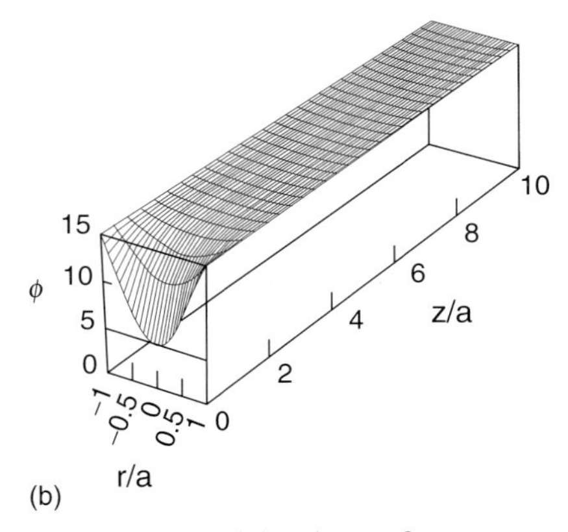

Fig. 5 (a) Schematic illustration of film surface, showing lines of constant potential. Reorganization of cations in the electrolyte allows the screening of electrons in the film, reducing their electrostatic effect. (b) Calculated electrostatic potential distribution for a columnar electrode with constant surface potential, showing how significant potential drops can be maintained only within a few tens of nm from the rear substrate (from Ref. 20).

processes involved in charge transport in dye-sensitized systems, we may now identify a number of relevant issues for device performance:

- *Recombination.* Because of the porous structure, the interfacial electron transfer processes discussed earlier compete with transport at all times. This limits both the cell voltage and the cell current at positive biases.
- *Resistive losses.* Current flow leads to gradients in the quasi Fermi levels. In the light this causes a drop in quasi Fermi level separation between interface and terminals. Increased resistance through charge trapping increases this loss.
- *Electrostatics.* Mobile electrolyte ions reorganize to effect charge screening and compensate charging of nanoparticles. This may improve conduction but at possible cost to the cell voltage.
- *Leakage.* This includes shunting of charges across the edges of the sample and charge transfer directly from working electrode to electrolyte.

5.3.3 Material Properties

5.3.3.1 Electron Conductor

The porous semiconducting film used in the DSSC should be transparent to visible light, n-type, stable, and have a large surface area. Many materials satisfy these requirements in principle, but in practice TiO_2 is by far the best studied and most widely used. Usually the porous films comprise colloidal particles of anatase TiO_2 some 10–20 nm in diameter assembled into a random porous network with a porosity of typically 50–60% [27, 28]. Particle and pore size can be varied by modifications to synthesis and film deposition. Films are usually deposited on glass substrates coated with a fluorine doped SnO_2 conducting layer ($SnO_2:F$), which is preferred to Indium-Tin Oxide (ITO) for the robustness of its conductive properties under heating. Alternative routes to porous TiO_2 films, more amenable to large-scale production, are being studied. The particle synthesis and film preparation is discussed elsewhere [14, 29, 27], as are the electronic and structural properties of the material [30–34]. Here we will focus only the material and morphological features relevant to charge transport.

1. TiO_2 is a *polar* material. Valence band states are associated with the valence orbitals of the O^{2-} ion, whereas conduction states are associated with the 3d orbitals of the Ti^{4+} ions. Band gap excitation corresponds to a shift in electron density from O to Ti sites.
2. Colloidal particles are crystalline and tend to adopt the *anatase* polymorph, rather than the more widespread and widely studied rutile (this has been confirmed by Raman and XRD studies [27, 28]). The preference for anatase at nanometer sizes has been attributed to surface tension arguments [35]. Anatase is less dense than rutile, with a different arrangement of TiO_6 octahedra in the unit cell, which leaves open channels parallel to the c-axis, and may assist intercalation [30, 34, 36]. Anatase has a band gap of 3.2 eV in the bulk, slightly larger than rutile, which makes it superior for dye-sensitized photovoltaic applications (less loss through absorption of blue light).
3. Random orientation of different crystallites leads to grain boundaries at the interfaces, which may reduce electronic mobility through enhanced scattering, small cross sectional area

and, possibly, through space charge induced potential barriers. However quantum confinement effects are not expected because the typical crystallite radius greatly exceeds the exciton radius (~2 nm [18]). Space charge effects are also expected to be small on account of the low doping density of the crystallites.

4. The material is nonstoichiometric and has a natural tendency to acquire *oxygen vacancies* through loss of O_2, in an oxygen-poor environment:

$$TiO_2 \longrightarrow TiO_{2-x} + xV_O + \tfrac{1}{2}xO_2 \qquad (12)$$

where V_O represents an oxygen vacancy and x the fraction of oxygen vacancies in the film. The rate of O_2 effusion can be controlled by the partial pressure of oxygen and temperature. Considered as a point defect, the oxygen vacancy contains two electrons in its ground state but is readily ionized to deliver one or both electrons to empty 3d orbitals on adjacent Ti sites, changing the Ti oxidation state from Ti^{4+} to Ti^{3+} and increasing the population of electrons available for conduction [30]. The electron in the Ti^{3+} defect state is localized but easily accessible to the conduction band. Calculations [37] and experiment [38] place its energy around 2–3 eV above the valence band edge in the case of rutile. Functionally, these states are important because electrons trapped in these defect levels are readily promoted to conduction band states, where they contribute to electrical conduction, whereas unoccupied states act as traps for conduction electrons. Oxygen vacancies thus effectively dope the material n-type. The control of conductivity through stoichiometry is well established for rutile, and exploited for applications in gas sensing [39].

5. The enormous surface area leads to a high-volume density of surface-related defects. Apart from oxygen vacancy induced Ti^{3+} states, these include defects due to surface adsorption of ions and intercalated species, particularly in the presence of an electrolyte. Surface adsorbed cations may lead to intraband gap states, similar in nature to the oxygen vacancy [40], and intercalated lithium leads to a similar intraband gap density of states, as observed by photoemission spectroscopy [41]. Such defects can trap electrons. The profusion of defects leads to the formation of a surface layer of different electronic structure to the bulk.

6. Under UV illumination, TiO_2 is highly hydrophilic and attracts a surface layer of hydrogen bonded H_2O molecules. In certain conditions [42, 43], the water dissociates to leave surface adsorbed OH groups – which may create electron traps – and itinerant protons. Protons intercalate easily and can strongly influence conductivity by assisting the formation of Ti^{3+} states [19, 44].

7. Ti^{3+} defect states have a characteristic red-infrared optical absorbance [45, 46, 19]. This can be used as a measure of the density of electrons accumulated in the film [47–50]. Although there is some debate about whether the states giving rise to the absorbance are delocalized conduction band states or intraband trap states, the evidence favors the latter [19, 51].

8. The nanocrystalline films are insulating in normal conditions, with a dark resistivity of 10^8–10^{10} Ω cm in air. This is substantially larger than

the dark resistivity of single crystal or polycrystalline anatase [32, 52] or rutile [53]. Conductivity can be increased by reducing the films by heating in vacuum (introducing oxygen vacancies) [32, 54] or by doping with lithium or other electron donors. There also appears to be some sensitivity to ambient water content, indicating a possible contribution from proton doping or protonic conductivity [44, 55]. Definitive results on the conductivity of "dry" films is rare, given the high resistivity and sensitivity to ambient conditions. However, it appears that conduction is thermally activated with an activation energy between 0.8 eV [56] and 1.6 eV [57]. Loss of thermal activation in strongly reduced polycrystalline anatase has been attributed to a metal-insulator like transition, which is specific to anatase [32], although the equivalent observation in nanocrystalline films is yet to be made. Because the material is polar, charge transport within crystallites is likely to be by small polaron hopping. In rutile, an activation energy of around 60 meV is observed and attributed to the hopping barrier [53]; similar observations for nanocrystalline films have not been reported.

9. Under band gap illumination the films become photoconducting, with DC conductivity increasing by several orders of magnitude [2, 57]. Sub-band gap illumination causes a small photocurrent [55] and photovoltaic effect [58] in dry films and a small photoconductivity in electrolytically contacted films [19], suggesting excitation of carriers out of trap states.
10. Transient photocurrent and dark current measurements are strongly dispersive, leading to difficulties in determining an electronic mobility. Dittrich and coworkers find a mobility of around 10^{-5} cm^2 V^{-1} s^{-1} at room temperature in an oxygenic atmosphere, using time of flight techniques [59]. Konenkamp and Hoyer [60] use junction recovery transients to derive a mobility lifetime product of around 10^{-10} cm^2 V^{-1}, which leads to injection dependent mobilities of 10^{-7}–10^{-5} cm^2 V^{-1} s^{-1}. These values are substantially smaller than the electronic mobility in crystalline anatase of around 1 cm^2 V^{-1}s^{-1} [52].

It is apparent that trap states influence conduction in several ways: through dispersive photocurrent and dark current transients; strongly space charge limited transport [59]; and sub–band gap photoconductive and photovoltaic effects.

5.3.3.1.1 **ZnO and Other Oxides** Alternative electron conductors to TiO_2 have been considered. For efficient function, the conduction band edge should lie between the conduction band of the substrate and the excited state potential energy of the dye. A wide range of oxides are potential materials but only a few (ZnO, SnO_2, and $SrTiO_3$) have been studied for dye-sensitized applications. Some oxides, such as $SrTiO_3$ offer the possibility of higher open-circuit voltages through closer alignment of sensitizer excited state levels with conduction band edge [61]. Others such as ZrO_2, are ruled out by a low electron affinity, which places the conduction band edge above the excited state potential of most dyes. This is illustrated in Fig. 6.

By far the best studied of these alternative materials is ZnO [18, 63–65]. Nanocrystalline ZnO can be prepared by sol-gel routes and sintered into porous

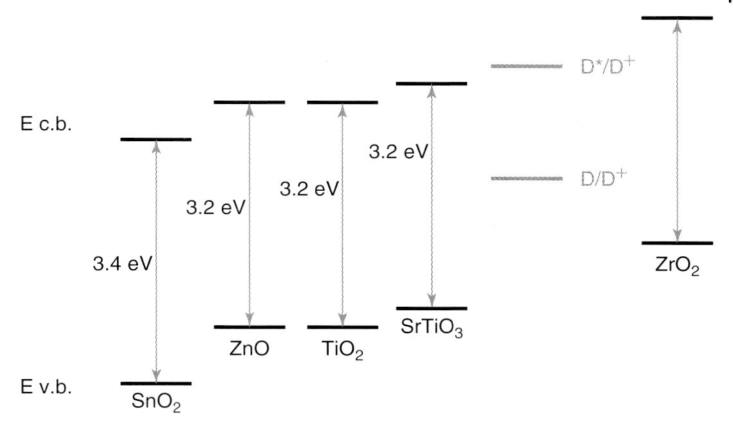

Fig. 6 Schematic energy band diagram of suitable and unsuitable metal oxides for dye sensitization. Electron injection from dye-excited state to conduction band is energetically favorable for SnO_2, ZnO, and $SrTiO_3$, as well as TiO_2, but forbidden for ZrO_2 [2, 62].

films on TCO substrates (usually $SnO_2:F$ or ITO) by similar procedures. It is similar in band gap (3.2 eV) and band edge position to TiO_2 [2] with similar (13 nm diameter) or smaller (5 nm) crystallite sizes than for typical TiO_2 films. As with TiO_2, it may be self doped n-type through oxygen deficiency.

Charge transport in nanocrystalline ZnO is less well studied than in TiO_2, but we may make a few relevant comments:

1. Nanocrystalline ZnO is highly nonstoichiometric with a defect density of up to 10^{18} cm^{-3}. This leads to a higher electronic mobility than TiO_2, of over 10^{-3} $cm^2\ V^{-1}\ s^{-1}$ [65].
2. The oxygen vacancy defects appear to function as deep traps for holes and recombination centers for electrons and shallow traps for electrons [66] (the limited dispersion of time constants in transient current experiments has been attributed to the absence of deep electron traps). No optical absorbance is associated with occupied defects but instead a bleaching of the absorption edge (Burstein shift) can be detected as an indicator of electron accumulation [18, 63].
3. The exciton radius is, at around 5 nm, larger than in TiO_2 (on account of a smaller effective mass), and may lead to quantum confinement effects in nanocrystalline films [18, 65, 67]. These affect the ease of transport between particles, leading to barriers between large and small particles for polydisperse particle size distributions.
4. There is some evidence for a barrier at the ZnO/ITO contact [67].
5. ZnO is not capable of intercalating Li ions.

5.3.3.1.2 ***p*-type Metal Oxides** Apart from the sensitization of n-type semiconductors with electron injecting dyes, there is, in principle, the possibility of sensitization of a p-type semiconductor with a hole-injecting dye. The requirement for hole injection is that the potential of the ground state of the dye lies below the valence band of the semiconductor. There has been some interest in nanoporous

Cu_2O, which can be sensitized with methylviologen [68], and NiO.

5.3.3.2 Hole Conductors

5.3.3.2.1 Liquid Electrolytes The usual hole-conducting medium is a polar solvent, such as propanol carbonate, ethylene glycol, or acetonitrile, containing the iodide–triiodide redox couple. In such liquid electrolytes the main issue for charge transport is the mobility of the iodide and triiodide ions through the pores of the structure. The relevant comparison is between the ionic conductivity of the electrolyte and the electronic conductivity of the TiO_2. Bulk diffusion coefficients for I^- and $I_3{}^-$ are typically 10^{-6} cm^2 V^{-1} s^{-1}, although the value for $I_3{}^-$ in particular is strongly dependent on electrolyte [69–71]. Diffusion coefficients in solid-state ion conducting electrolytes can be one to two orders of magnitude smaller [70]. The effect of the porous medium on triiodide diffusion has been studied, and is found to reduce the diffusion coefficient by an order of magnitude [72], which is still insufficient to make triiodide diffusion the limiting factor in current transport. Concentration of redox species in the electrolyte is typically 0.1 to 0.5 M, which mean a much higher charge density than in the semiconducting film at equilibrium. These values lead to ionic conductivities substantially larger than the electronic conductivity in the porous films and so ionic diffusion is normally disregarded in analysis of charge transport. However, at high injection levels where the electronic mobility improves, ionic conductivity may become rate limiting. Evidence for the effect of electrolyte on charge transport is inconclusive: it is concluded in Ref. 73 that ionic diffusion is not rate limiting, from studies showing no effect of electrolyte viscosity on current transients, whereas Solbrand and coworkers find, from similar experiments, that an effective diffusion coefficient derived from photocurrent transients increases with electrolyte conductivity [74]. It has been pointed out by Kopidakis and coworkers [75] that ambipolar diffusion is expected to apply in dye-sensitized systems, and although electron diffusion appears to dominate transport, the ambipolar model allows for some influence of electrolyte diffusion coefficient on transport.

Redox anions in the electrolyte must be charge balanced by cations. Typically, these are metal ions, such as lithium or sodium, or large organic cations such as tetrabutylammonium. Lithium is an important factor, as it appears to be required for very efficient charge injection and for stabilization of electrons in the TiO_2 film. For aqueous electrolytes, protons are also present. Although these cations do not take part in charge-transfer reactions and are electrochemically inert, they influence the electrostatics of the system, and are widely believed to neutralize the electric fields created by charge injection by reorganization at the surface. In the case of lithium and protons, intercalated cations are associated with the formation and stabilization of Ti^{3+} species in TiO_2 films.

5.3.3.2.2 Solid-state Hole Conductors Replacement of the liquid electrolyte with a solid-state analog is considered essential for widespread industrial application. However, at the time of writing, solid-state hole conductors in dye-sensitized devices have not been widely studied. In a solid state, hole-transporting material charge can be carried by a redox couple transported through an inert polymer, or by the hopping of charged excitations through the molecules of the hole-conducting material.

Mobile cations may or may not be present, depending on the material used. Key requirements are clearly the rapid regeneration of the dye cation, and slow recombination with electrons in the porous film. In addition, there are a number of physical (permeability into the pores of the film, stability, and optical transparency) and chemical (stability, redox potential consistency with TiO_2 film and dye) requirements.

Several approaches have been tried, including polymer gel electrolytes containing I^-/I_2 [76], inert ionic conducting polymers containing I^-/I_2 [77], hole-conducting polymers [78] and amorphous molecular hole conductors such as the chiral organic, OMeTAD [79]. Molecular hole conductors are preferred to ionic conductors because of the elimination of the volatile component, but with penalties for the charge transport. A peak short-circuit incident photon to current efficiency (IPCE) for OMeTAD of 33% indicates poor charge transport. It is not clear whether this results from reduced hole mobility relative to the ionic mobility of the liquid electrolyte, or from increased interfacial recombination because of reduced charge screening. In general, both are likely to matter. Although hole-transport properties are unimportant in the liquid cell relative to electron transport, the same will not be true of solid-state devices.

A limited number of inorganic solid-state hole conductors have been considered. These need to be *p*-type semiconductor materials, which can be grown within the pores of the TiO_2 film. The most successful materials reported to date are copper (I) compounds such as copper iodide and copper thiocyanate [80, 81].

5.3.3.3 Dyes

The most commonly used dye is a carboxylated ruthenium bipyridal dye complex. Under photoexcitation the charge density shifts from around the ruthenium ion to around the carboxylated bipyridal groups (a metal to ligand charge transfer reaction (MLCT)). This encourages rapid injection into acceptor states in the metal oxide. It is controversial whether MLCT is essential for rapid charge injection. Sub-ps charge injection has also been observed for porphyrin dyes in which MLCT does not occur [82]. Although the basic requirements for electronic structure of the dye are not clear, it has been pointed out [2] that electronic coupling between the π^* orbital of the dye complex and the conduction band of the semiconductor (a manifold of 3d orbitals in the case of TiO_2) may enable electron injection from dye to semiconductor with high yield.

Another much less well studied aspect of charge transport in dyes is the lateral transfer of excitations and charge between dye molecules anchored on the metal oxide surface. Some evidence for this "hole hopping" is provided by studies of quantum efficiency as a function of dye coverage, indicating that collection probability is affected by the density of dye molecules at low coverage [83]. Hole hopping in triarylamine dyes on nanocrystalline films has been inferred in Ref. 84 from studies of the redox activity of these dye molecules, which are energetically incapable of transferring charge to the metal oxide films.

A limited amount of work has been reported on inorganic sensitizers such as II-VI semiconductor QDs in dye-sensitized systems [10, 85, 86]. QD sensitizers offer the advantages of a tunable absorption edge and, in principle, tunable excited state energy level to optimize the matching of energy levels between sensitizer and electron conductor. However, charge separation is evidently poorer than for organic dyes, possibly due to poorer coupling between QD excited state levels and the semiconductor conduction band, and the injection yield is correspondingly less.

The remainder of this chapter is concerned primarily with electron transport in the porous semiconducting film. This is on the grounds that the high mobilities and concentrations of charge carriers in liquid electrolytes make ionic transport a secondary consideration, although it may affect the ease of electronic transport through effective parameters. Charge transport within solid-state electrolytes, on the other hand, cannot be neglected in considering solid-state systems, but, at the time of writing, the field is too young to make a fair review of the mechanisms and observations. Similarly, hole transport within the dye layer cannot be dismissed, but lies beyond the scope of the present work.

5.3.4 Theoretical Background

In this section we introduce the equations of device physics as normally applied to solar cells and outline the ways in which they must be modified for dye sensitized systems.

5.3.4.1 Basic Equations of Device Physics

In any electronic device, charge carriers must obey a set of continuity equations, which balance the generation and recombination of carriers [87]. For instance, the electron density $n(r, t)$ as a function of position r and time t must obey,

$$\frac{\partial n}{\partial t} = \frac{1}{q}\nabla \cdot J_{\mathrm{n}} + G - R \qquad (13)$$

where $J_{\mathrm{n}}(r, t)$ is the electron current density, q the electronic charge, G is the electron generation rate, and R is the electron recombination rate per unit volume. G and R may include capture and release of electrons by traps. A similar equation can be constructed for each type of charge carrier present in the system. If the form of J_{n}, G, and R are known, the continuity equations can be solved together with Poisson's equation and appropriate boundary conditions to yield the carrier density functions for particular physical situations. Carrier populations are related to the physically relevant measurables, current and voltage, through the definition of the current terms and the relationship between carrier density and electrochemical potential (Fermi level) at contacts.

For a solar cell, the most relevant quantity is the current density at the terminals of the device, J, delivered under steady state illumination when a given bias V is applied between the cell terminals, the $J(V)$ characteristic. In the steady state, the left hand side of Eq. (13) is zero, and equations are solved to determine $J(V)$ for a given incident spectrum and applied bias.

We can make some general comments about G and J_{n}. For a slab of material of uniform absorption oriented normal to the direction of incidence of the light, photogeneration at depth x below the surface is given by

$$G(x) = \int g(E, x)\, dE \qquad (14)$$

where

$$g(E, x) = \alpha(E) b_0(E)[1 - r(E)]\mathrm{e}^{-\alpha(E)x} \qquad (15)$$

where $\alpha(E)$ is the absorption coefficient of the material at photon energy E, $b_0(E)$ is the incident spectral photon flux, and $r(E)$ the surface reflectivity at energy E.

For a semiconductor, the electron current can usually be defined in terms of the electron quasi Fermi level, E_{Fn}, and the electron density, n

$$J_{\mathrm{n}} = \mu_{\mathrm{n}} n \nabla E_{\mathrm{Fn}} \tag{16}$$

where μ_{n} is the electron mobility. (This result is derived in the Boltzmann relaxation time approximation, where relaxation within a band is much faster than transitions between bands [88, 89]). The factor ∇E_{Fn} is sometimes called the *driving force* of the carrier population. n is related to E_{Fn} in the usual way through Fermi Dirac statistics:

$$n = \int g(E) \frac{1}{\mathrm{e}^{(E-E_{\mathrm{Fn}})/kT} + 1} dE \tag{17}$$

where $g(E)$ is the density of electron acceptor states at energy E, T is temperature, and k is Boltzmann's constant. For a nondegenerate semiconductor with no acceptor states in the band gap, the Boltzmann approximation is valid and the integral is taken up from the conduction band edge energy, E_{C}. Then

$$n = N_{\mathrm{C}}\, \mathrm{e}^{(E_{\mathrm{Fn}}-E_{\mathrm{C}})/kT} \tag{18}$$

where N_{C} is the effective density of states in the conduction band. In this limit, J_{n} can be broken down into contributions from diffusion and drift,

$$J_{\mathrm{n}} = qD_{\mathrm{n}} \nabla n + q\mu_{\mathrm{n}} E n \tag{19}$$

where D_{n} is the electron diffusion coefficient and E the electrostatic field.

5.3.4.2 **Charge Transport in a Conventional Photovoltaic Device**

First we consider the solution of the problem for a conventional p–n homojunction solar cell. Although a very different system, this is instructive, given that some of the terms and concepts have been applied to dye-sensitized devices.

In a p–n junction solar cell a number of simplifications can be made:

1. The system has one-dimensional symmetry,
2. Most of the photogeneration occurs in the doped, neutral regions adjacent to the field bearing p–n junction. In the neutral regions $E = 0$ and carriers move mainly by diffusion. For electrons in p-type material,

$$J_{\mathrm{n}} = qD_{\mathrm{n}} \frac{dn}{dx} \tag{20}$$

3. Most of the recombination occurs in the neutral regions and is dominated by the minority carriers, so that the recombination rate R varies linearly with the minority carrier density. For electrons in p-type material

$$R = \frac{n}{\tau_{\mathrm{n}}} \tag{21}$$

where τ_{n} is the electron lifetime. (In the Boltzmann relaxation approximation, τ_{n} is a constant).

Given these simplifications, the following linear differential equation can be constructed and solved for the (minority) electron density in the p region at steady state

$$D_{\mathrm{n}} \frac{d^2 n}{dx^2} - \frac{n}{\tau_{\mathrm{n}}} + g(E, x) = 0 \tag{22}$$

with $g(E, x)$ given by Eq. (15). Holes in the n-region obey an analogous equation. For constant D_{n} and τ_{n}, Eq. (22) can be solved exactly at a given photon energy E. The boundary conditions are (under current flow) that $n = n_0 \mathrm{e}^{qV/kT}$ at the edge of the depleted p–n junction region, where n_0 is the electron density in the dark and V the applied bias, or (in open circuit) that $\frac{dn}{dx} = 0$ at that plane. A second boundary condition relates $\frac{dn}{dx}$ at the surface to surface recombination velocity. In the steady state Eq. (22) yields hyperbolic profiles for both electron and current density, a dark current density that is proportional to $(e^{qV/kT} - 1)$, resembling that of an ideal diode. More refined models of p–n junction device physics allow for recombination in the junction region and lead to nonideal $J-V$ behavior. Evidently, the current voltage response can be described with knowledge of the transport parameters D_n and τ_n for either carrier, and the device structure. A useful derived parameter is the minority carrier diffusion length, L_n,

$$L_n = \sqrt{D_n \tau_n} \tag{23}$$

which determines the thickness of the neutral region from which photoexcited minority carriers can be collected effectively.

The transport parameters, although well known for common photovoltaic materials, are determined by transient measurements on doped semiconductor samples of known composition (for example, D_n from time of flight measurements of mobility, τ_n from time resolved photoluminescence).

Notice that although both electrons and holes are present, current in the neutral regions is dominated by the minority carrier because light and bias induced carrier gradients are larger. Majority carrier transport is described by similar diffusion equations but because the recombination probability is so low, the majority carrier density is effectively constant across the neutral region. Gradients in the carrier density may arise from resistive losses in low mobility materials.

5.3.4.3 Device Physics for Dye-sensitized Systems

Now we consider the case of a dye-sensitized device. On account of the complex geometry and multiple media, the system is in general more complicated. The following factors are important:

1. *More carrier types* A continuity equation should be constructed for each of the species involved in charge transfer: electrons, reduced and oxidized ions in the hole conductor. In addition, cations in the hole conductor are mobile and will affect electrostatics, and therefore they should be included, although these do not contribute to charge transfer.
2. *Three-dimensional geometry* Because of the structure of the junction, the geometry is three dimensional on the microscopic scale. The electrochemical junction gives rise to large local electric fields within the hole conductor at all points of the interface. These are typically a few nm in extent in the case of liquid electrolytes [23, 24].
3. *Boltzmann approximation invalid* Because the nanocrystalline film contains a high density of intraband gap states, the Boltzmann approximation for n (Eq. 18) may not be valid. Instead, n will be determined by some distribution of intraband gap states, with distribution $g(E)$, and Eq. (17) should apply.
4. *Nature of photogeneration* Photogeneration is a two-stage process that requires absorption of a photon by a sensitizer molecule and injection of an electron from dye to semiconductor.
5. *Nature of current* The nature of the driving force for the electron current is not definitely known. For example, the electron mobility is not expected to be a constant because the relaxation time approximation, which is used to derive Eq. (18), is not necessarily valid.
6. *Nature of recombination* The definition of the electron recombination rate is not known. In a dye-sensitized cell, unusually, two routes for electron

recombination exist: with hole conductor and with dye cation, as detailed earlier. Neither of them are necessarily expected to be linear with electron density.

7. *Trapping* Electron capture into and release from intraband gap "trap" states is expected to occur. In this case, it is useful to distinguish electrons that contribute to photocurrent from those that cannot. If we use n_C to represent the density of "free" electrons (those with energy above the conduction band, or mobility edge), then a net transfer of carriers to trap states should be included in the continuity equation for n_C:

$$\frac{\partial n_C}{\partial t} = \frac{1}{q}\nabla \cdot J_n + G - R - n_C \int_{E_V}^{E_C} g(E)\frac{df}{dt}\,dE \quad (24)$$

where $g(E)$ is the density of states at energy E in the intraband gap region, $E_{C/V}$ the conduction/valence band edge energy, and $f(E)$ is the (nonequilibrium) electron distribution function. The final term can be resolved into trapping and detrapping terms,

$$n_C \int_{E_V}^{E_C} \beta(E)g(E)(1 - f(E))\,dE - n_C \int_{E_V}^{E_C} \beta(E)e^{-(E-E_{Fn})/kT} \times g(E)f(E)\,dE \quad (25)$$

where $\beta(E)$ is the rate per unit volume of electron capture by a vacant trap of energy E, given by

$$\beta = \sigma\nu \quad (25)$$

where σ is the electron capture cross section and ν the electron thermal velocity. The volume rate of electron detrapping is given by

$$\beta e^{-(E-E_{Fn})/kT} \quad (26)$$

The energetic distribution and activity of such trap states is not known, and will be a function of film composition and morphology and chemical environment.

8. *Coulombic effects* Because of the small capacitance of the nanocrystals forming the porous film, the Coulombic energy required to charge a nanoparticle on addition of an electron may be significant. A change in electrical potential energy by application of bias, V, will not be translated solely into a shift in electron Fermi level, but will be divided between a shift in Fermi level ΔE_{Fn} and a shift in electrostatic potential energy ΔE_e,

$$qV = \Delta E_{Fn} + \Delta E_e \quad (27)$$

For a spherical nanoparticle of radius r,

$$\Delta E_e = \frac{q^2 n_1}{8\pi\varepsilon\varepsilon_0 r} \quad (28)$$

where n_1 is the number of electrons already contained in the particle, and ε the effective dielectric constant, which will in general depend on the chemical environment. This issue is relevant to both transport (Coulombic forces opposing further accumulation of charge on a nanoparticle) and cell photovoltage (not all of the electric potential of the electrons contributes to external voltage).

5.3.4.4 **Simplifications for Dye-sensitized Systems**

Although complex in general, the following simplifications can be made for dye-sensitized systems. Many of these are justified by experimental studies discussed in Sect. 5.3.5.

1. *Generation = absorption* In the case of bipyridal Ru dyes, electron injection from excited dye molecules occurs in subpicosecond time scales with unit efficiency [82, 90, 91]. Therefore light absorption by the dye can be considered equivalent to electron photogeneration.
2. *One dimensional geometry* Because the scale of phase separation in the junction is small compared to the device size, an effective medium approach is often applied, and one-dimensional symmetry assumed. This means neglecting the effect of local electric fields at the interface, and differences in behavior between electrons or ions that are near or further from the interface or grain boundaries. The effect of electrolyte on electron transport can be included through an effective diffusion coefficient.
3. *Neglect of hole transport* Because the density and mobility of charged species in the electrolyte is high, transport of ionic species is presumed to be facile and only *electron* transport needs to be considered to explain transient phenomena. This is analogous to the focus on minority carrier transport in the neutral regions of a p–n junction solar cell. Although electrons are majority carriers within the oxide film, in an effective medium that also contains electrolyte they are minority species. The solar cell contains an electron density of around 0.1 per nanoparticle in the dark, and a "hole" density of some tens (30–50 mM of I_3^-) [14, 20]. Transport in this system should more properly be regarded as ambipolar diffusion, but the relative ease of hole relative to electron transport means that the electron transport properties dominate [75].
4. *Neglect of electric fields* At least in the case of liquid electrolyte junctions, the electrolyte is concentrated enough and the ions mobile enough within the porous structure that electric fields are not expected to exist within the electrolyte over more than a few nm [12, 24, 92, 93]. The small size of the nanocrystals adds to the restriction that the porous film cannot sustain electric fields. Together these mean that no macroscopic electric fields are expected to exist within the film away from the interface with the SnO_2 substrate. (Local electric fields associated with individual charge carriers and the polarization of their environment are still expected, but are generally neglected.)
5. *Collection at SnO_2 interface* The higher electron affinity of doped SnO_2 than TiO_2 implies that a space charge region encouraging electron flow into the substrate should exist at this interface. This has been confirmed experimentally in both dry and wet films by the observation of directed photocharge transients [94] and direction dependent photocurrent efficiencies [73, 95]. The interface can be treated similarly to the depleted region of a p–n junction (although it requires electron accumulation, rather than depletion, in the TiO_2) in that all electrons reaching the interface are collected, and the excess photogenerated carrier density at that plane is zero.

These simplifications reduce the problem to that of electrons diffusing in a fieldless environment bordered by an absorbing boundary, that is,

$$\frac{\partial n}{\partial t} = \frac{\partial}{\partial x}\left(D_{\mathrm{n}}(n(x))\frac{\partial n}{\partial x}\right) + G - R \quad (29)$$

In the one-dimensional approximation, the boundary conditions are that $n(0) = n_{\mathrm{V}}(0)$ at the absorbing boundary $x = 0$,

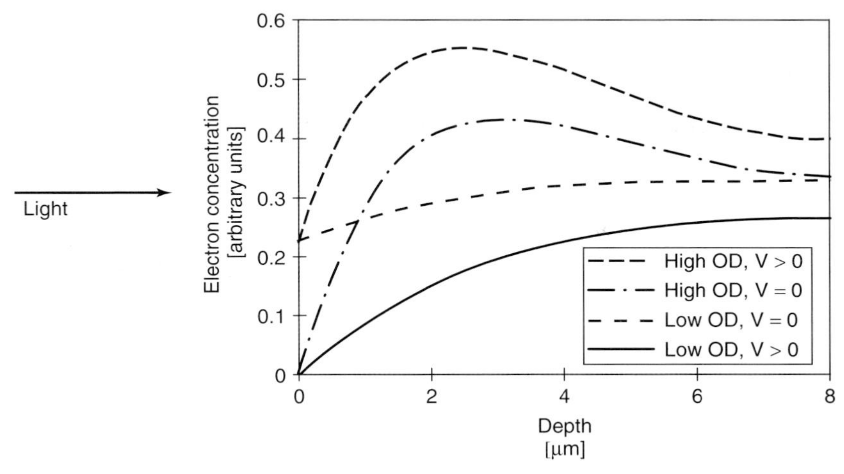

Fig. 7 Illustration of electron density satisfying the general diffusion equation for different optical densities and applied biases. It is assumed that light is incident from the negative x direction. In the diffusion picture the electron density is always depleted near to the interface with the TCO at $x = 0$, providing a statistical driving force for electron collection.

where $n_V(0)$ is the electron concentration in the dark when a bias V is applied across the terminals, and $\frac{dn}{dx} = 0$ at the extreme semiconductor–electrolyte interface. Equation (29) is more complex than the case of electrons in the p-type region of a p–n junction solar cell (Eq. 22) because of the generally nonlinear behavior of the recombination and diffusion, and the failure of the Boltzmann approximation. The diffusion coefficient is expressed as a function of $n(x)$ to reflect this. Some approaches to this problem are reviewed later. Elucidation of the mechanisms of recombination and diffusion to provide appropriate functional forms for D_n and R is a major experimental challenge, and is discussed in the following section.

5.3.5 Experimental Techniques and Results

We divide these into three groups: DC, time domain, and frequency domain. DC measurements under illumination are the most relevant to photovoltaic device performance, but time and frequency resolved measurements, usually on model systems, are the most revealing about the nature of charge transport.

5.3.5.1 DC Measurements

In the steady state the left-hand side of Eq. (13) is equal to 0. Measurables are the steady state photocurrent, photovoltage, light and dark current voltage characteristics, and the quantum efficiency spectrum (or IPCE). If the physical origin of J_n is known, the dependence of such DC measurements on variations in intensity, wavelength, and bias, deliver the parameters controlling J_n (for example, the values of the diffusion length and diffusion constant in the case of diffusion limited transport).

5.3.5.1.1 **Incident Photon to Current Efficiency** IPCE, also called external quantum efficiency, is the ratio of electrons collected at the terminals to incident photons under monochromatic illumination at short circuit. A high IPCE is a prerequisite for a useful photovoltaic device. At short circuit, recombination can usually be neglected and the continuity equation becomes, in the one-dimensional geometry,

$$\frac{1}{q}\frac{dJ_n(E,x)}{dx} + g(E,x) = 0 \qquad (30)$$

for light of a given photon energy, E. IPCE is defined as

$$\eta(E) = \frac{1}{q}\frac{J_n(E,0)}{b_0(E)} \qquad (31)$$

where J_n $(E,0)$ is the current passing through the SnO_2 interface at $x = 0$. Because the recombination rate, R, does not enter, this is a direct probe of the current mechanism. It is convenient to think of the IPCE as the product of the probabilities of (1) photon absorption, (2) electron injection, and (3) electron transport to the terminals. Because (1) is a known function of device thickness, sensitizer, and substrate and (2) is near unit efficient, (3) is the key quantity probed. We will refer to this probability as η_{coll}, the quantum efficiency for charge collection. IPCE's of around 80% at the absorption maximum of the dye have been reported for several Ru based sensitizer dyes on ~10 μm thick nanocrystalline TiO_2 films [14, 96, 97]. These are illustrated in Fig. 8.

Such high IPCE's correspond to almost 100% efficient collection because around 20% of the light is lost by absorption in the SnO_2 layer, by reflection and scattering. In terms of a simple diffusion model of electron transport, this indicates that the diffusion length of the electron is comparable to the optical depth of the cell at that wavelength. The very high IPCE's first reported for dye-sensitized nanocrystalline films were a surprising and promising feature.

More information about the diffusion mechanism can be extracted by studying the IPCE as a function of direction of

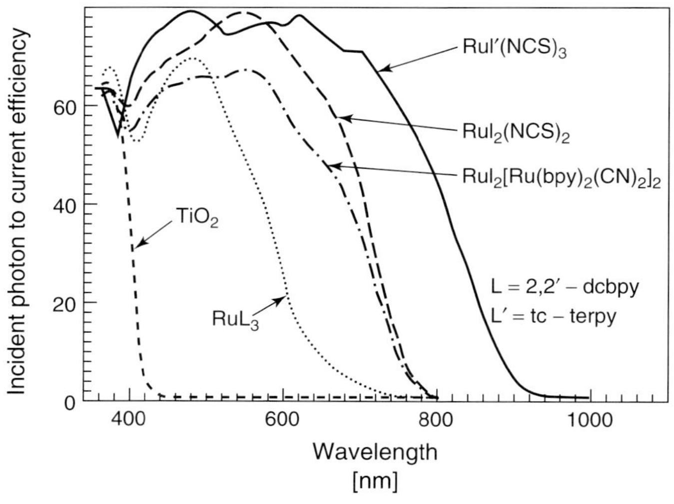

Fig. 8 IPCE's of some different sensitizer dyes, including the most widely used $Ru(bpy)_2(NCS)_2$, and the recently developed "black" dye, developed by Prof. Grätzel's group at EPFL [97].

illumination. In normal operation the device is illuminated from the glass substrate side, so that light is absorbed preferentially close to the $SnO_2 : TiO_2$ interface. If the diffusion length is comparable with the absorption depth and collection at the interface is easy, then high quantum efficiencies are expected for illumination from this direction. For illumination from the rear side (electrolyte side) of a film which is optically thick, most electrons have to cross the film to be collected, and study of the IPCE as a function of film thickness provides information on the electron diffusion length. Sodergren and coworkers [21, 98] have applied the diffusion solutions to fit the IPCE of a non-dye-sensitized device for illumination from electrolyte side and substrate side, and derive an electron diffusion length of 0.8 μm. Their results also confirm, importantly, that electrons generated close to the $SnO_2 : TiO_2$ interface are collected most efficiently, indicating the presence of an electric field driving collection at that interface. Measurements by Lindstrom and coworkers indicate that the electron diffusion length is longer in dye-sensitized devices [95], as would be expected from the absence of photogenerated holes.

The IPCE at the optimum wavelength can be interpreted as an indicator of the relative efficiency of electron transport in variations of the dye-sensitized TiO_2 device. For instance, the addition of electron acceptors such as O_2 and I_2 to the electrolyte lead to reduced IPCE at all wavelengths [95], which can be interpreted in terms of a shorter electron lifetime or shorter diffusion length.

Under forward bias (electrolyte at a more negative potential relative to the electrode potential) the electron density in the film increases and recombination becomes significant. At a bias that corresponds to the operating bias of the solar cell, IPCE at peak wavelength falls to about 45% of its saturation value [93]. This indicates competition between recombination and transport, although it is not trivial to disentangle the contributions from J_n and R.

The IPCE for sensitized ZnO films is smaller [64], although it is not clear whether this is because of poorer electron transport or poorer electron collection at the interface with substrate [67].

5.3.5.1.2 **Dark-current Voltage** In the dark, $G = 0$ and the steady state continuity Eq. (13) becomes

$$\frac{1}{q}\frac{dJ_n}{dx} - R = 0 \tag{32}$$

where R depends (usually strongly) on the applied bias V. The dark current $J_{dark}(V)$ is the current passing though $x = 0$ when a forward bias V is applied between the terminals. Note that J_{dark} acts in the opposite sense to the short circuit photocurrent. It is normally this process which limits the performance of a solar cell. Analysis of $J_{dark}(V)$ can in principle teach us something about R.

In a conventional solar cell, R results from the recombination of injected minority carriers. The dark current is rectifying and can be approximated by the form

$$J_{dark} = J_0(e^{qV/mkT} - 1) \tag{33}$$

where J_0 is a constant and the ideality factor m normally lies in the range $1 \leq m \leq 2$. m and J_0 are related to the mechanism for charge transport and recombination. The same parameters have been used to characterize dark currents of dye-sensitized cells, although their meaning is not clear.

In a dye-sensitized cell in the dark, R results mainly from the scavenging of electrically injected electrons by oxidized species (usually I_3^- ions) in the hole conductor. Dark currents typically reach current densities of several mA cm^{-2} at the operating bias of the solar cell, and exhibit rectifying behavior before being taken over by series resistance at current densities of ~10 mA cm^{-2} (Fig. 9a). Unlike solid-state junctions, the dark current does not saturate at negative biases because of the nonnegligible probability of electron transfer to the film at high reverse biases where the film is electron poor. The high series resistances are due to the series resistance of the conducting glass of some 10–15 ohm-square. Ideality factors of around 1 [99] and 2 [92] have been reported. In principle this number is related to the kinetics of the reduction reaction, but because interpretation in terms of transport parameters depends on the type of junction, for instance, p–n, p–i–n, Schottky barrier, it is not useful for DSSCs. Interpretations of dark-current behavior in terms of various barrier [99] or kinetic [92] models have been made but are speculative.

The magnitude of the dark current can be increased by various treatments that are expected to enhance electron recombination. These include the addition of electron scavengers such as oxygen [100] to the electrolyte, and pretreatment to increase the density of defects (oxygen vacancies) in the film [54]. The latter effect indicates that oxygen vacancies act as recombination centers. Any such

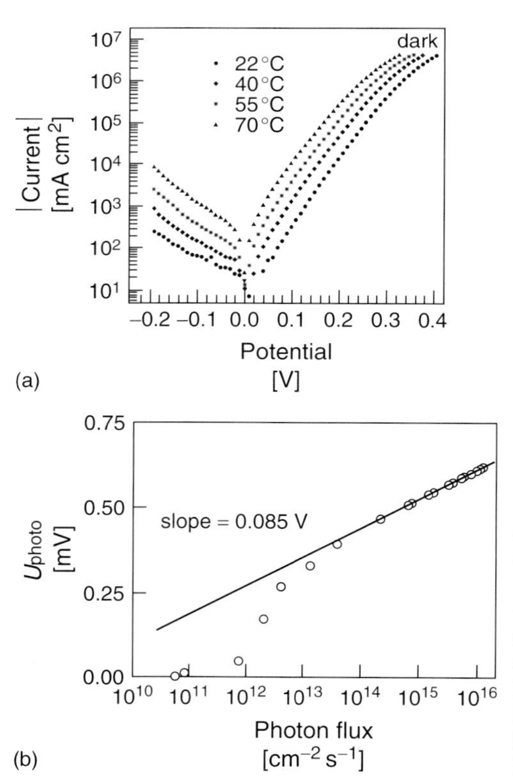

Fig. 9 (a) Dark current voltage characteristics of dye-sensitized cells at different temperatures, from Ref. 99. At high forward bias the current deviates from the exponential form on account of series resistance. Notice the bias dependence of the current at reverse bias. (b) Light intensity dependence of the open circuit voltage. At high intensities V_{oc} varies as expected from Eq. (36) (from Ref. 16).

treatment that enhances dark current will reduce the efficiency of charge collection.

5.3.5.1.3 **Light-current voltage** Although the most relevant measurement for solar energy conversion, steady state $J(V)$ is not the most useful for analyzing charge transport. The $J(V)$ curve for a typical dye-sensitized cell in standard solar illumination is presented in Fig. 10. The short-circuit photocurrent, $J_{sc} = J(0)$, corresponds to the IPCE integrated over the spectrum,

$$J_{sc} = \int \eta(E) b_0(E)\, dE \qquad (34)$$

In a good device, $J(V)$ remains constant into small forward bias, indicating that the photocurrent has indeed saturated at short circuit, and that recombination is negligible in this bias range.

At larger forward bias, the net photocurrent drops because of increased electron recombination. In a conventional solar cell, the process responsible for the recombination current is normally the same in light and dark, and it is common to make the approximation that $J(V) = J_{sc} - J_{dark}(V)$. Although this may be good for semiconductor p–n junction devices, it is not reliable for dye-sensitized photovoltaic devices. Ionization of the sensitizer opens up a new path for recombination in light, which is not active in dark (Fig. 11). There is some evidence [69] that the reverse current is larger under illumination than in the dark.

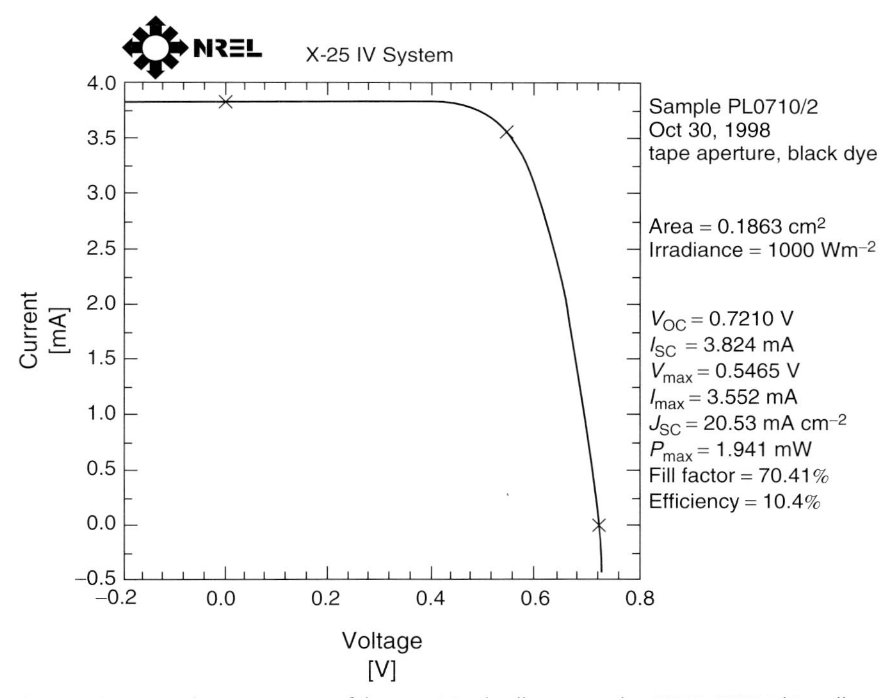

Fig. 10 Current voltage response of dye-sensitized cell measured at NREL [101]. This cell has a short-circuit current density of over 20 mA cm^{-2} and an open circuit voltage of over 0.7 V. The relatively low slope of I–V near V_{oc} indicates a significant series resistance, largely due to the sheet resistance of the conducting glass substrate.

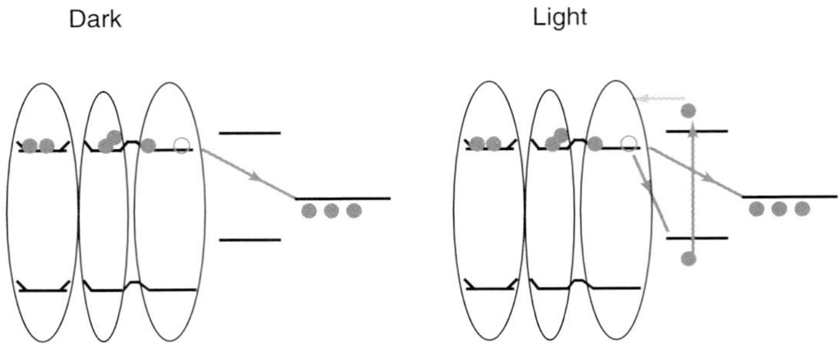

Fig. 11 Schematic illustration of charge recombination pathways in a dye-sensitized device in dark and light conditions. In the dark, electrons in the nanocrystalline semiconductor recombine only with the hole conductor. Under illumination, some of the dye molecules are ionized and electrons may recombine either with the hole conductor or with the dye cation. The second recombination route is often neglected, but may be significant at forward bias.

Study of the dark current is thus useful for understanding the reaction between electrons and redox couple, but not for understanding the limits to photovoltage of the DSSC. Nevertheless, simple models of $J(V)$ behavior derived from solid-state device physics can been applied to DSSC's to derive values of effective transport parameters. For example, Sodergren and coworkers [21] and Ferber and coworkers [102] have used a model, based on diffusion-controlled current, to fit the $J(V)$ behavior of DSSCs.

The open circuit photovoltage, V_{oc}, corresponds to the point, at which the recombination current and photocurrent are equal,

$$\frac{1}{q}\frac{dJ_n}{dx} - R_n(V_{oc}) + G(x) = 0 \qquad (35)$$

There is no *net* current at V_{oc}. If local currents are also small – for instance, if the generation rate is constant so that there are no internal diffusion currents – then the first term in Eq. (35) can be neglected and the variation of V_{oc} with light intensity tells us something about the form of R, as does the dark $J(V)$, but without the problem of series resistance. For a dye-sensitized cell it is a more relevant comparison as it is probing recombination under illumination, and not in the dark. Variation of V_{oc} with light intensity has been reported by various authors including Huang and coworkers [69] and Fisher and coworkers [16] for TiO_2 and by de Jongh and coworkers [67] for ZnO. For both materials the relationship tends to the form

$$V_{oc} = \frac{mkT}{q}\ln X + V_{oc}(1) \qquad (36)$$

at high light intensities, where X is the light concentration factor, $V_{oc}(1)$ is the open-circuit voltage at unit concentration, and m the apparent ideality factor. Values of m are in the range 1.3 to 1.5, but, as commented earlier, these values are not meaningful without a model for the recombination current.

V_{oc} is influenced by surface treatments and chemical environment. Huang and coworkers show that V_{oc} decreases with increasing concentration of triiodide, as expected for a recombination current which is dominated by the reaction of electrons with I_3^-. Several authors have reported the variation of V_{oc} by surface treatments of the TiO_2. Treatment with pyridine derivatives [69] and tetrabutylammonium phosphate [2] is observed to increase V_{oc}. The effect is attributed to the masking of electrons at the TiO_2 surface from electron scavengers in the electrolyte by inert adsorbed molecules. Treatment of TiO_2 surface by benzoic acid derivatives is observed to alter the V_{oc} of solid-state devices in a way that is consistent with modification of the surface dipole field by the adsorbed acids [103].

5.3.5.1.4 **Photoconductivity and Photovoltaic Properties** Relevant to the light response of dye-sensitized systems are the photoconductivity and photovoltaic properties of nanocrystalline films. Earlier it was noted that the conductivity of nanocrystalline TiO_2 is very small in the dark, around 10^{-10}–$10^{-8}\ \Omega^{-1}\ cm^{-1}$, rising by several orders of magnitude in UV light. Sensitized films are photoconductive in visible light, even in the absence of a hole conductor [10]. The relationship between light intensity and photoconductivity is superlinear, indicating increased electron mobility through trap filling [104]. It can be shown by simulation that this behavior result is consistent with dispersive transport (discussed below) in the presence of traps [105].

Steady state surface photovoltage (SPV) spectroscopy is useful for determining the nature of the junction between the nanocrystalline film and substrate. SPV studies of TiO_2 films on SnO_2 confirm that an electric field exists at the interface, driving electrons into the substrate. This is consistent with the values of the work functions of the two materials in vacuum (4.85 eV for $SnO_2:F$, 5.15 eV for TiO_2) [12, 103] and with evidence for directed photocurrents in dry films [94]. A comparison of the SPV spectra of an anatase single crystal and nanocrystalline film indicate the presence of a tail of states in the band gap below the conduction band edge [58].

5.3.5.2 Time Domain

Traditionally, time resolved measurements are used to study the mechanisms of charge transport and recombination in semiconductor materials.

5.3.5.2.1 **Photocurrent Transients** Time-resolved photocurrent at zero or reverse bias is, like IPCE, a probe of the forces driving conduction rather than recombination. If recombination can be neglected, the transient electron current resulting from an instantaneous or steady state light stimulus $G(x, t)$, is defined by

$$\frac{\partial n}{\partial t} = \frac{1}{q}\frac{\partial J_n}{\partial x} + G(x, t) \qquad (37)$$

in one dimension, where the form of J is not known.

Several configurations have been studied exhibiting some common features. Steady state and square-wave illumination of dye-sensitized electrochemical junctions produces a slow photocurrent, which rises to a saturation level [73, 106]. Photocurrent rise times are extremely long and intensity dependent (ms to s, depending on intensity). Both features are attributed to the trapping and release of photogenerated electrons. Schwarzburg and Willig [107] show with a simple rate equation treatment that the filling of a single trap level leads to intensity-dependent photocurrent kinetics, which are qualitatively similar to those observed. In their case of a single trap level Eq. (17) can be resolved into the sum of conduction (n_c) and trapped (n_t) electrons, and two continuity equations can be constructed for n_c and n_t with well defined rate constants for trapping and detrapping. Cao and coworkers [106] present transients at different light intensities and solve Eq. (37) for a light step ($G(x,t) = G_0$ for $t > 0$) and diffusive electron current ($J_n = qD_n \frac{dn}{dx}$). Although diffusion leads to a photocurrent which saturates at a level proportional to the light intensity, it predicts kinetics, which are intensity independent. This can be seen by normalizing n with respect to G_0 in Eq. (37): the solution for $n(x,t)/G_0$ is independent of G_0. In the data, however, a clear intensity dependence is seen (Fig. 12). The authors propose that the effective diffusion coefficient is n-dependent and demonstrate that the numerical solution of the diffusion equation when $D(n) = D_0 n$, leads to a rise time which decreases with increasing light intensity. This form of $D(n)$ is somewhat arbitrary, but is qualitatively consistent with trap filling: at higher injection levels, diffusion should be easier because fewer deep traps are available. Observations [73] of faster rise times under bias illumination and kinetics which depend on the direction of illumination (effectively, on the intensity reaching the SnO_2 interface) are consistent with intensity-dependent diffusion.

Transient photocurrents in response to a laser pulse have been studied in electrolyte supported and dry films [57, 60, 74, 108]. Such transient experiments are most informative when the optical depth of the film is high, and the light pulse effectively creates a sheet of photogenerated carriers, which travels through the film to the absorbing boundary. This is achieved by exciting in the UV where absorption of the TiO_2 is high. Solbrand and coworkers [74, 108] present laser pulse induced transient photocurrents for a TiO_2 electrochemical junction as a function of film thickness, electrolyte composition and direction of illumination (Fig. 13). The

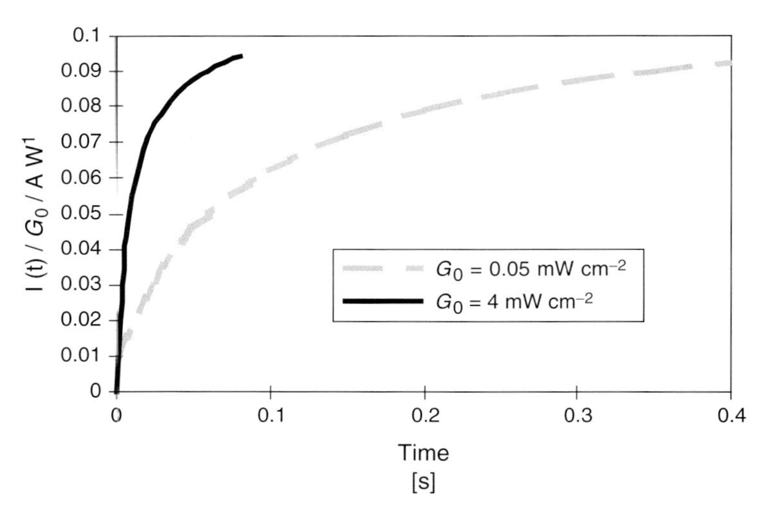

Fig. 12 Transient photocurrent in response to step illumination of different intensities normalized with respect to the light intensity. At higher intensities, the photocurrent rise time is faster, indicating an electron density dependent diffusion coefficient. Replotted from Ref. 106.

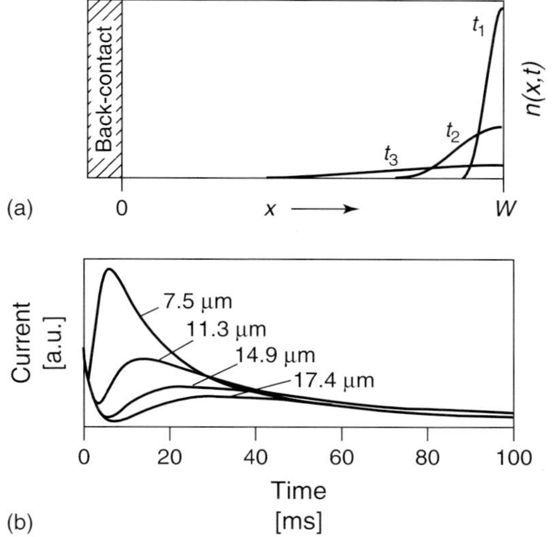

Fig. 13 In a transient photocurrent experiment, charges created near the TiO_2–electrolyte interface travel through the film to deliver a current at the collecting interface, from Ref. 74. The thickness dependence of the transient currents is consistent with diffusion, but contains some features of dispersive transport.

observed film thickness dependence is consistent with a diffusion picture, where the carrier profile generated near the semiconductor–electrolyte interface at $x = W$ spreads out over time to deliver a time-dependent flux of carriers through the absorbing boundary at $x = 0$. It is evident that the transit times are long (tens of ms to s) and the current signal becomes more dispersive with film thickness. They fit the slow portion of their transients (an initial fast decay is attributed to capacitive effects) to the solution of the diffusion equation for a delta function stimulus, and derive an effective diffusion coefficient, D, of the order of 10^{-6} cm^2 s^{-1}. The facts that (1) the experimental tail in the photocurrent is more persistent than predicted by simulation and (2) the kinetic shape of the normalized current, $J(0,t)/G_0$, is intensity dependent, indicate that simple diffusion is not a complete description. These features are, however, consistent with a model of "anomalous diffusion", a random walk through energetically distributed trap states, as discussed by Nelson [109]. A second and important observation by Solbrand and coworkers [74] is that the apparent diffusion coefficient is dependent on the conductivity and composition of the electrolyte. This is consistent with an activation energy for electron hopping which depends on the energy required to polarize its environment, a phenomenon which is familiar from conduction in doped disordered materials [110]. It is also a clear signal that the electrochemical interface is very important to transport, and consistent with the idea that electrons move along or close to the surface.

Laser pulse photocurrent transients in solid-state Schottky barrier junctions between dry TiO_2 and a platinum contact have been studied by Konenkamp and coworkers [57]. They observe a power law dependence of the collected photocurrent on time,

$$J(0,t) \propto t^{-\beta} \tag{38}$$

which is characteristic of "dispersive" transport. Such algebraic photocurrent transients are familiar from the physics of disordered semiconductors and have been shown to result from the anomalous diffusion of carriers within a medium containing a wide energetic dispersion of trap states. A continuous time random walk model due to Scher and Montroll [111] has been adapted by Nelson [109] for the case of nanocrystalline electrodes, and shown to be qualitatively consistent with observed photocurrent decays.

There is some evidence in the work cited earlier [57] for a dependence of the photocurrent kinetics on chemical environment, suggesting that the different environments or surface treatments (air, vacuum, and pH of electrolyte) influence the distribution of states acting as electron traps. Related evidence is provided by Hagfedlt and coworkers [112], who use time-resolved absorption at 900 nm to study the formation and disappearance of electrons (in absorbing Ti^{3+} states) following electron-hole pair generation by a UV laser pulse. They observe that in air, the signal initially decays rapidly (tens of ns), whereas in ethanol, which is an efficient hole scavenger, the signal persists for μs at least, implying that once holes have been removed the injected electron density decays very slowly. This suggests that it may be the presence of a hole-scavenging electrolyte, and not dye sensitization, which is the necessary condition for majority carrier injection in nanocrystalline films.

5.3.5.2.2 **Potential Step Transients** This refers to the current transient in response to a potential step. In solid-state devices a technique known as junction recovery is applied to barrier junctions as a probe of the charge density in localized states, particularly in amorphous materials [104]. A large negative potential step is applied to a forward biased junction and the recovered charge is measured as a function of forward current level. Konenkamp and coworkers [55, 60] have applied the technique to Schottky barrier nanocrystalline TiO_2 junctions, and extract an electron mobility-lifetime product of approximately 10^{-10} $cm^2\ V^{-1}$, by fitting their data with a simple model. They find that the mobility-lifetime product is apparently independent of injection level, whereas the recombination time, which is in the ms–μs range, falls with increasing injection. This behavior is typical of disordered materials such as amorphous semiconductors, where charge transport is dominated by the filling of localized states. The observation is explained in terms of trap filling by arguing that the mobility μ varies like

$$\mu = \mu_c \frac{n_c}{n} \quad (39)$$

where n_c is the free (conduction band) electron density, n the total including trapped electrons, and μ_c represents the mobility of conduction band electrons in the bulk crystal. The ratio of n_c to n is calculated using an exponential density of trap states, which appears to be consistent with observations of sub band gap photoconductivity. The authors show that the junction recovery current transients are both intensity dependent and tend toward a power law form, as shown in Fig. 14. They argue that this behavior is consistent with the thermal stimulation of electrons out of an exponential distribution of trap states.

A variant of the potential step technique, chronoamperometry, has been applied to electrochemical junctions in the dark to estimate the charge stored in the film. The transient current resulting from an applied potential step is measured and integrated to yield a stored charge. Hoyer and Weller used this technique to estimate the charge stored in nanocrystalline ZnO

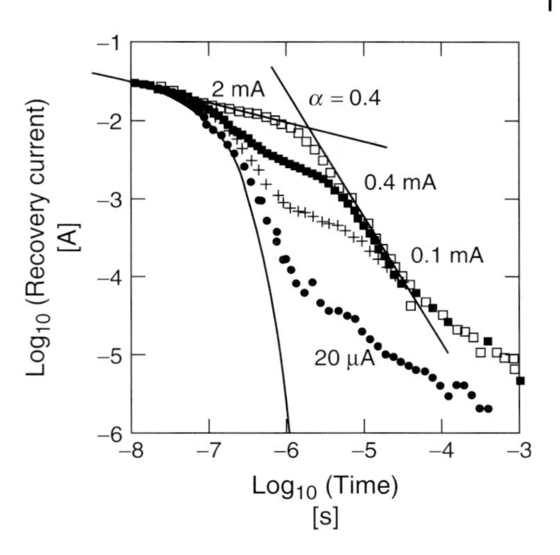

Fig. 14 Transient currents from a junction recovery experiment for a TiO_2 – Schottky barrier device [60]. The currents are dependent on injection level, and tend to a power law at long times, as expected for dispersive transport.

films [63] and others have used it to estimate charge stored in TiO_2 [113]. Results indicate an accumulated electron density of several electrons per ∼15 nm diameter nanoparticle, and are consistent with – but do not prove – an exponential density of surface states. A variant of the junction recovery technique, where recovered current from an illuminated electrochemical junction at open circuit is measured as a function of light intensity, has been applied to nanocrystalline TiO_2 by Duffy and coworkers [114]. These authors also infer an exponential density of states from their data and find that the recovered charge varies algebraically with time, which they attribute to a reaction of electrons with I_3^-, which is second order in n.

5.3.5.2.3 **Transient Absorption** Time-resolved absorption can be used to monitor the appearance or disappearance of the reduced or oxidized states of molecular species. Ultra fast time-resolved absorption was first used by Tachibana and coworkers to measure the rate of injection from dye-excited state into semiconductor, probing the absorption of the dye cation [90]. These and subsequent [82, 91] studies showed that injection occurs in sub picosecond time scales in a wide range of chemical environments, supporting the assertion made earlier that absorption is equivalent to photogeneration in a DSSC.

Although transient absorption is not a direct probe of transport, it can be used to probe the kinetics of recombination processes, which compete with transport. Haque and coworkers [115] studied the recombination between electrons in the TiO_2 electrode and photoionized dye molecules by monitoring the concentration of the dye cation following a laser pulse. By using a redox inactive environment, all other mechanisms to reduce the dye cation were suppressed. They found that this recombination reaction is slow (∼ms) at zero applied bias, but it is very strongly dependent on the bias applied to the junction, indicating that electrically injected electrons are capable of reducing the dye cation. Importantly, they find that this recombination reaction becomes sufficiently fast (ns-μs time scale) to compete with the reduction of dye cation by iodide at voltages near open circuit, and so should not be neglected in estimating the limits to performance of the DSSC. Simultaneous determination of the electron density in the film using the known optical absorbance of electrons trapped at Ti^{3+} states showed that the half life for decay, $t_{50\%}$, varies like

$$t_{50\%} \propto n^{-1/\alpha} \qquad (40)$$

where n is the total initial electron density including optically and electrically injected electrons, and α a fraction in the range from 0.25 to 0.5, which depends on chemical environment [11].

Neither the electron density dependence nor the shape (which is approximately stretched exponential) of the kinetics can be explained with second order reaction kinetics, where it is assumed that the reaction is controlled only by the concentrations of electrons and dye cations, nor are they consistent with simple electron transfer theory. An explanation was proposed by Nelson based on the continuous time random walk [109]. In the CTRW, electrons perform a random walk on a lattice, which contains trap sites distributed in energy, according to some distribution function, $g(E)$. In contrast to normal diffusion, where the mean time taken for each step is a constant, in the CTRW the time taken for each electron to move is determined by the time for thermal escape from the site currently occupied,

$$\tau_{\mathrm{hop}} \propto \beta \mathrm{e}^{(E_{\mathrm{C}}-E)/kT} \qquad (41)$$

where E_{C} is the conduction band edge energy, E is the energy of the trap site occupied, and β is the capture cross section defined in Eq. (26) earlier. The lattice is populated with a number of mobile electrons, n, and a number of static dye cations, $[D^+]$, and it is supposed that a dye cation is reduced whenever it meets an electron. If the trap energies are distributed according to an exponential density of states function with characteristic temperature T_0

$$g(E) = \frac{1}{kT_0} \mathrm{e}^{-(E_{\mathrm{C}}-E)/kT_0} \qquad (42)$$

then, in the limit where n exceeds $[D^+]$, the cation population evolves with time like a stretched exponential [116]

$$[\mathrm{D}^+] \propto \mathrm{e}^{-n(t/\tau)^{\alpha}} \qquad (43)$$

where τ and α are constants and

$$\alpha = \frac{T}{T_0}. \qquad (44)$$

It can be shown from Eq. (43) [117] that in these conditions the half life for cation decay, $t_{50\%}$, varies with n like

$$t_{50\%} \propto n^{-1/\alpha} \qquad (45)$$

as observed. Nelson and coworkers [11] have used Monte Carlo simulations of the CTRW to model observed recombination kinetics in different chemical environments with different values of α, (Fig. 15). If this is a valid description, it appears that recombination kinetics are strongly influenced by the presence and energetic distribution of trap states, and therefore may be controlled by controlling the film treatment and chemical environment. In a separate study of the kinetics of the recombination reaction between electrons in TiO_2 electrodes and rhenium sensitizer dyes, Hasselmann and Meyer [118] arrived at the conclusion that charge recombination is limited by the diffusional encounters of the electron with the dye cation, and not by the kinetics of the interfacial electron transfer step.

A less-used form of transient absorption is the monitoring of the electron density within the metal oxide electrode by measuring changes in the optical density in the red-IR, where Ti^{3+} states are

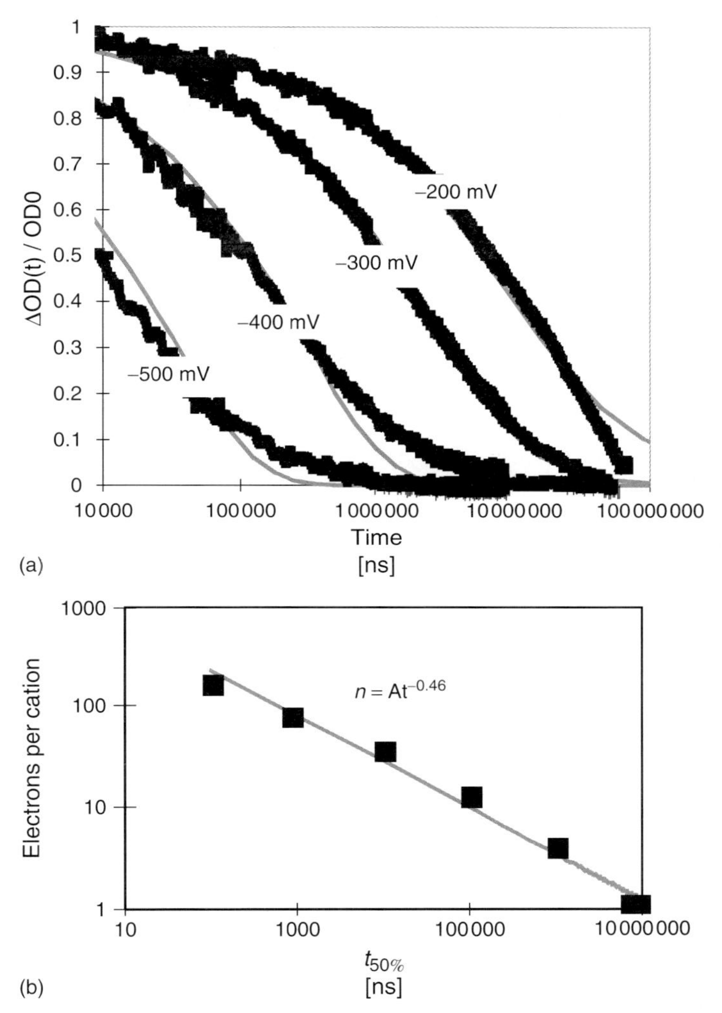

Fig. 15 (a) Kinetic decay curves for the rereduction of the dye cation by electrons in the TiO_2 film. The strong bias dependence indicates that electrically injected electrons are capable of reducing the dye. The smooth lines are fits using a continuous time random walk model of electron transport and assuming diffusion limited recombination. (b) Power law relationship between the electron density and the half life, $t_{50\%}$, for reduction of the dye cation. In the random walk model, the exponent α is related to the characteristic temperature of the density of trap states.

known to absorb. This has been used by Hagfeldt and coworkers to observe electron recombination kinetics following injection by a UV laser pulse [112] and by Olson and coworkers to observe the kinetics of electron accumulation following the application of a potential step in the dark [113]. Another novel probe of transient electron density is the use of time-resolved microwave conductivity to sense changes in electron density following application of light or bias. Salafsky has applied this technique to deduce an electron transit time of 100s of ms to seconds for a dye-sensitized device at short circuit [119].

5.3.5.3 Frequency Domain

Here we refer to a class of techniques where the response of a system to small, periodic modulations of the driving force is studied as a function of the frequency of the disturbance. These give the same information as time-domain studies but without the need for large amplitude disturbances. This is particularly useful for dye-sensitized – and for a wide range of complex systems – because the optical and electronic response is nonlinear in n, and therefore diffusion coefficient and recombination times are not constants. The use of small modulations permit linearization of the transport problem, and yields effective diffusion and recombination parameters, which can be related to the injection level through the underlying steady state driving force.

We will distinguish various modes. The techniques most widely used in DSSCs are: intensity modulated photocurrent spectroscopy (IMPS) which is, like IPCE, concerned with electron transport under short circuit conditions; Intensity modulated photovoltage spectroscopy IMVS, which probes the competition between transport and recombination at open circuit; and electrical impedance spectroscopy (EIS), which probes the bias-dependent electrical response, analogous to dark-current transients. Other variants, such as frequency-resolved transmittance, the frequency domain analogs of transient absorption, have been developed [50]. The techniques and their application to nanocrystalline systems are reviewed in Ref. 120.

In general, a driving force, which may be optical or electrical is applied and the phase and amplitude of the response, which may be a current, voltage, or optical density, is measured as a function of modulation frequency. For compactness, the results are often presented in the complex plane, as the locus of the complex response as frequency is varied, although phase and amplitude data plotted against frequency may be easier to interpret. For a simple system with a single time constant, τ, the response is a semicircle in the complex plane, and it is easy to show that the frequency at the point where the imaginary component of the response is a minimum, ω_{min}, is equal to $1/\tau$. Distorted semicircles imply multiphasic responses, which would be modeled with several time constants in the time domain.

5.3.5.3.1 Intensity Modulated Photocurrent Spectroscopy

IMPS was used on unsensitized nanocrystalline TiO_2 electrodes by de Jongh and Vanmaekelbergh [121]. In this configuration the driving force is a modulated light source and the response is the modulated short-circuit photocurrent. These authors showed that ω_{min} is dependent on light intensity, X, varying approximately like $\omega_{min} \propto X^{0.5}$ (as shown in Fig. 16) and were able to attribute this to the trapping and detrapping of electrons in a distribution of trap states which is wide compared to kT. They deduced a density of trap states of around 10^{17} cm^{-3} and from this value concluded that the traps are located on the large TiO_2–electrolyte interface and not at the TiO_2–TCO contact [122]. The trapping hypothesis is supported by studies on porous GaP electrodes, where ω_{min} is intensity independent and the trap density is known

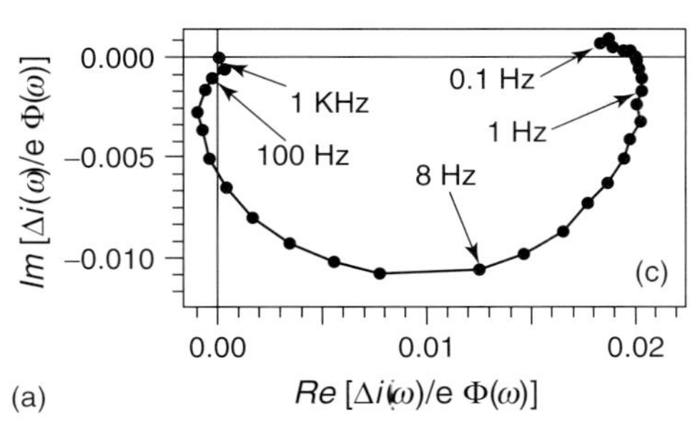

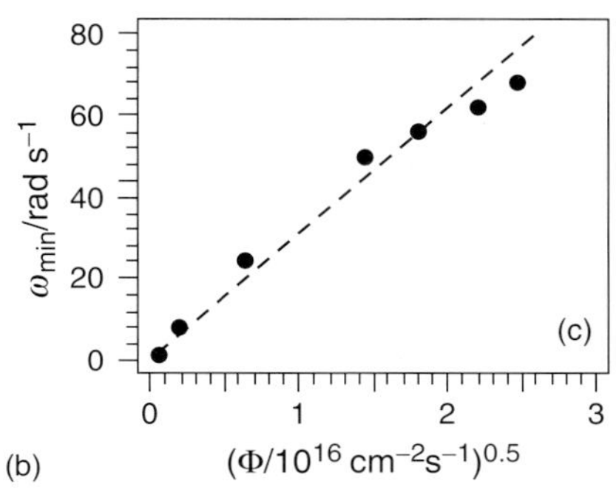

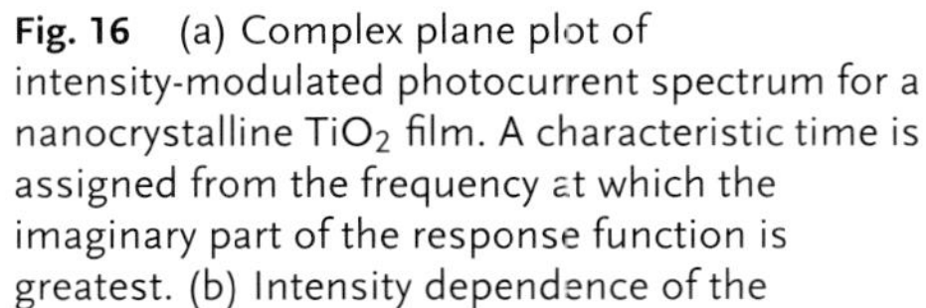

Fig. 16 (a) Complex plane plot of intensity-modulated photocurrent spectrum for a nanocrystalline TiO_2 film. A characteristic time is assigned from the frequency at which the imaginary part of the response function is greatest. (b) Intensity dependence of the characteristic frequency, which varies approximately like the square root of the light intensity. In a trap-free system this frequency would be independent of light intensity (from Ref. 121).

to be low; in this case multiple scattering is believed to dominate transport. The theory of trap limited IMPS response is developed in Ref. 123, where a distinction is made between multiple trapping and dispersive transport in the frequency domain, and in Ref. 124.

Dloczik and coworkers [92] later used IMPS to obtain an estimate of the electron diffusion coefficient in sensitized nanocrystalline TiO_2 electrodes. This is done by transforming the time dependent diffusion equation (Eq. 22) into frequency space and fitting the result for the photocurrent modulation to the measured values to extract values for an effective diffusion coefficient, D_n and electron lifetime, τ. This analysis does not explicitly account for the trapping and detrapping of electrons. However, it is pointed out that for the case of a single trap level, the effective diffusion coefficient is related to the true diffusion coefficient, D_c, for conduction band electrons through

$$D_n = D_c \frac{n_c}{n} \tag{46}$$

in analogy with Eq. (39). As injection level increases, the proportion of free electrons increases and so does D_n. Later work by the same authors [50, 16] showed that the diffusion coefficient derived this way varies with light intensity approximately like

$$D_n \propto X^{0.6} \tag{47}$$

Fisher and coworkers combine this observation with the observed (exponential) dependence of photovoltage on light intensity to derive a relationship between D_{eff} and electron Fermi level,

$$D_n \propto e^{(1-\beta)E_{Fn}/kT} \tag{48}$$

which is consistent with an exponential density of trap states of the form,

$$g(E) \propto e^{\beta E/kT} \tag{49}$$

as Eq. (42), with $\beta = \alpha$. They deduce a value of 0.57 for the exponent β from their data, which corresponds to a characteristic temperature T_0 of 526 K [16].

5.3.5.3.2 **IMVS** In IMVS, the driving force is again a modulated light source and the measured response is the modulation in open circuit photovoltage. The technique was first applied to dye-sensitized devices by Schlichthorl and coworkers [125].

The time constant, $1/\omega_{min}$, delivered by IMVS, corresponds to an effective lifetime for recombination, τ_{rec}, whereas the time constant delivered by IMPS results from a combination of recombination and charge collection,

$$\tau_{IMPS} = \left(\frac{1}{\tau_{rec}} + \frac{1}{\tau_{coll}}\right)^{-1} \tag{50}$$

although at short circuit the time constant for collection, τ_{coll}, often dominates. It is clearly important to be able to relate these experimentally accessible time constants to the functionally relevant quantity, the quantum efficiency η_{coll} for charge collection, but that is not trivial in a nonlinear system. Schlichthorl and coworkers [126] showed that although the linear approximation,

$$\eta_{coll} = \left(\frac{1 + \tau_{coll}}{\tau_{rec}}\right)^{-1} \tag{51}$$

is not adequate, a model based on electron diffusion with nonlinear relationships between electron density and recombination rate

$$R \propto n^{m_1} \tag{52}$$

where $m_1 \approx 2.2$, and between total and free electron density

$$n_{cb} \propto n^{m_2} \tag{53}$$

where $m_2 \approx 2.7$, is able to explain observed IMVS and IMPS behavior. The two power law relationships are deduced from observations of τ_{rec} versus light intensity using IMVS, and J_{sc} versus electron density using IMPS. The strong power law dependence of n_{cb} on n implied by the comparison is consistent with an exponential density of states, again of the form (40), but with $\alpha = 0.37$ or $T_0 = 810$ K [127].

The origin of the power law dependence of recombination time on light intensity, consistent with Eq. (52) is not clear. Such behavior has been observed elsewhere but attributed alternatively to a reaction between electrons and triiodide, which is second order in n [16], and to charge trapping [11, 128].

5.3.5.3.3 **Electrical Impedance Spectroscopy (EIS)** EIS measures the current response to a modulation in applied bias, and can be used in the light or dark. Traditionally, EIS results are interpreted in terms of an equivalent circuit and information extracted about relevant capacitive, resistive, and faradaic processes. The technique has been applied to nanocrystalline TiO_2 junctions in the dark by various groups [19, 22] and a film capacitance extracted and interpreted using a simple equivalent circuit. Results confirm that a potential drop does exist within the film in the dark. In reverse bias, where the film is depleted, the film capacitance appears to be bias independent and dominated by the potential drop at the TiO_2–TCO interface. In forward bias the film becomes populated and a high and a bias dependent capacitance develops at the large TiO_2–electrolyte interface, suggesting an increased charge storage capacity as surface or conduction band states become filled.

Van de Lagemaat and coworkers [93] apply EIS to illuminated nanocrystalline TiO_2 electrodes and again observe a capacitance which increases, approximately exponentially, with applied forward bias. They attribute this to the charging of surface states and show that the behavior is consistent with an exponential density of states. They further conclude that the potential drop across the TiO_2–TCO interface extends over a narrow region, and drops under illumination, which would be expected as the Fermi level of the TiO_2 is raised. EIS under illumination provides information about the time constant for recombination, as does IMVS, but is not limited to open circuit. This means that with an appropriate model for the extraction of η_{coll} from τ_{rec} and τ_{coll}, EIS and IMPS can provide an alternative measure of the bias dependent quantum efficiency of a dye-sensitized device. Those authors find $\eta_{coll} = 0.45$ at open circuit.

Frequency-resolved experiments in nanocrystalline electrodes often reveal a high degree of frequency dispersion in the underlying processes. In simple planar interface models this may be dealt with using a constant phase element (CPE) in the equivalent circuit. The CPE contains a degree of frequency dispersion but is hard to relate to physical processes. Bisquert [129] has taken the analysis of electrical impedance spectra one stage further by invoking a transmission line model of the interface. The model allows for two resistive channels representing charge flow in the two media, and distributed charge transfer at different points on the interface by means of a resistor-CPE element. They obtain excellent agreement with measured EIS in light and dark, and suggest that two characteristic relaxation frequencies can be resolved.

5.3.6 Summary

In summary, we can make the following observations about the nature of charge transport in dye-sensitized systems.

There is virtually no electric field within the porous film. There is an electric field driving charge collection at the interface with TCO, and an electric field at the Helmholtz layer. Under illumination, a photopotential is developed mainly across the semiconductor–electrolyte interface.

In electrolyte-supported systems, transport is rate limited by electron transport. This has some characteristics of diffusion, but is not well described by a simple diffusion equation. Both the apparent diffusion coefficient and the apparent recombination time are electron density dependent. The n-dependence of D is consistent with the concept of increasing ease of transport as deep traps are filled.

There is widespread evidence for electron traps in the nanocrystalline films, largely Ti^{3+} states because of oxygen vacancies and states because of adsorbed and intercalated species.

Nonlinear diffusion has been approached with a number of models: (1) n-dependent diffusion coefficient [106], which can be related to the fraction of free electrons in a medium containing traps, specifically, $D_n \propto n_c/n$ [16, 55, 75, 127]; (2) rate-equation approach involving a distribution of trap states [123]; (3) Monte Carlo simulations of anomalous diffusion in the presence of trap states [109]. Numerous authors have invoked an exponential distribution of trap states with a characteristic temperature of between 500 K and 810 K. There appears to be some consensus on this point.

Nonlinear recombination such as $\tau_{rec} \propto n^{-\beta}$ has been observed by several groups [11, 16, 60, 93, 115, 125]. This has been attributed to different origins: recombination reactions which are second order in n, and diffusion limited recombination, where the electron motion is limited by transport through traps.

The dispersion of time constants for conduction and recombination lead to dispersive features in both time-resolved and frequency-resolved measurements. These are not easily addressable with equivalent circuit models for planar interfaces.

The fact that recombination is slow and transport relatively efficient at short circuit, is qualitatively attributed to the effects of screening of electrons by cations in the electrolyte; and to the absence of mobile photogenerated holes. Recombination with the oxidized species in the hole conductor is normally assumed to be the main loss process for electrons, but recombination with dye cations cannot be ruled out.

Continuity equation approaches involving nonlinear forms for the electron diffusion and recombination terms are capable of reproducing some aspects of observed behavior [123, 126]. The challenge is to predictively relate the forms for recombination and diffusion to known material properties of the system.

Chemical environment strongly influences transport. Adsorption and intercalation influence the potential distribution at the interface, affecting the tendency for electrons to accumulate in the film and the photovoltage. Composition of electrolyte influences rate of electron scavenging. Film history influences the density of traps and recombination centers.

In nanoparticulate semiconductors other than TiO_2, quantum confinement may become important.

In solid-state systems charge transport within the electrolyte is expected to be rate limiting. Even in liquid electrolyte systems there is some evidence for an influence of electrolyte on electron transport. This is expected from the screening and reorganization energy. In solid-state systems the absence of macroscopic fields cannot be assumed and drift may be expected to become significant.

Most models applied to date use an effective medium in which electron transport is considered only in one dimension. Some consider transport within a single nanoparticle, or other highly simplified geometry. There is clearly a need to move on to models incorporating the interpenetrating features of the morphology.

Looking to the future, the following issues remain to be addressed quantitatively: the relationship between material properties and the apparent density of states; the effect of hole conductor composition and mobility on electron transport; Coulombic effects of nanoparticle charging and electron-hole interactions. An improved description of charge transport is needed to replace the simple diffusion model. Techniques developed to deal with disordered conductors in the past may prove useful here. Multiple-trapping models such as the continuous time random walk can explain certain dispersive features of current and recombination kinetics. Another tool that offers the potential to describe some features of the behavior of these systems is percolation theory. This has already been used to describe hole conduction in a polymeric hole conductor [130]; to describe charge hopping between quantum sized ZnO nanoparticles [65]; and to describe hole migration between dyes [83, 84].

Acknowledgment

The author is grateful to James Durrant, Sven Sodergren, Thierry Lutz, Carol Olson, Mili Biswas, and Ellen Moons for helpful discussions, and acknowledges the financial support of the Engineering and Physical Sciences Research Council and the Greenpeace Environmental Trust.

References

1. M. Grätzel, Nanocrystalline electronic junctions, *Stud. Surf. Sci. Catal.* **1997** *103*, 353–375.

2. A. Hagfeldt, M. Gratzel, Light-induced redox reactions in nanocrystalline systems, *Chem. Rev.* **1995**, *95*(1), 49–68.
3. B. O'Regan, M. Gratzel, A low-cost, high-efficiency solar-cell based on dye-sensitized colloidal TiO_2 films, *Nature* **1991**, *353*(6346), 737–740.
4. N. Vlachopoulos, P. Liska, J. Augustynski et al., Very efficient visible-light energy harvesting and conversion by spectral sensitization of high surface-area polycrystalline titanium-dioxide films, *J. Am. Chem. Soc.* **1988**, *110*(4), 1216–1220.
5. C. Bechinger, S. Ferrer, A. Zaban et al., Photoelectrochromic windows and displays, *Nature* **1996**, *383*(6601), 608–610.
6. J. Sotomayor, G. Will, D. Fitzmaurice, Photoelectrochromic heterosupramolecular assemblies, *J. Mater. Chem.* **2000**, *10*(3), 685–692.
7. E. Topoglidis, A. E. G. Cass, G. Gilardi et al., Protein adsorption on nanocrystalline TiO_2 films: an immobilization strategy for bioanalytical devices, *Anal. Chem.* **1998**, *70*(23), 5111–5113.
8. S. Y. Huang, L. Kavan, I. Exnar et al., Rocking chair lithium battery based on nanocrystalline Tio_2 (anatase), *J. Electrochem. Soc.* **1995**, *142*(9), L142–L144.
9. C. Bechinger, B. A. Gregg, Development of a new self-powered electrochromic device for light modulation without external power supply, *Sol. Energy Mater. Sol. Cells* **1998**, *54*(1–4), 405–410.
10. P. Hoyer, R. Konenkamp, Photoconduction in porous Tio_2 sensitized by Pbs quantum dots, *App. Phys. Lett.* **1995**, *66*(3), 349–351.
11. J. Nelson, S. A. Haque, D. R. Klug et al., Observation of trap limited recombination in dye-sensitized nanocrystalline metal oxide electrodes, *Phys. Rev. B* **2001**, *63*, 20 5321.
12. D. Cahen, G. Hodes, M. Gratzel et al., Nature of photovoltaic action in dye-sensitized solar cells, *J. Phys. Chem. B* **2000**, *104*(9), 2053–2059.
13. C. A. Kelly, F. Farzad, D. W. Thompson et al., Cation-controlled interfacial charge injection in sensitized nanocrystalline TiO_2, *Langmuir* **1999**, *15*(20), 7047–7054.
14. M. K. Nazeeruddin, A. Kay, I. Rodicio et al., Conversion of light to electricity by *cis*-X2bis(2,2′-Bipyridyl-4,4′-Dicarboxylate) Ruthenium(Ii) charge-transfer sensitizers (X = Cl-, Br-, I-, Cn-, and Scn-) on nanocrystalline Tio_2 electrodes, *J. Am. Chem. Soc.* **1993**, *115*(14), 6382–6390.
15. E. Topoglidis, T. Lutz, R. L. Willis et al., Protein adsorption on nanoporous TiO_2 films: a novel approach to studying photoinduced protein/electrode interactions, *Faraday Discuss.* **2000**, *116*, 35–46.
16. A. C. Fisher, L. M. Peter, E. A. Ponomarev et al., Intensity dependence of the back reaction and transport of electrons in dye-sensitized nanocrystalline TiO_2 solar cells, *J. Phys. Chem. B* **2000**, *104*(5), 949–958.
17. N. Myung, S. Licht, *J. Electrochem. Soc.* **1995**, *142*, L129–L132.
18. D. Fitzmaurice, Using spectroscopy to probe the band energetics of transparent nanocrystalline semiconductor-films, *Sol. Energy Mater. Sol. Cells* **1994**, *32*(3), 289–305.
19. F. Cao, G. Oskam, P. C. Searson et al., Electrical and optical-properties of porous nanocrystalline Tio_2 films, *J. Phys. Chem.* **1995**, *99*(31), 11974–11980.
20. J. Bisquert, G. GarciaBelmonte, F. Fabregat-Santiago, Modelling the electric potential distribution in the dark in nanoporous semiconductor electrodes, *J. Solid State Electrochem.* **1999**, *3*(6), 337–347.
21. S. Sodergren, A. Hagfeldt, J. Olsson et al., Theoretical-models for the action spectrum and the current-voltage characteristics of microporous semiconductor-films in photoelectrochemical cells, *J. Phys. Chem.* **1994**, *98*(21), 5552–5556.
22. A. Zaban, A. Meier, B. A. Gregg, Electric potential distribution and short-range screening in nanoporous TiO_2 electrodes, *J. Phys. Chem. B* **1997**, *101*(40), 7985–7990.
23. D. Matthews, P. Infelta, M. Gratzel, Calculation of the photocurrent-potential characteristic for regenerative, sensitized semiconductor electrodes, *Sol. Energy Mater. Sol. Cells* **1996**, *44*(2), 119–155.
24. B. A. Gregg, A. Zaban, S. Ferrere, Dye sensitized solar cells: energetic considerations and applications, Zeitschrift Fur Physikalische Chemie-International, *J. Res. Phys. Chem. Chem. Phys.* **1999**, *212*(Pt1), 11–22.
25. F. Pichot, B. A. Gregg, The photovoltage-determining mechanism in dye-sensitized solar cells, *J. Phys. Chem. B* **2000**, *104*(1), 6–10.

26. H. Lindstrom, S. Sodergren, A. Solbrand et al., Li^+ ion insertion in TiO_2 (anatase) chronoamperometry on CVD films and nanoporous films, *J. Phys. Chem. B* **1997**, *101*(39), 7710–7716.
27. C. J. Barbe, F. Arendse, P. Comte et al., Nanocrystalline titanium oxide electrodes for photovoltaic applications, *J. Am. Ceram. Soc.* **1997**, *80*(12), 3157–3171.
28. V. Shklover, M. K. Nazeeruddin, S. M. Zakeeruddin et al., Structure of nanocrystalline TiO_2 powders and precursor to their highly efficient photosensitizer, *Chem. Mater.* **1997**, *9*(2), 430–439.
29. A. Hagfeldt, B. Didriksson, T. Palmqvist et al., Verification of high efficiencies for the Gratzel-cell – a 7-percent efficient solar-cell based on dye-sensitized colloidal TiO_2 films, *Sol. Energy Mater. Sol. Cells* **1994**, *31*(4), 481–488.
30. P. A. Cox, Transition Metal Oxides, in *International Series of Monographs on Chemistry*, Oxford, Oxford University Press, 1992.
31. H. Tang, H. Berger, P. E. Schmid et al., Optical-properties of anatase (TiO_2), *Solid State Commun.* **1994**, *92*(3), 267–271.
32. H. Tang, K. Prasad, R. Sanjines et al., Electrical and optical-properties of TiO_2 anatase thin-films, *J. Appl. Phys.* **1994**, *75*(4), 2042–2047.
33. H. Tang, F. Levy, H. Berger et al., Urbach tail of anatase TiO_2, *Phys. Rev. B: Conden. Matter* **1995**, *52*(11), 7771–7774.
34. R. Sanjines, H. Tang, H. Berger et al., Electronic-structure of anatase TiO_2 oxide, *J. Appl. Phys.* **1994**, *75*(6), 2945–2951.
35. H. Z. Zhang, J. F. Banfield, Thermodynamic analysis of phase stability of nanocrystalline titania, *J. Mater. Chem.* **1998**, *8*(9), 2073–2076.
36. L. Kavan, M. Grätzel, S. E. Gilbert et al., Electrochemical and photoelectrochemical investigation of single-crystal anatase, *J. Amer. Chem. Soc.* **1996**, *118*(28), 6716–6723.
37. W. C. Mackrodt, E. A. Simson, N. M. Harrison, An ab initio Hartree-Fock study of the electron-excess gap states in oxygen-deficient rutile TiO_2, *Surf. Sci.* **1997**, *384*(1–3), 192–200.
38. R. G. Egdell, S. Eriksen, W. R. Flavell, Oxygen deficient SnO_2(110) and TiO_2(110) – a comparative-study by photoemission, *Solid State Commun.* **1986**, *60*(10), 835–838.
39. U. Kirner, K. D. Schierbaum, W. Gopel et al., Low and high-temperature TiO_2 oxygen sensors, *Sens. Actuators B-Chem.* **1990**, *1*(1–6), 103–107.
40. J. Muscat, N. M. Harrison, G. Thornton, First-principles study of potassium adsorption on TiO_2 surfaces, *Phys. Rev. B: Conden. Matter* **1999**, *59*(23), 15457–15463.
41. S. Sodergren, H. Siegbahn, H. Rensmo et al., Lithium intercalation in nanoporous anatase TiO_2 studied with XPS, *J. Phys. Chem. B* **1997**, *101*(16), 3087–3090.
42. A. Vittadini, A. Selloni, F. P. Rotzinger et al., Structure and energetics of water adsorbed at TiO_2 anatase (101) and (001) surfaces, *Phys. Rev. Lett.* **1998**, *81*(14), 2954–2957.
43. P. J. D. Lindan, N. M. Harrison, M. J. Gillan, Mixed dissociative and molecular adsorption of water on the rutile (110) surface, *Phys. Rev. Lett.* **1998**, *80*(4), 762–765.
44. H. Haerudin, S. Bertel, R. Kramer, Surface stoichiometry of "titanium suboxide" – Part I–Volumetric and FTIR study, *J. Chem. Soc., Faraday Trans.* **1998**, *94*(10), 1481–1487.
45. R. F. Howe, M. Grätzel, Electron-paramagnetic-res observation of trapped electrons in colloidal TiO_2, *J. of Phys. Chem.* **1985**, *89*(21), 4495–4499.
46. U. Kolle, J. Moser, M. Grätzel, Dynamics of interfacial charge-transfer reactions in semiconductor dispersions – reduction of cobaltoceniumdicarboxylate in colloidal TiO_2, *Inorg. Chem.* **1985**, *24*(14), 2253–2258.
47. B. O'Regan, M. Grätzel, D. Fitzmaurice, Optical electrochemistry steady state spectroscopy of conduction-band electrons in a metal-oxide semiconductor electrode, *Chem. Phys. Lett.* **1991**, *183*(1–2), 89–93.
48. G. Rothenberger, D. Fitzmaurice, M. Gratzel, Optical electrochemistry spectroscopy of conduction-band electrons in transparent metal-oxide semiconductor-films – optical determination of the flat-band potential of colloidal titanium-dioxide films, *J. Phys. Chem.* **1992**, *96*(14), 5983–5986.
49. G. Redmond, D. Fitzmaurice, Spectroscopic determination of flat-band potentials for polycrystalline Tio_2 electrodes in nonaqueous solvents, *J. Phys. Chem.* **1993**, *97*(7), 1426–1430.
50. G. Franco, J. Gehring, L. M. Peter et al., Frequency-resolved optical detection of photoinjected electrons in dye-sensitized

nanocrystalline photovoltaic cells, *J. Phys. Chem. B* **1999**, *103*(4), 692–698.
51. G. Boschloo, D. Fitzmaurice, Electron accumulation in nanostructured TiO_2 (anatase) electrodes, *J. Phys. Chem. B* **1999**, *103*(37), 7860–7868.
52. L. Forro, O. Chauvet, D. Emin et al., High-mobility *N*-type charge-carriers in large single-crystals of anatase (TiO_2), *J. Appl. Phys.* **1994**, *75*(1), 633–635.
53. B. Poumellec, J. F. Marucco, F. Lagnel, Electron-transport in TiO_2-X at intermediate temperatures 300-K less- than T less-than 1500-K, *Phys. Status Solidi A-Appl. Res.* **1985**, *89*(1), 375–382.
54. J. Weidmann, T. Dittrich, E. Konstantinova et al., Influence of oxygen and water related surface defects on the dye sensitized TiO_2 solar cell, *Sol. Energy Mater. Sol. Cells* **1999**, *56*(2), 153–165.
55. R. Konenkamp, Carrier transport in nanoporous TiO_2 films, *Phys. Rev. B* **2000**, *61*(16), 11057–11064.
56. T. Dittrich, J. Weidmann, V. Y. Timoshenko et al., Thermal activation of the electronic transport in porous titanium dioxides, *Mater. Sci. Eng., B-Solid State Materials For Advanced Technology* **2000**, *69*(SISI), 489–493.
57. R. Konenkamp, R. Henninger, P. Hoyer, Photocarrier transport in colloidal TiO_2 films, *J. Phys. Chem.* **1993**, *97*(28), 7328–7330.
58. E. Moons, M. Grätzel, *Photovoltage in Nanocrystalline TiO_2 Films* **1998**, Unpublished.
59. T. Dittrich, E. A. Lebedev, J. Weidmann, Electron drift mobility in porous TiO_2 (anatase), *Phys. Status Solidi A-Appl. Res.* **1998**, *165*(2), R5–R6.
60. R. Konenkamp, R. Henninger, Recombination in nanophase TiO_2 films, *Appl. Phys. A-Materials Science & Processing* **1994**, *58*(1), 87–90.
61. S. Burnside, J. E. Moser, K. Brooks et al., Nanocrystalline mesoporous strontium titanate as photoelectrode material for photosensitized solar devices: increasing photovoltage through flatband potential engineering, *J. Phys. Chem. B* **1999**, *103*(43), 9328–9332.
62. J. M. Stipkala, F. N. Castellano, T. A. Heimer et al., Light-induced charge separation at sensitized sol-gel processed semiconductors, *Chem. Mater.* **1997**, *9*(11), 2341–2353.
63. P. Hoyer, H. Weller, Potential-dependent electron injection in nanoporous colloidal ZnO films, *J. Phys. Chem.* **1995**, *99*(38), 14096–14100.
64. H. Rensmo, K. Keis, H. Lindstrom et al., High light-to-energy conversion efficiencies for solar cells based on nanostructured ZnO electrodes, *J. Phys. Chem. B* **1997**, *101*(14), 2598–2601.
65. E. A. Meulenkamp, Electron transport in nanoparticulate ZnO films, *J. Phys. Chem. B* **1999**, *103*(37), 7831–7838.
66. A. vanDijken, E. A. Meulenkamp, D. Vanmaekelbergh et al., The kinetics of the radiative and nonradiative processes in nanocrystalline ZnO particles upon photoexcitation, *J. Phys. Chem. B* **2000**, *104*(8), 1715–1723.
67. P. E. de Jongh, E. A. Meulenkamp, D. Vanmaekelbergh et al., Charge carrier dynamics in illuminated particulate ZnO electrodes, *J. Phys. Chem. B* **2000**, *104*, 7686–7693.
68. P. E. de Jongh, D. Vanmaekelbergh, J. J. Kelly, Photoelectrochemistry of electrodeposited Cu_2O, *J. Electrochem. Soc.* **2000**, *147*(2), 486–489.
69. S. Y. Huang, G. Schlichthorl, A. J. Nozik et al., Charge recombination in dye-sensitized nanocrystalline TiO_2 solar cells, *J. Phys. Chem. B* **1997**, *101*(14), 2576–2582.
70. A. Hauch, R. Kern, J. Ferber et al., Characterisation of the electrolyte-solid interfaces of dye-sensitized solar cells by means of impedance spectroscopy, *2nd World Conference on Photovoltaic Solar Energy Conversion*, Vienna, European Communities, 1998.
71. N. Papageorgiou, M. Gratzel, P. P. Infelta, On the relevance of mass transport in thin layer nanocrystalline photoelectrochemical solar cells, *Sol. Energy Mater. Sol. Cells* **1996**, *44*(4), 405–438.
72. Z. Kebede, S. E. Lindquist, The obstructed diffusion of the I-3(−) ion in mesoscopic TiO_2 membranes, *Sol. Energy Mater. Sol. Cells* **1998**, *51*(3–4), 291–303.
73. P. M. Sommeling, H. C. Rieffe, J. M. Kroon et al., Spectral response and response time of nanocrystalline dye-sensitized TiO_2 solar cells, *14th European Photovoltaic Solar Energy Conference*, Barcelona, H.S. Stephens and Associates, 1997.

74. A. Solbrand, H. Lindstrom, H. Rensmo et al., Electron transport in the nanostructured TiO_2-electrolyte system studied with time-resolved photocurrents, *J. Phys. Chem. B* **1997**, *101*(14), 2514–2518.
75. N. Kopidakis, E. A. Schiff, N.- G. Park et al., Ambipolar diffusion of carriers in electrolyte filled, nanoporous TiO_2, *J. Phys. Chem. B* **2000**, *104*(16), 3930–3936.
76. F. Cao, G. Oskam, P. C. Searson, A solid-state, dye-sensitized photoelectrochemical cell, *J. Phys. Chem.* **1995**, *99*(47), 17071–17073.
77. S. A. Haque, S. Sodergren, A. Holmes et al., Solid state dye-sensitized photovoltaic cells employing a polymer electrolyte, *International conference on photochemical conversion and storage of solar energy*, Colorado, 2000.
78. T. J. Savenije, J. M. Warman, A. Goossens, Visible light sensitization of titanium dioxide using a phenylene vinylene polymer, *Chem. Phys. Lett.* **1998**, *287*(1–2), 148–153.
79. U. Bach, D. Lupo, P. Comte et al., Solid-state dye-sensitized mesoporous TiO_2 solar cells with high photon-to-electron conversion efficiencies, *Nature* **1998**, *395*(6702), 583–585.
80. B. O'Regan, D. T. Schwartz, Efficient dye-sensitized charge separation in a wide-band-gap p–n heterojunction, *J. Appl. Phys.* **1996**, *80*(8), 4749–4754.
81. K. Tennakone, G. Kumara, A. R. Kumarasinghe et al., A Dye-sensitized nano-porous solid-state photovoltaic cell, *Semicond. Sci. Technol.* **1995**, *10*(12), 1689–1693.
82. Y. Tachibana, S. A. Haque, I. P. Mercer et al., Electron injection and recombination in dye sensitized nanocrystalline titanium dioxide films: a comparison of ruthenium bipyridyl and porphyrin sensitizer dyes, *J. Phys. Chem. B* **2000**, *104*(6), 1198–1205.
83. A. Fillinger, B. A. Parkinson, The adsorption behavior of a ruthenium-based sensitizing dye to nanocrystalline TiO_2 – coverage effects on the external and internal sensitization quantum yields, *J. Electrochem. Soc.* **1999**, *146*(12), 4559–4564.
84. P. Bonhote, E. Gogniat, S. Tingry et al., Efficient lateral electron transport inside a monolayer of aromatic amines anchored on nanocrystalline metal oxide films, *J. Phys. Chem. B* **1998**, *102*(9), 1498–1507.
85. R. Vogel, P. Hoyer, H. Weller, Quantum-sized Pbs, Cds, Ag2s, Sb2s3, and Bi2s3 particles as sensitizers for various nanoporous wide-bandgap semiconductors, *J. Phys. Chem.* **1994**, *98*(12), 3183–3188.
86. A. Zaban, O .I. Micic, B. A. Gregg et al., Photosensitization of nanoporous TiO_2 electrodes with InP quantum dots, *Langmuir* **1998**, *14*(12), 3153–3156.
87. M. Shur, in *Physics of Semiconductor Devices*, Prentice-Hall, Englewood Cliffs, 1990, pp. 86–102.
88. M. Lundstrom, in *Fundamentals of Carrier Transport*, Addison Wesley, Wokingham, 1990.
89. S. J. Fonash, in *Solar Cell Device Physics*, Academic Press, New York, 1981.
90. Y. Tachibana, J. E. Moser, M. Gratzel et al., Subpicosecond interfacial charge separation in dye-sensitized nanocrystalline titanium dioxide films, *J. Phys. Chem.* **1996**, *100*(51), 20056–20062.
91. T. Hannappel, B. Burfeindt, W. Storck et al., Measurement of ultrafast photoinduced electron transfer from chemically anchored Ru-dye molecules into empty electronic states in a colloidal anatase TiO_2 film, *J. Phys. Chem. B* **1997**, *101*(35), 6799–6802.
92. L. Dloczik, O. Ileperuma, I. Lauermann et al., Dynamic response of dye-sensitized nanocrystalline solar cells: characterization by intensity-modulated photocurrent spectroscopy, *J. Phys. Chem. B* **1997**, *101*(49), 10281–10289.
93. J. van de Lagemaat, N. G. Park, A. J. Frank, Influence of electrical potential distribution, charge transport, and recombination on the photopotential and photocurrent conversion efficiency of dye-sensitized nanocrystalline TiO_2 solar cells: a study by electrical impedance and optical modulation techniques, *J. Phys. Chem. B* **2000**, *104*(9), 2044–2052.
94. B. Levy, W. Liu, S. E. Gilbert, Directed photocurrents in nanostructured TiO_2/SnO_2 heterojunction diodes, *J. Phys. Chem. B* **1997**, *101*(10), 1810–1816.
95. H. Lindstrom, H. Rensmo, S. Sodergren et al., Electron transport properties in dye-sensitized nanoporous-nanocrystalline TiO_2 films, *J. Phys. Chem.* **1996**, *100*(8), 3084–3088.
96. M. K. Nazeeruddin, P. Pechy, M. Grätzel, Efficient panchromatic sensitization of

nanocrystalline TiO_2 films by a black dye based on a trithiocyanato-ruthenium complex, *Chem. Commun.* **1997**, *18*, 1705–1706.
97. M. Grätzel, Photoelectrochemical solar energy conversion by dye sensitization, *Future Generation Photovoltaic Technologies: First NREL Conference*, Denver: AIP, 1997.
98. A. Hagfeldt, U. Bjorksten, S. E. Lindquist, Photoelectrochemical studies of colloidal TiO_2-films – the charge separation process studied by means of action spectra in the UV region, *Sol. Energy Mater. Sol. Cells* **1992**, *27*(4), 293–304.
99. T. Dittrich, P. Beer, F. Koch et al., Barrier lowering in dye-sensitized porous-TiO_2 solar cells, *Appl. Phys. Lett.* **1998**, *73*(13), 1901–1903.
100. A. Stanley, B. Verity, D. Matthews, Minimizing the dark current at the dye-sensitized TiO_2 electrode, *Sol. Energy Mater. Sol. Cells* **1998**, *52*(1–2), 141–154.
101. A. Hagfedlt, M. Grätzel, *Acc. Chem. Res.* **2000**, *33*, 269–277.
102. J. Ferber, R. Stangl, J. Luther, An electrical model of the dye-sensitized solar cell, *Sol. Energy Mater. Sol. Cells* **1998**, *53*(1–2), 29–54.
103. J. Kruger, U. Bach, M. Grätzel, Modification of TiO_2 heterojunctions with benzoic acid derivatives in hybrid molecular solid-state devices, *Adv. Mater.* **2000**, *12*(6), 447 (6 pages).
104. R. A. Street, in *Hydrogenated Amorphous Silicon*, Cambridge University Press, Cambridge, 1991.
105. J. A. Anta, J. Nelson, N. Quirke, Charge transport model for disordered materials: application to sensitized TiO_2, *Phys. Rev. B* **2000**(Submitted).
106. F. Cao, G. Oskam, G. J. Meyer et al., Electron transport in porous nanocrystalline TiO_2 photoelectrochemical cells, *J. Phys. Chem.* **1996**, *100*(42), 17021–17027.
107. K. Schwarzburg, F. Willig, Influence of trap filling on photocurrent transients in polycrystalline TiO_2, *Appl. Phys. Lett.* **1991**, *58*(22), 2520–2522.
108. A. Solbrand, A. Henningsson, S. Sodergren et al., Charge transport properties in dye-sensitized nanostructured TiO_2 thin film electrodes studied by photoinduced current transients, *J. Phys. Chem. B* **1999**, *103*(7), 1078–1083.
109. J. Nelson, Continuous-time random-walk model of electron transport in nanocrystalline TiO_2 electrodes, *Phys. Rev. B: Condens. Matter* **1999**, *59*(23), 15374–15380.
110. N. F. Mott, Conduction in glasses containing transition metal ions, *J. Non-cryst. Solids* **1968**, *1*, 1–17.
111. H. Scher, E. W. Montroll, Anomalous transit time dispersion in amorphous solids, *Phys. Rev. B* **1975**, *12*, 2455.
112. A. Hagfeldt, S. E. Lindquist, M. Gratzel, Charge-carrier separation and charge-transport in nanocrystalline junctions, *Sol. Energy Mater. Sol. Cells* **1994**, *32*(3), 245–257.
113. C. L. Olson, R. L. Willis, T. Lutz et al. Current Transients, Optical Absorbance and Charge Recombination in Dye-sensitized Nanocrystalline Titanium Dioxide Electrodes, *13th International conference on photochemical conversion and storage of solar energy (IPS-2000)*, Snowmass, Colorado, 2000, Unpublished.
114. N. W. Duffy, L. M. Peter, R. M. G. Rajapakse et al., *Electrochemistry Communications 2*, **2000**, 658–662.
115. S. A. Haque, Y. Tachibana, D. R. Klug et al., Charge recombination kinetics in dye-sensitized nanocrystalline titanium dioxide films under externally applied bias, *J. Phys. Chem. B* **1998**, *102*(10), 1745–1749.
116. A. Blumen, G. Zumofen, J. Klafter, Target annihilation by random walkers, *Phys. Rev. B: Condens. Matter* **1984**, *30*(9), 5379–5382.
117. A. Blumen, J. Klafter, G. Zumofen, Models for reaction dynamics in glasses, in *Optical Spectroscopy of glasses*, (Ed.: I. Zschokke), D. Reidel, Dordrecht, The Netherlands, 1986, pp. 199–266.
118. G. M. Hasselmann, G.J. Meyer, Diffusion-limited interfacial electron transfer with large apparent driving forces, *J. Phys. Chem. B* **1999**, *103*(36), 7671–7675.
119. J. S. Salafsky, W. H. Lubberhuizen, E. van-Faassen et al., Charge dynamics following dye photoinjection into a TiO_2 nanocrystalline network, *J. Phys. Chem. B* **1998**, *102*(5), 766–769.
120. L. Peter, D. Vanmaekelbergh, Time and frequency resolved studies of photoelectrochemical kinetics, in *Advances in Electrochemical Science and Engineering*, (Eds.: R. Alkire, D. Kolb), Wiley WCH, Weinheim, 1999, pp. 77–163.

121. P. E. de Jongh, D. Vanmaekelbergh, Trap-limited electronic transport in assemblies of nanometer-size TiO_2 particles, *Phys. Rev. Lett.* **1996**, *77*(16), 3427–3430.
122. P. E. de Jongh, D. Vanmaekelbergh, Investigation of the electronic transport properties of nanocrystalline particulate TiO_2 electrodes by intensity-modulated photocurrent spectroscopy, *J. Phys. Chem. B* **1997**, *101*(14), 2716–2722.
123. D. Vanmaekelbergh, P. E. de Jongh, Electron transport in disordered semiconductors studied by a small harmonic modulation of the steady state, *Phys. Rev. B* **2000**, *61*(7), 4699–4704.
124. J. J. Kelly, D. Vanmaekelbergh, Charge carrier dynamics in nanoporous photoelectrodes, *Electrochim. Acta* **1998**, *43*(19–20), 2773–2780.
125. G. Schlichthorl, S. Y. Huang, J. Sprague et al., Band edge movement and recombination kinetics in dye-sensitized nanocrystalline TiO_2 solar cells: a study by intensity modulated photovoltage spectroscopy, *J. Phys. Chem. B* **1997**, *101*(41), 8141–8155.
126. G. Schlichthorl, N. G. Park, A. J. Frank, Evaluation of the charge-collection efficiency of dye-sensitized nanocrystalline TiO_2 solar cells, *J. Phys. Chem. B* **1999**, *103*(5), 782–791.
127. J. van de Lagemaat, A. J. Frank, Effect of the surface-state distribution on electron transport in dye-sensitized TiO_2 solar cells: Nonlinear electron-transport kinetics, *J. Phys. Chem. B* **2000**, *104*(18), 4292–4294.
128. N. W. Duffy, L. M. Peter, K. G. U. Wijayantha, Characterisation of electron transport and back reaction in dye-sensitized nanocrystalline solar cells by small amplitude laser pulse excitation, *Electrochem. Commun.* **2000**, *2*(4), 262–266.
129. J. Bisquert, G. GarciaBelmonte, F. Fabregat-Santiago et al., Doubling exponent models for the analysis of porous film electrodes by impedance. Relaxation of TiO_2 nanoporous in aqueous solution, *J. Phys. Chem. B* **2000**, *104*(10), 2287–2298.
130. J. S. Salafsky, Exciton dissociation, charge transport, and recombination in ultrathin, conjugated polymer-TiO_2 nanocrystal intermixed composites, *Phys. Rev. B: Condens. Matter* **1999**, *59*(16), 10885–10894.

5.4 Solid State Dye-sensitized Solar Cells – An Alternative Route Towards Low-Cost Photovoltaic Devices

Udo Bach
NTera Ltd., Dublin, Ireland

5.4.1 Introduction

To date, the field of applied photovoltaics is dominated exclusively by conventional inorganic semiconductor technologies such as silicon or thin-film solar cells. Since the invention of silicon solar cells in the 1940s, significant improvements in device manufacture and the introduction of novel thin-film techniques have led to a gradual increase in energy conversion efficiencies and made their production more energy- and cost-effective. However, classical inorganic p-n junction high-efficiency solar cells most probably will always demand the use of extremely pure starting materials and somewhat sophisticated deposition procedures.

A very promising alternative to classical, inorganic p-n junction solar cells is the concept of nanoporous, dye-sensitized photoelectrochemical solar cells, which was first introduced by Grätzel and coworkers [1]. In these cells, dye molecules, adsorbed onto a nanostructured TiO_2 electrode sensitize photoinduced charge separation across the interface. These cells exhibit external quantum efficiencies of up to 10% and can be made from less expensive starting materials and by simple printing techniques [2]. Therefore, they represent a promising alternative to classical inorganic p-n junction solar cells. Both industry and academia have research efforts toward development and commercialization of dye-sensitized solar cells. However, despite the very promising performance of these cells, the corrosive and volatile nature of the electrolyte represents an important drawback of this technology. Practical advantages might be gained by the replacement of the liquid electrolyte with a solid charge transport material (CTM). Organic materials and inorganic *p*-type semiconductors have been tested in this regard. The major focus of this chapter is to describe these concepts and to compare them to other recent ideas for combining new materials and concepts to solid-state photovoltaic junctions.

5.4.2 Solid-state Dye-sensitized Systems

When replacing the liquid electrolyte in dye-sensitized solar cells with a solid *p*-type conductor, a range of requirements have to be fulfilled to preserve the high external quantum efficiencies common to the original photoelectrochemical devices.

First, the *p*-type material needs to have electronic levels into which holes can be injected from the oxidized or excited state of the dye. The redox levels of the dye and the *p*-type material therefore have to be adapted carefully. An intimate contact between the sensitized metal oxide and the *p*-type material is vital to assure fast injection and regeneration processes (Fig. 1). This implies either the growth or deposition of one semiconductor inside a preformed, sensitized porous film of its counterpart or the in situ formation of the sensitized composite. Direct formation of the sensitized junction would be appreciable; however, charge collection within the two independent semiconductor networks, in which at least one semiconductor is formed from nanometer-sized inorganic semiconductor particles, demands intimate contact between the particles. Reduced

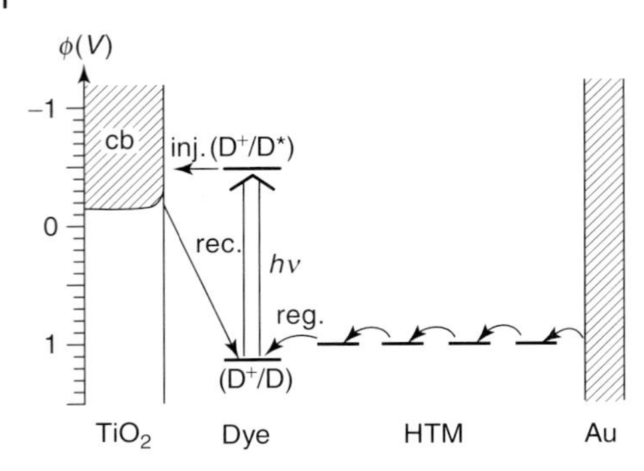

Fig. 1 Scheme for the electron transfer processes (injection, regeneration, and recapture) that occur in solid-state dye-sensitized heterojunctions.

interparticle contact was specified as a reason for the poor photovoltaic performance of early sensitized TiO_2/polymer composites [3–5]. However, Greenham and coworkers realized a CdSe/MEH-PPV (polymethoxyethyl hexyloxy-*p*-phenylene-vinylene) solar cell made via simple spin coating techniques, with a maximum incident photon-to-electron conversion efficiency (IPCE) of 12%, suggesting efficient electron percolation in the CdSe network [4]. Decapping of the nanoparticles before the formation of the composite was essential for the functioning of the device, a fact that has to be considered in future attempts to form composites with dye-derivatized nanometer-sized particles.

Once a positive charge is injected successfully into the *p*-type material, it has to be transported efficiently to the contacting electrode before interfacial recombination can occur. To allow for such efficient charge collection, both the *p*- and *n*-type material need sufficient conductivities so as not to impose high internal resistance and thereby limit charge collection. Screening of charges by the displacement of ions plays an important role in the charge collection process in photoelectrochemical solar cells. Solid-state dye-sensitized heterojunctions also exhibit increased photovoltaic performance, when ionic species introduced in the form of lithium salts were added to the phase of organic *p*-type conductors [6, 7]. It was postulated that the lithium salt might screen space charge layers, which are expected to build up in the illuminated heterojunction. The elimination of the space charge control of the photocurrent could thereby explain the observed increase in photovoltaic performance.

The contacting electrodes also have to be carefully chosen. The cathode material needs to form an ohmic contact to the *p*-type material so as to avoid injection barriers that would impede efficient charge collection. The anode needs to form a blocking contact to the *p*-type material so as to avoid internal shortcircuits. Generally, transparent conducting metal oxide layers such as FTO (fluorine doped tin(IV)oxide) or ITO (indium doped tin oxide) are used as the backcontact onto which the nanocrystalline film is applied. Iodine is unique, because it shows a very high overpotential at these materials. However, most solid *p*-type materials will form ohmic contacts with these transparent metal oxides causing internal short circuits at the points of contact [8]. This applies both to organic and inorganic materials. Consequently, whenever transparent conducting metal oxide layers are used in solid heterojunctions, care has to be taken to avoid direct contact between the p-type component of the sensitized heterojunction and the back contact of the photoanode.

5.4.2.1 Organic CTMs Applied to Nanoporous Dye-sensitized Metal Oxide Electrodes

Organic electronic materials nowadays are successfully replacing inorganic semiconductor components in a growing number of electronic devices such as xerographic ink drums or light emitting diodes (LEDs) [9, 10]. Further applications will soon follow including transistors [11], or entire integrated circuits fabricated on plastic [13]. Very recently, these materials also have been scrutinized with regard to their applicability in photovoltaic devices. In the following section, the application of organic materials in conjunction with dye-sensitized mesoporous metal oxide electrodes is reviewed, whereas alternative conceptual approaches toward organic solar cells that do not directly involve dye-sensitization are briefly discussed in the final section of this chapter.

Organic synthetic chemistry is a powerful tool used to develop materials with very specific electronic and optical properties. A large variety of electronically active materials with excellent, reproducible material properties became available over the last few years. These organic materials are still lagging their inorganic counterparts regarding their semiconducting properties such as conductivity and mobility. Organic materials, however, offer an almost unlimited potential to vary their specific properties. Furthermore, most of them can be prepared easily in their amorphous state. This is important because crystallization would most probably impair the formation of a good contact between the mesoporous surface of the sensitized metal oxide electrode and the CTM inside the sensitized heterojunction.

The application of organic CTMs in photovoltaic devices is a new research field, which still is in its infancy. Figure 2 overviews the organic charge transfer materials that were scrutinized with regard to their application in solid-state dye-sensitized solar cells. A much wider range of materials is mentioned in the patent literature [13–15].

The charge transfer events taking place in dye-sensitized heterojunctions are depicted in Fig. 1. Visible light is absorbed by the sensitizing dye. Electron injection from the excited state of the dye into the conduction band of TiO_2 (1) is followed by subsequent hole transfer from the photooxidized dye to the organic CTM, regenerating the dye's original ground state (2).

$$\mathrm{Dye}^* \longrightarrow \mathrm{Dye}^+ + \mathrm{e}^-(\mathrm{cb\ TiO_2}) \qquad (1)$$

$$\mathrm{Dye}^+ + \mathrm{CTM} \longrightarrow \mathrm{Dye} + \mathrm{CTM}^+ \qquad (2)$$

The latter process competes with recapture of the injected electron by the oxidized dye (3):

$$\mathrm{Dye}^+ + \mathrm{e}^-(\mathrm{cb\ TiO_2}) \longrightarrow \mathrm{Dye} \qquad (3)$$

The double injection process described by Eqs. (1) and (2) generates mobile majority carriers in both the TiO_2 and the hole conductor phases, which are subsequently collected at the contact electrodes. A second mechanism can be postulated that involves photoinduced hole transfer from the electronically excited dye to a neighboring charge transport molecule, followed by regeneration of the photoreduced state by electron injection into the TiO_2. Both mechanisms lead to the same initial charge separated state, which results in an electron in the conduction band of the TiO_2 and a positive charge, localized on a hole conductor molecule.

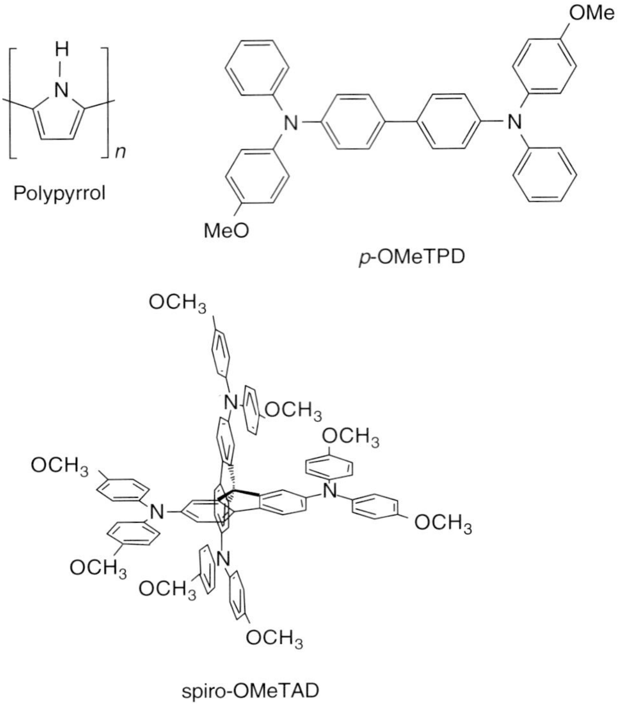

Polypyrrol

p-OMeTPD

spiro-OMeTAD

Fig. 2 Different CTMs used in solid-state dye-sensitized solar cells.

It is important to note, that hole conduction to the counter-electrode does not involve mass-transfer. When using low-molecular-weight CTMs, the underlying mechanism for charge transport generally is assumed to be a hopping mechanism as described by Bässler and coworkers [16], whereas charge transport in conjugated polymers is generally explained via a polaron mechanism [17].

Similar to photoelectrochemical solar cells, the concept of solid-state sensitized heterojunctions has the great advantage over conventional inorganic solar cells that only majority carriers are involved in the photoelectric conversion process. Photoelectric conversion in conventional inorganic p-n junction solar cells involves minority carriers whose lifetime is restricted because of recombination. Because they are generated throughout the semiconductor and away from the junction, expensive high-purity materials are required to maintain the minority carrier diffusion length at a level at which current losses caused by recombination are avoided.

The first to describe a solid-state device based on an organic CTM (*p*-OMeTPD, Fig. 2) were Hagen and coworkers [18]. The organic hole conductor was applied to a $Ru(dcbpy)_2(SCN)_2$-sensitized nanocrystalline TiO_2 electrode via thermal evaporation. A thin layer of gold, applied via vacuum deposition techniques, formed the back contact. IPCEs of these devices were still low (IPCE = 0.2%). Space charge effects were found to play a dominating role in this type of devices. The authors attribute the low mobility of electrons in the final device responsible for the poor performance. Incomplete pore filling and low conductivities might have been further reasons for the low efficiency, because thermal evaporation is not likely to transport organic material deep into the porous structure and because the charge carrier density inside the hole conductor matrix was low and merely influenced by accidental doping as a result of the presence of oxygen.

The first solid-state dye-sensitized heterojunction of TiO_2 and a semiconducting polymer was reported by Murakoshi and coworkers [19, 7]. They formed a solid-state heterojunction by photoelectrochemical polymerization of pyrrole into a nanoporous TiO_2 film, sensitized with $Ru(dcbpy)_2(SCN)_2$. Direct contact of polypyrrole and the FTO back contact was suppressed by electrodeposition of a thin TiO_2 film before the adsorption of the dye onto the nanocrystalline TiO_2 electrode. The doping density of the polymer was controlled electrochemically. Solar energy conversion efficiencies were somewhat higher than previously reported, but still did not exceed 0.1%.

Significant improvement in efficiency were achieved recently in a novel solar cell device applying the amorphous organic CTM 2,2′,7,7′-tetrakis(*N*,*N*-di-p-methoxyphenylamine)9,9′-spirobifluorene [7] (spiro-OMeTAD; Fig. 2). Photoinduced charge carrier generation was shown to be very efficient. A solar cell based on spiro-OMeTAD converts photons to electric current with a strikingly high yield of 50% [20]. The new hole conductor contained a spiro-center, which was introduced to improve the glass-forming properties and prevent crystallization of the material. Its glass-transition temperature of $T_g = 120\,^{\circ}C$, measured by differential scanning calorimetry, is much higher than that of comparable hole conductors. The methoxy groups were introduced to match the oxidation potential of the CTM to that of the applied sensitizer $Ru(dcbpy)_2(SCN)_2$.

Figure 3 shows a cross-section scheme of the sensitized TiO_2/spiro-OMeTAD heterojunction mentioned earlier. The

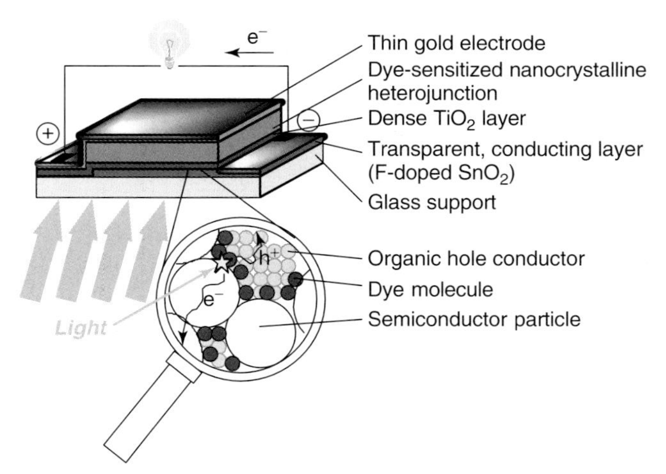

Fig. 3 Cross-section scheme for a solid-state dye-sensitized solar cell as described in by Bach and coworkers [6].

working electrode consisted of conducting glass (F-doped SnO_2) onto which a compact TiO_2 layer was deposited by spray pyrolysis [21]. This avoids direct contact between the hole conductor and the SnO_2, which would short-circuit the cell. A 4.2-μm thick mesoporous film of TiO_2 was deposited by screen printing onto the compact layer [22] and derivatized with $Ru(dcbpy)_2(SCN)_2$. The hole conductor was introduced into the mesopores by spin coating a concentrated solution of OMeTAD onto the TiO_2 film and subsequent evaporation of the solvent. A semitransparent gold back contact was evaporated on top of the hole conductor under vacuum.

The conductivity of the device was controlled by partial chemical oxidation of the hole conductor, before deposition onto the sensitized TiO_2 substrate to increase the number of free charge carriers [23]. Addition of the lithium salt $Li[(CF_3SO_2)_2N]$ to the spin coating solution of the hole conductor resulted in a strong performance increase in the final device. The underlying mechanism remained unidentified although charge screening due to partial ionic mobility inside the hole conductor matrix and/or the effect of the present lithium ions on the flat band potential of TiO_2 [24] were postulated as possible mechanisms.

Figure 4 shows the photocurrent action spectra of three typical cells under short circuit conditions [20]. Three different dyes were used to sensitize the heterojunction. In all three cases the action spectrum matches closely the absorptivity spectrum of the sensitizing dye. Solid-state heterojunctions sensitized with the merocyanine dyes (structure see Fig. 4) used in this study showed higher peak IPCE values when compared to $Ru(II)L_2(SCN)_2$-sensitized junctions. Strongly improved photovoltaic performances were reported recently for comparable systems [25].

Picosecond transient absorption laser spectroscopy was used to scrutinize the dynamics of the photoinduced charge separation process [26]. The dye regeneration process [2] could be time-resolved, and was shown to proceed primarily on the picosecond timescale with multiphasic kinetics. Therefore it is at least one order of magnitude faster than the dye regeneration process involving iodine or iodide electrolytes [2, 27]. The high hole injection rate insures that the quantum efficiency for the overall charge separation process across the junction is close to unity. This may be a key factor behind the high external quantum efficiency for charge carrier collection obtained for spiro-OMeTAD compared to other solid-state hole conducting materials studied to date. Transient absorption spectroscopy also gave indications that in the specific junction, electron injection into TiO_2 proceeds before hole injection into spiro-OMeTAD.

Analysis of the front and backside illumination photoresponse spectra proved to be a powerful tool to probe interfacial charge recombination during charge collection [20]. Figure 5 shows a typical example of a set of photoresponse spectra obtained on front and backside illumination, compared to the absorptivity of the sensitized TiO_2 layer. At the absorption maximum of the dye, the $IPCE^{front}$ (illumination through the conducting glass) exhibits a maximum, whereas the $IPCE^{back}$ spectrum (illumination through the semitransparent gold layer) exhibits a minimum. Even when corrected for the transmission losses of the conducting glass and gold electrode, respectively, the IPCE value for backside illumination mode is one order of magnitude smaller than the corresponding value for $IPCE^{front}$. This

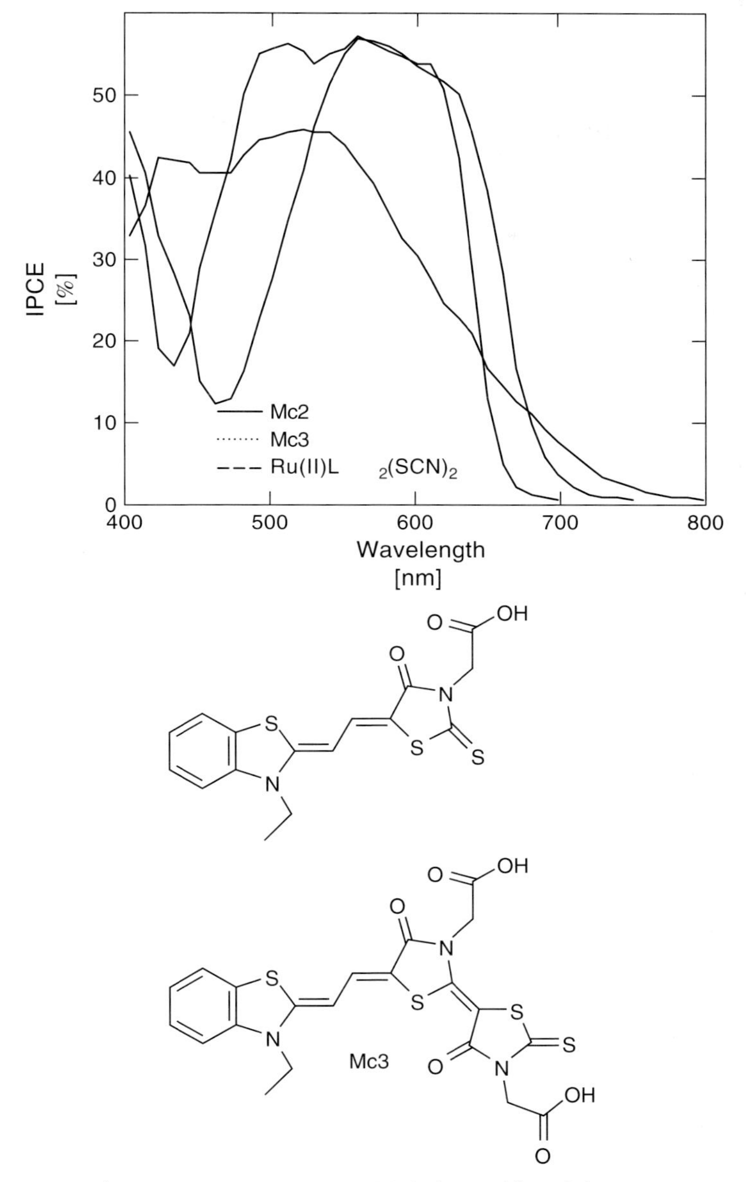

Fig. 4 Photocurrent action spectra (IPCE) obtained for solid TiO_2/spiro-OMeTAD heterojunctions, sensitized with three different sensitizers. (For device structure, see Fig. 3.) The dyes were absorbed from above solution for 48 hours at room temperature. All devices are based on 4-μm thick nanocrystalline TiO_2 electrodes and spiro-OMeTAD as hole conducting layer (0.070 mol% spiro-OMeTAD^{++}(PF_6) in respect to spiro-OMeTAD; the spin coating solution contained 15 mM $Li[(CF_3SO_2)_2N]$). The counter electrode was a 10-nm thick gold layer.

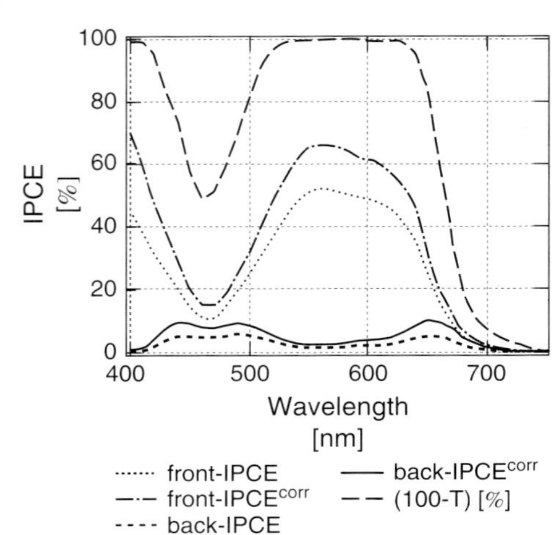

Fig. 5 Set of photoresponse spectra obtained under front and backside illumination conditions. The dotted curves show the IPCE spectra as measured, whereas the full curves corresponds to the corrected IPCE spectra, which take adsorption and reflection losses through the corresponding window materials (FTO glass or semitransparent gold layer, respectively). The top curve represents the absorptivity of the device, which corresponds to a calculated IPCE value assuming no reflection or recombination losses.

finding strongly suggests that recombination of photogenerated electrons while migrating toward the collecting electrode is a major loss mechanism. Applying a simple recombination model, a free electron diffusion length of about 1 μm could be determined for the observed junction. Charge recombination occurring during charge collection therefore could be identified as major loss mechanism of the cell.

A strong increase in energy conversion efficiency could be observed when tertiarybutylpyridine was introduced into the matrix of the organic hole conductor [20]. Although the typical open circuit voltage (V_{oc}) in absence of tertiarybutylpyridine reaches approximately 350 mV, it can easily exceed 700 mV in the presence of tertiarybutylpyridine. This leads to an increase in white light conversion efficiency from 0.74% to 1.75% (AM1.5, 10 mW cm^{-2}). Very similar effects were reported for classical sensitized I^-/I_3^- electrolyte/TiO_2 junctions [2]. The authors rationalized the effect in terms of 4-*tert*-butylpyridine adsorption at the TiO_2 surface blocking surface states that are active as intermediates in the interfacial charge recombination. They presumed that these sites are represented by Ti (IV) ions on the surface, which, because of their Lewis acidity, are prone to interact with the pyridine derivative. Similar results were obtained by Huang and coworkers [28], who attributed the observed increase in V_{oc} to either a charging of surface states or a shift of the conduction band edge.

Different techniques have been proposed to fill sensitized nanoporous structures with organic material. Evaporation [19], electrodeposition [7] and spin coating [6] techniques were described in this respect. Little is known regarding intimacy of the contact between the sensitized metal oxide surface and the organic CTM when simple spincoating procedures are used. Therefore complete pore filling still remains a critical problem. Although spin coating or dip coating techniques are very simple and inexpensive, electrodeposition might still be a more controlled technique. In situ polymerization of monomeric precursors appears to be the most reliable way to fill nanoporous electrodes with large-molecular-weight polymers. Very recently it was shown that low-molecular-weight organic semiconductors also can be electrodeposited into nanoporous TiO_2

electrodes [29], which thereby represents a very powerful technique to deposit poorly soluble materials sush as perylenes or phthalocyanines into porous structures.

Many questions regarding the function of these sensitized inorganic-organic hybrid materials still remain unsolved. Most importantly, the factors controlling the electronic properties of the junction have to be identified and understood. First steps toward a better control of the interface have been made. The dipole field, present at the interface because of the self-assembled monolayer of organic molecules proved to strongly affect the junction's electronic properties [30]. Further systematic studies will hopefully help to widen the understanding of such devices, leading the way to increased device performances.

5.4.2.2 Inorganic Nanoporous Dye-sensitized Heterojunctions

In dye-sensitized inorganic heterojunction solar cells, a monolayer of dye is sandwiched between two wide band gap semiconductors; one of them exhibits a *p*-type and the other an *n*-type conduction mechanism. Inorganic *p*-type semiconductors were successfully applied in an attempt to replace the liquid electrolyte in photoelectrochemical solar cells, and in fact the first solid-state dye-sensitized photovoltaic device described in the literature was based on a wide band gap *p*-type semiconductor material [31].

Such junctions mainly are of inorganic character, however, the dye present at the junction's interface still plays the central role in the initial formation of the charge separated state. The sensitizer takes on the role of both the light absorber and acts as the link between the two inorganic materials. The choice of inorganic wide band gap semiconductors applicable in such junctions is small, and only few combinations of wide band gap *p*-type and *n*-type semiconductors are energetically favorable. Organic and organo-metallic dyes easily can be adjusted with regard to their redox potentials and absorption characteristics to match the band positions of these materials. In principle, the dye also might be chosen to function as a flexible buffer between the two nanocrystalline phases, reducing strain effects, which arise as a result of the lattice mismatch of the two semiconductor materials. A potential advantage of inorganic semiconductors over their novel organic counterparts is their generally high charge mobility. However, the very limited choice of potentially interesting materials is a clear drawback, compared to the wide choice of organic CTMs.

Two mechanisms can contribute to the formation of the initial charge separated state. The electronic excitation of the dye either could be followed by electron injection into the *n*-type semiconductor or by hole injection into the *p*-type semiconductor. Both mechanisms have been shown to occur and both are likely to contribute to the charge separation mechanism [32]. However, no studies were undertaken to differentiate which of the two possible mechanisms mainly is responsible for the initial formation of the charge separated state.

Although dye sensitization of wide band gap *n*-type semiconductors has been comprehensively studied, little is known regarding the sensitization of *p*-type materials. Sensitized electron injection into an *n*-type semiconductor and hole injection into a *p*-type material are analogous processes. The misbalance might be explained simply in terms of the wider availability of stable wide band gap *n*-type materials, which allow for the preparation of nanoporous films with excellent electronic properties.

Only a few materials were studied concerning their applicability to dye-sensitized hole injection processes. Among those are different copper(I) compounds (e.g. Cu(I)SCN, Cu(I)I, Cu_2(I)O [33–35]) and nickel(II) oxide [36]. Photovoltaic performances of such devices are orders of magnitudes poorer than those of classical dye-sensitized photoelectrochemical solar cells based on *n*-type materials. Substantial advantages could arise if an efficient photo–hole injection process would be available. The formation of solid-state tandem solar cells would become feasible, and a quantum step in device efficiency of dye-sensitized solar cells could be at reach. However, because of the poor performance of all known photocathodes, a combination of available photoanodes and photocathodes to a tandem device always results in a device that is photovoltaically less efficient than the photoanode on its own. The concept for electrolyte-based tandem cells exists. However, it contains strong potential to improve the photovoltaic performance in both electrolytic and in solid-state, dye-sensitized solar cells.

Analogous to their photoelectrochemical counterparts, efficient dye-sensitized solid-state solar cells require a nanoporous junction structure. The first solar cell of this type was reported in 1995 [33] in form of a TiO_2/Cu(I)I heterojunction, sensitized with a cyanidine dye. When forming an inorganic interpenetrating network on the nanometer scale, the crystal lattice mismatch of the two inorganic semiconductors is likely to impair the formation of a good contact between the two nanocrystalline structures. Recently, O'Regan and coworkers reported on a sensitized nanocomposite of ZnO and CuSCN forming a bicontinuous network heterojunction [37]. Nanoporous ZnO films with a columnar morphology, oriented perpendicular to the substrate surface could be realized, using an electrodeposition technique. These highly anisotropic dye-sensitized films were shown to form a complete inner surface electrical contact when *p*-type CuSCN was electrodeposited into the pores (Fig. 6). These junctions exhibited external quantum efficiencies of up to 50%, and incident energy conversion efficiencies of 1.5%. Dye-sensitized heterojunctions made from colloidal ZnO particles on the other hand, showed maximum external quantum efficiencies of only 20%. It could thereby be shown that lattice mismatch is not an intrinsic problem of dye-sensitized inorganic heterojunctions but can be overcome by choosing appropriate device assembly techniques and materials. The sensitizing dye might also exhibit a certain buffer function between the two crystal structures, especially if the dye is substituted with two sets of attachment groups, each of them strongly interacting with one of the semiconductor components.

Solid-state dye-sensitized TiO_2/CuI heterojunction were recently realized by Tennakone and coworkers [38]. The authors reported on surprisingly high solar energy conversion efficiencies, showing that in the future it might be feasible to form solid-state dye-sensitized solar cells with similar efficiencies as their photoelectrochemical counterparts.

5.4.3 Alternative Organic Solid-state Approaches

5.4.3.1 Organic Solar Cells

Thin films of organic pigments are known to exhibit semiconducting properties from their use in xerography; Morel, Ghosh, Feng and coworkers [39, 40] reported on a solar cell in 1978 based on the Schottky

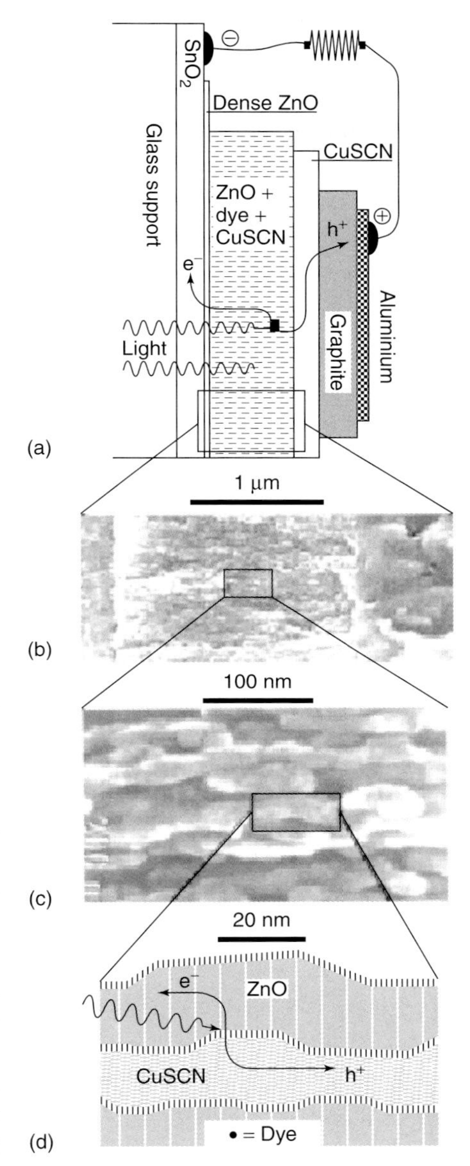

Fig. 6 Scheme for a sensitized nanocomposite of ZnO and CuSCN, as described by O'Regan and coworkers [37]. (a) Layered structure of the dye-sensitized heterojunction. (b) Fractured edge of a complete DSH, showing the SnO_2 layer, the porous ZnO/dye/CuSCN layer, and some of the overlying CuSCN. (c) Fractured edge of a porous ZnO layer, without CuSCN, showing the columnar structure perpendicular to the substrate. (d) Schematic of the ZnO/dye/CuSCN interface with dye molecules shown to scale, but at low packing density.

junction of aluminum and a merocyanine dye, which showed an overall sunlight conversion efficiency of 0.7% – about 35 times higher than any efficiency that had been reported on organic films before. Following that, interest in organic solar cells strongly increased, particularly driven by the research effort undertaken by companies such as Exxon, Xerox, Kodak, and Shell in the 1980s. However, white light conversion efficiencies never exceeded 1%, so that the general interest in this technology decreased after a number of years. Several comprehensive reviews describing organic solar cells in great in-depth were published since [41–44]. A short summary follows, to allow for the comparison with alternative technologies.

In thin-film organic solar cells, light is absorbed by a dye layer, typically of a thickness of several tens of nanometers. Light absorption leads to the formation of an exciton (an electron/hole pair, which is strongly bound). These excitons can give rise to free charge carriers by dissociation under the influence of a strong electric field. In general, such fields develop close to the interface of two materials exhibiting different work functions because of the formation of a charge carrier depletion layer. According to the origin of this built-in field, one can distinguish three different types of organic solar cells. Most organic photodiodes are based on the rectifying junction of an organic semiconductor and a metal, forming a Schottky type cell such as the cell reported by Morel and coworkers. A rectifying contact can also be formed by the formation of a p-n junction between the light-absorbing, semiconducting dye and a second semiconductor of opposite conduction type. This second semiconductor either can be an inorganic wide band gap semiconductor or a second light-absorbing, semiconducting dye, giving rise to an *organic/inorganic* or *organic/organic* p-n heterojunction.

The main limitation of all three device types is the limited diffusion length of excitons in organic materials, which usually does not exceed several tens of nanometers. Therefore the photovoltaically active region in a thin-film solar cell is limited to a very thin layer close to the junction. This usually is too thin to assure good light harvesting over a wide spectral range. Very similar to conventional inorganic p-n junction solar cells, the exciton lifetime strongly decreases in the presence of bulk impurities, demanding highly purified dyes. Heterojunction formation by two different organic dyes can improve the light harvesting, especially if their absorption spectra are complementary to each other [45, 46]. However, layer thicknesses have to be optimized to prevent optical filter effects caused by photovoltaically inactive areas away from the junction. Attempts were made to widen the thickness of the photovoltaically active area by incorporation of a coevaporated layer of the two organic, semiconducting device components, resulting in a significant increase in device efficiency [47].

Another limitation of organic thin-film solar cells is the generally poor conductivity of the organic layer. Organic devices commonly show strong sublinearity of the photocurrent at illumination intensities close to AM1.5 (100 mW cm^{-2}). However, control of the charge carrier density via chemical doping with electron acceptors such as iodine for *p*-type materials [48] or with donors such as H_2 for *n*-type materials [49] can improve drastically the conductivity and device performance.

An interesting approach partially overcoming the problem of low conductivity and limited exciton diffusion widths is to combine two organic p-n junctions in series, leading to a tandem cell with a doubled photovoltage, compared to the simple p-n structure device [50].

5.4.3.2 **Interpenetrating Networks**

As described in the previous section, the recombination of an exciton before it reaches the built-in field region is the main loss mechanism in organic thin-film solar cells. This problem can be overcome by the formation of an interpenetrating p-n heterojunction in which the average width of the *p*-type and *n*-type domains are smaller than the exciton diffusion length, thereby guaranteeing that every exciton can reach the interface before it decays. However, the role of the electric field in these devices is questionable because the theoretical depletion layer width generally is much wider than the dimension of phase segregation. It might be more exact to assume the totality of the interpenetrating network to be depleted with exciton splitting occurring via an interfacial electron transfer mechanism. Similar to thin-film solar cells, the rectifying junction can be formed by two organic semiconductors, or by a blend of an organic and an inorganic semiconductor.

Semiconducting polymers are superior to their low-molecular-weight analogs, with regard to their applicability in interpenetrating junctions. This is due to the fact that charge transport in semiconducting polymers mainly proceeds along the polymer chain, so that percolation starts already at low concentration thresholds (for a review on polymer solar cells see Ref. 51).

Poly(*p*-phenylenevinylene) derivatives (PPVs) are the most prominent examples of polymers used in organic solar cells. Depending on their substituents they either can exhibit *p*-type or *n*-type behavior. The first junction of this type was reported in 1995, using a phase-segregated thin film of MEH-PPV as *p*-type material and CN-PPV (cyano-*p*-phenylenevinylene) as *n*-type material [52]. The phase segregation was on the scale of 10–100 nm and IPCE values as high as 6% were recorded. Recently, significantly improved device performances were reported based on a polythiophene/PPV-derivative heterojunction, assembled via a lamination technique with IPCE values reaching 29% [53]. The cross-section scheme and the molecular structures can be seen in Fig. 7.

A different approach to form organic interpenetrating heterojunctions is to dope strongly a *p*-type conducting polymer with a low-molecular-weight acceptor until charge percolation occurs inside the acceptor phase. The blends attracting most attention are those of PPV- or thiophene derivatives doped with fullerenes as acceptor, reaching 23% IPCE for polythiophene/C_{60} junctions [54] and 29% for PPV/C_{60} junctions [55]. Perylene nanocrystals were also used as doping acceptor [56]. Different theoretical models were developed to describe the functioning principals of these devices [57, 58].

5.4.3.3 Nonsensitized Inorganic/Organic Junctions

Recently, increased interest was raised in the photovoltaic properties of composites of semiconducting polymers and nanosized semiconductor particles, in which either the polymer or the semiconductor *and* the polymer take over the function of the light absorber. Among those, nanocrystalline heterojunctions of TiO_2 either with PPV [59] or polythiophene [60] derivatives have received particular interest. Several different techniques were applied to form the heterojunctions. Commonly the first step was the formation of a nanoporous thin film of TiO_2 via spray pyrolysis or via deposition of a colloidal paste followed by a sintering step while the polymer was applied via a spincoating technique or (in the case of polythiophene) via electrodeposition. IPCE values reported on these systems are still very low (below 3%). However, just recently, highly efficient TiO_2/polythiophene solar cells exhibiting peak IPCE values of up to 10% were accomplished [61]. Photovoltaic cells of comparable efficiency were obtained when polycrystalline films of low band gap semiconductors such as CdSe, CdS, or $CuInSe_2$ are contacted with a thin film of polythiophene via electrodeposition [62]. Overall white light conversion efficiencies of 1.3% at 56 mW cm^{-2} were reported for a nanostructured CdSe/poly(3-methylethiophene) junction [63].

One particularly interesting aspect of nanoparticle/polymer solar cells is that charge "hopping" from one semiconductor particle to another takes place even in blends prepared from spin coating solutions containing both the polymer and the colloidal nanoparticles. Usually, those devices are not exposed to successive heat treatments. This is surprising because nanocrystalline semiconductor films used in photoelectrochemical solar cells generally are believed to need such a heat treatment to induce interparticle contact and "necking." Less expensive, flexible dye-sensitized solar cells based on plastic substrates could be realized if this heat treatment could be made redundant.

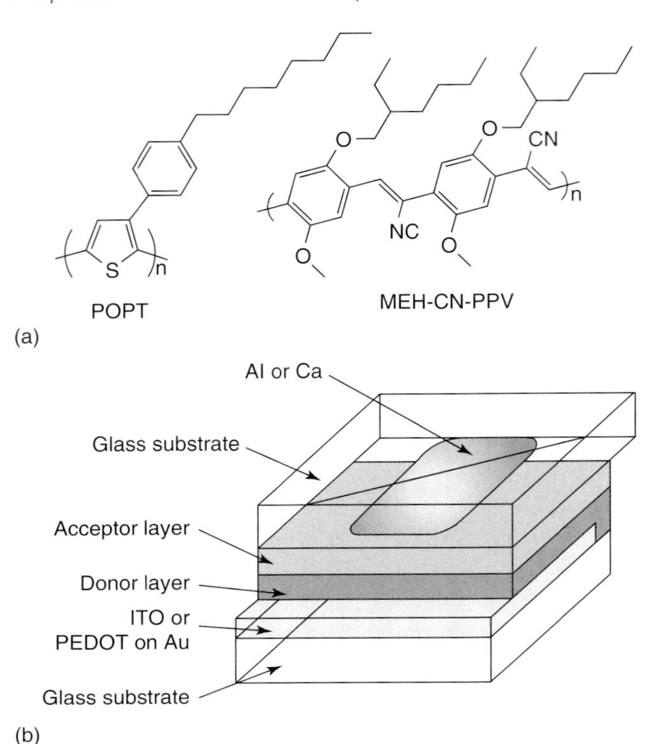

Fig. 7 Chemical structures and device structure for an organic polymer photodiode as described by Granström and coworkers [53]. (a) Chemical structures of the polymers used in the devices. POPT is a regioregular phenyl-octyl substituted polythiophene. MEH/CN-PPV is a cyano-derivative of poly(*p*-phenylene vinylene), with a large electron affinity as a result of the electron-withdrawing cyano groups. (b) Device structure. Both polymers (see (a) for chemical structures) were dissolved in chloroform or toluene (5 mg ml^{-1}). For the top half of the device, aluminum or calcium contacts were thermally evaporated on glass substrates and the acceptor material, MEH/CN-PPV (and a small amount of POPT, usually 5%), was spin-coated on top of the metal electrode. The other half of the device comprises the donor material, POPT (and a small amount of MEH/CN-PPV, usually 5%), either spin-coated on ITO substrates or glass-coated with poly(ethylene dioxythiophene) doped with polystyrene sulfonic acid. To ensure a low contact resistance, a thin layer of gold ($\sim$10 nm) was thermally evaporated on the glass slide before spin-coating the PEDOT film from a water solution. The thickness of the PEDOT layer was 100 nm. To get the desired structure of the POPT, this half of the device was heated to 200 °C under vacuum before the device was laminated together by applying a light pressure while one half was still at elevated temperature. The total thickness of the organic semiconductive layer was 70–80 nm, and the active area 2.5 mm^2.

Various examples of composite nanoparticle or polymer films made via spin coating can be found in the literature [3, 4, 64]. Most examples actually show that charge percolation inside the nanoparticle network is reduced strongly, compared to nanocrystalline, sintered films. Arango and coworkers [3] reported a decrease in photoconductivity of 2 orders of magnitude when comparing a sintered polymer-coated TiO_2 film, to a composite made via spincoating of a TiO_2-colloid/polymer solution. The reduced interparticle contact also leads to low photovoltaic performances. An exception to these observations is the CdSe/PPV composite solar cell reported by Greenham and coworkers [38]. Peak IPCE values of 12% were obtained from spin-coated composite thin films with a weight ratio of CdSe (ϕ 5 nm): MEH-PPV = 9 : 1 (film thickness = 300–500 nm). Removal of surface-adsorbed species from the colloid before the cell assembly was shown to be crucial for interparticle electron transfer to occur.

5.4.4 Conclusion and Outlook

A large number of different approaches toward solid-state dye-sensitized or organic solar cells were recently described in the literature, and some of these are briefly described in this chapter. The efficiencies of these novel prototype devices are constantly rising, some of them showing energy conversion efficiencies of up to 2%. Such concepts not only are of scientific but also of potentially economic interest, indicating that the realization of less expensive, efficient plastic solar cells might soon be at hand. These recent developments strongly benefit from the continued success in the synthesis of novel organic electronic components. Those organic materials have found their way into fields, which until recently were purely dominated by inorganic device physics. Organic LEDs, for example, are already outperforming their inorganic counterparts in certain types of applications. Very often the pioneering work that was done to realize those novel applications finally gave access to materials that also showed very interesting photovoltaic properties. Examples are PPV derivatives developed for their use in organic LEDs [51] and pentacene single crystals studied by Lucent Technologies for use in organic transistors [11]. It was only at a later stage that interesting photovoltaic properties of such materials were scrutinized [52–55, 65]. The organic compounds used in these early studies were poorly optimized with regard to their photovoltaic properties; nonetheless, those initial prototypes showed impressive performances.

Triggered by these initial promising results, new efforts currently are under way to develop tailor-made organic materials with improved optical and electronic properties for their use in solar cells. These second-generation organic photovoltaic materials almost necessarily will lead to significant improvements in device performances. Along with the development of novel materials, there is the need to gain a better understanding of the different concepts described in this chapter. Most of these are only poorly understood, when compared to well-established systems such as photoelectrochemical solar cells. Nevertheless, very exciting advancements were realized over the last years and the field of organic photovoltaics is steadily progressing. Organic solar cells have the potential for very low production costs at reasonable high efficiencies. These properties make them interesting for their application in mobile electronic consumables, in which high efficiencies and long-term stability in extreme environments are not an ultimate requisite. Therefore, the predescribed concepts emerge as a very interesting and viable option for future low-cost plastic solar cells.

Acknowledgment

Greg Smestad is gratefully acknowledged for proofreading this chapter.

References

1. B. O'Regan, M. Grätzel, *Nature* **1991**, *353*, 737–739.
2. M. K. Nazeeruddin et al., *J. Am. Chem. Soc.* **1993**, *115*, 6382–6390.
3. A. C. Arango, S. A. Carter, P. Brook, *J. Appl. Phys. Chem.* **1999**, *74*, 1698–1700.
4. N. C. Greenham, X. Peng, A. P. Alivisatos, *Phys. Rev. B* **1996**, *54*, 17628.
5. M. Kocher, T. K. Däubler, E. Harth et al., *Appl. Phys. Lett.* **1998**, *72*, 650–652.
6. U. Bach, D. Lupo, P. Compte et al., *Nature* **1998**, *395*, 583–585.
7. K. Murakoshi, R. Kogure, Y. Wada et al., *Sol. Energy Mater. Sol. cells* **1998**, *55*, 113–125.
8. U. Bach, Solid-state dye-sensitized mesoporous TiO_2 solar cells, 2000, Ph.D. thesis N° 2187, EPFL Lausanne, Switzerland.
9. G. Hadziioannou, P. F. van Hutten in *Semiconducting polymers*, Wiley-VCH, Weinheim, 2000.
10. K. Müllen, G. Wegner in *Electronic Materials: The Oligomer Approach*, Wiley-VCH, Weinheim, 1998.
11. C. D. Dimitrakopoulos, S. Purushothaman, J. Kymissis et al., *Science* **1999**, *283*, 822–824.
12. B. Crone, A. Dodabalapur, Y.-Y. Lin et al., *Nature* **2000**, *403*, 521–523.
13. D. Lupo, J. Salbeck 1999, US5885368.
14. L. Hausslinger et al., 1997, US5683833.
15. K. Shiratsuchi, H. Takizawa 2000, US 6084176.
16. P. M. Borsenberger, L. Pautmeier, H. Bässler, *J. Chem. Phys.* **1991**, *94*, 5447–5454.
17. M. N. Bussac, L. Zuppiroli, *Phys. Rev.* **1996**, *B54*, 4674.
18. J. Hagen, W. Schaffrath, P. Otschik et al., *Synth. Met.* **1997**, *89*, 215–220.
19. K. Murakoshi, R. Kogure, S. Yanagida, *Chem. Lett.* **1997**, *5*, 471–472.
20. U. Bach, J. Krüger, M. Grätzel, *SPIE proceedings San Diego*, San Diego, Calif., 2000, in print.
21. L. Kavan, M. Grätzel, *Electrochim. Acta* **1995**, *40*, 643–652.
22. C. J. Barbé et al., *J. Am. Ceram. Soc.* **1997**, *80*, 3157–3171.
23. M. Abkowitz, D. M. Pai, *Philos Mag. B* **1986**, *53*, 193–216.
24. B. Enright, G. Redmond, D. Fitzmaurice, *J. Phys. Chem.* **1994**, *97*, 1426–1430.
25. J. Krüger, R. Plass, L. Cavey et al., *Appl. Phys. Lett.* **2001**, *79*, 2085–2087.
26. U. Bach, Y. Tachibana, J.-E. Moser et al., *J. Am. Chem. Soc.* **1998**, *121*, 7445–7446.
27. S. A. Haque, Y. Tachibana, D. R. Klug et al., *J. Phys. Chem. B* **1998**, *102*, 1745–1749.
28. S. Y. Huang, G. Schlichthörl, A. J. Notzik et al., *J. Phys. Chem. B* **1997**, *101*, 2576–2582.
29. A. Zaban, Y. Diamant, *J. Phys. Chem. B* **2000**, *104*, 10043–10046.
30. J. Krüger, U. Bach, M. Grätzel, *Adv. Mater.* **2000**, *12*, 447–451.
31. K. Tennakone, K. P. Hewaparakkrama, M. Dewasurendra et al., *Semicond. Sci. Technol.* **1988**, *3*, 382–387.
32. B. O'Regan, D. T. Schwartz, *Semicond. Chem. Mater.* **1998**, *10*, 1501–1509.
33. K. Tennakone, G. R. R. A. Kumara, A. R. Kumarasinghe et al., *Semicond. Sci. Technol.* **1995**, *10*, 1689–1693.
34. C. A. N. Fernando, *Sol. Energy. Mater.* **1993**, *30*, 211–220.
35. C. R. N. Fernando, *Sol. Energy. Mater.* **1992**, *28*, 255–271.
36. J. He, H. Lindström, A. Hagfeldt et al., *J. Phys. Chem. B* **1999**, *103*, 8940–8943.
37. B. O'Regan, D. T. Schwartz, M. S. Zakeerudin et al., *Adv. Mater.* **2000**, *12*, 1263–67.
38. K. Tennakone, G. R. A. A. Kumara, I. R. M. Kottegoda et al., *J. Phys. D: Appl. Phys.* **1998**, *31*, 1492–1496.
39. A. K. Ghosh, T. Feng, *J. Appl. Phys.* **1978**, *49*, 5982–5989.
40. D. L. Morel, A. K. Ghosh, T. Feng et al., *Appl. Phys. Lett.* **1978**, *32*, 495–497.
41. G. A. Chamberlain, *Sol. Cells* **1983**, *8*, 47.
42. G. Horowitz, *Adv. Mater.* **1990**, *2*, 287.
43. D. Wöhrle, D. Meissner, *Adv. Mater.* **1991**, *3*, 129.

44. J. Simon, J. J. André, in *Molecular Semiconductors*, Springer-Verlag, Berlin, 1985.
45. C. W. Tang, *Appl. Phys Lett.* **1986**, *48*, 183.
46. K. Kudo, T. Moriizumi, *Jpn. J. Appl. Phys.* **1981**, *20*, L553–L556.
47. M. Hiramoto, H. Fujiwara, M. Yokoyama, *J. Appl. Phys.* **1992**, *72*, 3781–3787.
48. K. Yamashita, Y. Harima, T. Matsubayashi, *J. Phys. Chem.* **1989**, *93*, 5311–5315.
49. M. Hiramoto, Y. Kishigami, M. Yokoyama, *Chem. Lett.* **1990**, 119–122.
50. M. Hiramoto, M. Suezaki, M. Yokoyama, *Chem. Lett.* **1990**, 327–330.
51. G. Hadziioannou, P. F. van Hutten, in *Semiconducting Polymers*, Wiley-VCH, Weinheim, 2000.
52. J. J. M. Halls, C. A. Walsh, N. C. Greenham et al., *Nature* **1995**, *376*, 498–500.
53. M. Granström, K. Petrisch, A. C. Arias et al., *Nature* **1998**, *395*, 257–260.
54. L. S. Roman, W. Mammo, L. A. A. Pettersson et al., *Adv. Mater.* **1998**, *10*, 774–777.
55. G. Yu, J. Gao, J. C. Hummelen et al., *Science* **1995**, *270*, 1789.
56. J. J. Dittmer et al., *Sol. Energy Mater. Sol. Cells* **2000**, *61*, 53–61.
57. A. Köhler, D. A. D. Santos, D. Beljonne et al., *Nature* **1998**, *392*, 903–906.
58. L. A. A. Pettersson, L. S. Roman, O. Inganäs, *J. Appl. Phys.* **1999**, *86*, 487–496.
59. T. J. Savenije, J. M. Warman, A. Goossens, *Chem. Phys. Lett.* **1998**, *287*, 148–153.
60. K. Kajihara, K. Tanaka, K. Hirao et al., *Jpn. J. Appl. Phys.* **1997**, *36*, 5537–5542.
61. S. Spiekermann, S. Mayer, W. Blau et al., unpublished results.
62. S. A. Gamboa, H. Nguyen-Cong, P. Chartier et al., *Sol. Energy Mater. Sol. Cells* **1998**, *55*, 95–104.
63. P. Chartier, H. N. Cong, C. Sene, *Sol. Energy Mater. Sol. Cells* **1998**, *52*, 413–421.
64. J. S. Salafsky, W. H. Lubbrthuizen, R. E. I. Schropp, *Chem. Phys. Lett.* **1998**, *290*, 297–303.
65. J. H. Schoen, C. Kloc, E. Bucher et al., *Nature* **2000**, *403*, 408–410.

6
Nonsolar Energy Applications

6.1 Fundamentals of Photocatalysis

Akira Fujishima and Donald A. Tryk
The University of Tokyo, Tokyo, Japan

6.1.1 Introduction

Remarkable progress has been made in the area of photocatalysis in almost 25 years since the earliest work appeared [1–3]. It was recognized that TiO_2 is a highly suitable material for photocatalysis because of its complete resistance to photocorrosion. This virtue of TiO_2 had been recognized only a few years earlier in the work on photoelectrochemical (PEC) water splitting [4–7]. Other virtues include the low cost and nontoxicity of TiO_2. These have allowed TiO_2 to be used in an increasing number of applications involving both indoor and outdoor air purification, water purification, self-cleaning surfaces, and antifogging surfaces. This introductory chapter focuses on the basic aspects of photocatalysis.

6.1.2 Background

As outlined by Heller [8], the extensive work that was carried out in the field of photoelectrochemistry during the 1960s, 1970s, and 1980s provided an excellent basis for work on photocatalysis. Each of the basic principles related to photoelectrochemistry is also involved in photocatalysis. To a first approximation, the framework of photocatalysis diverges from other PEC effects in two aspects: (1) scale, that is, micrometer or nanometer instead of millimeter; and (2) in most cases, lack of an external electrical circuit. The latter situation is analogous to that of electrochemical corrosion, in which anodic and cathodic electrochemical reactions short-circuit locally on a conductive metal surface. In the case of photocatalysis, a PEC reaction, usually an oxidation reaction, is short-circuited with a purely electrochemical reduction reaction, for example, that of molecular oxygen, as discussed later in detail in this chapter.

In a solar PEC cell, photogenerated electron-hole pairs are driven efficiently in opposite directions by an electric field existing at the interface between a semiconductor and an electrolyte solution. The same processes can occur on a single semiconductor particle, although the electric field may be vanishingly small [9, 10]. Thus, the carriers must be separated by means of diffusion, with nanoscale sources and sinks. In a system where semiconductor particles are suspended in a liquid solution, excitation of the semiconductor can lead to redox processes in the interfacial region around each particle. These types of systems have drawn the attention of a large number of investigators over the past twenty years. For example, in the area of photocatalytic water splitting, there has been a very active effort, as reviewed recently [11, 12].

In the early work on PEC solar cells for water electrolysis, one of the first types of electrode materials examined was TiO_2, partly because it has a sufficiently positive valence band edge to oxidize water to oxygen [4–7]. It is also an extremely stable material in the presence of aqueous electrolyte solutions. The possibility of solar photoelectrolysis was demonstrated for the first time with a system in which an immersed n-type TiO_2 semiconductor electrode, connected through an electrical load to an immersed platinum black counter electrode, was exposed to near-UV (ultraviolet) light. When the surface of the TiO_2 electrode was irradiated with light consisting of wavelengths shorter than ~415 nm, photocurrent flowed from the platinum counter electrode to the TiO_2 electrode through the external circuit. The direction of the current revealed that the oxidation reaction (oxygen evolution) occurs at the TiO_2 electrode and the reduction reaction (hydrogen evolution) at the Pt electrode.

When the conduction band energy, E_{CB}, is higher (i.e. closer to the vacuum level or more negative on the electrochemical scale) than the hydrogen evolution potential, photogenerated electrons can flow to the counter electrode and drive the reduction of protons, resulting in hydrogen gas evolution without an applied potential. Hence, for this to occur, E_{CB} should be at least as negative as -0.4 V [standard hydrogen electrode (SHE)] in acid solution or -1.2 V (SHE) in alkaline solution. Among the oxide semiconductors, TiO_2, $SrTiO_3$, $CaTiO_3$, $KTaO_3$, Ta_2O_5, and ZrO_2 satisfy this requirement, although in most cases the band gaps are relatively large (>3.0 eV), and thus the efficiency of solar energy conversion is very low.

Following early TiO_2 -induced water splitting studies, it was discovered that TiO_2 particularly excels in photocatalytically breaking down organic compounds and other pollutants to innocuous products, neutralizing foul odors, and killing bacteria. Frank and Bard in 1977 were among the first to examine the possibilities of using TiO_2 to decompose pollutants; in their case, cyanide, in water [2, 3]. Since then, there has been an increasing interest in environmental applications. These authors quite correctly pointed out the implications of their result in the field of environmental purification. Their prediction has indeed been borne out, as evidenced by the extensive international efforts in this area [8, 12–23].

6.1.3 TiO_2 Characteristics

6.1.3.1 General

Titanium dioxide occurs in nature in three crystalline forms, rutile, anatase, and brookite. The crystal structures are shown in Fig. 1 [24]. Some of the physical properties are given in Table 1. Much of the fundamental work on TiO_2 has been carried out with rutile single crystals because these are relatively straightforward to produce. Anatase single crystals are more difficult to produce, and some work has been carried out with natural crystals but it is possible to grow such crystals with the use of the vapor-phase transport method [25]. Photocatalytic reactions

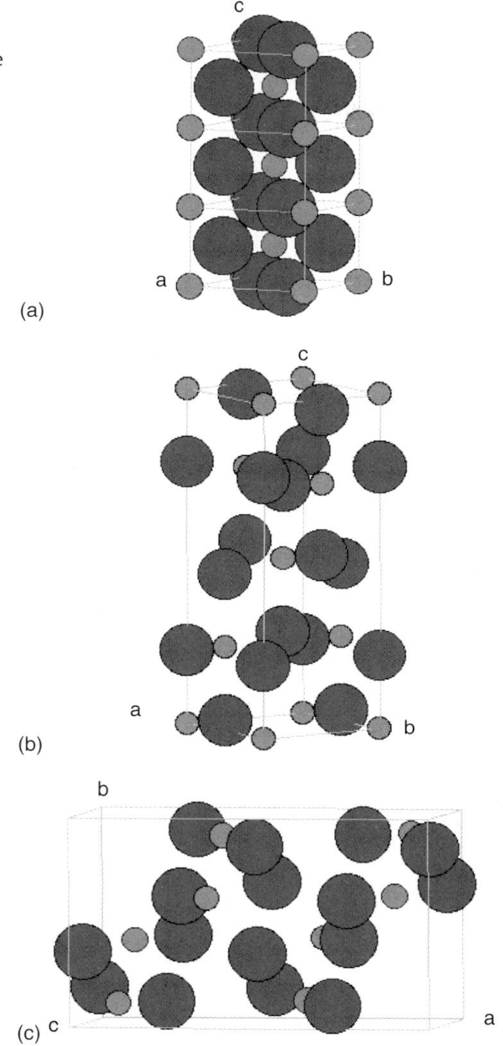

Fig. 1 TiO_2 crystal structures: (a) rutile; (b) anatase; (c) brookite. Structure data were taken from Wyckoff [24].

have been shown to exhibit larger rates on anatase compared to rutile, and it will be very useful to carry out further comparative studies with single crystals.

6.1.3.2 Band Edge Energies

The energies of the valence and conduction bands of TiO_2 are determined by the chemical termination of the material, together with the interactions of the surface with gas or liquid environment. The electrochemical flat band potentials, E_{fb}, can be used to estimate the band edge energies for titanium dioxide. E_{fb} values for rutile single crystals have been reported by several authors [26]. In acid solution (pH 2.1), a value of −0.05 V versus SHE has been proposed by Morrison, based on the results of several workers [27]. In alkaline solution, values of −0.6 [27] and

Tab. 1 Selected properties of TiO_2

Property	***Rutile***	***Anatase***
Crystal structure	Tetragonal	Tetragonal
Space group	P42/mnm (136)	I41/amd (141)
Density, g cm^{-3}	4.250	3.894
Band gap, eV	3.02	3.2
Type	Indirect	Indirect

Note: Data taken from Finklea [24] and Wyckoff [26]

−0.76 V [28] have been given. Recently, Kavan and coworkers have measured E_{fb} values for both rutile and synthetically grown anatase single crystals in 1 M H_2SO_4 and obtained values of +0.04 and −0.16 V against SHE, respectively [25]. The authors point out that the latter value is sufficiently negative so that no external bias needs to be applied to photoelectrochemically evolve hydrogen.

On the basis of the flat band potential, the actual location of the conduction band edge energy, E_{CB}, then depends on the doping level, according to the following equation:

$$E_{CB} - E_F = \xi = -kT \ln \frac{n}{N_C} \qquad (1)$$

where E_F is the Fermi level energy, n is the occupation of conduction band states with electrons, and N_C is the density of conduction band states at the band edge [29]. For normal semiconductor doping levels, ξ could be in the range of 0.1 to 0.2 V [27]. In subsequent discussions we will refer to the E_{fb} values, which should then be considered to be lower (energy) limits for the E_{CB} values.

As already stated, the band positions depend on the environment external to the solid surface, for example, the hydrogen ion concentration or pH. This is a situation common to all oxide semiconductors and arises from the fact that the surface oxygens of the oxide interact strongly with protons and hydroxide ions [27]. To understand the processes underlying photocatalysis on TiO_2, it is useful to take the pH dependencies of the band edge energies into perspective. The relationships of these energies to those corresponding to various redox couples, for example, those for various forms of reduced oxygen (see next section) or for various organic oxidations, then become apparent (Fig. 2). Here we have adapted the commonly used Pourbaix representation in a modified form (i.e. with negative potentials uppermost), to be consistent with the prevailing practice in energy band diagrams [30]. Therefore, the ubiquitous lines for the H^+/H_2 and O_2/H_2O redox couples are inverted and slope upward with increasing pH (more negative electrochemical potentials). Likewise, the valence band (VB) and conduction band (CB) energies for TiO_2 slope upward (higher electron energies) with increasing pH.

The VB energies lie approximately 1.8 eV below those necessary to evolve molecular oxygen from water over the entire pH range. Electrons in the CB for rutile are very close in potential to that necessary to evolve molecular hydrogen from water but not at appreciable rates. In early TiO_2-based PEC solar cells, this problem was circumvented through the use of a two-compartment cell. The TiO_2 electrode was immersed in a more alkaline half-cell and a Pt counter electrode was immersed in a more acid electrolyte in the second half-cell. The electrons generated at the photoelectrode diffuse at the back of the electrode and then toward the counter electrode. In this case, electrons photogenerated at the alkaline TiO_2 electrode have a relatively high electrochemical energy, which is sufficient to evolve hydrogen in

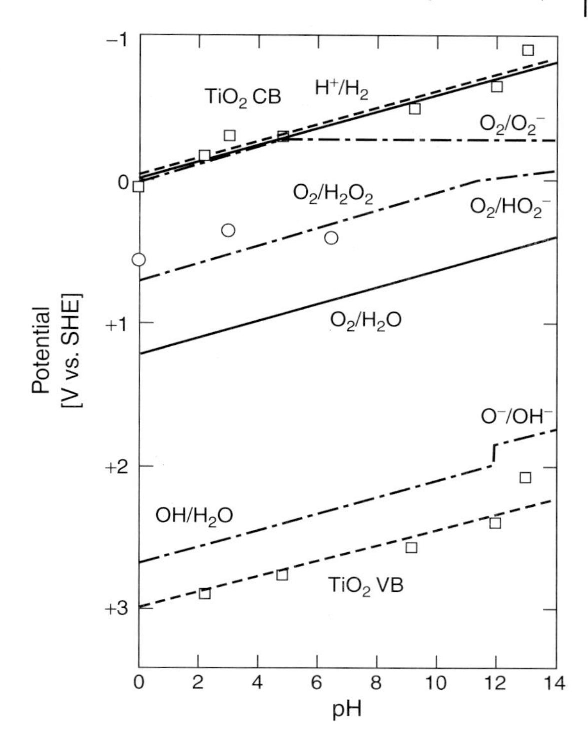

Fig. 2 TiO_2 band edge energies. Data were taken from Finklea [26] and from the data given in Table 2.

the more acidic solution. In the case of photocatalysis, the same type of effect may also be operative: specifically, if the local pH is high, for example, at a reducing site at a particular location on a TiO_2 particle, additional reducing power is available for the reduction of molecular oxygen to superoxide.

In contrast, the oxidizing power of photogenerated holes decreases with increasing pH but the standard potential for the one-electron oxidation of any given organic compound RH to its radical cation $^{\bullet}RH^+$ is not pH-dependent. For example, the standard potential for this reaction for *n*-octane in acetonitrile is +2.71 V against Ag/Ag^+ [34], which can be converted to +2.84 V against SHE. This value may not be sufficiently positive because solvent reactions make elucidation of the reversible potential problematic. At present, the topic of the actual energetics of various oxidation reactions pertinent to photocatalysis is beginning to be explored in detail [35]. For example, because of the presence of water, the actual reaction could be considered to be an electron transfer followed by an extremely fast chemical reaction with water.

$$R \longrightarrow {}^{\bullet}R^+ + e^- \tag{2}$$

$${}^{\bullet}R^+ + H_2O \longrightarrow {}^{\bullet}RH + OH^- \tag{3}$$

The effective redox potential of this reaction would be shifted significantly in the negative direction because of consumption of the oxidized product. Thus, it is quite possible that trapped holes with redox potentials of +1.5 V [36] to +1.75 [37], have sufficient oxidizing power to induce the production of the neutral radical, which would involve Eqs. (2) and (3).

Tab. 2 Standard redox potentials for various electrochemical reactions involving molecular oxygen and its reduced forms

Reaction			*pH*	E^0 *[V]*	*G* *[eV]*	pK_a	*References*
Acid dissociations							
$HO_2^\bullet$	$\longleftrightarrow$	$^\bullet O_2^- + H^+$				4.88	31
H_2O_2	$\longleftrightarrow$	$HO_2^- + H^+$				11.63	31
$^\bullet OH$	$\longleftrightarrow$	$^\bullet O^- + H^+$				11.9	32
Redox reactions							
$O_2 + e^-$	$\longleftrightarrow$	$^\bullet O_2^-$	all	−0.284	+0.284		33
$O_2 + H^+ + e^-$	$\longleftrightarrow$	$HO_2^\bullet$	0	+0.005	−0.005		c
$O_2 + H^+ + e^-$	$\longleftrightarrow$	$HO_2^\bullet$	7	−0.409	+0.409		c
$O_2 + 2H^+ + 2e^-$	$\longleftrightarrow$	H_2O_2	0	+0.693	−1.386[a]		c
$O_2 + 2H^+ + 2e^-$	$\longleftrightarrow$	H_2O_2	7	+0.279	−0.558[a]		c
$O_2 + H^+ + 2e^-$	$\longleftrightarrow$	HO_2^-	14	−0.065	+0.130[a]		33
$O_2 + H^+ + 2e^-$	$\longleftrightarrow$	HO_2^-	7	+0.142	−0.284[a]		c
$O_2 + 4H^+ + 4e^-$	$\longleftrightarrow$	$2H_2O$	0	+1.229	−4.916[a]		33
$O_2 + 4H^+ + 4e^-$	$\longleftrightarrow$	$2H_2O$	7	+0.815	−3.260[a]		c
$O_2 + 4H^+ + 4e^-$	$\longleftrightarrow$	$2H_2O$	14	+0.401	−1.603[a]		c
$^\bullet O_2^- + 2H^+ + e^-$	$\longleftrightarrow$	H_2O_2	7	+0.842	−0.558[a]		c
$HO_2^\bullet + e^-$	$\longleftrightarrow$	HO_2^-	all	+0.694			c
$HO_2^\bullet + H^+ + e^-$	$\longleftrightarrow$	H_2O_2	7	+0.967	−0.558[a]		c
$H_2O_2 + H^+ + e^-$	$\longleftrightarrow$	$^\bullet OH + H_2O$	7	+0.433	−0.991[a]		c
$H_2O_2 + 2H^+ + 2e^-$	$\longleftrightarrow$	$2H_2O$	7	+1.351	−3.260[a]		c
$^\bullet OH + H^+ + e^-$	$\longleftrightarrow$	H_2O	0	+2.683	−2.233[b]		32
$^\bullet OH + H^+ + e^-$	$\longleftrightarrow$	H_2O	7	+2.269	−0.991[b]		32
$^\bullet OH + e^-$	$\longleftrightarrow$	OH^-	14	+1.857	+0.252[b]		32
$^\bullet O^- + 2H^+ + e^-$	$\longleftrightarrow$	H_2O	7	+2.559	−0.701[b]		c
$^\bullet O^- + 2H^+ + e^-$	$\longleftrightarrow$	H_2O	14	+1.733	+0.128[b]		c

[a] The free energy is for the product.
[b] The free energy is for $^\bullet OH$ or $^\bullet O^-$.
[c] Calculated.
Note: Data were taken from Hoare [31], Buxton and coworkers [32], and Tarasevich and coworkers [33].

Solely considering Eq. (2), it would be expected that the reaction of a hole in the TiO_2 valence band with the organic compound would become less favorable at high pH. However, considering Eq. (3), a normal pH dependence, that is, about 60 mV/decade at 25 °C should exist. Overall, it might be expected that there should not be significant pH effects in terms of changes in excess oxidizing power of valence band holes or trapped holes. In fact, the rates of photocatalytic oxidation reactions are often higher at higher pH [18]. For organic oxidation reactions in aqueous solution, not involving radical species, standard redox potentials are known (Table 3) and many exhibit pH dependencies of about 60 mV pH^{-1}, although the behavior can be somewhat more complex, with at least one change in slope [38] because of involvement of acid-base equilibria.

The majority of the VB and CB values (the latter being approximated by E_{fb} as

Tab. 3 Selected free energies and standard potentials for CO_2 reduction reactions

Reaction			*pH*	*ΔG^0 [kJ mol^{-1}]*	*E^0, V vs. SHE*	*References*
$CO_2(g) + 2H^+ + 2e^-$	$\longleftrightarrow$	$CO(g) + H_2O$	0	+20.06	−0.104	39
$CO_2(g) + 2H^+ + 2e^-$	$\longleftrightarrow$	$HCOOH(aq)$	0	+38.32	−0.199	38, 40
$CO_2(g) + 4H^+ + 4e^-$	$\longleftrightarrow$	$HCOH(aq) + H_2O$	0	+27.49	−0.071	38, 40
$CO_2(g) + 6H^+ + 6e^-$	$\longleftrightarrow$	$CH_3OH(aq) + H_2O$	0	−17.24	+0.030	38, 40
$CO_2(g) + 8H^+ + 8e^-$	$\longleftrightarrow$	$CH_4(g) + 2H_2O$	0	−130.62	+0.169	39
$HCOOH(aq) + 4H^+ + 4e^-$	$\longleftrightarrow$	$CH_3OH(aq) + H_2O$	0	−55.96	+0.145	40
$HCHO(aq) + 2H^+ + 2e^-$	$\longleftrightarrow$	$CH_3OH(aq)$	0	−44.77	+0.232	40
$HCOOH(aq) + 2H^+ + 2e^-$	$\longleftrightarrow$	$HCHO(aq) + H_2O$	0	−10.81	+0.056	40
$CH_3OH^{\bullet+}(CH_3CN) + e^-$	$\longleftrightarrow$	$CH_3OH(CH_3CN)$	NA	> −268.23[a]	> +2.78[b]	41
$CH_3CH_2OH^{\bullet+}(CH_3CN) + e^-$	$\longleftrightarrow$	$CH_3CH_2OH(CH_3CN)$	NA	> −256.65[a]	> +2.66[b]	41
2-propanol$^{\bullet+}(CH_3CN) + e^-$	$\longleftrightarrow$	2-propanol(CH_3CN)	NA	> −246.04[a]	> +2.55[b]	41
n-pentane$^{\bullet+}(CH_3CN) + e^-$	$\longleftrightarrow$	n-pentane(CH_3CN)	NA	> −281.74[a]	> +2.92[c]	34
n-hexane$^{\bullet+}(CH_3CN) + e^-$	$\longleftrightarrow$	n-hexane(CH_3CN)	NA	> −277.88[a]	> +2.88[c]	34
n-heptane$^{\bullet+}(CH_3CN) + e^-$	$\longleftrightarrow$	n-heptane(CH_3CN)	NA	> −274.02[a]	> +2.84[c]	34
n-octane$^{\bullet+}(CH_3CN) + e^-$	$\longleftrightarrow$	n-octane(CH_3CN)	NA	> −274.02[a]	> +2.84[c]	34

[a] Indicates that the actual ΔG^0 values are more negative than those given.
[b] converted from values versus ferrocene/ferrocinium [135].
[c] converted from values versus Ag/Ag+ (0.1 M) [135].

already discussed) shown in Fig. 2 are for rutile single crystals. It has already been pointed out that the crystalline form more commonly used in photocatalysis, anatase, has a higher CB energy (more negative electrochemical potential). A point is included for single crystal anatase, approximately 200 mV above the line drawn through the points for rutile, and a point is also shown for a nanocrystalline anatase thin film: its CB energy is elevated approximately 100 mV with respect to that for rutile.

Lying approximately 0.6 to 0.8 eV below the CB energy, there are surface states on TiO_2; three experimental points are included in Fig. 2 at specific pH values but additional experimental evidence exists for such surface states under conditions with no pH control, for example, in the gas phase, and therefore such data cannot be included in the figure. A reasonable characterization for this type of surface state is that it involves a trapped electron with an accompanying proton, viz.:

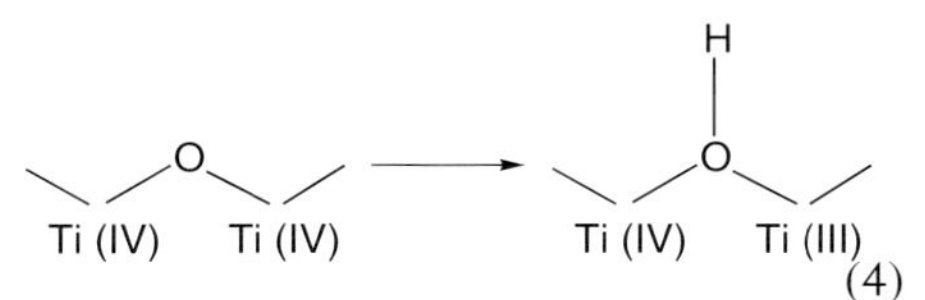

(4)

This type of species is important for photocatalysis because it provides a surface sink for photogenerated electrons. Additionally, this type of species is particularly important with respect to the wetting properties of TiO_2, which is of importance to self-cleaning surfaces.

6.1.4 Photocatalytic Reactions

6.1.4.1 Oxygen Electrochemistry

The active chemical species that are involved in photocatalytic reactions are generally forms of reduced oxygen, including superoxide $^{\bullet}O_2^-$, the hydroperoxyl radical $^{\bullet}O_2H$, hydrogen peroxide H_2O_2, the peroxide anion HO_2^-, the hydroxyl radical $^{\bullet}OH$, and the oxyl radical anion $^{\bullet}O^-$. The relevant reactions involving these species are shown in Table 2. The pH dependences of these reactions are shown in the same figure with the TiO_2 band edge energies (Fig. 2). It is interesting that the reducing power of the conduction band electrons is barely sufficient to reduce O_2 to $^{\bullet}O_2^-$ at pH values below the pK_a for $^{\bullet}O_2H$ (4.88). With increasing pH, however, there is an increasing excess of reducing power for this reaction.

The pH dependence of the oxidizing power of the $^{\bullet}OH$ radical parallels that for valence band holes, except that there is a discontinuity at pH 11.9, which is the pK for the equilibrium

$$^{\bullet}OH \Longleftrightarrow H^+ + O^- \quad (5)$$

At pH 0, the potential for the one-electron reduction of $^{\bullet}OH$ is +2.683 V [32] (Table 2). As already mentioned, the redox potential for trapped holes has been estimated to be significantly less positive, +1.50 to +1.75 V against SHE.

The standard potentials for the various redox processes, involving molecular oxygen, can be conveniently visualized in a Frost diagram, given here for pH 7 (Fig. 3) [33, 43, 44]. In this type of diagram, the relative state of reduction of the oxygen is plotted along the horizontal axis, with O_2 itself being taken as zero and H_2O being taken as four. The vertical axis is the free energy of formation in units of electron volts. The slopes of the lines between species of differing oxidation states are the negative of the potential for the corresponding redox process. The free energy change for the process is the slope multiplied by the number of electrons involved

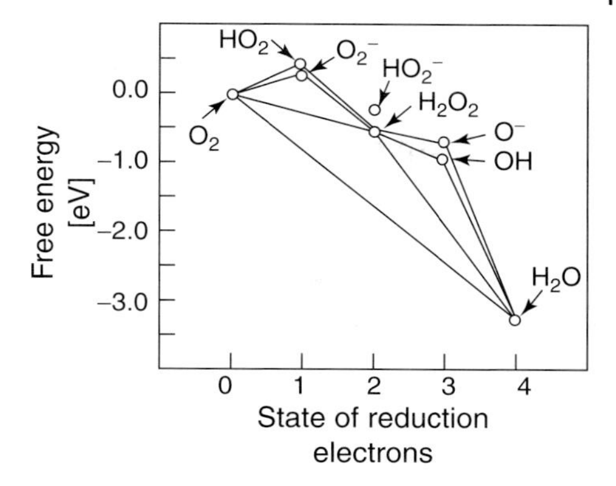

Fig. 3 Frost diagram for O_2 reduction at pH 7. Data were taken from Table 2.

in the oxidation state change. At any particular oxidation state, there can be more than one state of protonation depending on the difference of pH 7 and the pK_a value for a particular equilibrium. The free energy change is obtained from the relationship

$$\frac{\Delta G}{F} = \frac{2.303RT}{F}(\text{pH} - \text{pK}_\text{a}) \qquad (6)$$

where $\Delta G/F$ gives the free energy change directly in eV. This diagram shows that each of the species with oxidation states −1, −2 and −3 are unstable with respect to disproportionation. It shows why the hydroxyl radical is such a powerful oxidant: it is simply because it gains so much stability by being reduced to water. It also shows that superoxide is not a very strong one-electron oxidant and that peroxide is a moderately strong two-electron oxidant. The redox potentials are given in Table 2.

6.1.4.2 Basic Reactions, Kinetics, and Active Species

The basic photocatalytic reactions have been reviewed by several authors. A generic set of reaction steps for a given organic compound RH can be written as follows:

$$h\nu \longrightarrow \text{h}^+ + \text{e}^- \qquad (7)$$

$$\text{h}^+ + \text{RH} \longrightarrow \text{RH}^+ \qquad (8)$$

$$\text{RH}^+ \longrightarrow \text{R}^\bullet + \text{H}^+ \qquad (9)$$

$$\text{h}^+ + \text{H}_2\text{O} \longrightarrow {}^\bullet\text{OH} + \text{H}^+ \qquad (10)$$

$$\text{RH} + {}^\bullet\text{OH} \longrightarrow \text{R}^\bullet + \text{H}_2\text{O} \qquad (11)$$

As shown here, the initial attack on the organic compound can be carried out either by the hole (h^+) or the hydroxyl radical ($^\bullet OH$). The question of which species is more important has been a subject of much debate. One reason for the uncertainty is that it is often difficult to distinguish between products of reactions with holes and products of reactions with hydroxyl radicals. Similar to other aspects of photocatalysis, the answer must be given in very specific terms. One must specify the precise conditions under which a particular result is obtained. For example, one must consider the surface coverage of the organic compound when discussing the active species. If the coverage is high, direct attack by holes appears to be possible. For relatively low coverages, it appears that $^\bullet OH$ is the active species.

Lawless and coworkers argued that an adsorbed $^\bullet OH$ radical is identical to a trapped hole [36]. By reacting colloidal TiO_2 (13 nm) in aqueous suspension with radiolytically generated $^{\bullet}OH$, they observed a broad absorption band (ca. 350 nm). The intensity of this signal exhibited first-order kinetics, with a rate that was proportional to the $^{\bullet}OH$ concentration as a result of the self-reaction of $^{\bullet}OH$ to produce hydrogen peroxide:

$$^{\bullet}OH + {^{\bullet}OH} \longrightarrow H_2O_2 \qquad (12)$$

with a time constant of about 50 ms. In homogeneous solution, this reaction is second-order (Table 4) but becomes first-order on the two-dimensional surface, as discussed later. The trapped hole can be represented as $Ti^{IV} - O^{-}$. On the basis of the reaction with SCN^{-}, the E^0 for this species was estimated to be about 1.5 V (pH 3), which is approximately 1.3 eV above the valence band. These workers conclude that the existence of adsorbed $^{\bullet}OH$ (surface-trapped holes) can explain most of the observed results, including the production of H_2O_2 and hydroxylated products of organic compounds. The desorption of free $^{\bullet}OH$ is considered to be unlikely because of the extremely large rate constant for the reverse reaction.

A later study by Goldstein and coworkers involved phenol as the reactant [45]. The photocatalytic reaction products were compared with those of radiolytically generated $^{\bullet}OH$ on one hand and $SO_4^{\bullet -}$ on the other. The latter is a strong one-electron oxidant and would mimic the effect of a hole (see section on mechanisms). At low phenol concentrations, the oxidation products of the photocatalytic reaction are similar to those for $^{\bullet}OH$, whereas at high phenol concentration, they are similar to those for reaction with $SO_4^{\bullet -}$, indicating involvement of holes. This study shows that the trapped hole acts as a $^{\bullet}OH$ transfer agent. At high phenol concentrations, phenol adsorbs on the TiO_2 surface and competes for mobile holes, decreasing the rates of hole localization and electron-hole recombination.

Tab. 4 Homogeneous reaction rate constants for selected reactions involving hydroxyl radicals

Reactants		*Products*	*Rate constant* $[M^{-1}\ s^{-1}]$
$HO_2^{\bullet} + {^{\bullet}OH}$	$\longrightarrow$	$H_2O + O_2$	1×10^{10}
$^{\bullet}O_2^{-} + {^{\bullet}OH}$	$\longrightarrow$	$OH^{-} + O_2$	1×10^{10}
$^{\bullet}OH + {^{\bullet}OH}$	$\longrightarrow$	H_2O_2	4.2×10^9
$CH_3CHO + {^{\bullet}OH}$	$\longrightarrow$	$H_2O + CH_3CO$	3.6×10^9
$CH_3COCH_3 + {^{\bullet}OH}$	$\longrightarrow$	Products	1.3×10^8
$CH_3CH_2CH_3 + {^{\bullet}OH}$	$\longrightarrow$	Products	2.3×10^9
2-propanol + $^{\bullet}OH$	$\longrightarrow$	Products	1.6×10^9
1,4-$C_6H_4(CO_2H^{-}) + {^{\bullet}OH}$	$\longrightarrow$	Products	3.3×10^9
Coumarin + $^{\bullet}OH$	$\longrightarrow$	Products	2×10^9
$C_6H_5OH + {^{\bullet}OH}$	$\longrightarrow$	dihydroxycyclohexadienyl	6.6×10^9
DMPO + $^{\bullet}OH$	$\longrightarrow$	DMPO-OH	4.3×10^9

Note: Values were taken from A. B. Ross, NDRL/NIST Solution Kinetics Database, v. 2.0, Standard Reference Data, NIST, Gaithersburg, MD 20899, USA (http://allen.rad.nd.edu), also found in Ref. 32. DMPO stands for 5,5-dimethyl-1-pyrroline N-oxide.

Bahnemann and coworkers have distinguished between shallow and deeply trapped holes [46]. They point out that the absorption features commonly used to indicate trapped electrons (650 nm, 1.7 eV) and holes (430 nm, 2.7 eV), could not be the actual ones involved with photocatalytic reactions because they would correspond to energy levels near the middle of the band gap, for example, 1.7 eV below the CB or 2.7 eV above the VB. The latter is assigned to an excitation of a surface peroxide. The active oxidizing species is thought to be a shallow trapped hole, which can equilibrate with VB holes and would have a similar oxidizing power. These authors showed that strongly adsorbing species, such as citrate and acetate, react with free holes, whereas more weakly adsorbed species, such as ethanol, could react with deeply trapped holes (i.e. surface-bound $^{\bullet}OH$). They also showed that dichloroacetate and thiocyanate adsorb strongly and react with either freely or shallowly trapped holes but not deeply trapped holes.

Cermenati and coworkers used a very interesting approach to assess the relative importance of holes and hydroxyl radicals [47]. For a model reactant, they employed quinoline, which has two aromatic rings with differing electron density, the nitrogen-containing ring with less. Because $^{\bullet}OH$ is electrophilic, the attack should occur on the benzene ring. On the basis of a detailed analysis of products as a function of time, the major mechanistic pathways involving $^{\bullet}OH$ were discounted. Instead, a mechanism involving oxidation with holes, together with superoxide, was proposed (see section on mechanisms). A third mechanism was proposed in which initial oxidation by either holes or $^{\bullet}OH$ is followed by reaction with neutral molecular oxygen O_2. The effect of raising the pH from 3 to 6 was consistent with the second pathway because both the $^{\bullet}Q^-$ and $^{\bullet}O_2{}^-$ would be stabilized at the higher pH.

A different approach was taken by Kesselman and coworkers, who examined the oxidation of some organic anions (chloroacetates) at heavily Nb-doped TiO_2 electrodes. Because the selectivity for reaction of these species with $^{\bullet}OH$ in homogeneous solution is known, as is the ordering of the electrochemical redox potentials, it is possible to distinguish between pathways. In addition, the extent to which the oxidation current depends on the reactant concentration is a measure of the involvement of the direct-hole pathway. If there is no dependence on concentration, this indicates that $^{\bullet}OH$ is produced and then reacts with the organic species. These authors found a full range of behavior from 12% to 97% direct-hole involvement for formate and 4-chlorocatechol, respectively. The results were found to vary significantly from one electrode to another (15% vs. 65% direct-hole involvement for acetate).

From the work of Nosaka and coworkers with electron paramagnetic resonance (EPR) spectroscopy, the proportion of the direct hole and $^{\bullet}OH$-mediated pathways can be determined for various types of TiO_2 samples. These workers used the photocatalytic oxidation of acetate as a probe reaction and found mixtures of EPR-active products. They have proposed the following schemes:

$$CH_3COOH + H^+ \longrightarrow {}^{\bullet}CH_3 + CO_2 + H^+ \tag{13}$$

$$CH_3COOH + {}^{\bullet}OH \longrightarrow {}^{\bullet}CH_3 + CO_2 + H_2O \tag{14}$$

$$H_2O + h^+ \longrightarrow {}^{\bullet}OH + H^+ \quad (15)$$

$$CH_3COOH + {}^{\bullet}OH \longrightarrow {}^{\bullet}CH_2COOH + H_2O \quad (16)$$

The methyl radical ${}^{\bullet}CH_3$ can be produced by reaction with either h^+ or ${}^{\bullet}OH$, whereas the carboxymethyl radical ${}^{\bullet}CH_2COOH$ can be produced only by reaction with ${}^{\bullet}OH$. Thus the ratio of ${}^{\bullet}CH_3$ to ${}^{\bullet}CH_2COOH$ radical yields the proportion of the direct h^+ path against the ${}^{\bullet}OH$ path. It was found that the ${}^{\bullet}OH$ path was favored for TiO_2 samples with higher coverages of OH groups. On such surfaces, the type of active surface species has been proposed by Howe and Grätzel [48], based on EPR measurements, to be

$$Ti^{4+}O^{-\bullet}Ti^{4+}OH^{-} \quad (S1)$$

On TiO_2 surfaces that contain fewer OH groups, perhaps as a result of heat treatment, the active species may instead be [49, 50]:

$$Ti^{4+}O_2{}^{-}Ti^{4+}O^{-\bullet} \quad (S2)$$

Hirakawa and coworkers propose that the reaction

$$Ti^{4+}O^{-\bullet}Ti^{4+}OH^{-} + H_2O \longrightarrow Ti^{4+}O_2{}^{-}Ti^{4+}OH_2 + {}^{\bullet}OH \quad (17)$$

takes place more easily than the reaction

$$Ti^{4+}O_2{}^{-}Ti^{4+}O^{-\bullet} + H_2O \longrightarrow Ti^{4+}O_2{}^{-}Ti^{4+}OH^{-} + {}^{\bullet}OH \quad (18)$$

The structural arrangements of these species are shown in Fig. 4.

From work in our laboratory, it was found that the yields of fluorescent products from the reactions of ${}^{\bullet}OH$ with coumarin and 1,4-phthalate could be measured as a function of time, with a linear dependence being found for fluorescence intensity versus illumination time [51]. It was concluded that the ${}^{\bullet}OH$ pathway predominates for the low reactant concentrations used (10^{-4} to 10^{-3} M) in aqueous solution.

In later work, it was found that the quantum yield for the reaction of ${}^{\bullet}OH$ with 1,4-phthalate was relatively small (7×10^{-5}) under the experimental conditions used (4×10^{-4} M, 2×10^{-3} M NaOH; UV intensity, $1\ mW\ cm^{-2}$) [52]. In contrast,

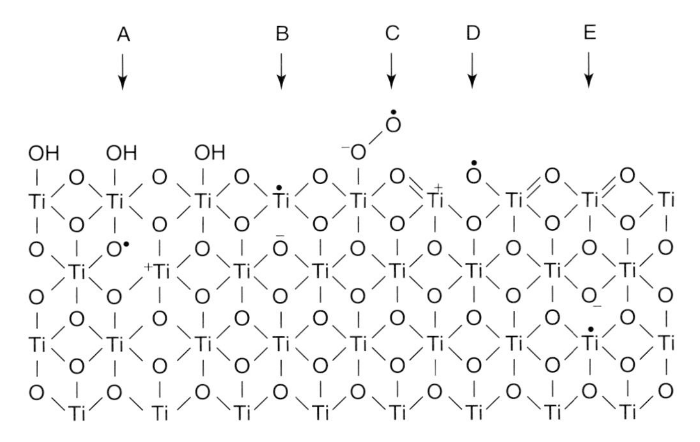

Fig. 4 TiO_2 surface species, from Nakaoka and coworkers [45].

the quantum yield for the photocatalytic oxidation of iodide I^- to iodine I_2 was found to be relatively high (5.7×10^{-2}) for the conditions used (10^{-2} M KI; UV intensity, 1 mW cm^{-2}). The quantum yields usually observed for photocatalytic reactions at similar light intensities are more in line with the latter, and therefore it was concluded that the hole is the active species in most cases.

As already mentioned, the desorption of active oxygen species, particularly $^{\bullet}OH$, from the TiO_2 surface, with subsequent reaction in homogeneous aqueous solution, has been downplayed as a significant pathway for TiO_2 photocatalysis, although not by direct experiments. Nevertheless, our group has found it possible to observe the oxidative degradation of organic films, for example, methylene blue, that are separated from the illuminated TiO_2 surface by air gaps of up to 125 μm [53]. In an effort to identify the active species, various types of organic reactant films were examined [54]. Changes in mass were also monitored by use of the quartz-crystal-microbalance (QCM) technique. In addition to methylene blue, a polyaniline derivative, poly(2-methoxyaniline-5-sulfonic acid) (PMAS), polystyrene (PS), octadecyltriethoxysilane (ODS)-coated glass, and polyvinyl alcohol were examined. With the ODS-coated glass separated from the illuminated TiO_2 film by 50 μm, oxygen-containing functional groups were produced, as observed with X-ray photoelectron spectroscopy, creating a hydrophilic surface. For PMAS, PS, and ODS, the QCM results showed decreases in frequency, that is, increases in mass due to oxygenation. The fact that aliphatic hydrocarbons are oxidized suggests that the active oxidizing agent may not be $HO_2^{\bullet}$, H_2O_2, or $^1O_2$. Thus the most likely candidate is $^{\bullet}OH$. However, it may also be possible that $HO_2^{\bullet}$ or H_2O_2 could be the diffusing species and could then be converted to $^{\bullet}OH$ on reaching the organic film, as in the following reactions:

$$2HO_2^{\bullet} \longrightarrow H_2O_2 + O_2 \qquad (19)$$

$$H_2O_2 \xrightarrow{h\nu} 2^{\bullet}OH \qquad (20)$$

In experiments with chemically generated $^{\bullet}OH$ (Fenton reagent), similar effects were observed in water contact angle for PS and ODS.

In earlier work, it was found that *Escherichia coli* cells were killed in aqueous suspensions in contact with illuminated TiO_2 films, even when separated by a 50-μm-thick porous membrane [55]. On the basis of the inhibitory effect of catalase, it was concluded that the active species is hydrogen peroxide but there appeared to be a cooperative effect, possibly due to superoxide, which significantly enhanced the lethal effect.

Now that many of the important photogenerated species have been identified, it is useful to consider the kinetics of formation and decay of some of these species. The absorption of a photon and creation of an electron-hole pair in a TiO_2 particle occurs in a time range that has, thus far, been too short to measure, that is, a few femtoseconds. These charge carriers are initially considered to exist in the conduction band and valence band, respectively but are then quickly trapped. Skinner and coworkers have determined the trapping time for electrons to approximately 180 fs based on measurements on TiO_2 sols in acetonitrile, which was used to avoid the absorbance of hydrated electrons [56]. The nanoparticle surface consists of many different types of local geometries and corresponding trap energies, leading to a broad range for the absorbance spectrum.

Holes have been estimated to be trapped within 1–10 ps by Serpone and coworkers, and these workers find that 90–100% of all electron-hole (e-h) pairs have undergone recombination

$$e^- + h^+ \longrightarrow \text{heat} \tag{21}$$

within 10 ns for colloidal TiO_2 particles from 2–27 nm in diameter [57]. However, Colombo and coworkers estimate the half-life of an e-h pair to be approximately 30 ps [58]. This is just one example of the wide variety of kinetic results that have been reported (Table 5). Second-order kinetics are observed particularly at short times and high light intensities, that is, when there are multiple e-h pairs existing simultaneously in a single particle [57, 59].

Results for time-resolved microwave conductivity measurements, which reflect the number of charge carriers, indicate that the decay kinetics span many orders of magnitude in time, for example, from ns to s [60, 61, 63]. Striking differences were observed, comparing anatase and rutile, with the latter exhibiting much faster decay kinetics [60]. Major differences were observed among various types of colloidal TiO_2 preparations and in comparison to Degussa P25 in terms of recombination kinetics [63]. However, Colombo and coworkers, using diffuse reflectance flash photolysis experiments, have found the behavior of TiO_2 colloidal solutions versus powder form, rutile versus anatase, and nanoparticle versus P25, all to be similar [58, 65]. These workers point out that the diffuse reflectivity results are specific for trapped electrons, rather than for carriers in general and assign the long-lived species observed in this and other studies to such electrons [65]. They also show that the effect of a hole scavenger, such as thiocyanate, is to greatly increase the lifetime of the trapped electrons [58].

It had already been observed in early work with flash photolysis, albeit on much longer timescales, that recombination could be greatly inhibited in the presence of either hole scavengers or electron scavengers [67]. Lifetimes of seconds or even hours have been reported for trapped electrons in the presence of hole scavengers such as 2-propanol [68]. Lifetimes of milliseconds (in acid) or microseconds (in base) have been reported for trapped holes in the presence of an electron scavenger such as methylviologen [67].

The decay process is usually assumed to involve recombination of electrons and holes in various states of trapping or detrapping. Because at short time intervals and relatively high light intensities, there may be approximately equal numbers of each type of carrier in a single particle, the recombination process can be expected to be second-order. When there is on the order of only a single e-h pair in a particle, the decay process is known to become first-order. This has been explained in terms of stochastic processes [59, 69–71], that is, because of the discrete nature of the species and corresponding events, they must be treated statistically rather than continuously. The simple fact that reactions are taking place on a two-dimensional surface is also important [72]. However, it is well known in the field of solid-state electronics that direct recombination in a doped semiconductor behaves as a first-order process for low injection rates, simply because the magnitude of the squared term in the rate expression becomes small [73]. This is the analog of pseudo-first-order kinetics. In the case of indirect recombination (i.e. mediated by a band gap

Tab. 5 Kinetics of photogenerated charge carrier trapping and decay

Reaction	*TiO_2 type*	*Particle size, nm*	*Medium*	*pH*	*Gas*	*e-h/ particle*	*Technique*[a]	*Half-life*[b]	*Comments*	*Year*	*Author*	*References*
e^--trapping	Sol	–	aqueous	2.7	Inert	67	LFP	<30 ps		1985	Rothenberger	59
h^+-trapping	Sol	–	aqueous	2.7	Inert	67	LFP	~250 ns		1985	Rothenberger	59
e^--decay	Sol	–	aqueous	2.7	Inert	67	LFP	200–500 ps		1985	Rothenberger	59
e^--decay	Sol	–	aqueous	2.7	Inert	6	LFP	20–50 ns		1985	Rothenberger	59
e^--decay	Sol	–	aqueous	2.7	Inert	0.85	LFP	~100 ns		1985	Rothenberger	59
e^--decay	Anatase powder		dry	–	–	–	TRMC	1 μs to 10 s		1990	Schindler	60
e^--decay	Rutile powder		dry	–	–	–	TRMC	1–5 μs		1990	Schindler	60
e^--decay	P25 powder	29 nm	dry	–	–	–	TRMC	1 μs–10 s		1990	Schindler	60
•OH decay	Sol	13 nm	aqueous	Acid	Inert	–	PR	20 ns	•OH self-reaction	1991	Lawless	36
e^--decay	P25	29 nm	dry	–	–	0.8	PR-TRMC	100 ns		1991	Warman	61
e^--decay	P25	29 nm	2-propanol	–	–	0.8	PR-TRMC	25 μs		1991	Warman	61
e, h-trapping	P25	29 nm (124 nm)	aqueous	2.5	O_2	–	LFP	0.25–1 ns		1993	Lepore	62
e, h-decay	P25	10 ns	aqueous	2.5	O_2	–	LFP	~10 ns		1993	Lepore	62
e^--decay	P25	29 nm	paste	–	–	–	TRMC	100 ns to 10 ms	trans-decalin	1994	Martin	63
e^--decay	S13	–	paste	–	–	–	TRMC	>60 ms	trans-decalin	1994	Martin	63
e^--decay	P25	29 nm	dry	–	–	–	TRMC	230 ns to 1.6 ms	Various adsorbates	1994	Martin	64
e^--decay	Q-size	2–4 nm	dry	–	–	–	TRMC	1.6 μs to 11 ms	Various adsorbates	1994	Martin	64

(continued overleaf)

Tab. 5 *(continued)*

Reaction	*TiO_2 type*	*Particle size, nm*	*Medium*	*pH*	*Gas*	*e-h/ particle*	*Technique*[a]	*Half-life*[b]	*Comments*	*Year*	*Author*	*References*
e^--decay	Q-size	–	aqueous	–	–	–	TRDRS	~5 ps		1995	Colombo	65
e^--decay	Q-size	–	dry	–	–	–	TRDRS	~20 ps		1995	Colombo	65
e^--decay	Rutile	–	aqueous	–	–	–	TRDRS	~20 ps		1995	Colombo	65
e^--decay	Anatase	–	aqueous	–	–	–	TRDRS	~5 ps		1995	Colombo	65
e^--decay	P25	–	aqueous	–	–	–	TRDRS	~10 ps		1995	Colombo	65
e^--trapping	Q-size	2 nm	aqueous	2.7	–	–	TRDRS	~500 fs		1995	Colombo	66
e^--decay	Q-size	2 nm	aqueous	2.7	–	17.3	TRDRS	~10 ps		1995	Colombo	66
e^--decay	Q-size	2 nm	aqueous	2.7	–	2.9	TRDRS	~20 ps		1995	Colombo	66
e, h-trapping	Sol	2.1–26.7 nm	aqueous	–	Air	0.2–1504	LFP	1–10 ps		1995	Serpone	57
e^--decay	Sol	2.1 nm	aqueous	–	Air	0.2	LFP	483 ps		1995	Serpone	57
e^--decay	Sol	13.3 nm	aqueous	–	Air	157	LFP	120 ps		1995	Serpone	57
e^--decay	Sol	26.7 nm	aqueous	–	Air	1504	LFP	38 ps		1995	Serpone	57
e-trapping	Sol	2 nm	CH_3CN	–	–	–	LFP	180 fs		1995	Skinner	56
e^--decay	Sol	–	aqueous	4.5	–	–	TRDRS	~3 ps		1996	Colombo	58
e^--decay	P25	29 nm	dry	–	–	–	TRDRS	~10 ps		1996	Colombo	58
e^--decay	P25	29 nm	wet	–	–	–	TRDRS	~7 ps		1996	Colombo	58
e^--decay	P25	29 nm	aqueous	–	–	–	TRDRS	~7 ps		1996	Colombo	58
e^--decay	P25	29 nm	aqueous	–	–	–	TRDRS	>100 ps	adsorbed SCN^-	1996	Colombo	58
e^--trapping	Sol	2.4 nm	aqueous	2.3–2.6	Air	<1	LFP	<20 ns		1996	Bahnemann	46
e^--decay	Sol	2.4 nm	aqueous	2.3–2.6	Air	<1	LFP	~200 ns		1996	Bahnemann	46
h^+-decay	Sol	2.4 nm	aqueous	2.3–2.6	Air	<1	LFP	~500 ns		1996	Bahnemann	46

[a] LFP = laser flash photolysis; TRMC = time-resolved microwave conductivity; PR = pulse radiolysis; time-resolved diffuse reflectance spectroscopy.
[b] Values in parentheses are times for which the intensities decrease by one-half.

recombination level), first-order kinetics may also be observed.

Although e-h recombination is certainly important, particularly for large numbers of e-h pairs per particle, an additional decay channel that must not be overlooked is the hole-hole process:

$$h^+ + h^+ \longrightarrow (h^+)_2 \qquad (22)$$

or the related process with adsorbed $^\bullet OH$:

$$^\bullet OH + {}^\bullet OH \longrightarrow H_2O_2 \qquad (12)$$

These processes could also, in principle, be either second-order or first-order. For example, as already mentioned, Lawless and coworkers generated $^\bullet OH$ radicals radiolytically and allowed them to react with colloidal TiO_2 particles. There were no excess electrons present in this case, and the decay of the absorbance at 350 nm was first-order, assigned to Eq. (12). The existence of peroxide-like species on illuminated TiO_2 surfaces has been reported in cases in which molecular oxygen was absent [74, 75]. As already mentioned, Bahnemann and coworkers have pointed out that the optical absorbance signal at 430 nm (2.7 eV) could not correspond to a trapped hole because it was too deep within the band gap and have instead assigned this absorbance to a peroxide-like species [46]. In an EPR study by Howe and Grätzel of illuminated TiO_2 particles in vacuo at 4.2 K, Eq. (12) was proposed to occur on warming to 77 K, followed by

$$H_2O_2 + h^+ \longrightarrow {}^\bullet O_2{}^- + 2H^+ \qquad (23)$$

to explain the existence of $^\bullet O_2{}^-$ in the absence of O_2. We should also note that these decay channels, that is, hole-hole or hydroxyl-hydroxyl, help to explain the fact that trapped holes, in general, are not as long-lived as trapped electrons. As exemplified, in the presence of appropriate scavengers, there is no corresponding pathway for trapped electrons to self-react.

6.1.4.3 Quantum Yield

6.1.4.3.1 General There is an important need to assess the rates of photocatalytic reactions in terms of various other parameters, such as the amount of catalyst, the catalyst surface area, the number of active catalytic sites, and the illumination intensity. This is based on the desire to compare results for various substrates, catalysts, photocatalytic reactors, light sources, and other experimental conditions [76–79]. Progress in the photocatalytic field is facilitated by a clear set of standard definitions and methodologies, and several ambiguities need to be clarified or resolved in this area.

For example, the quantum yield (QY or Φ) is a fundamental parameter in heterogeneous photocatalysis, whose definition should be carefully considered. A standard definition has been given in terms of a particular reaction, with a defined reactant or product, at a given wavelength λ [78]:

$$\Phi_\lambda = \frac{\text{moles of reactant or product}}{\text{moles of photons (einsteins) absorbed}} \qquad (24)$$

This definition raises a number of issues. For example, commonly used light sources are rarely monochromatic. However, it is also possible to use an averaged value for a given wavelength range or it is also possible to use a close approximation of a monochromatic source. A second issue is that the measurement of the number of photons absorbed by the photocatalyst is a nontrivial task, as has been discussed for the cases of liquid suspensions [76, 79] and immobilized films in gas phase [77].

For photocatalytic reactions in aqueous suspensions, it has been proposed that a standard reactant (phenol) and standard photocatalyst (Degussa P25) be used to calibrate the results from a particular laboratory or a particular experimental arrangement in terms of quantum yield [78]. This proposal appears to have some merit, in that it is not difficult to implement, requiring no specialized equipment. Specifically, one would measure the effective quantum yield for any arbitrary reactant and photocatalyst, normalized to that for phenol and P25 TiO_2.

For gas-phase photocatalytic reactions, we would like to propose that isopropanol be used as a standard reactant because it degrades relatively rapidly to acetone in a pathway that involves only the hydroxyl radical, with a much slower decomposition of acetone [80, 81]:

$$H_3C{-}CHOH{-}CH_3 + {}^{\bullet}OH \longrightarrow H_3C{-}{}^{\bullet}COH{-}CH_3 + H_2O \qquad (25)$$

Before getting into the details of the oxidation reaction for 2-propanol, we will first mention the use of TiO_2 films on various solid substrates, which are now being used in many gas-phase applications. Initially, to avoid the difficulties of using TiO_2 powder for water purification, which entails a separation step, various researchers began to work on ways of immobilizing TiO_2 particles, for example, in thin-film form. One of the first reports on the preparation of TiO_2 films was that of Matthews [82]. Such films have also been developed by Anderson [77], by Heller [8], and by our group [83–88]. Our group has been developing ways to put photocatalytic TiO_2 coatings on various types of support materials, for example, ceramic tiles.

6.1.4.3.2 **Factors Determining Quantum Yield** Several groups have found that Φ values tend to approach a maximum at low light intensities and/or high reactant concentrations. Early work by Egerton and King, with TiO_2 suspensions in liquid 2-propanol, found Φ values between 0.5 and 1.0 for light intensities of 2×10^{13} to 5×10^{15} quanta s^{-1} for 2-propanol oxidation to acetone [89]. Above this range, a square-root dependence of rate on intensity was observed. Kormann and coworkers measured a limiting value of 0.56 for the oxidation of chloroform in aqueous TiO_2 suspension at a light intensity of 2.8×10^{-6} einstein L^{-1} min^{-1} [90]. Lepore and coworkers found an experimental value of approximately 0.6 for gas-phase propanol oxidation, with an extrapolated value of 1.0 for high concentration [62]. In EPR spin-trapping experiments, in transparent aqueous TiO_2 sols, Grela and coworkers measured a limiting value of 0.54 for high spin-trap concentrations [71].

In gas-phase experiments with TiO_2 films on glass, we have found that under specific conditions (i.e. high 2-propanol coverage and low light intensity (36 nW cm^{-2} to 45 uW cm^{-2}), the quantum yield Φ tends to approach a maximum value that is intrinsic to a particular TiO_2 film [81]. Even films made from the same starting materials can have small differences in this maximum value. It was proposed that this value depends on such factors as the levels of impurities in the bulk of the TiO_2 particles and the degree of crystallinity [12], which would control the rate of bulk recombination. Another idea has been advanced that suggests that the limiting Φ value at low intensity is due to particle-size effects, with lower Φ values exhibited by small particles [69].

Furthermore, we found that the Φ values are constant for a particular ratio of absorbed photons to adsorbed molecules [81]:

$$I_{\text{norm}} = \frac{J_{h\nu}}{\Gamma_{2-\text{p}}} \quad (26)$$

where I_{norm} is termed the normalized light intensity, $J_{h\nu}$ is the light flux in number of moles per unit apparent area per unit time (mol cm^{-2} s^{-1}), and $\Gamma_{2-\text{p}}$ is the number of moles of adsorbed 2-propanol per unit apparent area per unit time (mol cm^{-2} s^{-1}). For any given initial value for $\Gamma_{2-\text{p}}$, we propose that the shape of the curve (Fig. 5) is governed by the competition between reactions of either trapped holes or hydroxyl radicals with isopropanol (Eq. 25) and one of the following: either e-h recombination (Eq. 21), hole-hole annihilation (Eq. 22), or the hydroxyl radical self-reaction (Eq. 12). Assuming that $^{\bullet}$OH is the active species, one can write an equation for the branching of the light flux into these two types of processes, based on the assumption that a steady state surface concentration of hydroxyl radicals or trapped holes is reached.

$$J_{h\nu} = k_{25}\Gamma_{2-\text{p}}\Gamma_{\bullet\text{OH}} + k_{12}\Gamma^2_{\bullet\text{OH}} \quad (27)$$

$$\Phi = \frac{k_{25}\Gamma_{2-\text{p}}\Gamma_{\bullet\text{OH}}}{J_{h\nu}} = \frac{k_{25}\Gamma_{2-\text{p}}\Gamma_{\bullet\text{OH}}}{k_{25}\Gamma_{2-\text{p}}\Gamma_{\bullet\text{OH}} + k_{12}\Gamma^2_{\bullet\text{OH}}} \quad (28)$$

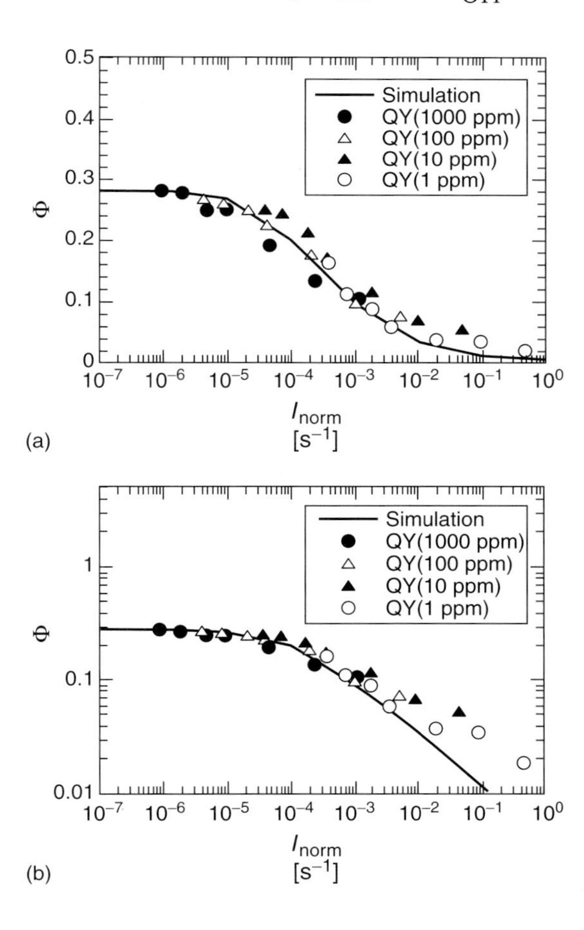

Fig. 5 Quantum yield as a function of normalized light intensity: 2-propanol, taken from Ohko and coworkers [81].

This equation can be rearranged to the following form:

$$\Phi = \frac{2}{1 + \sqrt{1 + \dfrac{4J_{h\nu}k_{12}}{k_{25}^2 \Gamma_{2-p}^2}}} \qquad (29)$$

It can be seen that this function approaches unity at low light intensity and zero at high intensity. This type of curve was used to fit the data in Fig. 5. Of course, it is also possible that such a relationship could also result from a scheme, based on e-h recombination instead of the $^{\bullet}$OH self-reaction. It is highly intriguing that this type of behavior could be exhibited over such a wide range of initial reactant surface concentrations (three orders of magnitude) and light intensities (seven orders of magnitude).

We have proposed a simple model to rationalize the fact that the quantum yield versus light intensity behavior tends to remain constant for a given ratio of adsorbed photons to adsorbed molecules. Basically, this model involves the assumption that the diffusion of $^{\bullet}$OH on the surface is fast enough, that it is not a limiting factor, and every $^{\bullet}$OH generated has an opportunity to react with all of the adsorbed 2-propanol molecules in its vicinity. Let us imagine for a moment that the reactions are taking place on a perfectly flat, nonporous surface with a certain number of adsorbed 2-propanol molecules, say 3000. For a low initial gas-phase 2-propanol concentration, this number of molecules will occupy a certain surface area A_{low}. For a high initial concentration, the molecules will occupy a smaller surface area A_{high}, simply due to the adsorption equilibrium. For each surface area, we now assume that a certain number of photons is absorbed, for example one every second, so that the ratio of absorbed photons to adsorbed molecules is the same, 1 : 3000. We find that Φ is the same in both cases, 0.14, or half the limiting value of 0.28. Hence, each $^{\bullet}$OH is approximately as likely to react with a 2-propanol molecule, as it is to react with another $^{\bullet}$OH. Thus, the Φ value is determined only by the probabilities that the $^{\bullet}$OH radicals will either react with a 2-propanol molecule or with each other.

However, this simple model requires further modification because the Φ values tend to deviate from the theoretical curve at higher light intensities; this becomes apparent when $\log \Phi$ is plotted (Fig. 5b), which indicates a mixing of first and second-order kinetics. Second-order behavior is approached for the lowest 2-propanol concentration but deviates for higher concentrations. For the latter, trapped holes would become increasingly favored as the active species. Additionally, the adsorbed molecules may block sites at which holes can self-react, and therefore another decay channel might become important, for example, trapping at a different type of site at which the h-h reaction is favored or e-h recombination. In any case, the results show clearly that 2-propanol decomposition can be used as a type of standard for quantum yield, if high reactant concentrations and low light intensities are used to estimate the maximum, limiting value. However, it is still necessary to estimate the amount of absorbed light, for example, using the methodology proposed by Anderson and coworkers [77].

The idea of $^{\bullet}$OH radicals self-reacting, has also been proposed for the decomposition of chloroform at illuminated TiO_2 aqueous suspension by Kormann and coworkers [18, 90]. These authors have given a limiting form of Eq. (29) at high light intensity:

$$\Phi = \frac{k_{rds}\Gamma_{chloro}}{\sqrt{k_{12}J_{h\nu}}} \quad (30)$$

$$k_{obs} = \frac{k_{rds}}{\sqrt{k_{12}}} = 2.0 \times 10^{-3}(\text{M s})^{-1/2} \quad (31)$$

where k_{obs} is the observed rate constant, k_{rds} is the rate constant for the reaction between a surface-bound $^{\bullet}OH$ radical and a chloroform molecule, corresponding to our k_{25}, and k_{12} is the rate constant for the $^{\bullet}OH$ self-reaction. From the value of k_{obs}, they conclude that the rds is slow compared to Eq. (10). This conclusion is consistent with that of Ohko and coworkers for 2-propanol oxidation on TiO_2 films [81].

It is an important challenge to resolve this issue of whether e-h recombination or another reaction, such as 12 or 25, is more important in determining Φ at higher light intensities. The answer could help to guide further efforts to improve Φ values.

We continue to emphasize that low light intensities are not only interesting from a fundamental standpoint but are also highly important for indoor environmental applications, in which intense UV light sources cannot be used [12, 20]. For these applications, the most important point is that the photocatalytic reaction can be quite efficient, even at UV light intensities that are compatible with safe human exposure levels.

6.1.4.4 Reaction Mechanisms

6.1.4.4.1 General The reaction steps already given are just the beginning of a complex mix of reactions that can be proposed to explain the experimental photocatalysis results. These results often involve the complete mineralization of the starting organic compound, that is, complete conversion to inorganic compounds such as CO_2 and H_2O. In some cases, it is possible to monitor specific intermediate compounds, and this approach can be quite powerful in elucidating the mechanistic pathways, as discussed in the following sections.

6.1.4.4.2 Active Species The first set of reactions that follow those already given, involve either molecular oxygen or superoxide radical anion $^{\bullet}O_2^-$:

$$R^{\bullet} + O_2 \longrightarrow ROO^{\bullet} \quad (32)$$

In many cases, this radical can then react with superoxide, which has been experimentally demonstrated to be produced on illuminated-TiO_2 [91–93]:

$$ROO^{\bullet} + {}^{\bullet}O_2^- \longrightarrow ROOOO^{\bullet} \quad (33)$$

The latter species is termed a tetroxide anion. It can become protonated to the $^{\bullet}$ROOOOH form, a tetroxide radical, which can take on a cyclic structure, depending on the nature of the reactant. These will be discussed further in the section on hydrocarbons and related compounds. We shall now proceed to describe the reactions that have been proposed for various specific types of reactants.

6.1.4.4.3 Aldehydes For aldehydes, it has been known for quite some time that radical initiators can induce chain-type oxidation in the presence of oxygen [94–97].

$$^{\bullet}OH + RH \longrightarrow R^{\bullet} + H_2O \quad \text{(initiation)} \quad (11)$$

$$R^{\bullet} + O_2 \longrightarrow ROO^{\bullet} \quad \text{(propogation)} \quad (34)$$

$$ROO^{\bullet} + RH \longrightarrow ROOH + R^{\bullet} \quad (35)$$

$$ROO^{\bullet} + ROO^{\bullet} \longrightarrow \text{nonradical products} \quad \text{(termination)} \quad (36)$$

where we have inserted •OH as the radical initiator.

We were able to observe clear evidence for the chain-type mechanism in experiments, involving acetaldehyde decomposition in the gas-phase [98], similar to those already discussed for 2-propanol. With acetaldehyde, the Φ values exceeded the maximum value obtained for a similar film for 2-propanol oxidation (0.28) (Fig. 6). As already discussed, the latter value may be considered to be an intrinsic maximum Φ value for this particular film. Therefore, if Φ exceeds the intrinsic maximum value, it indicates that radical chain reactions are important, that is, a single photon can cause more than one photodecomposition reaction.

Other groups have also obtained relatively high Φ values for chain-type reactions. As already mentioned, Lepore and coworkers found values, approaching unity for propanol oxidation [62]. Raupp and Junio found values exceeding unity, as large as 3.0, for the oxidation of acetone and methyl-*t*-butyl ether [99]. Stark and Rabani measured values exceeding unity, as high as 1.4, for the photocatalytic dechlorination of carbon tetrachloride [100].

6.1.4.4.4 **Hydrocarbons and Related Compounds** Reaction pathways involving superoxide have been treated in detail by Schwitzgebel and coworkers [101], as shown in Fig. 7. These reactions can operate on alkanes and their oxidation products so that with the exception of aldehydes, as noted in the earlier section, any of the main types of CH and C−H−O compounds (i.e. alkanes, alcohols, ketones, and carboxylic acids) can be completely decomposed in this fashion to CO_2 and H_2O. These reactions involve cyclic tetroxide intermediates, similar to those originally proposed by Russell [102]. It is quite interesting that TiO_2 is so well suited to the decomposition of organic compounds, with a combined attack of oxidatively and reductively produced radical species.

In Fig. 7(b), a possible three-dimensional structure for the tetroxide intermediate for the photocatalytic oxidation of *n*-octane

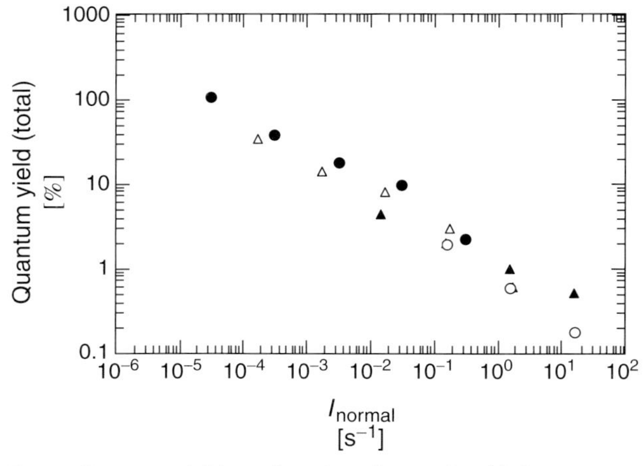

Fig. 6 Quantum yield as a function of normalized light intensity: acetaldehyde, taken from Ohko and coworkers [98].

n-octane:	$RCH(OOOOH)R' \longrightarrow RCOR' + O_2 + H_2O$
3-octanol:	$RCOH(OOOOH)R' \longrightarrow RCOR' + O_2 + H_2O_2$
3-octanone:	$RCH_2COCH(OOOOH)R' \longrightarrow RCHO + R'CHO + CO_2 + H_2O$
n-octanoic acid:	$RCH_2COOH + h^+ (^{\bullet}OH) \longrightarrow RCH_2^{\bullet} + CO_2 + H^+(H_2O)$
	$RCH_2^{\bullet} + O_2 \longrightarrow RCH_2OO^{\bullet}$
	$RCH_2OO^{\bullet} + {}^{\bullet}OOH \longrightarrow RCH_2OOOOH$
	$RCH_2OOOOH + RCH_2COOH \longrightarrow RCH_2OH + RCHO + O_2 + CO_2$

(a)

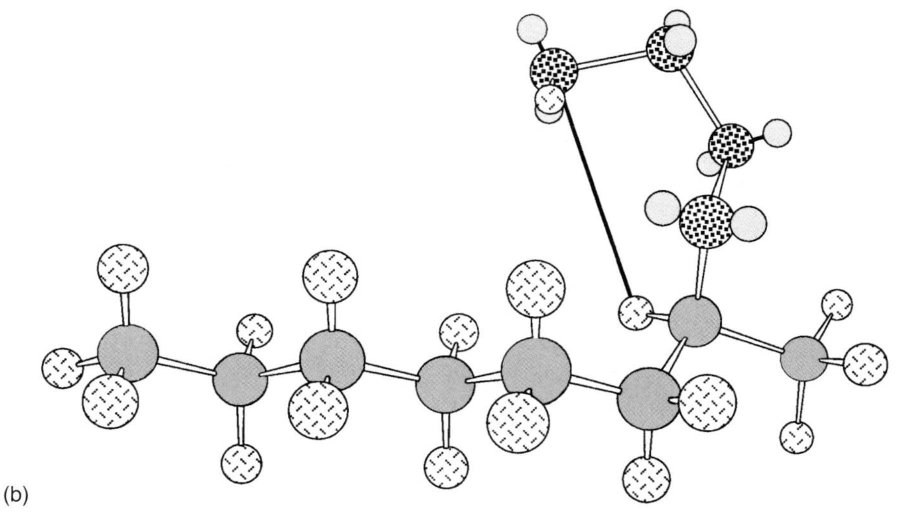

(b)

Fig. 7 (a) Photocatalytic reactions involving tetroxides, taken from Schwitzgebel and coworkers [101]. (b) Proposed structure for the tetroxide intermediate in the oxidation of octane, with the hydrogen bond shown as a solid rod. Carbon atoms are gray, hydrogen atoms are a light stipple, oxygen atoms are a dark stipple, and oxygen lone pairs are white.

is shown. Very similar intermediates can be drawn for the corresponding alcohol and ketone. For the alcohol (3-octanol), the cyclic tetroxide structure can form at C-2, similar to the structure shown for *n*-octane but the intermediate given in the equation in Fig. 7(a) actually corresponds to the cyclic structure, forming at C-3. In that case, the alcohol −OH group becomes part of the ring. For 3-octanone, the terminal oxygen of the tetroxide may hydrogen-bond across to the hydrogens on C-4, which could facilitate the breakage of the chain, with the formation of two aldehydes and a CO_2 molecule.

The pathways delineated in the preceding paragraph are probably also important in the decomposition of long-chain hydrocarbons, carboxylic acids, fats, oils, and even long-chain polymers [103–105]. In the latter case, we have direct infrared evidence that the hydrocarbon chains can be attacked at any point along their length, with C−C bond scission, presumably with subsequent degradation

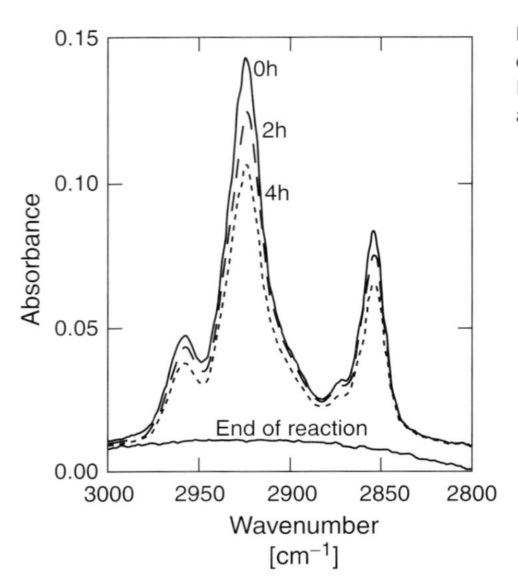

Fig. 8 Photocatalytic decomposition of octadecane: IR spectra, taken from Minabe and coworkers [105].

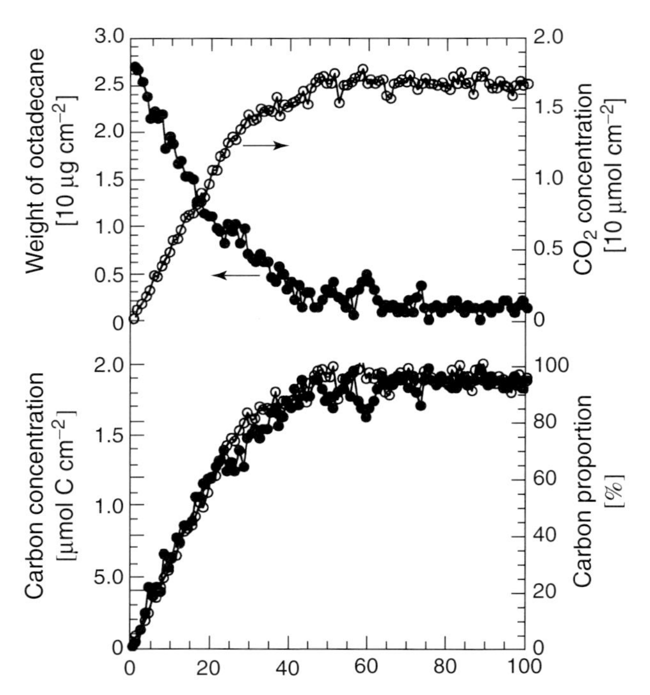

Fig. 9 Photocatalytic decomposition of octadecane: weight loss and CO_2 generation, taken from Minabe and coworkers [105].

via the resulting aldehydes (Fig. 8). The aldehydes may be unstable enough to account for the fact that they are not detected by IR on the TiO_2 film. Also, it is important to note that no gas-phase products, other than CO_2, are detected in the decomposition of octadecane, stearic acid, and glycerol trioleate, demonstrating the safety of the photocatalytic approach to self-cleaning technology. For example, a 36-μm-thick film of octadecane was completely converted to CO_2 after approximately 50 hours of illumination at 0.8 mW cm^{-2} (Fig. 9).

6.1.4.4.5 **Aromatics** As already briefly mentioned, the photocatalytic oxidation of phenol was studied by Goldstein and coworkers [45]. On the basis of the analysis of the possible reactions (Fig. 10), these authors were able to conclude that either $^{\bullet}OH$ or h^+ can be the principal active species. Also, the use of quinoline as a probe molecule was already mentioned. The full scheme developed by Cermenati and coworkers is shown in Fig. 11 [47]. This scheme shows that the combined action of holes and superoxide is important.

6.1.4.5 Kinetic and Mass Transport Considerations

6.1.4.5.1 **Adsorption** For moderate-intensity illumination, over a wide range of experimental conditions, photocatalytic decomposition follows first-order kinetics in the adsorbed concentration of the organic compound and one-half-order in light intensity due to recombination.

$$\text{rate} = k_{\text{oxidation}}\Gamma_{\text{reactant}}I^{1/2} \qquad (37)$$

Where $k_{\text{oxidation}}$ is a first-order rate constant, Γ_{reactant} is concentration per unit real surface area, and I is the light intensity. As pointed out by Emeline and coworkers, however, this equation is a special case of a more general one [106].

$$\text{rate} = k_{\text{oxidation}}\Gamma^{n}_{\text{reactant}}I^{m} \qquad (38)$$

where the n and m values can vary with I and Γ_{reactant}, respectively.

Many organic compounds and noxious gases, such as H_2S, follow Langmuir adsorption behavior so that Eq. (37) can be converted into the familiar Langmuir–Hinshelwood form by substituting the adsorption equilibrium expression for

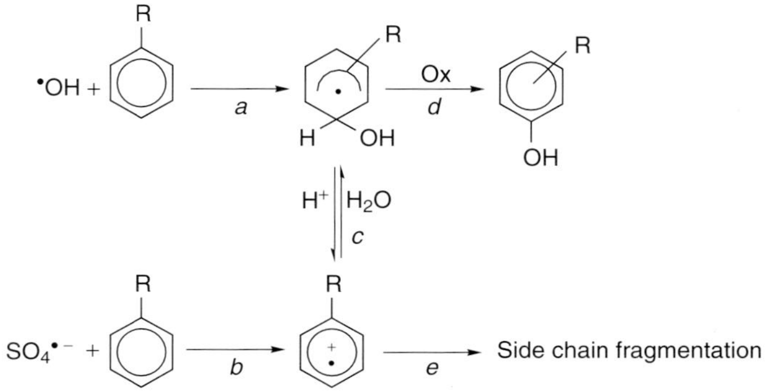

Fig. 10 Photocatalytic decomposition of phenol: scheme, based on Goldstein and coworkers [45].

Γ_{reactant}:

$$\Gamma_{\text{reactant}}^{-1} = \Gamma_{\text{reactant},0}^{-1} + (K_{\text{eq}} c_{\text{reactant}})^{-1} \quad (39)$$

where $\Gamma_{\text{reactant},0}$ is the surface concentration at full coverage, K_{eq} is the adsorption equilibrium constant, and c_{reactant} is the gas-phase or liquid-phase concentration.

An example of this type of behavior is provided by our recent work with the gas-phase decomposition of formaldehyde and acetaldehyde, which have been implicated in the "sick-building" syndrome [107]. These two compounds were found to obey Langmuir adsorption and Langmuir–Hinshelwood kinetics were followed (Fig. 12). It was also found that formaldehyde adsorbs relatively strongly on the TiO_2 surface, showing that TiO_2

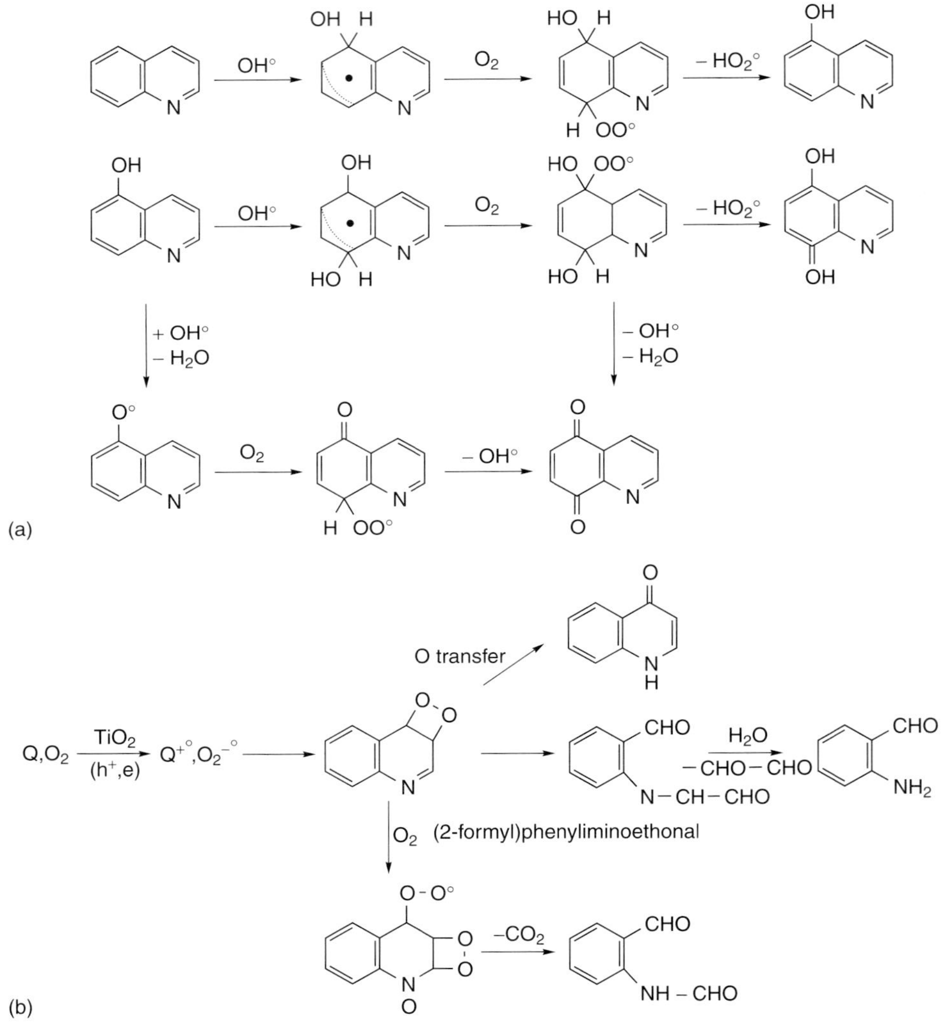

Fig. 11 Photocatalytic decomposition of quinoline: scheme, based on [47].

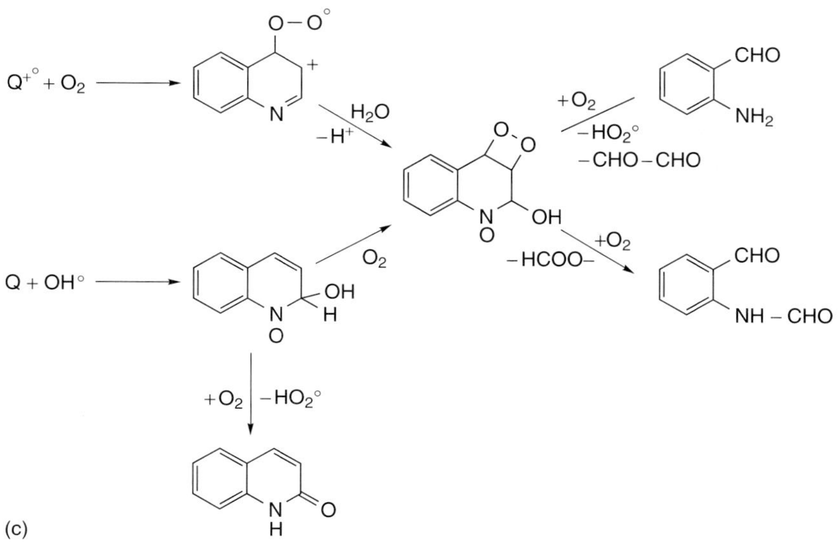

(c)

Fig. 11 (*Continued*)

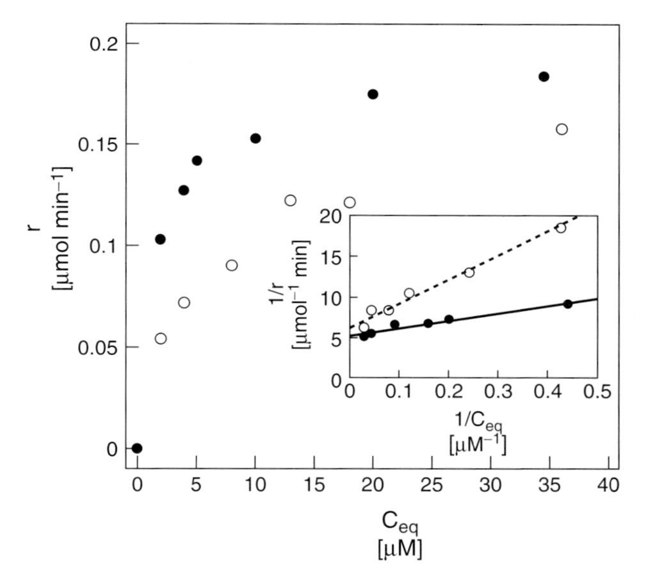

Fig. 12 Photocatalytic decomposition of acetaldehyde: Langmuir–Hinshelwood plot, based on Noguchi and coworkers [107].

can be effective even at low formaldehyde concentrations. The adsorption characteristics of TiO_2 in this case are similar, if not better than those of activated carbon.

Many groups have obtained results that are consistent with Langmuir–Hinshelwood kinetics. However, Emeline and coworkers have emphasized the well-known difficulties in distinguishing between various mechanisms purely based on the observed kinetic rate laws [106].

6.1.4.5.2 **Mass Transport** Under conditions of high light intensity and low reactant concentration, it is possible to reach a point at which the mass transport of the organic compound is no longer sufficient to maintain the adsorption equilibrium. Such diffusion-controlled conditions have been explored in further work with 2-propanol as a model reactant [108]. These results show that it is possible to clearly distinguish the experimental conditions under which various types of kinetics are operative (e.g. light-limited conditions, a middle range of light intensity in which either Eqs. (37) or (38) are valid, and a high light intensity range, in which mass-transport limitations become important). These types of conditions can be seen in a plot of light intensity versus reactant concentration (Fig. 13). Such plots must also take into account the Langmuir isotherm characteristics of the particular reactant of interest on the specific type of TiO_2 being considered.

6.1.4.6 Special Techniques

In addition to the techniques that have already been mentioned, with which to investigate photocatalytic reactions, including laser flash photolysis and EPR, there are several others that can yield very useful information. These include the slurry electrode technique, the use of biased electrodes or photoelectrodes, the scanning electrochemical microscopic (SECM) technique, and the Langmuir–Blodgett (LB) film technique. These will be discussed briefly.

6.1.4.6.1 **Slurry Electrodes** The slurry electrode technique enables one to carry out electrochemical measurements on a particulate suspension or slurry. If the suspension is illuminated, then one has a PEC slurry cell, a technique introduced by Bard and coworkers [109–111] and later extended by other groups [112–114], including our group [93]. For example, a TiO_2 suspension in aqueous electrolyte is stirred and current is measured at a potentiostated platinum electrode. With illuminated TiO_2 in a deaerated solution with no hole scavenger, the results show a net anodic current due to oxidation of trapped electrons. As soon as the illumination is stopped, the anodic current decays over a period of up to 250 ms [112]. The negative charge build-up on the illuminated particle is due to the need for a pathway by which photogenerated holes can be consumed and the product desorbed. In the work of Peterson and coworkers and that of our group, this was assumed to be the hydroxyl radical itself (Fig. 14). However, based on our results with photocatalytic effects at a distance from an illuminated TiO_2 film, it appears to be more likely that the hydroxyl radicals self-react (Eq. 12) and hydrogen peroxide desorbs [53–55]. With the addition of hole scavengers, such as formate, acetate or methanol, the anodic photocurrent is enhanced greatly and decays much more slowly when illumination is stopped [112–114].

In related work, we have found that illuminated suspensions of TiO_2 particles in silicone oil exhibit a tendency to form bridges between biased electrodes that varies, depending on the water content of the TiO_2 [115, 116]. For relatively dry types of TiO_2, such as P25 (ca. 1%), the bridge-forming tendency is larger because the photogenerated charges can be polarized at opposite ends of the particles. With TiO_2 that contains greater amounts of water,

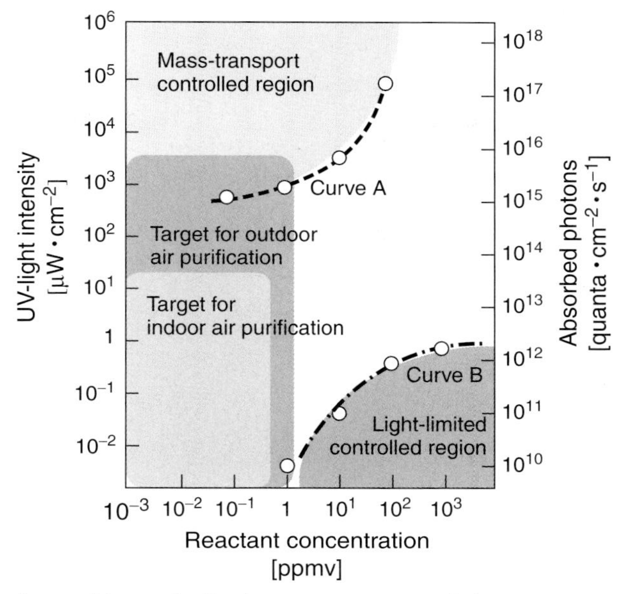

Fig. 13 Master plot for determining operating light intensity as a function of reactant concentration, based on Ohko and coworkers [64].

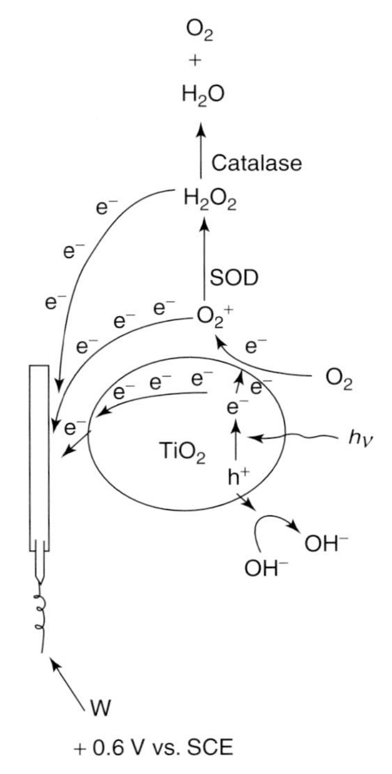

Fig. 14 Reactions taking place between an illuminated TiO_2 particle in aqueous solution, based on Cai and coworkers [93].

there is a tendency for the particles to develop excess charge when they make contact with the electrode, so that they move under the influence of the electric field and circulate electrophoretically back and forth between the electrodes. This is because the photogenerated carriers can be effectively trapped via reactions involving hydroxide ions or water. Because of the absence of bulk water surrounding each particle, excess charge cannot be carried away, for example, by the production of solution-phase peroxide.

6.1.4.6.2 **Biased Electrodes** It is possible to significantly enhance the oxidation of organic compounds, such as 4-chlorophenol, at a nanoparticulate photoelectrode by applying a positive bias potential [117, 118]. This result provides convincing evidence that recombination is important because it seems clear that the effect of the electric field is to enhance the separation of electrons and holes.

It was mentioned at the beginning of this chapter that the photocatalytic process is closely related to the corrosion of metals, in that both anodic and cathodic processes are occurring on the same surface. Indeed, for an electrically isolated material, the anodic and cathodic currents must exactly match to maintain charge neutrality. One can measure the separate anodic and cathodic current-potential curves for an appropriate electrode to understand the overall behavior. There is a potential at which the anodic and cathodic currents are equal in magnitude and this potential is termed the "mixed potential." The basic aspects of metal corrosion have been treated by Vetter [119] and Sato [120].

A similar approach has been developed for the photocatalytic process by Kesselman and coworkers [121]. These workers, using a single-crystal rutile rotating-disk electrode, have shown that the mixed potential, and more importantly the magnitude of the current, is limited by the rate of the cathodic process, that is, O_2 reduction (Fig. 15). In this case, the magnitudes of both currents at open circuit were much lower than the maximum anodic current, which is due to photo-assisted O_2 evolution and is limited only by the rate of photogeneration of holes. It was found that the deposition of ~2 Å of Pt was effective in increasing the current density for O_2 reduction, such that the magnitudes of the anodic and cathodic reactions at open circuit were then limited by the anodic reaction. These results imply that photocatalytic reaction rates on illuminated TiO_2 could be significantly accelerated by the use of catalytic metals, as discussed later. One point in which the current-matching experiments appear to deviate from the situation with practical photocatalytic phenomena is that the anodic process in the former case is found to be almost purely O_2 evolution. This cannot be true, in general, for TiO_2 photocatalysis because the quantum yields would then be very low.

It is interesting to note that, in Vetter's treatment of corrosion, there are three different special cases: (1) chemically and physically homogeneous surfaces; (2) chemically inhomogeneous surfaces;

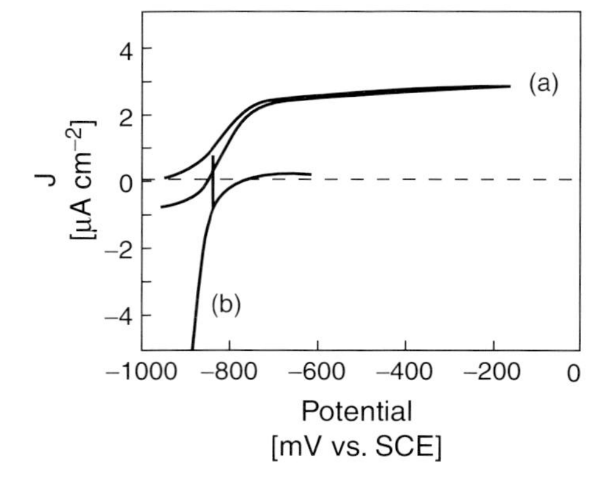

Fig. 15 Current–potential curves for an illuminated TiO_2 photoelectrode, taken from Kesselman and coworkers [121].

and (3) chemically homogeneous, physically inhomogeneous surfaces. An example of case 2 in a metal would be a Zn-Cu alloy, in which Zn undergoes corrosion but Cu does not. A corresponding example in photocatalysis would be TiO_2, with deposited Pt, in which the anodic process might occur predominantly on TiO_2 and O_2 reduction might occur predominantly on Pt. An example of case 3 in terms of corrosion would be any polycrystalline metal, in which different crystallographic faces are exposed to the electrolyte, such that the relative proportions of the anodic and cathodic processes are different, although both reactions occur on both types of surfaces. In terms of photocatalysis, this is the type of situation that could be expected for a pure polycrystalline TiO_2 sample. A direct comparison of the anodic and cathodic reactions on the various single crystal faces of either anatase or rutile does not yet appear to have been studied.

6.1.4.6.3 **Scanning Electrochemical Microscopy** Another way in which it can be confirmed that oxidative and reductive photocatalytic reactions simultaneously occur on TiO_2 particles is to set up a simple heterogeneous model electrode system. This can be used to simulate model individual particles and can be compatible with microelectrode detection of reaction products [122–126]. Such a system, shown in

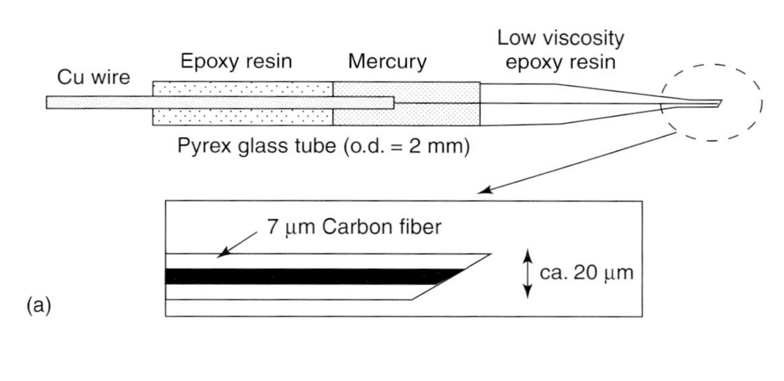

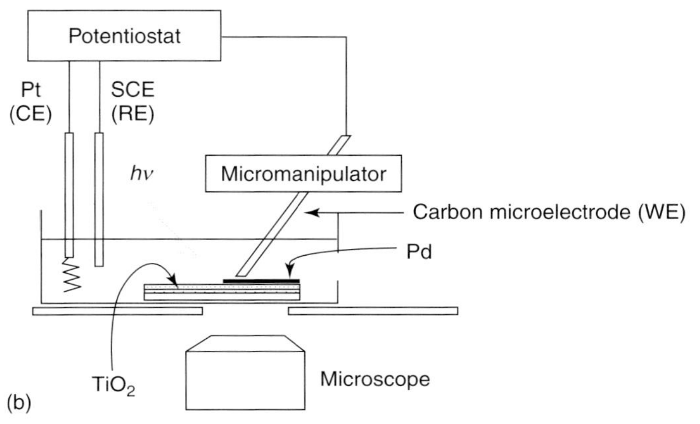

Fig. 16 Schematic diagram of a scanning electrochemical microelectrode system, taken from Fujishima and coworkers [12].

Fig. 16, involves a film that is half-covered with TiO_2 and half-covered with metallic regions, for example, Pd or ITO (In-Sn oxide). A scanning microelectrode can be positioned close to the surface, as close as 50 μm. Its potential can be set at values at which, for example, either O_2 or H_2O_2 can be monitored as a function of time.

The oxidation and reduction reactions can be monitored separately to obtain information regarding the mechanism of the photocatalytic reaction. In photocatalytic reactions with metal-deposited TiO_2 particles and films, the deposited metal acts as a reduction site, thereby increasing the efficiency of photogenerated charge separation. Thus, a TiO_2-ITO composite film can act as a simple model for the metal-deposited photocatalyst. A carbon microelectrode is employed to detect the reaction products. By positioning this electrode close to either the TiO_2 or the ITO surface (50–500 μm), the oxidation and reduction reactions, respectively, can be monitored on these regions. This method was used to detect hydrogen peroxide at the two types of regions, separately, by use of a "wired" horseradish peroxidase microsensor fabricated on the carbon microelectrode surface [122]. Later, it was found that peroxide could also be detected adequately by setting the microelectrode potential at a very positive potential (+1.0 V versus SCE) [124]. Additionally, the carbon microelectrode is able to detect O_2 by setting the potential at a negative potential (−1.0 V versus SCE).

In one of the early studies, it was found that H_2O_2 is produced largely at the ITO portion of the illuminated TiO_2-ITO composite surface due to O_2 reduction [122]. In a subsequent study on the separate detection of dissolved oxygen in aqueous solution at reducing regions (in this case, Pd) and oxidizing regions (TiO_2), using a carbon microelectrode, we found an increase in O_2 concentration near the TiO_2 surface and a decrease in the O_2 concentration near the TiO_2 surface under illumination (Fig. 17) [124]. Oxygen is produced via water oxidation on TiO_2 and is then consumed at the metallic electrode (ITO or Pd) via reduction. Principally, reduction to peroxide occurs although reduction to superoxide or to water may also contribute on an actual TiO_2 microparticle. There can also be varying contributions by these three reduction reactions, depending on whether there is a deposited catalyst. As already discussed, some photocatalytic reaction pathways can involve $^{\bullet}O_2^-$; this is a desirable product, and metallic catalysts may in fact be counterproductive, if they promote the two- or four-electron O_2 reduction reactions.

In the presence of ethanol, although dissolved oxygen was consumed at both TiO_2 and Pd sites, the consumption of oxygen was larger at the TiO_2 surface [122]. A possible scheme that is consistent with O_2 consumption at TiO_2 is:

$$^{\bullet}OH + CH_3CH_2OH \longrightarrow CH_3C^{\bullet}HOH + H_2O \quad (40)$$

$$CH_3C^{\bullet}HOH + O_2 \longrightarrow CH_3CH(OH)OO^{\bullet} \quad (41)$$

producing an organoperoxyl radical, which can either participate in a chain-type process

$$CH_3CH(OH)OO^{\bullet} + CH_3CH_2OH \longrightarrow CH_3CH(OH)OOH + CH_3C^{\bullet}HOH \quad (42)$$

$$CH_3CH(OH)OOH \longrightarrow CH_3CHO + H_2O_2 \quad (43)$$

or can react further with $^{\bullet}O_2^-$ or $HO_2^{\bullet}$ to produce a tetroxide intermediate, as already discussed. The importance of radical intermediates was found to be great, as

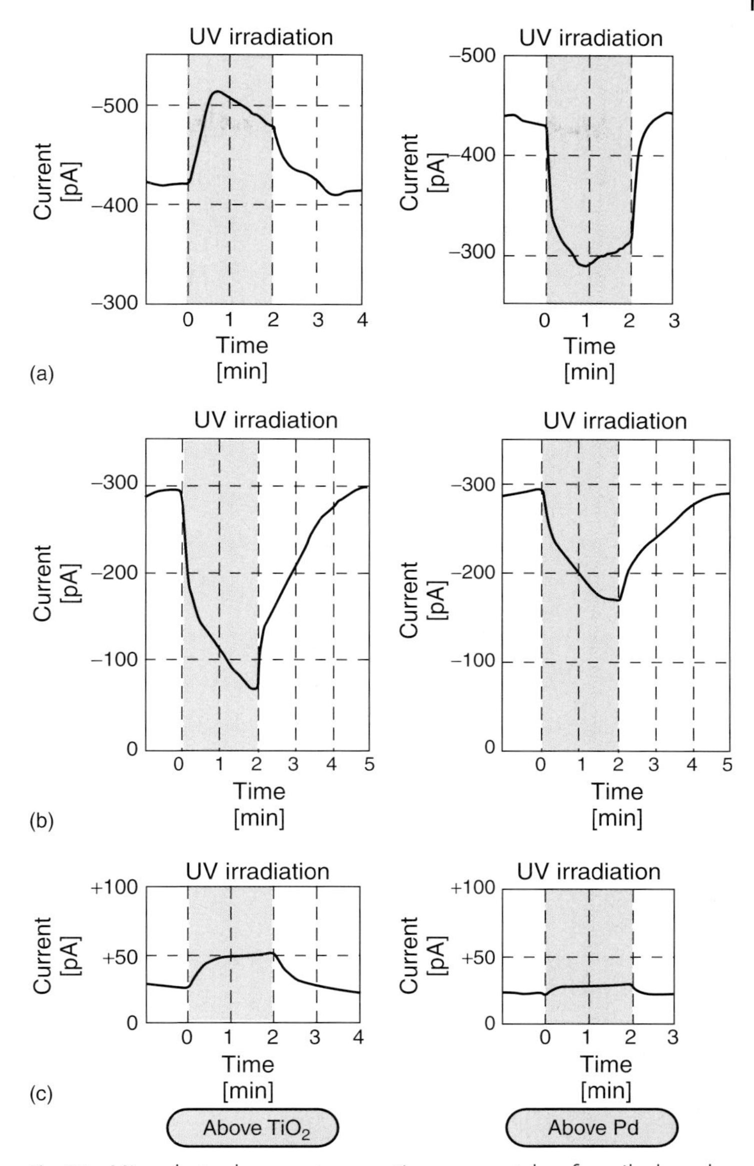

Fig. 17 Microelectrode current versus time curves, taken from Ikeda and coworkers [124].

evidenced by the strong decrease in O_2 consumption in the presence of phenol, a radical scavenger.

It was also found that the consumption of O_2 over the illuminated TiO_2 portion is faster for the photodecomposition of acetaldehyde compared to ethanol because of the strong contribution of radical chain reactions for the former [125]. In the presence of superoxide dismutase, the

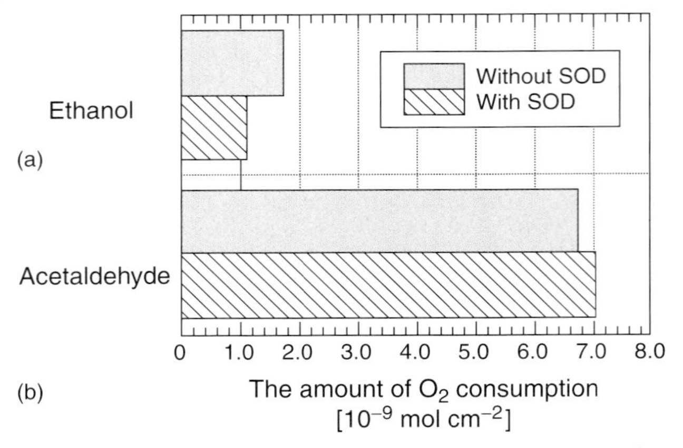

Fig. 18 The effect of SOD on the rate of oxygen consumption for ethanol versus acetaldehyde [125].

O_2 consumption decreased significantly in the case of ethanol but negligibly in the case of acetaldehyde (Fig. 18). Thus it appears that the tetroxide pathway may be important for ethanol. However, aldehydes can decompose via radical-initiated chain reactions, consuming O_2 but not $^{\bullet}O_2^-$ [95–97].

6.1.4.6.4 **LB Techniques** Fundamental information concerning the photocatalytic decomposition of stearic acid LB films on both polycrystalline anatase films and single crystal rutile can be obtained with the AFM technique [127, 128]. Such studies have shown that, for less than full monolayer coverage, the photocatalytic decomposition process takes place at random locations on the film. In the case of full monolayers, the film structure is disturbed very quickly, with molecules aggregating into relatively tall ridges. It appears that there are randomly located active sites, which can become passivated by reaction products. New sites can also form. In contrast to results obtained with thicker layers (100 nm), for which zero-order kinetics were obtained, the LB film results obeyed first-order kinetics, which is reasonable because of the extremely limited amount of material.

6.1.4.7 New Photocatalysts

6.1.4.7.1 **General** The results presented thus far certainly demonstrate the superior characteristics of TiO_2 as a photocatalyst. However, there are various characteristics of TiO_2 that could, in principle, be improved. For example, various researchers have probed the idea of improving the charge separation through catalysis of electron transfer to electron acceptors, such as O_2, by means of deposited metal particles [22]. TiO_2 has a drawback, in that the proportion of ambient light as either sunlight or even indoor light that can be utilized to induce photocatalytic reactions is still quite small. The value for full sunlight, which has a total power of approximately 100 mW cm^{-2}, is just about 1% or 1 mW cm^{-2}. For indoor light, the values are much lower, on the order of 1 μW cm^{-2} under a study lamp or 0.1 μw cm^{-2} on the walls or floor of a room [20]. In addition, for photocatalytic air or water treatment, it is necessary to use relatively intense UV light sources. It may be possible to greatly increase the light-harvesting ability of TiO_2 through band gap engineering.

6.1.4.7.2 **Metal-modified Photocatalysts**

The area of metal-modified TiO_2 photocatalysts is an extensive one but some aspects have been reviewed recently [22]. We will only cite a few examples here. We have already introduced two of the main themes: (1) improvement of electron transfer; and (2) improvement of light harvesting. One extremely important application, the photocatalytic deposition of metals, has also been treated in depth by Litter.

In the area of photocatalytic water splitting, the use of deposited noble metals has been studied for many years [11]. The objective in this case is to enhance the transfer of photogenerated electrons from the TiO_2 conduction band to water to produce hydrogen gas. This approach does indeed lead to improvements in H_2 production due to the well-known electrocatalytic activity of platinum for this reaction. In the area of photocatalytic decomposition of organic compounds, as already discussed, it is often desirable to produce superoxide from O_2 to assist in the formation of tetroxides, for example, Ref. 101. The energetic requirement for the production of superoxide is substantially less than that for hydrogen, so that in principle, an electrocatalyst is not needed. In fact, as already mentioned, the use of a metallic catalyst may be counterproductive if it leads to a multielectron product such as H_2O_2 or H_2O. In terms of simply increasing the electron transfer rate for oxygen reduction a metallic catalyst, such as Pt or Pd, could be beneficial as shown by Gerischer, Heller, and coworkers [113, 129].

In addition, metal ions can enhance oxidation rates by acting as electron acceptors in themselves. Litter points out that metal ions can be either beneficial or detrimental [22]. Examples are Cu^{2+}, Fe^{3+}, and Ag^{+}, all of which can have beneficial effects in certain concentration ranges and detrimental effects at higher concentrations. Metal ions can be present in water that is being photocatalytically treated and the effects must be taken into account. Finally, Litter points out that the addition of metal ions is generally too expensive in large-scale applications.

In terms of the modification of the light absorption properties of TiO_2, there have been several notable efforts. Anpo and coworkers have used the ion implantation technique extensively to extend the absorption spectrum into the visible range [130]. With the use of implanted Cr, the spectrum was shifted by as much as about 80 nm, with significant absorption at 450 nm. One challenge is to try to devise methods of preparing such photocatalysts with less technology-intensive approaches.

Zang and coworkers have prepared Pt(IV) chloride-modified amorphous TiO_2 that is effective at visible wavelengths [131]. These authors have shown that the Pt is present in a complex ion form, still surrounded by chloride. This constitutes a highly effective use of the precious metal. Photocatalytic activity was found at wavelengths as high as 546 nm. Hamerski and coworkers have reported the improvement of photocatalytic activity for oil decomposition under sunlight with calcium-modified TiO_2 [132].

Murata and coworkers have thermally prepared rutile films on alloys of Ti with various 4d and 5d transition metals [133]. With the Ti-Ta and Ti-W alloys, visible light activity was found. These results are promising and should be followed up with efforts to produce practical materials.

6.1.4.7.3 **Mixed Oxides** In the area of modification of TiO_2 with other oxides, one of the rationales is to increase the surface acidity and thereby modify the adsorption and catalytic properties. Fu and coworkers have reviewed the previous literature in this area and have prepared TiO_2/SiO_2 and TiO_2/ZrO_2 materials via sol-gel techniques [134]. Improvements by a factor of three were observed for the gas-phase oxidation of ethylene.

An important area of activity in terms of modification of TiO_2 with other oxides is that of adsorption enhancement. This area has been pursued by the Bard group [135, 136] and by the Yoneyama group [137, 138]. The rationale is to use a metal oxide, with optimized adsorption properties for a particular target pollutant, to capture the pollutant more effectively than would be possible on TiO_2 by itself. If a microheterogeneous material is used, the pollutant can be captured and delivered to active TiO_2 sites over distances that are as short as possible. Both groups reported improvements in the performance of photocatalyst systems using this approach.

6.1.5
Conclusion

Although a great deal of progress has been made in the fundamental aspects of photocatalysis in the last 25 years, there are still many areas that are incompletely understood. One example is the understanding of the nature of the photocatalytically active sites on TiO_2. Another is a detailed understanding of the factors that determine quantum yield. A third might be the understanding of the nanoscale details of the oxidation and reduction reactions, occurring simultaneously on small TiO_2 particles. A fourth might be a thorough understanding of the electronic effects of the doping of TiO_2 with other elements. There are certainly a host of other topics worth intensive study.

Even so, the number of applications has advanced, proving that complete understanding is never a necessity in developing new technology. However, we hope that there will continue to be a healthy synergy between fundamental and applied work in photocatalysis.

References

1. J. H. Carey, J. Lawrence, H. M. Tosine, *Bull. Environ. Contam. Toxicol.* **1976**, *16*, 697.
2. S. N. Frank, A. J. Bard, *J. Am. Chem. Soc.* **1977**, *99*, 303–304.
3. S. N. Frank, A. J. Bard, *J. Phys. Chem.* **1977**, *81*, 1484–1488.
4. A. Fujishima, K. Honda, S. Kikuchi, *Kogyo Kagaku Zasshi* **1969**, *72*, 108–113.
5. A. Fujishima, K. Honda, *Nature* **1972**, *238*, 37–38.
6. A. Fujishima, K. Kobayakawa, K. Honda, *J. Electrochem. Soc.* **1975**, *122*, 1487–1489.
7. A. Fujishima, K. Kobayakawa, K. Honda, *Bull. Chem. Soc. Jpn.* **1975**, *48*, 1041–1042.
8. A. Heller, *Acc. Chem. Res.* **1995**, *28*, 503–508.
9. B. O'Regan, J. Moser, M. Anderson et al., *J. Phys. Chem.* **1990**, *94*, 9820–8726.
10. L. Kavan, T. Stoto, M. Grätzel et al., *J. Phys. Chem.* **1993**, *97*, 9493–9498.
11. D. A. Tryk, A. Fujishima, K. Honda, *Electrochim. Acta* **2000**, *45*, 2363–2376.
12. A. Fujishima, T. N. Rao, D. A. Tryk, *J. Photochem. Photobiol. C, Photochem. Rev.* **2000**, *1*, 1–21.
13. E. Pelizzetti, N. Serpone, Eds., *Homogeneous and Heterogeneous Photocatalysis*, D. Reidel Publishing Company, Dordrecht, The Netherlands, 1986.
14. N. Serpone, E. Pelizzetti, Eds., *Photocatalysis – Fundamentals and Applications*, John Wiley & Sons, New York, 1989.
15. E. Pelizzetti, M. Schiavello, Eds., *Photochemical Conversion and Storage of Solar*

Energy, Kluwer Academic Publishers, Dordrecht, The Netherlands, 1991.
16. M. A. Fox, M. T. Dulay, *Chem. Rev.* **1993**, *93*, 341–357.
17. D. F. Ollis, H. Al-Ekabi, Eds., *Photocatalytic Purification and Treatment of Water and Air*, Elsevier, Amsterdam, The Netherlands, 1993.
18. M. R. Hoffmann, S. T. Martin, W. Choi et al., *Chem. Rev.* **1995**, *95*, 69–96.
19. A. L. Linsebigler, G. Lu, J. J. T. Yates, *Chem. Rev.* **1995**, *95*, 735–758.
20. A. Fujishima, K. Hashimoto, T. Watanabe, in *TiO2 Photocatalysis: Fundamentals and Applications*, BKC, Inc., Tokyo, 1999.
21. J. -M. Herrmann, *Catal. Today* **1999**, *53*, 115–129.
22. M. I. Litter, *Appl. Catal. B: Environ.* **1999**, *23*, 89–114.
23. D. F. Ollis, *C. R. Acad. Sci. Paris, Serie IIC, Chim.* **2000**, *3*, 405–411.
24. R. W. G. Wyckoff, in *Crystal Structures*, Interscience Publishers, New York, 1963, p. 205.
25. L. Kavan, M. Grätzel, S. E. Gilbert et al., *J. Am. Chem. Soc.* **1996**, *118*, 6716–6723.
26. H. O. Finklea, in *Semiconductor Electrodes*, (Eds.: H. O. Finklea), Elsevier, Amsterdam, 1988, pp. 43–145, Vol. 55.
27. S. R. Morrison, in *Electrochemistry at Semiconductor and Oxidized Metal Electrodes*, Plenum Press, New York, 1980, p. 183.
28. A. J. Nozik, in *Annual Review of Physical Chemistry*, (Eds.: B. S. Rabinovitch, J. M. Schurr, H. L. Strauss), 1978, pp. 189–222.
29. W. Jaegermann, in *Modern Aspects of Electrochemistry*, (Eds.: R. E. White, B. E. Conway, J. O. M. Bockris), Plenum Press, New York, 1996, pp. 1–185, Vol. 30.
30. M. R. Prairie, B. M. Stange, L. R. Evans, in *Photocatalytic Purification and Treatment of Water and Air*, (Eds.: D. F. Ollis, H. Al-Ekabi), Elsevier Science, Amsterdam, The Netherlands, 1993, pp. 353–363.
31. M. Herlem, F. Bobilliart, A. Thiebault, in *Encyclopedia of Electrochemistry of the Elements, Organic Section*, (Eds.: A. J. Bard, H. Lund), Marcel Dekker, Inc., New York, 1978, pp. 5–16, Vol. XI.
32. D. Lawless, N. Serpone, D. Meisel, *J. Phys. Chem.* **1991**, *95*, 5166–5170.
33. Y. Nakato, A. Tsumura, H. Tsubomura, *J. Phys. Chem.* **1983**, *87*, 2402.
34. A. Taghizadeh, M. F. Lawrence, L. Miller et al., *J. Photochem. Photobiol. A: Chem.* **2000**, *130*, 145–156.
35. F. R. Keene, in *Electrochemical and Electrocatalytic Reactions of Carbon Dioxide*, (Eds.: B. P. Sullivan, K. Krist, H. E. Guard), Elsevier, Amsterdam, The Netherlands, 1993, pp. 1–18.
36. G. V. Buxton, C. L. Greenstock, W. P. Helman et al., *J. Phys. Chem. Ref. Data* **1988**, *17*, 513–535.
37. A. A. Frost, *J. Am. Chem. Soc.* **1951**, *73*, 2680–2682.
38. M. R. Tarasevich, A. Sadkowski, E. Yeager, in Comprehensive Treatise of Electrochemistry, *Kinetics and Mechanisms of Electrode Processes*, (Eds.: B. E. Conway, J. O. M. Bockris, E. Yeager et al.), Plenum Press, New York, 1983, pp. 301–398, Vol. 7.
39. D. F. Shriver, P. W. Atkins, C. H. Langford, in *Inorganic Chemistry*, Oxford University Press, Oxford, UK, 1995, pp. 300–304.
40. S. Goldstein, G. Czapski, J. Rabani, *J. Phys. Chem.* **1994**, *98*, 6586–6591.
41. D. W. Bahnemann, M. Hilgendorff, R. Memming, *J. Phys. Chem. B* **1997**, *101*, 4265–4275.
42. L. Cermenati, P. Pichat, C. Guillard et al., *J. Phys. Chem. B* **1997**, *101*, 2650–2658.
43. R. F. Howe, M. Grätzel, *J. Phys. Chem.* **1987**, *91*, 3906–3909.
44. O. I. Micic, Y. Zhang, K. R. Cromack et al., *J. Phys. Chem.* **1993**, *97*, 7277–7283.
45. Y. Nakaoka, Y. Nosaka, *J. Photochem. Photobiol. A: Chem.* **1997**, *110*, 299–305.
46. K. Ishibashi, Y. Nosaka, K. Hashimoto et al., *J. Phys. Chem. B* **1998**, *102*, 2117–2120.
47. K. Ishibashi, A. Fujishima, T. Watanabe et al., *J. Photochem. Photobiol A: Chem.* **2000**, *134*, 139–142.
48. T. Tatsuma, S. Tachibana, T. Miwa et al., *J. Phys. Chem. B* **1999**, *103*, 8033–8035 .
49. T. Tatsuma, S. Tachibana, A. Fujishima, submitted.
50. Y. Kikuchi, K. Sunada, T. Iyoda et al., *J. Photochem. Photobiol. A: Chem.* **1997**, *106*, 51–56.
51. D. E. Skinner, D. P. Colombo, Jr., J. J. Cavaleri et al., *J. Phys. Chem.* **1995**, *99*, 7853–7856.
52. N. Serpone, D. Lawless, R. Khairutdinov et al., *J. Phys. Chem.* **1995**, *99*, 16655–16661.
53. D. P. Colombo, Jr., R. M. Bowman, *J. Phys. Chem.* **1996**, *100*, 18445–18449.

54. G. Rothenberger, J. Moser, M. Grätzel et al., *J. Am. Chem. Soc.* **1985**, *107*, 8054–8059.
55. K. -M. Schindler, M. Kunst, *J. Phys. Chem.* **1990**, *94*, 8222–8226.
56. J. M. Warman, M. P. Haas, P. Pichat et al., *J. Phys. Chem.* **1991**, *95*, 8858–8861.
57. S. T. Martin, H. Herrmann, W. Choi et al., *J. Chem. Soc. Faraday Trans.* **1994**, *90*, 3315–3322.
58. D. P. Colombo, Jr., R. M. Bowman, *J. Phys. Chem.* **1995**, *99*, 11752–11756.
59. D. Bahnemann, A. Henglein, J. Lilie et al., *J. Phys. Chem.* **1984**, *88*, 709–711.
60. A. Henglein, *Pure & Appl. Chem.* **1984**, *56*, 1215–1224.
61. Y. Nosaka, *J. Phys. Chem.* **1990**, *94*, 3752–3755.
62. M. O'Neil, J. Marohn, G. McLendon, *J. Phys. Chem.* **1990**, *94*, 4356–4363.
63. M. A. Grela, M. E. J. Coronel, A. J. Colussi, *J. Phys. Chem. B* **1996**, *100*, 16940–16946.
64. R. Kopelman, *Science* **1988**, *241*, 1620–1626.
65. B. G. Streetman, in *Solid State Electronic Devices*, Prentice-Hall International, Englewood Cliffs, New Jersey, USA, 1995, pp. 103–106.
66. D. Duonghong, M. Grätzel, *J. Chem. Soc., Chem. Commun.*, **1984**, 1597–1599.
67. Y. Nosaka, Y. Yamashita, H. Fukuyama, *J. Phys. Chem. B* **1997**, *101*, 5822–5827.
68. V. Augugliaro, L. Palmisano, M. Schiavello, *AIChE J.* **1991**, *37*, 1096–1100.
69. M. A. Aguado, M. A. Anderson, C. G. Hill, Jr., *J. Mol. Catal.* **1994**, *89*, 165–178.
70. N. Serpone, *J. Photochem. Photobiol. A: Chem.* **1997**, *104*, 1–12.
71. L. Davydov, P. G. Smirniotis, S. E. Pratsinis, *Indust. Eng. Chem. Res.* **1999**, *38*, 1376–1383.
72. R. I. Bickley, *J. Catal.* **1973**, *31*, 398–407.
73. Y. Ohko, K. Hashimoto, A. Fujishima, *J. Phys. Chem. A* **1997**, *101*, 8057–8062.
74. R. W. Matthews, *J. Phys. Chem.* **1987**, *91*, 3328–3333.
75. I. Sopyan, S. Murasawa, K. Hashimoto et al., *Chem. Lett.* **1994**, 723–726.
76. N. Negishi, T. Iyoda, K. Hashimoto et al., *Chem. Lett.* **1995**, 841–842.
77. I. Sopyan, M. Watanabe, S. Murasawa et al., *J. Photochem. Photobiol. A: Chem.* **1996**, *98*, 79–86.
78. I. Sopyan, M. Watanabe, S. Marasawa et al., *J. Electroanal. Chem.* **1996**, *415*, 183–186.
79. S. Matsushita, T. Miwa, A. Fujishima, *Chem. Lett.* **1997**, 925–926.
80. S. I. Matsushita, T. Miwa, D. A. Tryk et al., *Langmuir* **1998**, *14*, 6441–6447.
81. T. A. Egerton, C. J. King, *J. Oil Col. Chem. Assoc.* **1979**, *62*, 386–391.
82. C. Kormann, D. W. Bahnemann, M. R. Hoffmann, *Environ. Sci. Technol.* **1991**, *25*, 494–500.
83. G. P. Lepore, C. H. Langford, J. Vichova et al., *J. Photochem. Photobiol. A: Chem.* **1993**, *75*, 67–75.
84. M. Anpo, N. Aikawa, Y. Kubokawa et al., *J. Phys. Chem.* **1985**, *89*, 5689.
85. R. F. Howe, M. Grätzel, *J. Phys. Chem.* **1985**, *89*, 4495–4499.
86. R. Cai, R. Baba, K. Hashimoto et al., *J. Electroanal. Chem.* **1993**, *360*, 237–245.
87. K. U. Ingold, *Acc. Chem. Res.* **1969**, *2*, 1–9.
88. N. A. Clinton, R. A. Kenley, T. G. Traylor, *J. Am. Chem. Soc.* **1975**, *97*, 3746–3751.
89. N. A. Clinton, R. A. Kenley, T. G. Traylor, *J. Am. Chem. Soc.* **1975**, *97*, 3752–3757.
90. N. A. Clinton, R. A. Kenley, T. G. Traylor, *J. Am. Chem. Soc.* **1975**, *97*, 3757–3762.
91. Y. Ohko, D. A. Tryk, K. Hashimoto et al., *J. Phys. Chem. B* **1998**, *102*, 2699–2704.
92. G. B. Raupp, C. T. Junio, *Appl. Surf. Sci.* **1993**, *72*, 321–327.
93. J. Stark, J. Rabani, *J. Phys. Chem. B* **1999**, *103*, 8524–8531.
94. J. Schwitzgebel, J. G. Ekerdt, H. Gerischer et al., *J. Phys. Chem.* **1995**, *99*, 5633–5638.
95. G. A. Russell, *J. Am. Chem. Soc.* **1957**, *79*, 3871–3877.
96. S. Sitkiewitz, A. Heller, *Nouv. J. Chem.* **1996**, *20*, 233–241.
97. T. Minabe, P. Swanyama, Y. Kikuchi et al., *Electrochem.* **1999**, *67*, 1132–1134.
98. T. Minabe, D. A. Tryk, P. Sawunyama et al., *J. Photochem. Photobiol. A: Chem.* **2000**, *137*, 53–62.
99. A. V. Emiline, V. Ryabchuk, N. Serpone, *J. Photochem. Photobiol. A: Chem.* **2000**, *133*, 89–97.
100. T. Noguchi, A. Fujishima, P. Sawunyama et al., *Environ. Sci. Technol.* **1998**, *32*, 3831–3833.
101. Y. Ohko, A. Fujishima, K. Hashimoto, *J. Phys. Chem. B* **1998**, *102*, 1724–1729.
102. W. W. Dunn, Y. Aikawa, A. J. Bard, *J. Am. Chem. Soc.* **1981**, *103*, 3456–3459.
103. W. W. Dunn, Y. Aikawa, A. J. Bard, *J. Electrochem. Soc.* **1981**, *128*, 222–224.
104. M. D. Ward, A. J. Bard, *J. Phys. Chem.* **1982**, *86*, 3599–3605.

105. M. W. Peterson, J. A. Turner, A. F. Nozik, *J. Phys. Chem.* **1991**, *95*, 221–225.
106. C. -M. Wang, A. Heller, H. Gerischer, *J. Am. Chem. Soc.* **1992**, *114*, 5230–5234.
107. W. -Y. Lin, N. R. de Tacconi, R. L. Smith et al., *J. Electrochem. Soc.* **1997**, *144*, 497–502.
108. Y. Komoda, N. Sakai, T. N. Rao et al., *Langmuir* **1998**, *14*, 1081–1091.
109. Y. Komoda, T. N. Rao, D. A. Tryk et al., *J. Electroanal. Chem.* **1998**, *459*, 155–165.
110. K. Vinodgopal, S. Hotchandani, P. V. Kamat, *J. Phys. Chem.* **1993**, *97*, 9040–9044.
111. K. Vinodgopal, U. Stafford, K. A. Gray et al., *J. Phys. Chem.* **1994**, *98*, 6797–6803.
112. K. J. Vetter, in *Electrochemical Kinetics: Theoretical and Experimental Aspects,* Academic Press, New York, 1967, pp. 406–417.
113. N. Sato, in *Electrochemistry at Metal and Semiconductor Electrodes,* Elsevier, Amsterdam, 1998, p. 400.
114. J. M. Kesselman, G. A. Shreve, M. R. Hoffmann et al., *J. Phys. Chem.* **1994**, *98*, 13385–13395.
115. H. Sakai, R. Baba, K. Hashimoto et al., *J. Phys. Chem.* **1995**, *99*, 11896–11900.
116. K. Ikeda, H. Sakai, R. Baba et al., *Chem. Lett.* **1995**, 979–980.
117. K. Ikeda, H. Sakai, R. Baba et al., *J. Phys. Chem. B* **1997**, *101*, 2617–2620.
118. K. Ikeda, K. Hashimoto, A. Fujishima, *J. Electroanal. Chem.* **1997**, *437*, 241–244.
119. H. Maeda, K. Ikeda, K. Hashimoto et al., *J. Phys. Chem. B* **1999**, *103*, 3213–3217.
120. P. Sawunyama, L. Jiang, A. Fujishima et al., *J. Phys. Chem. B* **1997**, *101*, 11000–11003.
121. P. Sawunyama, A. Fujishima, K. Hashimoto, *Langmuir* **1999**, *15*, 3551–3556.
122. H. Gerischer, A. Heller, *J. Phys. Chem.* **1991**, *95*, 5261–5267.
123. M. Anpo, Y. Ichihashi, M. Takeuchi et al., *Res. Chem. Intermed.* **1998**, *24*, 143–149.
124. L. Zang, C. Lange, I. Abraham et al., *J. Phys. Chem. B* **1998**, *102*, 10765–10771.
125. M. Hamerski, J. Grzechulska, A. W. Morawski, *Sol. Energy* **1999**, *66*, 395–399.
126. Y. Murata, S. Fukuta, S. Ishikawa et al., *Sol. Energy Mater. Sol. Cells* **2000**, *62*, 157–165.
127. X. Fu, l. A. Clark, Q. Yang et al., *Environ. Sci. Technol.* **1996**, *30*, 647–653.
128. C. Anderson, A. J. Bard, *J. Phys. Chem.* **1995**, *99*, 9882–9885.
129. C. Anderson, A. J. Bard, *J. Phys. Chem. B* **1997**, *101*, 2611–2616.
130. N. Takeda, T. Torimoto, S. Sampath et al., *J. Phys. Chem.* **1995**, *99*, 9986–9991.
131. N. Takeda, M. Ohtani, T. Torimotoet al., *J. Phys. Chem. B* **1997**, *101*, 2644–2649.
132. P. W. Atkins, in *Physical Chemistry,* Oxford University Press, Oxford, 1998.
133. J. VanMuylder, M. Pourbaix, in *Atlas of Electrochemical Equilibria in Aqueous Solutions,* (Eds.: M. Pourbaix), Pergamon, Oxford, 1966, pp. 449–457.
134. V. D. Parker, G. Sundholm, U. Svanholm et al., in *Encyclopedia of Electrochemistry of the Elements, Organic Section,* (Eds.: A. J. Bard, H. Lund), Marcel Dekker, New York, 1978, pp. 181–225, Vol. XI.
135. J. N. Butler, in *Advances in Electrochemistry and Electrochemical Engineering* (Eds.: P. Delahay), Interscience Publishers, New York, 1970, pp. 77–139, Vol. 7.
136. J. P. Hoare, in *Encyclopedia of Electrochemistry of the Elements,* (Eds.: A. J. Bard), Vol. 2, Marcel Dekker, New York, 1974, p. 220.
137. Y. Ohko, K. Ikeda, T. N. Rao et al., *Z. phys. Chem.* **1998**, *213*, 33–42.

6.2 Applications of TiO_2 Photocatalysis

Tata N. Rao, Donald A. Tryk, and Akira Fujishima
University of Tokyo, Tokyo, Japan

6.2.1 Introduction

The applications of titanium dioxide (TiO_2) photocatalysis date from the mid-1970s, with the solar-assisted purification of contaminated water, but now include a staggering array of diverse ideas from the purification and sterilization of indoor air to self-cleaning ceramics and glass, to antifogging mirrors, and to light-induced cancer treatment. These technologies are becoming more attractive because of the environmentally benign nature of TiO_2, its low cost, nontoxicity, and ability to self-regenerate combined with the possibility of using only solar light or ambient indoor light together with ambient oxygen (O). Additional attractive features include its unique chemical properties, for example, its resistance to photocorrosion, its ability to simultaneously produce at least two types of highly reactive oxygen radical species, and its ability to modify its wetting properties under illumination. Part of the appeal stems from the recent emphasis on indoor environments in which photocatalysis is expected to play a significant role. Another appealing aspect is the ability of TiO_2 to maintain sterile conditions on illuminated surfaces without the use of antiseptics.

The growth in the number of publications on photocatalysis has been accelerating steadily over the past quarter century, as seen in Fig. 1. The total number of papers that have appeared up to September 2000 is slightly more than 5000 and the number of patents is just more than 1700 [1–4]. Besides a rapid increase in the number of publications, the number of patents is also growing as a result of high commercial interest in applications, many of these being developed in Japan. The topics covered are shown in Fig. 2, together with the number of publications in each area [5]. In addition to the comprehensive listings of publications that have been assembled by Blake, there have been a number of reviews of the photocatalysis area [6–32].

The Section "Fundamentals of Catalysis" provides an overview of the fundamental aspects of TiO_2 photocatalysis, and the present chapter is devoted to the presentation of some of the myriad applications (Table 1). The intent of this chapter is to present an overview of the state of the art in the applications of TiO_2 photocatalysis for sterilization, disinfection, air and water purification, and self-cleaning and antifogging surfaces. To indicate the broad range of applications, a brief overview of selected applications is given, including discussion of their practicability and future prospects. Emphasis is given to photocatalytic systems that either are already being commercialized or are ready for commercialization.

6.2.2 Photocatalytic Sterilization and Disinfection

In the medical field, the development of ways to control outbreaks of disease caused by new strains of deadly bacteria and viruses has been a topic of serious discussion. Sterilization and disinfection are seen to be increasingly important, even in environments other than hospitals, such as bathrooms and kitchens. The photocatalytic action of TiO_2 can be extremely useful in preventing disease, particularly

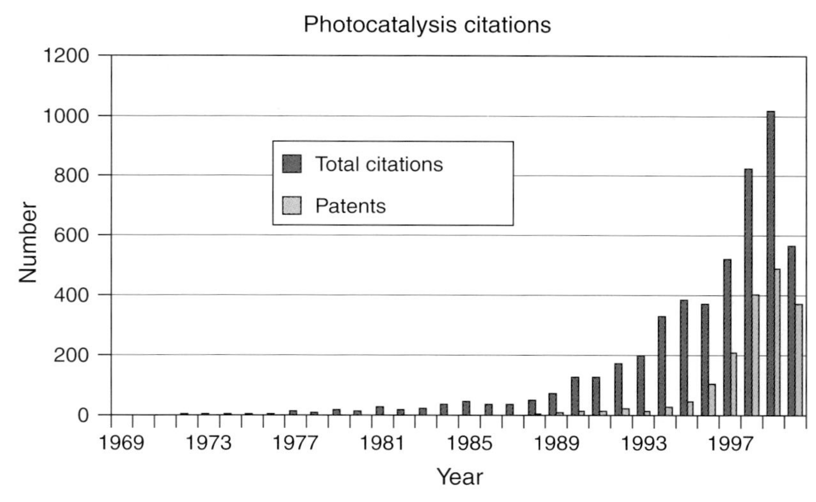

Fig. 1 Numbers of papers and patents related to photocatalysis up to September 2000 [4].

Tab. 1 Selected applications of photocatalysis

Property	***Category***	***Application***
Self-cleaning	Materials for residential and office buildings	Exterior tiles, kitchen and bathroom components, interior furnishings, plastic surfaces, aluminum siding, building stone and curtains, paper window blinds
	Indoor and outdoor lamps and related systems	Translucent paper for indoor lamp covers, coatings on fluorescent lamps and highway tunnel lamp cover glass
	Materials for roads	Tunnel wall, soundproofed wall, traffic signs and reflectors
	Others	Tent material, cloth for hospital garments and uniforms and spray coating for cars
Air cleaning	Indoor air cleaners	Room air cleaner, photocatalyst-equipped air conditioners, and interior air cleaner for factories
	Outdoor air purifiers	Concrete for highways, roadways and footpaths, tunnel walls, soundproofed walls, and building walls
Water purification	Drinking water	River water, groundwater, lakes, and water-storage tanks
	Others	Fish feeding tanks, drainage water, and industrial wastewater
Antitumor activity	Cancer therapy	Endoscopic-like instruments
Self-sterilizing	Hospital	Tiles to cover the floor and walls of operating rooms, silicone rubber for medical catheters, and hospital garments and uniforms
	Others	Public rest rooms, bathrooms, and rat-breeding rooms

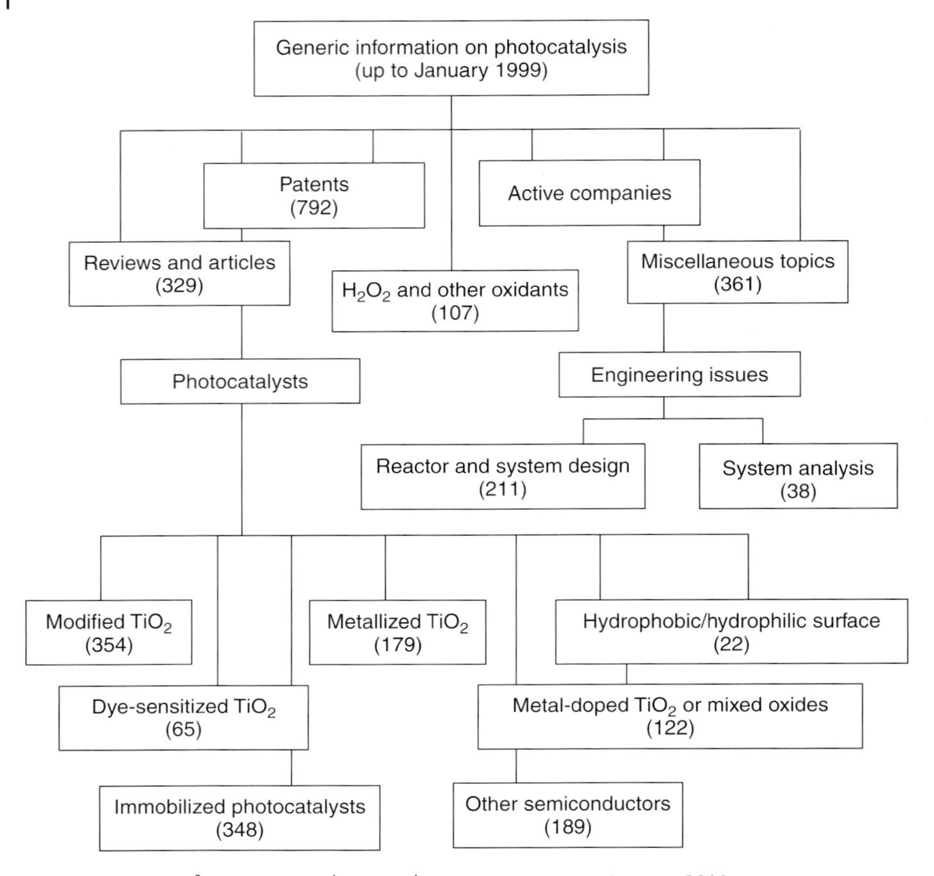

Fig. 2 Generic information on photocatalytic processes up to January 1999.

because it operates in a passive fashion, that is, without the need for electrical power or chemical reagents, only light and oxygen being required. For these reasons photocatalysis recently has attracted the attention not only of the medical world but also of the construction industry in Japan.

Most of the work reported on photocatalytic disinfection has been carried out in aqueous suspensions, in view of applications in water sterilization. In 1985, for the first time, Matsunaga and coworkers reported the possibility of killing microbial cells in water using a TiO_2-Pt (platinum) photocatalyst under ultraviolet (UV) illumination [33]. Since then there have been several reports and patents appearing on this topic [5, 34–44]. Blake and coworkers have published a comprehensive review of this, including a list of patents on photocatalytic disinfection and killing of cancer cells using TiO_2 [5]. Many of the commercial products available in the market at present are based on self-sterilizing photocatalytic surfaces and disinfection of indoor air.

Bacteria find places in bathrooms, kitchens, and hospital operating rooms where they can multiply. Once they find a suitable environment, they can increase in number at an exponential rate. The key point regarding photocatalysis is that it can be very effective in controlling bacteria while they still are relatively small in number, before they start multiplying out of control.

6.2.2.1 Photocatalytic Antibacterial Tiles

In general, the effective use of photocatalytic technology depends heavily on the specific way in which the TiO_2 powder is used. One of the simplest approaches, often used in model studies on water treatment, is to mix the powder with the water and expose the suspension to UV light or sunlight. Although the separation of the catalyst powder from the water is not an insurmountable problem, an alternate approach is to immobilize the powder on a substrate [25].

In view of indoor applications, various kinds of substrates such as glass plates and tiles have been examined as substrates. It has been found that even very thin films (less than a few micrometers in thickness) are photocatalytically active [45, 46]. Moreover, (~1 μm) TiO_2 films on the surfaces of ceramic tiles have been found to be effective for antibacterial functions. These films can be prepared by spraying a liquid suspension containing TiO_2 on a glazed ceramic tile, followed by heating at 800 °C [25, 47, 48]. Films prepared this way are attached to the tiles so strongly that they are difficult to scrape off.

In the case of indoor antibacterial tiles, one has to keep in mind the possibility of accumulation of bacteria in areas either temporarily or permanently inaccessible to light. To overcome this problem, silver or copper metal particles have been deposited photocatalytically on catalyst surfaces [25]. These metals are widely used as antimicrobial agents. Either of the metal solutions is sprayed on the TiO_2-coated tile and, when the surface is irradiated with UV light, the metal cations are photocatalytically converted to the metallic form as ultrafine particles that are firmly attached to the surface. The photocatalytic deposition of various metals on TiO_2 is also discussed in a later section.

The present method of metal deposition was found to have an advantage over the conventional method in which the metal powders are mixed with the glaze and fired on a tile [25]. With the conventional method, most of the metal particles are submerged in the glaze, only a few being exposed on the surface. Thus, the particles in the glaze are not effective for the antibacterial effect. The new photocatalytic approach allows the formation of a high-density layer of antibacterial metal particles on the TiO_2 layer, which then exhibits a strong antibacterial activity. The photocatalytic antibacterial tiles prepared this way are highly effective and durable.

6.2.2.2 Sterilizing Effect of TiO_2

As already indicated, one of the promising applications of TiO_2 is photocatalytic sterilization. *Escherichia coli* has been studied widely, but work on other organisms also has been reported. As recently reviewed, various organisms have been examined for photocatalytic sterilization [5]. Although the feasibility of sterilization was realized from the beginning, the mechanism of cell death was not understood completely. Initially, the hydroxyl radical was thought to be the main species responsible for cell death. However, peroxide currently is considered to be one of the most active agents. Evidence against the involvement of hydroxyl radicals is that there is little effect of mannitol, a hydroxyl radical scavenger [39].

A typical experiment involves placing 150 μL of an *E. coli* suspension containing 3×10^4 cells on an illuminated TiO_2-coated glass plate (1 mW cm^{-2} UV light) [39]. Under these conditions, there were no surviving cells after one hour of illumination. By contrast, in the absence of a TiO_2 film, it took four hours for the destruction of 50% of the cells. In another experiment, the *E. coli* suspension was separated from the TiO_2 surface by a porous PTFE membrane (pore size, 0.4 μm). The function of this membrane is that it prevents the passage of *E. coli* cells through its pores. Because of its thickness (50 μm), relatively reactive radical species such as $^{\bullet}OH$ generated at the irradiated TiO_2 surface are expected to be deactivated during passage through it. Under these conditions, the complete destruction of cells was observed in four hours. This result tends to rule out the involvement of $^{\bullet}OH$, but not that of peroxide. However, the latter by itself is not as effective, so the additional presence of other oxygen species must be considered. Further experiments are necessary to understand this mechanism in more detail. Regardless of the oxidative species involved, the cell membrane damage was attributed to oxidative damage caused by peroxidation of lipids in the cell membrane [42]. After eliminating the protection of the cell wall, the oxidative damage takes place on the underlying cytoplasmic membrane [44]. Photocatalytic action progressively increases the cell permeability and subsequently allows the free efflux of intracellular contents, eventually leading to cell death. Free TiO_2 particles also are expected to gain access into membrane-damaged cells, where they can attack the intracellular components, accelerating cell death.

In view of practical applications, photocatalytic antibacterial tiles mentioned earlier also were tested by placing live bacteria on them. The bacteria on these tiles were killed completely in an hour under illumination equal to the brightness of the surface of a study desk (~4 μW cm^{-2}) (Fig. 3). On the basis of these results, such tiles were tested on the floor and walls of a hospital operating room where sterile conditions are crucial. After installing the tiles, the bacterial counts on the walls decreased to negligible levels in a period of one hour. Surprisingly, the bacterial counts in the surrounding air also decreased significantly (Fig. 4).

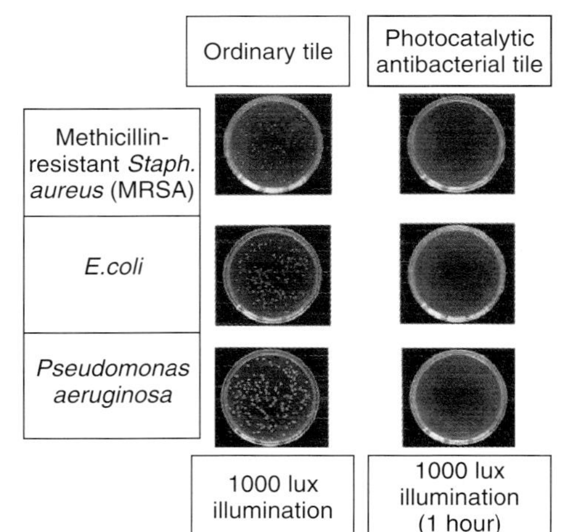

Fig. 3 Demonstration of the self-sterilizing action of photocatalytic tiles before and after illumination [25].

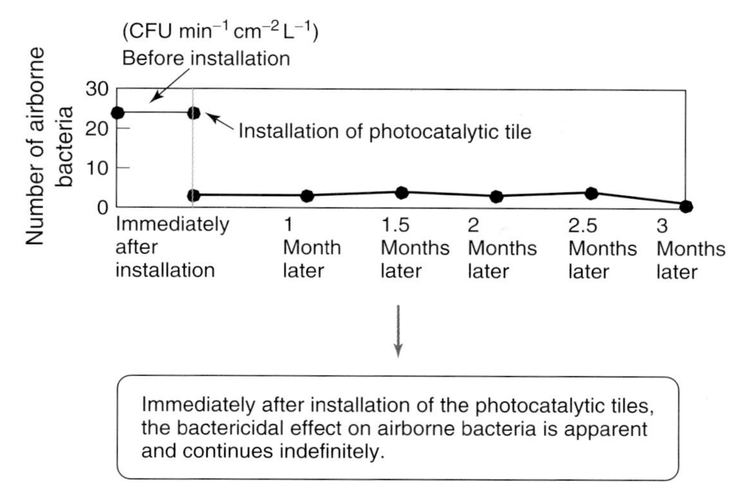

Fig. 4 Sterilizing effect of illuminated photocatalytic tiles on airborne bacteria as a function of time [25].

The applications of these tiles can be extended further to public toilet facilities, shower rooms, and biological experimental facilities such as breeding rooms for experimental rats. The tests conducted in these facilities have indicated that the antibacterial tiles are effective not only in killing the bacteria but also in significantly reducing the odor caused by ammonia, which is common in such facilities. The deodorizing functions of TiO_2 are discussed in a later section.

6.2.2.3 **Destruction of Microbial Toxins**

A major advantage of photocatalytic antibacterial tiles over conventional antibacterial tiles is that they not only kill bacterial cells but also completely decompose them. Conventional self-sterilizing surfaces, such as those with silver coatings, are not useful for long-term use because dead cells can eventually cover the surface, rendering it ineffective. The advantage of TiO_2 is that it continues to work well even when there are cells covering the surface and while the bacteria are actively propagating.

TiO_2 is capable of decomposing not only dead cells but also the toxins released during cell death. Jacoby and coworkers have demonstrated the complete oxidation of *E. coli* cells to carbon dioxide (CO_2) [40]. In the case of *E. coli*, TiO_2 decomposes both the living cells and the endotoxin released from these cells at the time of death [49]. This unique function of TiO_2 has attracted tremendous attention in Japan. Endotoxin is a cell wall constituent of bacteria that consists of a sugar chain, an O-antigen, and a complex lipid referred to as lipid A. This endotoxin is toxic and can cause critical problems in medical facilities and in factories manufacturing pharmaceuticals and medical devices, as well as in food-processing factories. In the summer of 1996 in western Japan, there was a serious outbreak of food poisoning, in which nearly 1800 people, including many children, were hospitalized, and twelve died. The poisoning was caused by the O-157 endotoxin.

Figure 5 shows the concentration change of endotoxin and the survival ratio

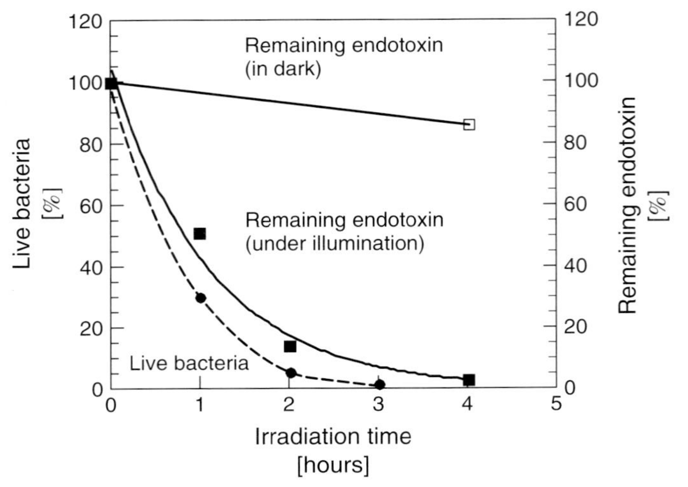

Fig. 5 Sterilization of *E. coli* and endotoxin decomposition (endotoxin amount: 3–8 EU cc^{-1} = 100%). Both sterilization and endotoxin decomposition commence immediately [49].

of *E. coli* as a function of illumination time on a TiO_2-coated glass plate. It was found that two hours were required to decompose most of the toxin and four hours were required to decompose it completely under UV illumination (0.4 mW cm^{-2}) on a TiO_2 film [49]. This can be contrasted with other techniques that have been developed, for example, thermal treatment at 250 °C for 30 minutes or chemical treatment in ethanol containing NaOH. These results clearly indicate that the antibacterial effects of TiO_2-coated materials involve not only the nullification of the viability of the bacteria but also the destruction of the bacterial cell.

6.2.2.4 **Photocatalytic Cancer Therapy**

Cancer treatment is carried out mainly by means of radiation and chemical therapies, which may generate a wide range of side effects. Photocatalysis offers a new approach in cancer treatment. Cancer arises from a loss of normal growth control. In normal tissues, the rates of new cell growth and old cell death are kept in balance. In cancer, this balance is disrupted. This disruption can result in the formation of a malignant tumor. The selective killing of cancer cells in vivo by photocatalysis is a challenging task.

The possibility of applying photocatalytic techniques to the photodynamic therapy (PDT) of cancer led to early efforts in this area. It was found that the multiplication of human malignant cells was suppressed remarkably both in vitro [50–54] and in vivo [55, 56] in the presence of photoirradiated TiO_2. Initially, TiO_2 was found to be effective in killing HeLe cells under UV irradiation [50]. Later, various experimental conditions were examined, including the effect of superoxide dismutase, which enhances the effect, caused by the production of peroxide [51–53]. The possibility of selectively killing a cancer cell was demonstrated using a polarized, illuminated TiO_2 microelectrode [54]. Photocatalytic killing of human U937 monocytic leukemia cells also was reported by Chinese researchers using colloidal TiO_2 [57]. A complete photocatalytic destruction of these cells was observed after 30 minutes of UV illumination. The destruction process involved membrane blistering and DNA fragmentation. All of these effects are characteristics of apoptosis (programmed cell death) that is known to occur when cancer cells are treated with antitumor agents, radiation, or high temperature.

In collaboration with urologists at Yokohama City University, animal experiments also have been conducted. Cancer cells were implanted under the skin of mice to cause tumors to form. When the size of the tumors grew to about 0.5 cm, a solution containing fine particles of TiO_2 was injected. After two or three days, the skin was cut open to expose the tumor, which then was irradiated. This treatment clearly inhibited tumor growth (Fig. 6). After 13 more days, the treatment was repeated with TiO_2 photocatalyst and a further marked antineoplastic effect was observed [55].

Although this technique offers some promise for cancer therapy, it may not be effective in treating large tumors. To make this technique more useful, a device (a

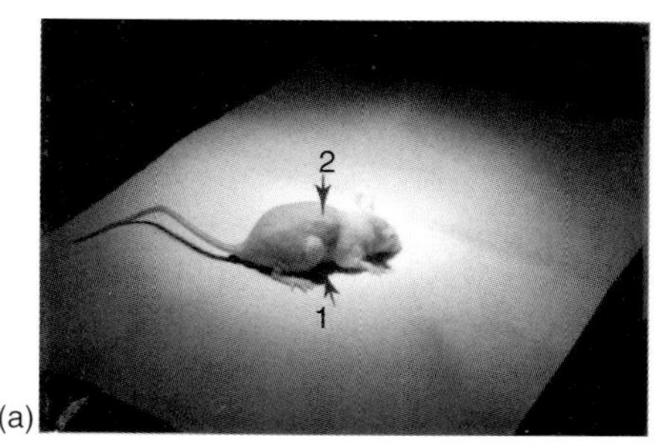

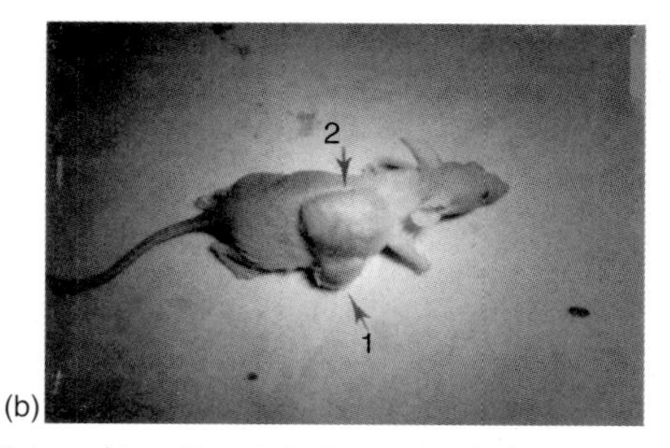

Fig. 6 Animal test of photocatalytic cancer therapy; photograph of nude mouse just after initial treatment (a) and four weeks after treatment (b). TiO_2 powder (0.4 mg) was injected into Tumor 1. Tumor 2, which was not injected with TiO_2 particles, was also opened surgically. Both tumors were irradiated with a mercury lamp for one hour [25].

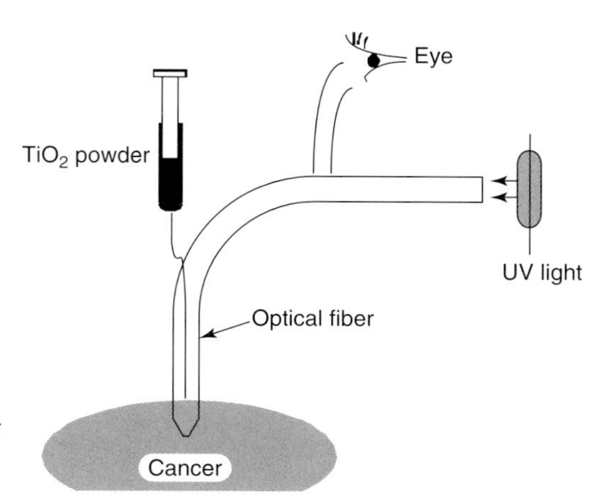

Fig. 7 Proposed instrument for photocatalytic cancer therapy [25].

modified endoscope) has been developed to allow the tumor to be exposed to light while TiO_2 powder is being added to the tumor (Fig. 7). Because a photocatalytic reaction occurs at the illuminated areas, it is possible to make an attack on cancer cells alone. There are special types of dyes that have been developed that selectively bind to cancer cells. They can be used along with TiO_2 powder to trace the location of cancer cells and treat with the new endoscope-like instrument. Thus it is possible to avoid damage to normal cells.

Obviously, the excitation light should not cause mutations in normal cells. The results of animal experiments have shown that the near UV light that is used in photocatalytic reactions, with wavelengths of 300–400 nm, is safe.

6.2.3 Photocatalytic Air Purification

Photocatalytic oxidation of various air pollutants currently is a rapidly developing field of heterogeneous photocatalysis. Volatile organic compounds (VOCs), nitrogen oxides, and sulfur oxides are the most important anthropogenic pollutants generated in urban and industrial areas. For example, volatile chlorinated organic compounds such as trichloroethylene (TCE), tetrachloroethylene, 1,3-dichlorobenzene, and dichloromethane, which are widely used as extracting solvents in industrial processes, are released into the environment in large quantities. Air stripping followed by adsorption on activated carbon is a common method employed to treat water and soil that is contaminated with such volatile organics. Such a treatment method, merely transferring toxic materials from one medium to another, is not a long-term solution to the problem of hazardous waste disposal. TiO_2 photocatalysis is advantageous in handling such problems because of its ability to completely and efficiently mineralize a wide range of pollutants. Purification of indoor air is another area where photocatalysis plays a dominant role. Photocatalytic treatment of various kinds of gaseous pollutants is summarized in this section.

6.2.3.1 Deodorization of Indoor Air

TiO_2 photocatalysis is very successful for the purification of indoor air in which the air pollutants are relatively low in concentration. Experimental results have indicated that the quantum efficiency of photocatalytic reactions is high at low light intensities (Fig. 8) [25]. Thus, the

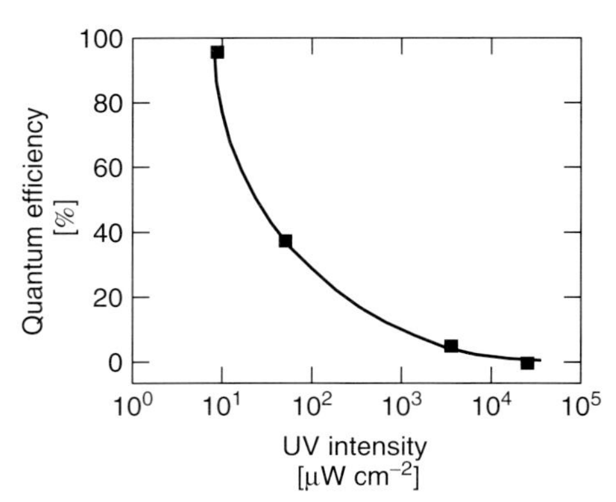

Fig. 8 Inverse relation between quantum efficiency and UV light intensity for low-pollutant concentrations [25].

most appropriate applications for the photocatalytic approach are those that involve low concentrations of pollutants but that also involve serious risks to health or comfort. A malodorous pollutant is a good example because the actual quantity of the chemical may be very small. The photocatalytic sterilization discussed in the previous section is another example.

The Japanese government regulates the concentrations of major malodorous substances as ammonia, hydrogen sulfide, methyl mercaptan, and acetaldehyde in the environment. Methyl mercaptan, for example, is regulated to concentrations lower than 0.002–0.01 ppm. The human sense of smell, however, senses it at a concentration as low as 0.00012 ppm. A conventional fluorescent lamp can be used as a light source for a photocatalytic reactor to control this substance [25]. Regarding acetaldehyde, which is a highly malodorous compound, a fluorescent lamp is also sufficient to decompose it at an air concentration of 5 ppm, which is 10–100 times higher than that at which it becomes offensive. Several studies have been conducted on acetaldehyde decomposition under various light intensities and concentrations to understand the reaction mechanism [45, 58–60].

Another interesting application of TiO_2 photocatalysis was demonstrated using TiO_2-containing paper. In collaboration with the Gifu Prefectural Paper Research Institute, TiO_2-containing papers possessing a high catalytic activity toward the decomposition of acetaldehyde and cigarette smoke residues were prepared. The TiO_2-containing paper was prepared from a mixture of softwood kraft pulp and TiO_2 aqueous solution. The details of the preparation were reported earlier [61]. The amount of added TiO_2 was varied from 2 to 10% based on the weight of the pulp. The important point in this preparation is that TiO_2 particles should be aggregated before they are added to the paper pulp to maintain the paper strength. If the TiO_2 particles are dispersed in the paper, there is photocatalytic attack on the fibers and loss of strength. Acetaldehyde degradation was seen with the TiO_2-containing papers even under very weak UV irradiation. The decomposition rate was found to increase with increasing amounts of TiO_2 contained in the paper [61]. Furthermore, the papers showed high stability under illumination for about 600 hours. These papers also were tested for the decomposition of cigarette smoke residue. Smoke from one cigarette was adsorbed on a 5 × 5-cm sheet of TiO_2-containing paper. TiO_2-free paper was also used for control experiments. The smoke residue on the TiO_2 paper disappeared after only two hours under black light (1 mW cm^{-2}) illumination and after one week under regular fluorescent light (10 $\mu W cm^{-2}$).

On the basis of these findings, various kinds of commercial photocatalytic air cleaners currently are being produced in Japan by Daikin Industries and are being marketed worldwide (Fig. 9). The Ishihara Techno Corporation has developed photocatalytic filter elements for air cleaners. They contain TiO_2 and activated carbon (C), the latter trapping pollutants for subsequent destruction by the irradiated TiO_2. These filters feature a honeycomb-type construction for minimum pressure drop. The air cleaners are available in various sizes, which have been developed to clean the air in a car or in an entire factory or hospital.

6.2.3.2 **Treatment of Industrial Gaseous Effluents**

Because gas-phase photocatalysis is in general more than 10 times more effective

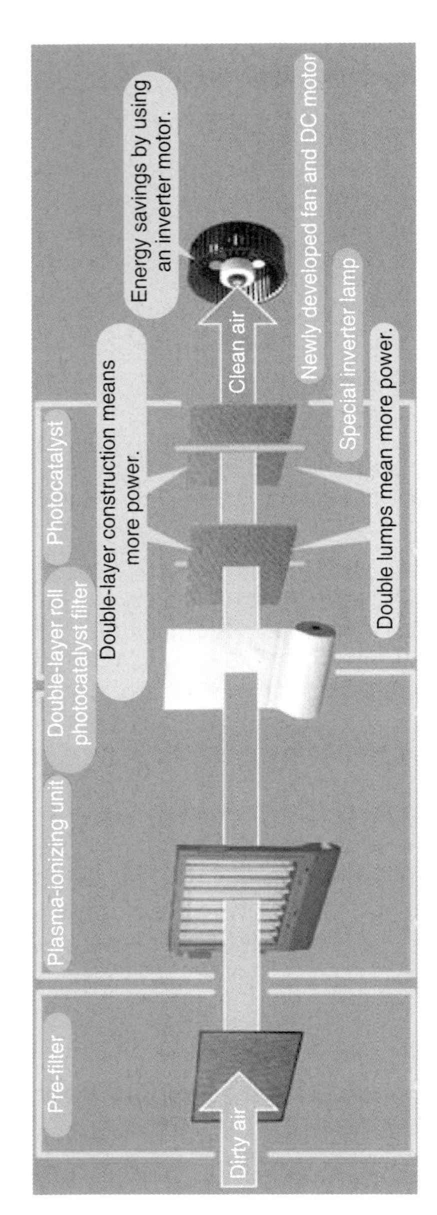

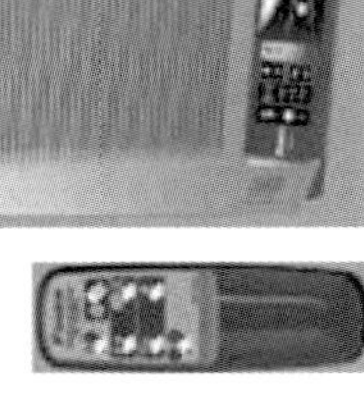

Fig. 9 Schematic diagram of the components of an indoor room air-purification unit (Courtesy of Daikin Industries).

than that in liquid phase [22] the use of photocatalytic purification methods for industrial gas effluents has gained much attention. Almost all kinds of organic compounds are photocatalytically degradable on TiO_2. Gas-phase carbon tetrachloride (CCl_4) appears to be an exception, in that it is resistant to photocatalytic degradation [62, 63]. The presence of methanol helps in slightly increasing the photocatalytic reduction of CCl_4, the effect being higher in aqueous solutions [63, 64]. There have been several reports on the photocatalytic oxidation of individual gaseous substrates such as halogenated organics [62, 65–67], alkanes [68], alcohols [62, 69], aldehydes [60, 70], aromatics [62, 66, 71, 72], and heterocycles [73]. Among these organic substrates, only a few compounds such as TCE, methanol, and ethanol exhibited quantum efficiencies between 80 and 100% [22, 62]. In contrast, the quantum efficiencies for the degradation of nonhalogenated organic substrates such as benzene, toluene, and xylene were found to be below 15% [62]. However, a drastic enhancement in the efficiencies for the degradation of some of these compounds has been reported to be possible by mixing with chlorocarbons such as TCE that cause radical chain reactions [62].

The presence of water vapor plays an important role in the gas-phase photocatalysis. For example, moderate levels of water vapor promote the photocatalytic degradation of toluene and *m*-xylene [66, 74], whereas they inhibit the degradation of ethylene [75]. In some cases, such as TCE [62], deactivation of the photocatalyst was observed in the absence of water vapor, which probably is due to the exhaustion of surface hydroxyls or to the formation of an intermediate species that blocks the active catalyst sites. However, high levels of water vapor (>4000 ppm) were found to decrease the degradation efficiency for acetone [76]. The role of water vapor in gas-phase photocatalysis is not completely understood. Even for the same compound, divergent results can be obtained by different groups [77, 78]. This may result from subtle differences in the film microstructure. For example, one film may have relatively small pores and may become flooded easily in the presence of high humidity, but such a film may be more effective in dry conditions because water that is produced during the photocatalytic reaction may be trapped more effectively [78]. This is an area that requires further study.

The design of suitable reactors is very crucial for the achievement of maximum conversion efficiencies. Ollis has made comparisons of various types of photoreactors for air purification [9]. Although monolithic photocatalytic reactors are well suited for air handling in buildings and exhausts, fluidized-bed reactors are advantageous for large-scale operations. The monolithic catalysts allow a drastic reduction in the pressure drop produced by the passage of the gas through the catalyst. Monolithic structures increase the illumination area of the catalyst in a three-dimensional way. Although fluidized-bed reactors offer good catalyst-light and catalyst-gas contacts, bubble-phase formation is a disadvantage. Recently, Vorontsov and coworkers [76], have demonstrated photocatalytic oxidation of acetone using a vibrofluidized-bed reactor. In this type of reactor, fluidization is achieved by vibration. The advantage is that it can operate under low-gas flow rates compared to conventional fluidized-bed reactors.

6.2.3.3 Outdoor Air Purification (NO_x Removal)

Air pollution, especially nitrogen oxides (NO_x) contamination from the combustion of hydrocarbons, is a particularly serious problem in urban areas. Despite serious efforts toward emission control, the concentrations of NO_x often exceed the air-quality standard, especially in large cities. TiO_2 photocatalysis appears to be a promising technology for the removal of low concentrations of NO_x from ambient air [79]. Daikin Industries has demonstrated the efficient removal of NO_x from indoor air using a photocatalyst coated on activated carbon [25]. According to their results, the indoor NO_x concentration decreases from 0.1 ppm to 0.06 ppm (the air-pollution standard) in 25 minutes.

The mechanism of nitrogen oxide (NO) removal by a photocatalyst is somewhat complicated. It is assumed that the NO is photocatalytically oxidized to NO_2, which is finally converted to nitric acid (HNO_3) (Fig. 10) [25]. At the NO_2 stage, part of the gas may escape from the photocatalyst surface but with an adsorbent such as activated carbon mixed with the catalyst, this gas may be trapped. The accumulation of HNO_3 on the catalyst surface retards the photocatalytic action. Therefore it is necessary to wash the catalyst after certain time intervals. However, if the photocatalyst is used outdoors, the products may be washed off by rainfall. A similar mechanism is expected for SO_2 removal as well.

The range of concentrations of air pollutants that can be removed efficiently is from 0.01 ppm to 10 ppm. These concentrations range from those present typically in the atmosphere to those in highway tunnels. However, at the end of an automobile's exhaust pipe, the concentration of pollutants is higher than 100 ppm. Therefore, attaching TiO_2 directly to the car exhaust system does not seem to be a good idea. This is better handled by a catalytic converter.

Further efforts are being made to improve the catalytic response of TiO_2 for NO_x removal. For example, Nakamura and coworkers [80], have used hydrogen plasma–treated TiO_2 to sensitize the catalytic response in the visible wavelength region to make use of sunlight more efficiently. Anpo and coworkers have used chromium (Cr) ion–implanted

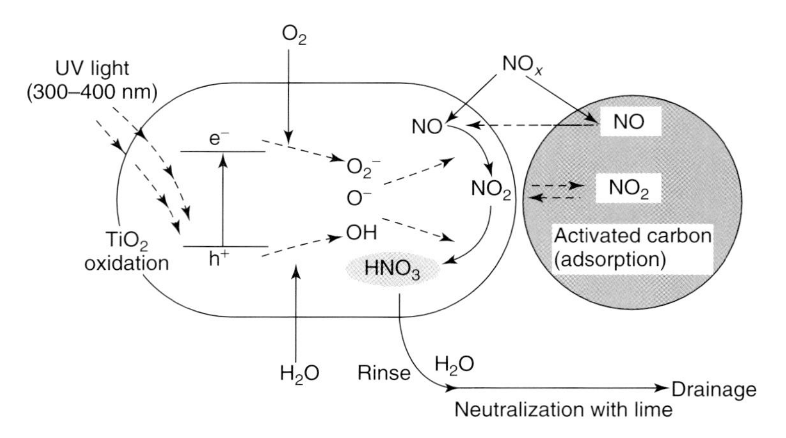

Fig. 10 NO_x removal by TiO_2 photocatalyst (courtesy of Dr. Koji Takeuchi).

TiO_2 to obtain enhanced response for decomposition in the visible region [81].

Because photocatalytic reactions are not thermochemical, it is not necessary to supply heat. They proceed at temperatures available in the ambient environment. The required UV light intensity is not more than about 0.1 mW cm^{-2}, which corresponds to the intensity level of outdoor sunlight that can be found on a cloudy winter day. Outdoor experiments have confirmed that photocatalytic effects can be observed from sunrise to sunset. Photocatalysts, however, are by no means a cure-all. Further efforts are necessary to use fully their special characteristics.

6.2.4 Photocatalytic Water Purification

6.2.4.1 Wastewater Purification

In a recent article, Herrmann has shown the feasibility of photocatalytic treatment of an actual highly loaded industrial wastewater (Fig. 11) [26]. In that experiment, the foul-smelling black-colored waste was diluted 1000 times before treatment. The treatment started after an adsorption period of one hour in the dark. After four hours of photocatalytic treatment under UV light, the chemical oxygen demand (COD) decreased by 95%, the water turning clear and odorless.

The advantage of photocatalysis in water purification is the complete mineralization of organics caused by the photogeneration of •OH radicals originating from water via the OH groups of the TiO_2 surface. In the case of gas-phase catalysis, water vapor must be present to achieve complete mineralization for most compounds. However, a disadvantage with regard to the photocatalytic purification of water is the lower quantum efficiencies in comparison to gas-phase photocatalysis. The principal reason for this is the smaller availability of oxygen in liquid water compared to air.

Several reports have appeared in the literature on the photocatalytic treatment of industrial effluents such as textile effluents [82–84], rinse waters of pesticide containers [85], wastewater from a phenolic resin factory [86], distillation effluents from a pharmaceutical company [86], waters that are produced during recovery of natural gas and crude

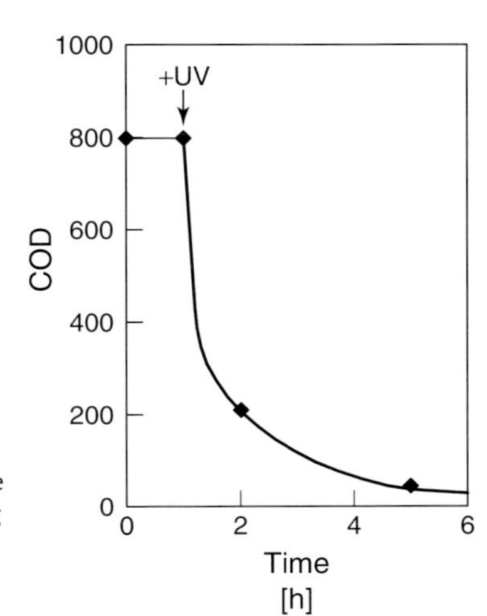

Fig. 11 Kinetics of the decrease of the COD (in mg kg^{-1}) during the photocatalytic treatment of a real industrial wastewater (diluted 1000 times) [26].

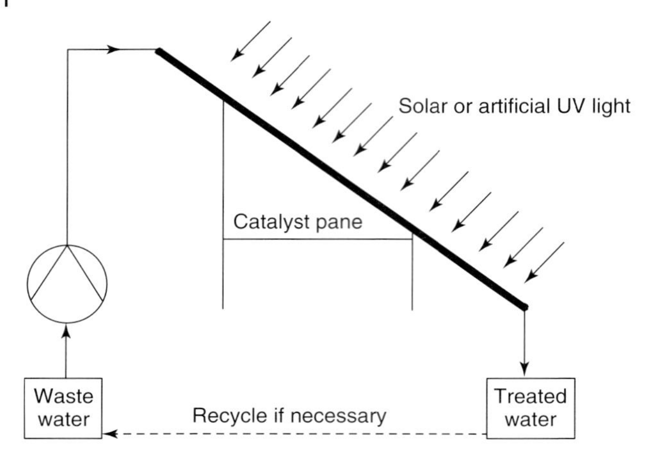

Fig. 12 Flow sheet of the thin-film fixed-bed reactor [91].

oil [87], explosive-contaminated water [29, 88, 89], and oil-contaminated water [90]. Various reactor configurations and their applications for wastewater treatment have been summarized recently [29]. A comparative study made on different solar reactors indicated that the thin-film fixed-bed reactor (Fig. 12) offers the advantages of a high degree of degradation, lower-cost construction, and reduced operating costs [91].

Although photocatalysis is attractive because of its ability to break down organic compounds completely to CO_2 and H_2O, that is, mineralization, industrial applications of this technology for solar wastewater purification still involve some fundamental questions, such as low hydroxyl radical–production efficiency and slow kinetics. These two points are barriers to marketing the technology. However, solar photocatalytic technology may become practical for the purification of low-level contaminants, similar to the case with indoor air purification. Freudenhammer and coworkers have indicated the suitability of photocatalysis for final-stage purification of wastewater after treatment with low-cost conventional techniques such as biological treatment [91]. Photocatalytic purification of drinking water is also promising. However, there are other questions related to equipment design, for example, how to increase the contact area between the photocatalyst and water and how to improve the light-irradiation efficiency.

6.2.4.2 Drinking Water Purification

Disinfection and decontamination of public drinking water supplies is an extremely important environmental matter because a large percentage of the world's population relies on these waters, especially in developing countries. As pointed out by Matthews [92], one of the most promising applications of photocatalysis is for domestic photocatalytic water purification, which would supply sufficient water for drinking and cooking requirements. A 15-W to 40-W photocatalytic water purifier that could process 10–20 liters per day was estimated to be less expensive than an activated carbon filter. Application of traditional methods to water treatment appears to be impractical for the low concentrations of organic pollutants in drinking water. The complete mineralization of many

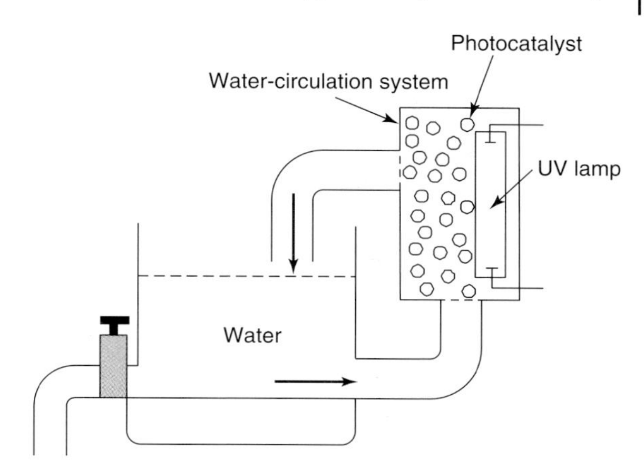

Fig. 13 Drinking water treatment with TiO_2 photocatalyst [25].

types of organic compounds is possible by photocatalysis. Matthews has demonstrated the photocatalytic degradation of several organic compounds that are typical water contaminants [93].

Several research groups recently have demonstrated the feasibility of drinking water purification by photocatalysis. Platinized TiO_2 was used to photocatalyze the destruction of bromate ions in tap water, at the ppb levels typically found in ozonated water [94]. Chen and Jenq have examined the degradation of dissolved organic carbon in drinking water sources [95]. Vidal has conducted pilot-plant studies to examine the effectiveness of the solar photocatalytic process for the destruction of low-level contaminants such as lindane (an organochlorine pesticide) in drinking water [96]. He has conducted tests with the municipal water supply of Madrid spiked with lindane and found that the lindane concentration was reduced from 200–500 $\mu g\ L^{-1}$ to maximum permitted levels (0.1 $\mu g\ L^{-1}$) during a 30-min solar exposure. These studies have indicated that inorganic substances found in municipal water slightly reduced the oxidation rates expected from related laboratory data. Shephard and coworkers demonstrated the rapid removal of cyanobacterial microcystins in lake waters [97]. However, humic substances in drinking water could be removed only partially with the formation of fluorescent intermediates [98]. Overall, the photocatalytic process seems to be very promising for drinking water purification, especially on the small scale required for domestic purposes (Fig. 13). Recently, the photocatalytic destruction of endocrine disruptors, such as bisphenol A [99], and that of natural and synthetic estrogens [100] has been demonstrated in water.

6.2.4.3 **Remediation of Metal Contamination**

Precious metals are released into waters by industrial processes such as hydrometallurgy, plating, and photography. Silver contamination from photographic processes, Cr from electroplating, and lead (Pb) from various biological and industrial sources are typical examples. Many of the metals are highly toxic, and accumulation of such metals in effluents and industrial wastes is of great environmental concern. Photocatalysis again helps in either the transformation or deposition of metals from aqueous solutions.

These metallic species can be recovered from the photocatalyst by mechanical or chemical methods.

Dissolved metals in the form of ions can affect the photocatalytic reaction rates and efficiencies of organic degradation reactions to a significant extent. Recently, Litter has published an extensive review on this subject [27]. Both increases [101, 102] and decreases [101, 103] in the photocatalytic reaction rates were reported, depending on the metal ion concentrations in the aqueous solutions. At low metal concentrations, trapping of photogenerated electrons by the metal ions, with resulting inhibition of electron-hole recombination, was proposed to be one of the reasons for the enhanced reaction rates [101, 102]. Photo-Fenton-type reactions may also take place in the presence of metal ions by reaction with photogenerated H_2O_2 [104]. Photogeneration of H_2O_2 caused by oxygen reduction has been demonstrated [105].

The detrimental effect on the oxidation rate of the presence of high metal ion concentrations was attributed to the oxidation of the reduced metal ions, either by $^{\bullet}OH$ radicals or directly by photogenerated holes [104]. Other reasons include the filter effect as a result of absorption of UV light by the metal species (e.g. Fe^{II}, Cu^{II}, and Ni^{II}) and precipitation of dissolved metals as hydroxides on the photocatalyst [103].

Photocatalytic removal of metals is an important application in its own right. There are three pathways for metal removal, two direct and one indirect. Direct reduction of metal ions by conduction band electrons is the simplest way. However, the redox potential of the metallic couple (M^{n+}/M^0) should be more positive than the conduction band edge (Fig. 14). Experimental results have indicated that only metal ions with potentials more positive than ~0.4 V can be reduced on TiO_2. Prairie and coworkers demonstrated the reduction of Cr^{VI}, Au^{III}, Hg^{II}, Pd^{II}, Pt^{IV}, and Ag^{I} in deaerated conditions in the presence of salicylic acid [106]. Insignificant metal deposition is observed for platinum (Pt), silver (Ag), and gold (Au) in the absence of a hole scavenger [107]. With an alcohol as a hole scavenger, metal deposition may be an indirect reduction process involving $^{\bullet}ROH$ radicals produced from alcohol degradation by $OH^{\bullet}$ radicals, rather than a direct capture of conduction band electrons. These are the two possibilities for the direct photocatalytic deposition of metals. A third, indirect pathway involves metal species such as Pb^{2+}, Mn^{2+}, and Tl^{+}, with relatively negative redox potentials, which may follow an oxidative route through hole attack because of the existence of stable higher oxidation states [27]. Litter has summarized the pathways for the photocatalytic deposition of several metals [27].

6.2.5 Self-cleaning and Antifogging Surfaces

In 1997, it was reported that TiO_2 surfaces can become highly hydrophilic (superhydrophilic) under UV illumination [108, 109]. Although the mechanism for this effect still is not completely understood, great progress has been made, and it is clear that it involves reactions that are distinct from those involved in the photocatalytic decomposition of organic compounds [109–111]. In fact, the possibility that the conversion of the TiO_2 surface from hydrophobic to hydrophilic was due to the photocatalytic removal of a thin layer of organic contamination was considered, but mounting evidence has shown this not to be true. For example,

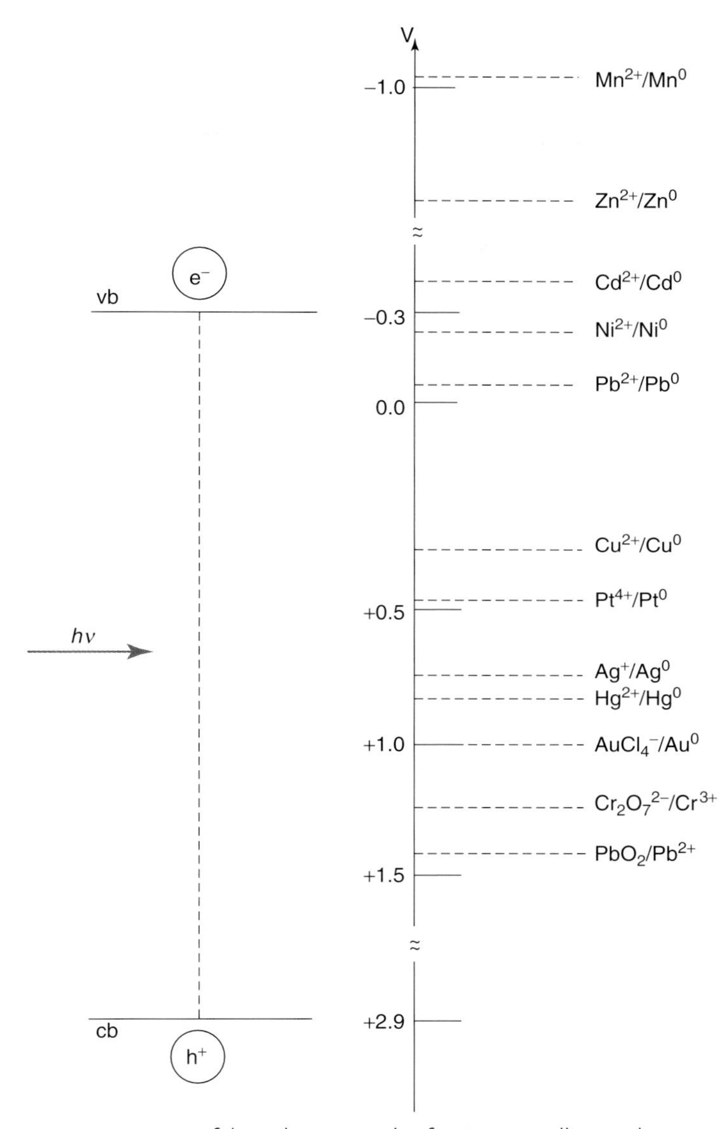

Fig. 14 Positions of the redox potentials of various metallic couples related to the energy levels of the conduction and valence bands of TiO_2 [27].

the surface wettability, which increases under relatively low intensity UV illumination over the span of about one hour (Fig. 15), can be converted back to the hydrophobic state by ultrasonic treatment. The latter is thought to oxidize reduced surface states (Ti^{3+}) by $^{\bullet}OH$ radicals or hydrogen peroxide. The surface can then be reconverted to the hydrophilic state by illumination [110].

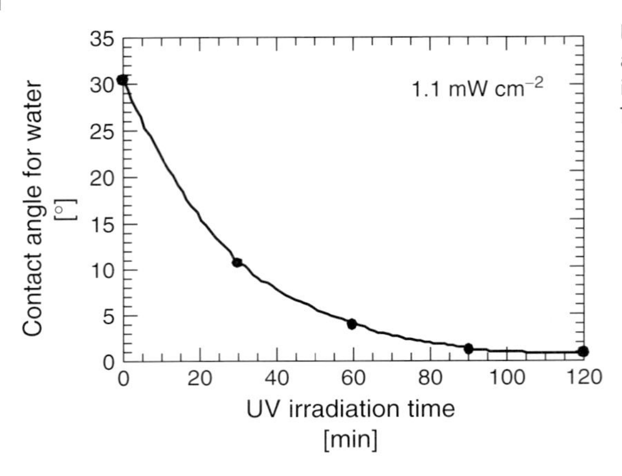

Fig. 15 Water contact angle as a function of time under UV illumination for a polycrystalline TiO_2 film on glass.

The same fundamental phenomenon, that is, the great increase in wettability on UV illumination, can be used in two broad areas of applications, that is, self-cleaning surfaces and antifogging or antibeading surfaces. Both of these areas are treated briefly.

6.2.5.1 Self-cleaning Surfaces

The concept of the *superhydrophilic* self-cleaning surface or easily washable surface is a very appealing one and in a functional sense is very closely related to that of the *photocatalytic* self-cleaning surface. However, the mechanisms are different. What makes this situation particularly interesting is that any particular TiO_2 surface possesses both types of properties simultaneously. Here the focus is on superhydrophilic properties.

The mechanism for superhydrophilic self-cleaning principally involves the very

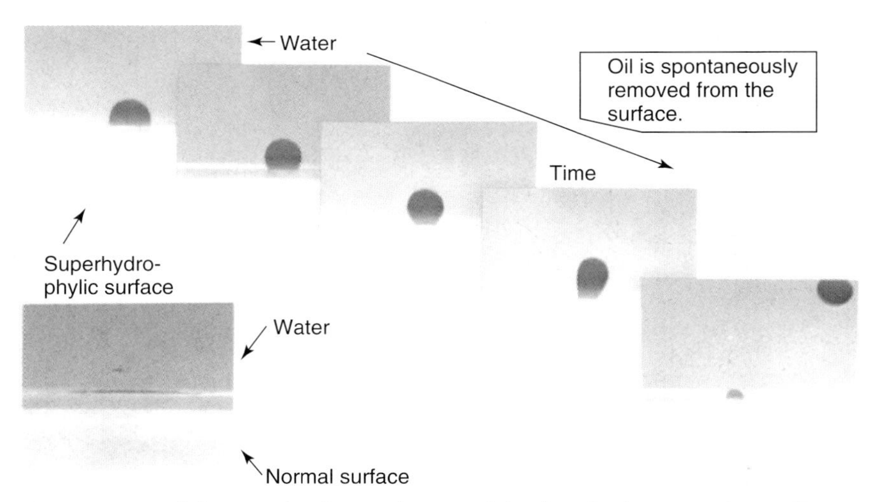

Fig. 16 Series of photographs showing how an oil droplet adhering to a polycrystalline TiO_2 film on glass is gradually removed from the surface by water as a result of a UV-induced increase in hydrophilicity [25].

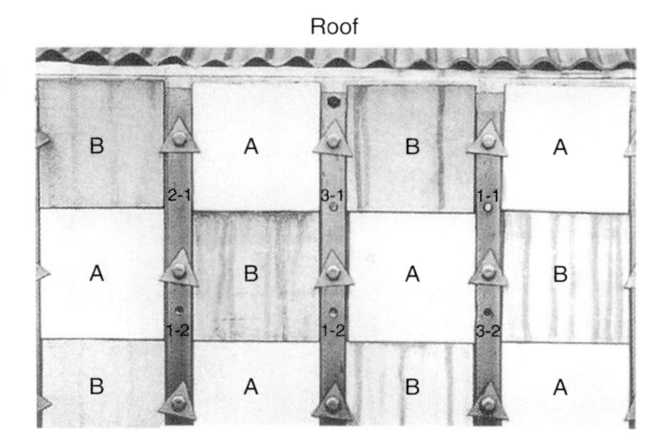

Fig. 17 Photograph showing alternating photocatalytic (A) and nonphotocatalytic (B) exterior wall tiles that have been placed so that water from a corrugated metal roof runs off onto the wall [25].

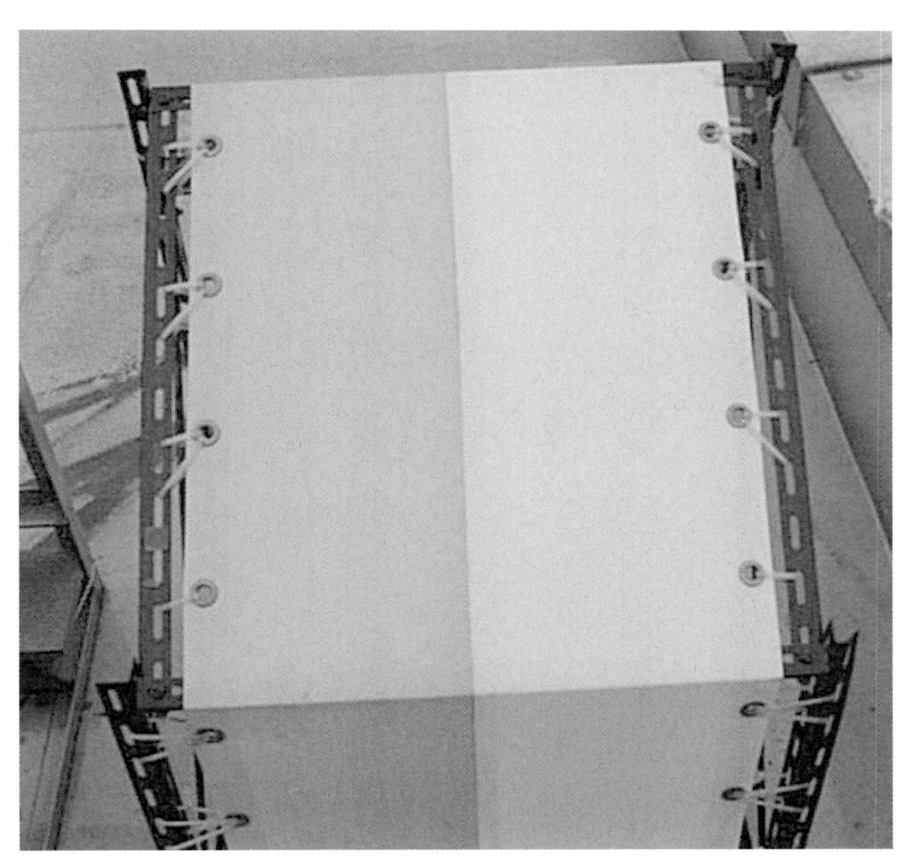

Fig. 18 Photograph showing heavy tarpaulin material, noncoated on the left and TiO_2-coated on the right, after exposure to urban air for several months.

strong adsorption of water on the photochemically activated TiO_2 surface. Organic compounds that are attached or absorbed on the surface are simply lifted off due to the fact that water is more strongly adsorbed. For example, if an oil droplet is attached to the surface, water simply begins to undercut it until it dislodges (Fig. 16). This effect can be extremely useful for construction materials to be used in the urban environment because pollutants can be easily washed off by rainwater or with a stream of water [25, 112, 113]. For example, siding material that has been treated with TiO_2 is resistant to staining from water that runs off a roof (Fig. 17). Another example is the heavy cloth material (tarpaulin, Fig. 18) that is being used for temporary storage structures or shelters (Fig. 19). The corresponding conventional materials are quite difficult to clean because of their flexible nature.

It again should be stressed that it is possible for the same TiO_2 surface to support both the photocatalytic decomposition reactions and the superhydrophilic reactions. The proportion of the two pathways can be controlled through the variation of the film composition, for example, by SiO_2 addition to increase the proportion of the hydrophilic pathway [112–114]. For given applications and ambient conditions, for example, light intensity, humidity, and so forth, it may be necessary to fine-tune the proportion of the two pathways.

6.2.5.2 **Antifogging, Antibeading Surfaces**

Because one of the critical attributes of glass is its extreme visual clarity, there is a host of possible applications for glass surfaces that do not fog or on which water does not bead. Some of these are listed in Table 2, together with those that are related to hydrophilic self-cleaning surfaces. Water, in the form

Fig. 19 Photograph of a temporary storage structure made from TiO_2-treated tarpaulin (courtesy of Taiyo Kogyo Corp.).

Tab. 2 Applications of superhydrophilic technology

Property	*Category*	*Application*
Self-cleaning	Roads	Tunnel lighting, tunnel walls, traffic signs, and sound-proof walls
	Houses	Tiles on kitchen walls and bathrooms, exterior tiles, roofs, and windows
	Buildings	Aluminum panels, tiles, building stone, crystallized glass, and glass film
	Agriculture	Plastic and glass greenhouses
	Electric and electronic environment	Computer displays and cover glass for solar cells
	Vehicles	Paint work, coatings for exterior surfaces of windows, and headlights
	Daily necessities and consumer products	Tableware, kitchenware, and spray-on antifouling coatings
	Paint	General-purpose paints and coatings
Antifogging property	Roads	Road mirrors
	Houses	Mirrors for bathrooms and dressers
	Stores	Refrigerated showcases
	Electric and electronic environment	Heat exchangers for air conditioners, high-voltage electric transmission equipment
	Vehicles	Inside surfaces of windows, glass films, rear-view mirrors, and windshields
	Daily necessities and consumer products	Spray-on antifogging coatings and films
	Paint	General-purpose paints and coatings
	Optical instruments	Optical lenses
Biocompatibility	Medical instruments and supplies	Contact lenses and catheters

of droplets ranging from micrometer to millimeter size, tends either to scatter light or simply to reflect or refract it randomly. In both cases, visual clarity is impaired drastically.

In a serendipitous discovery, Watanabe and coworkers found that a TiO_2-SiO_2 surface could become extremely hydrophilic under UV illumination [114]. The result of this property is that water spreads evenly across the surface (Fig. 20). If the amount of water is relatively small, the water layer becomes very thin and evaporates quickly. If the amount of water is larger, it forms a sheetlike layer that also has high visual clarity. The first commercial application of this phenomenon has been for automobile side-view mirrors. The TOTO Company in Japan currently is supplying TiO_2-treated glass for such mirrors to major automobile manufacturers. In addition, it is marketing adhesive TiO_2–coated plastic films for the conversion of existing mirrors.

As the properties of superhydrophilic films continue to be improved, there will be many other possible applications of antifogging, antibeading glass. For example, one of the principal challenges is to develop films that can be activated at longer

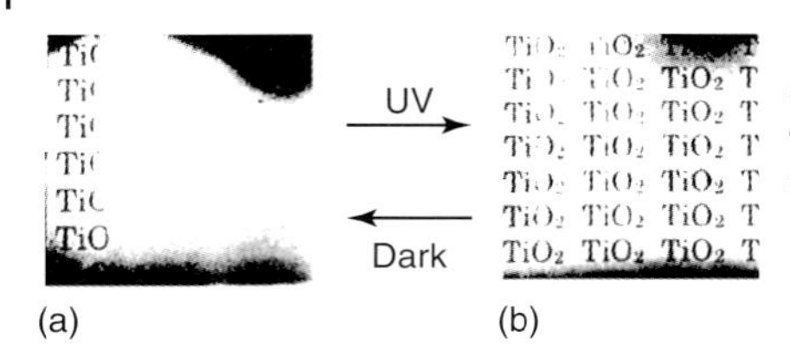

Fig. 20 Photograph of the effect of UV illumination on water droplets covering a sheet of TiO_2-treated glass [108].

wavelengths, principally by decreasing the magnitude of the band gap. Efforts along these lines were discussed briefly in the previous chapter. However, even with the band gap of unmodified anatase, the number of possible applications is substantial, particularly those in which solar light can be used, for example, for window glass for building exteriors or vehicles. These promise to be important applications.

6.2.6
Conclusion

Photocatalysis has become an increasingly promising technology for air and water purification. Although there are obstacles to overcome on the way to commercialization of this technology for water purification, commercial products for air purification already have appeared on the market in Japan and recently in other countries. Photocatalytic systems for the purification of room air and car interiors already have been developed and commercialized. TiO_2-coated tiles have been installed in a number of hospital operating rooms and have successfully eliminated bacteria from the walls and from the air. TiO_2-coated concrete paving blocks have been installed on busy roads in Osaka by the Mitsubishi Company in Japan. The success of these developments is mainly because of the selection of environments where the target gas-phase pollutant levels are low and thus the efficiency of TiO_2 is high under the low available levels of UV irradiation. Regarding the development of photocatalytic water–purification technology, some of the specific challenges are more complete mineralization, higher quantum yields, better regeneration of the photocatalyst, and lower costs. Photocatalysis may play a dominant role in the purification of pretreated wastewater, and this may be more economical than treating raw wastewater directly. In addition, domestic photocatalytic water purifiers may be an attractive new prospect for commercial development.

The ability of TiO_2 surfaces to undergo photoinduced changes in wettability, specifically, with large increases under UV illumination, leads to a large number of possible applications, including self-cleaning construction materials, self-cleaning glass, and antifogging, antibeading glass for windows and mirrors. These applications are being commercialized at present, and the range of applications should continue to broaden as ways are found to enhance the ability to become activated by visible light.

Acknowledgment

The authors would like to express appreciation to a number of coworkers who have contributed heavily to this work, including Professors Kazuhito Hashimoto, Toshiya Watanabe, and Rong Wang, Dr. Yoshihisa Ohko, and Mr. Nobuyuki Sakai. Dr. Bulusu V. Sarada helped in the preparation of the figures. The authors appreciate Dr. D.M. Blake for providing copies of his comprehensive bibliography on photocatalysis.

References

1. D. M. Blake, in *Bibliography of Work on the Photocatalytic Removal of Hazardous Compounds from Water and Air*, NREL/TP-430-6084, National Renewable Energy Laboratory, Golden, Colorado, 1994.
2. D. M. Blake, in *Bibliography of Work on the Photocatalytic Removal of Hazardous Compounds from Water and Air*, NREL/TP-473-20300, Update No. 1 (to June 1995), National Renewable Energy Laboratory, Golden, Colorado, 1995.
3. D. M. Blake, in *Bibliography of Work on the Photocatalytic Removal of Hazardous Compounds from Water and Air*, NREL/TP-430-22197, Update No. 2 (to October 1996), National Renewable Energy Laboratory, Golden, Colorado, 1996.
4. D. M. Blake, in *Bibliography of Work on the Photocatalytic Removal of Hazardous Compounds from Water and Air*, NREL/TP-570-26797, Update No. 3 (to January 1999), National Renewable Energy Laboratory, Golden, Colorado, 1999, available at: *http://www.ott.doe.gov/coolcar/pubs.html*.
5. D. M. Blake, P. Maness, Z. Huang et al., *Sep. Purif. Methods.* **1999**, *28*, 1–50.
6. M. A. Fox, M. T. Dulay, *Chem. Rev.* **1993**, *93*, 341–357.
7. P. V. Kamat, *Chem. Rev.* **1993**, *93*, 267.
8. A. Mills, R. H. Davies, D. Worsley, *Chem. Soc. Rev.* **1993**, *22*, 417.
9. D. F. Ollis, H. Al-Ekabi, Eds. in *Photocatalytic Purification and Treatment of Water and Air*, Elsevier, Amsterdam, 1993.
10. H. Yoneyama, *Crit. Rev. Solid State* **1993**, *18*, 69–111.
11. W. A. Zeltner, C. G. Hill, M. A. Anderson, *CHEMTECH* **1993**, *23*, 21–28.
12. N. Serpone, *Res. Chem. Intermed.* **1994**, *20*, 953–992.
13. K. I. Zamaraev, M. I. Khramov, V. N. Parmon, *Catal. Rev.* **1994**, *36*, 617.
14. A. Heller, *Acc. Chem. Res.* **1995**, *28*, 503–508.
15. M. R. Hoffmann, S. T. Martin, W. Choi et al., *Chem. Rev.* **1995**, *95*, 69–96.
16. A. L. Linsebigler, G. Lu, J. J. T. Yates, *Chem. Rev.* **1995**, *95*, 735–758.
17. K. Rajeshwar, *J. Appl. Electrochem.* **1995**, *25*, 1067–1082.
18. Y. Parent, D. Blake, K. M. Bair et al., *Solar Energy* **1996**, *56*, 429–437.
19. U. Stafford, K. A. Gray, P. V. Kamat, *Heterogen. Chem. Rev.* **1996**, *3*, 77–104.
20. K. Vinodgopal, P. V. Kamat, *CHEMTECH* **1996**, *26*, 18–22.
21. A. Fujishima, T. N. Rao, *Proc. Indian Acad. Sci. (Chem. Sci.)* **1997**, *109*, 471–486.
22. A. Mills, S. LeHunte, *J. Photochem. Photobiol. A: Chem.* **1997**, *108*, 1–35.
23. J. Peral, X. Domenech, D. F. Ollis, *J. Chem. Tech. Biotechnol.* **1997**, *70*, 117–140.
24. N. Serpone, *J. Photochem. Photobiol. A: Chem.* **1997**, *104*, 1–12.
25. A. Fujishima, K. Hashimoto, T. Watanabe, in *TiO_2 Photocatalysis: Fundamentals and Applications*, BKC, Inc., Tokyo, Japan, 1999.
26. J.-M. Hermann, *Catal. Today* **1999**, *53*, 115–129.
27. M. I. Litter, *Appl. Catal. B: Environ.* **1999**, *23*, 89–114.
28. M. Romero, J. Blanco, B. Sanchez et al., *Solar Energy* **1999**, *66*, 169–182.
29. O. M. Alfano, D. Bahnemann, A. E. Cassano et al., *Catalysis Today* **2000**, *58*, 199–230.
30. A. Fujishima, T. N. Rao, D. A. Tryk, *J. Photochem. Photobiol. C* **2000**, *1*, 1–21.
31. M. Tomkiewicz, *Catal. Today* **2000**, *58*, 115–123.
32. D. A. Tryk, A. Fujishima, K. Honda, *Electrochim. Acta* **2000**, *45*, 2363–2376.
33. T. Matsunaga, R. Tomada, T. Nakajima et al., *FEMS Microbiol. Lett.* **1985**, *29*, 211–14.
34. C. Wei, W.-Y. Lin, Z. Zainal et al., *Environ Sci. Technol.* **1994**, *28*, 934.
35. T. Matsunaga, M. Okochi, *Environ. Sci. Technol.* **1995**, *29*, 501–505.
36. R. J. Watts, S. Kong, M. P. Orr et al., *Water Res.* **1995**, *29*, 95–100.
37. M. Bekbolet, C. V. Araz, *Chemosphere* **1996**, *32*, 959–965.
38. M. Bekbolet, *Water Sci. Technol.* **1997**, *35*, 95–100.
39. Y. Kikuchi, K. Sunada, T. Iyoda et al., *J. Photochem. Photobiol. A: Chem.* **1997**, *106*, 51–56.
40. W. A. Jacoby, B. C. Maness, E. J. Wolfrum et al., *Environ. Sci. Technol.* **1998**, *32*, 2650–2653.

41. K. Sunada, Y. Kikuchi, K. Hashimoto et al., *Environ. Sci. Technol.* **1998**, *32*, 726–728.
42. P.-C. Maness, S. Smolinski, D. M. Blake et al., *Appl. Environ. Microbiol.* **1999**, *65*, 4094–4098.
43. A. Vidal, A. I. Diaz, A. E. Hraiki et al., *Catal. Today* **1999**, *54*, 283–290.
44. Z. Huang, P.-C. Maness, D. M. Blake et al., *J. Photochem. Photobiol. A: Chem.* **2000**, *130*, 163–172.
45. I. Sopyan, M. Watanabe, S. Murasawa et al., *J. Photochem. Photobiol. A: Chem.* **1996**, *98*, 79–86.
46. Y. Ohko, K. Hashimoto, A. Fujishima, *J. Phys. Chem. A* **1997**, *101*, 8057–8062.
47. T. Watanabe, A. Kitamura, E. Kojima et al., U.S. Patent 5,874,701, February 23, 1999, in *U. S. Patent & Trademark Office (Patent Full Text and Image Database)*, Toto Co., Ltd., USA, 1999.
48. T. Watanabe, A. Kitamura, E. Kojima et al., U.S. Patent 6,139,803, October 31, 2000, in *U. S. Patent & Trademark Office (Patent Full Text and Image Database)*, Toto Co., Ltd., USA, 2000.
49. K. Sunada, Y. Kikuchi, K. Hashimoto et al., *Environ. Sci. Technol.* **1998**, *32*, 726–728.
50. A. Fujishima, J. Ootsuki, T. Yamashita et al., *Photomed. Photobiol.* **1986**, *8*, 45–46.
51. R. Cai, K. Hashimoto, K. Itoh et al., *Bull. Chem. Soc. Jpn.* **1991**, *64*, 1268–1273.
52. R. Cai, K. Hashimoto, Y. Kubota et al., *Chem. Lett.* **1992**, 427–430.
53. R. Cai, H. Sakai, K. Hashimoto, Y. Kubota et al., *Denki Kagaku* **1992**, *60*, 314–321.
54. H. Sakai, R. Baba, K. Hashimoto et al., *Chem. Lett.* **1995**, 185–186.
55. R. Cai, Y. Kubota, T. Shuin et al., *Cancer Res.* **1992**, *52*, 2346–2348.
56. A. Fujishima, R. X. Cai, J. Otsuki et al., *Electrochim. Acta* **1993**, *38*, 153–157.
57. N. Huang, M. Xu, C. Yuan et al., *J. Photochem. Photobiol. A: Chem* **1997**, *108*, 229–233.
58. I. Sopyan, S. Murasawa, K. Hashimoto et al., *Chem. Lett.* **1994**, 723–726.
59. T. Noguchi, A. Fujishima, P. Sawunyama et al., *Environ. Sci. Technol.* **1998**, *32*, 3831–3833.
60. Y. Ohko, D. A. Tryk, K. Hashimoto et al., *J. Phys. Chem. B* **1998**, *102*, 2699–2704.
61. H. Matsubara, M. Takada, S. Koyama et al., *Chem. Lett.* **1995**, 767–768.
62. M. L. Sauer, M. A. Hale, D. F. Ollis, *J. Photochem. Photobiol. A: Chem.* **1995**, *88*, 169–178.
63. R. M. Alberici, W. F. Jardim, *Appl. Catal. B: Environ.* **1997**, *14*, 55–68.
64. J. Stark, J. Rabani, *J. Phys. Chem. B* **1999**, *103*, 8524–8531.
65. W. A. Jacoby, M. R. Nimlos, D. M. Blake et al., *Environ. Sci. Technol.* **1994**, *28*, 1661–1668.
66. Y. Luo, D. F. Ollis, *J. Catal.* **1996**, *163*, 1–11.
67. O. Hennezel, D. F. Ollis, *J. Catal.* **1997**, *167*, 118–126.
68. K. Wada, K. Yoshida, Y. Watanabe et al., *J. Chem. Soc.*, Faraday Trans. **1996**, *92*, 685–691.
69. S. A. Larson, J. A. Widegren, J. L. Falconer, *J. Catal.* **1995**, *157*, 611–625.
70. N. Takeda, M. Ohtani, T. Torimoto et al., *J. Phys. Chem. B* **1997**, *101*, 2644–2649.
71. T. N. Obee, R. T. Brown, *Environ. Sci. Technol.* **1995**, *29*, 1223–1231.
72. N. N. Lichtin, M. Sedeghi, *J. Photochem. Photobiol. A: Chem.* **1998**, *113*, 81–88.
73. P. Kopf, E. Gilbert, S. H. Eberle, *J. Photochem. Photobiol. A: Chem.* **2000**, *136*, 163–168.
74. J. Peral, D. F. Ollis, *J. Catal.* **1992**, *136*, 554–565.
75. S. Yamazaki, S. Tanaka, H. Tsukamoto, *J. Photochem. Photobiol. A: Chem.* **1999**, *121*, 55–61.
76. A. V. Vorontsov, E. N. Savinov, P. G. Smirniotis, *Chem. Eng. Sci.* **2000**, *55*, 5089–5098.
77. S. Sitkiewitz, A. Heller, *Nouv. J. Chim.* **1996**, *20*, 233–241.
78. T. Minabe, D. A. Tryk, P. Sawunyama et al., *J. Photochem. Photobiol. A: Chem.* **2000**, *137*, 53–62.
79. T. Ibusuki, K. Takeuchi, *J. Mol. Catal.* **1994**, *88*, 93–102.
80. I. Nakamura, N. Negishi, S. Kutsuna et al., *J. Mol. Catal. A: Chem.* **2000**, *162*, 205–212.
81. M. Anpo, Y. Ichihashi, M. Takeuchi et al., *Res. Chem. Intermed.* **1998**, *24*, 143–149.
82. H. Q. Zhan, H. Tian, *Dye Pigment.* **1998**, *37*, 231–239.
83. S. G. D. Moraes, R. S. Freire, N. Duran, *Chemosphere* **2000**, *40*, 369–373.
84. K. Tanaka, K. Padermpole, T. Hisanaga, *Wat. Res.* **2000**, *34*, 327–333.
85. S. Malato, J. Blanco, C. Richter et al., *Appl. Catal. B: Environ.* **2000**, *25*, 31–38.

86. S. M. Rodriguez, C. Richter, J. B. Galvez et al., *Solar Energy* **1996**, *56*, 401–410.
87. E. Bessa, G. L. Sant' Anna, M. Dezotti, *J. Adv. Oxid. Technol.* **1999**, *4*, 196–202.
88. N. Z. Muradov, *Solar Energy* **1994**, *52*, 283–288.
89. R. Alnaizy, A. Akgerman, *Water Res.* **1999**, *33*, 2021–2030.
90. M. Hamerski, J. Grzechulska, A. W. Morawski, *Solar Energy* **1999**, *66*, 395–399.
91. H. Freudenhammer, D. Bahnemann, L. Bousselmi et al., *Wat. Sci. Technol.* **1997**, *35*, 149–156.
92. R. W. Matthews, in *Photocatalytic Purification and Treatment of Water and Air* (Eds.: D. F. Ollis, H. Al-Ekabi), Elsevier, Amsterdam, 1993, pp. 121–138.
93. R. W. Matthews, *Water Res.* **1986**, *20*, 569–578.
94. A. Mills, A. Belghazi, D. Rodman, *Water Res.* **1996**, *30*, 1973–1978.
95. P. H. Chen, C. H. Jenq, *Environ. Internat.* **1998**, *24*, 871–879.
96. A. Vidal, *Chemosphere* **1998**, *37*, 387.
97. G. S. Shephard, S. Stockenstrom, D. D. Villiers et al., *Toxicon* **1998**, *36*, 1895–1901.
98. A. R. Eggins, F. L. Palmer, J. A. Byrne, *Water Res.* **1997**, *31*, 1223–1226.
99. Y. Ohko, I. Ando, C. Niwa et al., *Environ. Sci. Technol.* **2001**, *35*, 2365–2368.
100. Y. Kubota, C. Niwa, T. Iguchi et al., in preparation.
101. K. Okamoto, Y. Yamamoto, H. Tanaka et al., *Bull. Chem. Soc. Jpn.* **1985**, *58*, 2023–2028.
102. E. Pelizzetti, M. Borgarello, E. Borgarello et al., *Chemosphere* **1988**, *17*, 499–510.
103. A. Sclafani, L. Palmisano, E. Davi, *J. Photochem. Photobiol. A: Chem.* **1991**, *56*, 113–123.
104. J. Sykora, *Coord. Chem. Rev.* **1997**, *159*, 95–108.
105. H. Sakai, R. Baba, K. Hashimoto et al., *J. Phys. Chem.* **1995**, *99*, 11896–11900.
106. M. R. Prairie, L. R. Evens, B. M. Stange et al., *Environ. Sci. Technol.* **1993**, *27*, 1776–1782.
107. R. Baba, R. Honda, A. Fujishima et al., *Chem. Lett.* **1986**, 1307.
108. R. Wang, K. Hashimoto, A. Fujishima et al., *Nature* **1997**, *388*, 431–432.
109. R. Wang, K. Hashimoto, A. Fujishima et al., *Adv. Mater.* **1998**, *10*, 135–138.
110. N. Sakai, R. Wang, A. Fujishima et al., *Langmuir* **1998**, *14*, 5918–5920.
111. R. Wang, N. Sakai, A. Fujishima et al., *J. Phys. Chem. B* **1999**, *103*, 2188–2194.
112. M. Chikuni, M. Hayakawa, T. Watanabe et al., U. S. Patent 5,755,867, May 26, 1998, in *U. S. Patent & Trademark Office (Patent Full Text and Image Database)*, Shin-Etsu Chemical Co., Ltd. and Toto Ltd., USA, 1998.
113. M. Hayakawa, E. Kojima, K. Norimoto et al., U. S. Patent 6,013,372, January 11, 2000, in *U. S. Patent & Trademark Office (Patent Full Text and Image Database)*, Toto, Ltd., USA, 2000.
114. A. Fujishima, D. A. Tryk, T. Watanabe et al., *Internat. Glass Rev.* **1998**, 114–116.

6.3
Silverless Photography – Optical Image Recordings by Means of Photoelectrochemical Processes

Hiroshi Yoneyama
Anan National College of Technology, Tokushima, Japan

6.3.1
Introduction

The term *Silverless photography* is used for photoimage-formation processes using a variety of photochemical reactions including those using photoelectrochemical systems [1]. Silverless photography provides a photochemical means of image processing [2]. Light-sensitive chelate coordination compounds containing divalent elements are often used for such purposes [3]. For example, the photochemical cleavage of polymeric films of coordination compounds such as $[Fe^{II}(CN)_6\text{-}Pt^{IV}(NH_3)_4]_n$, followed by reactions of the resulting complexes with 3D transition metal cations in solution, affords image formation in the polymeric films [4]. Photoisomerization of molecules such as azobenzene derivatives [5] and spirooxazine dyes [6] are also useful in optical image recordings. Furthermore, the use of photochromic molecules with a combination of liquid crystals can provide the optical image storage [7, 8]. This article, however, is restricted to electrochemical systems in which semiconductive materials and/or redox dye molecules are used as the photosensitizers.

Photoelectrochemical imaging systems may be roughly classified into two classes; one is concerned with photoinduced electrochemical reaction systems using various electrode configurations and the other with light-induced heterogeneous reaction systems. The former may be further divided into three categories: area-selective deposition onto semiconductor electrodes, area-selective electrochemical reactions in the surface layer of semiconductor electrodes, and area-selective reactions in electrochromic thin-film-covered semiconductor electrodes. The latter may be classified into two groups: area-selective reactions of electrochromic films containing dispersed semiconductor particles as the photosensitizer and area-selective reactions of dye films with use of a molecular photosensitizer.

6.3.2
Area-selective Photoinduced Deposition on Semiconductor Electrodes

6.3.2.1 Deposition on Semiconductor Electrodes with External Bias

The most significant feature of the semiconductor electrode is that photosensitized electrochemical reactions that are induced to occur using photogenerated minority charge carriers take place only on the irradiated electrode surfaces. Therefore, if *n*-type semiconductor electrodes are irradiated under anodic bias, photosensitized oxidation reactions are induced to take place at the irradiated electrode surfaces by the photogenerated valence band holes (as described in earlier chapters). Similarly, if *p*-type semiconductor electrodes are irradiated under cathodic bias, photosensitized reduction reactions are induced to take place on the irradiated electrode surfaces with the use of photogenerated conduction band electrons. Oxidative electrochemical deposition reactions are then used in photoimage formations on *n*-type semiconductor electrodes and electrochemical reductive depositions are used on *p*-type semiconductor electrodes. If light scattering at the irradiated semiconductor electrode surfaces were ignored, the highest resolution of the area-selective deposition would be twice as large as the mean diffusion length of the photogenerated minority charge carriers.

Area-selective photoinduced reductive depositions of copper (Cu), nickel (Ni), palladium (Pd), and gold (Au) on cathodically biased *p*-Si and *p*-GaAs [9] from aqueous solutions containing the corresponding metal ions were shown to take place at a resolution of less than 10 μm. Area-selective reductive deposition of heptylviologen radical cation salt ($HV^{+\cdot}Br^-$) from a heptylviologen bromide solution onto *p*-GaAs electrode under a cathodic bias was useful in photoinduced electrochromic displays [10]. Optical images due to the deposited $HV^{+\cdot}Br^-$ were easily erased by applying the reverse bias (i.e. anodic bias) to the electrode. The area-selective deposition was also reported for photooxidative deposition of PbO_2 from a Pb^{2+} solution onto *n*-type ZnO, TiO_2, and $SrTiO_3$ [11], and of Tl_2O_3 on these semiconductor electrodes from a Tl^+ solution [12]. Using these photodeposition reactions, the photoelectrochemical imaging was successfully achieved at the anodically biased semiconductor electrodes. One example of the photoimages produced is given in Fig. 1. The deposited Tl_2O_3 do not grow beyond the irradiated part of the electrodes because the photoinduced oxidation reaction took place on the irradiated surfaces only. For example, the rate of the deposition reaction of Tl_2O_3 is proportional to the irradiation intensities at the electrode surfaces. Here, irradiation of the *n*-type semiconductor electrodes through a negative film of photography resulted in the deposition of the brown oxide of various contrasts in reflection to the images of the negative film, allowing the formation of a portrait. Attempts to form photoimages on *n*-Si using the Tl_2O_3 deposition were also reported [13]. The produced images are easily erased by applying the reverse bias to the electrodes in the dark. Area-selective photooxidative deposition of polypyrrole from a pyrrole solution onto TiO_2 was used to produce a conducting polymer pattern [14, 15]. The photoelectrochemical oxidation of water-soluble aromatic compounds such as *O*-toluidine on TiO_2 electrodes resulted in hydrophobic deposits on the irradiated surface only, which was developed by applying oily color ink [16]. Photo-patterning was also reported with self-assembled monolayer (SAM) of bis[11-[(4-azidobenzoyl)oxy]-1-undecyl]disulfide [17]. Irradiation of the SAM-covered Au electrode through a photomask in the presence of several amines such as $HNEt_2$, $HN(n\text{-}Bu)_2$ or $HN(CH_2CH_2OH)_2$ resulted in the attachments of the amine, and the amine-attached regions showed little activity for electrochemical deposition of polyaniline from an acidic aniline solution, giving selective deposition of polyaniline on the nonirradiated SAM surfaces on Au. The area-selectivity in the photoelectrochemical reactions finds another application – the micromachining of ZnO surfaces by anodic photoetching [18]. Micropatterns having 130-μm line width were successfully produced using this technique.

6.3.2.2 Deposition on Semiconductor Surfaces with no External Bias

When a photooxidative deposition occurs on semiconductor surfaces without any applications of external bias, a reduction reaction takes place simultaneously as the counterpart reaction (see Chapter 8). Irradiation of *p*-Si wafers in a zinc (Zn) or cadmium (Cd) salt solution resulted in area-selective deposition of metals on the irradiated part if the back surface of

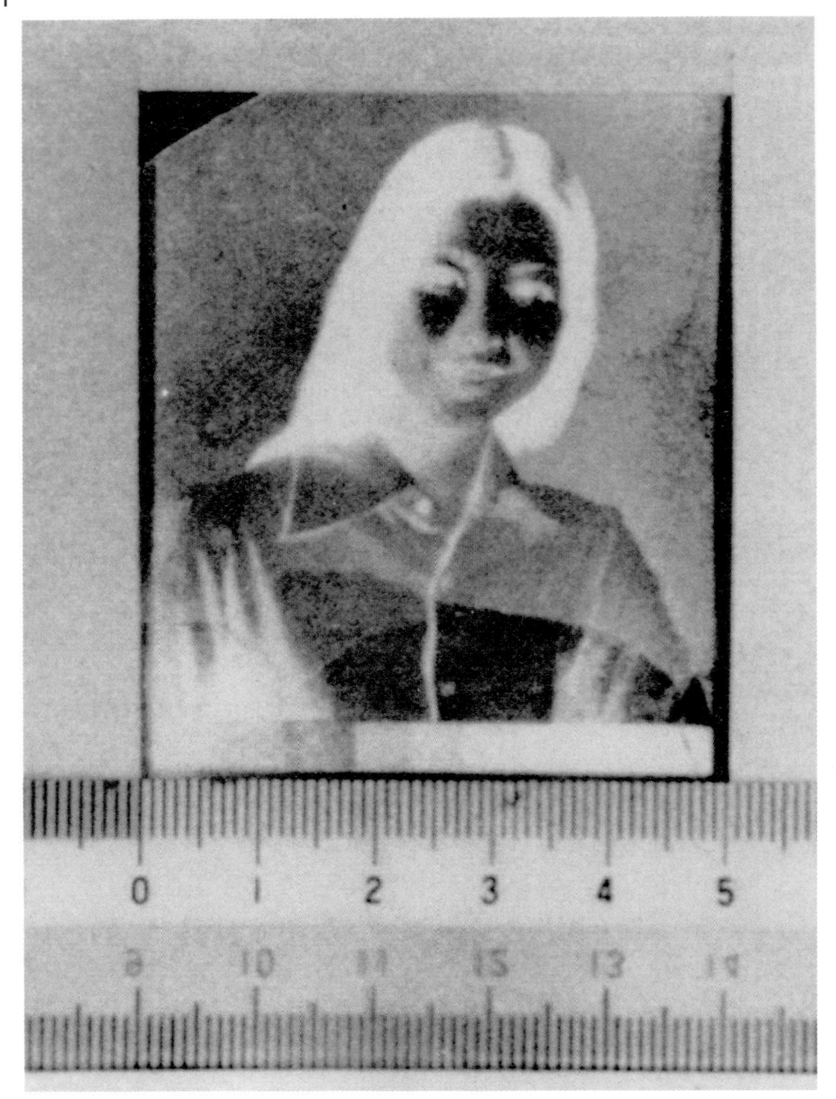

Fig. 1 Image recorded by using photoinduced deposition of metallic oxides on *n*-type semiconductor electrodes. This example shows Tl_2O_3 deposition on ZnO electrode from an aqueous solution containing a Tl(I) salt [12]. (With permission by the Electrochemical Society, Inc.)

the wafers was coated by metallic films of Zn or Cd [19]. The photoinduced reaction in that case is thought to consist of the photoinduced cathodic deposition of the metal in the irradiated surface and anodic dissolution of the coated metal in the back surface. The metal film on the back surface was useful in enhancing the band bending with which the rate of the transfer of photogenerated electrons to the solution phase was enhanced. Irradiation of TiO_2 single crystal plates in a solution containing a Pd salt caused selective deposition of Pd at surface flaws [20]. The surface flaws seem to provide electron-transfer channels from the TiO_2 to the electrolyte solution. In that case, the oxidation reaction as the counterpart reaction is thought to be photoinduced oxidation of water on the flaw-free irradiated surfaces. Such reaction selectivity of semiconductor surfaces was confirmed by observing which part of the semiconductor surfaces, flaws in the surfaces, or flaw-free well-etched surfaces is more favored for oxidative or reductive deposition reactions [21]. There seems to be a general trend that when n-type semiconductor such as TiO_2 is used, photoinduced oxidation reactions take place on the flaw-free well-etched surfaces, whereas reduction reactions occur on the flaws especially in the dark. Similarly, if p-type semiconductors are used, photoinduced reduction reactions occur on the well-etched flaw-free surfaces, whereas oxidation reactions occur on the flaws especially in the dark. Therefore, it may be said that electrochemical reactions in which the photogenerated minority charge carriers are involved, take place on the irradiated defect-free surfaces and the counterpart reactions that are induced to occur by the majority charge carriers take place on the defects in the nonirradiated surfaces.

Using such reaction selectivity of the semiconductor surfaces, line-shaped patterns of polypyrrole were successfully produced on well-etched n-type Si wafers, which had mechanically damaged back surfaces, by scanning He-Ne laser beam on the front surfaces in aqueous solution containing pyrrole and Ag^+ ions [22]. The deposition of silver (Ag) took place selectively on the whole damaged surfaces in the dark as the counterpart reaction. Using a similar technique, polythiophene patterns of high resolution were produced on n-Si wafers [23] from a solution containing α-terthienyl and Ag^+ ions. Irradiation of a TiO_2-adhered alumina plate in an electroless Ni plating solution containing citric acid as the hole scavenger resulted in area-selective deposition of Ni on the irradiated parts [24]. The mechanism for this event would be such that electrons and holes are generated in the irradiated surface, and the positive holes are quickly captured by the hole scavenger, leaving the electrons at the irradiated surface; then the reductive deposition of the metals takes place at the irradiated surface.

6.3.3 Image Formations Using Photoinduced Electrochromism

6.3.3.1 Electrochromic Materials

Electrochromic materials are often used as photoimage-recording materials. However, the process for making photoimages in the electrochromic materials is different depending on whether the electrochromic materials have semiconductivity or not. The electrochromic materials having semiconductivity work both as the photosensitizer and as the substrates in which the photoimages are produced, whereas those having metallic or insulating properties need another photosensitizer to produce photoimages in the electrochromic substrates.

Table 1 gives popular electrochromic materials useful in photoinduced image formations [25–28]. The list includes several semiconductor electrochromic oxides such as MoO_3, WO_3, and Nb_2O_5.

6.3.3.2 Photoinduced Electrochromism of Semiconductive Oxides

WO_3, MoO_3, and Nb_2O_5 have n-type semiconductivity and are colored when

Tab. 1 Electrochromic materials

Material	*Color*	
	oxidized	*reduced*
WO_3	Trans.[a]	Blue
IrO_x	Trans.	Blue black
phosphotungstic acid	White	Blue
MoO_3	Yellow	Blue
Rh_2O_3	Yellow	Green
V_2O_5	Yellow	Black
Nb_2O_5	Trans.	Dark blue
NiO_x	Trans.	Dark bronze
CoO_x	Red purple	Gray black
Prussian	Blue	White
Lutechium diphtalocyanine	Green	Red violet
Heptylviologen	Trans.	Blue purple
Benzylviologen	Trans.	Blue
Tetrathiafulvalene	Yellow	Purple
Polypyrrole	Brown black	Yellow
Polyaniline	Green	Yellow
Polythiophene	Green	red

[a] trans.: transparent.

reduced. The photoinduced coloring of a WO_3 film without any applied bias in air atmosphere has been known since 1979 [25]. WO_3 particles suspended in water or in formic acid solution changes in color to blue with photon absorption [29]. The coloring occurs as a result of reduction of WO_3 to hydrogen tungsten bronze. The photooxidation of water or formic acid occurs simultaneously as the counterpart reaction. The same is true for a WO_3 film and WO_3-TiO_2 composite film [30]. Detailed studies of the photoinduced color change of WO_3 films caused by irradiation were carried out mostly in air atmosphere containing aliphatic alcohol vapors such as ethanol and propanol as the hole scavenger [31–34]. When the hole scavengers are oxidized by the photogenerated positive holes, protons are released, which are involved in hydrogen tungsten bronze formation. Because of the generation of photogenerated positive holes and because the production of protons are restricted to the irradiated surface only, the color change occurs only at the irradiated surfaces. The colored images on the oxide films are easily erased by anodically polarizing the oxide films in the image-forming solution.

Similar photoinduced electrochromism has been published on MoO_3 films prepared by electrodeposition [35] and vacuum evaporation [36–39]. MoO_3 needs irradiation with near ultraviolet (UV) light for its photoexcitation. However, it was reported that a MoO_3 film prepared by vacuum evaporation has visible light sensitivity by cathodic prepolarization in a propylene carbonate solution containing $LiClO_4$ [36, 37, 39]. The cathodic prepolarization may form visible light-sensitive Li_xMoO_3 in the MoO_3 lattice. Both UV- and visible light–induced electrochromic response are enhanced by coating the oxide surface with a thin porous Au or platinum (Pt) layer (20 nm) [39]. The formation of a large Schottky barrier at the coated metal–MoO_3 interface may assist in the separation of photogenerated positive holes and electrons, resulting in the enhancement of the coloring of the oxide. The photoinduced coloration of MoO_3 was found to be enhanced by mixing with WO_3, and the highest enhancement was achieved on a film having 92% WO_3 and 8% MoO_3 [40]. As in the case of WO_3, the images formed in the MoO_3 surfaces are erased by anodically polarizing the oxide electrodes in the image-forming baths. A Nb_2O_5 film prepared by vacuum evaporation showed similar photoinduced electrochromism if ethanol vapor was used as a hole scavenger in nitrogen atmosphere [41].

6.3.3.3 Combined Systems of Electrochromic Materials and Semiconductor Electrodes

Attempts were made to form area-selective images on electrochromic thin films coated on semiconductor electrodes. The first reports on this category were on Prussian blue film-coated TiO_2 [42, 43] electrodes. Later, similar systems were constructed using $SrTiO_3$ films as the photosensitive substrate [44, 45]. The coating with Prussian blue films on semiconductor electrodes was carried out by electrochemical synthesis of the Prussian blue films using the semiconductor as the electrode. To produce area-selective photoimages, the entire Prussian blue-coated film was first reduced to Prussian white. The irradiation was then carried out with a xenon lamp through a photomask under anodic bias to allow area-selective photooxidation of the coated film, resulting in blue images in the white film. The same principle was applied to poly(*N*-methylpyrrole) film–coated *n*-Si [46] and polypyrrole film–coated Si electrodes of both *n*-type and *p*-type conductivity [47]. These conducting films were deposited on the Si electrode surfaces by electrochemical polymerization of the corresponding monomer using the Si as the electrode. When *n*-type Si electrodes are used, the entire polypyrrole-coated film is first reduced by cathodic polarization to give yellow, and then the polypyrrole-coated Si electrodes are anodically polarized and irradiated through a mask. The area-selective photooxidation of the polypyrrole films is then induced to take place to give brown-black images in the irradiated surfaces. In the case of using *p*-type Si electrodes, the as-deposited brown black polypyrrole films changed in color to yellow at the irradiated part under cathodic bias. However, it was found that the images produced in the coated polypyrrole film had very poor resolution. With an increase in the irradiation time, the produced images easily spread beyond the irradiated area because of high electrical conductivity of polypyrrole.

Polyacrylamide gel films containing Fe(III) dipyridinium complex as an electrochromic component were useful as the substrate for the formation of photographic images when coated on a CdSe film [48]. Poly-$Re(CO)_3$(vbpy)Cl-coated *n*-$MoSe_2$ also allowed photoinduced electrochromic recordings in the coated polymer layer under anodic bias [49]. The images produced are easily erased by applying the reverse bias to the polymer-coated semiconductor electrodes.

6.3.3.4 Composites Made of Electrochromic Materials and Semiconductor Powders

Photoimages of high resolution were successfully produced in polyaniline films containing dispersed WO_3 [50, 52], and TiO_2 [51, 53, 54] powder particles in the films. The polyaniline/semiconductor particles composite films were synthesized by electrochemical polymerization of aniline in aqueous acidic solution in the presence of the suspended semiconductor particles. The prepared composite films showed the color of polyaniline even if a high amount of the oxide particles was used because WO_3 and TiO_2 are colorless. One example of a photoimage formed in a polyaniline/TiO_2 composite film is presented in Fig. 2. As shown in the figure, a portrait of high resolution was obtained. Complex electrochemical reactions are involved in the formation of such good photoimages. Figure 3 shows the reaction scheme for the photoimage formation in the composite film, derived from analyses of changes in absorption spectra of polyaniline films obtained during

Fig. 2 Photoimage formed on a composite film of polyaniline and dispersed TiO_2 particles by projecting the positive imaged on the film that was immersed in 0.5 mol dm^{-3} phosphate buffer (pH 7) containing 20 wt% methanol. The illumination was carried out with a 55 Xe lamp for 1 minute [49]. (With permission by the Royal Society of Chemistry).

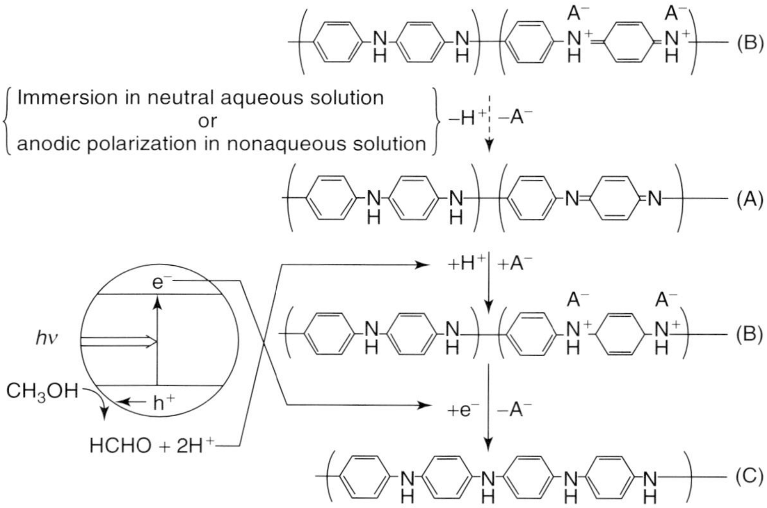

(B)

Immersion in neutral aqueous solution
or
anodic polarization in nonaqueous solution

$-H^+$ $-A^-$

(A)

$+H^+$ $+A^-$

e^-

$h\nu$

CH_3OH

h^+

$HCHO + 2H^+$

(B)

$+e^-$ $-A^-$

(C)

Fig. 3 Schematic illustration of compositional changes in polyaniline caused by photoreduction. (A) deprotonated blue film, (B) as-grown and/or green film, and (C) reduced yellow film. A^- denotes a monovalent electrolyte anion [51]. (With permission by American Chemical Society).

the course of the image formation [53] and from studies using a quartz crystal microbalance [54]. When the composite films are prepared, they exhibited vivid green, which is characteristic of highly conducting polyaniline having high electrochemical activities. Polymer films of high electrical conductivity and of high electrochemical reactivity do not allow the formation of photoimage of high resolution, as already described earlier for polypyrrole-coated Si electrodes. To produce photoimage of high resolution, the electrical conductivity of polyaniline of the composites has to be changed to be as low as possible. Fortunately, polyaniline films of high electrical conductivity are easily changed into electrically insulating ones either by anodically polarizing in alcohol solutions such as methanol and ethanol or in aqueous neutral solutions or by immersion in concentrated alkaline solutions [55]. With the use of either one of these procedures, the deprotonation occurs at polyaniline in an oxidized state, and the color changes from vivid green to blue. The polyaniline/TiO_2composite films having an electrically insulating blue film prepared in such a manner is then immersed in ethanol solution, and irradiated under no applied bias through a photomask with lights whose energy is high enough to photoexcite the incorporated TiO_2. Yellow images, which are characteristic of the reduced polyaniline, are then formed in the blue films with high resolution. An important phenomenon inherent in the photoimage formation-processes is that the electrochemically inactive insulating polyaniline is changed into electrochemically active one at the irradiated parts only. As shown in Fig. 3 for the reaction scheme, if the blue composite film is irradiated, electrons and positive holes are generated in the TiO_2 particles in the film. The generated positive holes cause oxidation of methanol at the irradiated TiO_2 surfaces, resulting in release of protons there. The produced protons are instantly picked up by the deprotonated polyaniline, and the electrochemically inactive blue polyaniline is then converted to electrochemically active green polyaniline, which is then reduced by the photogenerated electrons in the TiO_2 to form yellow images. Because the production of protons and their attachments to the deprotonated polyaniline

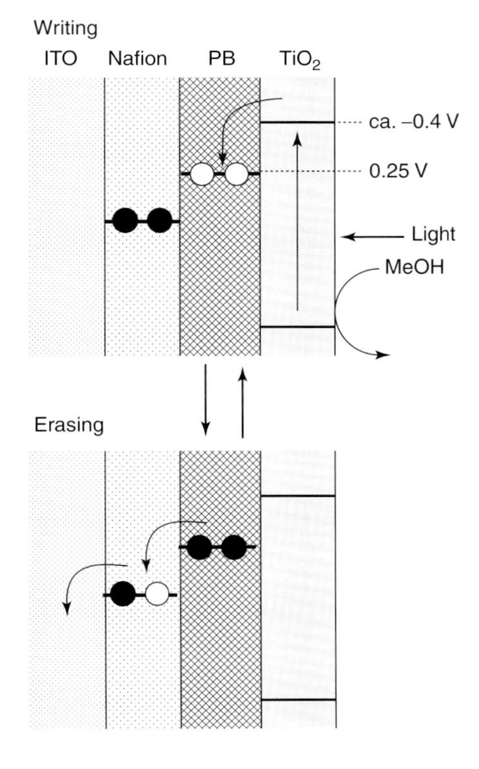

Fig. 4 Energy level diagram for electron-transfer events at the bilayer electrode consisting of a Nafion film containing ferrocenylmethyltrimethylamnnonium ions and a Prussian-blue film containing 14 wt% TiO_2 particles [55]. (With permission by the Electrochemical Society, Inc.).

occurs only in the irradiated area, image formations of high resolution take place at the irradiated part only.

The same idea was used in the image formation in polyaniline/CdS composite films [56], but because of the poor stability of CdS in the strong acidic medium in which the composite is prepared, results obtained were not so good as those obtained with the polyaniline/TiO_2 composites. The composite films of Prussian blue and TiO_2 fine particles

(a)

(b)

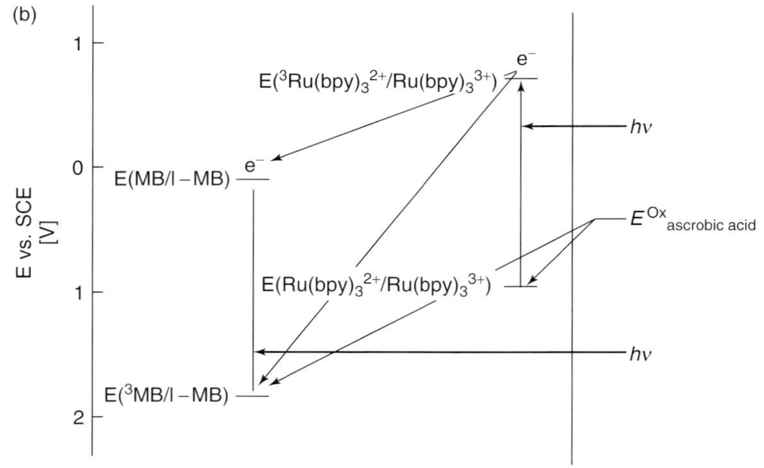

Fig. 5 (a) Photoimage formed in polyaniline film containing methylene blue- and $Ru(bpy)_3{}^{2+}$-bound Nafion by illumination with a 500-W Xenon lamp for 30 seconds in 1 mol dm^{-3} HCl containing 0.1 mol dm^{-3} ascorbic acid, and (b) Energetic correlation of photoexcited methylene blue and $Ru(bpy)_3{}^{2+}$ to ascorbic acid used as an electron donor. (With permission by the Electrochemical Society, Inc.).

prepared on ITO electrodes also showed good properties for photoimage formations with high resolution [57]. However, it was found that the photoimages produced grew beyond the irradiated area with increasing the irradiation time. The electron transport from the produced Prussian white to the surrounding Prussian blue through the conducting ITO was found to be responsible. This serious problem was solved by inserting a thin Nafion film containing ferrocenylmethyltrimethylammonium (FA) as an electron relay between the ITO substrate and the Prussian blue/TiO_2 composite film. The energy scheme for the image formation is given in Fig. 4.

6.3.3.5 Image Formation with the Use of a Dye as the Photosensitizer

Photoimages could be formed if the semiconductor powder dispersed in the electrochromic film were replaced by a redox dye. The utility of the idea was demonstrated by immobilizing in electrically conductive polyaniline film, with redox active methylene blue used as an image-forming molecule and $Ru(bpy)_3^{3+/4+}$ as a photosensitizer [58]. If aniline is electrochemically polymerized in the presence of methylene blue and Nafion, methylene blue–bound Nafion is incorporated in the resulting polyaniline films, and the film exhibits electrochromic properties as a result of the redox reaction of methylene blue [59]. The polyaniline worked as a conducting matrix in that case. A photosensitive film with an ability for recording photoimages was prepared by electrochemical polymerization of aniline in the presence of methylene blue–bound Nafion and $Ru(bpy)_3^{3+/4+}$. The reaction scheme of the image-forming processes is illustrated in Fig. 5. A solid-state device using the same components was constructed using a composite made of a cation exchange polymer and polyethylene glycol as a solid electrolyte [60].

6.3.4 Summary

So far, various optical image recordings with the use of photoelectrochemical reactions have been published, as described in the preceding sections. Many systems use a couple of a photosensitizer and an electrochromic material. The key for construction of recording systems in these cases is to appropriately choose the photosensitizer and the electrochromic material to obtain good photoimages of high resolution. Because there are a lot of electrochromic materials and photosensitizers available for photoinduced electrochromic reactions, the development of novel recording systems with the use of a variety of combinations would still be of active concern.

References

1. Yu. A. Cherkasov, *J. Opt. Technol.* **1997**, *64*, 872–879.
2. M. R. V. Sahyun, *J. Chem. Educ.* **1973**, *50*, 88–93.
3. O. V. Mikhailov, *J. Coord. Chem.* **1999**, *47*, 31–58.
4. Y. Wu, B. W. Pfenning, A. Bocarsly et al., *Inorg. Chem.* **1995**, *34*, 4262–4267.
5. Z.-F. Liu, B. H. Loo, R. Baba et al., *Chem. Lett.* **1990**, 1023–1026.
6. W. Du, H. Y. Tam, *Optics Commun.* **1996**, *126*, 223–229.
7. T. Ikeda, O. Tsutumi, T. Sasaki, *Synth. Met.* **1996**, *81*, 289–296.
8. M. V. Kozlovsky, V. P. Shibaev, A. I. Stakhanovs et al., *Liq. Cryst.* **1998**, *24*, 759–767.
9. R. H. Micheels, A. D. Darrow, II, R. D. Rauch, *Appl. Phys. Lett.* **1981**, *39*, 418–420.
10. B. Reichman, F.-R. F. Fan, A. J. Bard, *J. Elecrochem. Soc.* **1980**, *127*, 333–338.
11. T. Inoue, A. Fujishima, K. Honda, *Jpn. J. Appl. Phys.* **1979**, *18*, 2177–2178.

12. T. Inoue, A. Fujishima, K. Honda, *J. Electrochem. Soc.* **1980**, *127*, 1582–1588.
13. R. J. Phillips, M. J. Shane, J. A. Switzer, *J. Mater. Res.* **1989**, *4*, 923–929.
14. M. Okano, K. Itoh, A. Fujishima et al., *Chem. Lett.* **1986**, 469–472.
15. M. Okano, K. Itoh, A. Fujishima et al., *J. Electrochem. Soc.* **1987**, *134*, 837–841.
16. H. Masuda, N. Shimidzu, S. Ohno, *Chem. Lett.* **1984**, 1701–1704.
17. L. F. Rozsnyai, M. S. Wrighton, *J. Am. Chem. Soc.* **1994**, *116*, 5993–5994.
18. M. Futsuhara, K. Yoshioka, Y. Ishida et al., *J.Electrochem. Soc.* **1996**, *143*, 3743–3746.
19. T. L. Rose, D. H. Longendorfer, R. D. Rauh, *Appl. Phys. Lett.* **1983**, *42*, 193–195.
20. H. Yoneyama, H. Shiotani, N. Nishimura et al., *Chem. Lett.* **1981**, 157–160.
21. T. Kobayashi, Y. Taniguchi, H. Yoneyama et al., *J. Phys. Chem.* **1983**, *87*, 768–775.
22. H. Yoneyama, M. Kitayama, *Chem. Lett.* **1986**, 657–660.
23. H. Yoneyama, K. Kawai, S. Kuwabata, *J. Electrochem. Soc.* **1988**, *135*, 1699–1702.
24. S. Morishita, K. Suzuki, *Bull. Chem. Soc. Jpn.* **1994**, *67*, 843–846.
25. R. J. Colton, A. M. Guzman, J. W. Rabalais, *Acc. Chem. Res.* **1978**, *11*, 170–176.
26. C. M. Lampert, *Sol. Energy Mater.* **1984**, *11*, 1–27.
27. K. Hyodo, *Electrochim. Acta* **1994**, *39*, 265–272.
28. R. J. Mortimer, *Electrochim. Acta* **1999**, *44*, 2971–2981.
29. M. Fujii, T. Kawai, H. Nakamatsu et al., *J. Chem. Soc. Chem. Commun.* **1983**, 1428–1429.
30. B. Ohtani, T. Atsumi, S. Nishimoto et al., *Chem. Lett.* **1988**, 295–298.
31. E. Kikuchi, K. Iida, A. Fujishima, *J. Electroanal. Chem.* **1993**, *351*, 105–114.
32. B. H. Loo, J. N. Yao, H. D. Coble et al., *Appl. Surf. Sci.* **1994**, *81*, 175–181.
33. E. Kikuchi, N. Hirota, A. Fujishima et al., *J. Electroanal. Chem.* **1995**, *381*, 15–19.
34. Y. A. Yang, Y. W. Cao, P. Chen et al., *J. Phys. Chem. Solids* **1998**, *59*, 1667–1670.
35. J. N. Yao, B. H. Loo, K. Hashimoto et al., *J. Electroanal. Chem.* **1990**, *290*, 263–267.
36. J. N. Yao, B. H. Loo, A. Fujishima, *Ber. Bunsen-Ges. Phys. Chem.* **1990**, *94*, 13–17.
37. J. N. Yao, B. H. Loo, A. Fujishima, *Nature* **1992**, *355*, 624–626.
38. K. Ajito, L. A. Nagahara, D. A. Tryk et al., *J. Phys. Chem.* **1995**, *99*, 16383–16388.
39. J. N. Yao, Y. A. Yang, B. H. Loo, *J. Phys. Chem. B* **1998**, *102*, 1856–1860.
40. J. N. Yao, B. H. Loo, K. Hashimoto et al., *Ber. Bunsen-Ges. Phys.Chem.* **1991**, *95*, 554–556.
41. J. N. Yao, B. H. Loo, K. Hashimoto et al., *Ber. Bunsen-Ges. Phys. Chem.* **1992**, *96*, 699–701.
42. D. W. DeBerry, A. Viehbeck, *J. Electrochem. Soc.* **1983**, *130*, 249–251.
43. K. Itaya, I. Uchida, S. Toshima et al., *J. Electrochem. Soc.* **1984**, *131*, 2086–2091.
44. J. P. Ziegler, J. C. Hemminger, *J. Electrochem. Soc.* **1988**, *134*, 358–363.
45. J. P. Ziegler, E. K. Lesniewski, J. C. Hemminger, *J. Appl. Phys.* **1987**, *61*, 3099–3104.
46. O. Inganas, I. Lundstrom, *J. Electrochem. Soc.* **1984**, *131*, 1129–1132.
47. H. Yoneyama, K. Wakamoto, H. Tamura, *J. Electrochem. Soc.* **1985**, *132*, 2414–2417.
48. A. A. Nekrasov, V. F. Ivano, A. V. Vannikov, *J. Photogr. Sci.* **1993**, *41*, 43–47.
49. H.-T. Zhang, P. Subaramanian, O. Fussarydel et al., *Sol. Energy Mater.* **1992**, *25*, 315–325.
50. H. Yoneyama, S. Hirao, S. Kuwabata, *J. Electrochem. Soc.* **1992**, *139*, 3141–3146.
51. H. Yoneyama, N. Takahashi, S. Kuwabata, *J. Chem. Soc. Chem. Commun.* **1992**, 716–717.
52. S. Kuwabata, S. Hirao, H. Yoneyama, *Denki Kagaku oyobi kogyo Butsuri Kagaku* **1992**, *60*, 1097–1100.
53. S. Kuwabata, N. Takahashi, S. Hirao et al., *Chem. Mater.* **1993**, *5*, 437–441.
54. S. Kuwabata, A. Kishimoto, H. Yoneyama, *J. Electroanal. Chem.* **1994**, *377*, 261–268.
55. S. Kuwabata, N. Kihira, H. Yoneyama, *Chem. Mater.* **1993**, *5*, 716–719.
56. H. Yoneyama, M. Tokuda, S. Kuwabata, *Electrochim. Acta* **1994**, *39*, 1315–1320.
57. M. Nishizawa, S. Kuwabata, H. Yoneyama, *J. Electrochem. Soc.* **1996**, *143*, 3462–3465.
58. S. Kuwabata, K. Mitsui, H. Yoneyama, *J. Electrochem. Soc.* **1992**, *139*, 1824–1830.
59. S. Kuwabata, K. Mitsui, H. Yoneyama, *J. Electroanal. Chem.* **1990**, *281*, 97–107.
60. N. Kobayashi, T. Yano, K. Teshima et al., *Electrochim. Acta* **1998**, *43*, 1645–1649.

6.4
Photoelectrochemical Etching

Hideki Minoura and Takashi Sugiura
Gifu University, Yanagido, Japan

6.4.1
Introduction

Since the earliest stage of the study of semiconductor photoelectrochemistry, much attention has been drawn to the instability of semiconductors toward their photocorrosion in solution (see Chapter 1). However, this photocorrosion reaction has also been effectively utilized as a unique technique for semiconductor surface processing, which is called "*photoelectrochemical etching*" or simply "photoetching." This term is analogous to that used for the metal corrosion applied to surface treatment, which is often called *chemical etching*. In chemical etching processes, the etching reaction is caused by an oxidizing agent in solution. In photoetching, the presence of a photogenerated hole in an n-type semiconductor at the electrolyte interface is analogous to a broken chemical bond and allows the ionic dissolution of the semiconductor itself. Photoetching technique has several unique features. The etching rate is usually proportional to the rate of photogeneration of a minority carrier, that is, light intensity. This can be monitored by measuring the photocurrent flowing throughout the external circuit, and likewise, the etching quantity can be monitored during the etching process. One more important feature is spatial selectivity at the etching site. Etching proceeds only at the irradiated site of the surface. It also depends on the crystallographic orientation, the presence of lattice defects, and the energy band structure at semiconductor–electrolyte interface.

In the 1970s, the idea of photoelectrochemical cells for light to electrical and/or chemical energy conversion attracted scientists in various fields as a novel energy conversion device. Photoetching has been widely used to modify the semiconductor surface and consequently improve the cell efficiency [1, 2]. This improvement is attributed to several effects, such as a decrease in reflection losses [3], an increase in effective electrode area, a removal of surface defects acting as recombination centers [4–7], and change in the chemical composition of surface [8, 9]. The photoetching works well for typically photocorrosive semiconductors, such as CdS, CdSe, and CdTe. The photocorrosion of other compound semiconductors have been also examined, for example, $CuIn_5S_8$, some other ternary semiconductors, InSe, $MoSe_2$, and WSe_2, and so forth [10–23]. The significant effect of photoetching is usually seen with a polycrystalline sample, which has a high defect density at the surface. Photoetching can be effectively applied, often better than chemical etching to the surface processing of thin film samples because of its better controllability of etching rate and etching quantity by measuring the photocurrent.

Photoetching is known to be a powerful technique for the characterization of semiconductor materials, especially III–V compound semiconductors [24–37], which is useful not only to reveal and decorate dislocations, defects, and precipitates but also to determine the local conductivity type at the surface.

Photoetching has been also applied to fabricate microstructures on the surface of semiconductor. Because the photoetching reaction occurs selectively at the irradiated part of the surface, one can etch the semiconductor to form three-dimensional structures by spatial modulation of the light intensity or by providing physical masking. An example is the production of gratings on the surface of several semiconductors, for example, CdS [38], GaAs [39, 40], and InP [41–43], by the use of a holographic set-up. This technique allows fast generation of gratings with submicron meter size without using photoresist mask. An excellent review article covers the photoelectrochemical surface processing of III–V semiconductors [44].

Another example of photoetching application is the preparation of monodisperse CdS nanoparticles by the use of size-selective photoetching technique, which has been recently reported by Yoneyama and his coworkers [45–49]. Meijerink and his coworkers applied this technique to other compound semiconductor particles, such as ZnS, PbS, and ZnO [50, 51]. It is based on the fact that as the band gap of size-quantized semiconductor nanoparticles increases with a decrease in their size, larger CdS nanoparticles can be selectively photoexcited and photocorroded under irradiation by light.

Photoetching has also been applied to make three-dimensional processing of silicon surfaces. Porous structured silicon is of great interest for electronic and micromechanical devices (see Chapter 10).

Photoanodic dissolution does not proceed homogeneously over the surface. Preferential dissolution of specified sites on the surface of semiconductors results in the generation of a unique etching pattern. In what follows, we shall describe some examples that we have found in recent years, especially focusing on the surface tailoring of cadmium chalcogenide and TiO_2 semiconductors.

6.4.2 Surface-tailoring of Semiconductor

6.4.2.1 CdX(X = S, Se)

In this section, we review the characterization of photoetching patterns evolved on the surface of polycrystalline CdX(X = S, Se), which are one of the most extensively studied semiconductors as photoelectrodes in photoelectrochemical cells, photoconductors, and other optoelectronic usages. Anodic photocurrent flowing through the electrode–electrolyte interface is entirely due to the oxidative dissolution of the semiconductors themselves to form Cd^{2+} ion and elemental chalcogen, except in certain redox solutions, such as aqueous polysulfide, polyselenide, and alkanolamine solutions. For example, in an aqueous NaCl solution, CdSe is photoanodically decomposed to form Cd^{2+} ion and elemental Se. After the photoelectrolysis of a CdSe electrode under potentiostatic conditions and a subsequent

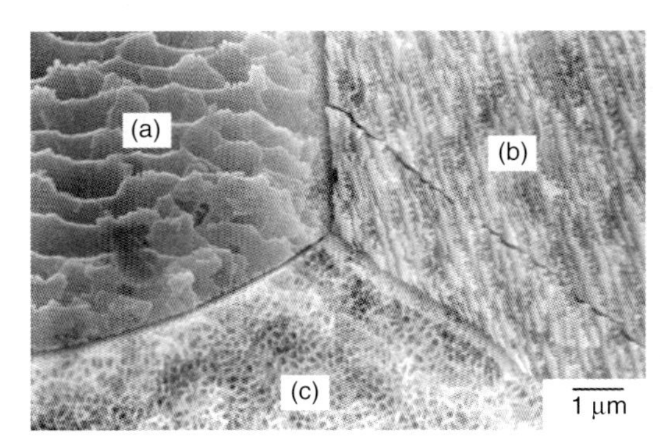

Fig. 1 SEM photograph of a polycrystalline CdSe surface after photoetching at +0.5 V versus saturated calomel electrode (SCE) in NaCl solution.

dipping in a polysulfide solution to remove surface selenium layer, a fine etching pattern appears on the surface [52–56]. A scanning electron microscope (SEM) photograph shown in Fig. 1 is a typical etching pattern of the surface of a polycrystalline CdSe, photoanodized at +0.5 V versus SCE in NaCl solution. It is apparent that the grain boundary is selectively dissolved and characteristic etching patterns, which depend on the crystallographic orientation, appear on the surface of each grain.

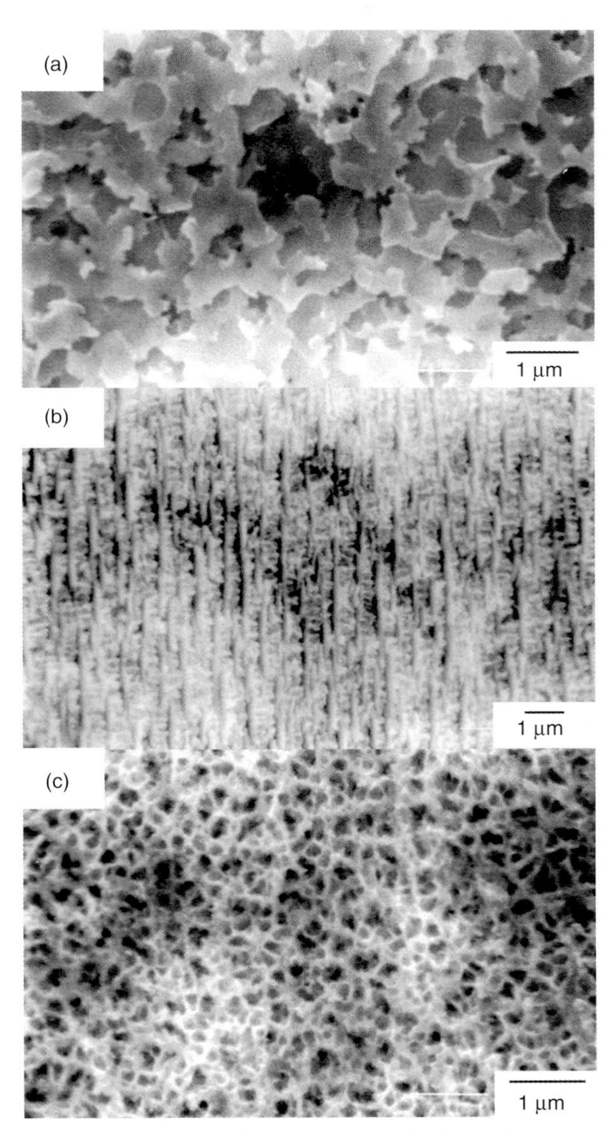

Fig. 2 SEM photographs of single crystal CdSe surfaces with different crystallographic plane exposed after photoetching.

Figure 2 shows SEM photographs of single-crystal CdSe surface with different crystallographic planes exposed after photoetching under the same condition as Fig. 1. By comparing these SEM photographs one can determine the crystallographic orientation of each grain of polycrystalline samples. It turns out that an exposed surface of the grain (a), which has terrace-like etching pattern, corresponds to (0002) face, terminated with Cd atoms. Likewise, an exposed surface of the grain (b), which has layered etching pattern, corresponds to the face parallel to c-axis, such as $(10\bar{1}0)$ or $(11\bar{2}0)$, and that of the grain (c), which has fine pit pattern, corresponds to $(000\bar{2})$ face, terminated with Se atoms. II–VI compound crystals with wurtzite structure lack the inversion symmetry along their polar axis (c-axis) and many workers have reported the difference in their properties [57, 58]. The photoetching site selectivity also shows a significant difference between the two polar surfaces.

One of the important findings in our studies on photoetching pattern observation is that the photoetching pattern differs by varying the photoetching potential. Quite interestingly, under weak anodic polarization, the grain boundaries are left undissolved and the bulk of each grain are selectively dissolved, as shown in Fig. 3, which is in contrast to Fig. 1. An evolution of such a unique skeleton structure, consisting of grain boundaries should be noted in the following points; one is that such a structure does not evolve by other techniques and the other is that grain boundaries thus left undissolved are subject to closer examinations of grain boundary.

Such a potential-dependent etching site selectivity is explained by using an energy band model. Figure 4 is a schematic energy band model of the polycrystalline *n*-type semiconductor with a grain boundary in contact with an electrolyte. It is well known that there is a buildup of a potential barrier because of the formation of electron acceptor levels at the grain boundary. The space charge layers are formed not only at the semiconductor–electrolyte interface but also on both sides of the grain boundary. In the case of weak anodic polarization, namely, near the flat band potential, almost all photogenerated holes that reach the grain boundary are likely to recombine with the trapped electrons, resulting in little or no photocurrent flowing at the grain boundary. It means that the grain boundary has less tendency to be photoanodically dissolved. More anodic polarization brings about the following two effects: one is to reduce the number of trapped electrons by lowering the Fermi level and the other is to generate stronger electric field in the space

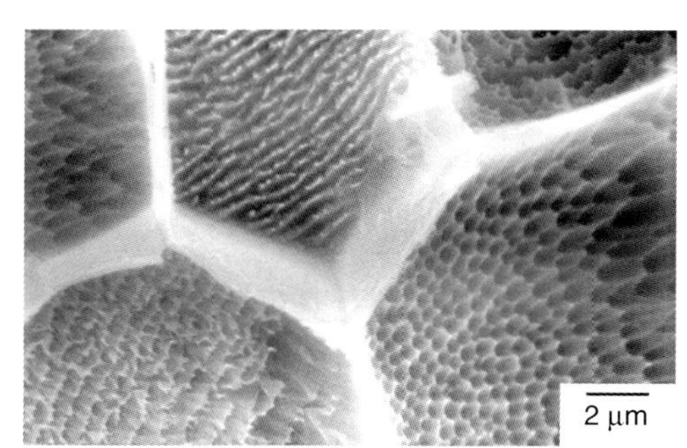

Fig. 3 SEM photograph of a polycrystalline CdSe surface after photoetching at −0.3 V versus SCE in NaCl solution.

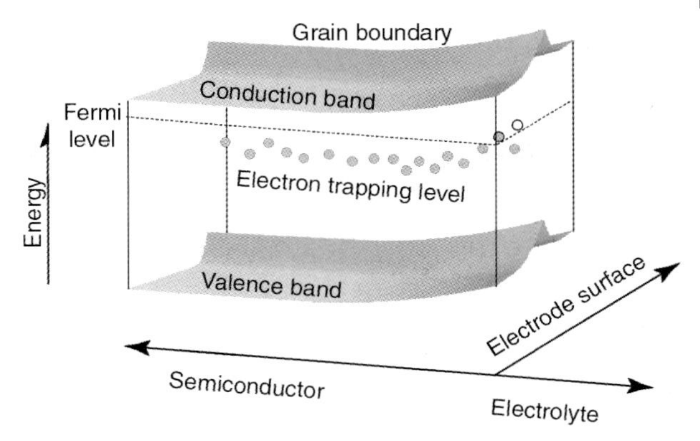

Fig. 4 Schematic energy band model of the polycrystalline *n*-type semiconductor, having grain boundary at the solid–electrolyte interface.

charge layer, especially at a grain boundary, where an additional electric field exists due to the grain boundary potential. This situation explains the preferential dissolution of grain boundary.

CdS, CdSe-CdS mixed crystals, and CdS_xSe_{1-x} solid solution electrodes also exhibit the same etching site selectivity, depending on the crystallographic orientation and the etching potential [54]. It is well known that the grain boundaries play an important role in the electroceramic semiconductor devices, such as a photoelectrochemical cell, a varistor, and a positive temperature coefficient (PTC) thermistor [59–62]. More detailed studies on the microstructure of the grain boundary by examining the species, thus formed, and separated from semiconductor pellets, would contribute further understanding of the grain boundary and will be described in Sect. 6.4.3.

6.4.2.2 TiO_2

TiO_2 has been playing a leading role in active recent research for the utilization of photoenergy. Because of its high oxidative power, stability, and nontoxicity, it promises a broad range of uses as a photocatalyst to decompose harmful organic compounds in the environment (see Chapters 21, 22). The TiO_2-based dye-sensitized photoelectrochemical cell is also receiving great attention as a possible candidate for converting solar energy into electricity on a large scale (see Chapter 16–19). TiO_2 with a large surface area and a high crystallinity is preferable for these applications. It has been found that the photoetching helps satisfy both of these requirements [63–65]. Because the photoanodic dissolution of TiO_2 takes place only in an aqueous solution of sulfuric acid in competition with water oxidation [66–70], the photoetching should be carried out in this solution.

Figure 5 shows the surface morphology of a polycrystalline TiO_2 after photoetching at +1.0 V versus SCE, which is about 1.2 V anodic with respect to photocurrent onset at the electrode. The electrode used is a pellet prepared by sintering at 1300 °C for six hours in N_2 atmosphere and subsequently reduced at 700 °C for four hours in 10% H_2/N_2 gas to get *n*-type semiconductivity. These SEM pictures show that the characteristic etching pattern

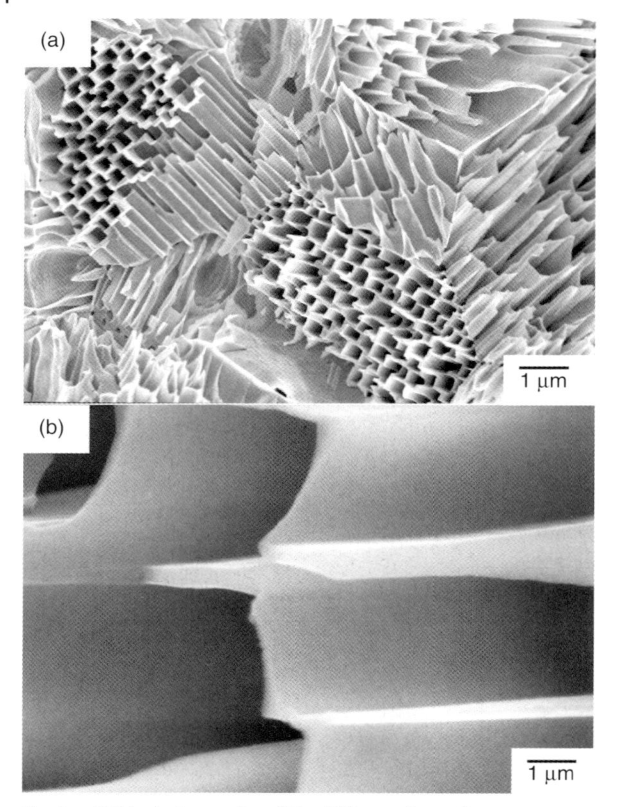

Fig. 5 SEM photographs of the TiO_2 surface after photoelectrochemical etching under strong anodic polarization (+1.0 V versus SCE).

appeared on the surface. The platelets are regularly ordered and interconnected thin walls of TiO_2, creating quadrangular cells with uniform size. This suggests that the etching reaction proceeds along a specific crystallographic orientation. This morphology was named as the nano-honeycomb structure. High magnification observation (b) reveals that the thickness of the platelets is only 20–30 nm and the surface is very smooth. Faradic efficiency for photoelectrochemical dissolution of TiO_2 was determined to be approximately 3% by assuming this reaction to be a two-hole process as expressed by Eq. (1).

$$TiO_2 + SO_4{}^{2-} + 2h^+ \longrightarrow TiO \cdot SO_4 + \tfrac{1}{2}O_2 \qquad (1)$$

A transmission electron microscope (TEM) photograph (a) and its complementary selected area electron diffraction (SAD) pattern (b) of the thin walls of TiO_2, collected from the photoetched surface are shown in Fig. 6. These are thin and long platelets, having a uniform thickness of several tens of nanometers. The well-regulated lattice fringe of 0.25 nm is observed in the TEM photograph (a) and it corresponds to the spacing of (101) planes (0.2487 nm). The SAD pattern (b) of

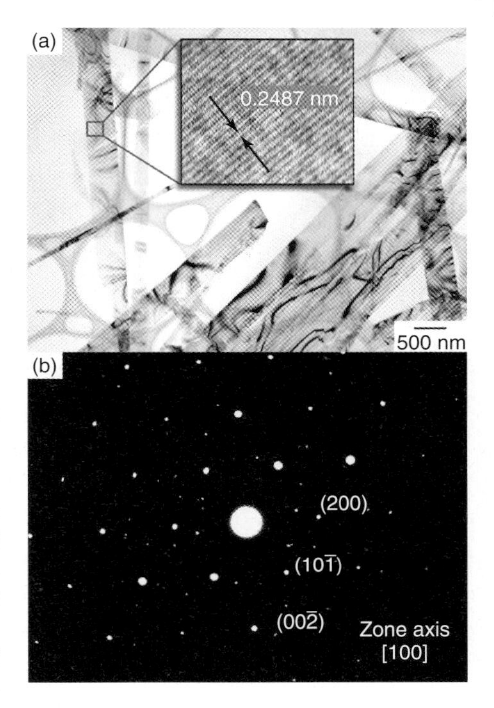

Fig. 6 TEM photograph (a) and its complementary SAD pattern (b) of specimens collected from the photoetched TiO_2 surface. The high-magnification image of lattice fringe is inserted in (a).

the same part of the specimen has a (100) zone axis of TiO_2 with a rutile structure. The direction of (10$\bar{1}$) diffraction spot is perpendicular to that of the lattice fringe shown in the inset of (a), indicating that the platelet has its c-axis parallel with its long side. This result shows that the photoetching reaction proceeds along the c-axis to construct the deep square honeycomb seen in Fig. 5. Detailed analysis of many platelet samples revealed that the surface of all the walls that construct nano-haneycomb structure has (100) crystallographic face.

It is interesting to note that the surface of the platelet is very smooth and the roughness is less than several nanometers, as is seen in atomic force microscopy (AFM) photograph (Fig. 7a). Some irregularities in this picture are probably caused by physically attached particles of dusts. Figure 7 suggests that this platelet is 200 nm in width, 1 μm in length, and 25 nm in thickness. These observations are in good agreement with the SEM pictures (Fig. 5). It is likely that the photoetching proceeds by the layer-by-layer dissolution mechanism to create atomically flat surfaces of TiO_2.

To find useful applications of the nanostructured TiO_2, the size of each honeycomb should be controlled. According to our recent experimental results, the size is independent of the photoetching conditions and is varied by changing donor density (N_d) of TiO_2. Figure 8 shows SEM photographs of polycrystalline TiO_2, having different donor densities after photoetching at +1.0 V versus SCE. Donor density was increased by raising the reducing temperature. The size of the honeycomb of the TiO_2 sample with higher donor density (B in Fig. 8) is obviously smaller than that of a sample with lower donor density (A in Fig. 8). Figure 9 shows a SEM photograph of single crystal rutile TiO_2 surface ((001) face) ($Nd = 10^{20}$ cm^{-3}) after photoetching at +1.0 V versus SCE. Regularly ordered submicron porous structure, consisting of a thin wall with (100) crystal faces, evolved over the surface. Such a structure apparently satisfies the requirements for a large specific surface area and high crystallinity.

In analogy with cadmium chalcogenides, mentioned in the previous section, the formation of a "skeleton" structure, consisting of only grain boundaries is also possible by anodizing a polycrystalline TiO_2 electrode at weak anodic polarization under illumination. A typical SEM photograph of the TiO_2 surface

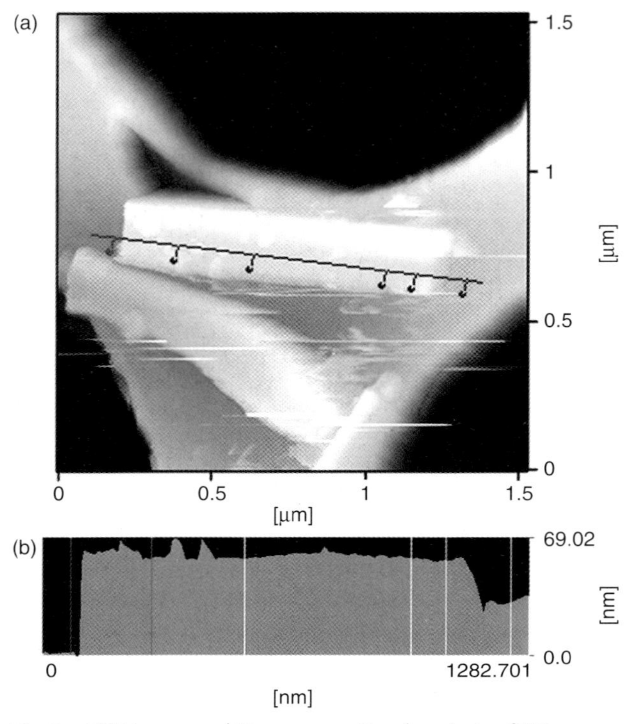

Fig. 7 AFM image and it's cross-sectional analysis of TiO_2 specimen collected from the photoetched surface.

photoetched at +0.1 V versus SCE is shown in Fig. 10. Interestingly, the grains are selectively dissolved, whereas the grain boundaries were left undissolved. The height of the grain boundary wall is several microns and the thickness is $10 \sim 20$ nm. Such a potential dependence of the photoetching pattern can be explained in the same manner shown in Fig. 4.

6.4.3 Analyses of Grain Boundaries

Now that one can separate and isolate grain boundaries from a photoetched polycrystalline TiO_2 surface, these samples are subjected to closer characterization. A TEM photograph of a grain boundary taken out from the surface of a photoetched TiO_2 is shown in Fig. 11. A grain boundary has a uniform thickness of about several tens of nanometers. Detailed analyses of grain boundary is described as follows. Figure 12 shows its high-resolution TEM photograph and corresponding SAD pattern. The incident direction of the electron beam is parallel to the (110) zone axis. The well-regulated lattice fringe with a distance of 0.32 nm, which corresponds to the interplanar spacing of (110) plane (0.3247 nm). This result shows that the grain boundary is not amorphous but crystalline.

A TEM photograph and its corresponding SAD pattern of another part of the same specimen are shown in Fig. 13. The moiré pattern of the 0.7 nm width is seen in the TEM image (a). Such a moiré pattern is produced by the interference of crystal

Fig. 8 SEM photographs of polycrystalline TiO_2 having different donor density after photoetching at +1.0 V versus SCE. Samples reduced at (a) 700 °C, four hours and (b) 1000 °C, one hour, respectively, in the stream of 10% H_2/N_2 gas.

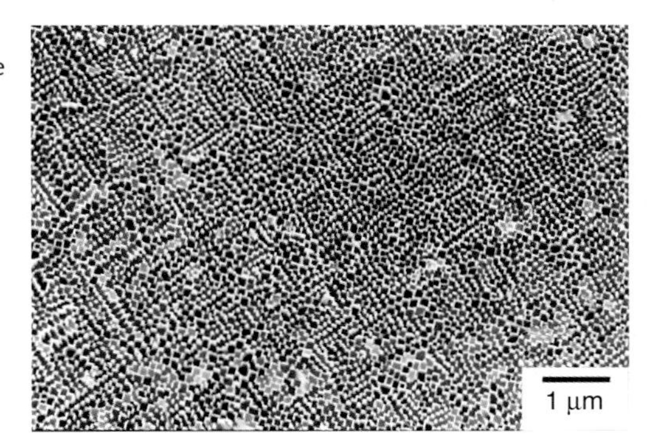

Fig. 9 SEM photograph of single crystal rutile TiO_2 surface ((001) face) after photoetching at +1.0 V versus SCE.

Fig. 10 SEM photograph of skeleton-structured TiO_2 surface after photoetching at +0.1 V versus SCE.

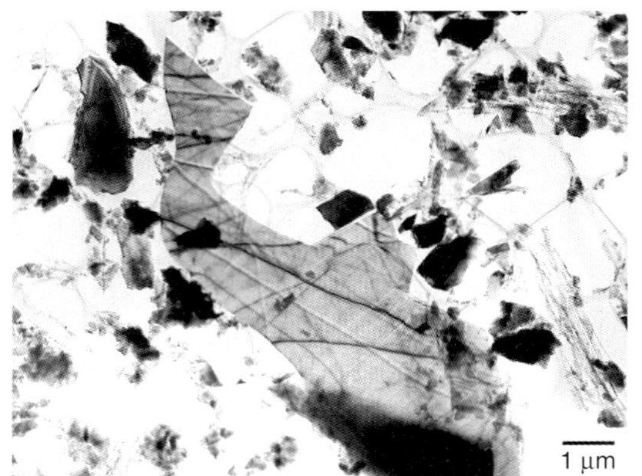

Fig. 11 TEM photograph of the grain boundary specimen picked up from the TiO_2 surface after photoetching.

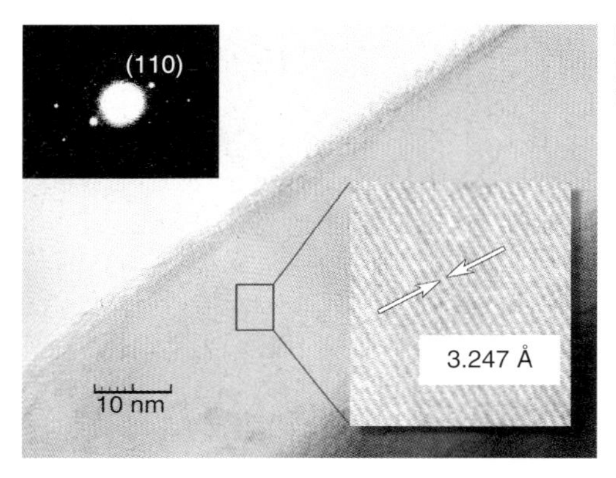

Fig. 12 High-resolution TEM photograph and SAD pattern of the grain boundary.

lattices of two twisted thin crystals. In the SAD pattern (b) an unexpected diffraction spot, C, emerges in addition to two sets of (111) lattice reflections (A & B) and (110) lattice reflections, D, of a rutile structure. It was interpreted that the spot C originates from double electron diffraction at the grain boundary, as illustrated in Fig. 14. Considering that the grain boundary has a sandwich structure that consists of two thin crystals (a, b) of TiO_2, having different crystal orientations, is quite reasonable. One of the (111) lattice diffraction (spot B in Fig. 13b) and (110) lattice diffraction (spot D in Fig. 13b) are assigned to the diffraction spots, arising from the crystal (b) with the (110) zone axis parallel to the electron beam. Another (111) lattice diffraction (spot A) arises from the crystal (a) and is superimposed with a twist of 18°. When an incident electron beam is reflected two times by each (111) lattice planes of crystals (a) and (b), it causes double diffraction of the electron beam to create spot C. Overlapping of two (111) lattice planes twisted with a small angle produces a moiré fringe. Its spacing is calculated by Eq. (2).

$$d_M = \frac{d_1 d_2}{(d_1^2 + d_2^2 - 2d_1 d_2 \cos\alpha)^{1/2}} \tag{2}$$

where d_M is the spacing of moiré fringe, d_1, d_2 refers to the spacing of each lattice, and α is the angle between the two lattices. By substituting 0.2188 nm (lattice spacing

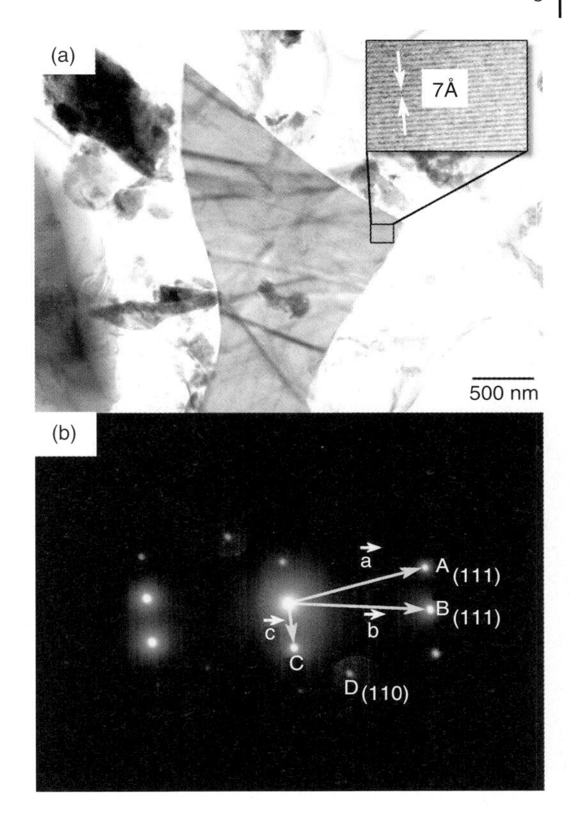

Fig. 13 TEM photograph (a) and its complementary SAD pattern (b) of the grain boundary.

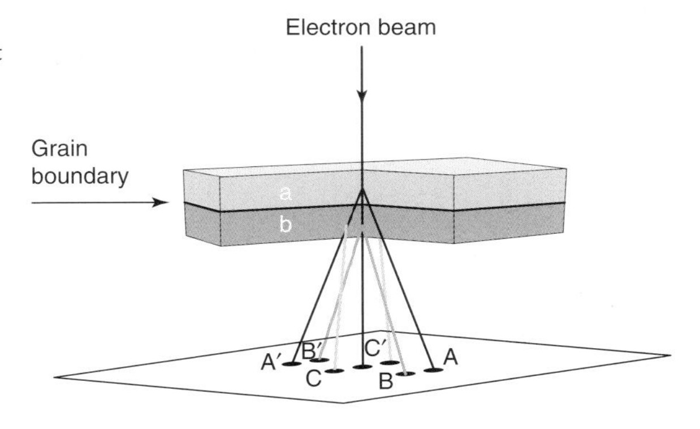

Fig. 14 Schematic illustration of double-electron diffraction at the grain boundary.

of (111) plane) for d_1 and d_2, and 18° (the angle between two electron diffraction spots, A and B) for α, d_M is calculated to be 0.7 nm, which is consistent with the spacing of moiré fringe in Fig. 13(a). The direction of the moiré fringe is perpendicular to the vector **c** of the electron diffraction spot C, which is the sum of the vectors **a** and **b** of diffraction spots of two (111) planes (A and B).

On the basis of the above information the structure of the polycrystalline TiO_2

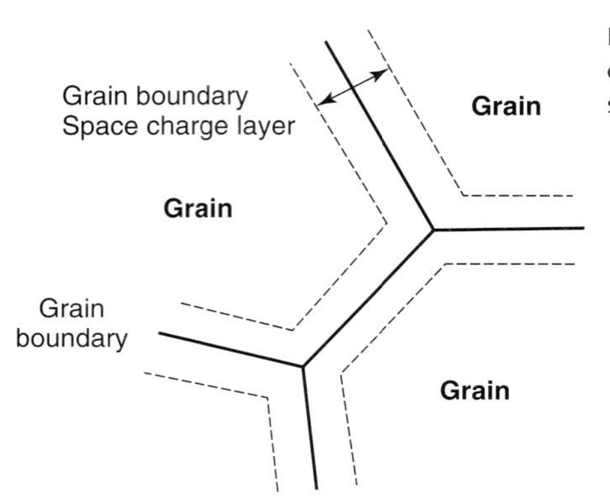

Fig. 15 Schematic illustration of the grain boundary model of semiconductor.

surface can be schematically illustrated as Fig. 15. The grain boundaries are represented by a solid line and the space charge layer that is formed near the grain boundary is represented by a dotted line. At the grain boundary region in contact with an electrolyte, the photogenerated holes arrives at the surface recombine with the electrons trapped in the grain boundary region, whereas the dissolution of the grain bulk in contact with the electrolyte takes place in competition with the water oxidation. Consequently, it is understandable that the grain boundary thus left undissolved is a thin plate, consisting of two platelets with different crystallographic orientations attached to each other. It seems that the thickness of the grain boundary thus left undissolved, depends on the thickness of the space charge layer formed at the grain boundary. This model is consistent with our results of the grain boundary observation of CdX (X = S, Se) [54, 55], where SAD patterns of the grain boundary obtained from the photoetched polycrystalline CdSe and CdSe–CdS mixed crystal has diffraction spots having multiple zone axes. Interestingly, diffraction spots by both CdS and CdSe are also observed for grain boundary samples of a photoetched CdSe–CdS mixed crystal. More detailed analyses of grain boundary by using this technique must contribute to further understanding of the grain boundary of ceramics.

6.4.4 Conclusion and Prospect

In conclusion, the photoetching technique has been found to have an advantage of unique site-selective dissolution and to contribute not only to microtailoring of semiconductor surface but also to further characterization of grain structure. Particularly, realization of TiO_2 with a large specific surface area and high crystallinity seems to be very effective for improving sensor devices, a photocatalyst, and a dye-sensitized solar cell.

References

1. G. Hodes, J. Manassen, D. Cahen, *US Patent 4* **1983**, *386*, 142.
2. R. N. Pandey, K. S. Babu, K. S. Chandra, O. N. Srivastava, *Prog. Surf. Sci.* **1996**, *52*, 125–192.
3. D. Cahen, G. Hodes, J. Manassen, R. Tenne, in *Photoeffects at Semiconductor – Electrolyte Interface* (Eds.: A. J. Nozik), The American Chemical Society, Washington, D.C., 1981, pp. 369–385.
4. K. Matsushita, R. Chiba, S. Okuyama et al., *J. Electrochem. Soc.* **1993**, *140*, 2097–2100.
5. C. R. Elliott, J. C. Regnault, *J. Electrochem. Soc.* **1980**, *127*, 1557–1562.

6. T. Saitoh, S. Matsubara, S. Minagawa, *J. Electrochem. Soc.* **1975**, *122*, 670–674.
7. M. M. Faktor, J. L. Stevenson, *J. Electrochem. Soc.* **1978**, *125*, 621–629.
8. E. Galun, G. Hodes, M. Peisach et al., *J. Cryst. Growth* **1992**, *117*, 666–671.
9. R. Khare, D. B. Young, E. L. Hu, *J. Electrochem. Soc.* **1993**, *140*, L117–L118.
10. H. H. Streckert, J. Tong, A. B. Ellis, *J. Am. Chem. Soc.* **1982**, *104*, 581–588.
11. R. Tenne, *Appl. Phys. Lett.* **1983**, *43*, 201–203.
12. G. Hodes, D. Cahen, H. J. Leamy, *J. Appl. Phys.* **1983**, *54*, 4676–4678.
13. M. A. Russak, C. Creter, *J. Electrochem. Soc.* **1985**, *132*, 1741–1745.
14. G. Hodes, J. Manassen, D. Cahen, *J. Electrochem. Soc.* **1981**, *128*, 2325–2330.
15. R. Tenne, G. Hodes, *Appl. Phys. Lett.* **1980**, *37*, 428–430.
16. N. Mueller, R. Tenne, *Appl. Phys. Lett.* **1981**, *39*, 283–285.
17. G. Dagan, S. Endo, G. Hodes et al., *Sol. Energy. Mater.* **1984**, *11*, 57–74.
18. G. Dagan, D. Cahen, *J. Electrochem. Soc.* **1987**, *134*, 592–600.
19. R. Tenne, B. Theys, J. Rioux et al., *J. Appl. Phys.* **1985**, *57*, 141–145.
20. R. Tenne, A. Wold, *Appl. Phys. Lett.* **1985**, *47*, 707–709.
21. A. M. Chaparro, P. Salvador, A. Mir, *J. Electroanal. Chem.* **1997**, *422*, 35–44.
22. V. D. Das, L. Damodare, *J. Appl. Phys.* **1997**, *81*, 1522–1530.
23. H. Minoura, Y. Ueno, H. Kaigawa et al., *J. Electrochem. Soc.* **1989**, *136*, 1392–1395.
24. L. Hollan, J. C. Tranchart, R. Memming, *J. Electrochem. Soc.* **1979**, *126*, 855–859.
25. A. Yamamoto, S. Yano, *J. Electrochem. Soc.* **1975**, *122*, 260–267.
26. A. Yamamoto, S. Tohno, C. Umemura, *J. Electrochem. Soc.* **1981**, *128*, 1095–1100.
27. T. Saito, S. Matsubara, S. Minagawa, *J. Electrochem. Soc.* **1975**, *122*, 670–674.
28. F. Kuhn-Kuhnenfeld, *J. Electrochem. Soc.* **1972**, *119*, 1063–1068.
29. K. Takahashi, *Jap. J. Appl. Phys.* **1979**, *18*, 1741.
30. P. A. Kohl, *IBM J. Res. Dev.* **1998**, *42*, 629–637.
31. E. A. Miller, G. L. Richmond, *J. Phys. Chem. B* **1997**, *101*, 2669–2677.
32. S. Mueller, J. L. Weyher, K. Koehler et al., *Int. Phys. Conf. Ser.* **1996**, *145*, 279–284.
33. J. L. Weyher, *Int. Phys. Conf. Ser.* **1995**, *146*, 399–408.
34. D. Rotter, J. Uffmann, J. Ackermann et al., *Mater. Res. Soc. Symp. Proc.* **1998**, *482*, 1003–1008.
35. C. Youtsey, G. Bulman, I. Adesida, *Mater. Res. Soc. Symp. Proc.* **1997**, *468*, 349–354.
36. C. Youtsey, I. Adesida, G. Bulman, *Electron. Lett.* **1997**, *33*, 245–246.
37. D. Soltz, M. A. D. Paoli, L. Cescato, *J. Vac. Sci. Technol. B* **1996**, *14*, 1784–1790.
38. V. A. Tyagai, V. A. Sterligov, Ya. Kolbasov, *Electrochim. Acta* **1977**, *22*, 819.
39. D. V. Podlesnik, H. H. Gilgen, R. M. Osgood et al., *Appl. Phys. Lett.* **1983**, *43*, 1083–1085.
40. R. M. Lum, A. M. Glass, F. W. Ostermayer et al., *J. Appl. Phys.* **1985**, *57*, 39–44.
41. R. Matz, J. Zirrgiebel, *J. Appl. Phys.* **1988**, *64*, 3402–3406.
42. R. M. Lum, F. W. Ostermayer, P. A. Kohl et al., *Appl. Phys. Lett.* **1985**, *47*, 269–271.
43. F. Decker, D. A. Soltz, L. Cescato, *Electrochim. Acta* **1993**, *38*, 95–99.
44. R. David Rauh, in *Electrochemistry of Semiconductor and Electronics* (Eds.: J. McHardy, F. Ludwig), Park Ridge, N.J., 1992, pp. 177–216.
45. H. Matsumoto, T. Sakata, H. Mori et al., *Chem. Lett.* **1995**, 595–596.
46. H. Matsumoto, T. Sakata, H. Mori et al., *J. Phys.Chem.* **1996**, *100*, 13781.
47. T. Torimoto, H. Nishiyama, T. Sakata et al., *J. Electrochem. Soc.* **1998**, *145*, 1964–1968.
48. M. Miyake, T. Torimoto, T. Sakata et al., *Langmuir* **1999**, *15*, 1503–1507.
49. T. Torimoto, H. Kontani, T. Sakata et al., *Chem. Lett.* **1999**, 379–380.
50. A. van Dijken, D. Vanmaekelbergh, A. Meijerink, *Chem. Phys. Lett.* **1997**, *269*, 494–499.
51. A. van Dijken, A. H. Janssen, M. H. P. Smitsmans et al., *Chem. Mater.* **1998**, *10*, 3513–3522.
52. T. Sugiura, H. Minoura, Y. Ueno, *Denki Kagaku* **1989**, *57*, 1133–1136.
53. T. Sugiura, H. Tomimatsu, H. Minoura et al., *Denki Kagaku* **1993**, *61*, 661–665.
54. T. Sugiura, R. Tenpaku, H. Minoura et al., *Electrochim. Acta* **1993**, *38*, 593–569.
55. T. Sugiura, M. Hida, H. Minoura et al., *Appl. Phys. Lett.* **1990**, *56*, 1954–1956.
56. T. Sugiura, M. Hida, M. Ohnishi et al., *Electrochim. Acta* **1992**, *37*, 1429–1432.

57. H. Iwanaga, T. Yoshiie, T. Yamaguchi et al., *J. Crystal Growth* **1979**, *47*, 703–711.
58. T. Yoshiie, H. Iwanaga, T. Yamaguchi et al., *J. Crystal Growth* **1981**, *51*, 624–626.
59. L. M. Levinson, S. Hirano, Eds., Grain Boundaries and Interfacial Phenomena in Electro Ceramics, in *Ceramic Transactions*, The American Ceramic Soc., Westerville, Ohio, 1993, Vol. 41.
60. T. R. Gupta, *J. Am. Ceram. Soc.* **1990**, *73*, 1817–1840.
61. J. A. S. Ikeda, Y.-M. Chiang, *J. Am. Ceram. Soc.* **1993**, *76*, 2437–2446.
62. J. A. S. Ikeda, Y.-M. Chiang, A. J. Garratt-Reed et al., *J. Am. Ceram. Soc.* **1993**, *76*, 2447–2459.
63. T. Sugiura, T. Yoshida, H. Minoura, *Electrochem. and Solid-state Lett.* **1998**, *1*, 175–177.
64. T. Sugiura, S. Itoh, T. Ooi et al., *J. Electroanal. Chem.* **1999**, *473*, 204–208.
65. T. Sugiura, T. Yoshida, H. Minoura, *Electrochemistry* **1999**, *67*, 1234–1236.
66. L. A. Harris, R. H. Wilson, *J. Electrochem. Soc.* **1976**, *123*, 1010–1015.
67. L. A. Harris, D. R. Cross, M. E. Gerstner, *J. Electrochem. Soc.* **1977**, *124*, 839–844.
68. M. E. Gerstner, *J. Electrochem. Soc.* **1979**, *126*, 944–949.
69. Y. Nakato, H. Akanuma, J. Shimizu et al., *J. Electroanal. Chem.* **1995**, *396*, 35–39.
70. Y. Nakato, H. Akanuma, Y. Magari et al., *J. Phys. Chem. B* **1997**, *101*, 4934.

Index

d

i

l

m

t